History of Liquid Propellant Rocket Engines

History of Liquid Propellant Rocket Engines

George P. Sutton

American Institute of Aeronautics and Astronautics
1801 Alexander Bell Drive
Reston, Virginia 20191-4344
Publishers since 1930

American Institute of Aeronautics and Astronautics, Inc., Reston, Virginia

Library of Congress Cataloging-in-Publication Data on file.   1 2 3 4 5

Cover design by David C. Milam

Copyright © 2006 by the American Institute of Aeronautics and Astronautics, Inc. All rights reserved. Printed in the United States of America. No part of this publication may be reproduced, distributed, or transmitted, in any form or by any means, or stored in a database or retrieval system, without the prior written permission of the publisher.

Data and information appearing in this book are for informational purposes only. AIAA is not responsible for any injury or damage resulting from use or reliance, nor does AIAA warrant that use or reliance will be free from privately owned rights.

CONTENTS

Preface **ix**

Acknowledgments **xiii**

Glossary of Abbreviations and Symbols **xv**

1. Introduction **1**

2. Merits of Liquid Propellant Rocket Engines and Their Applications **5**
   2.1 Why Were Liquid Propellant Rocket Engines Used? **5**
   2.2 Early Applications **6**
   2.3 Current Applications **16**

3. Technology Trends and Historical Changes **23**
   3.1 Expanding the Range of the Thrust, 0.01 to 1,800,000 lbf **23**
   3.2 Increasing the Chamber Pressure **25**
   3.3 So Many Liquid Propellants **25**
   3.4 Engines Can Come in Families **27**
   3.5 Thrust-to-Engine Weight Ratio Has Gone Up **28**
   3.6 Costs Have Gone Down **29**
   3.7 Extra Functions Have Increased Engine Complexity **29**
   3.8 Reliable Operations Were Achieved **32**

4. Technology and Hardware **33**
   4.1 Propellants, Past and Present **33**
   4.2 Engine Systems **54**
   4.3 Large Thrust Chambers **74**
   4.4 Turbopumps **108**
   4.5 Gas Generators, Preburners, and Chemical Tank Pressurization **131**
   4.6 Small Thrusters for Attitude Control and Trajectory Corrections **147**
   4.7 Controls, Valves, and Interconnecting Components **182**
   4.8 Starting and Ignition **201**
   4.9 Steering or Flight Trajectory Control **218**
   4.10 Combustion and Vibrations **228**

5. The Early Years, 1903 to the 1940s **241**
   5.1 Konstantin E. Tsiolkowsky **241**
   5.2 Robert H. Goddard **247**

      5.3   Hermann Oberth 271
      5.4   Other Pioneers 276
      5.5   Amateur Rocket Societies 281
      5.6   Early Team Efforts 289

6. Liquid Propellant Rocket–Engine Organizations, Worldwide, 1932 to 2003 **293**

7. Liquid Propellant Rocket Engines in the United States (Summary) **303**
      7.1   Liquid Propellant Rocket Engine Developers and Manufacturers in the United States 307
      7.2   Reaction Motors, Inc. 311
      7.3   General Electric Company 327
      7.4   Curtiss–Wright Corporation 346
      7.5   M.W. Kellogg Company 351
      7.6   Walter Kidde and Company 356
      7.7   Aerojet Propulsion Company, a Subsidiary Unit of Gen Corp, Inc. 359
      7.8   The Boeing Company, Rocketdyne Propulsion and Power 404
      7.9   Propulsion Products Center, Northrop Grumman Corporation 474
      7.10 Pratt & Whitney, a United Technologies Company 491
      7.11 Atlantic Research Corporation (ARC), Liquid Rocket Division 510

8. Liquid Propellant Rocket Engines in Russia, Ukraine, and the former Soviet Union **531**
      8.1   Early History (1929–1944) 532
      8.2   Rocket Engines for Piloted Aircraft 559
      8.3   Organizations Working on Liquid Propellant Rocket Engines 577
      8.4   NPO Energomash 585
      8.5   KB Khimautomatiki or Chemical Automatics Design Bureau 629
      8.6   KB Khimmash or Chemical Machinery Design Bureau 661
      8.7   NII Mashinostroeniya or the R&D Institute of Mechanical Engineering 684
      8.8   NPO Saturn, formerly OKB Lyulka 693
      8.9   OKB Kuznetsov, Reorganized as NPO Samara 698
      8.10 NPO Youzhnoye 710
      8.11 Korolev's Design Bureau, Later NPO Energiya 721
      8.12 OKB Fakel 725
      8.13 R&D Institutes 726
      8.14 Summary of Soviet or Russian Efforts in Liquid Propellant Rocket Engines 728

9. Liquid Propellant Rocket Engines in Germany **737**
      9.1   Early Efforts and Early Propellant Evaluations 738
      9.2   The Army Research Station at Peenemünde 740
      9.3   Hellmuth Walter Corporation 754
      9.4   Bayrische Motoren Werke (Bavarian Motor Works) 763
      9.5   German Liquid Propellant Rocket Engines Since 1945 769

10. France's Liquid Propellant Rocket–Engine History **785**

11. Japan's Liquid Propellant Rocket–Engine History **815**

12. Liquid Propellant Rocket Engines in the United Kingdom or Britain **843**

13. Liquid Propellant Rocket Engines in the People's Republic of China **865**

14. Liquid Propellant Rocket Engines in India **881**

15. General Findings, Comments, and Conclusions **887**

About the Author **895**

Index **897**

Supporting Materials **911**

PREFACE

The idea of writing a historical book did not occur to me until I had a phone call in 2001 from Vigor Yang, the Editor-in-Chief of the *Journal of Propulsion and Power*. He asked me to write this book, because he believed that a written history was needed, and I was one of the few people still alive, who had personal experience with early liquid propellant rocket engines (LPRE) in the United States, and who was still active in this field after retirement.

At that time I really was not aware of the large amount of effort needed to research this subject, find old reference documents, obtain permissions of copyrighted material, or contact and obtain information from others, who had contributed to this field. Perhaps the most time-consuming work was to verify and validate the historical data on specific rocket engines in several countries, early pioneers in this endeavor, and the key organizations who were dedicated to developing new rocket engines. The book covers principal events in the United States as well as in other countries that had a significant effort in several types of LPREs.

The book has been written from the perspective of an engineer in this field with emphasis on LPRE technology; engine developments; design, ground or flight testing; and flight applications. It covers the timing of the introduction of key technical features in the history of LPREs and the impact of key developments primarily on engines that have flown successfully. A number of unusual and technically interesting engines, which were not flown, are also discussed. The book covers most of the larger (higher thrust) rocket engines which have actually flown and many of the numerous smaller rocket engines or thrusters. While the book also gives the names and some details about key personalities and a few of the early pioneers, the emphasis is on the evolution of the technology, the key rocket engine characteristics, and the major development/manufacturing organizations in this field.

There are many sources of information for this book. These sources include technical papers presented at technical meetings, company brochures, company data sheets, photos, patents, official and unofficial reports, internet sites, press releases, and other written documents. Some of the best sources were discussions and correspondence with people, who were key investigators or designers and had done some of the early work in this field. A good number of them were eager to provide information. Their contributions were indeed appreciated and their names are given in the text and the acknowledgment section.

A number of documents or technical reports on pieces of this history of LPREs have already been published and most are listed in the references. All seem to cover a narrow, limited part of this history, such as the history of a specific rocket engine family, books or articles about outstanding individuals in this field, developments in a few countries, or the LPRE history in a few specific organizations or a few key companies. What seems to be different and new in this book is putting the history of all organizations and countries into perspective—filling many historical voids—which had not been described in earlier literature, discussing how the successful efforts in one company or one country has influenced the development of LPREs in another country or another company or in a government agency. Comparisons are made of the timing for introducing several key technical events or engine feature innovations in one country and how they were thereafter used or improved by other development organizations in other countries.

My own experience in developing early engine components or complete engines, visits to other LPRE organizations, both domestic and foreign, and work through technical societies are also reflected in this book. I started in this field as a young engineer out of graduate school in 1943 and worked for two prominent U.S. LPRE development companies. My perspective was broadened, because I worked not only in the commercial industrial profit-motivated world, but also for several years as an academician and professor and several years as an appointed government official making decisions and allocating resources. For example, while working for the Department of Defense (DOD) Advanced Research Project Agency in the Pentagon, I participated in initiating the development of several large LPREs and in transferring several DOD rocket engine developments to the NACA, which subsequently became NASA.

My book treats the history of LPREs from several different points of view. After an introduction, Chapters 2 and 3 deal with questions such as, "Why LPREs were developed?" "What were their merits?" or "What were their successful flight applications?." Some of these applications are obsolete today. Chapters, 4.1–4.11 cover specific technical trends in the development history of LPREs as seen by me. This section includes tendencies in liquid propellants, combustion, starting, and cooling of combustion devices. Chapters 5–5.6 are about the early pioneers in several countries and some of their analytical and experimental investigations, which laid the basis for future developments. Also included are chapters about the early rocket societies in different countries and an example of an early team effort. The bulk of the Chapters (6–14) are concerned with the technical work and/or production achievements of specific LPRE organizations, descriptions of a good number of historical rocket engines developed by these organizations, their novelties, and flight applications. The chapter on Rocketdyne is the longest, because I had many years of personal experience at this company and have a more intimate knowledge of its engines. These chapters are organized by the eight countries, where a variety of different types of LPREs have been actively pursued. Finally, Chapter 15 gives the general findings, comments, and conclusions.

The field of LPREs is relatively young when compared to solid propellant rocket motors. The decisive early work was done in three countries, the United States, pre-war Germany, and the Soviet Union. In the United States the first experimental work was performed by Robert H. Goddard in the early 1920s and the first flight with a LPRE was accomplished by him in 1926. Experimental work on large LPREs in the United States started in the late 1940s and early 1950s. Germany started serious work in about 1930 and flew the first large LPRE (V-2) in the early 1940s. The Soviet Union started also in about 1930 and produced large LPREs in the late 1940s and early 1950s. In Germany and in the Soviet Union the work was done in government organizations and was kept secret until the end of World War II in 1945. Within a few decades, the development of LPREs had proliferated and spread to a number of other countries, which reinvented and/or copied much of the technology. This book describes some of the accomplishments of eight of these countries, each of which produced and flew several types of LPREs. There are additional countries, where only one or two types of LPREs were developed and/or where some limited development and/or production had occurred or is still in progress; they are metioned, but their rocket engines are not described in this book. The total number of countries is about 30 and they are listed.

This book should be of interest to practicing engineers in LPRE design or development organizations and in the flight vehicle design or integration field. It will give clues on the step-wise improvements or the maturing of the technology and it will also show which ideas or features have worked and which have been tried, but abandoned or replaced. It should also be of interest to the historians or educators, who follow the technical progress in the aerospace industry and to organization management, who want to review the history of these products from the perspective of their own company or agency.

George P. Sutton
September 2005

ACKNOWLEDGMENTS

It is with gratitude and a sense of having been blessed that I acknowledge the good help from friends, associates, and correspondents, who made this book a better publication. The early version was kindly reviewed by Vince Wheelock, Bob Kraemer, Mark Fisher, Tony Springer, Jim Morehart, Fred Durant, and an anonymous reviewer. Their comments helped to shape the reorganization and style of the book, identify errors, and add to the content. Reviews of individual chapters were done by Jim Morehart (Russian chapters), Vince Wheelock with help from John Halchek, Charles McKeon, and Jeff Kincaid (Rocketdyne chapter), Kenji Kishimoto (Japan), Vladimir Ratchuk (CADB-Russia), Mayne Marvin (ARC), Gordon Dressler (Northrop Grumman), and George Suzuki (formerly of Rocketdyne, reviewed three chapters). With the help of Vigor Yang of Pennsylvania State University a chapter review and inputs were obtained from several of his colleagues in the Peoples' Republic of China, mainly Minchu Gu, but with inputs from Pinxin Hu, Ningchang Zhu, Baojiong Zhang, and Guogiu Liu.

Many of the contributors provided data, technical reports, photos, or brochures on different liquid propellant rocket engines or on organizations, which developed and built these engines. In particular Jim Morehart (The Aerospace Corporation), Vince Wheelock (Rocketdyne's unofficial historian), and Mark Coleman (Chemical Propulsion Information Agency) provided a lot of useful historical information, photos, and suggestions. Good contributions were received from Charles Ehresman (Purdue University, formerly of Aerojet), Jiro Nakamichi (JAXA-Japan), Fred Baroody (formerly of ARC), Asif Siddiqi (data on Soviet Union), Frank Winter (Smithsonian Institute – data on General Electric), Mike Yost, Ed Shuster, John Halchak, Scot Claflin, Bob Dillaway (all provided Rocketdyne information), Olwen Morgan, Carl Stechman, and Judith Bauer (all of Aerojet), Carl Pignoli and Patricia Mills (Pratt & Whitney), Mayne Marvin (ARC), Lt. Col. Dennis Lileikis (U.S. Air Force), Andrew Kubica (formerly of General Electric), Ye. G. Larin (chief designer of NII Mash, Russia), Christian Lardier (France, data on KB Khimmash), Gerald Hagemann (specialist at EADS in Germany), Vladimir Ratchuk (chief designer and general manager of CADB, Russia), V. K. Chvanov (first deputy of general director and general designer of NPO Energomash), Eckart Schmidt (formerly of Rocket Research Company), Sam Smyth (retired from Lockheed Martin, historical aircraft expert), Douglas Millard (Science Museum, London), Tom Moore (CPIA), and Narajam Agappa (India).

Several people, who have passed away, unwittingly contributed information to this book. These are individuals with whom I had various technical discussions between 1947 and 1960, but at the time I did not know that some of this technical information could be historically useful years later. I want to gratefully acknowledge inputs from the late Richard Porter (head of rocket organization at General Electric), Charles Chillson (top rocket engineer at Curtiss Wright), Dave Calhoun (Aerojet Chief Engineer), Val Cleaver (leader of rocket engineering at Rolls Royce, UK), and Walter Riedel (head of V-2 development – war-time Germany).

Let me thank Mark Fisher and several of the graphic designers of the NASA Marshall Space Flight Center for making clear reproducible line drawings or annotated illustrations out of my crude sketches. The assistance from the AIAA offices is indeed appreciated. Jeanne Trimble cheerfully conducted computer searches of databases of technical publications, Chris Grady supplied some technical papers, Justin Opoku obtained some copyright permissions, Roger Williams gave me advice, and Meredith Perkins made a neat looking professional book out of a coarse manuscript.

GLOSSARY

ACS attitude control system
Aerozine 50 or A-50 fuel consisting of 50% hydrazine and 50% UDMH
AF Air Force
ARC Atlantic Research Corporation
CADB Chemical Automatics Design Bureau (Russia)
CPIA Chemical Propulsion Information Agency
DB design bureau (Russia); see KB
EADS European Aeronautics Defence and Space
GALCIT Guggenheim Aeronautical Laboratory/California Institute of Technology
GG gas generator
ICBM intercontinental ballistic missile
IRFNA see RFNA
I_s specific impulse, thrust force divided by propellant mass flow per second
ISRO India Space Research Organization
JATO jet assisted takeoff
JPL Jet Propulsion Laboratory (Pasadena, California) and also one in St. Petersburg (Leningrad) in the former Soviet Union
KB konstructorskoye buro (design or construction bureau); see DB
L^* characterisic length of combustion chamber, which is chamber volume divided by the throat area
LH_2 liquid hydrogen
LOX liquid oxygen
LPRE liquid propellant rocket engine (or sometimes just engine)
MMH monomethyl hydrazine
MRBM medium-range ballistic missile
NASA National Aeronautics and Space Administration
NII Nauchno-Technichsekey Kompleks (Science and Technical Complex)
NPO Nauchno-Proizvidstvennoye Obedinenie (Science and Production Association)
NTO nitrogen tetroxide (N_2O_4)
OKB opitnoye konstructorskoye buro (experimental design bureau)
PB preburner
RCS reaction control system
RFNA red-fuming nitric acid (also IRFNA, for inhibited RFNA)
RNII Rossiyskiy Nauchno – Isseldovatelskiy Institut (Russian Science and Research Institute)
SAM surface-to-air missile
SL sea level or sea-level condition
SLBM submarine-launched ballistic missile
SLV space launch vehicle
SRBM short-range ballistic missile
SSME space shuttle main engine
TC thrust chamber or thruster
TP turbopump

TVC thrust vector control
UDMH unsymmetrical dimethyl hydrazine
U.S. United States
USAF U.S. Air Force
USSR Union of Soviet Socialist Republics or Soviet Union
Vac or vac vacuum of space or vacuum condition
WFNA white-fuming nitric

Introduction

Today the liquid propellant rocket engine (LPRE) is a proven means of flight propulsion. It was conceived over 100 years ago, but its first actual design was in 1921; the first static test firing took place in 1923, and the first flight of a LPRE took place in 1926. This technology is now sufficiently well developed and proven, so that a broad range of LPRE can be designed, built, and flown with confidence. In 1930 there were only three countries, where a small group was struggling with an early research and development effort on LPREs, namely, the Soviet Union, Germany, and the United States of America. The LPRE capability has proliferated and grown, and today there are at least eight countries that have or have had a mature broad LPRE technical base and perhaps 10 more countries that have a limited capability in one or two specific types of LPREs. The total number of countries involved in this field is a small percentage of all of the countries that exist. Other means of propulsion, such as turbojets or electrical rocket propulsion, also thrive only in a few countries. This is not surprising because only a few countries have the resources, interest and technical capability to be active in this endeavor.

Many people in different countries made important contributions to the analysis, design, materials, fabrication, or operation of LPREs. Only a few names of these key people are included in this book, and little detail is presented about them. An estimated total of approximately 1300 different LPREs have been designed, built, and tested. It is impossible to collect information or even to make a list of all of them or to identify all that might be historic in value. Therefore this book will cover only selected pieces of this history, which have been documented and are available. If a significant event, important engine, or major historical accomplishment was omitted, it was not by intent, but by the limitation of the information known or available to the author.

There is no single LPRE concept or type, but rather several that are related. All of them have one or more thrust chambers. There are significant differences between large (high-thrust) and small (low-thrust) LPREs, between those using cryogenic vs storable propellants, monopropellants or bipropellants, single vs multiple starts during the flight, steady vs random variable thrust, single operation or reusable for several flight operations, and those with pumps or gas pressure expulsion of propellants in their feed systems. The history of all of these types will be discussed.

The development and production of a LPRE has always been a team effort. Chapters 6–14 identify the groups, laboratories and companies, who played key roles in the history of LPREs in eight different countries. These chapters describe key historical engines for each organization and some of their technical accomplishments. The exceptions to the team effort are few, such as some the early visionaries and pioneers in this field, who worked alone. Their activities will be briefly identified and described in Chapter 5. The names or organizational structure of the teams or companies often changed. They will be identified by the recent name of the organization and by the country, but sometimes also by the name of a few of their leading people. Many important people, who were leaders or strong voices on these teams and some of the previous organization names or parent companies, will not be mentioned.

In this book "a successful LPRE" is defined as one that 1) has been put into serial production and/or 2) has flown satisfactorily more than once. Research and design (R&D) efforts are for naught, if they do not culminate in a flight application. There have been many experimental and prototype LPREs, engine components, and propellants that were conceived, designed, built, and tested, but for various reasons were never successful, and most fell by the wayside. Some important lessons were gleaned from some of them. This history concentrates on the successful LPREs. However, some of the other developments that might not fit this definition of success but have interesting technology or some historic significance will also be discussed.

It is not the purpose of this historical account to present detailed information on specific LPREs or to include all of the world's LPREs. There are too many of them. Only a few have been selected for this book. For each of those, only a few pieces of data, some characteristic parameters, and/or figures will be given in this history. For more detailed parameters and for other engines, the reader is referred to the references or to the developer's organization. However the references are not inclusive or complete. Although some of the flight vehicles driven by a LPRE (airplanes, missiles, space launch vehicles, spacecraft, etc.) are mentioned here briefly by name or identification number, the emphasis in this work is on the rocket engines and not the vehicles. Gaseous-propellant engine systems are included because they are usually grouped with the LPREs. Solid-propellant rocket motors, electrical propulsion, hybrid propulsion, and combination rocket-airbreathing engines are excluded.

The numerical designation for LPREs is different in each company and each country, and usually there is more than one designation. For example the 80,000-lb-thrust engine built by Aerojet, a GenCorp Company, has a company designation AJ23-131, a military or government designation of LR91-AJ-3, and an application designation, namely, the engine for the second stage of the Titan I missile. In this book engines will be identified by the application or flight vehicle and in some cases by the company's designation or by another designation.

The general references and classical textbooks that follow are for those readers who want to acquire some background or pursue a general study of this book's subject.

Bibliography

Barrére, M., Jaumotte, A., de Veubeke, B. F., and Vandenkerckhove, J., *Rocket Propulsion*, Elsevier, 1960.

Gruntman, M., *Blazing the Trail, The Early History of Spacecraft and Rocketry*, AIAA, Reston, VA, 2004.

Herrick, J., and Burgess, E. (eds.), *Rocket Encyclopedia, Illustrated*, Aeropublishers, Inc., Los Angeles, 1959.

Huzel, D. K., and Huang, D. H., *Modern Engineering for Design of Liquid Propellant Rocket Engines* (rev. ed.), Progress in Astronautics and Aeronautics, Vol. 147, AIAA, Washington, DC, 1992.

Isakovitz, S. J., Hopkins, J. B., and Hopkins, J. P., *International Reference Guide to Space Launch Systems*, 4th ed., AIAA, Reston, VA, 2004.

Jane's Space Directory, Jane's Information Group, Coulsdon, Surrey, England, U.K. (Revised about every two years, comprehensive information on liquid propellant rocket engines, space launch vehicles, and missiles), 1995–1996, 1997–1998, 1999–2000.

Kölle, H. H. (ed.), "Liquid Propellant Propulsion Systems," *Handbook of Astronautical Engineering*, McGraw–Hill, New York, 1961, Chap. 20.

Sutton, G. P., and Biblarz, O., *Rocket Propulsion Elements*, 7th ed., Wiley, Hoboken, NJ, 2001.

Merits of Liquid Propellant Rocket Engines and Their Applications

2.1 Why Were Liquid Propellant Rocket Engines Used?

Why were LPREs used? Because they propelled certain military and space vehicles better than any other type of chemical propulsion and because they provided some operating characteristics that could not be duplicated at the time by any other means of propulsion.[1] LPREs made it possible to build sounding rockets (1926–1973); they propelled military aircraft and assisted with their takeoff (1940–1970). They went into production for several early tactical missiles (1946–1973) because solid-propellant rocket motors could not meet the operating temperature limit or size requirements during the 1940s and 1950s. LPREs were selected for all of the initial ballistic missiles and helped to build up the military missile inventory needed urgently by the U.S. and USSR governments in the 1950s to 1970s. Since 1960, LPREs propelled all of the large space launch vehicles and just about all of the U.S. spacecraft and satellites. They constitute the propulsion machinery that drove us into the space age.

The *features and performance characteristics of LPREs*, which allowed their selection for the missions already mentioned, were unique and are briefly reviewed next.[1,2] Liquid bipropellants generally give a *higher specific impulse* than other chemical propulsion means, such as monopropellants or those using solid or hybrid propellants. Cryogenic propellants give the highest specific impulse. LPREs can be designed over a very wide *range of thrust values* to fit specific applications (by a factor of 10^8). They are the only form of chemical propulsion that can be designed for *quick restart, fast pulsing, and ready reuse*. They can be designed for a *random thrust variation upon command*. They have been uniquely suitable for controlling quick *attitude changes (pitch, yaw, or roll)* and minor *velocity changes* of individual stages of missiles, space launch vehicles, spacecraft, space stations, and satellites. A precise repeatable *thrust termination* permits an accurate terminal flight velocity, which is important in ballistic missiles, precise orbit attainment, and certain defensive missiles. Many LPREs can be functionally *checked out and even fully tested* before they are used or flown. An *engine-out capability* can be designed into engine clusters of several individual engines (4 to 30). A remarkably high *reliability* has been achieved in production LPREs. Lightweight LPRE hardware has allowed flight vehicles to achieve a *high propellant fraction* and a high *vehicle mass ratio*; these are critical parameters for high vehicle flight performance. Instant *readiness* has been achieved with storable propellants. These propellants have been *stored for more than 20 years* in a vehicle. All common propellants used today can discharge a *clean transparent exhaust* gas without smoke. However gas from a few of the storable propellants can give a trace of smoke. The exhaust gas from common propellants does not contain some condensing species of chemicals, which can form a noticeable deposit on sensitive vehicle surfaces, such as spacecraft windows or mirrors. Most of the exhaust gases of LPREs using modern common propellants are *not toxic* and *environmentally friendly*.

2.2 Early Applications

It is usually the application or the mission that determines the type and the specification of the particular propulsion system. The following LPRE applications or missions were very important before 1975, but are no longer needed. However, they were the first to fly with LPREs, and they represent a significant chapter in the history of LPREs. The specific LPREs and/or their developers, which are mentioned here as examples of applications, will be found again in Chapters 7–14 together with some additional references. Engines were selected for each mission because of some of the unique LPRE characteristics mentioned in the preceding paragraph.

Sounding Rockets

Sounding rockets was the original application for which the development of the first few LPREs was undertaken.[3] The early flying engines of Goddard (USA), Tsander (Russia), Veronique (France) or Jet Propulsion Laboratory (JPL, USA) were developed for sounding rockets. Many of the rocket vehicles built and flown by amateur societies in several countries (Chapter 5.2) were for this application. These engines were useful for simple vehicles doing meteorological investigations between 1926 and about 1975. They had simple gas-pressurized feed systems suitable for a single start. LPREs were selected because solid-propellant motors were then not available in the sizes needed and could, at this early stage of development, not be stored if the ambient temperature fluctuated widely (–65 to +160°F). An early sounding rocket vehicle with a LPRE is shown in Fig. 2-1. As our knowledge about the Earth's atmosphere increased, the need to explore by sounding rocket has greatly diminished, and the prediction of weather seems today to rely on data other than those gathered by sounding rockets. The few sounding rockets used today now have mostly solid-propellant rocket motors with modern capable solid propellants.

Fig. 2-1 WAC Corporal sounding rocket developed at the Jet Propulsion Laboratory in 1943–1946. Courtesy of JPL; copied with permission from second edition of Ref. 2.

Jet-Assisted Takeoff Systems

Jet-assisted takeoff or JATO systems were needed between 1935 and 1970 to assist heavily loaded aircraft in taking off from marginal airports, particularly high-altitude airports.[4-6] Many liquid-propellant JATOs were developed and flight tested with various experimental aircraft. A few years later solid-propellant JATO motors were also produced and used in several countries. A takeoff with a LPRE JATO can be seen in Fig. 2-2. The instant

Fig. 2-2 View of underside of a modified Boeing B-47 bomber aircraft taking off with the assistance of two Aerojet LPRE JATO units, each with two TCs. Courtesy of U.S. Air Force; photo furnished by the author of Ref. 5.

readiness, the reuse of the same JATO (after it was parachuted to the ground, cleaned, and reloaded), the wide storage temperature limits, the ability to clean and check the unit prior to reuse, and the relatively nontoxic clean exhausts were some of the engine characteristics that caused LPREs to be chosen. The first JATO was developed in Germany around 1935, in the Soviet Union a few years later, in Britain in 1944, and in France in 1945. Goddard developed and tested a JATO engine that assisted the takeoff of a PBY flying boat in the early 1940s. The first and only mass production of LPREs for JATO use was in Germany beginning in the late 1930s by the Hellmut Walter Company. The German Air Force deployed these JATO units and made extensive use of them in their bomber fleet during World War II. JATO units increased the range and/or the bomb load that could be carried in a bomber aircraft. The air forces of several other countries supported their development and flight testing, but they were not deployed or used operationally. The first limited production of a LPRE JATO in the United States was in 1943 by Aerojet. More information on JATO developments is found in Chapters 5.4, 7.5, 7.7, 8.1, 9.3, 10, and 12. JATO units became largely unnecessary when jet engines became more powerful and used afterburners.

Aircraft-Assist Rocket Engines or Superperformance Aircraft Rocket Engines

Aircraft-assist rocket engines or superperformance aircraft rocket engines were used with military fighter aircraft to enhance the power of turbojet engines for rapidly climbing to intercept altitude and/or for higher-speed flight maneuvers.[3-6] Many different aircraft-assist rocket engines were developed, and most were flight tested in different military airplanes in five countries: Germany, USSR, France, Britain, and United States. These LPREs were developed and flight tested and applied between 1937 and 1970. The early aircraft LPREs used gas-pressurized feed systems, which were very heavy and severely limited the range of the aircraft. Later versions used a turbopump feed system, which improved the range. Two French LPREs became operational, each in a group of military French fighter aircraft, and they were in the military service in France until 1984. France exported a few of the rocket-propelled fighter airplanes to Switzerland, Pakistan, Lybia, and South Africa. The last rocket-assisted flight was in Switzerland in about 1994. This application used unique LPREs that had multiple restarts, were reusable, man-rated, could be stored or flown in hot or cold weather, allowed ground check-out at full thrust, and many featured throttling, a commanded reduction of the thrust. Thrust levels of these superperformance engines were between 200 and 7000 lbf. The French initially used nitric acid with a hypergolic fuel, but later switched to kerosene or jet fuel, which was taken from the aircraft's fuel tanks. The pumps were driven by a jack shaft from the aircraft's principal engine. Figure 2-3. shows an operational French fighter aircraft being propelled by turbojets and a superperformance rocket engine. The Soviets developed about 12 such rocket superperformance rocket engines and flight tested most of them in various experimental fighter aircraft. In the United States about a dozen different aircraft-assist engines were developed. Aerojet developed several, some with pump feed systems, using storable propellants (nitric acid) and flew them in

Fig. 2-3 French Trident fighter aircraft flying with an early SEPR 631 auxiliary superperformance LPRE flew first in 1957. Courtesy of SEPR/EADS; copied with the publisher's permission from second edition of Ref. 2.

modified fighter airplanes, which were adapted for this use. Rocketdyne developed four different superperformance engines using hydrogen peroxide and jet fuel and flew two of them in three different experimental aircraft. The Rocketdyne AR-2-3 LPRE (turbopump feed, hydrogen peroxide, and jet fuel) was installed in several U.S. F-104 fighter military aircraft, and these airplanes have been used for astronaut training. See Chapters 7.7, 7.8, 9.3, 10, and 12. None of the aircraft equipped with superperformance LPREs were used in actual combat. The development of more powerful jet engines with afterburners obviated this application.

Rocket-Propelled Aircraft

A rocket-propelled aircraft used a rocket engine as its only power plant. Several experimental aircraft, which were intended to explore supersonic flight and high-altitude operation, were driven by different experimental LPREs in the United States, Germany, and Russia, and their engines are described in this book.[4,5] The very first engine to be flight tested was a LPRE from the Hellmuth Walter Company (Germany) using a monopropellant hydrogen-peroxide propellant with a liquid catalyst, a thrust of about 600 kg or 1320 lbf, and a turbopump feed system. In 1939 it took off on rocket power and propelled a Heinkel (He 176) aircraft, which was specifically designed for a rocket-powered engine and is shown in Fig. 2-4. The first Soviet LPRE for driving an aircraft was the RDA-1-150, and it was first flown about three years later in 1940 in an experimental RP-318 glider-type aircraft, which can be seen in Fig. 8.1-13. The aircraft was towed to altitude by another airplane and released for flying under rocket power. A higher thrust 1400 kgf or 3080 lbf LPRE was developed in the Soviet Union, and it could take off on rocket power, first in 1942. See Chapters 8.1 and 9.3 for more information. The first U.S. aircraft powered by a small LPRE was a small experimental flying wing shown in Fig. 2-5, and it had a small Aerojet engine, which is shown in Fig. 7.7-10 (Refs. 4–6). Your author assisted in the development of this Aerojet engine. It flew in 1943, which was later than the German and Russian airplanes. The aircraft was so small that the pilot had to lie down with his head in a plastic bubble at the nose of the airplane. The three early pioneering rocket flights were performed with LPREs, whose thrust was too low to permit a high-speed or a fast climb.

Larger LPREs and better R&D airplanes for investigating supersonic flight were developed next. In the United States the first such aircraft LPRE was the RMI 6000C engine (6000 lbf thrust), shown in Fig. 7.2-3, which flew first in the X-1 supersonic experimental airplane in 1947. The RMI XRL-99 engine at 50,000 lbf thrust, described in Chapter 7.2, flew in the X-15 experimental aircraft in 1961. A perspective view of the X-15 research aircraft, developed by North American Aviation, is shown in Fig. 2-6. It also had an attitude control system for flight control above the atmosphere with multiple small thrusters using the catalytic decomposition products of hydrogen peroxide as a monopropellant. These propellants are described in Chapter 4.1. The German Messerschmitt

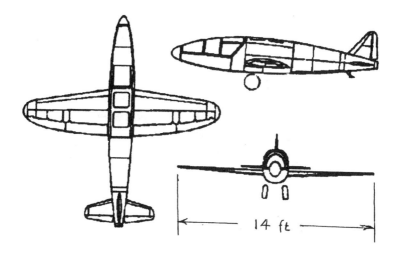

Fig. 2-4 This small German aircraft, the Heinkel Kadett (He-176), was the first airplane in 1939 to have a LPRE as its only powerplant for takeoff and flight propulsion. It used a Walter monopropellant hydrogen-peroxide LPRE (not shown). Copy provided by Sam Smyth (personal communication, 2004).

Me-163 interceptor aircraft, which is shown in Fig. 2-7, was propelled by a couple of versions of the German Walter 109-509 rocket engine, and it is shown in Fig. 9-7. Flights started in 1940. The final engine version had about 15 kN or almost 3400 lbf thrust, could be throttled and restarted, and used hydrogen peroxide and a hypergolic fuel. It is the only rocket-propelled fighter aircraft in the history of LPREs, which became operational and was used in combat by the German Air Force toward the end of World War II (1944 and 1945).

Because these aircraft rocket engines and their propellants were heavy, the range and operational capability of the fighter aircraft were severely limited. This mission of defense with fighter aircraft has been superseded by defensive guided missiles beginning in the late 1940s and 1950s.

Guided Tactical Missiles with Prepackaged LPREs

A series of guided tactical missiles with prepackaged LPREs were developed between 1946 and 1970.[7,8] In the United States this included the Bullpup (air-to-surface missile), shown in Fig. 7.2-7, Lance (surface-to-surface short-range Army transportable missile) described in Chapter 7.8, or Terrier (ship-defense Navy missile), which were all deployed with the military forces. The Nike antiaircraft missile, shown in Fig. 2-8, had a solid-propellant booster motor and an Aerojet LPRE in its upper stage. It had 2600 lbf thrust and used nitric acid and aniline as propellants. Engine development started in the late 1940s. The Germans developed the Wasserfall and the Taifun antiaircraft missile at the end of World War II,

12 History of Liquid Propellant Rocket Engines

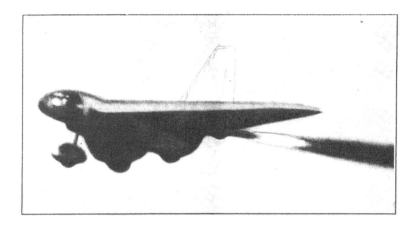

Fig. 2-5 Enlarged motion picture frame of the MX 324 model aircraft, a very small experimental flying wing developed by Northrop Corporation for aerodynamic research. It was the first U.S. aircraft to be propelled only by a LPRE, namely, a small Aerojet engine in 1943. Courtesy of Northrop Grumman; copied with permission from third edition of Ref. 2.

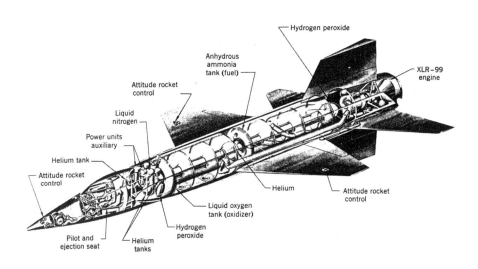

Fig. 2-6 The X-15 was an airplane for supersonic aerodynamic research with an unusual LPRE and small attitude control thrusters for operation at high altitude. Courtesy of NASA; copied with permission from third edition of Ref. 2.

Fig. 2-7 German Messerschmitt Me-163 fighter aircraft was powered 100% by a Walter LPRE and was used in combat toward the end of World War II (1944–1945). Copy from Sam Smyth (personal communication, 2004), and Ref. 5.

but their development had not been completed before the war ended. These engines are discussed in Chapter 9.4. The United States, USSR, Britain, Germany, France, and others developed tactical missiles driven by LPREs with sealed, prepackaged tanks for storable propellants, and they all had pressure-fed engine systems. The Soviets had several operational surface-to-air missiles with LPREs. Contrary to other countries, where gas pressure feed systems were used, the Soviets used engines with turbopump feed systems on the upper stages of their air defense missiles, as described in Chapters 8.2, 8.5, and 8.6. These were produced, and the antiaircraft missiles were deployed in the USSR in very large quantities. These LPREs for tactical applications were needed because at that time the solid-propellant rocket motors could not operate over wide ambient temperature ranges. In many of these tactical applications, the LPREs have since been replaced by improved solid-propellant motors. The hazards associated with spilled propellants and the relative complexity of the LPREs made them less attractive.

Ballistic Surface-to-Surface Missiles

Large ballistic surface-to-surface missiles capable of delivering a warhead over long distances of 1500 to 7000 miles were very important as weapons in the Cold War period between the USSR and the United States. Large LPREs were in strong demand for these applications during the 1950s and 1960s in the United States, Soviet Union, France, China, and Britain.[8,9] A launch of the Atlas

14 History of Liquid Propellant Rocket Engines

Fig. 2-8 Nike-Ajax, a tactical U.S. antiaircraft missile had a LPRE in its upper stage. Courtesy of U.S. Army; copied with permission from second edition of Ref. 2.

intercontinental ballistic missile is shown in Fig. 2-9. In this picture the cold oxygen tank had condensed atmospheric moisture (mist and fluffy small ice particles) on the outside of the tank wall, and this condensate can be seen as it is being sluffed off during the launch. The bright clouds at the bottom of the missile are burning exhaust gases reflected from the flame bucket of the launch stand. Typical LPREs are depicted in Fig. 7.7-18 and 8.4-7. Many large ballistic missiles have already been taken out of service by treaty or by becoming obsolete, and some more are scheduled to be decommissioned. In the United States this included the Redstone, Atlas, Thor, Jupiter, and Titan II missiles and in Russia the R-1, R-2, R-3, R-12, and R-14. Today several countries,

Fig. 2-9 The Atlas was the first U.S. intercontinental ballistic missile. Courtesy of U.S. Air Force; copied with permission from second edition of Ref. 2.

such as India, Russia, or China, still have some of these ballistic missiles with LPREs in service as a deterrent.

There were other applications of LPREs, such as the propelling of reusable rocket sleds[10] (to conduct experiments under high deceleration) or the short-term generation of hot gas for power, as was used in the launching ramps of the German V-! missiles. They did not fly and are not discussed further.

2.3 Current Applications

Today there are several principal areas of current LPRE applications, and they are all concerned with space flight.

Space Launch Vehicles

Space launch vehicles (SLV) for space missions and exploration require single or multiple LPREs with relatively high thrust, typically between 7 and 450 tons (15,000 and 1,750,000 lbf) for each engine.[1,2,9] The launch-vehicle booster or lowest stage usually has the highest total thrust value. High specific impulse and high reliability were very important in this application, but accurate cut-off, clean exhaust, modest throttling, environmental compatibility, or engine-out capability also were significant. Boosters typically use liquid oxygen as oxidizer and a hydrocarbon or liquid hydrogen as fuels. The upper stages have the higher nozzle exit area ratio. They give the best vehicle performance when liquid hydrogen and liquid oxygen are used as propellants, but other propellants have occasionally also been used. Figure 2-10 shows a 2002 night launch of Boeing's Delta IV space launch vehicle. This version has two solid-propellant strap-on booster rocket motors, and its principal booster engine is the Rocketdyne RS-68 LPRE.

Some SLVs use storable propellants, usually because they originally were are ballistic missiles, that have been decommissioned and converted for space use. Almost all large engines use pump feed systems and condition monitoring systems. The Space Shuttle Orbiter has been the only reusable SLV, and its several LPREs are designed for multiple flights. It is shown in Fig. 2-11, and all of its LPREs are identified and also can be found in Chapters 7.7 and 7.8. LPREs for SLVs are discussed in general terms in Chapter 4.2, 4.3, and 4.6, their propellants in Chapter 4.1, and specific engines in Chapters 7-14.

The largest space launch vehicle with six stages was the U.S. Apollo/Saturn V, which successfully completed several moon missions, and the Soviet N-1 space launch vehicle, also intended for a moon mission. These large multistage vehicles are a propulsion man's dream. For example, the Apollo/Saturn has 83 propulsion systems; 66 were LPREs, and 17 were solid-propellant rocket motors. The N-1 program did not reach the moon, and the program was canceled after four unsatisfactory flights. The LPREs used in both these large vehicles are discussed in this book

The former Soviet Union has developed multiple *booster strap-on stages with LPREs* for augmenting the booster stage of a ballistic missile and of several of their space launch vehicles. These strap-on stages were first used in the first Soviet ICBM, identified as R-7. The strap-on boosters increase the payload and overall vehicle performance, and they are dropped off first during the ascent. China and India have copied the concept and developed liquid-propellant strap-on boosters for their SLVs. The United States has preferred large solid-propellant strap-on boosters, such as those shown in Fig. 2-10.

Merits of Liquid Propellant Rocket Engines and Their Applications 17

Fig. 2-10 One version of the Delta SLV with two solid-propellant strap-on boosters during takeoff. Courtesy of The Boeing Company copied from a company brochure.

In the United States and in the former Soviet Union, large surface-to-surface military ballistic missiles were developed in the 1950s. Ballistic missiles usually use storable propellants. After they were decommissioned, they were converted and slightly modified to serve for the early space-launch-vehicle missions. Some of these converted launch vehicles, such as the Titan II, were still being used at the time of this writing, but they are gradually being replaced by more modern space launch vehicles. For small payloads and low

18   History of Liquid Propellant Rocket Engines

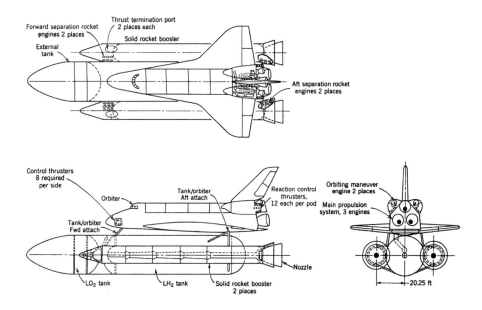

Fig. 2-11 Space shuttle with its reusable orbiter vehicle, solid-propellant boosters, and external tank. The orbiter is the size of a medium-sized commercial aircraft and was designed for 100 flight missions with maintenance and parts replacement. Courtesy of NASA; copied with permission from Ref. 2.

Earth orbits some storable LPREs and some solid-propellant rocket motors have also been used in multistaged vehicles. For heavy space payloads and deep-space missions the higher performing cryogenic propellant engines are universally preferred.

Spacecraft or Satellite Propulsion for Major Space Maneuvers

Spacecraft or satellite propulsion for major space maneuvers, such as orbit injection, retroaction while approaching the moon or a planet, correcting the thrust misalignment of the main engines, and some orbit transfers of satellites, requires LPREs with medium thrust levels, typically 40 to 1500 kg (88 to 3300 lbf), high specific impulse, and for some missions the top stage engine needs to be restarted in flight.[2,9] The higher thrust levels (up to 4500 kg or 15,000 lbf) are for planetary or lunar landings, and they also will need deep throttling or variable thrust. The Apollo moon landing LPRE of Fig. 7.9-7 had approximately 10,000 lbf thrust and a 10-to-1 throttling capability. If the spacecraft is flying for more than perhaps a day, then storable propellants are necessary because cryogenic propellants would suffer from excessive evaporation losses. The preferred storable oxidizer was initially nitric acid, but it was replaced later by nitrogen tetroxide (NTO). The preferred fuels were hypergolic (self-igniting), such as hydrazine, monomethylhydrazine (MMH), unsymmetrical dimethylhydrazine (UDMH), or a mixture of these. Some of these major maneuver engines developed in the former Soviet Union use a turbopump feed system, whereas the rest of the world seems to prefer pressurized gas feed systems.

Reaction Control and Minor Maneuvers

The reaction control and minor maneuvers (flight trajectory adjustments) of satellites, spacecraft, space stations, and upper stages of launch vehicles are usually accomplished by low thrust propulsion systems.[2,9,11] Each LPRE for such a reaction control system (RCS) has a series of small thrust chambers (in a size typically between 0.1 to 1000 lbf thrust), and all usually use pressurized gas feed systems. The thrust size for pitch, yaw, and roll maneuvers depends on the vehicle mass to be maneuvered and the rate of rotary acceleration. These RCS engines are discussed in Chapter 4.6 and their propellants in Chapter 4.1.

This category includes engines for attitude control (pitch, yaw, and/or roll) before, during, and after the operation of a larger principal LPRE on the same vehicle or stage. It also includes maneuvers for propellant settling, station-keeping for overcoming perturbing forces (such as the attraction of the sun or the moon), rendezvous or docking maneuvers, minor flight-path corrections, deorbit maneuvers, momentum wheel desaturation, or spacecraft orientation, such as keeping a solar-cell array, telescope, or antenna pointed in a specific direction. Repeated and random starting, stopping, and restarting, often many thousands of thrust pulses, are needed for some of the maneuvers just listed. Storable bipropellants are usually preferred (because they are stored for long periods and give an adequate performance). However monopropellants (hydrazine) have also been popular because they allow a simpler system with fewer components. Cold-gas propulsion systems (using nitrogen, air, or helium) were used on older systems (extensively in the 1950s and 1960s, but a few have been flown as late as 1985), and they are even simpler, but their performance is low, and their gas tanks are so heavy.

These systems of small thrusters have also been used in controlling the attitude of one or more stages of launch vehicles or missiles, particularly with upper stages. Here they supplement one or more large LPREs with fixed or movable thrust chambers (gimbaled or hinged). In these applications the reaction control systems perform some of the functions mentioned earlier in this subchapter, but they also can perform other duties. This includes compensating for the misalignment of the thrust of the large thrust chambers during operation and also aligning the direction of the thrust of the large thrust chambers prior to their altitude start. It can also include the settling of liquid propellants in their tanks prior to altitude start of the large engines. This prevents the pressurizing gas from entering into the tank outlets and the aspirating of gas bubbles into the feed pipe to the large thrust chambers.

In recent years electrical propulsion systems have been used successfully in several of these applications. However for many maneuvers, such as attitude control during powered flight or rendezvous and docking, LPREs are preferred. Some types of future spacecraft will most likely have both electrical and small liquid propellant thruster systems.

There are two special categories of these reaction control LPREs, namely, for post boost control systems (for achieving an accurate terminal velocity of each of several warheads in a ballistic missile) and for terminal maneuvering systems of defensive missiles.[2] The first system imparts a final velocity (in both magnitude and direction) to each of several warheads of the same long-range ballistic missile. It has a series of small attitude control system (ACS) TCs for pitch, yaw, and roll control and one larger TC for increasing the axial velocity of the stage. They operate with storable propellants, in a pulsing mode, and one Soviet system uses two levels of thrust in the same thrust chambers. An example of such a system is in the fourth stage of the Peacekeeper missile (designed in 1978) shown in Fig. 4.6-20. The second system is found in the homing upper stage of an antiaircraft or an antimissile vehicle (1980s). It has a series of small pulsing ACS thrusters and at least one larger pulsing TC for providing a side force or "divert" force to the terminal stage.

More small thrusters in reaction control systems have been flown than the higher thrust larger LPREs, which usually have turbopumps. The production quantities of small thrusters have been substantially larger. Also the number of organizations worldwide would seem to be larger.

Certain Types of Long-Range Missiles

Certain kinds of long-range ballistic missiles with LPREs are still in service in several different countries.[9] Their LPREs still need support services and replacement parts. For example, Russia still has submarine-launched missiles with LPREs, and China has LPREs in their ballistic missiles. Russia still has some ballistic missiles with storable propellants, but several are being decommissioned. This same application category of propulsion for ballistic missiles with LPREs was also included in Sec. 2.2, Ballistic Surface-to-surface Missiles, in this chapter on obsolete early flight applications; in this chapter it concerned ballistic missiles, which are no longer operational and have been destroyed or the vehicles have been converted to SLVs.

References

[1]Sutton, G. P., "History of Liquid Propellant Engines in the United States," *Journal of Propulsion and Power*, Vol. 19, No. 6, Nov.-Dec. 2003, pp. 978–1007.

[2]Sutton, G. P., and Biblarz, O. *Rocket Propulsion Elements*, 7th ed., Wiley, Hoboken, NJ. 2001.

[3]Newell, H. J., *Sounding Rockets*, McGraw–Hill, New York, 1959.

[4]Schnare, C. W., "The Development of ATO and Engines for Manned Rocket Aircraft," American Rocket Society Preprint No. 2088-61, 1961.

[5]Ehresman, C. M., "Liquid Rocket Propulsion as Applied to Manned Aircraft in Historical Perspective," AIAA Paper 91-2554, 1991.

[6]Ehresman, C. M., "The M.W. Kellogg Company's Liquid Propellant Rocket Venture," AIAA paper 2002-3584, July 2002.

[7]Jensen, G. E., and Netzer, D. W., *Tactical Missile Propulsion*, Progress in Astronautics and Aeronautics, Vol. 170, AIAA, Reston, VA, 1996, pp. 33–55.

[8]Gatland, K. W., *Development of the Guided Missile*, A Flight Publication, London, 1952.

[9]*Jane's Space Directory*, Jane's Information Group, Coulsdon, Surrey, England, U.K. (revised about every two years, comprehensive information on liquid-propellant rocket engines, space launch vehicles, and missiles), 1995–1996, 1997–1998, 1999–2000.

[10]Emery, E. M., and Fletcher, R. M., "New Rocket Sled Developments," *Missile Design and Development*, Vol. 4, Feb. 1958, pp. 10, 11.

[11]Sutton, G. P., "History of Small Liquid Propellant Rocket Thrusters," JANNAF, May 2004 (available through the Chemical Propulsion Information Agency).

Technology Trends and Historical Changes

Throughout this LPRE history one can discern some technical trends and growth patterns. Eight such trends, as seen by the author, are briefly listed next, and their order has no significance. There are no references listed in this chapter because all of the items mentioned here, such as specific LPREs or particular technical subjects, are described in another chapter, where suitable references are listed.

3.1 Expanding the Range of the Thrust, 0.01 to 1,800,000 lbf

The thrust magnitude is dictated by the application. The first thrust chambers (Goddard 1921–1925) had between 40 and 100 lbf thrust. Some of his early small TCs are shown in Fig. 5.2-1. Historically the thrust levels went both up and down. Between 1944 and 1950 a series of hydrogen-peroxide monopropellant thrust chambers (for reaction control) became available with thrust values as low as 0.1 lbf. With inert gas or with stored heated gas as the propellant, the thrust levels went even below 0.01 lbf. It took 40 years to increase of the thrust from about 100 to 1,500,000 lbf (at sea level) with the F-1 LPRE as seen in Table 3.1-1. The highest thrust for a flying single LPRE is 1.64 million lb (sea level), and it came about a dozen years later with the RD-170 engine (Russia), which can be seen in Fig. 8.4-15 and 8.4-16. The highest thrust pump-fed LPRE, which has been ground tested, is the F-1A at 1.8 million lb thrust, but it never flew, and the program was not continued.

The highest vehicle takeoff thrust in a cluster of engines was with the ill-fated Soviet N-1 moon vehicle with 30 Kutznetsov NK-15 booster engines at a total thrust of 10.2 million lbf. Launches began in 1969. This is the largest number of large LPRE and the highest total thrust in any booster stage. In the United States the highest takeoff thrust was with five F-1 engines at total of approximately 7.6 million lb for the booster stage of the Saturn V SLV, and the largest number of main engines in a stage was eight (Saturn I or Black Arrow from the United Kingdom). There were no application requirements for higher thrusts since about 1970.

Table 3.1-1 Historical Increases in Thrust Level of LPREs, which have Flown

Initial ground test year	No. of TCs per engine	No. of LPREs in cluster	Thrust, lbf per LPRE (sea level)	Engine ID and vehicle application	Developer
1923	1	1	40–100	Experimental	Goddard
Up to 1939	1	1	150–2,000	Sounding rockets	Goddard
1936	1	1	3,000	A-3, A-5 test vehicle	German Army
1939	1	1	54,300	German V-2 engine and missile	German Army
1947	1	1	57,800	Soviet RD-100/R-1 SRBM	Energomash
1948	1	1	81,900	Soviet RD-101/R-2 SRBM	Energomash
1948	1	1	99,000	Soviet RD-103M/ R5-M MRBM	Energomash
1953	1	2	120,000	Navaho G-26/ Boost-glide missile	Rocketdyne
1955	1	2	135,000	Early Atlas booster/ICBM	Rocketdyne
1955	4	4	184,000	Soviet RD-107/R-7 ICBM	Energomash
1960	4	4	320,000	Soviet RD-111/R-9A ICBM	Energomash
1963	1	5	1,520,000	U.S. F-1/ Saturn V Booster	Rocketdyne
1977	4	4	1,640,000	Soviet RD-170/ Energiya	Energomash

3.2 Increasing the Chamber Pressure

The historical trend has been to raise the chamber pressure in pump-fed LPREs. This makes it is possible to increase specific impulse between 4 and 8%; the values depend on the specific design, chamber pressure, nozzle configuration, and application. Goddard started (1920s) with relatively low chamber pressure, typically 50 to 100 psi, but he later went up to 350 psi with a pump feed system in 1939. In the 1940s some gas-pressurized feed systems allowed increases to 400 psi. With pumped feed systems these pressures reached 1000 psi in the 1950s. There were some exceptions. Some small experimental TCs were tested at more than 5000 psi in the 1970s. The highest chamber pressure of a flying U.S. engine was 3319 psi in Block I of the space shuttle main engine (SSME), whose development started in 1972. The USSR had several engines with chamber pressures between 3300 and 3800 psi and one that was higher than 4000 psi. The higher pressure allows the TC to be smaller, which makes it easier to place into a vehicle. For the same engine length the smaller TC allows a higher nozzle-area ratio. Both the increased pressure and the larger nozzle-area ratio increase the specific impulse. The higher pressure also requires a higher nozzle-area ratio for nozzle flow separation to occur at sea level. This flow separation phenomena will be discussed later. There were some disadvantages, which prevented going to even higher values. Because heat transfer increases approximately linearly with the chamber pressure, the cooling of TCs becomes much more difficult at higher chamber pressure. If considerable supplementary film cooling is required to keep the TC walls relatively cool, the engine performance increase can be small or negative. The power of the turbopump (TP) and the amount of gas flow needed to drive the turbines increase with chamber pressure, as does the mass of the TP. Also pipes, pump housings, or valves need thicker walls and the engines become heavier. The optimum chamber pressure for a given application is usually a compromise between the higher performance, the increases in inert engine mass, and the possible development problems.

3.3 So Many Liquid Propellants

In the first two decades of this history, a very large number of widely different chemicals were evaluated for becoming a potential propellant, and in 80 years the selections were narrowed down to just a few practical propellant combinations. The effort to seek better liquid propellants was large and conducted in several countries. More than an estimated 1800 different liquid propellants and more than 2000 bipropellant combinations have undergone laboratory evaluations, and about 300 were also tested in small TCs. More than 40 different propellant combinations have been flown, each at least once. The analytical and laboratory evaluations of different propellants centered around several themes: determining the physical, chemical, and combustion properties of promising propellants; finding or synthesizing propellants that will give the highest possible engine performance; finding or synthesizing propellants with a high density (results in smaller and lighter vehicle tanks); compatibility with materials of construction; achieving the fastest hypergolic ignition over a wide range of temperatures and ambient pressures; lowering the freezing point or raising the boiling point to conform to the operational temperature limits; enhancing storage stability, or making the exhaust gases environmentally harmless. The effort involved many organizations, such as most of the LPRE companies, some R&D organizations, and some chemical or oil companies. More information on propellants is given in Chapter 4.1.

Before and during War II (1935–1945), the Germans investigated more than 1100 different propellant combinations for hypergolic ignition as described in Chapters 4.1 and 9.1. Other countries also looked for some of the same hypergolic propellant combinations and also some new compounds, but the effort was with fewer combinations and occurred well after 1945. The intensive search for better hypergolic fuels or self-igniting storable fuel mixtures did not result in compounds that were significantly better than hydrazine and its organic derivatives MMH or UDMH.

A big international effort was made to discover propellants that would give higher specific impulse and/or higher density. The United States and the Soviet Union investigated at least 100 novel toxic energetic and exotic chemicals as potential high-energy propellants for upper stages. Theoretically fluorine or compounds of fluorine, boron, and chlorine gave very high performances, and therefore they were investigated. New propellant formulations were synthesized. Several of these high-energy propellant investigations went beyond the small TC test stage and were ground fired in larger experimental TCs. Complete engines have been operated and ground tested with fluorine as the oxidizer, but they did not fly. The large effort on high-energy chemicals was abandoned (around 1980) because of the toxicity and handling difficulties of these energetic materials and their potential drastic effects on people, machinery, or the environment in case of an accident, such as a tank car derailment, failure of a LPRE during ascending flight, or a major spill.

The only high-energy propellant combination that was accepted to be practical was liquid oxygen/liquid hydrogen (LOX/LH$_2$), and it has been used extensively to this day. It has been investigated experimentally since 1945. High specific impulse is very significant in space missions, where the cumulative mission flight velocity is high. Here even a small increase in specific impulse leads to major increases in payload or orbit height. Vacuum specific impulse values between 410 and 467 s have been achieved with different flying engines. It has been preferred for upper stages of SLVs with engines, such as the Pratt and Whitney RL-10 flying since 1963 and discussed in Chapter 7.10 or the Rocketdyne J-2 LPRE used in Saturn V shown in Fig. 7.8-13. The extra performance usually overcomes the disadvantage of the low density of LH$_2$, which means very large insulated fuel tanks, extra tank weight, and more drag.

There was no discernible trend in the early propellant selections, and each project team, company, or government agency picked the propellants they thought most suitable for their application. All of this effort did not lead to a universally acceptable single liquid propellant combination. All propellant selections for flying vehicles were a compromise between good qualities (high performance, high density, high specific heat for good cooling, easy start, low cost, low vapor pressure, stable combustion, or stable, long time storage) and bad qualities [corrosive, flammable, toxic, smoky exhaust plumes, prone to decay during storage, high vapor pressures (difficult to pump), high freezing point, deposits from nozzle exhaust plume on sensitive spacecraft surfaces, such as mirrors, or a tendency toward combustion instability].

Over the decades much experience was accumulated about the logistics or the handling of the propellants, their behavior during flight, or the consequences of accidents. Gradually the industry began to concentrate on certain propellant combinations for certain applications. After years of operational experience, a few propellant combinations seem to have emerged as being practical to use with current space applications and they are listed in Table 3.3-1.

There are a few exceptions to these five propellant combinations. For example, some organizations believe there is a good future for hydrogen peroxide as a storable oxidizer and are spending some of their resources on it. Some of the older propellant combinations, such as nitric acid, are still used in some countries with old designs of vehicles and engines. Also there are dual-propellant reaction control systems, which combine hydrazine monopropellant thrusters of very low thrust with a few larger thrust bipropellant thrusters using NTO/hydrazine for the major space vehicle maneuvers. One small company and some amateurs prefer to use LOX/alcohol, which was a favorite combination in the first four decades of this history.

3.4 Engines Can Come in Families

There has been a trend for many of the LPREs to be a part of an engine family. Once a good and proven LPRE has flown satisfactorily, it is often common practice to develop variations or modifications of this engine. It has always been easier and usually also faster and cheaper to change or up-rate an

Table 3.3-1 Practical Current Propellants and their Applications

Propellant	Application
LOX/kerosene (RP-1)	Some SLV booster stages
LOX/LH$_2$	Most SLV upper stages and some SLV booster stages
NTO/MMH (preferred in United States), or NTO/UDMH (preferred in Russia)	Attitude or reaction control systems, (for orbit change, reentry, or space rendezvous); planetary missions, terminal maneuvers, reentry, or post boost control systems, some missiles
Hydrazine monopropellant	Some reaction control systems

engine and keep the basic proven engine design concept and key components, rather than to make a new engine. These engine changes are usually triggered by changes in the vehicle (e.g., more payload) or the mission. A modification of an existing engine allows substantial savings in development, proof testing or qualification testing, in manufacturing operations or tooling, in using qualified vendors, or the inventory of spare parts. There also is usually less uncertainty in achieving the performance, reliability, or cost targets. A modification of a proven engine is also usually more acceptable to the vehicle developer company or the customer. The following kind of changes are typical: higher thrust, higher chamber pressure, include deeper throttling, change the nozzle-area ratio, use modified or alternate propellants, reduce the design complexity, or reduce the cost. Certainly only some and not all of the LPREs are in families.

For example, the Titan booster and sustainer engines went from the Titan I version (300,000 lbf nominal thrust at sea level (SL) with LOX/kerosene), to the Titan II (430,000 lbf and NTO/50% N_2H_4/50% UDMH), to two versions of the Titan III (450,000 lbf), and to the Titan IV [478,000 lbf thrust at SL and 548,000 lbf vacuum condition (vac)]. The Soviet RD-107 and RD-108 engine family has been used for an ICBM and a SLV. About 17 known versions exist with slightly different performances and for different vehicle applications. The RD-4 TC developed by Marquardt (today Aerojet) of about 100 lbf thrust has a number of different versions, with different materials, nozzle-area ratios, insulations, slightly different thrusts, or alternate propellants. The family relationships of the large booster LPREs used for ballistic missiles in the Soviet Union are explained in Chapter 8.4.

3.5 Thrust-to-Engine-Weight Ratio Has Gone Up

This parameter is the nominal thrust divided by the dry weight of the engine. An increase in this number means that the engine dry mass has gone down or that the engine can produce more thrust for a given engine mass. It has been a widely accepted measure of the effectiveness of the design. Over the 80 years of history, this ratio has changed from a range of about 0.2 to 20 in the 1930s to about 50 to 170 units of thrust per unit sea level weight. Other parameters have also been used. The thrust-to-weight ratio is not an accurate parameter because the definition of what is the nominal or average thrust or what is included in the engine weight varies greatly between manufacturers and between applications. For example, some experts define the engine mass to include the mass of residual propellants, unused pressurization gas, gimbal actuators, instrumentation, and even some parts of the vehicle structure. Also it is very difficult to make a valid and accurate comparison between two engines, because the engine parameters (nozzle-area ratio, mixture ratio, chamber pressure, etc.) are usually different. Nevertheless, there has been a noticeable trend to reduce engine weight for a given thrust. Lower engine weights have resulted in improved vehicle mass ratios and thus in improved vehicle performances.

The improvement in this parameter has been because of better and stronger materials (allowing higher stresses and thus lighter-weight components). There has been good progress over the years in finding or developing better and stronger materials for specific engine components. Improvements also came with clever design, fewer parts, less variation of physical properties between material batches, which allows a reduced safety factor, or a better understanding of the failure modes for each of the key engine components. Contributing to the reduction of inert engine mass has been a trend toward lighter turbopumps, with better suction characteristics. In turn, this allows a higher pump and turbine shaft speeds, which leads to smaller diameter and lighter TP housings and rotary assemblies. The application of inducer impellers, booster pumps, or ejectors has avoided the debilitating cavitation and helped to increase the shaft speed of the main high-pressure TPs. Similarly, other engine components have been reduced in mass. This improvement over the decades has been truly amazing.

3.6 Costs Have Gone Down

It is very difficult to make rigorous comparison of engine costs. There are many factors that have to be considered when comparing the cost of an older engine with a newer engine, even when at the same thrust, chamber pressure or propellants. It includes factors such as the quantity of engines fabricated in a production run, the experience of the manufacturing organization, inflation, number of acceptance tests, the use of experienced suppliers, or variations in the material or labor costs. Therefore no specific data will be quoted here. The author, after discussions with accountants and others in this business, believes that there has been a trend for engine costs to decrease. Certain general measures such as sales per employee show the same trend.

It is reasonable to expect costs to go down because the people doing the work are more experienced (learning curve); the use of computers in the development, testing, design, fabrication, or field service has led to real savings, improved manufacturing processes, better tooling; or inspection techniques have reduced fabrication man-hours, or excess material has been trimmed off.

3.7 Extra Functions Have Increased Engine Complexity

Whenever the requirements of a mission or application demand extra capability or a new function in an engine, the engine has usually become more complex, and this implies more hardware, more testing, and/or more software. There is a noticeable trend to add capabilities in each category of LPREs. The early pioneers in the 1930s in the United States, Germany, or the Soviet Union ran simple basic LPREs, which operated steadily at essentially constant thrust for a single operation [jet-assisted takeoff (JATO), sounding rockets]. They typically had pressurized feed systems, a single thrust chamber, valves for start and stop, some instrumentation to measure a few engine parameters, and provisions for filling and draining propellant tanks, for filling and venting the high-pressure gas, and for cleaning or flushing the propellant system after a ground test.

The primary function of any propulsion system is to provide thrust or acceleration to a vehicle for a certain time. Most of these extra capabilities listed next were not aimed at providing thrust. Following is a list of extra engine functions that make the engine more complex, and the items are not in any particular order. Each will be discussed again later in this book:

1) Restart (at low altitude) requires some control changes and probably some flushing or purging of residual propellants in larger engines. Restarts were introduced with aircraft rocket engines beginning in the 1930s.

2) Restart in the gravity-free environment of space flight requires the settling of liquid propellants in the tank so that gas would not be allowed to enter the feed pipes. This became an important issue in the 1950s.

3) Many, many restarts or pulsing of small thrust chambers for attitude control means a different design with multiple TCs. These LPREs first flew in the 1950s.

4) Using pump feed, rather than a pressure feed system, will make the engine more complex and heavier, but often really reduces the tank weights and usually results in an overall performance improvement of the vehicle. Pumps are usually for larger engines and/or longer cumulative operating durations, and they require a power source of hot gas for driving the turbine, and this makes an engine even more complex.

5) Thrust-vector-control devices can control the flight path during the operation of the principal engine. It usually needs actuators, extra power, and more hardware and software.

6) Random variable thrust upon command is needed only in a few applications, such as lunar landing. Controls are more complex, particularly with pump feed systems. It was first implemented with some aircraft rocket engines in the 1940s.

7) Thrust control or keeping the thrust within predetermined, often narrow limits is relatively easy with a pressurized gas system, but more complex with a TP feed system. It was first considered by Goddard in the 1920s, but was not incorporated into flying pump-fed engines until the 1940s. It allows a predetermined flight trajectory. It usually results in more complex vehicle controls.

8) Minimizing the amount of residual or unused propellant avoids a penalty of a heavier final engine mass and a loss in vehicle payload or performance. It is commonly called mixture ratio control. It requires the measurement of

the propellants remaining in the tanks at any one time and at least one throttle valve and some software.

9) Reusing the engine for another flight implies extra postrun inspections, cleaning, maintenance, and check-out procedures. The aircraft engines of the 1940s and the reusable JATO units needed these features.

10) Condition monitoring or health monitoring before and during operation adds extra sensors, but can allow the correcting of abnormal unsafe operating conditions or cause a safe engine shutdown. This was first done properly in the 1960s and 1970s.

11) Side accelerations during a flight require provisions in the propellant tanks (such as baffles) to minimize sloshing and prevent gas from entering the feed line to the engine. The first baffles were installed by Goddard in the 1920s, and several positive expulsion devices were developed a little later.

12) Man rating of an engine can include some component redundancy, extra sensors, sophisticated control logic, and some manual overrides of some of the engine controls.

13) By providing "warm" gas for tank pressurization (instead of a cold stored pressurized gas), the heavy separate high-pressure inert gas tank and supply system can be largely eliminated, thus allowing a reduction in inert weight. Examples would be gas tapped off the exhaust of a turbine or directly from a gas generator or a preburner and subsequently cooled. Gasified cryogenic propellant, which has been heated in a TC cooling jacket or in a heat exchanger, has also been used. It requires some additional engine hardware and controls.

14) To achieve a precise vehicle flight terminal velocity, the engine must have accurate, predictable thrust termination with a known repeatable total impulse during the thrust decay period. This is needed only in some applications, such as in the terminal stage of a SLV or ballistic missile. Some larger engines have used a reduced thrust phase for a short time during thrust cutoff, which reduces the uncertainty in the cutoff total impulse. This usually leads to a more sophisticated shut-off control.

15) In an engine cluster (two or more engines) one can provide an engine-out capability, by prematurely shutting off one of the engines, which has gone outside its normal operating parameters, and completing the flight with the remaining engines. This was first implemented in the 1960s and 1970s.

16) For propellants that are not hypergolic, an ignition system is needed. Substantial improvements in ignition systems were made in the 1920s and 1930s.

There might have been other functions that could have been added to this listing, such as the amount and type of measurements or instrumentation. With each added function the engines usually became more complex, and there were, of course, also needs for additional tests during development or qualification, more inspections, and more engineering effort. Most of the current engines incorporate some of these extra features or functions, but there is no known engine that incorporates all or nearly all of the functions just mentioned. The extra functions are no longer novel and have become common in LPREs, particularly with the larger thrust engines.

3.8 Reliable Operations Were Achieved

The early investigators (1925–1955) had many engine failures, and they were pleased if their engine would operate for the planned duration and at the desired performance without malfunction (outside of expected operating limits) or failure, such as leaks, faulty valves, burnouts, or explosion. This author remembers a series of runs in the 1940s on a small TC, where about every third run had some problem that would have caused the mission not to be completed. By the 1960s the state of the art of the technology had advanced far enough so that some engines would fail only once during 500 runs. Most of the random failures had been encountered and remedied, and the reliability of the components has also been improved tremendously. Although this was a vast improvement, it was not good enough. Large flight vehicles are so expensive that one flight failure would pay for a lot of engine R&D, and therefore a better reliability was justified. Today many engines have had a 100% reliable flight record and an estimated ground-test reliability of 0.999 (1 failure in 1000 operations). Of course in order to achieve this reliability in the old days with full confidence, it would be necessary to test this engine or a set of identical engines for 1000 runs with only one malfunction, and this is expensive. So the imputed reliability is estimated with a certain degree of confidence, based on data from fewer actual engine runs, prior data from similar engines or components, and from R&D engine test data. Even with these statistical shortcuts, the reliability of LPREs has greatly improved.

There were other trends, such as in cooling methods, nozzle shapes, or specific design features, and they are treated in later chapters.

Technology and Hardware

4.1 Propellants, Past and Present

This chapter is an expansion and continuation of Sec. 3.3 "So Many Liquid Propellants." The liquid rocket propellants are the working fluids of LPREs. All categories of propellants have been used, namely, *bipropellants* (liquid oxidizer and liquid fuel), *monopropellants* (single energetic liquid or a mix of liquids that is decomposed and gasified by a catalyst), *self-igniting hypergolic propellants* or *nonhypergolic propellants*, which require a separate means of ignition, *cryogenic propellants* (liquid only at low temperature), and *storable propellants* (liquid at ambient temperature).[1] So many different liquid propellant combinations have been investigated in the past 80 years that it is just not possible to discuss very many of the propellants here.

In the early decades of the LPRE history, there did not seem to be any pattern or trend in the propellants used by different organizations or countries. Early investigators and/or their government sponsors selected those propellants that were known to them, were believed to be most suitable for the specific applications involved, were available, and had some favorable background experience. As already mentioned in Chapter 3.3, there is no single ideal liquid rocket propellant that fits all applications. All propellant selections for flying vehicles were a compromise between the known propellant *good qualities* and the *bad qualities* and depended on the application. Most liquid propellants have unique qualities (described in Chapter 3.3) and give a higher performance than solid propellants, but all liquid propellants have some inadequacy, undesirable physical properties, or potential hazards. Many were either toxic, had low density, were corrosive, flammable, potentially explosive, had undesirable ignition characteristics, too high a freezing point, limited heat-absorption capacities, high vapor pressure (poor pumping), poor storage stability, or were prone to combustion instability. Because rocket propellants by their very nature are highly energetic substances, it is impossible to make them totally harmless materials. The selection of the liquid propellants for specific applications has therefore always involved compromises of the risks, performance, costs, and the desirable qualities needed by the requirements of the application.[1-4]

Early Propellant Investigations (1921–1940)

The early investigators used a variety of different propellants, and there does not seem to be a pattern to their selections. In the Soviet Union the first few LPREs used NTO/toluene, nitric acid/gasoline, and LOX/gasoline as described in Chapters 8.1 and 8.2. In 1935 the Germans selected LOX/75% ethyl alcohol for the V-2 as described in Chapter 9.1. This propellant combination was then used after the war for the early large LPREs in America and in the Soviet Union. In the United States, Goddard used LOX/ether for his early 1923 tests but then switched to LOX/gasoline; Reaction Motors, Inc. (RMI) built its first engine with LOX/alcohol. GALCIT (Graduate Aeronautical Laboratory of the California Institute of Technology) used nitric acid with either hydrocarbon or alcohol in the 1930s and subsequently aniline as a fuel. Most of the propellants mentioned in this paragraph are no longer being used for reasons that will be explained.

As already stated in Chapter 3.3, more than *1800 different liquid propellants* have undergone laboratory investigations. Probably more than 300 bipropellant combinations have been test fired in small trust chambers, and all of the LPRE organizations have been involved to some degree. Most of these propellant investigations were performed between 1923 and 1965. More than 40 different propellant combinations have been flown worldwide, each in at least one LPRE. After all of these investigations and after many decades of operating experience, less than a dozen propellant combinations have emerged as being practical to use in specific applications. They are listed in Table 3.3-1.

A good number of propellants that were popular during the early days of LPREs are today no longer used. Here are two examples: nitric acid, and particularly red-fuming nitric acid, which was a favorite storable oxidizer in the 1930s and 1940s, is not commonly used in the United States, Europe, or Russia today. This is partly because the evaporation of the nitric oxide and the corrosion of the tank materials (formation of metal nitrates) cause the density and energy content to vary in an unpredictable way, making it difficult to meet tight performance requirements.[5] However inhibited red-fuming nitric acid (IRFNA) is still being used in China. Hydrogen peroxide as a monopropellant was very popular for small thrusters (ACS) and used actively in the 1940s and 1950s.[6] Because of its poor performance and its gradual self-decomposition during storage, it was replaced by more powerful propellants. Some of the propellants selected by early pioneers, just mentioned, such as ether, toluene, or ordinary gasoline, are no longer in use.

Extensive laboratory investigations were done early and mostly in three countries. In Germany (1935–1945) this encompassed an estimated 1800 to 2000 different propellants,[7] in the Soviet Union (number unknown) it was between 1945 to 1970, and the United States about 1300 mostly between 1936 and 1970. In the Soviet Union R&D on liquid propellants was done in part by the State Institute of Applied Chemistry, which was started in 1948. Its work is briefly described in Chapter 8.13. All sorts of strange chemicals or mixtures of two or more ingredients were tried.[5] Some good work was also performed in France, Britain, and other countries. It involved not only LPRE companies and government laboratories, but also academic research groups, chemical companies, petroleum companies, and commercial laboratories.[4] The properties of some of the more relevant propellants were collected, measured if missing, and identified.[1-3,5,7,8]

The objectives of this work were several.[2,3,5,7] Many projects were aimed at increasing the combustion energy release, reducing the average molecular mass of the exhaust gases, or better combustion efficiency; any of these would improve the vehicle performance. Other projects were to reduce the freezing point of certain propellants (field operations require liquids down to −65°F initial temperature) or to raise the boiling point of other propellants (above 160°F). Searches for increasing the density (smaller lighter-propellant tanks) included the synthesis of new compounds. It was also important to select the most suitable materials of construction and determine and improve the compatibility of the propellants with the hardware materials of LPREs and tanks. Some propellants needed better storability or minimizing their deterioration or self-decomposition when stored for long periods with variations in the operating temperature range. Furthermore there was a big effort to minimize the time delay of self-ignition for hypergolic propellant combinations, particularly at low temperatures. Considerable experimental work was done to determine or confirm the physical and chemical properties of propellants and the effect of impurities on these properties. Some investigations were aimed at improving the process for manufacturing some of the propellants. Other efforts were aimed at enhancing the ability of the coolant to absorb more heat without vigorous boiling of self-decomposition or at reducing the vapor pressure to avoid excessive cavitation in the pumps. For monopropellants the investigations had additional objectives, such as desensitizing certain liquid propellants during storage (minimize self-decomposition), improving the safety of handling, and avoiding unscheduled explosions.

Some of the conclusions of these investigations were these: the materials (common metals or plastics) compatible (or incompatible) with specific propellants were identified. The important chemical and physical properties of propellants were identified and validated.[7] Several pieces of data in old literature were found to be incorrect. On several occasions the author has been involved in incidents where (by error or ignorance) the wrong material was allowed to come in contact with certain propellants. For example, the wrong gasket material was assembled in the discharge flange of a RFNA pump during a test. During the lunch period, the acid had eaten out the gasket, but this was then not known. When the pump was turned on again after lunch, a spray of nitric acid squirted through the test bay, and the three-man crew, including this author, was sprayed with RFNA. Fortunately the crew stood in a location, where the spray was in droplets and not in jets. Quick stopping of the pump and dousing ourselves with water prevented more severe consequences. The company paid for new pants and new shirts to replace those that were ruined by acid.

Here are more conclusions: all of the many high-energy storable propellants that were investigated or synthesized did not give a vehicle performance or physical/chemical/handling properties substantially better than those propellants that were commonly used in the 1960s and 1970s. Many compounds were discovered to give acceptable hypergolic ignition with storable oxidizers, but none resulted in shorter ignition delays than the popular fuels, namely, hydrazine, UDMH, or MMH.[7] The addition of dense solid suspended particles (such as small perchlorate crystals) to an oxidizer has generally not been successful. All monopropellants that were formulated and evaluated would explode, if stimulated with enough energy (heat, shock, or impact) or with contaminants that usually acted as catalysts. The exceptions were hydrazine and hydrogen-peroxide monopropellants, where it was learned to contain the risk of detonation to an acceptably low level. Both are described briefly in this chapter.

Water was used as an additive or as a separate third propellant in some of the early experimental engines. This was done in part because cooling with the propellants alone was then not satisfactory with the early cooling jacket designs of that period. Although the addition of water decreases the specific impulse, it did increase the propellant mass flow and thus the thrust. Alcohol was diluted with 25% water in the German V-2 (1940) and also in the U.S. Redstone engine (1950) and the RD-100 Soviet engine (1948). A few years later the water content was reduced and the ethyl alcohol content raised to 92.5%.

The French used water-diluted hydrazine or hydrazine hydrate instead of pure hydrazine in some of their early LPREs. The French Viking engines of the Ariane 4 SLV (Chapter 7.4) continued to use a storable propellant consisting mostly of UDMH, but it also included some water. The Germans in 1936 diluted their "C" Stoff fuel with water (mix of hydrazine and alcohol with 13% water) in order to limit the maximum combustion temperature to a tolerable level. All of these engines can be found in later chapters. Several experimenters (including Goddard of the United States) used water directly as a coolant and then injected the warmed water into the combustion chamber. It increased the thrust, but reduced the specific impulse to an unacceptably low level. The Soviets and Goddard added water to their first gas-generator (GG) propellant flow to reduce the gas temperature. Eugen Sänger used to evaporate the cooling water of his early TCs, and the resulting steam was used to drive the turbine. Because of the performance penalty, the use of water as an additive or as a third propellant has since been discontinued.

Self-Igniting or Hypergolic Propellant Combinations

The discovery of fuels, which ignited spontaneously upon contact with an oxidizer, (called hypergolic propellant combinations) was a milestone in the early history of LPREs, and it led to several new propellants. This was first discovered in Russia in about 1933, independently in Germany about 1936, and independently in the United States about 1940. The word "hypergolic" was mentioned in the literature in 1936, and it was coined by Dr. Noeggerath in Germany in 1935 (Ref.5). It allowed ignition without a separate pyrotechnic igniter, a simpler starting system, and a simpler multiple restart operation.[5] In the United States GALCIT and the Navy Annapolis researchers discovered aniline to be hypergolic with nitric acid in 1940. A large variety of other propellants was identified as being hypergolic, was investigated in a number of different countries between 1936 and 1970, and some propellants were used in LPREs.[5,8]

Between 1935 and 1945 a department of the German Aeronautical Institute headed by O. Lutz performed a thorough investigation of about 1100 different propellants for their self-ignition properties.[7] A large variety of candidate hypergolic fuels was tested in this laboratory. This included various optoles, aldehydes, cyclopentadiene, or furfural alcohol. None ignited as rapidly as hydrazine or hydrazine hydrate. Many of the chemicals were then not readily available in war-time Germany, and some had undesirable physical properties. It was found that some mixtures of hypergolic fuels with nonhypergolic fuels, such as alcohol or certain hydrocarbons, would also allow auto-ignition, often at a lower cost. Concentrated 80% hydrogen peroxide was the oxidizer for the German Walter series of rocket engines, and it needed a hypergolic fuel. Early German experiments were with 50% hydrogen peroxide with 50% water and a fuel mix of 50% hydrazine hydrate and 50% methyl alcohol. A small amount of copper catalyst was added to the fuel. This mix failed at low temperatures at or below −25 C or −13 F. Finally the Laboratory researchers and the Walter Company arrived at C-Stoff (57% alcohol with 30% hydrazine hydrate and 13% water), which was a successful hypergolic fuel used in operational LPREs.[7] The water was added because engineers at the Walter Company, the engine developer, stated that the combustion temperature had to be below 1750° C or 3182° F to prevent engine overheating. More discussion of this effort is in Chapters 9.1 and 9.3.

The German Government Laboratory also tested many hypergolic fuels for igniting nitric acid and also nitric acid with dissolved nitric oxide, which was used by BMW's aircraft rocket engine and by several tactical missiles under development in the early 1940s. Most of the fuels were mixtures of two or more ingredients in order to optimize the ignition characteristics over the range of operating conditions. Several kinds of nitric acid were tested and flown. They were concentrated nitric acid, also called white fuming nitric acid, with 2% water (simplest, lowest cost); red-fuming nitric acid (higher density and higher vapor pressure) with 5 to 30% NO_2, which is a reddish gas that evaporates; and a mix of concentrated nitric acid with 10% sulfuric acid (faster ignition).[5,7] All three acids were very corrosive and toxic. These nitric-acid oxidizers are discussed with specific LPREs in Chapters 7.7 (Aerojet), Chapter 9.4 (BMW), and Chapter 13 (China). Most of the fuels being investigated at that time were mixtures of two or more ingredients in order to optimize the ignition characteristics over the range of operating conditions. A typical result with different proportions of two different fuels is shown in Fig. 4.1-1. Fuels included aliphatic amino compounds (such as di-ethylene tri-amine, xylidine or tri-ethyl amine), unsaturated compounds (such as vinyl ether, diacetylone, or diketenes), and furan compounds (such as furfuryl alcohol) and hydrazine and its organic

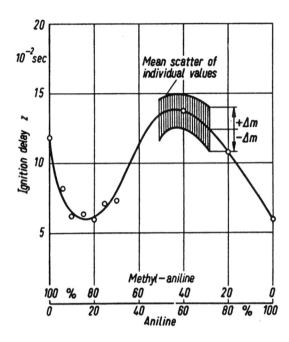

Fig. 4.1-1 Typical variation of ignition delay of a different mixes of two fuels reacting with nitric acid. Copied with American Astronautical Society permission from Ref. 7.

derivatives. The best results or shortest ignition delays were obtained with pure hydrazine, hydrazine hydrate, furfural alcohol, MMH, and UDMH.[7] Most of the propellants investigated for being hypergolic did not give as fast an ignition as these five. All of these hypergolic propellants are toxic materials, and most of them are carcinogenic, that is, cancer-causing chemicals.

It has also been demonstrated (in several countries during the 1940s) that any nonhypergolic propellant combination (such as oxygen/kerosene or nitric acid/gasoline) could be ignited by injecting a small initial amount of hypergolic propellant into the combustion chamber at the start of rocket operation. Once the combustion was started, the propellants continued to burn by thermal ignition from the hot flame, and the further addition of a hypergolic propellant was not necessary. Ignition generally worked well by injecting an initial quantity of some of those five propellants just mentioned. As the ambient temperature is reduced, the ignition lag time becomes longer, it becomes more difficult to obtain reliable ignition, and there is more danger of an explosion of the unignited mixture of propellants in the combustion chamber.

Hypergolic ignition will work with cold liquid oxygen, if one injects a very reactive liquid chemical, such as tri-ethyl aluminum or zinc-di-ethyl, which will burn with air or oxygen. These chemicals are toxic, highly reactive with air, and must be kept sealed. Some are light sensitive, and most are difficult to transfer or handle.[8,*] In spite of these disadvantages, many LOX/kerosene LPREs and some LOX/LH$_2$ engines have used and are still using this method for starting.

Tailoring Propellants with Additives

The tailoring of liquid propellants[1] with additives for specific applications started in the 1930s in Germany. In part it was the result of the intensive laboratory effort already mentioned. The density of nitric acid was increased by adding up to 30% NO$_2$ or nitric oxide. The chemists say that NO$_2$ is in chemical equilibrium with N$_2$O$_4$ (NTO) in the liquid acid, and the addition of either of these is equally effective. This mixture gives off toxic reddish fumes of nitric oxide, and it thus became red-fuming nitric acid or RFNA. It was first achieved by Friedrich Wilhelm Sanders, a German in 1931 and first applied to propellants in Germany in the 1930s.[5] Another example is Aerozine 50, a mixture of 50% hydrazine and 50% UDMH; it was developed at Aerojet as a compromise on freezing point, density, and performance and used on the Titan II, III, and IV first- and second-stage rocket engines and the Apollo service and command module engines. The liquid temperature range of NTO was extended by adding between 2 and 20% nitrogen oxide (NO). This was called mixed oxides of nitrogen; the freezing point was lower, and many of them had an unfavorable high vapor pressure.

Beginning around 1950 a small amount of fluorine ion (in the form of 1/2 to 1% of HF or hydrofluoric acid or alternatively a soluble fluorine salt) was added to nitric acid or RFNA to reduce corrosion of the walls of tanks, pipes, or valves.[5,*] This additive forms a thin fluoride coating, which inhibits further corrosion, on the walls of aluminum and steel tanks. This acid has since been called inhibited red-fuming nitric acid or IRFNA, and it has been used extensively in acid propellants in most countries that have used nitric acid oxidizers. Prior to the fluorine addition, the concentrated nitric acids would react with the stainless-steel metal of the tank wall material and form an unknown amount of iron nitrate, chromium nitrate, and nickel nitrate; often these nitrates appeared as jelly-like constituents of the acid, and the amount of these nitrates increased with storage time. These nitrates changed the physical properties of the acid and reduced the performance or the specific impulse. This addition of fluorine ion to the acid was discovered by a chemist at the Naval Air Rocket Test Station in Dover, New Jersey, in 1951[5] and reported in a contract progress report, which was not circulated to the LPRE industry and was probably not read by many people. Ammonium bifluoride was added to the acid, and a substantial reduction in the corrosion of the tank material and a reduction in forming dissolved metal nitrates was obtained. It remained an exciting laboratory discovery, but it did not find its way into a production

---

*Personal communications, V. Wheelock and B. Thompson, The Boeing Company, Rocketdyne Propulsion and Power, 2000.

propellant tank of a LPRE. This additive was also discovered independently in 1952 or 1953 by Irving Kanarek, a chemist at Rocketdyne. It was quickly implemented in various LPREs both in the United States and in other countries.* With some steel tanks iodine has also been used as a corrosion inhibitor with nitric acid, but this remedy did not find its way into very many oxidizer tanks in the United States.

Theoretical Analysis of Performance

The initial crude theoretical thermochemical analyses of the performance of different propellants were undertaken by several of the pioneers in the 1920s and 1930s. These early analyses were only approximate (theory was then not adequate) and were hampered by the lack of good data on thermodynamic properties of the propellants, their chemical ingredients, and of the reaction product gases. More precise analytical work started in Germany in the early 1940s and was also conducted in the United States and presumably also in the Soviet Union. For any given propellant combination it was possible to determine the specific impulse and the combustion temperature at a selected chamber pressure, nozzle-area ratio and mixture ratio.[1] By 1950 it was possible to calculate (by thermochemical analysis) reasonably good values of the theoretical performance of different propellant combinations for different operating conditions. Laboratory work on getting good data on specific heats, enthalpy, heat of vaporization and other properties at different temperatures and pressures was very useful and was continued.[9] A satisfactory analysis method and computer program became available in the 1960s and 1970s in the United States. Investigators at the NASA Lewis Center (today NASA Glenn Center) have developed a good theory, which was implemented by computer programs; it has been adopted by the U.S. industry.[1,10,11] Refinements in the analysis have been made since to include various losses, such as wall friction, chemical equilibrium, or shock waves. Shifting and frozen equilibrium, assumptions are no longer needed.

An example of early results of the theoretical performance analysis is given in Table 4.1-1, which has data excerpted from a Rocketdyne chart prepared sometime during the 1950s covering 54 different propellant combinations.* It applies to optimum expansion going from 1000-psia chamber pressure to the atmospheric nozzle-exit pressure of 1 atm. Most of the data in this table come very close compared to more modern analysis techniques. These calculations were based on adiabatic combustion, isentropic expansion, and one-dimensional flow. There are two values for each propellant combination in the table. The top line of numbers is obtained with a shifting equilibrium assumption (the composition of the exhaust gases varies during the expansion in the nozzle) and the bottom line with a frozen equilibrium assumption. (The composition of the exhaust gases in the nozzle is the same as in the chamber.) The

---

*Personal communications, B. Thompson, The Boeing Company, Rocketdyne Propulsion and Power, 2000.

Table 4.1-1 Sample Data of Early Performance Analysis

Oxidizer	Fuel	I_s	r	T_c °F	d
LOX	LH_2	389.5	4.05	4913	0.28
		365.3	2.5	4503	0.26
LOX	RP-1	300.1	2.60	6160	1.02
		285.4	2.28	5968	1.00
F_2	LH_2	411.1	7.90	6650	0.46
		389.0	4.60	5084	0.33
N_2O_4	MMH	288.7	2.15	5653	1.20
		277.8	1.64	5408	1.16

I_s – theoretical maximum specific impulse.
r – mixture ratio (oxidizer flow rate/fuel flow rate).
T_c – chamber combustion temperature.
d – bulk density (sum of propellant masses divided by sum of volumes), grams/cm^3.

actual value was between these two numbers. The more modern analysis include the change in chemical equilibrium in the nozzle, two-dimensional flow, and sometimes shock phenomena and boundary-layer effects.

These analyses helped to reduce the amount of testing that would have been needed to determine these values experimentally and also helped to eliminate a lot of propellant combinations from further consideration. Furthermore it helped greatly to determine the effectiveness of the design of any particular LPRE by knowing how close the actual measured performance came to the calculated theoretical performance value. Some of the early LPRE achieved only about 75 to 90% of theoretical performance. The steady design improvements have raised this, and today's LPREs routinely achieve between 96 and 99.5% of theoretical performance. Since about 1975, sophisticated software for this performance analysis has been available to the U.S. LPRE industry.

Hydrocarbon Fuels

A large variety of hydrocarbons and related chemicals has been investigated as potential liquid rocket fuels, such as propane, heptane, methane, hexane, octane, various alcohols, different kinds of gasoline, several kinds of kerosene, ether, toluene, benzene, acetylene, cyclohexane, turpentine, or vinyl ether.[2-5] Kerosene and gasoline are readily available hydrocarbons and became popular fuels for early LPREs in several countries. They are hydrocarbon refinery products, and their composition and properties vary, depending on the oil field from which the crude oil was obtained and on the refining process.[1,*] The

---

*Personal communications, B. Thompson, The Boeing Company, Rocketdyne Propulsion and Power, 1990.

same gasoline might have nearly the same heating value, but slightly different densities or viscosities. Therefore early firing tests with these two groups of hydrocarbon fuels had inconsistent and somewhat variable results. Between 1950 and 1958 the use of gasoline was phased out because of this uncertainty in properties and performance. Switching to jet fuel (JP-4) did not help because the specification for this fuel allowed relatively large variations in viscosity, density, and fractions of olefins and aromatics. When JP-4 (jet fuel, a type of kerosene) was run with nitric acid, the engine performance results were variable or unpredictable and did not always meet the narrow limits of the specified performance, which is needed for a predictable flight path of many a vehicle. To limit the variations in the properties of kerosene, a new specification was agreed upon with the U.S. petroleum industry, and it imposed narrow limits on the fuel's density, viscosity, distillation fraction, and other properties. A narrow-cut kerosene for rocket use, called *RP-1 (rocket propellant # 1)*, was thus created by government specification, which had more consistent properties from batch to batch. Refining and blending processes were then modified to obtain this blend. RP-1 has been used in many U.S. rocket engines up to the present time The Soviets and other countries created specifications for a similar narrow-cut hydrocarbon kerosene type fuel. Gasoline, which was used by early investigators and amateurs, and JP-4 have largely been replaced by RP-1 in the United States in the early 1950s and later with some similar types of narrow-cut kerosene in some of the other countries. Since the late 1950s, the preferred oxidizer to burn with RP-1 kerosene has been liquid oxygen because it gives somewhat better specific impulse than storable liquid propellants and has an environmentally acceptable exhaust gas.

Around 1970 the Soviets devised a special synthetic kerosene that they call *Syntin*.* It had a cyclopropane-type hydrocarbon, had somewhat more energy content and contained more hydrogen. It gave a small increase of specific impulse compared to RP-1 or ordinary kerosenes and was used in the upper stages of the large Proton SLV and upper stage of the Soyuz-U2 SLV. With the 11D58M LPRE in the Proton SLV, Syntin provided about 10 s more specific impulse than kerosene and that translated into 200 kg of extra payload. The source of Syntin was in one of the former republics of the Soviet Union and was outside of Russia. After the breakup of the Soviet Union, the Russians no longer controlled Syntin production. Apparently the price of this fuel had become so high that the Russians looked for other means for increasing payload and no longer purchased this fuel. Syntin is mentioned again in Chapters 8.4, 8.5, 8.6, and 8.13. Rocketdyne synthesized a hydrocarbon fuel that gave a few seconds more performance than RP-1. It was diethylcyclohexane, a highly reproducible mixture of isomers, which could be readily manufactured.[5,†] However the U.S. government had just settled on standardizing RP-1 kerosene at that time and was not interested to change to a somewhat better

---

*Personal communications, B. Thompson, The Boeing Company, Rocketdyne Propulsion and Power, 2002.
†Personal Communication, Jim Morehart, Aerospace Corporation, 2002.

hydrocarbon fuel. The Soviets also used *supercooled oxygen*, that is, liquid oxygen which is colder, denser (better mass ratio), and more cavitation resistant than ordinary oxygen.* It also gave a small performance increase and an extra margin for the payload. In fact the Russians used Syntin fuel together with supercooled LOX for an upper-stage mission, that would otherwise be marginal.

Liquid methane has the simplest molecular structure of any hydrocarbon, contains a lot of hydrogen, is readily available, and was suggested by several of the early visionaries and pioneers to be a very good rocket fuel. It has predictable properties and is usually burned with LOX in LPREs. The exhaust products are environmentally friendly.[2,3,8] Methane's specific impulse, combustion temperature, and average propellant density are intermediate between hydrogen and kerosene. The following estimates were calculated (at 1000-psi chamber pressure with expansion to sea level and with shifting thermochemical equilibrium in the nozzle) for the specific impulse and the specific gravity: LOX/methane 310.7s and 0.81; LOX/LH_2 389.5s and 0.28; LOX/RP-1 300.1 s and 1.02 (Ref. 12). Tests of LOX/methane with small TCs were conducted in the United States. Aerojet started the development of an LOX/methane attitude control system (with multiple small TCS) for the X-33 flight-test vehicle, but this effort was canceled. The Soviet Union developed at least five complete experimental engines with pump feed systems using this propellant combination, and they are mentioned in Chapter 8.4. This author does not know of an engine with LOX/methane that has ever flown (up to 2003) or become operational. In the United States the benefits of a methane-propellant LPRE were not considered to be good enough (compared to other common propellants) to warrant the effort of engine qualification, adapting launch systems to a new propellant, and setting up a suitable supply system for it.

An unexpected *solid carbon formation layer problem* has occurred in TCs using a fuel that contains carbon atoms. This can occur with all hydrocarbons, including RP-1, but under some circumstances it can include MMH and UDMH. A thin solid layer of carbon deposits was occasionally formed on the inside of the inner wall of cooling passages of a cooling jacket. It typically occurs at locations of the highest heat transfer, such as just upstream of the nozzle throat. This thin layer (usually 0.002 to 0.04 in.) consists mostly of carbon particles with some complex fuel derivatives, acts as a thermal insulator, impedes the heat transfer, raises the inner wall temperature, and has caused failures, when the wall became too hot. Of course the formation of such a layer must be avoided. Studies have revealed that the forming of such a carbon-type layer depends critically on the temperature of the fuel, but also on the wall material, wall temperature, heat-transfer rate, and the propellant itself.[13] Modern TCs are designed to avoid these conditions. Solid layers of carbon formation in the turbine nozzles and blades have occurred under some operating circumstances with fuel-rich gas generators, and good GG designs have avoided this condition.

---

*Personal communications, J Morehart, Aerospace Corp., 2004.

Modern Storable Propellants

In the more recent spacecraft applications and attitude control applications of rocket engines, storable propellants (NTO as an oxidizer and either UDMH or MMH as fuels) have been popular.[1-3,8,12] These propellants had also been used for tactical missile propulsion and for ballistic missiles in several countries. They have largely replaced nitric acid and hydrocarbon fuels since the 1970s. Although MMH or UDMH or Aerozine 50 have a slightly lower performance (specific impulse) than pure hydrazine, they have a lower freezing point, are also hypergolic, and seem to be preferred today in most countries. NTO does have a somewhat lower density than RFNA, but it has more predictable properties, is noncorrosive (unless mixed with water), can be stored in ordinary steel tanks, and permits a more predictable engine performance. The United States has preferred MMH fuel for small thrusters and in some applications Aerozine 50 because it gives good performance and lowers the freezing point (when compared to hydrazine), thus avoiding the need to heat valves and pipe lines. Russia or the former Soviet Union has preferred UDMH in most applications for the same reasons, has believed it allows good stability, and the lower combustion temperature causes less thermal stress, particularly on small thruster cooling systems. Storable propellants for small thrusters are discussed in Chapter 4.6.

High-Performance Propellants

Getting high performance was the driving motivation for many propellant investigations between 1947 and 1980. All sorts of high-energy liquid propellants were investigated in the United States, the Soviet Union and other countries.[2,3,5,12] The lure was the higher specific impulse and higher density, which means either more payload, more range, or less propellant for a particular mission. Both a cryogenic oxidizer with the highest possible performance (fluorine) and several new types of storable propellants were investigated analytically (for upper-stage engines as well as for attitude control systems) and also experimentally in laboratories and in firings of small thrust chambers. This included fluorine type propellants (oxygen difluoride, mixtures of liquid oxygen and liquid fluorine), chlorine compounds (chlorine trifluoride, chlorine pentafluoride, and also compounds of oxygen, fluorine and chlorine), and boron compounds. Tests of diborane and pentaborane, using concentrated hydrogen peroxide as an oxidizer, showed that boron-hydrogen compounds had very poor combustion efficiencies (see Chapter 7.3). Liquid ozone (O_3) was also investigated, but abandoned, because it was unstable. Beryllium hydride theoretically made an excellent high-energy fuel, but was abandoned because it was not possible to stop the gradual self-decomposing of this material during storage and because beryllium (as a powder) is very toxic. These boron-type propellants were investigated and abandoned both in the Soviet Union and the United States. Some new high-energy storable propellants were synthesized, such as bromium pentafluoride, which had a relatively very high specific gravity of 2.48, but a poor performance.[5] The properties of these high-energy propellant and their compatibility with other materials were investigated in several laboratories.

Several U.S. and foreign (German and French) companies ran different high-energy propellant combinations in small (typically up to 1000 lbf) thrust chambers, but Curtiss–Wright tested liquid fluorine with hydrazine at 3750 lbf thrust. Low thrust firings were also made of chlorine pentafluoride (ClF_5), chlorine triflouride (ClF_3), and pentaborane (Bo_5H_9). With some high-energy propellant combinations the experimental investigations went beyond merely firing small TCs. Beginning in 1958, Rocketdyne developed the NOMAD experimental upper stage (liquid fluorine and hydrazine) with a 12,000 lbf engine, cryogenic fluorine oxidizer and hydrazine fuel, and a helium-gas pressurized feed system. To prevent the highly toxic fluorine gas from evaporating and escaping from the liquid fluorine tank prior to operation, a jacket with liquid nitrogen was built over the ullage space in the tank in order to recondense the fluorine vapors, which otherwise would have been vented overboard. This can be seen in Fig. 7.8-23. In 1962 ARC ran an Agena gas-generator cycle engine with a TP using liquid fluorine. The Germans and the French also tested small TCs with liquid fluorine. Between 1969 and 1975 Enerogomash (Russia) built and tested a completely new pump-fed 22,000-lb-thrust upper-stage experimental turbopump-fed engine RD-301 using liquid fluorine as oxidizer with ammonia as a fuel. These engines are described in Chapter 8.4. Earlier (1960 to 1966) Energomash built an engine using pentaborane with hydrogen peroxide as propellants (RD-502). Both of these high-energy engines were originally intended to become an upper-stage engine for a SLV, but they were never flown.

An alternate approach was to add liquid fluorine to liquid oxygen, and the mixture was called FLOX. It was tested in the United States using simple pressurized feed systems by TRW (today part of Northrop Grumman) and Rocketdyne (today part of the Boeing Company), and this can be found in Chapters 7.8 and 7.9.

All these and other related efforts proved that high-energy liquid propellant could be harnessed in rocket engines, but the safety precautions and safety rules were onerous, took a lot of time, were expensive, and were not fool proof. When Rocketdyne managed to fire its first small thrust chambers with liquid fluorine and oxygen/fluorine mixtures as the oxidizer, a lot of extra safety rules had to be obeyed, and a variety of personal safety gear had to be used.* Also the exhaust gas had to be trapped and rinsed with water containing an aqueous calcium solution in order to form calcium fluoride, which was insoluble and could be precipitated out of the rinsing liquor. With this technique of scrubbing the exhaust gas with a chemical solution, very little fluorine from the exhaust gas escaped into the local atmosphere.

Discussions about these exotic propellants were held with many vehicle designers and with safety officers of the launch sites. These people indicated

---

*Personal communications, V. Wheelock and B. Thompson, The Boeing Company, Rocketdyne Propulsion and Power, 2000.

that they really did not want to have a highly toxic propellant in their vehicles or on their launch stations and that design costs and launch costs would increase substantially.

None of these high-energy propellants discussed in this section were adopted for a flight application in any of the countries. Further efforts were stopped starting about 1980. The reasons were obvious: very high toxicity, difficult handling procedures, more hazardous, time consuming, and expensive test or launch preparations, or problems with compatible materials for construction. The most compelling reason was the potential effects on the crew, vehicle, equipment, nearby communities or the environment in case of certain types of failures of the engine, spills of the toxic propellants, escape of a lot of vaporized propellant, or accidents of the flight vehicle at the launch facility or during flight. However there was one propellant combination that was accepted universally, namely, oxygen and hydrogen. It is discussed in another part of this chapter.

Monopropellants

Monopropellants allow a simpler engine than a bipropellant, but their performance is inferior to bipropellants by about 15 to 35%. Monopropellants were initially investigated in the 1920s to 1940s.[2,3,5,8] For example, Goddard and others worked with nitromethane in the 1920s. Although it had a reasonable performance (I_s = 225 s), there were several explosions (or violent self-decompositions) with this material, and today it is no longer used. Experimental investigations have been done with nitromethane, ethylene oxide, various synthetic chemicals, and mixtures of nitroglycerine and alcohol. The early Soviets investigators mixed some liquid oxidizers with some fuels to create their own monopropellants. They too had some explosions and abandoned this approach. Warmke, who worked at Penemünde in Germany during the war, tried a mix of 80% hydrogen peroxide and 20% alcohol. It exploded and killed him. An estimated 50 different monopropellants have been analyzed or investigated in the laboratory or small TC firings, but they were all abandoned because a monopropellant really is a liquid explosive. With some stimuli, such as energy inputs (heat, impact, friction, etc.) or with certain impurities, a monopropellant is very likely going to explode.

Only two monopropellants were found to be safe under certain conditions, namely, *hydrogen peroxide* and *hydrazine*.[14-16] Historically hydrogen peroxide (H_2O_2) came first. It was safe to use, if there were no excessive energy input, such as impact or heating to a high temperature and if proper containment materials were used. The self-decomposition during storage could be minimized only if the pipes, valves, or tanks were "superclean" and properly passivated and if the propellant were free of impurities. During World War II, the Germans successfully deployed several LPRE systems (JATOs, torpedoes, aircraft engines) using 80% concentrated hydrogen peroxide with a liquid or a solid catalyst for decomposition as mentioned in Chapter 9.3. This monopropellant was also used with a solid catalyst bed in early ACS engines in the

United States, as described in Chapters 7.3, 7.6, and 7.11 *(GE, Kidde, and ARC)*. This monopropellant worked reliably for attitude control applications of spacecraft and upper stages (1934–1965), and more than 1200 TCs were delivered and flown on U.S. vehicles. This monopropellant was also used successfully in GG for large LPREs with turbopumps. It was used in the GG of the LPRE of the V-2 German missile (80% H_2O_2), the U.S. engine for the Redstone missile, and the USSR engines for the first two Soviet ballistic missiles. This propellant has a relatively high density (specific gravity of 1.4), but its specific impulse is low, as can be seen in Table 4.1-2.[12]

One of the merits of this monopropellant is that a gas generator will never have a high enough gas temperature (Shown in Table 4.1-2) to overheat a turbine. In contrast a bipropellant gas generator is operated at either a fuel-rich or an oxidizer-rich mixture ratio; if it malfunctions, it can operate at off-mixture ratio and generate very hot gas that will melt the turbine blades. Hydrogen peroxide was replaced by other more powerful propellants in the 1960s. Hydrogen peroxide (90% concentration) can also be an oxidizer in a bipropellant combination, gives good performance, and this is discussed later in this chapter.

Hydrazine (N_2H_4), the second acceptable monopropellant, gives a better performance than hydrogen peroxide or most of the other monopropellants. Its typical altitude specific impulse is between 220 and 245s (depending on pressure, nozzle-area ratio, and design), and its density is relatively high when compared to other fuels.[1–3,16] To prevent freezing of the hydrazine (freezes at 1°C or 34°F), the valves, catalysts, and other critical engine components usually have to be heated. Several unsuccessful efforts were made to find a suitable freezing point depressant as an additive; it was found that it takes a large amount of additive to reduce the freezing point below −20°C, but this caused an unacceptable decrease in performance and ignition qualities.[5] The first flight of hydrazine as a monopropellant in small LPREs for reaction control occurred in the United States in a Pioneer spacecraft in 1959 with an engine from TRW, the predecessor of Northrop Grumman Corporation. Because a suitable catalyst was not available at that time, an initial slug of NTO oxidizer was used for starting. After a good solid catalyst became available (in the United States it was developed at JPL in the 1950s), a suitable catalyst bed for decomposing the hydrazine was developed.[16] The first flight with a catalyst in

Table 4.1-2 Performance of Hydrogen Peroxide with Water

H_2O_2, %	Temperature, °F	Specific impulse, s
100	1856	163
95	1637	156
90	1372	148
80	909	129
75	678	117

the United States took place in about 1961. Since then, an estimated 30,000 hydrazine monopropellant thrusters (worldwide for attitude control and trajectory maneuvers) have been built and flown. They have been preferred for applications where the performance is not critical (bipropellant thrusters give better performance), where environmentally clean exhausts are desired, and where simplicity of the engine system is appreciated. These hydrazine monopropellant thrusters are still being used today in many spaceflight vehicles.

There is now a relatively new, synthetic monopropellant called *hydroxyl ammonium nitrate*. It has the potential to replace the highly toxic hydrazine in some applications. It has a low freezing point, and its exhaust gas is also environmentally acceptable.[15] It has been investigated for about seven years, as reported in Chapter 7.7. This seems to be one of the few new potentially useful liquid propellants in the last 25 years.

Gelled Propellants

Serious investigations of *gelled propellants* were undertaken beginning about 30 years ago. This work appears to have been done primarily in the United States. By means of suitable gelling additives, almost any liquid propellant can become thixotropic like a jelly.[17] The gelled propellant is stable (will not separate or precipitate in storage) and can be pumped or expelled by a simple pressurized feed system. The gelling agents have to be carefully selected so that the changes in the gel properties (viscosity and density) with temperature are similar for the fuel and the oxidizer; this allows maintaining a constant mixture ratio over a reasonable range of tank storage temperatures. Enhanced safety is the principal benefit, because it is less likely to leak or be spilled or to react violently to impact. By adding fine aluminum powder to a fuel, the density and the energy content can be increased substantially. Its principal demerits are its somewhat poorer atomization and combustion, causing a slight performance decrease (−2 to −4%) and a slightly higher amount of residual (unusable) propellant. Some very clever LPREs using gelled propellants have been successfully tested, and at least one has flown in an experimental munition (see Chapter 7.9). None have been put into production in the United States at the time of this writing. During World War II, the Germans also investigated suspending solid particles in some fuels, but they did not pursue this potential improvement. They also tried to suspend solid oxidizers (such as small nitrate particles) in an oxidizer liquid, but they did not continue this approach.

Silicon-Dioxide Thermal Insulation

In the 1950s the General Electric Company (GE) conceived and tested an internal self-depositing thermal insulation layer to reduce the heat transfer from the hot gases to the TC walls as mentioned in Chapter 7.3. By adding a small amount of silicon oil to the fuel, a fluffy porous deposit of small silicon oxide (SiO_2) particles would form on the inner walls of the chamber, nozzle, and injector. If a piece of the insulating layer would be come loose and be ejected from the nozzle, a new layer would form immediately over the exposed wall. This insulating layer was very effective in reducing the heat transfer to the wall by 20 to 50%, and in turn this would reduce the critical wall temperature. This concept was tested not only at GE with LOX/kerosene and LOX/alcohol, but also at RMI, at Rocketdyne (LOX/RP-1), and at Bell Aerospace/ARC in a full-scale Agena engine using heavy IRFNA (with 44% NO_2) and UDMH. All fuels contained about 1% silicon oil. ARC reported a 33% reduction in heat transfer, and Rocketdyne had a little higher number.

There were some problems that are discussed in Chapters 7.8 and 7.11. This method was not fully reliable and was abandoned at GE, RMI, and Rocketdyne. However it apparently was successful at ARC, where the Agena engine (16,000 lbf) ran successfully with a 1% silicon oil addition to the UDMH fuel in 1971 and 1972.*

Hydrogen Peroxide as an Oxidizer

Concentrated liquid *hydrogen peroxide* as an oxidizer can yield good performance with one of several fuels (such as C-Stoff in Germany or kerosene in the United States). It was first used in a production rocket vehicle by the Germans and later by the United States, Soviets, and the British. Historically the firm of Hellmuth Walter Works in Kiel (see Chapter 9.3) was the first to use it in applications as aircraft JATO units, LPREs for experimental flight-test vehicles, and fighter aircraft (as the primary engine of the Messerschmitt 163 fighter aircraft). This was between 1936 and 1945. Great Britain built and flew several LPREs with H_2O_2 as the oxidizer (see Chapter 12) in the 1946 to 1972 period. The aircraft auxiliary rocket engine in the U.S. F104 fighter airplane (Rocketdyne AR-2 engine, 1960, see Chapter 7.8) also used this powerful oxidizer. In the Soviet Union an engine was developed (Chapter 8.4) with hydrogen peroxide as the oxidizer and Bo_5H_9 or pentaborane as the fuel. This oxidizer's density is high, its performance is comparable to other storable oxidizers, and the engine exhaust gases are environmentally benign and nontoxic. Early uses in Germany were initially with 50%, then 70%, and shortly thereafter with 80% hydrogen peroxide (for production LPREs), which gave reasonably good performance. In the United States and Britain it was 90%. This performance and the reaction temperature can be increased, by using higher concentrations such as 95% and even 98% hydrogen peroxide, and they were experimentally investigated. Concerns over storage stability (even today up to 1% of 90% hydrogen peroxide self-decomposes yearly during storage) and potential violent self-decompositions have detracted further applications of concentrations over 90%. Today it is still an attractive propellant combination, and some work on it continues (at the time of the writing of the manuscript) at a low level in at least two countries.

The United States developed and successfully flew (in 1961) the XLR-99 LPRE using *liquid ammonia* and LOX in the X-15 research aircraft (Chapter

---
*Personal communication, Fred Baroody, ARC, 2003.

7.2). The aircraft with its propellant tanks is shown in Fig. 2-6. The complexity of handling this propellant, the modest performance improvement, its relatively high vapor pressure for a fuel, and its toxicity seem to have prevented additional applications.

Liquid Oxygen and Liquid Hydrogen

All of the theoretical investigations of this propellant combination (LOX/LH$_2$), including those conducted by the early pioneers, show that this is the best practical high-energy propellant combination.[1-4,8] Many studies have indicated superior vehicle flight performance with these two propellants. LOX/LH$_2$ was experimentally investigated since the 1940s and subsequently developed in LPREs for various space-launch-vehicle applications, but only after plants were built to produce liquid hydrogen in quantity. The high specific impulse of 409 to 467 s (depending on the design details) makes this propellant combination the choice for upper stages of launch vehicles and some first stages of SLVs. The cryogenic nature of these propellants causes handling and launch complications, and the low liquid-hydrogen density results in very large cryogenic fuel tanks and large vehicles. Condensation of moisture from the air forms ice on the cold surfaces of the vehicles (tanks, pipes), and the ice causes extra unwanted inert vehicle mass. Also the loosening and falling off of chunks of ice has caused damage (by impact) to wings and structures of the space shuttle. The propellants and their combustion products are environmentally benign.

Work on small thrusters started at Aerojet in 1945 and was also pursued at the NASA Lewis Laboratory. The very first flight with this high-energy cryogenic propellant combination was in 1963 with the Pratt & Whitney RL 10 engine, and this is explained in Chapter 7.10. LOX/LH$_2$ has become a common propellant for many upper-stage engines, and such LPREs have flown in seven different countries. The Rocketdyne J-2 engine with 230,000 lbf vacuum thrust was used for the two upper stages of the Saturn V launch vehicle and was first launched in 1966. It can be found in Chapter 7.8. As described in Chapter 7.7, Aerojet had a contract for a large experimental engine (M-1), with 1,500,000 lbf thrust, but work did not progress beyond the testing of several components. Beginning in 1972, the space shuttle main engine with LOX/LH$_2$ was then developed at Rocketdyne. It flew first in 1981. In the Soviet Union at least two experimental engines (D-57 and KVD-1) were developed and ground tested in the late 1960s and early 1970s. Compared to other countries the Soviets were late in flying a LOX/LH$_2$ engine, namely, the RD-0120 in the Energiya vehicle in 1987; details are in Chapter 8.5. The Japanese flew their first LE-5 engine beginning in 1977, the French first flew their HM-7 in 1979, and the Chinese also developed and flew their YF-73 and YF-75 oxygen/hydrogen engines as described in Chapters 11, 10, and 13. At the time of this writing, the Indians had flown a Russian KVD-1 engine and also were developing a LOX/LH$_2$ upper-stage engine of their own (see Chapter 14).

In the 1960s some work was done on "slush" hydrogen, which is a mixture of solidified small particles of very cold solid hydrogen and liquid hydrogen.

The objective was to increase the very low density or specific gravity of this cryogenic fuel (specific gravity of 0.071 and boiling point of 20.4°K or − 423°F) and contain it in smaller tanks. There were difficulties in obtaining and keeping hydrogen as a suspended solid. Even if the problems would have been solved, the improvement in bulk density was small; a mix of 50% solid particles and 50% liquid had a specific gravity of 0.082 as compared to 0.071 for the pure liquid hydrogen. Slush hydrogen has not been pursued and has not been adopted for a flying space vehicle.

Tripropellant Rocket Engines

Several countries sponsored studies and limited experimental efforts of a tripropellant rocket engine for a potential single-stage-to-orbit application. Here liquid oxygen is the oxidizer of choice. A hydrocarbon fuel (such as RP-1) together with a small amount of LH_2 is the initial fuel (at takeoff and for the flight through the atmosphere) with relatively high thrust. Liquid hydrogen is the only fuel during the second operating mode (final ascent portion of the flight and orbit maneuvers), but at about 20 to 30% of the takeoff thrust. The Soviets not only studied several vehicles, but they actually built and tested heavy duty versions of at least three such engines, each based on an existing engine. The RD-701 and the RD-704 are from the Energomash design bureaus, and the other is a variant of the RD-0120 from the Chemical Automatics Design Bureau. They are discussed briefly in Chapters 8.4 and 8.5. A good tripropellant LPRE does offer a rather small increase in vehicle performance, but the pump-fed engines (typically six turbopumps including three booster pumps) become quite complex and expensive. Enthusiasm for such an engine seems to have diminished, and there are no known plans to use it in an actual launch vehicle.

Surface Contamination

In the late 1960s it was a surprise to find that spacecraft windows fogged up during orbital flights. This phenomenon was traced to exhaust gases from TCs using storable propellants (such as NTO/MMH or NTO/UDMH) containing gas species that can (in the vacuum of space) condense on sensitive spacecraft surfaces. This surface contamination degrades the function of mirrors, solar cells, optical lenses, or windows of space vehicles or space stations. Condensed hydrazinium nitrate is one species that condensed on spacecraft windows; it is a material present in small quantities in the exhaust from several storable propellant combinations at certain mixture ratios. Small solid carbon particles from incomplete combustion of hydrocarbons can also form such deposits. Propellant combinations that form only gas products, such as LOX/LH_2, LOX/hydrazine, or monopropellant hydrazine, do not cause deposits on spacecraft surfaces. With complete combustion and judicious placement of the ACS thrusters on the vehicle, this deposit of contaminants with storable propellants is minimal or less likely to occur.[18]

Who Did All of the Work on Propellants?

It was a relatively large number of rather different organizations during the first several decades of this historical account.[5] Most of the LPRE companies have some propellant specialists or a propellant R&D department, and of course they were involved. For example, Rocketdyne and Allied Chemical tried to synthesize ONF_3, Rocketdyne tested ClF_5, and Aerojet ran Bo_5H_9 (pentaborane) in small TCs. Furthermore there were various government laboratories, private R&D organizations, chemical companies, and oil companies.

In Germany the chemical manufacturer of I. G. Farben worked with a LPRE company and the German Government Laboratory to discover, investigate, and proof test various fuels, which were considered to react with nitric acid as the oxidizer. They might have evaluated more than 2000 chemicals for their potential to become a rocket propellant. In the Soviet Union the propellant work was done by a State R&D Institute (see Chapter 8.13), by the several LPRE design bureaus, and a few chemical companies.

As an example, some of the U.S. organizations working on propellants in the 1960s and 1970s are cited here, but the list is not complete.[5] It included these government laboratories: Naval Ordnance Test Station, Inyokern, California; Naval Air Rocket Test Station, Dover, New Jersey (has since been closed); the NASA Jet Propulsion Laboratory, Pasadena, California; and the Air Force Philips Laboratory (formerly Rocket Propulsion Laboratory), Edwards, California. Callery Chemical tried to minimize the formation of nitrates in nitric acid tanks; Standard Oil of California through its California Research organization in Richmond, California, measured ignition delays; Standard Oil of Ohio measured physical properties; Air Reduction worked on cryogenic processes, and American Cyanamid did synthesis. There were also contributors from Academia and the R&D community, such as the chemical engineering department of New York University (additives to kerosene), Purdue University (effects of high pressure), or the Stanford Research Institute (SRI), working on propellant synthesis. Harshaw Chemicals tried to synthesize $HClF_4$, and Stauffer Chemical tried to synthesize NF_3. Today only a few of these organizations are still involved in liquid propellant R&D, and work seems to be sporadic.

References

[1]Sutton, G. P., and Biblarz, O. (eds.), *Rocket Propulsion Elements*, 7th ed., Wiley, New York, 2001, Chaps. 5, 7.

[2]Penner, S. S., *Chemistry Problems in Jet Propulsion*, Pergamon, London, 1957.

[3]Bollinger, L. E., Goldsmith, M., and Lemmon, A. W., Jr., *Liquid Rockets and Propellants*, Vol. 2, Progress in Astronautics and Aeronautics, AIAA, New York, 1960.

[4]Ehricke, K. A., "A Comparison of Propellants and Working Fluids in Rocket Propulsion," *Journal of the American Rocket Society*, Vol. 23, No. 5, Sept.–Oct. 1953.

[5]Clark, J. D., *Ignition*, Rutgers Univ. Press, NJ, 1972.

[6]Bloom, R., Davis, N. S., and Levine, S. D., "Hydrogen Peroxide as a Propellant," *Journal of the American Rocket Society*, No. 80, March 1950.

[7]Lutz, O., "A Historical Review of Developments in Propellants and Materials for Rocket Engines," *First Steps Toward Space*, American Astronautical Society History Series, edited by F. C. Durant, and G. S. James, Vol. 6, Univelt, Inc., San Diego, CA, 1985, pp. 103–112.

[8]*Liquid Propellant Manual M-4*, Chemical Propulsion Information Agency, John Hopkins Univ., Columbia, MD, 1970–2003.

[9]Wagman, D. D., Evans, W. H., Parker, V. B., Schumm, R. H., Halow, I., Bailey, S. M., Churney, K. L., and Nutall, R. L., "The NBS Tables of Chemical Thermodynamic Properties," *Journal of Physical and Chemical Reference Data*, Vol. 11, Supplement 2, 1982.

[10]Gordon, S., and McBride, B."Computer Program for Calculation of Complex Chemical Equilibrium Compositions, Rocket Performance, Incident and Reflected Shocks, and Chapman-Jouquet Detonations," *NASA–SP-273*, 1971.

[11]Gordon, S., and McBride, B. J,. "Computer Programs for Calcuation of Complex Chemical Equilibrium Compositions and Applications," Vol. 1: *Analysis*, Vol. 2: *User Manual and Program Description NASA Reference Publication 1311*, Oct. 1994 and June 1996.

[12]Koelle, H. H. (ed.) *Handbook of Astronautical Engineering*, McGraw–Hill, New York, 1961, Chap. 20, pp. 19–51.

[13]Liang, K., Yang, B., and Zhang, Z., "Investigations of Heat Transfer and Coking Characteristics of Hydrocarbon Fuel," *Journal of Propulsion and Power*, Vol. 14, No. 5, Sept–Oct, 1998, pp. 789–796.

[14]Zube, D. M., Wucherer, E. J., and Reed, B., "Evaluation of HAN Based Propellant Blends," AIAA Paper 2003-4643, July 2003.

[15]Morgan, O. M., and Meinhardt, D. S., "Monopropellant Selection Criteria—Hydrazine and Other Options," AIAA Paper 95-2595, June 1995.

[16]Schmidt, E. W., *Hydrazine and its Derivatives, Preparation, Properties, and Applications*, Wiley, New York, 1984.

[17]Natan, B., and Rahami, S., "The Status of Jel Propellants in the Year 2000," *Combustion of Energetic Materials*, edited by K. K. Kuo, and L. deLuca, Begel House, Boca Raton, FL, 2001.

[18]Hoffman, R. J., English, W. D., Oeding, R. G., and Webber, W. T., "Plume Contamination Effects Prediction," Air Force Rocket Propulsion Lab., Report, Edwards, CA, Dec. 1971.

4.2 Engine Systems

Introduction and Definition

This chapter describes complete engines, their feed systems, engine cycles, and engine analyses. In this book a *liquid propellant rocket engine system* is defined as containing the following hardware subsystems: 1) one or more thrust chambers; 2) propellant tanks; 3) a feed system to transfer the propellants from the tanks into the thrust chamber(s) either by gas pressurization or by pumps, which raise the pressure of the propellants; 4) pipes or propellant feed lines, often with flexible joints; 5) controls (valves, regulators, venturis, switches, orifices, electric controllers, etc.) to start and stop the operation and achieve the intended propellant flows, sometimes also to calibrate or regulate the trust level and occasionally to calibrate or regulate the mixture ratio (oxidizer to fuel flow rates), and an electrical control subsystem to provide electric or pneumatic power and command signals; 6) a structure to hold the components together and transmit the thrust to the vehicle; and 7) some engines can include mechanisms for changing the direction of the thrust vector during flight.

An engine system has been described simply as a LPRE with tanks, and the most obvious difference between a LPRE and an engine system is therefore primarily the propellant tanks and pressurizing gas tanks. Large thrust chambers (TCs) and their histories are discussed in Chapter 4.3 and small TCs in Chapter 4.6. Feed systems are briefly discussed in this chapter, but two of the feed systems components, namely, turbopumps and gas generators, are discussed in Chapters 4.4 and 4.5, respectively. Starting and ignition of LPREs is in Chapter 4.8. Vehicle steering and flight-path controls are in Chapter 4.9.

The early pioneers and the early developers, including Robert Goddard, had to investigate, design, and build most or all seven of these hardware items just listed in order to build a usable engine system. The early JATO units, and some tactical missiles, had pressurized feed systems and were delivered by the LPRE maker with their tanks and controls as a complete autonomous engine system. As the business grew, it became more specialized, and the vehicle contractors needed to integrate the tanks with their vehicle structure; therefore, the vehicle or integrating contractors then developed and built the tanks and sometimes also the pressurizing system and some of the controls. So beginning in approximately the late 1940s, the LPRE developers/manufacturers of larger engines in the United States, Britain, or France have delivered a partial engine system, without tanks and without all of the controls to pressurize the tanks. Since about 1950, most makers of small thrusters have been delivering only individual thrust chambers with propellant valves, and the vehicle contractor then usually develops and builds the tanks, pressurizing system, controls, and structure. There were, however, a few small engine systems, where the LPRE company was and still is responsible for the whole engine system such as in the Minuteman post boost control system described in Chapter 7.11. When LPREs with pumped feed systems were developed beginning in the late 1940s, the LPRE makers/suppliers delivered only the engine with engine controls, engine structure, and engine piping, but the tanks were then usually not developed by the LPRE manufacturers.

Liquid Propellant Rocket Engine Systems

LPRE systems have been classified in several different ways, such as by propellant, first flight, thrust magnitude, application, dates of development, or manufacturer.[1,2] One common way is to divide the LPREs into two thrust classes. These two classes overlap. Most of the low thrust engines have pressurized gas feed systems and most of the high thrust engines use a pump feed system. A comparison of these two classes is shown in Table 4.2-1. There are some exceptions to the statements in the table 4.2-1.

Historically the small engines came first. As can be seen from Table 4.2-1, the two classes of engine systems are quite different. The high-thrust LPRE has components (cooling jacket, turbopump, gas generator, and more valves) and

Table 4.2-1 Characteristics of Two Classes of LPREs[a]

Characteristic	Large LPREs	Small LPREs
Thrust size, typical	High (above 2000 lbf) Highest is 1,628,000 (SL)	Low (0.5 to 1300 N, or 300 lbf) Range is 0.01–1,000 lbf
Typical purpose or mission	Accelerate vehicle along its flight path	Attitude control, trajectory maneuvers or corrections
Typical application	Launch vehicle, first, second, or third stage	Satellite ACS, orbit correction, rendezvous maneuvers
Total impulse, lbf-s	Above 300,000	Below 400,000
Number of thrust chambers/engine	Usually 1; sometimes 2, 3, or 4, with axial orientation	Between 4 and 24, usually oriented in different directions
Chamber pressure	350–3,600 psia	30–300 psia
Feed system	Turbopump with low tank pressure	Expulsion by pressurized gas
Propellant tank pressure	Low (20–60 psia)	High (150–900 psia)
TC cooling method	Typically regenerative, can be augmented by film cooling	Radiation cooled, film cooled, ablative; occasionally regenerative
Number of restarts	Usually no restart; sometimes one or two for upper stage only	1000 restarts are typical, for medium thrust less than 10 restarts
First ground tests	1937–1942	1921–1925
First flight	Germany, 1939, with turbopump	Goddard, 1926

[a]Adapted from Ref. 1.

design arrangements (engine cycles, structure) not found in the small-thrust LPREs. The small LPREs have characteristics not found in the larger version, namely, multiple thrust chambers, pulsing thrust operations (many restarts), relatively low chamber pressures, very fast starts, and nozzles oriented into different directions. Typical chamber pressures or propellant tank pressures are very different. In fact a number of companies in this business specialized in one or the other, and only some companies do both. The technology for the large- or high-thrust LPREs encompasses a greater number of additional specialties, the development time is longer, and the cost per engine or cost per unit thrust is usually much higher.

Feed Systems

Figure 4.2-1 shows one way for classifying feed systems for LPREs. For each major class in this figure, some general or historic comments will be made, and examples of specific actual engines, which fit into that class, will be cited in the next few paragraphs. The figure also shows several different engine cycles, which will be discussed later in this chapter.

Historically the first feed system was the *pressurized-gas feed system*, where high-pressure gas expels the propellants from their propellant tanks.[1,2] Goddard's first thrust chamber tests (1923) used it, and all pioneers and early investigators used it. Figure 4.2-2 is an examples of a pressurized-gas feed system with most of the hardware for test firings, filling propellants and

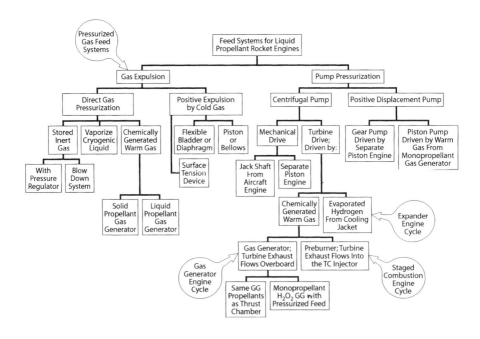

Fig. 4.2-1 Major feed system categories for LPREs during the past 65 years.

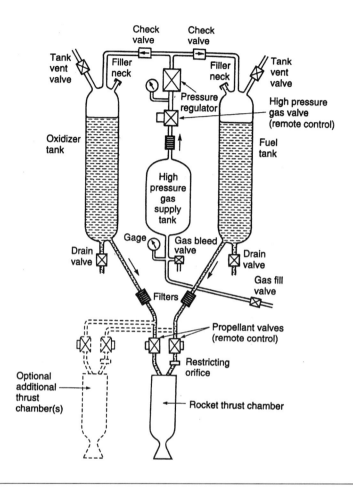

Fig. 4.2-2 Schematic flow diagram of a typical pressurized-gas feed system with call-outs of detailed features. Copied with permission from Ref. 1

pressurizing gases, emptying or draining residual propellants, and flushing out remaining propellants. However it does not show all of the possible sensors or measuring instruments. Simpler versions, with burst diaphragms instead of valves, have also been developed. This type of feed system is still used extensively today in many applications, mostly for small LPREs, but there are a few larger-thrust LPREs. It is the preferred feed system, when thrust levels are low and the total impulse is relatively low. Also for 80 years a pressurized-gas feed system has been the preferred for testing large TCs as a component in test facilities, prior to assembling it into a pump-fed system. It requires heavy propellant tanks and a heavy thick-walled high-pressure gas tank with higher tank pressures than those listed in Table 4.2-1. Every LPRE development organization started out with this type of feed system using a stored high-pressure inert gas as the pressurizing fluid.

When the pressurizing gas is stored as a cryogenic liquid, such as liquid nitrogen or liquid ammonia, the storage tank is much smaller and lighter in mass than with an equivalent high-pressure gas. It is vaporized by electric heat or by heat convection and used for tank pressurization in a gaseous condition. An example is the vaporized ammonia pressurizing system in Chapter 7.9, and it was flown in satellites.

Two categories of pressurized feed systems with stored inert gas have been developed and flown, as shown in Fig. 4.2-1. The first feed systems used a *pressure regulator* to reduce the pressure from the high-pressure gas storage tank to the pressure of the propellant tank. An example is shown in Fig. 4.2-2, and it permits constant chamber pressure and nearly constant thrust. All of the early LPRE pioneers used a version of this system. The other category is called a *blowdown system*, was invented in the late 1940s, used extensively beginning in 1960, and it does not have a pressure regulator.[1,3,4] Here the gas is stored under pressure inside the propellant tank together with the propellants. As the gas expands during operation, the tank pressure, the thrust, and chamber pressure decrease steadily down to perhaps 25 to 45% of the initial values. This system is simpler (no separate gas tank and no regulator or gas valve), but the mixture ratio can change slightly during the operation, the total volume occupied by the feed system is larger, and the average specific impulse is somewhat lower. Both of these systems are still used.

Because the high-pressure gas tank was usually a very heavy component of a LPRE, investigators in the period of the 1940s–1960s turned to schemes that were lighter. As already mentioned, they flew systems, where a liquefied gas in a small tank is evaporated by controlled heat addition (1950s). Then they investigated a chemically generated "warm" gas for tank pressurization because this allowed a substantial reduction in the inert mass of an engine with a pressure feed system. The term warm gas (typically 500–1800°F) was used to distinguish it from the hot combustion chamber gas (typically 5000–6000°F). Both liquid propellant and solid-propellant gas generators were developed. An example is the RMI's engine for the Bullpup missile. It used a solid-propellant GG to generate a high-pressure warm gas, which expelled the propellants from their tanks. This project started in 1958 and is described in Chapter 7.2 and in Fig. 7.2-6. The Bomarc guided missile (started in 1951) used a LPRE with two different chemically generated gases to expel the propellants from their tanks. It generated these two gases at high pressure with two small liquid propellant gas generators, one with a fuel-rich gas output for pressurizing the fuel tank and one with an oxidizer-rich gas output to pressurize the oxidizer tank. Figure 7.7-16 shows that there was a small high-pressure inert gas tank just for pressurizing the two small propellant tanks for feeding the two gas generators. The three small tanks were much lighter than a single larger inert gas tank, which would have to pressurize the two large propellant tanks.

The largest LPRE with a gas-pressurized feed system was developed by Société d´ Études de la Propulsion par Réaction (SEPR) in France for the Diamant B Missile, mentioned in Chapter 10. It had a thrust of 84,000 lbf, a chamber pressure of 284 psia, a tank pressure of about 350 psi, and used storable propellants. The tank walls were thick and very heavy. This engine really should have used a turbopump feed system, which would have allowed a much smaller and lighter and less expensive missile. However the French engine experts at the time of the design (1960) believed that their TP technology was not ready for an urgent ballistic missile production program.

Positive expulsion tanks or special surface-tension devices were needed in the beginning of the 1940s, to prevent the pressurizing gas from entering into the high-pressure propellant feed pipes and forming gas bubbles in the propellant. As already mentioned elsewhere in this book, gas bubbles in the propellant cause erratic thrust behavior, lower performance, and a possible initiation of unstable combustion; bubbles must be avoided. Positive expulsion is needed for flight applications with side accelerations, where the tank outlet could become uncovered during propellant sloshing and also for a start or restart in 0-g space, where the lack of gravity would cause the propellant to float around in the tank, thus again uncovering the outlet pipe of the tank. Five different positive expulsion devices, which have been flown successfully, are shown by simple sketches in Fig. 4.2-3. There are other designs of positive expulsion, but they are not shown here. The first four have a barrier or a physical separation of the pressurizing gas from the liquid propellant, thus ensuring a bubble-free feeding of propellant to the thrust chamber. Historically the flexible bladder or diaphragm type came first for multiple restarts in space. These positive expulsion tanks were flown first in the later 1950s and early 1960s in several applications. Bladders and diaphragms were used first, and they worked well at moderate accelerations, had an excellent volumetric expulsion efficiency, but plastic bladders can deteriorate with time or suffer fatigue damage by the daily variations in ambient temperature. Bellows-type tanks are usually a little heavier, but better for long storage. Pistons have worked well, but some had problems with seals, if the tank is stored for long durations.

The Agena LPRE used small bellows-type expulsion devices on the propellant start tanks, which can be seen in Fig. 7.11-8. Pistons and bellows usually work well in a flight mission with high side accelerations, such as with maneuvering antiaircraft missiles.

The fifth device in the figure, which is a set of shaped fine-mesh screens, works on the principle of surface tension and holds enough propellants near the tank outlet to ensure a bubble-free start. The acceleration from the engine's thrust will then automatically orient the propellant in the tank and continue to cover the tank outlet with liquid. However surface-tension devices work only for relatively small vehicle accelerations and do not have a barrier between gas and liquids. A large variety of different bladders, diaphragms, bellows, pistons, or surface-tension devices has been successful.[5]

All of the early investigators understood that a *pump-fed engine* system would greatly reduce the pressure in the propellant tanks and thus allow a much thinner tank wall, which of course would be much lighter and allow improved vehicle performance, such as more payload or more range. At the

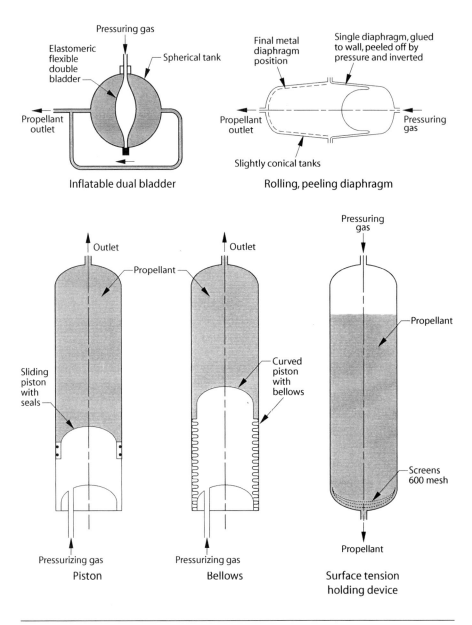

Fig. 4.2-3 Simple sketches of several positive expulsion devices used with pressurized-cold-gas feed systems.

beginning of this history, it was not clear what kind of a pump should be used. The space visionary Konstantin Tsiolkowsky visualized a piston pump in 1903, and Robert H. Goddard in his 1914 patent also showed a piston pump. Crude sketches of these are seen in Figs. 5-2 and 5-5. Goddard tried several types of positive displacement pumps (gear pumps, piston pumps, vane pumps) in the 1920s and 1930s because he correctly believed that these pumps would ensure a good mixture-ratio control. However, he was not successful because evaporation of liquid oxygen caused gas bubbles, and the density of the two-phase oxygen mix varied in an unpredictable manner. Later Goddard developed centrifugal pumps, which actually have flown. Valentin P. Glushko, a key man for LPREs in Russia (his efforts are discussed in Chapters 8.2 and 8.4), also investigated several types of positive displacement pumps some years later, and he was successful with gear pumps feeding storable propellants to a thrust chamber. The gear pumps of the RD-1 aircraft rocket engine and its follow-on versions were driven by a jack shaft and coupling from the aircraft's main engine. During the 1940s, the rocket engines with the gear pumps have flown successfully as an auxiliary power plant or as a superperformance rocket engine in several Soviet experimental aircraft. These engines and the gear pump are described in Chapter 8.2. Another example of small gear pumps feeding a pair of small thrusters with an external power source (small electric motor) was used with the U.S. Agena engine in the 1950s and 1960s, and it is shown in Fig. 4.4-10. An example of a LPRE, where the power for driving the pumps is provided by a separate small airbreathing piston engine is the Aerojet 40XAL-4000 LPRE for a JATO of the PB2Y-3 flying boat. The French used centrifugal pumps to raise the pressure of the propellants in their auxiliary aircraft rocket engines SEPR 825 and SEPR 844, which are shown in Figs. 10-5 and 10-6. The power for driving the pumps was provided from the aircraft's main jet engine through a jack shaft and not through a turbine. All of these various engine system schemes are seen in Fig. 4.2-1. Using power from the aircraft engine as tapped-off compressor gas was not a totally satisfactory solution because the rocket caused the engine to have less power available, particularly for maneuvers at altitude.

A turbine-driven centrifugal propellant pump, which today is called a *turbopump* (TP), was visualized by several of the early LPRE pioneers. Its turbine was driven by gas generated by the LPRE itself, independent of the main aircraft engine or an external power source. Turbopumps were a logical solution to allow a significant savings in inert vehicle mass, and they allowed higher pump discharge pressures, which in turn allowed a little more engine performance. Figure 4.2-4 shows a simplified diagram of a typical LPRE with a turbopump, propellant tanks, and a system for pressurizing the tanks to a low tank pressure. Not shown are the components necessary for controlling the engine operation, filling, draining, or flushing propellants with helium. This TP has a gear case, which allows a high-speed turbine to drive lower-speed pumps more efficiently.

The first ground tests of centrifugal pumps driven by turbines were performed by Goddard in 1934, but the early model had problems and was not suitable for flight. Later a LPRE with two improved TPs was tested successfully with a gas generator in 1938 and was flown by Goddard for the first time in 1939. This is described in Chapter 5.2. In Germany a single-shaft TP was first operated on the ground in about 1938 and first flown in a LPRE in 1939.

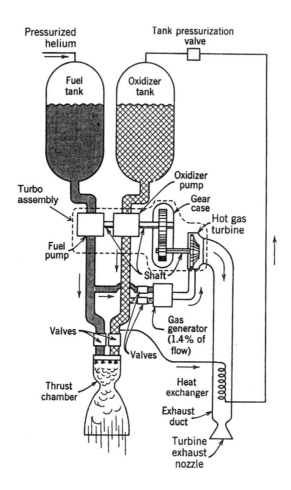

Fig. 4.2-4 Typical flow diagram of a LPRE operating on a gas generator engine with a gas generator, a geared TP, an overboard turbine exhaust, and gasified oxygen for pressurizing the LOX tank.

An improved model was the first to be put into production and used as an aircraft engine for a fighter airplane during World War II (Chapter 9.3). In the Soviet Union the first complete and successful turbopump ground test was in 1945, and the first flight of an auxiliary engine with a TP was in an aircraft in 1947. Chapter 4.4 discusses TPs in detail.

For the small propellant flow regimes needed by engines with small thrusters, centrifugal pumps are not generally used because they are very inefficient at low flows. However positive displacement pumps are efficient with small flows. A set of piston pumps was developed at the Lawrence Livermore National Laboratory in the 1980s and 1990s, and one set of four was flown in an experimental vehicle 1994. Figure 4.4-11 shows an early model of such a piston pump, and it was driven by a gas piston using warm gas from the catalytic decomposition of hydrazine, which was the fluid being pumped. The flight LPRE had a set of four such piston pumps, and these pumps were timed to operate in sequence and deliver a reasonably steady flow of liquid hydrazine to four small thrusters, which propelled the flight vehicle. This piston pump is described with a reference in Chapter 4.4

Engine Cycles

Engines with TPs in their feed systems have become the favorite approach for almost all larger LPREs. There are several different designs whereby a turbine can be integrated into a LPRE, and this was identified as different engine cycles.[1,2,6] An engine cycle describes the propellant flowpaths through the major engine components, the method of providing hot gas to one or more turbines, and the method of handing and discharging the turbine exhaust gas. A closed cycle has all of the propellants go through the combustion chamber, where they are burned efficiently, whereas an open cycle has most of the gasified propellant go through the combustion chamber, but a small flow is dumped overboard or dumped into the nozzle exit at a pressure lower than the combustion chamber pressure. The Russians also refer to these as engines with incomplete or complete combustion.

The concepts for engine cycles were introduced in the Soviet Union and in the United States in the 1950s and later in other countries. Of the several possible *engine cycles*, only four were actually developed and flown, and they are shown schematically (together with one other cycle) in Fig. 4.2-5. The spirals in the figure indicate the hydraulic resistance of an axisymmetric cooling jacket. Three of these cycles are also identified in Fig. 4.2-1.

Historically the *gas-generator cycle* was the first. It has a separate gas generator, where fuel and oxidizer are burned at a mixture ratio that results in low enough temperature for the turbine inlet gases (typically between 1200 and 1600°F) to allow uncooled turbines. This cycle is the simplest, often the lowest in cost, gives a low engine inert (empty) mass, but gives somewhat lower performance (1.5–7% less engine specific impulse depending on the design details) than the expander or the staged combustion cycles. Also the engines with the GG cycle usually have lower internal pressures. Figure 4.2-6 shows a diagram of a large LPRE (205,000 lbf thrust) with a typical GG engine cycle. The engine has a TP, GG, solid-propellant start cartridge, gear lubricant supply with an oil-fuel mix, and overboard turbine exhaust. Flow diagrams of other engines with this GG cycle are shown for a Rocketdyne aircraft engine in Fig. 7.8-34, an Energomash SLV booster engine in Fig. 8.4-8, and for the General Electric Vanguard booster engine in Fig. 7.3-5. The German V-2 missile engine (design about 1937, in Fig. 9-3), and the German Hellmuth Walter Company's aircraft rocket engines in World War II (design 1936–1938, in Fig. 9-7) used a GG engine cycle, were flight tested between 1939 and 1942, and were the first to become operational with the German military forces in 1944 and 1945. These early LPRE used monopropellant hydrogen-peroxide gas generators. In the United States Goddard flew experimental sounding rockets with a TP-fed

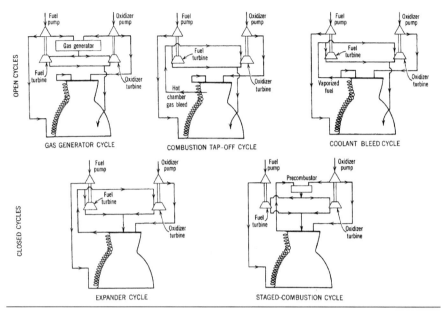

Fig. 4.2-5 Simplified diagrams of five engine cycles for LPREs with a turbopump feed system. The spirals represent the flow resistance of an axisymmetrical cooling jacket. Adapted and copied with permission from Ref. 1.

engine in 1939 and again in 1940 as described in Chapter 5.2. The date for the first ground test of a Soviet experimental engine with a gas-generator cycle was around 1945 and the first flight in 1947.

Most designs since about 1955 have a bipropellant gas generator, with the same propellants as the main engine, but at a different mixture ratio. The GG cycle is still used today in some of the newer engines such as the French Vulcain of Fig. 10-13 upper-stage engine and the Rocketdyne RS-68 shown in Fig. 7.8-20. History tells us that there were a number of successful LPREs that started to fly with a gas-pressurized feed system, which was heavy, and were later converted to use a turbopump feed system with a gas-generator cycle, and this reduced the inert mass of the vehicle and improved the vehicle's performance. This includes the 6000-C4 aircraft LPRE (Reaction Motors, Inc., late 1940s, see Fig. 7.2-1) and two aircraft LPREs of the former Soviet Union (see Chapter 8.2).

The *expander-engine cycle* relies on using a cryogenic fuel, which is gasified and heated in the TC cooling jacket, to drive the turbine(s).[1,2,6,7] The relatively cool turbine exhaust gas of evaporated hydrogen is subsequently fed into the combustion chamber. There are no GGs or preburners. Figure 4.2-7 shows the simple flow diagram of the first such LPRE with an expander cycle. The performance of such an engine is slightly better (2–7% depending on design details) than the gas-generator cycle, but the internal fuel pressures and inert engine mass are somewhat higher than an engine with an equivalent GG cycle. There is no real performance penalty when compared to the

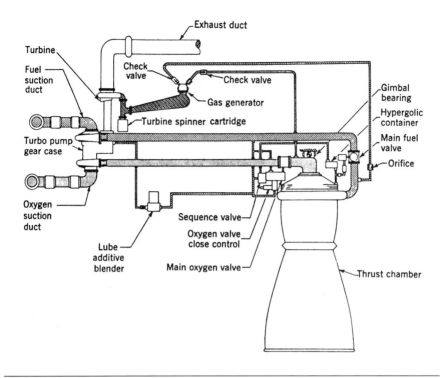

Fig. 4.2-6 Simplified schematic diagram of the H-1 LPRE with a gas-generator cycle. Eight of these engines were used in the booster stage of the Saturn C-I SLV. Courtesy of The Boeing Company, Rocketdyne Propulsion and Power; copied with permission from an earlier edition of Ref. 1.

staged combustion cycle, which is discussed next. This expander cycle works only with a cryogenic fuel that can be evaporated, such as hydrogen. It would not work with storable fuels, such as kerosene or UDMH. To date all LPREs with an expander engine cycle have used LOX/LH$_2$. The RL-10 engine of Pratt & Whitney was the first in the world to use the expander cycle for an upper-stage launch-vehicle application. Its flow diagram is in Fig. 4.2-7. It was first tested on the ground in 1959 and flew first in 1963. Improved and uprated versions of the RL-10 continue to fly using this cycle. The Russians built and tested an experimental engine with an expander cycle in 1998, and Rocketdyne developed one between 1981 to 1983, but none of these LPREs have flown.

A variation of this expander cycle is the coolant bleed cycle, and it is shown in Fig. 4.2-5. The Japanese called it the nozzle-expander cycle. The turbine exhaust flow is dumped into the nozzle exit, and this gas flow contributes some to the nozzle gas expansion. The engine with this cycle is not quite as efficient as one with a pure expander cycle, but its performance is better than an engine with a GG cycle. It was used by the Japan's Mitsubishi Heavy

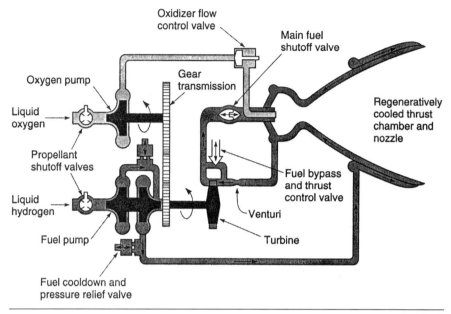

Fig. 4.2-7 Simplified schematic flow sheet of the RL-10B upper-stage LOX/LH$_2$ engine, the first to fly with an expander engine cycle in 1963. Courtesy of Pratt & Whitney, a United Technologies Company. Photo from Pratt & Whitney.

Industries LE-5A engine, which flew first in 1994. A flow diagram of this LPRE is shown in Figs. 11-6 and 11-7.

The *staged combustion cycle* allows a higher performance LPRE (2 to 7% better than a gas-generator cycle LPRE, depending on the design) and about the same performance as a LPRE with an expander-engine cycle. It usually has higher internal propellant pressures and a somewhat higher inert engine mass, when compared to a LPRE with an equivalent GG cycle or expander cycle.[1,2,6,8] It has one or sometimes two preburners that deliver warm gas to the turbines at temperatures typically between 900 and 1350°F (480 to 730°C), and a few have been higher. The hot turbine exhaust is at a higher pressure than in the other two main cycles and is fed through the injector into the combustion chamber, where it is completely burned with all of the other propellants at the optimum mixture ratio. The Soviets were the first to work on a LPRE with this cycle beginning in about 1958. This was done at Nauchno-Technichsekey Kumpleks (NII) Reaction Propulsion Research Institute initially with heavy duty hardware. The first Soviet engine with this dual combustion cycle was first flown in 1961. It was the S1-5400 LPRE designed by V.M. Melnikov in Korolev's vehicle design bureau, and the engine is shown in Fig. 8.11-1. It was developed between 1958 and 1960 and has flown first in an upper stage of the Soviet Venera SLV. The Soviet Union has since developed at least 30 LPRE with this cycle, more than all of the other countries together, and most of them are identified in this book. The Soviets used both storable propellants (NTO/UDMH) and cryogenic liquid oxygen/kerosene with their staged combustion cycle engines. Examples of large LPREs with this cycle are the RD-253 and RD-170 from Energomash (Chapter 8.4). A flow diagram of the RD-253, which was the first large LPRE with storable propellants using this cycle, can be seen in Fig. 4.2-8. A photo of this engine can be seen in Fig. 8.4-25. A flow diagram for the RD-170 is shown in Fig. 8.4-17. Most of the Soviets engines, designed with a staged combustion cycle, had an oxidizer-rich preburner, but their LOX/LH$_2$ engines and a couple of others from the Yushnoye Design Bureau used a fuel-rich preburners.

The U.S. space shuttle main engine was the first engine with a staged combustion cycle that has flown outside the Soviet Union. Its flow diagram is described in Fig. 7.8-18, and the engine is discussed in Chapter 7.8. Its design started in 1972, and it flew for the first time in 1981, which was 20 years after the first Soviet flight with this engine cycle. It uses oxygen/hydrogen propellants, has two fuel-rich preburners, and the turbine exhaust gas drives two of the four turbines. The fuel booster pump is driven by hydrogen gas tapped off the cooling jacket, and the oxidizer booster pump is driven by a liquid turbine using liquid oxygen tapped off the main oxygen pump. The Japanese LE-7 booster engine was also designed with a staged combustion engine cycle, ran on LOX/LH$_2$, and was developed between 1984 and 1990. It is described in Chapter 11 and shown in Fig. 11-9.

The *combustion tap-off engine cycle* is shown schematically in a simplified form in Fig. 4.2-1. It also has been called a topping cycle or a chamber bleed cycle, and it used a bleed or tap-off of a small quantity of combustion gas, which is cooled to a warm gas temperature and used to drive the turbine. The turbine exhaust is either dumped overboard or into the lower part of the diverging nozzle. Reaction Motors, Inc., developed an experimental engine in the early 1960s at a thrust level of about 50,000 lbf, where small flows of combustion gas were tapped off at the injector, and the gas was diluted and cooled with extra fuel. Control of the flow of gas and fuel was by orifices and a venturi flow devices. It eliminated a gas generator and its controls. No other data about this engine were available. Rocketdyne operated a modified version of the J-2 LOX/LH$_2$ engine (used in the second and third stage of the Saturn V SLV) with a tap-off engine cycle in the late 1960s. This experimental engine, called J-2S, had 15% more thrust (265,000 lbf in vacuum) and also featured a number of unrelated improvements and simplifications compared to the J-2 engine. Both engines are described in Chapter 7.8. Hot gas from the injector was bled off symmetrically in cooled passages in the injector and diluted with hydrogen fuel flow from the cooling jacket, and the cooled gas was used to drive the turbine of the fuel TP and the LOX TP. The turbine exhaust gases were injected into the nozzle diverging section and provided nozzle wall cooling and a little more thrust. The engine performance was about the same as an engine with a gas-generator cycle, but the investigators believed that it could be improved. Neither one of these two tap-off engines was considered seriously for a flight application.

One of the problems that has occurred in many LPRE feed systems is *water hammer*.[9] It manifests itself as strong pressure surges, which occur when the flow in a propellant feed system is suddenly stopped or in rapid priming of evacuated propellant lines. It can happen in large and small LPREs. The surge pressures can be high enough to break hardware and cause the engine to fail. Between 1950 and 1970 designers learned how to analyze water hammer in feed systems with branched piping and to include features to mitigate or eliminate this problem. An example of a remedy is a fast-acting relief valve that will open at high pressure and prevent the remainder of the system from experiencing undue pressure. If the LPRE is fired only once, then a burst diaphragm in combination with an explosive shut-off valve has been used. One of these, developed in the Soviet Union, is shown in Fig. 4.7-8. Water hammer also can be alleviated by making the sudden flow changes more gradual.

Engine Systems with Small Thrusters

Small LPRE systems have many of the same basic hardware components listed at the beginning of this chapter, and some of their characteristics have been identified in Table 4.2-1. These LPREs usually had typically four to 16 or more small thrusters, each with a relatively low thrust level, and many operated in a pulse mode (typically 4 to 20 cycles per second).[1] A sample flow diagram of such a system is shown in Fig. 4.2-9 for the INSAT-1 satellite propulsion system.[10] Here the higher thrust TC of 100 lbf or 445 N is used for orbit insertion or orbit changes, and the other thrusters at 5 lbf or 1.1 lbf each are for attitude control and are supplied in two redundant sets of six. This particular set of thrusters was delivered by the Marquardt Corporation, which has since become a part of Aerojet. Although the thrusters are shown next to each other, they are really mounted in different parts of the flying vehicle. A pair of thrusters firing simultaneously provides roll torque. All of these systems used storable propellants, such as NTO/UDMH or NTO/MMH, fast-acting, no-leak propellant valves, and almost all had a pressurized feed system.

Although experimental work on the engine's principal or main thrust chamber started in the 1920s, the work on small thrusters and their reaction control systems did not really start until the 1950s. At that time it was understood how the flight paths and the angular orientation of launch vehicles and satellites had to be controlled. Small multiple thrusters (typically between 0.2 and 200 lbf thrust) appeared to be suitable for this purpose. The development of larger TCs and larger LPRE evolved from the work done earlier by pioneers and amateurs, and the satellite missions came from studies done by early pioneers. However, the requirements for engine systems with small thrusters came as the result of earnest system studies (of various space flight missions) performed by experts in several countries between 1940 and 1950. The Marquardt Company (now part of Aerojet) did such a study in 1956, and the R&D Institute for Mechanical Engineering in the Soviet Union did it also. There were several options for controlling the flight path and the vehicle attitude, as discussed in Chapter 4.9. One of the options was to use multiple small TCs or thrusters. These thrusters would be arranged in pairs in such locations on the vehicle that moments or torques could be applied to the vehicle for pitch, yaw,

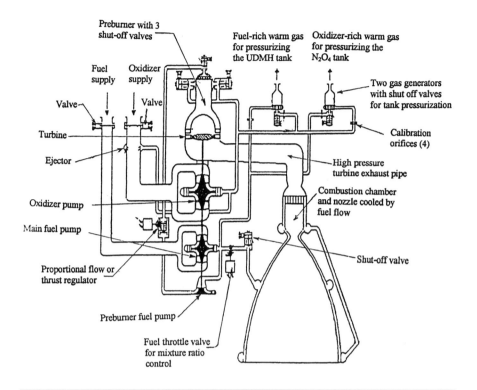

Fig. 4.2-8 Simplified flow diagram of the RD-253 LPRE with a staged combustion engine cycle using NTO/UDMH. Six of these engines propelled the first stage of the Russian Proton SLV. Courtesy of NPO Energomash. Personal communication, V. K. Chvanov, first deputy to general director and general designer, NPO Energomash.

and roll maneuvers.[1,2] A pulsing mode with short burn times and many restarts was found to be preferable over continuous burns or longer periods of firing. Furthermore it allowed attitude control and minor vehicle maneuvers at a time when the main propulsion system was not operating. Other requirements also emerged from these studies; for example, the engine system for some space missions had to be available for firing after long periods of coast time (in some flights it is months or years) in orbit. The key component in these small engine systems is the thrust chamber or thrusters, which can be supplied with a bipropellant or a monopropellant, and they are discussed in Chapter 4.6.

Almost all early small LPREs have used cold-gas-pressurizing systems, and they are categorized in Fig. 4.2-1. For missions where there are no large side accelerations and no restart in gravity-free space a simple cold-gas feed system is used. Here the high-pressure pressurizing gas is in direct contact with the propellant in the tank. The pressurizing gas has to be selected, so that little or none of it will dissolve in the liquid propellant and dilute the propellant. Dissolved gas can come out of solution and form small bubbles, when the pressure is lowered, for example, when the propellant enters the combustion chamber. There have been a few instances when these bubbles caused combustion problems. For missions with major fast maneuvers (major side accelerations), with pulsing (multiple restarts), and in gravity free space, a positive expulsion tank must be used. Several devices for positive expulsion are shown in Fig. 4.7-3 and were discussed earlier.

The Soviet Union is the only country where a turbopump system has been used to feed small thrusters. A common Soviet concept for flight-path control is to use a large LPRE, which is fixed (not gimbaled), and four smaller hinge-mounted vernier thrust chambers. These four thrust chambers serve to provide not only pitch, yaw, and roll control, but they also provide some of the axial principal thrust, typically between 10 and 20% of the total thrust. In some of the Soviet LPREs, these four vernier thrusters are supplied with propellants by their own small turbopump feed system. Depending on the vehicle, the engine, and the specific application, the thrust of a vernier thrusters can vary from about 400 to 8500 lbf thrust each. Most of these have too high a thrust to be considered a small thruster, but a couple of the vernier engines do qualify to be small thrusters.

Engine System Analyses

The early analyses (1930s and 1940s) of a complete engine system were essentially calculations of the amount of propellants and high-pressure gas needed, the propellant flows and pressure drops through the various major components (through pipes, valves, injector, cooling jacket, or orifices), an analysis of the gas pressure losses and flows (pressurizing gas), and the pressure drops of the combustion gases in the supersonic nozzle. These calculations helped to establish key parameters such as the sizes of the propellant tanks (for a given ullage), sizes of calibrating orifices, sizes of injection holes, gas regulator settings, or pressures across the inner wall of nozzles and thus the thickness of this wall. The purpose was to obtain the desired mixture ratio and propellant flows in order to achieve the specified thrust and specific impulse. Early heat-transfer analyses were then not accurate, used an approximate analysis in just a few locations, were hampered by the lack of good data on the coolants or the materials of construction, and were enlightening, but not really useful. A further purpose was to compare the actual measurements with the calculated data, to refine the analysis, and to attempt predictions of future LPREs. The author remembers that there were no calculators then available; these calculations had to be done by pencil, paper, and a slide rule.

As calculating machines and later computers became available, it became easier to make these analyses and to obtain more data for plotting better curves. The engines and the analyses became more complicated as turbopumps, gas generators, and additional engine functions (such as throttling or restart in flight) were added. As the engine specifications required tighter tolerances on the thrust, or the specific impulse, or the mixture ratio, the accuracy of the engine system analysis had to be improved. So various refinements were added to the analyses between 1955 and 1975. As more test results became available, it was possible to improve some of the estimated pressure drops, transients, or evaporation losses. A better understanding of the flow through impellers and turbine blades helped to form a better set of algorithms for turbines and pumps. The development of finite element analyses was a major milestone in the analysis of LPRE parameters, such as stress analysis, heat transfer, complex flow passages, and three-dimensional profiles of pump impellers or turbines blades. Transient behavior during start, stop, or thrust change became a separate set of system analyses.

Around the 1950s the thermochemical analysis of combustion reactions was added, and later the changes in thermochemical equilibrium of the nozzle flow were added. This allowed better estimates of the combustion temperature, the gas flow kinetic behavior, the reaction products and their predicted properties, and estimates of specific impulses for different chamber pressures and nozzle expansion area ratios. More recently two- and three-dimensional simulations of the kinetic changes in nozzles have become available. A heat-transfer analysis can allow estimates of certain key wall temperatures. For several decades it has been possible to simulate the engine operation and behavior by appropriate algorithms in computer programs. The two papers cited in the references are just a small sample of the many papers that have been written on the subject.[11,12] New applications of these analyses

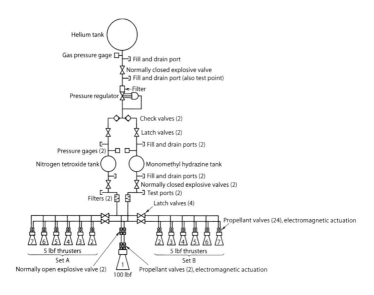

Fig. 4.2-9 Simplified schematic flow diagram of the INSAT-1 bipropellant propulsion system (developed in 1980) with a pressurized-gas feed system, one 100-lbf thruster (for apogee and other maneuvers), and 12 5-lbf thrusters for attitude control and stationkeeping. Courtesy of the Marquardt Company, which today is part of Aerojet; copied with permission from Ref. 10.

have emerged. For example, data from the system analysis can be used in preliminary design, in actual design, in comparing actual test results of selected engine parameters with the calculated predicted results, or in determining certain detail design changes, if the engine thrust is to be uprated. It is possible to determine the influence of potential design changes or design tolerance changes upon the performance of the LPRE. The analysis of rocket exhaust gas plumes (open flames outside of the nozzle) has made big advances in the last several decades.

A key improvement in the accuracy of the analyses was caused by a better understanding of the loads (forces, pressures) imposed on the various pieces of hardware under different operating or thermal conditions. Another key improvement over the last 40 years was to obtain more accurate data on the various materials and their properties at different temperatures and conditions. Improvements in all of these various data and analytical methods have been gradual and have continued to this day.

For the last three decades it has been possible to develop health monitoring systems. They monitor key engine parameters, such as positions of key valves, turbine or pump speeds, various pressures, and temperatures. Among other things, they can perform, in real time, a comparison between the actual set of engine parameters and the calculated parameters; if there is a major difference between these two quantities, it can generate a signal for a computer to determine, what corrective action, if any, should be taken. These systems have been useful in ground tests by quickly taking a remedial action, such as engine shutdown, before there is serious engine damage, and this has saved some hardware. There have been very few opportunities to use the outputs of a health monitoring system in flight and to take rapid remedial action during flight; the author knows of only one such instance, where a faulty engine had to be shut down, but the mission was completed.

Computer-aided design, computer-assisted manufacturing, and computer-aided inventory control, and computer parts histories have now been used in some of the LPRE companies. They have been useful in saving effort and hardware. Only in the last decade has it been possible to design a complete LPRE on a computer.

These system analysis programs and the key people, who know how to use these analysis programs, have become a very important asset to organizations in this business. Each major LPRE organization has its own sets of computer programs, and some of these have been considered to be proprietary or a trade secret.

References

[1]Sutton, G. P., and Biblarz, O., *Rocket Propulsion Elements*, 7th ed., Wiley, New York, 2001, Chap. 6, pp. 197–241.

[2]Huzel, D. K., and Huang, D. H., *Design of Liquid Propellant Rocket Engines*, AIAA, Washington, DC 1992, pp. 224, 226.

[3]Hearn, H. C., "Design and Development of Large Bipropellant Blowdown Propulsion Systems," *Journal of Propulsion and Power*, Vol. 11, No. 5, Sept.–Oct. 1995.

[4]Hearn, H. C., "Thruster Requirements and Concerns for Bipropellant Blowdown Systems," *Journal of Propulsion and Power*, Vol. 4, No. 1, Jan.–Feb. 1988, pp. 47–52.

[5]Rollins, J. R., Grove, R. K., and Jaeckle, D. E., Jr., "Twenty-Five Years of Surface Tension Propellant Management Systems Design, Development, Manufacture, Test, and Operation," AIAA Paper 85-1199, 1985.

[6]Manski, D., Goertz, C., Saβnick, H-D., Hulka, J. R., Goracke, B. D., and Levack, D. J. H., "Cycles for Earth-to-Orbit Propulsion," *Journal of Propulsion and Power*, Vol. 14, No. 5, Sept.–Oct. 1998.

[7]Hill, P. G., and Peterson, C. R., *Mechanics and Thermodynamics of Propulsion*, 2nd ed., Addison Wesley Longman, Reading, MA, 1992, p. 543.

[8]Hawk, C. W., "Lessons Learned in the Development of Staged Combustion Liquid Rocket Engines," *Proceedings of the International Symposium on Space Technology and Science*, Agne Publishing, Tokyo, 1994, pp. 20–40.

[9]Prickett, R. P., Mayer, E., and Hermel, J., "Water Hammer in Spacecraft Propellant Feed Systems," *Journal of Propulsion and Power*, Vol. 8, No. 3, May–June 1992, pp. 592–596.

[10]Yeh, T. P., "Bipropellant Propulsion Performance and Propellants Remaining, Prediction for INSAT-1B," *Journal of Propulsion and Power*, Vol. 8, No. 1, Jan.–Feb. 1992, pp. 74–80.

[11]Iffly, A., and Brixhe, M., "Performance Model of the Vulcain Ariane 5 Main Engine," AIAA Paper 99-2472, 1999.

[12]Kalnin, V. M., and Sherstiannikov, V. A., "Hydrodynamic Modelling of the Starting Process in Liquid Propellant Engines," International Astronautical Federation, Sept.–Oct. 1977; also *Acta Astronautica*, Vol. 8, 1981, pp. 231–242.

4.3 Large Thrust Chambers

The most significant component of a LPRE is its *thrust chamber* (TC).[1,2] It consists of three key components, namely, 1) a *combustion chamber*, where the propellants burn and undergo a set of chemical reactions at elevated pressure creating a very hot gas; 2) a *supersonic nozzle*, where the hot gas is thermodynamically accelerated to supersonic velocities; and 3) an *injector*, where the propellants are introduced into the chamber, metered, broken up into small liquid droplets, and evenly distributed (in flow and in mixture ratio) over the cross section of the chamber. Furthermore many TCs have a cooling jacket, and many are mounted on a gimbal or hinge joint to allow thrust vector control (TVC). Many of the LPREs have more than one thrust chamber per engine.

Chamber Configuration

The chamber configuration has changed since the early decades. Several of the early investigators had a different concept of the combustion process and injector functions than are commonly accepted today. As a result, their early chamber designs were different in shape or configuration, and they were usually larger in volume. A TC with a cone-shaped forward end was used by early German investigators. A sketch of one of Hermann Oberth's TCs can be seen in Fig. 5-20. They believed a cone shape would enhance the mixing and atomization processes; it was assumed that the gases would not be very hot at the cone. In many of the early designs, the injection holes are located on the side of the chamber, just upstream of the converging nozzle section, and the injection holes or injection spray elements are often pointed toward the center or partly forward toward the vehicle's nose. An example of an early thrust chambers is shown in Fig. 8.1-6. Later the conical chambers and side injection near the nozzle turned out to be inefficient. Some of the designers of the 1930s used spherical chambers or nearly spherical chambers, like the German V-2 shown in Fig. 9-4. By testing different TC configurations, it was learned that a cylinder-shaped chamber gives good combustion and is easier to manufacture. It was first conceived by Goddard in the 1920s and used by the Germans as well as the Soviets in the 1930s. A cylindrically shaped chamber (or a slightly tapered cylinder) are today well accepted and used for all thrust chambers.

Many early TCs had a *relatively large chamber volume*. In a smaller chamber volume the combustion process would not be complete. The chamber volume had to be large enough to allow reasonable complete atomization into small droplets and good mixing and burning with the design of injectors then available. A smaller chamber volume would have resulted in poor combustion efficiency and low performance. With improved injection, better mixing, and hotter combustion, it has been possible to substantially reduce the chamber volume by a factor of two or three by about 1950. This allowed a smaller and lighter chamber, which still gave very good combustion efficiency.

Cooling of Thrust Chambers

The cooling of TCs has truly been a major challenge to the early LPRE investigators. Because the gas temperature usually exceeds the melting point of the TC's metal wall material by perhaps a factor of two, there were frequent burnout or meltdown failures. Over the years a variety of different cooling methods have been conceived, developed, and flown in TCs.[1,2] Table 4.3-1 lists some of the more common types, and each will be discussed.

With uncooled TCs the time of operations is limited by the heat-absorbing capacity of its wall, typically 0.5 to 3 s. All of the early TCs had many burnout failures, and most occurred in the nozzle region. The early investigators of the 1930s and 1940s understood that some form of chamber wall insulation and/or cooling would be necessary. *Uncooled thrust chambers* with a single wall have been used and are still being used when the burning time is just a few seconds, such as in certain guided missiles and as alternate TCs in short development tests, such as for combustion instability tests or engine igniter development. Several attempts were made to use air cooling in the 1930s, and cooling fins were put on TCs by some U.S. amateurs and by several early Soviet investigators, but this approach failed because air cooling does not remove anywhere near enough heat.

Film cooling was the first viable solution to cool TCs, and it allowed a longer operating duration. It was first applied by Goddard in the late 1920s as described in Chapter 5.2. Usually some fuel is injected at a low velocity along the chamber walls, and a film or boundary layer of evaporated cooling fluid protects the inner TC wall from direct exposure to the high combustion gas temperatures.[1-3] In one of his early film-cooled TCs, Goddard used 100% of the fuel for film cooling (injected through slots at the injector) to allow operation for more than a few seconds (see Chapter 5.2). He later was able to run with 25 to 50% of the fuel for film cooling. Although this prevented burnouts, the early film-cooled tests caused a relatively large (probably 8 to 25%) loss in specific impulse as a result of incomplete mixing and incomplete combustion. One of Goddard's injectors, which used only about half of the fuel for film cooling through holes at the periphery of the injector face is shown in Fig. 5-7. Several early Soviet engines (1933) also depended on film cooling as the sole method of heat protection for the chamber and nozzle walls as mentioned in Chapter 8.1. With further development and more effective boundary layers it was possible to reduce the performance penalty of the cooler boundary layer to less than 3% by the 1960s. The French Viking engine, which propels the first and second stages of the Ariane SLV, uses fuel film cooling as its sole cooling method, has a single chamber wall made of a high temperature alloy, and a composite material nozzle throat insert. It is run on NTO/UDMH with 25% hydrazine hydrate. The initial version had an initial sea level thrust of 617 kN or 138,700 lbf and is shown in Figs. 10-7 and 10-8. This engine was originally designed in the 1970s, and a modified uprated version is still used today relying on film cooling.

Table 4.3-1 Different Methods of Cooling Chambers and Nozzles of LPREs

Method	Typical features	First implementation, example, and comments
Thrust chamber and nozzle throat region		
1) No cooling	Thick, single wall	1923, Goddard, USA; 1931, Glushko, USSR
2) Film cooling as only cooling method	Boundary layer of cool gas at walls	Late 1920s, Goddard, USA; French TCs for Ariane SLV, 1965–2003
3) Regenerative cooling	a) Double wall construction	1934/1935, Eugen Sänger, Austria; 1933, USSR, nozzle cooling jacket only; circa 1934, USSR, chamber and nozzle; 1938, James Wyld, USA
	b) Formed tubes, brazed together	USA, 1947–1955, RMI, Aerojet, and Rocketdyne
	c) Milled channel construction	1953, USSR RD-107/108 throat region only; 1958, Bölkow; 1972, Rocketdyne SSME;
	c) Corrugated intermediate sheet	1949, USSR
4) Ablative liner (large TC)	Has supplementary film cooling	1970, TRW Lunar descent engine, Aerojet Apollo Service Module LPRE
5) Interregen cooling	Uses beryllium	Rocketdyne, circa 1972; for small thrusters
Nozzle extension or lower nozzle region		
1) Regenerative cooling	a) Unidirectional	1958, RMI, X-15;
	b) Upflow in every other tube	1955, Rocketdyne Atlas, Thor, and Navaho LPREs
2) Radiation cooled or ablative cooling	Made of stainless steel, niobium, ablative, or carbon fibers in a carbon matrix, or ceramic matrix composite	Many examples of upper-stage engines and small thrusters with nozzle extension; for larger LPREs; it can be designed to be extendible during flight
3) Dump cooling	Turbine exhaust gas forms boundary layer	1963 Rocketdyne F-1 engine, nozzle-exit section only

In the late 1930s during the development of the V-2 thrust chamber, the Germans were the first to use film cooling only as a *supplementary cooling method* (by slowly injecting a little fuel at three different stations in the chamber); this augmented the regenerative cooling, which was then considered to be marginal. The three film-cooling manifolds and the rows of low-flow injection holes can be seen in Fig. 9-4. One of these V-2 engine film-cooling manifolds is actually in the diverging section of the nozzle, and the coolant from this location might not have been needed. Subsequent TCs did not have film-coolant injection in the diverging nozzle section. Supplementary film cooling was also used by the Soviets in the beginning in the late 1940s, for example, in the thrust chambers of the RD 170 seen in Fig. 8.4-18 and the RD-0110 seen in Fig. 8.5-6; the improved models of these LPREs are still flying today. The liquid film coolant is usually injected from separate manifolds at one or more stations along the chamber wall or the converging nozzle section.

In the United States this supplementary cooling method for a large engine was first used on the Redstone TC (1949) as described in Chapter 7.8. In 1950 the first film-cooling injection from the injector was used in this engine by drilling extra fuel holes at the periphery of the flat injector face of this LPRE. In this TC the film-cooling design concept of injecting film-cooling fuel from the injector was first proposed by the author, but it was then not known to the Rocketdyne designers that Goddard had done this more than 20 years earlier. This injection of a little extra fuel at the injector periphery (typically up to 3%) has been used ever since that time on many large TCs. These extra injection holes are still used effectively today on small bipropellant thrust chambers (5–300 lbf thrust). The TCs in the United States have used the fuel as the film coolant, but the Soviets use the oxidizer as the film coolant in most of their larger TCs. At the time there was concern in the United States about using the oxidizer as a film coolant because an oxidizer-rich boundary layer was believed to cause potential oxidation of the hot wall material.

Transpiration cooling or *sweat cooling* is really a special variation of film cooling, where a porous wall or a wall with many film-cooling holes is used to inject a small flow of coolant (usually a part of the fuel flow) at a low velocity. This coolant absorbs heat from the metal wall, also heat from the reaction gas, forms a protective gaseous boundary layer of relatively cool gas, and this reduces the heat transfer and the wall temperatures. Since the 1960s, a number of organizations have tried transpiration-cooled combustion chambers, but none have been fully satisfactory, and none are known to have flown. One such transpiration-cooling concept for a chamber and nozzle was conceived and tested at the 50,000 lbf thrust level by Pratt and Whitney; it used wafers with fuel passages and is explained in Figs. 7.10-5 and 7.10-6. However transpiration cooling has been successful for use on the faces of large injectors. This was first accomplished by Pratt & Whitney in the late 1950s as described in Chapter 7.10.

Regenerative cooling is historically the second method. Early attempts were with a stagnant (nonflowing) water, but the coolant did not absorb enough heat, and the operating duration was limited to a few seconds. It would fail at longer durations by burning a hole into the nozzle throat or chamber inner wall. An example of a thrust chamber, which was immersed in a bath of water, is the first Soviet TC (ORM-1) shown in Fig. 8.1-1. Some of the TCs of the amateur rocket societies also had stagnant water or low-velocity water in a container around the TC. It was soon learned by the early pioneers that the coolant needed to flow at a good velocity. Today the cooling velocity is typically between 5 and 40 ft/s, depending on the coolant, chamber pressure, mixture ratio, and other parameters. One of the propellants, usually fuel, is routed or circulated through a cooling jacket around the chamber and the nozzle walls, before this fuel is injected into the combustion chamber.[1,2] The early regeneratively cooled TCs were made of steel sheets and used a double-wall construction; the cooling passages were between the inner hot wall and the outer cooler wall. An example of a double-wall design is shown in Fig. 4.3-1. This TC was developed by General Electric in the 1940s. The fuel flows through spiral wound cooling passages between the two walls. The pitch and the height of the spiral change, so that the cooling velocity is highest in the throat region, where the heat transfer is the highest.

The term regenerative cooling was borrowed from steam power plant practice. The heat absorbed by the propellant in the cooling jacket is not lost; it raises the temperature of the cooling fluid, and this heat is regenerated; this extra thermal energy slightly augments the heat of combustion and increases the specific impulse very slightly. It was logical to introduce a cooling jacket around the nozzle, where the burnouts occurred most frequently. Therefore in the early Soviet TCs (circa 1933) only the nozzle was regeneratively cooled, and the chamber was uncooled or film cooled as shown in Fig. 4.3-2. The

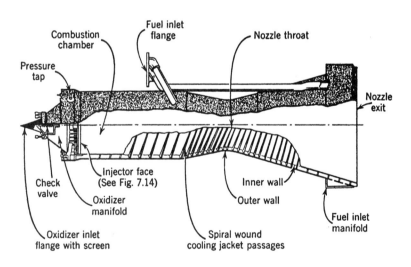

Fig. 4.3-1 Partial section of the thrust chamber of the LPRE of the Hermes A-3 flight vehicle with a double-wall construction (around 1950). Courtesy of General Electric Company.

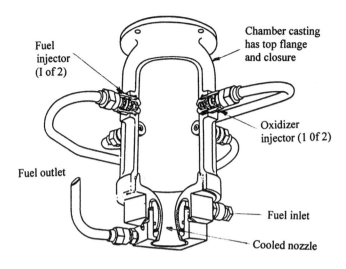

Fig. 4.3-2 Only the nozzle is regeneratively cooled in this simple sketch of the ORM-50 thrust chamber developed and tested in the Soviet Union in 1933. Copied from Ref. 3 in Chapter 8.1.

injector spray elements are located at about halfway down the chamber wall. By 1934 the Soviets tested a design with two cooling jackets: one around the nozzle (fuel cooled) and one along the chamber (oxidizer cooled). It required considerable external piping, the joint between the two jackets had problems, and the forward end of the chamber was still uncooled. The first truly regenerative cooling (circa 1935) in the USSR had a single cooling jacket around both the chamber and the nozzle and also had cooling passages behind the injector face. Several types of regenerative cooling schemes were tested in Germany around 1936 (see Chapters 9.1 and 9.2) and in Russia between 1935 and 1937 (see Chapter 8.1). These early regenerative cooling accomplishments in Germany and the USSR were kept secret at the time and were unknown in other countries. The Austrian pioneer Eugen Sänger published an article about his demonstrating regenerative cooling in 1936, after he had accomplished cooling with water in a similar cooling jacket beginning in 1934. Sänger's efforts are mentioned in Chapter 5.4. The American James Wyld, who was an amateur rocketeer at the time, designed and tested the first regeneratively cooled thrust chamber in the United States in 1938. His sketch is shown in Fig. 4.3-3. It was of the double-wall construction without any spiral ribs. His first attempts were with aluminum TCs, and they were unsatisfactory and failed. When he switched to monel, an alloy of nickel and copper, he was able to run for more than a minute, a major achievement at the time. His TC is mentioned again in Chapters 5.5 and 7.2.

It was not until the late 1930s and early 1940s that creditable heat-transfer analyses were made in several countries.[2,4,5] These studies gave heat-transfer

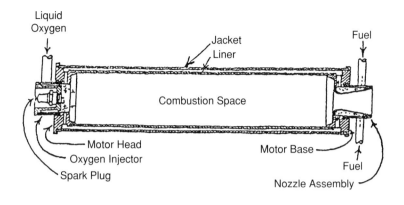

Fig. 4.3-3 Sketch by James Wyld of the first American regeneratively cooled double-wall thrust chamber. Copied with permission from the *Journal of the American Rocket Society*. Circa 1939.

data, coolant temperature rise, and wall temperatures at different TC locations. The studies showed that the highest heat transfer and wall temperatures are at the nozzle throat region and the second highest normally in the converging nozzle section and the chamber. As a result, the cooling jacket passages at the nozzle throat were made smaller, thus increasing the local cooling velocity and improving the local heat absorption. A higher cooling velocity can be obtained by putting in spiral internal guide vanes into the cooling passages as seen in Fig. 4.3-1. The allowable stresses in the inner wall material decrease as the wall temperatures rise and the materials become weaker. Enough heat has to be absorbed by the coolant to keep these walls at a sufficiently low temperature, so that the wall material is strong enough to withstand the stresses imposed by the chamber gas pressure, coolant pressure, thermal gradients, and other loads.

Figure 4.3-4 shows simple sketches of several of the common cooling schemes, each for a piece of an arc, which is cut out from the cooling jacket at the combustion chamber region. It also shows the dramatic progress that has happened in the design of regenerative cooling jackets, and each design will be discussed subsequently in this section.

In the double-wall construction there are two walls and sometimes spiral ribs to form a spiral flow passage as exemplified by Fig. 4.3-1. Some of the early double-wall TCs did not have adequate provisions for the thermal expansion of the inner wall, which gets hot and expands relative to the cooler outer wall. The material of the inner wall would locally exceed its yield point during the firing; upon cooldown after the run it can wrinkle or form local cracks, most often in the throat region. With each restart the inner wall would yield more, and the cracks would grow until the chamber failed. Different clever expansion provisions were then designed into restartable TCs to allow repeated operation without deformation or cracks.

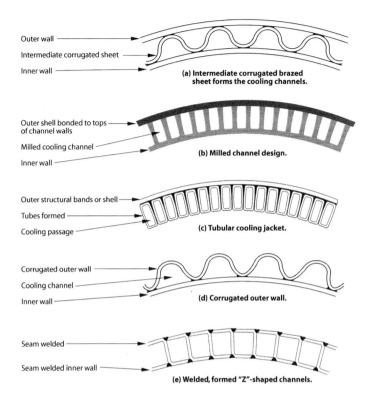

Fig. 4.3-4 Simplified sketches of sections through a portion of the cooling jacket of several different cooling schemes in regeneratively cooled thrust chambers.

This double-wall design is still used effectively today in some thrust chambers with relatively low chamber pressures and thrusts between approximately 400 and 8000 lbf. The exact values depend on chamber pressure, mixture ratio, propellants, nozzle-area ratio, and other factors. For higher thrust the walls became too thick, too heavy, and too difficult to cool, so that some other cooling technique had to be found.

The heat-transfer analyses indicated that the inner wall temperatures can be reduced if the wall is thin, and this led to *the tubular cooling jacket* idea in the United States. The principle of this cooling scheme is also shown in Fig. 4.3-4. In the late 1940s three U.S. companies (Reaction Motors, Aerojet, and Rocketdyne) came up with the idea to use a bundle of flattened and shaped tubes for the passages of a cooling jacket. Reaction Motors, Inc., claimed to have conceived of this idea in 1946 or 1947 and was granted a U.S. patent on the tube bundle idea in 1965, about 19 years after they conceived the idea (U.S. patent #3,190,070 filed on 5 April 1950). RMI tested such a chamber in 1948 or 1949. The RMI LPRE for the X-15 research aircraft had a tubular cooling jacket at the 50,000 lbf thrust level in the 1950s, and it flew successfully for the first time in 1961. In 1948 Aerojet built and tested a small TC with flattened aluminum tubes, which were welded together. This aluminum TC design was not pursued, but Aerojet built LPREs with stainless-steel tubes beginning in 1955 for the Titan TC and flew them successfully beginning in 1960. In 1947 Rocketdyne tested various tubes and their fabrication method. In 1948 Rocketdyne began to design large thrust chambers that used a brazed-together bundle of formed and shaped thin-walled stainless-steel tubes for circulating the fuel around the combustion chamber and nozzle. It was first used with a large thrust chamber (120,000 lbf) of the NAVAHO missile, which first flew in 1956. This was the first flight of a tubular cooling jacket. Figure 4.3-5 shows an early version of the Thor engine, which used the same tubular design. The tube bundle of the Thor LPRE flew first in 1957. This drawing was originally prepared by Rolls Royce in England, after that company had obtained a U.S. license to copy and build this tubular TC. Two of these engines propelled the British Blue Streak vehicle as discussed in Chapter 12. More details about tubular cooling jackets can be found in a section on tubular TCs in Chapter 7.2.

The flow through the tubes is either unidirectional (usually from the nozzle exit to the injector), as in RMI X-15 LPRE just mentioned, or it is in different flow directions in adjacent tubes. This unidirectional flow toward the injector is used in most of the nozzle and chamber regions of the thrust chambers in order to obtain a high cooling flow velocity in the critical heat-transfer regions of the TC. In the lower part of the nozzle, the heat transfer is very much lower, and a low cooling velocity is usually adequate. Thus a low velocity can be used. The coolant flow area increases with the nozzle diameter, and therefore the flow can often be in different directions in adjacent tubes. The flow in one tube is down toward the nozzle exit, and the return flow is in the opposite flow direction in the adjacent tube. This avoids a heavy large coolant inlet manifold at the nozzle exit, results in lower coolant velocities, and reduces the pressure drop in the cooling jacket.

The fabrication process of shaping the tubes started with long pieces of straight tubes. The tube was swaged or made smaller in diameter in the middle of the cut tube piece (in order to increase the flow velocity in the throat region), but remained full diameter at both ends. The tapering was hydraulically formed or rolled by a manufacturer of tapered golf shaft shanks. Next the tapered tubes were bent to the chamber/nozzle contour. Finally the cross section of the tube was changed to a more rectangular shape by hydraulic presses.[7] The tubes were cut to the final length and assembled in a bundle. Instead of a separate outer wall, some designs use a wire wrap or bands of high-strength steel to contain the gas pressure loads as seen in Fig. 4.3-5. The tubular method did away with a separate relatively thick inner thrust chamber wall, saved considerable inert mass, improved the heat transfer, and reduced the temperature drop across the inner wall. Near the exit of nozzles with a high nozzle-area ratio, the designer applied stiffening bands around the brazed tubes, in order to keep the nozzle from deviating from a circular shape as it was subjected to pressures, particularly when the nozzle gas flow is overexpanded. TCs using the following tube materials were manufactured: ductile

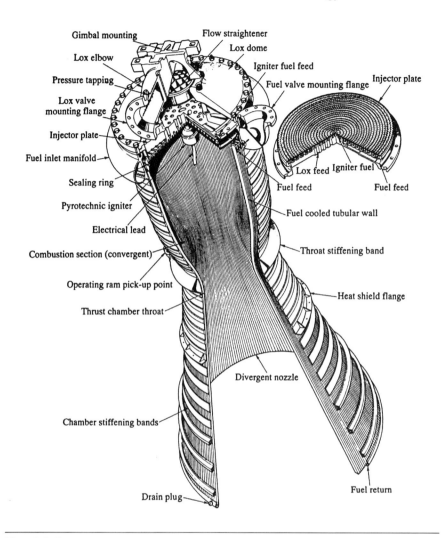

Fig. 4.3-5 Sectioned perspective drawing by Rolls Royce of the British Blue Streak RZ-2 thrust chamber. It uses LOX/kerosene and is a modified copy of the early version of the Rocketdyne Thor TC. The nozzle-throat diameter is about 15 in. Courtesy of Rolls–Royce (U.K) and The Boeing Company, Rocketdyne Propulsion and Power.

stainless steels, copper alloyed with small amounts of other metals, aluminum alloys, titanium, and nickel.

The brazing of the assembled tube bundle was done in a special furnace with a reducing atmosphere and special fixtures to hold the tubes in place. These fixtures had a special oxide coating, so that the brazing material would not bond the fixture to the TC. The selection of the brazing material was critical, and several were found to be acceptable. Making the tubes out of nickel is desirable because it had a better thermal conductivity and ductility. This was tried by Rocketdyne and found to be very difficult because of problems with hot nickel deteriorating in the presence of certain small impurities (sulfur) in the fuel or the brazing material.* Nevertheless, RMI learned in 1950 how to build a tubular TC with nickel tubes for the X-15 LPRE. The Japanese and others also learned how to make TCs with nickel tubes.

Tubular TC walls have been used successfully ever since 1950 in many LPREs. This includes those used by Pratt & Whitney in their RL-10 engines, the Rocketdyne engines for the Atlas, and Thor missiles, and the Saturn space launch vehicles, the Aerojet engines for the Titan vehicle series; the French Ariane upper-stage LPRE; or the Mitsubishi engines for the Japanese SLVs.

The Soviets did not use tapered tubes, but a clever different construction for achieving thin inner chamber walls. It has a precisely shaped relatively thin inner wall of a high-conductivity material and a strong outer wall, held together by a thin corrugated sheet metal, thus forming (nearly triangular) channels for the cooling flow. These three walls were brazed together as illustrated in item a) of Fig. 4.3-4. The outer wall, which was relatively cool and carried the loads, was made of a strong stainless-type steel, and the hot inner wall and the corrugated sheets were made of a high conductivity material, such as a bronze or copper alloy. The higher conductivity reduced the inner wall's temperature gradient. The connection between the walls was made by a corrugated (wave-shaped) thin sheet, which was brazed to the other two. The brazing material was manganese based and was coated in selected locations to the inner side of the outer wall, the outer side of the inner wall, and on both sides of the corrugated wall. The brazing was performed in a special furnace with a vacuum atmosphere (or perhaps an inert gas atmosphere), and this bonded the three walls together. This method has been used successfully in most Soviet and Russian large- and medium-sized TCs. This design was devel-oped in the late 1940s and is discussed in Chapters 8.4 and 8.6. This cooling jacket configuration can in many cases be somewhat lower in inert mass than an equivalent tubular cooling jacket.

Another method of regenerative cooling for very high heat fluxes was developed probably in the late 1940s in the Soviet Union and again independently in the 1960s in Germany and the United States. It is item b) in Fig. 4.3-4. It has essentially rectangular *channels* (of different cross sections in the chamber, nozzle throat, or nozzle exit) machined or milled into a solid inner wall in a straight (and not spiral wound) pattern.[1,2] The outer wall is built over the channels as shown in a section in Fig. 4.3-4. The milling operation of the inner wall of the chamber and nozzle throat area of the Rocketdyne space shuttle main engine is shown in Fig. 4.3-6. This design is used, for example, on some upper-stage engines developed by Aerojet and in the throat region of a number of Soviet engines. In the United States the outer wall was electroformed, in the Soviet Union the outer shells were brazed to the milled channel inner piece.

Some people have experimented with a cold isostatic pressing of powdered metals to form the outer wall. This author believes that the Soviets used a pres-

---

*Personal communication, John Holchek, The Boeing Company, Rocketdyne Propulsion and Power.

Fig. 4.3-6 View of the milling operation that cuts vertical cooling passages into the outside of the inner copper-alloy wall of the thrust chamber of the space shuttle main engine. The coolant channels are cut to variable depth in order to vary the flow velocity of the coolant. Courtesy of The Boeing Company, Rocketdyne Propulsion and Power, Canoga Park, CA.

sure brazing process to obtain a good fit and small enough gaps, so that the brazing material will flow, bond, and fill the gaps. The milled channel construction can have very thin inner walls, and it can absorb more heat and keep the inner walls cooler than the other types of regenerative cooling. The Soviet designers have used channel wall construction only in the critical nozzle throat regions of some of their TCs; a corrugated sheet construction was used in the chamber region and the exit section of the nozzle. This idea of using two different coolant passage designs in the same TC cooling jacket has been used in the USSR RD-107, RD-0110, and the RD-170. It allows the use of high-conductivity material in thin inner walls, and this promotes a low temperature gradient across the wall and thus low thermal stresses with good heat transfer.

An alternate to channel design discussed in the preceding paragraph is an all-welded design illustrated in item d) of Fig. 4.3-4. It avoids the brazing or electrodepositing processes used in the fabrication of milled channels or corrugated intermediate sheets, which can have unbonded areas that cannot be easily inspected by visual means. Furthermore the brazing material can become weak at a lower temperature than the wall material. The machine-made welds can withstand a somewhat higher temperature, can usually be visually inspected, and require special welding machinery. Using the same material for the inner and outer walls, this design compromises the properties of the wall (high conductivity for inner wall, high strength of outer wall). This welded construction has been developed in Europe and has been used on some nozzle-exit sections of LPREs in upper-stage engines for the Ariane SLV.

The most *critical material* in a TC is the *inner wall* in the region of highest heat transfer, namely, in the nozzle throat region and the adjacent converging nozzle section. In a tubular cooling jacket it is that part of the tube which is exposed to the hot gases. The hot wall material requires a good strength at high temperature and high thermal conductivity. When the cooling liquid is NTO, kerosene, UDMH, or MMH, a stainless steel or a special alloy steel are used today. The RMI XLR-99 (1958) flown with the experimental X-15 aircraft initially used nickel tubes, and the Soviets used titanium in the 1960s. These two wall materials are no longer used today.

A copper alloy is used for the inner wall or the tubes of a LPREs, when hydrogen fuel is the cooling fluid. Copper is an excellent heat conductor, but it becomes weak at the high temperatures, typical of inner walls of the cooling jacket. Therefore several copper alloys with small additives have been formulated especially for this application beginning in about 1960. They are much stronger at elevated temperatures than pure copper, still have an acceptably high conductivity, and are still used today. NARloy-Z was formulated at Rocketdyne in the 1960s, and it consists of pure copper with small amounts of silver and zirconium.* It was based on some related German alloys. The USSR has used a similar copper material with small hardening additives in the inner walls of their LOX/LH$_2$ engines. These are examples of new materials that were especially formulated for LPREs.

For smaller thrust levels the heat rejected by the hot gas to all of the internal walls is greater than the capacity of the cooling propellant to absorb this heat without excessive boiling and therefore regenerative cooling is not usually applied to small thrusters. This is discussed further in Chapter 4.6.

An unusual cooling jacket is shown in Fig. 4.3-7, and it corresponds to sketch d) of the regenerative cooling designs shown in Fig. 4.3-4. It was first developed at the Jet Propulsion Laboratory and later used on a JATO LPRE engineered by the M. W. Kellogg Company in the late 1940s. It requires special tooling to fabricate. Contrary to the double-wall concept, this design puts all of the loads and stresses into the inner wall, which has to operate at elevated temperature. The maximum stress that can be accepted is less than that of a cooler outer wall. This convolution cooling jacket is satisfactory for applications with moderate heat transfer. A very similar concept seems to have been invented in the Soviet Union at KB Khimautomatiki (CADB) in the late 1950s with a single inner wall and a corrugated outer wall to confine the cooling passages. These passages were straight and not of spiral configuration as in the Kellogg design. This construction was used by CADB in regions of low heat transfer, such as on the exit portion of the nozzle of the RD-0110 as mentioned in Chapter 8.5.

Ablative materials have also been used for chambers and nozzles in some large LPREs beginning in the 1960s. These materials contain oriented strong

---

*Personal communications with J. Halchak and V. Wheelock, Rocketdyne Propulsion and Power, The Boeing Company, Canaga Park, CA, 2001–2004.

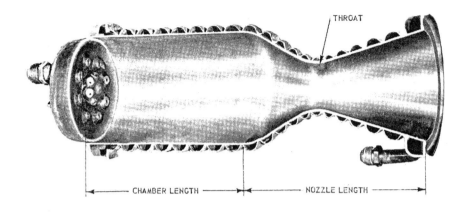

Fig. 4.3-7 This convoluted (spiral path) outer wall for a TC cooling jacket was conceived and tested by the Jet Propulsion Laboratory around 1944 and used about 1946 by the M. W. Kellogg Company in a JATO application. This cooling configuration is not used today. Courtesy of the M. W. Kellogg Company, from Ref. 6.

fibers suitable for a high-temperature environment (glass, silica, or carbon fibers) usually in a matrix of phenolic plastics compounds. The earlier success of ablative materials in solid-propellant rocket motors inspired their use in LPREs. The uncooled chamber and/or nozzle walls, exposed to the hot gases, are lined with a relatively thick layer of ablative material, and there is no cooling jacket. The chemical cracking (decomposition), the melting, and the evaporation of the ablative material absorb energy, and the low thermal conductivity helps to provide a relatively cool surface gas layer over the walls. These ablative materials have also been used on small TC, and this cooling method is described in more detail in Chapter 4.6 and some of its figures. Examples of large TCs with ablative cooling are the Lance engine (Rocketdyne designed in 1964, Chapter 7.8), the lunar descent engine (TRW, design in 1963, Chapter 7.9), and the Apollo service module engine (Aerojet, 1965, Chapter 7.7). All had supplementary film cooling. Ablative liners are also used in nozzle-exit skirts, such as with Aerojet's Titan upper-stage thrust chambers with storable propellants or the Rocketdyne's RS-68 LOX/LH$_2$ engine

The prediction of the life of a reusable thrust chamber has always been difficult, but some investigators have made estimates.[8] One approach is concerned with the progressive yielding of the inner wall, when the stress (mostly a thermally induced stress) exceeds the yield point of the hot wall material. After several start cycles, the yielded material starts to form cracks during cooldown, and when the cracks become progressively deeper and larger the cracks reach the cooling passage, the coolant will leak, and the TC will fail.

Nozzles

The Germans were the first (late 1930s) to recognize that the best performance of a nozzle is obtained when the nozzle-exit-area ratio (exit area divided by throat area) is just big enough to expand the combustion gases to the local atmospheric pressure. This has been called optimum nozzle expansion. When the nozzle area ratio is too small, then the gases are *underexpanded* and can expand some more outside the nozzle, but at a slight loss in specific impulse or performance.[1,2] This is illustrated in Fig. 4.3-8, with simplified sketches of the behavior of the exhaust gases of three typical rocket nozzles of a three-stage flight vehicle for both high-altitude and sea-level conditions. The first stage has the biggest chamber and the highest thrust, but usually the lowest nozzle-exit-area ratio, and it is usually underexpanded. When the nozzle exit is too large, then the nozzle exit pressure is below the atmospheric pressure, the gases are *overexpanded*, and the rise of the gas pressure at the nozzle exit to the atmospheric pressure causes shock waves and a contraction of the plume diameter and also a performance loss. The third stage in the preceding sketches is greatly overexpanded, if operated at sea level, such as during ground tests. Of course, a high area ratio nozzle is very desirable for high per-

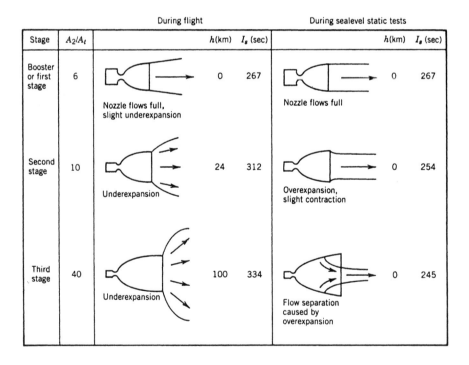

Fig. 4.3-8 Exhaust flame or plume can expand outside the nozzle, contract in diameter outside the nozzle or the flow can separate inside the nozzle. Adapted from Ref. 2.

formance at high altitude. It is normally started and operated only at altitude. When the pressure difference between the expanded gas at the nozzle exit and the atmosphere rises, then the flow will separate upstream from the nozzle-exit plane, and the losses become higher.

Overexpansion can have unpleasant effects, as discussed in a following separate section, and generally overexpansion conditions are to be avoided. So the Germans avoided overexpansion and used a conservative nozzle area ratio for the V-2 LPRE. The nozzle exit was smaller (underexpanded) than the optimum nozzle area ratio at sea level. In the United States the phenomena of over- and underexpansion were first analyzed in the late 1930s by GALCIT, and some useful design criteria were formulated. Propulsion engineers in other countries learned how to pick an optimum nozzle area ratio a few years later in the 1940s.

Several kinds of supersonic nozzles have been used on TCs.[1,2,9] Some are pictured in Fig. 4.3-9 together with their typical flow patterns at sea level and high altitude. In this section the *conical nozzle exit* and the *curved nozzle exit* will be discussed. Aerospike nozzles and expansion/deflection nozzles will be found later in this chapter. The earliest TCs (1921–1936) had a straight conical diverging nozzle section with a small half-angle (about 5 deg) as preferred by early investigators. The thermodynamic analysis of the flow (local pressure, temperature, and velocity) through bow-tie-shaped supersonic nozzles was understood very early, when the first concepts of LPRE were sketched. However the exact nozzle wall contour and the half-angle of the diverging nozzle-exit cone were then not optimized. As can be seen in Fig. 5-5, 5-6, 5-22, and 5-27, the nozzle-exit half-angle for these nozzle-exit cones in the 1920s and 1930s was typically 4 or 5 deg. Although such a small angle minimized the flow divergence losses (which the early rocket experimenters wanted to do), it increased the wall friction forces, increased the inert nozzle mass, and caused the engine to be very long and thus sometimes awkward to install in a vehicle. It was not until the late 1930s that analyses were performed optimizing the exit cone angle. GALCIT did one of these analyses (see Chapter 5.4). It was found that a shorter nozzle with a cone half-angle between 12 and 18 deg was best, and this angle depended on the application and the design. By 1940 a half-angle of about 15 deg appeared in many designs, such as those shown in Figs. 7.3-7, 7.8-3, or 7.11-4.

The Soviets were the first to adopt a *curved nozzle-exit profile*, which allows a small performance gain (typically 0.4 to 1.5%). The subject of a curved nozzle-exit contour was supposedly mentioned in a letter from V. P Glushko, the top man in the USSR on large LPREs, written to K. E. Tsiolkowsky, Russia's father of rocketry, in 1930. A few of the TCs developed in the Soviet Union in the 1930s and 1940s had curved nozzle exits on several experimental TCs, but most used conical exit sections. This subject is discussed in Chapter 8.4. It is not known if they used analysis, as the United States did, to recognize the benefit of a small increase in performance by using a bell-shaped nozzle exit. Apparently they relied on a clever simple experimental device to optimize the diverging nozzle shape. They mounted a TC with two opposed nozzle

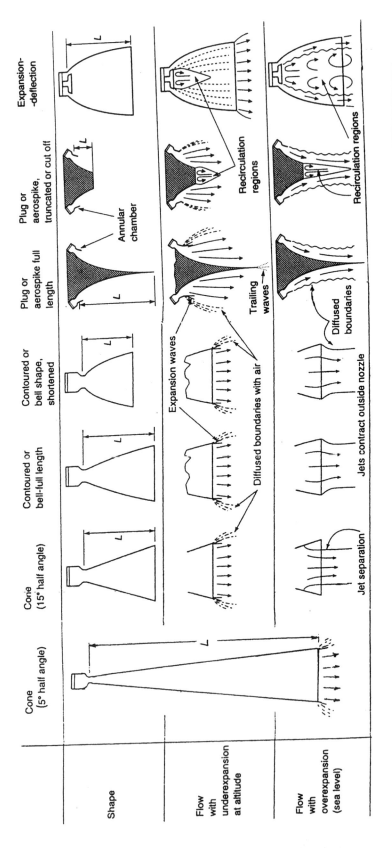

Fig. 4.3-9 Simplified diagrams of several types of contours for the diverging portion of a supersonic nozzle. Adapted from a similar figure in Ref. 2.

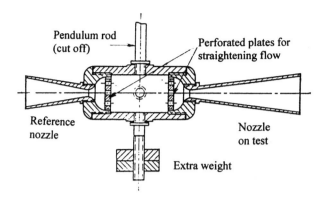

Fig. 4.3-10 Pendulum apparatus used to compare the performance of two nozzles. Courtesy of NPO Energomash, Russia.*

outlets, but with the same nozzle throat diameter, on a pendulum as shown in Fig. 4.3-10. One nozzle had the novel curved contour or different nozzle divergence angle, and the other had a standard contour as a reference.* In this way they determined that simple curved contours (such as a circular arc) gave a little more performance (typically 0.5 to 1.5%) than a straight cone of the same length. Initially the contours were arcs of circles, which give much, but not all of the possible performance gain. Some of the early flying Soviet engines (such as the ORM-65, RD-103, or D-1A-1100) appeared with a conical nozzle exit. Beginning about 1955 all of the Soviet engines were built with curved contours.

In the United States an in-house analysis was performed at Rocketdyne in 1955 or 1956 to determine the benefit of a curved nozzle exit using the method of characteristics for the analysis. This was first suggested and analyzed by Robert B. Dillaway, who worked in the author's section at Rocketdyne at that time. It was documented in an internal company memo. In 1957 or 1958 a pendulum balance was built and tested to confirm the analysis. This apparatus was very similar to the concept shown in Fig. 4.3-10 and used at an earlier date by the USSR, but this USSR work was not known in the United States at that time. The angular deflection of the pendulum was a measurement of the thrust differences between the two opposed nozzles with identical throat areas. The pendulum tests were first performed with an inert gas and later with actual hot rocket combustion gas. One nozzle served as a reference (typically a cone with a 15-deg half-angle), and the other nozzle had different kinds of contours. The test data confirmed the calculated

---

*Personal communications, V. K. Chvanov, first deputy to general director and general designer, NPO Energomash, Khimki, Moscow Region, Russia, 2002.

performance gain of up to about 1% more specific impulse, if compared to an equivalent conical nozzle with a 15-deg half-angle. In 1958 G. V. R. Rao, who worked at that time with Dillaway and the author at Rocketdyne, published the analysis of optimum curved nozzle-exit contours.[9,10] He analytically derived that wall contour by the method of characteristics, but today this can also done by a two-dimensional finite element code with thermodynamic changes in the gas and chemical reactions during the nozzle flow. This new contour minimizes the losses of internal shock waves in the supersonic flow. The words bell-shaped nozzle were coined at that time. The ideal calculated new bell shape was relatively very long, would be heavy, when built, and is shown schematically in Fig. 4.3-9.[2,9–11] A new shortened bell nozzle shape was also determined by this analysis method; it gave slightly higher losses than a long bell-shaped ideal nozzle, but it had a shorter nozzle length, which could be even shorter than an equivalent 15-deg cone.

This bell shape or curved exit contour is used almost universally today for nozzles designed since about 1960 for large as well as small thrust chamber nozzles and also for liquid propellant as well as solid-propellant rocket nozzles. Several large LPREs, which were designed originally with a straight conical nozzle-exit section and had flown, were then modified to have a bell-shaped nozzle. In the United States this happened, for example, with the Atlas and the Thor rocket engines.

Nozzle Extension

The diverging nozzle exit segment of a rocket nozzle has a relatively very *low heat transfer* compared to the chamber or nozzle throat sections.[1,2,11] The early cooled TC also had regenerative cooling in the nozzle. It is not necessary to cool this nozzle-exit segment with propellants in a cooling jacket; therefore, simpler, lower-cost, single-wall concepts can be used. One of the following three approaches has been used for this nozzle-exit segment. A low-density ablative material at the nozzle-exit section in a thin metal outer wall was used on Aerojet's Titan upper-stage engines (1962, see Fig. 7.7-20) and more recently on the Rocketdyne RS-68 (designed 1997, Fig. 7.8-20). Usually this was the lowest-cost approach. Alternatively a radiation-cooled nozzle-exit segment with a thin refractory metal wall of niobium or carbon has been effective. It has been used in several applications including the space shuttle orbiting maneuver engine of Aerojet seen in Fig. 7.7-13. The third design, called dump cooling, has a single wall usually made of stainless steel, is gas cooled by the warm turbine exhaust gases (typically 700 to 900°F), which are injected or dumped into the diverging nozzle section at a location of a low area ratio. It can be considered a type of film cooling with warm gas, which increases in temperature as it flows in a protective layer inside the nozzle-exit wall. The gases are then dumped at the nozzle exit and augment the thrust by a very small amount (usually 0.5% or less). Dump cooling was performed in the United States with the Rocketdyne F-1 LPRE in 1961 as shown in an early version in Fig. 7.8-10 and with one version of the French Ariane second stage

Viking 4 engine (described in Chapter 10) in the 1990s. An alternate form of dump cooling uses tubes for the flow of the turbine exhaust gases in the nozzle-exit section. These tubes are open at the nozzle exit, and the warmed gas is dumped from the tubes along with the exhaust gas flow from the thrust chamber. This was used in the French Vulcain LOX/LH$_2$ engine discussed and shown in Chapter 10.

Nozzle Flow Separation

Problems were encountered (beginning about 1945 in the United States[2,8,11] and possibly earlier in the Soviet Union) with the separation of the overexpanded supersonic flow from the inside walls of the diverging nozzle segment. The phenomena of overexpansion and flow separation are illustrated in Fig. 4.3-8 and were analytically investigated in the 1940s. Overexpansion occurs when a nozzle with a large area ratio is operated at a lower altitude than the design altitude for optimum expansion. This usually happens when an upper-stage TC or engine, which usually has a large nozzle area ratio, is tested in a static test facility at or near sea-level conditions. Overexpansion causes a significant loss in performance and is not desirable.[2,11] During normal operation with overexpansion, the flow usually separates symmetrically in the nozzle. When the flow does not separate symmetrically, as can happen in the startup transit or shutdown transition, then large internal aerodynamic side forces can occur inside the nozzle-exit section, and these forces are sometimes so large that they have broken nozzle hardware or actuators of gimbal-mounted TCs and caused an engine failure. Some progress in understanding the phenomena of unsymmetrical separation has been made in the last decade.[9,12]

Historically there were three approaches to test thrust chambers with high area ratio nozzle exits in a ground-based test facility. New altitude test facilities were built in the USSR, United States, and later also in Japan, and these allowed full simulation of altitude conditions and the testing of LPREs with nozzles of high nozzle area ratio without the concern of potential nozzle separation. Since about 1960, the United States has had such a facility at the AF Arnold Engineering Development Center in Tullahoma, Tennessee, where large LPREs can be started, restarted, and operated in a vacuum equal to an altitude of about 34,500 m or 100,000 ft. This vacuum is normally sufficient to avoid separation. There are also other facilities in the United States and in other countries, where steam ejectors are used to simulate altitude condition. A lower-cost solution was the use of an ejector, sometimes called diffuser or jet diffuser, where ambient air can be aspirated into the nozzle-exit high-speed gas flow of the TC's exhaust plume. They are used widely, and they create a vacuum around the nozzle exit with the high nozzle area ratio during full thrust operation.

However this method does not simulate altitude conditions during the start or the restart. The third solution was to build a separate essentially identical TC specifically for the sea-level ground tests of an upper-stage engine or TC, but with a cutoff or shortened nozzle at a lower area ratio. The nozzle is then underexpanded and will not cause flow separation. The measured performance is then corrected to the larger area ratio and altitude conditions. This allows firing tests in a ground-test facility without the possibility of overexpansion and unsymmetrical flow separation. Shortened nozzles have been used for ground tests by all countries since the 1950s. However this solution again does not simulate an altitude start, shutdown, or vacuum restart.

Special Nozzles

Special configurations of a nozzle are represented by the *aerospike nozzle* (also called *plug nozzle*) and the *expansion/deflection (E/D) nozzle*; both are sketched in Fig. 4.3-9. Their principal merit is to provide altitude compensation (the effective nozzle expansion area ratio is at optimum value at all altitudes) and thus to give somewhat better performance than a fixed area ratio nozzle, which would give optimum expansion at only one altitude. Also the nozzle length (and thus the engine length) can be very short with a cutoff or truncated nozzle, which usually implies a shorter, lighter vehicle. It is possible to start and run these LPREs at sea level without flow separation. The aerospike or plug nozzle was first proposed by General Electric Company in the late 1950s and then promoted by Rocketdyne beginning in the early 1960s. A test firing at 40,000 to 50,000 lbf thrust is shown in Fig. 7.3-12 and 7.8-35. It has been described as an inside-out nozzle with an annular combustion chamber.[2,13,14] Alternatively the chamber consisted of a multiplicity of small thrust chambers (also called thrust cells) with low nozzle area ratios arranged in a ring. There is a curved inner-wall contour (shaped like a spike or a cutoff spike) and an outer virtual wall formed by the interface between the expanding nozzle exhaust gases and the external airflow. The pressure distribution on the spike surface and the flow pattern of an aerospike nozzle are shown in Fig. 4.3-9 and 4.3-11 for low- and high-altitude operation. Tests of this special nozzle have been performed in the United States at General Electric at thrusts of 16,000 and 50,000 lbf in the early 1960s and at Rocketdyne with several versions with thrusts between a few thousand pounds and up to 250,000 lbf in the late 1960s and 1970s. Small-scale ground tests were also done later in Japan and Germany.

A *linear version of an aerospike* was developed and test fired by Rocketdyne initially in 1969 and 1970 at 50,000 lbf and then later at thrust levels up to 100,000 and then 200,000 lbf. An improved version was developed between 1998 and 2001 for an experimental single-stage-to-orbit vehicle with two linear aerospike engines, each at 200,000 lb thrust. One engine is shown in Fig. 7.8-37. All used LOX/LH$_2$. This LPRE is described near the end of Chapter 7.8. The linear aerospike engine again can attain optimum performance at all altitudes and is easier to integrate with certain blended wing-body vehicles than an engine with a round aerospike or round (axisymmetric) conventional nozzle.[2,13] It also allowed simple thrust vector control in pitch, yaw, or roll by throttling of selected banks of thrust cells. This principle of TVC is explained in Fig. 4.9-6 and has been used in engine clusters or multiple large LPREs in a vehicle stage. Tests of two linear aerospike engines coupled together were performed at the 400,000 lbf thrust level, confirming the design as well as the thrust-vector capability by selective throttling of certain groups of thrust cells. The engine was successfully developed and statically tested, but the project was stopped in 2001 because the vehicle program was canceled. There have been flights of an experimental small aerospike engine by rocket enthusiasts in 2004.

There are additional small losses (estimated to be less than 0.3%) in all aerospike and E/D nozzles with multiple thrust cells, caused by shock losses when the supersonic exhaust jets from adjacent thrust cells meet and impinge against each other and caused by the flow going from a round nozzle throat to a rectangular nozzle exit. The linear aerospike also has more wall friction loss than a round aerospike. However these losses are less than the gain of altitude compensation and probably could have been overcome with further development. Because the large LPREs with large area ratio nozzles are long and awkward to install, people have looked for a way to make them shorter. One way has been to use the cutoff or shortened bell nozzle, mentioned earlier in this chapter; however, it has a slight performance loss.

Several LPREs with experimental *expansion/deflection nozzles* have been developed and ground tested, but they have not flown. The benefit again is altitude compensation (the engine can run at optimum expansion at all altitudes) and the possibility of a very short engine. The RD-0126 (tested in 1998) of Chemical Automatics Design Bureau (CADB) in Russia, running on LOX/LH$_2$ with a thrust of about 39 kN or almost 9000 lbf, had such a nozzle and tested it successfully as reported in Chapter 8.5. Three experimental thrust chambers of Rocketdyne used this E/D nozzle. Two operated at 50,000- and 10,000-lb nominal thrust using NTO/Aerozine 50. A smaller E/D TC was tested with LOX/LH$_2$. All three were developed and tested in the 1960s.

The aerospike or plug nozzle and the expansion deflection nozzle offer definite merits. However the performance gain of the flight vehicle is relatively small, when compared to existing proven engines with conventional nozzles. Apparently the advantages were not considered attractive enough to spend the effort to fully qualify and produce these engines with special nozzles and to build the vehicles for them.

Another way to shorten the engine length is with a concept called an *extendible nozzle*.[2] It is shown in a simplified sketch in Fig. 4.3-12 and in more detail in Fig. 7.10-5. for the RL10-B2 LPRE of Pratt & Whitney. Its first flight was in 2000. By making the nozzle (for an upper-vehicle-stage engine) in annular segments, these segments can be stored, one inside the other (like a boy scout cup) and then extended during unpowered flight. It is suitable only for upper-stage propulsion systems with high area ratio nozzles. After dropping off the lower vehicle stage, the two position nozzle segment is assembled (extended into operating position), thus automatically forming a complete large nozzle-exit section. This is done prior to the start of this upper-stage

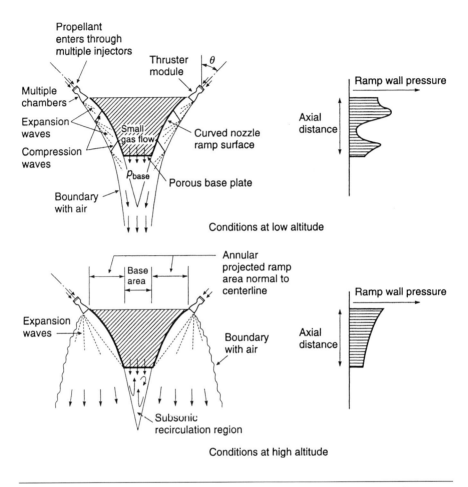

Fig. 4.3-11 Flow patterns and pressure profile on the curved surface of an aerospike nozzle at low and high altitude. Copied with permission from Ref. 2.

engine. It makes it possible to reduce the length of the vehicle and thus save some vehicle structure and inert mass. This concept was first implemented on solid-propellant upper-stage rocket motors (first flown in 1990) and then adapted for LPREs. The Pratt & Whitney engine RL10B-4 uses a large extendible radiation-cooled nozzle segment made of carbon fibers in a carbon matrix, which was developed by SEP in France. The altitude area ratio of the extended nozzle is 285 to 1.0 and it gives a very high performance, namely, a specific impulse of about 466 s. As described in Chapter 7.10, this is the highest known specific impulse of any flying LPRE. The Soviets reportedly have also built at least two of these extendible nozzles, one for the D-57 LOX/LH$_2$ engine, but they have not flown as yet. The Chinese are reported to have developed an extendible nozzle, which is deployed or put into its operating position while the engine is firing, but at 50% thrust. It has flown.

Technology and Hardware   97

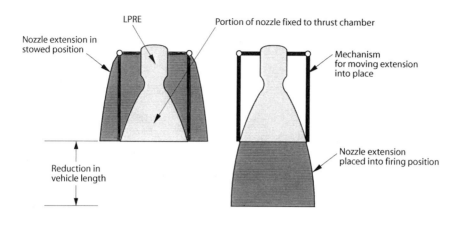

Fig. 4.3-12 Simplified sketch of the extendable nozzle concept.

Injectors

An injector consist of a *structure* to withstand the pressures; an *inner face or wall*, which is usually cooled by propellant to limit the heat transfer from the hot combustion chamber gas to the injector; a series of *injector elements* (a set of holes, coaxial elements, or sprays) arranged in a logical orderly pattern over the face of the injector; and *manifolds* or feed passages to supply propellants to all of the injection elements. Small thrusters can have injectors with only one injection element, such as a single spray element or a single doublet impinging set of holes. Figures 4.3-13 show typical simplified sketches of multiple injector elements made of individual propellant streams emanating from injection holes.[2] When like-on-like (fuel on fuel or oxidizer on oxidizer) injection streams impinge, they form a thin fan-shaped liquid sheet, which then breaks up into droplets, which evaporate, and the chemical reaction then occurs mostly in the gas phase. When unlike streams of hypergolic propellants impinge, they partly start to burn almost immediately, and the hot-gas evolution breaks the remaining liquids into small droplets, which mix, evaporate, and burn. There are others injection elements with holes, which are not shown here, such as a single fuel jet or stream impinging with three or four smaller diameter jets of oxidizer (used in a French orbit maneuver engine of 1350 lbf thrust) or the V-2 oxidizer brass injector insert with many holes for direct injection into the chamber without intended impingement.

The early pioneers and investigators used mostly nonimpinging injector designs (also called shower head injector), with poor breakup of injected propellant streams into small droplets and with poor, somewhat random mixing. Examples were Goddard's early TCs, such as the one on Fig. 5-7, where much of the fuel was in the boundary film-cooling layer. In the V-2 injector there is no direct impingement as seen in Fig. 9-4. The oxidizer enters the chamber of each of the 18 injector heads or domes from a single brass fitting through many radial and axial holes, and the fuel enters through three rows of

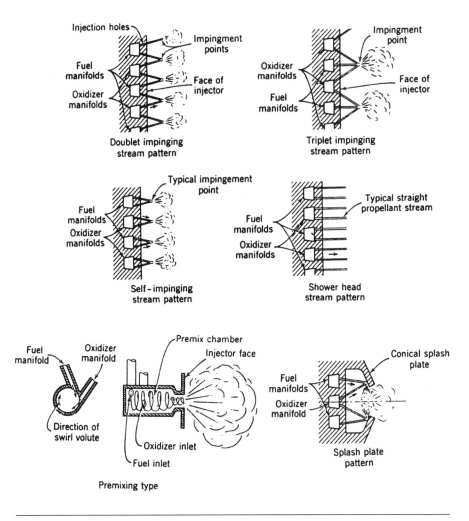

Fig. 4.3-13 Simplified cross sections of several common types of injection hole patterns or injection hole elements. Copied in part with permission from Ref. 2.

swirl-type fuel inserts and also through rows of holes. There are wide variations in local mixture ratio. The combustion efficiency of early TC was poor, and it required relatively large chamber volumes to attain attain a combustion efficiency of more than 90%. Specific impulse values were 7 to 12 below the theoretical values. With impingement or coaxial injection elements it is possible to reach between 92 and 98%.

Figure 4.3-14 shows various common spray-type injection elements.[1] In some of the older TCs, these spray elements were separate parts or inserts, which were screwed or fastened to the injector face. A simple single spray element forms a conical thin sheet of one of the liquid propellants; the sheet breaks up into small droplets, and these evaporate and burn. Sketches **a** and

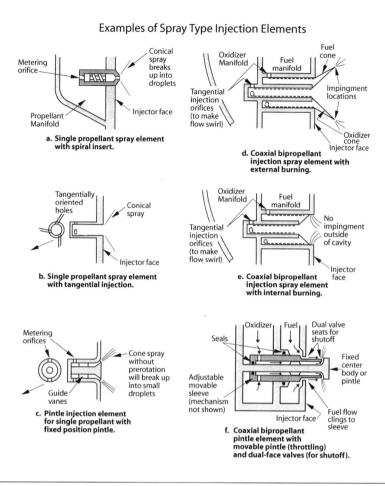

Fig. 4.3-14 Simplified cross sections of several common types of spray injection elements.

b are representative of single-propellant sprays. By means of tangential injection into each posts or insert, or by using a spiral insert, the liquid propellants experience a rotation or swirling motion inside the injection element cavity, and by judicious design the angle of the spray cone can be controlled. With bipropellant coaxial injection elements, two conical sheet sprays of propellant are formed, as seen in sketches **d** or **e** of Fig. 4.3-14. In one the impingement occurs outside the cylinder element, and in the other it occurs inside the bottom cavity of the cylinder element, which means that a major energy release occurs within this cavity. Some are recessed from the injector face. There are many variations of these injection spray elements, and each usually has some merit.

The development of injectors has been largely empirical, and many different designs have been built and tested.[1,2] In addition to its basic functions of introducing the propellant into the chamber and distributing them evenly over the cross section of the chamber, the injector configuration determines or influences the atomization and evaporation of the propellant and many of the combustion characteristics. This includes the required chamber volume for good combustion efficiency, the combustion stability, and the heat transfer to the chamber and nozzle walls.[2] The injector design also affects the starting characteristics. The operating behavior of the injector is affected by chamber pressure, mixture ratio, propellant, initial propellant temperatures, and by the design of the injection element. All of these influences and effects of injector performance on the characteristics listed in this paragraph were not known at the beginning of this LPRE history, but were discovered gradually decade by decade. Many rocket-engine R&D groups, in the United States and other countries, have undertaken systematic experimental injector investigations. There is extensive literature on injectors, injection phenomena, and injection elements, but it is not well organized. Only two references are cited here.[15,16]

In the 1930s several of the early investigators did not have the injector at the forward end of the TC. They placed the injection spray elements midway on the side of the chamber or just upstream of the nozzle converging section, partly because they had an erroneous concept of the combustion process. Examples of this type of injection can be seen in Figs. 4.3-2 and 5-20. This was tried by investigators in the USSR, Germany (Hermann Oberth and the VfR or German Rocket Society), and some amateurs in the United States. This approach was abandoned within 10 years. A few investigators started with the more effective concept, namely, to place the injection elements at the forward end or head end of a cylindrical TC, where it still is today. This included the TCs of the pioneers Goddard and Sanger, which are discussed in Chapter 5.

In the 1940s it was found that minor variations in hole size, hole entry conditions (chamfer or radius), impingement geometry, or hole distribution over the injector face can have significant effects on one or more of the injector's characteristics. Small burrs or sharp edges at the hole entries can affect the flow, the local mixture ratio, combustion, and the performance. The cleaning of the hole surfaces, precise angular positioning, and having a consistent and good entry radius on every hole were found to be very important. The original fabrication method was to drill the holes in the desired location, with the correct angular orientation, and surface condition. At times it was difficult to drill accurate holes. The drills cutting edge would become dull, would no longer make a good hole, and the drill had to be replaced often. Many times drills broke during the drilling, particularly when the hole diameter was small. Some companies still use drilling effectively today. An alternate and more recent fabrication technique uses electrodischarge machining or EDM. The work piece is immersed in an EDM fluid or oil. A hollow copper tube is used as the electrode, and sparks at the end of the electrode remove the metal in small pieces. The copper electrode is positioned at the proper angle, rotated to ensure even metal removal and a straight hole, and slowly move forward into the hole. The debris of metal particles is sucked out through the hole in the hollow electrode. The sparks are generated through pulsing electric current from an adjustable special power supply.

For good propellant injection and combustion with spray elements or coaxial injection elements the following are critical: certain dimensions changes or certain tolerances, the entry shape and size of the tangential holes for admitting the propellants, the liquid cone angle, the impingement location of the two cones, and the surface finish.

Each organization or country seems to have its favorite injection elements for specific applications. The USSR has used coaxial swirling spray-type injection elements since the 1930s for most of their major engines, with all of their common propellant combinations and for the current Russian engines. Typical coaxial bipropellant elements are shown in Figures 8.4-20 and 8.5-8. For film cooling the fuel-only spray elements at the periphery of the injector face have been effective. Also a number of the large LPREs used single-propellant spray elements arranged in an orderly pattern. China also uses coaxial swirling injection elements with various storable propellant combinations. One of the tricks in bipropellant spray injectors was to not use a continuous single conical sheet, but to have gaps in these sheets, so that hot-gas reaction products can go from one side of the cone to the other without bending the cone out of shape or causing flutter. The U.S. practice has favored discrete impinging propellant jets admitted through precisely oriented holes, and these were used in most larger U.S. injectors with storable propellants between 1950 and the 1980s. For LOX/RP-1 and storable propellants a triplet or doublet impinging stream pattern seems to have been effective and was used in U.S. designed injectors up through the present time. However patterns other than impinging streams and swirling sprays will often also be satisfactory. For example in the V-2 injector the Germans used basically a nonimpinging multiple stream pattern. For liquid oxygen with heated, evaporated gaseous hydrogen fuel, hollow posts or coaxial spray elements have been successful, and this type is used in LOX/LH$_2$ engines in all of the different countries. In this case it is a relatively cold gaseous hydrogen fuel (it was gasified by absorbing heat in the cooling jacket) at a relatively high injection velocity being swirled and mixed with cryogenic liquid oxygen at a much lower injection velocity, but much higher density.[15] There have been some LOX/LH$_2$ injectors with doublet or triplet impinging streams, and they have also given a high performance and good stability, but as far as can be determined, they have not been flown.

A *pintle injector* is basically a spray element with a pintle or central rod, which produces a sheet spray. Sketches **c** and **f** in Fig. 4.3-14 show a fixed pintle for one propellant and a movable pintle for a bipropellant. Pintle injectors for bipropellants can provide two injection sheets that impinge. There is usually no swirl device, but there are slots through which a liquid sheet is injected. The Northrop-Grumman organization and its predecessor TRW have successfully used this pintle injector concept extensively with a variety of different propellants and thrust sizes. There are simple pintle injectors without moving parts as shown in Fig. 4.3-14, pintle injectors with a two position pintle that allows face shutoff, and pintle injectors with a moving, adjustable position pintle to provide variable thrust. Variable thrust is obtained by varying the width of the two injected sheets of propellant as shown in Fig. 4.3-14. The pintle can be moved to close off the two slots, and this acts as propellant valves, which are placed directly on the injector face; this minimizes dribble volume (afterburning) and facilitates restart. It was successfully developed by TRW (see Chapter 7.9) for the lunar descent and landing throttling rocket engine in the late 1960s. By changing the gap or thickness of the spray sheet, it was possible to maintain about the same injection pressure drop and the same injection velocity over a 10 to 1.0 thrust range. A single pintle injector was used in this LPRE, and it requires a relatively large chamber volume for good combustion efficiency and a reasonably high performance.

Many of the early injector designs for storable propellants and oxygen/hydrocarbon had relatively large injection holes, and they were not always very effective. Later smaller hole sizes and more holes were found to usually give better mixing and a more uniform mixture ratio across the chamber and gave better performance with a smaller chamber volume. The use of many very small holes, such as those that can be built with the patented Aerojet platelet design of Fig. 7.7-15, are likely to give a more uniform flow and mixture ratio across the chamber, smaller droplets and thus more rapid evaporation, and thus more complete combustion or performance. This figure shows two different ways for constructing a platelet injector from brazed together stamped or cut out multiple metal plates or sheets. Several of Aerojet's upper-stage engines and the Aerojet Shuttle orbital maneuver engine, shown in Fig. 7.7-13; used a platelet-type injector with many small injection holes. Platelet injectors were developed beginning in 1965, have given good injection, and have been used principally by Aerojet

There were many injector configurations that were developed and tested, but they were often deficient in fulfilling all of the requirements. Figure 4.3-15 shows three U.S. injectors with different injection elements, which were unsatisfactory and were abandoned in the 1940s or 1950s.[6] The splash plate concept was based on atomizing the propellant streams into small droplets by impact against a metal surface, and it worked well for some injectors with storable propellants. However, this particular injector had combustion stability problems. The idea of a precombustion chamber with swirling propellants was to achieve good premixing, which was expected to result in a good combustion efficiency and a small volume of the main combustion chamber. However the premixed propellants would at times explode. The rosette-type injector at the bottom had unlike doublet impinging stream elements. The orientation of the doublets and their liquid fans depended on manufacturing tolerances. It experienced occasional burn-throughs of the chamber inner wall and nozzle inner wall, believed to have been caused by misaligned injectors holes and burrs.

The scaling of injectors has been largely empirical and is still not fully understood. It was a common practice in the early decades of the LPRE history to test a new injector pattern with fewer, but identical injector elements on a small-scale experimental TC (usually smaller diameter and lower thrust). It was hoped to reduce the cost of development of the full-scale TC. Some injectors, which performed well on a small scale, experienced problems, once they were scaled up in thrust (to full scale and/or a larger diameter). Combustion

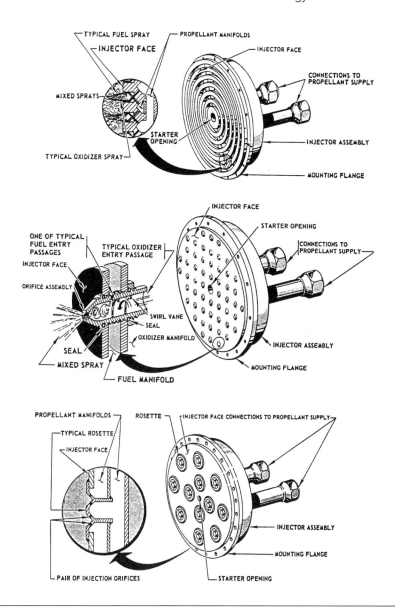

Fig. 4.3-15 Three injector designs that were tried in the 1940s and 1950s, but were not satisfactory. The splash plate type (top) had occasional unstable burning, the premix type (center) had explosions, and the rosette type (bottom) caused occasional TC burnouts. From Ref. 6.

phenomena are strongly influenced by the injector design, and this subject is discussed in Chapter 4.10.

The *injector's structure* has to withstand the loads imposed by the chamber pressure and the propellant feed pressures, accept the stresses imposed by

104 History of Liquid Propellant Rocket Engines

thermal gradients, prevent internal leaks, which could allow one propellant to enter into the flow passage or manifold of the other propellant (forming a combustible or explosive mixture in the manifold), and distribute the flow of both propellants evenly to all of the injector elements. In the first three or four decades of thrust chamber development, several designs for the injector structure for large diameter injectors became prominent. Four of the common injector structures are sketched in Fig. 4.3-16. All had a flat face or nearly flat face, and some provided for film cooling by having fuel-only injection elements at the periphery.

The first and probably one of the oldest schemes is shown in the upper left of the figure and was similar to the one used by GALCIT, some amateur rocket societies, and some early Soviet investigators between the 1920s and the early 1940s. It was built from machined billets welded together, with no potential leaks between the fuel and the oxidizer passages, a single circle of unlike impinging doublet injection elements. It is still used in small thrust chambers, such as the R4 100-lbf bipropellant Apollo ACS thruster of Fig. 7.7-26 or the RS-14 reaction control thruster shown in Fig. 7.8-28.

The machined forging design with a separate oxidizer dome, radial internal fuel feed holes, and a series of concentric annular grooves was conceived at General Electric in the United States in about 1944 as mentioned in Chapter

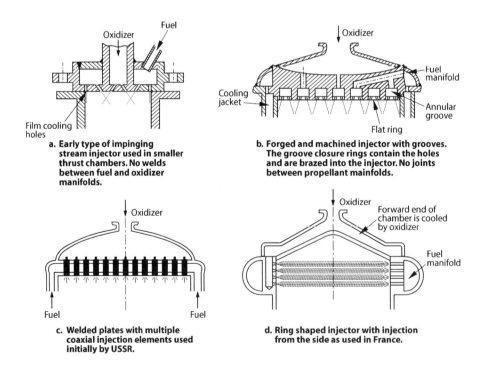

Fig. 4.3-16 Simplified sketches of the cross sections of four injector assembly structural configurations.

7.3. As depicted in the upper-right sketch of Fig. 4.3-14, it featured grooves with face rings; the injection holes were predrilled into the rings, which were then soldered or brazed into the flat injector face as shown in Fig. 7.3-3. The odd-numbered grooves contained liquid oxidizer, and the even-numbered grooves were supplied with fuel. The propellants flowing in the grooves and some of the internal manifolds or passages cooled the hot injector face. It had no internal brazed or welded joints between oxidizer and fuel passages inside the injector assembly, and a smooth radius at the entrance to each hole could be machined into the rings prior to brazing. This construction was pioneered by General Electric Company in the 1940s, copied and adapted by Rocketdyne, and later by Aerojet and others. This injector construction with a forging was scaled up by using a larger number of rings, and improved versions with more rings were produced for the engines of the Redstone, Titan, Atlas, and Thor missiles and the booster engines of the Saturn I and Saturn V space launch vehicles.

One of the cardinal design rules in the United States up to about 1972 was that there shall be no welds, brazed joints, or possible leak paths from a fuel flow passage or fuel manifold to an oxidizer manifold. If there were a leak, it could result in local gas formation (bubbles) or sometimes in an explosive mixture of propellants inside the injector. Later it was found that some other countries did not abide by this rule and instead made sure that these joints were leakproof.

In the late 1940s or early 1950s the USSR designed a two-plate welded structure with a dome and a number of hollow posts or injection spray elements. As shown in the lower left of Fig. 4.3-16, the injector face has many posts or injector elements, which became part of the structure and are brazed to the face plate and intermediate plate. Some are single-propellant spray elements such as in the RD-216 shown in Fig. 8.4-22, and some are coaxial bipropellant elements, such as for the RD-107 and RD-0110 in Fig. 8.4-10 and 8.5-6. The space between the chamber face metal plate and the intermediate plate provides the passage for distributing one of the propellants. The other propellant flows from the top manifold (covered by the top plate or dome) through the inner tubes of the coaxial injector elements. In some designs the top dome-shaped plate is structurally connected to the other two plates. This structural three-plate design is usually lighter in mass than the forged, machined design. There are brazed joints and welds between fuel and oxidizer passages in these injectors. The author is not aware of a problem caused by a leak. The engineers developed reliable brazing and welding techniques and appropriate inspection methods for avoiding potential internal leaks. The Soviet Union used this metal plate design in most of their larger LPREs. The Germans in the Aestus engine shown in Fig. 9-16 and the Chinese in their YF-1 engines seen in Fig. 13-1, all with storable propellants, also use a similar design, but with a heavier machined plate and brazed coaxial posts in their injectors.

The initial thrusters burning LOX/LH$_2$ in the United States used impinging propellant streams injected through holes in their injectors. At the time the U.S. designers were not aware of the Soviet spray element developments. The

NASA Research Center at Cleveland, Ohio, independently developed and demonstrated an injector with coaxial injection spray elements. Beginning about 1958 this type of spray element was used for the RL-10 (see Fig. 7.10-1 and 7.10-2; it flew in 1963) and the J-2 LOX/LH$_2$ (flew first in 1966) for upper stages of Saturn I and Saturn V and later in the main engine of the space shuttle (first flew 1980). This concept has also been selected by several Japanese, Chinese, Soviet, and French LOX/LH$_2$ engines. There are some exceptions to the use of coaxial injection elements for LOX/LH$_2$ LPREs. For example, in 1958 to 1960 Aerojet operated its Titan I LPRE with LOX/LH$_2$. Even though this LPRE was originally designed for and operated with LOX/RP-1 (kerosene) and had impinging propellant streams in the injector, the tests showed good performance with H$_2$ as the alternate fuel. In U.S. LOX/LH$_2$ engines the injector face has a sheet of pressed metal shavings (a porous material), through which a small amount of fuel flows to transpiration cool this hot face. This was a contribution to the state of the art of LPRE technology by Pratt & Whitney with their RL 10 LPRE. In engines with staged combustion cycles, such as in the Rocketdyne SSME shown in Fig. 4.10-2, the coaxial elements are extended inside the injector, and the hot turbine exhaust gas is used to heat the liquid propellants prior to injection. An extension of selected coaxial injection elements into the chamber is used to form baffles for the control of certain types of combustion instability.

Sketch **d** in Fig. 4.3-16 shows a simplified version of a side injection design, where the propellants are injected in a radial direction through rows of holes forming unlike doublets. The fuel enters from a manifold around the injector, and the oxidizer flows through a cooling passage of the top of the chamber and enters the injector through vertical holes in the annular injector structure. There are multiple rows of injection holes on top of each other. This design originated in France in the 1950s, has been used in about five LPREs of several French flight vehicles, beginning with the Coralie in 1960, and is still used on the uprated Viking engines of the most recent version of the Ariane SLV. This injector type is shown as part of the TCs in Fig. 10-8 and 10-4.

The placement of the injection elements on the face of the injector is usually in some kind of orderly geometric pattern, usually axisymmetrical. Figures 4.10-2, 8.4-10, 8.4-19, or 8.4-22 show examples of the placement of injection elements on the face of an injector. The original intent has been to distribute the flow of propellants evenly, which means that the mass flow per unit area and the local mixture ratio is the same across the face. Several of these injectors showed high-frequency instabilities, presumably because the impingement distance from the face was identical for each element and the zone of the most intensive energy release was essentially in a plane. Combustion stability was often achieved by changing to an uneven distribution of mass flow and mixture ratio or by changing injection angles of some elements, which changes the location of the impingement and greatest energy release. The number of injection elements per unit cross-section area also has an influence on the performance. A single element is usually easier to build and lower in cost, but the atomization and mixing is not as good, and it requires a large chamber volume (more inert mass) for attaining essentially complete combustion. A multitude of smaller elements gives usually good atomization and mixing and a high combustion efficiency, but is more delicate to build and often more prone to instability.

References

[1]Huzel, D. K., and Huang, D. H., *Design of Liquid Propellant Rocket Engines*, Vol. 147, Progress in Astronautics and Aeronautics, AIAA, Washington, DC, Chap. 4, 1992, pp. 67–133.

[2]Sutton, G. P., and Biblarz, O., *Rocket Propulsion Elements*, 7th ed., Wiley, New York, 2001, Chap. 8, pp. 268–341.

[3]Knuth, E. L., "The Mechanics of Film Cooling," *Journal of the American Rocket Society*, Nov.–Dec. 1954.

[4]Bartz, D. R., "A Simple Equation for Rapid Estimation of Rocket Nozzle Convective Heat Transfer Coefficients," *Jet Propulsion*, Vol. 27, No. 1, Jan. 1957, pp. 49–51.

[5]Bartz, D. R., "Survey of the Relationship Between Theory and Experiment for Convective Heat Transfer in Rocket Combustion Gases," *Advances in Tactical Rocket Propulsion*, edited by S.S. Penner, AGARD, Paris, 1968.

[6]Herrick, J., and Burgess, E. (eds.), *Rocket Encyclopedia, Ilustrated*, Aeropublishers, Inc., Los Angeles, 1959.

[7]Kazaroff, J. M., and Pavli, A. J., "Advanced Tube Bundle Rocket Thrust Chambers," *Journal of Propulsion and Power*, Vol. 8, No. 4, July-Aug. 1992, pp. 786–791.

[8]Porowski, J. S., O'Donnell, W. J., Badlani, M. L., and Kasraie, B., "Simplified Design and Life Predictions of Rocket Thrustchambers," *Journal of Spacecraft and Rockets*, Vol. 22, No. 2, Mar.–Apr. 1985, pp. 181–187.

[9]Hageman, G., Immich, H., Nguyen, T. V., and Dumnov, D. E., "Advanced Rocket Nozzles," *Journal of Propulsion and Power*, Vol. 14, No. 5, 1998, pp. 620–634.

[10]Rao, G. V. R., "Exhaust Nozzle Contours for Optimum Thrust," *Jet Propulsion*, Vol. 28, No. 6, 1958, pp. 377–382.

[11]Rao, G. V. R., "Recent Developments in Nozzle Configurations," *Journal of the ARS*, Vol. 31, No. 11, 1961, pp. 1488–1492.

[12]Summerfield, M., Foster, C. R., and Swan, W. C., "Flow Separation in Overexpanded Supersonic Exhaust Nozzles," Inst. of Fluid Mechanics and Heat Transfer, June 1948.

[13]Berman, K., "The Plug Nozzle: A New Approach to Engine Design," *Astronautics*, No. 5, 1960, p. 5.

[14]Kinkaid, J. S., "Aerospike Evolution," *Threshold*, No. 18, Spring 2000, pp. 4–12.

[15]Stehling, K. R., "Injector Sprays and Hydraulic Factors in Rocket Motor Analysis," *Journal of the American Rocket Society*, May–June 1952.

[16]Zhou, J., Hu, X., Huang, Y., and Wang, Zh., "Flowrate and Acoustics Characteristics of Coaxial Swirling Injector of Hydrogen/Oxygen Rocket Engine," AIAA Paper 96-3135, 1996.

4.4 Turbopumps

The turbopump (TP) is a key component of a pump-fed LPRE. It is really a high-precision, high-speed, sophisticated piece of rotating machinery. Its fabrication is expensive, and its design is engineering intensive.[1–3] The technology of TP has changed and improved throughout the history of pump-fed LPREs. A typical TP has these subassemblies: 1) one or two principal centrifugal propellant pumps; 2) turbine(s) to provide the rotary power to the pumps; 3) usually only one, but sometimes two or more shafts; 4) bearings and means to lubricate and cool them; 5) shaft seals to limit leakage and to prevent mixing of fuel with oxidizer, or turbine gas with a propellant; and 6) most have also inducer impellers upstream of the main propellant impellers

Some but not all of the turbopumps also have a gear case, usually to reduce the higher shaft speed of a turbine to a lower shaft speed of a pump. They have provisions for lubrication and cooling of the gear case and its bearings. Some TPs have provisions for driving accessory equipment, such as a lubricating oil pump or a hydraulic oil pump

Turbopump designs for a LPRE can have *different physical arrangement* or placements of turbines, pumps, and bearings relative to each other.[4] Several common versions are seen schematically in Fig. 4.4-1. The top two sketches **a** and **b** are representative of the early TPs in the 1940s, but there were some exceptions. When the two propellants have nearly similar pump discharge heads and roughly the same density (within +/− about 35%, as with most storable propellants and also with LOX/kerosene), then the fuel and oxidizer pumps have about the same diameter and speed, and they can then both be designed to be on the same shaft as the turbine and have reasonably good pump efficiencies. This single shaft type is represented by sketches **a, b, g,** and the main TP in **h**.

Two in-line shafts connected by a coupling are shown in sketch **g**; this was first introduced in the late 1930s by the Germans for their V-2 LPRE, and it is discussed later in this chapter. In many TPs the turbine has to run at a higher shaft speed to be efficient, and therefore gear cases were introduced with LOX/kerosene and storable propellants, mostly in the United States in the 1950s. Sketches **d** and **f** of Fig. 4.4-1 are representative of the different arrangements of geared TPs, and Fig. 4.4-2 shows a typical example with a high power gear case for transmitting power to the pumps and a small gear case for driving auxiliaries, such as a hydraulic pump for hydraulic gimbal actuators and a lubricant pump. With LOX/LH$_2$ there is a large difference in densities and discharge heads. The low-density hydrogen has a large pump head, and therefore the hydrogen pump usually has two or more pump stages and runs best at higher speed than the LOX pump. The initial solution was a gear case. However it often is best to have a separate fuel turbopump and a separate oxidizer turbopump, each running at a different speed and at near optimum efficiency, as shown in sketch **c** or **e** of Fig. 4.4-1. To provide an extra margin against cavitation in the main pump impellers, booster pumps were added as shown schematically in sketches **c** and **h**. Although the sketch

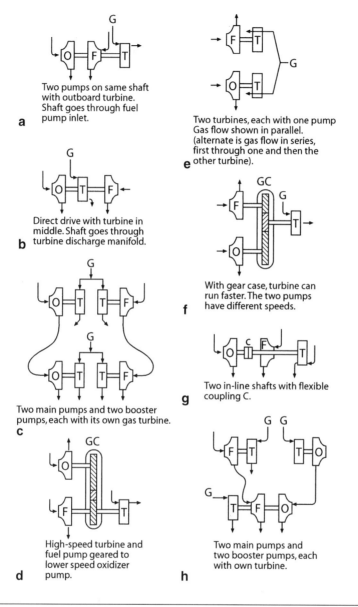

Fig. 4.4-1 Simplified diagrams of several different component arrangements of turbopumps (TP); F, fuel pump; O, oxidizer pump; T, turbine; G, hot gas; C, shaft coupling; and GC, gear case.

suggests that a gas G (usually warm gas) be used to drive the turbines of booster pumps, some oxidizer booster pumps are driven by liquid oxidizer (tapped off the main oxidizer pump discharge) in a liquid turbine. This avoids putting fuel species from the warm gas into an oxidizer pump. All the different

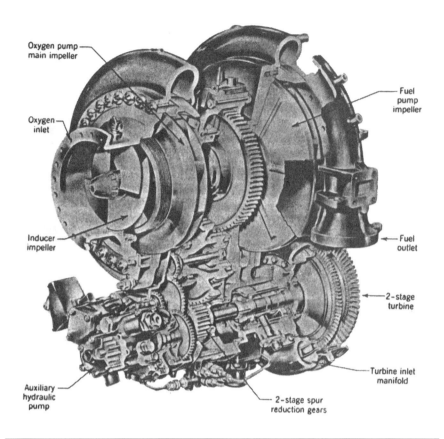

Fig. 4.4-2 Geared Rocketdyne turbopump used in early version of the Atlas booster engine. The turbine rotates at a higher speed than the two pumps (on same shaft), and the auxiliary hydraulic pump runs at the slowest shaft speed. Courtesy of The Boeing Company, Rocketdyne Propulsion and Power; copied with permission from Ref. 4.

TP configurations shown in Fig. 4.4-1 have been tested succesfully and most have flown. However the more recent designs seem to favor TPs without gears and without dual shafts.

Early Turbopumps

The first few TPs were developed in three different countries between 1934 to 1947. The Germans were the first to use and fly a TP. The first turbopump for a LPRE was developed by the Hellmuth Walter Company in Kiel for a monopropellant airplane rocket engine pumping only one propellant, namely, 80% hydrogen peroxide (1936). The hydrogen peroxide was decomposed by catalysts to form steam, which contained some hot oxygen, and this gas produced the thrust. A small portion of the flow (typically 2%) is decomposed separately in a gas generator, and the resulting gas drives the turbine. The two pumps, the turbine, and the two inducer impellers were all on a single shaft. This monopropellant engine first flew in 1939 and was the first LPRE with a TP to fly in an aircraft. A few years later a bipropellant engine versions of this same basic engine was developed (identified as the 109-509 LPRE), and there was an additional fuel pump (for C-Stoff) in the TP assembly. Figure 4.4-3 shows some of the key components, drains, and construction details of this TP with two sets of pumps. Chapter 9.3 give further information on this historical TP and its engine. It was the very first TP with inducer impellers, which were added and first tested in 1940. The bipropellant engine with this TP and its inducers went into production as the power plant for the Me 163 fighter aircraft. The Germans developed the very first large TP for the high-thrust LPRE of the V-2 missile in the late 1930s and the early 1940s. This historical TP is treated in a separate section of this chapter.

The first TPs in the United States were developed by Robert H. Goddard between about 1934 and 1936. They will be discussed in Chapter 5.2 and shown in Fig. 5-11 and 5-12. There were two TPs, each with its separate

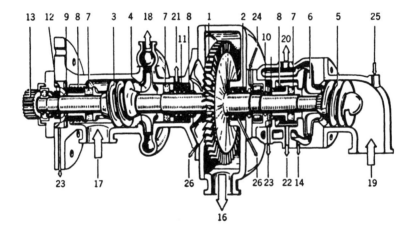

Fig. 4.4-3 Section drawing of a TP designed around 1940 by Hellmuth Walter Company, (Kiel, Germany) for the 109-509 LPRE used in a German Me-163 fighter aircraft: copied from Ref. 7, courtesy of HWK. 1, turbine wheel; 2, carbon seal rings; 3, oxidizer inducer impeller; 4, oxidizer pump impeller; 5, fuel inducer impeller; 6, fuel pump impeller; 7, carbon face seal; 8, seal spring; 9, synthetic rubber membrane; 10, felt seal ring; 11, metal bellows; 12, oil seal plate; 13, starter drive spur gear; 14, fuel discharge pressure tap; 15, steam inlet to turbine (not shown); 16, turbine discharge; 17, oxidizer pump inlet; 18, oxidizer pump discharge; 19, fuel pump inlet; 20, fuel pump discharge; 21, oxidizer leakage return line to tank; 22, fuel leakage return line to tank; 23, leakage drain; 24, grease passage; 25, fuel return drain from control unit; and 26, steam leakage drain.

turbine, one driving a small fuel pump and one a small oxidizer pump. Goddard's first set of turbines was driven by the hot exhaust gas from the thrust chamber, but were only partly immersed into the exhaust flow emerging at supersonic speed from the nozzle. The placement of the TPs just outside the nozzle can be seen in Fig. 5-13; only a small sector of the turbine blades were immersed in the exhaust jet at any one time. These TPs were small and relatively inefficient because the flow of gas reached only a few of the turbine blades at any one time. Because the stagnation temperature of the exhaust gas (over 4000°F) was much higher than the melting point of the turbine blades, Goddard experienced frequent turbine failures. None of these early Goddard TPs were installed in a flight vehicle. In 1938 Goddard developed a separate gas generator (GG), which burned at a lower combustion temperature (600–750°F) than the melting point of the blade material. He then successfully used this GG to drive two improved larger, but similar TPs in ground tests and vehicle tie-down tests in 1939. The engine with the two TPs flew in a Goddard sounding rocket in 1939 and again in 1941. These were the world's first vertically launched flights with a TP-fed LPRE

In the Soviet Union the design and component testing of a pseudoturbopump was initiated at RNII (Reaction Propulsion Research Institute) in 1933 using heavy-duty hardware. The program was aimed at an engine for an aircraft superperformance at 300 kg thrust or 660 lbf (Ref.6). They did not design a turbine, but obtained one from a small existing marine engine and ran it with combustion gas (ambient air and fuel burning at about 930°F). It was really not a rocket TP, but a test setup for a rocket engine pump. The pump tests showed a high discharge pressure (75 atm). The design provided for a single shaft for the turbine, a small nitric acid pump, and a small kerosene pump. Neither the engine nor the TP was fully developed. The author could not find any further information on this pump.

There was little or no work done in the Soviet Union on TPs for more than 10 years. This delay happened, in part because the Soviet Union's engineers had developed a very satisfactory set of gear pumps, which were driven by a jack shaft from the aircraft's piston engine.[6] As discussed in Chapter 8.2, the rocket engines with the gear pumps flew satisfactorily in several experimental Soviet aircraft, and there was no urgency to develop a TP or a turbine to drive these gear pumps. Gear pumps, a viable method for raising the pressure of the propellants, are discussed later in this chapter.

The RNII team headed by Leonid. S. Dushkin developed two single-shaft simple turbopumps, one with a two-stage turbine, in the 1944 to 1949 period.[6] Their LPREs were aimed at aircraft applications and are described further in Chapter 8.2. One TP was first developed and tested for the RD-2M engine using nitric acid and kerosene propellants. Its gas generator used the catalytic decomposition of hydrogen peroxide as the gas to drive the two-stage turbine. The pump-fed engine is reported to have had a maximum thrust of 1400 kg or 3080 lbf and could be throttled to 25% of rated thrust. After 40 to 45 starts the combustion chamber had to be replaced. The turbopump was reported to have run well for a 1.5-h endurance test. No other information was found on this TP. It was the first Soviet TP in an aircraft-type LPRE to fly in 1947. This happened nine years after the Germans performed such a flight with a TP, but a couple of years earlier than a similar accomplishment in the United States.

The RD-2M-3V LPRE, also developed by Dushkin at the Soviet RNII R&D organization, used a larger turbopump. This engine was designed and built with a TP in 1945 and 1946 and was rated at 2000 kg or 4400 lbf thrust using nitric acid and kerosene as propellants. In the single-shaft TP design the turbine drove three pumps, one each for nitric acid, kerosene, and 80% hydrogen peroxide.[6] The peroxide was decomposed by a solid catalyst, and the resulting gas drove the turbine. Ground testing of the turbopump and engine occurred in 1947 and 1948. At about the same time another TP was being developed by RNII for an aircraft rocket engine, which operated with LOX/alcohol and could reportedly be throttled from 1500 kg (3300 lbf) to 300 kg thrust. It also used decomposed 80% hydrogen peroxide for driving its turbine and also was ground tested in 1947.[6] The author could not find any photos or drawings of these two USSR engines or their TPs, and there is no record that they have flown. Later they operated an engine with a bipropellant gas generator, which had been developed as a separate component earlier (1934), and this GG is described in the next chapter. Work on LPREs with gear pumps, but without a turbine, was successful; its engines have flown, and the programs are described in Chapter 8.2. The United States and the Soviet Union later developed several small more sophisticated TPs for jet-assisted-takeoff (JATO) units and experimental aircraft (thrusts of 3000 to 10,000 lbf) between 1942 and 1949. Several are described in Chapters 7.7, 7.8, and 8.2. All used a simple TP with a single shaft, a single turbine, and two propellant pumps.

Turbopump for the German V-2 Engine

This is a historically significant TP. In 1938 or 1939 the Germans had a design of the large turbopump in their V-2 missile engine as seen in Fig. 4.4-4. It was the most powerful (465 hp) and most efficient turbopump (about 70% pump efficiency) of its day, and it astonished the LPRE industry, when the TP information became available in 1945 after World War II. It was larger (in thrust and power) by a factor of almost 10 than any prior TP. This engine used liquid oxygen and 75% alcohol fuel, and the engine is shown in Fig. 9-2, and engine data are in Table 9-1. The turbine was driven by a small flow of 80% hydrogen peroxide into a gas generator, where it was catalyzed and decomposed into steam and some hot oxygen. The turbine inlet gas temperature was about 840°F, and this made it possible to use aluminum for the turbine material. The two propellant pumps were also made out of aluminum castings. All of the early German TPs used ball bearings and had mechanical shaft seals cooled and lubricated by a very small flow of propellant; any small leakage past the seals was dumped overboard. Photos of the parts of the disassembled two pumps and the turbine can be found in Ref. 7.

114 History of Liquid Propellant Rocket Engines

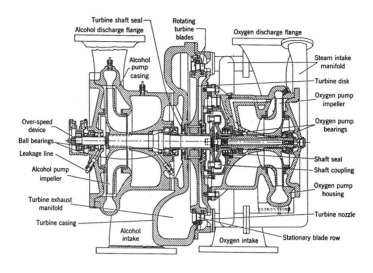

Figure 4.4-4 Turbopump for the LPRE of the German V-2 missile with two in-line shafts that are coupled. Copied with permission from Ref. 7.

This TP for the V-2 LPRE was the first with *two in-line shafts*, a type that is shown schematically in sketch **g** of Fig. 4.4-1 and as a section in Fig. 4.4-4. The fuel pump and the turbine are on one shaft, and the oxidizer pump is on the second shaft. The two shafts are connected by a flexible coupling, which allows for some shaft misalignment and minimizes the flow of heat through the shaft from the hot turbine assembly to the cold cryogenic LOX pump. The four bearings (two on each of the two shafts) and a coupling can be seen in the figure. With the smaller span between each set of bearings, this arrangement results in substantially smaller shaft diameters, which in turn reduces the mass, diameter, and inertia of the rotary assembly, the bearing loads, sizes and costs, and also the mass of the pump impellers and pump housings. If the turbine and both pumps would have been on the same shaft, the TP shaft and assembly would have been much larger in diameter (to have the same stiffness for minimizing shaft deflection), be heavier, and probably run at a lower speed. This two in-line shaft TP design requires often more parts in the TP, has perhaps more cost, and results in a longer TP assembly, which can be more difficult to package into an engine.

This German V-2 TP had a major influence on TPs developed in the Soviet Union. The Soviets used this design concept of two in-line coupled shafts in several of their larger TPs in the late 1940s and 1950s. It does not seem to have been used in other countries. The Soviets must have learned about this double-shaft concept from the Germans in 1945 and adopted it and improved it. For example they replaced the large flange-type coupling with a smaller, lighter, and simpler curvic coupling. Several early large Soviet LPREs had TPs with two in-line shafts. Example are shown in Fig. 8.4-9, which is the TP for the RD-107 LOX/kerosene LPRE and in Fig. 4.4-5, which is TP for the RD-119

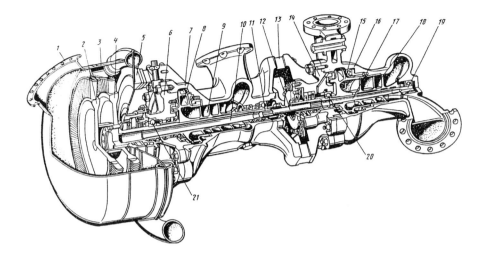

Fig. 4.4-5 Turbopump for the RD-119 is long and has two separate in-line shafts connected through an internal splined shaft coupling. Copied from texbook *Construction and Design of Liquid Rocket Engines* (in Russian) Gahun, G. G., editor, Mashinostroenie, Moscow, 1989; 1, turbine exhaust flange; 2, 4, turbine disks with blades; 3, stator housing (stator vanes are not shown); 5, housing with gas inlet duct; 6, 14, bearing and seal housings; 7, 16, pump housings; 8, 15, pump main impellers; 9, 18, inducer pump impellers; 10, 17, shafts; 11, spline coupling connection; 12, internal jack shaft; 13, connecting spacer and housing; and 19, 20, 21, bearings.

LOX/UDMH upper-stage LPRE. In more recent LPRE designs the Soviets seem to have preferred single-shaft TPs.

Turbopump Development

The design of TPs is engineering intensive. It is not the purpose of this book to explain how turbopumps are designed or what criteria have been used to select the best design. The literature contains a number of reports on design, selection of TP parameters, testing, development issues, and problems encountered with TPs.[8-11] A couple are listed in the references, but they are not comprehensive.[1-3] A number of development problems have had to be solved. For example, several kinds of vibration problems were common in the early developments.[9-11] The nature of the development issues were usually different with different TPs, and there are a number of reports on the development of specific TPs.[12-15] The methods for designing TPs have changed and improved over the last 60 years. For example, the analytical models or tools for performance, hydraulic flow through impellers, stress analysis, turbine blade contours, or thermal analysis are better, more versatile, and more sophisticated. The available materials of construction are stronger and better in other ways. The testing of TP components, TP assemblies, and TPs operating within an engine is today more pertinent, usually better instrumented, and less costly than it used to be. Special test equipment and test facilities had to be developed and installed. This included the driving of pumps with a variable-speed electric motor at shaft speeds and power levels, which at the time, were beyond the state of the art of electric motors, novel facilities for destructive testing of impellers or turbine wheels by running them at higher than rated speed under some simulated operating conditions, bearing and seal test machines using real propellants, hydraulic testing of pumps, and facilities for running TP with gas generators or preburners, but without TCs.

It is easier and usually faster and less costly to modify or uprate an existing, proven TP, than to design, develop, qualify, and produce a brand new TP. Therefore there are at least a dozen different groups of related LPREs that use basically the same design concept for the TP. For example the several versions of the Titan rocket engine described in Chapter 7.7 use the same geared turbopump, except that the propellants were changed, the flows and pressures were increased, and several minor changes were incorporated. Some of the engine changes are described in Chapter 7.7.

Different types of turbines were used.[1-4] With gas generator engine cycles the turbine flow is small, and the pressure drop across the turbine is large, but with expander cycles or the staged combustion engine cycle the warm gas flow is large, and the pressure drop is small. With higher power two wheels of blades are common. The turbines are driven by hot gas, and the source of this gas depends on the engine cycle as discussed in Chapter 4.2. When the turbines for the first TPs were designed, the engineers borrowed from the existing technology and materials of steam turbines, aircraft engines, and stationary gas turbines. In a rocket-engine TP the thermal stresses are more severe than in commercial turbines or pumps because of the rapid transition to full power and the high-temperature gradients during the brief start sequence. Furthermore the pressures are higher, and the gas composition is very different.

As already mentioned in Chapters 4.1 and 4.5, the early turbines were usually supplied with warm gas from a hydrogen-peroxide monopropellant gas source, and the maximum gas temperature was therefore limited as defined in Table 4.1-2. Because of the relatively low inlet gas temperature, the early turbines could be made of aluminum. This was the material of choice for turbines of LPREs developed in Germany, Soviet Union, Britain, and the United States in the period between 1935 and 1955. Between 1945 and 1955 the Soviet Union and the United States developed bipropellant GG and put into production several engines with TPs for JATO units or aircraft superperformance applications. The TPs of these engines used alloy steel turbines and had turbine inlet temperatures of 1000–1450°F. The higher turbine inlet temperature reduced the gas-generator flow and permitted a small engine performance increase. When the turbines were made out of one of several specialty nickel-based superalloys, the GG gas temperatures could be raised up to about 1650°F and still retain enough strength at the high gas temperatures to accept the centrifugal forces and other loads. However there was more risk for turbine damage or turbine failure, if the GG should inadvertently be operated at an off-design mixture ratio and thus the gas should become hotter. Some turbines used blades with fir tree roots, a technology originally developed for jet engines and steam turbines. Fig. 4.4-6 shows such blades. They can be fabricated out of a different and more appropriate material than the turbine disk. A common construction has been to machine the blades and the turbine disk out of one piece of high-temperature metal, and this approach was used in the early TPs and is still popular today.

The design parameters for the *propellant pumps*[1–4,16–18] are derived from the engine parameters, the general engine requirements, and the need to avoid cavitation. The selection of the pump impeller type and size and the volute configuration depends on required propellant flow rates, discharge pressures, the pump inlet conditions, the throttling requirements, if any, and other parameters. Certain types of centrifugal impellers are most effective for particular regimes of flow and head (fluid pressure divide by fluid density). Most of the

Fig. 4.4-6 Blades with fir-tree roots from the first-stage turbine of the Rocketdyne TP for the space shuttle main engine. It uses a high nickel-chromium alloy that solidifies directionally. Courtesy of The Boeing Company, Rocketdyne Propulsion and Power. Copied from Rocketdyne photo with permission.

pumps for larger LPREs (for engines with 10,000 lbf thrust or more) have used Frances-type or mixed-flow-type centrifugal impellers.[1-3,18,19] The various impeller configurations are defined in Refs. 4 and 18. For low flow application, radial flow impellers or straight vane-type pumps seem to have been used. For very low thrust engines and thus for very low propellant flows, the centrifugal pumps become very small in size, have a relatively large leakage around the impeller, and are very inefficient (7–30%). Positive displacement pumps, discussed at the end of this chapter, give very much better efficiencies (70–90%) at low propellant flows.

Aluminum castings and/or forgings have been used for pump impellers and housings, and these were used for most of the early pumps. Now with stronger aluminum alloys they are used today up to discharge pressures of about 2000 psi. Later other materials, including alloy steels and nickel-based superalloys, have been employed for pumps with higher discharge pressures and higher impeller tip speeds. The Rocketdyne SSME uses Inconel 718 extensively. A few years ago Pratt & Whitney used a single titanium piece for a rotor with machined pump impellers as well as turbine blades in an experimental hydrogen TP suitable for a 50,000-lb-thrust LPRE. It is shown in Fig. 7.10-10. More recently, powder metallurgy was used to construct impellers allowing higher discharge pressures.[20]

The first large U.S. liquid-hydrogen pump for the Rocketdyne J-2 LPRE was an exception. It is shown in Fig. 4.4-7 and used seven stages of axial-flow impellers, which were based on an investigation of high solidity inducer impellers.[20,21] The axial-flow impellers allowed a higher pump efficiency at the design conditions than a centrifugal pump, but were not as adaptable to off-design operating conditions and might have had a higher risk of possible impeller stall. This axial-flow pump was developed between 1960 and 1966 and was operated satisfactorily in the 1960s and 1970s. No other multi-stage axial-flow pumps are known to have been developed and flown since that time.

Most impellers have a single inlet on one side. An inlet manifold with an internal rotating shaft is more likely to have cavitation problems than an inlet at the end of the TP (without a protruding shaft). This can be visualized by examining sketches **a** and **b** of Fig. 4.4-1. Double- or dual-inlet impellers have been used in some of the early TPs in order to balance the axial hydraulic loads on the rotating assembly and to reduce impeller inlet velocity, which makes the impeller less likely to cavitate. Such dual-inlet impellers can be seen in the SSME high-pressure oxidizer pump, which is assembled to the SSME power head in Fig. 7.8-19 and the Soviet RD-111 oxidizer pump in Fig. 8.4-12.

The early investigators learned in the 1930s and 1940s that it was not easy to pump liquid oxygen because the cold cryogenic propellant would evaporate or partly evaporate as the initial flow was fed into the ambient temperature hardware. The average density of the initial flow of oxygen was unpredictable and usually unknown. This caused combustion problems and the thrust to be very low. So they learned how to cool the TP, valves, and pipes by bleeding some cold oxygen through them just prior to the engine start. This

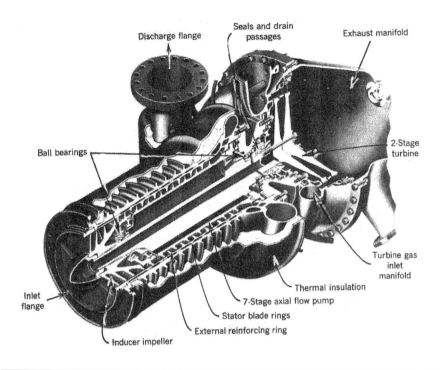

Fig. 4.4-7 Liquid-hydrogen turbopump used in the Rocketdyne J-2 LPRE. It has an axial inducer impeller and seven stages of axial-flow impellers. Courtesy of The Boeing Company, Rocketdyne Propulsion and Power; copied with permission from fourth edition of Ref. 4.

procedure, called "chill down," cools the hardware and allows a predictable start with liquid oxygen. Today it is used with LOX as well as LH_2. It is more difficult to chill down the hardware of an upper-stage engine prior to a start during the flight in space. Various techniques have been used to minimize the amount of propellant used for chill down because this propellant is not available for propulsion and detracts from the vehicle's performance. Precooling (just prior to takeoff) the TP and other hardware on the launch stand with cold propellant supplied from the ground was effective and was started in the 1960s. An alternate procedure was to replenish the propellant used for chill down just prior to launch.

In the United States or Japan the flight-path control, during the operation of a main LPRE, is usually achieved by a multithruster reaction control system (RCS) using a pressurized gas feed system and multiple thrusters of low thrust (1–1000 lbf depending on vehicle size and mission). The Soviet Union has used a TP for supplying propellants to their hinged vernier engines, which control the flight path and they usually have a much higher thrust. There are typically four hinge-mounted vernier TCs arranged around the larger fixed (not gimbaled) main engine, which has its own larger turbopump. Examples are the Russian RD-0212 LPRE from CADB or the RD-0243 submarine missile engine, also from CADB, and both are depicted and described in Chapter 8.5. Typically the four hinge-mounted vernier TCs provide 10–20% of the total thrust. With a thrust of typically 1000–4000 lbf for each vernier TC, they have a high enough total propellant flow for the four vernier thrust chambers to use their own separate small TP with reasonably good pump efficiencies, such as more than 40%. Using a separate TP for the vernier thrusters allows a saving of inert propulsion system mass and also allows vernier operation before, during, and after the operation of the main engine. The low efficiency is one of the reasons why there are usually no TPs for attitude flight control systems with low thrusts of typically 100 lbf.

Inducers and Booster Pumps

An *inducer* is axial-flow-type pump, and it typically has a spiral shape as shown in Fig. 4.4-8. Typically the inducer is on the same shaft as the main impeller, but just upstream of the main pump impellers.[22] Most of the turbopumps before 1950 did not have inducer impellers. They required relatively high propellant tank pressures (40 psi or more) and relatively heavy tanks to suppress pump cavitation. As stated elsewhere, cavitation is the formation of vapor bubbles in low-pressure regions of the flow, and it usually occurs at the leading edge of the pump impeller vanes. Cavitation bubbles can cause volumetric changes in the propellant flow, which in turn can cause combustion

Fig. 4.4-8 Inducer pump impellers usually have axial flow (flow parallel to axis), and most are of the spiral type shown here. Courtesy of the Boeing Company, Rocketdyne Propulsion and Power. Copied with permission from Ref. 4.

problems, mixture-ratio changes (which can cause major gas temperature changes), and decreases in thrust and total impulse. It also can cause the TP to overspeed and thus to fail.[23,24] In about 1939 the Germans realized that cavitation had occurred in one of their TPs and that the design must provide enough pump suction pressure to avoid cavitation. The pressure rise in the inducer is typically 6–12% of the pressure rise of the main impeller, and this is usually enough to prevent cavitation.

The first inducers were put into the turbopumps of the Hellmut Walter Company's LPREs developed at Kiel, Germany. This was done by Walter because an earlier TP (without inducers) experienced cavitation and operated at greatly reduced thrust. An example of the early use of inducers is the TP for the bipropellant 109-509 rocket engine in the early 1940s. One version can be seen in Fig. 4.4-3. Inducers were introduced in the Soviet Union in the RD-111 about 1950 and in the United States for the Atlas and Thor engines (around the mid-1950s). Even though the Soviet Union and the United States learned about the German inducers in 1945 after the end of World War II, both countries were slow in using this technology. Inducers allowed an extra margin for cavitation at the main pump impellers, a reduction in tank pressure (20 or 25 psia) and thus in tank inert mass and a lower risk of cavitation. In turn this allows a higher speed of the main pump, which in turn can result in a smaller, lighter main TP. In the 1950s some of the early existing U.S. TPs had to be redesigned and requalified to include new inducer impellers. Modern TPs today generally have inducers ahead of the main impellers.

Booster pumps were included in some engine designs (particularly those with staged combustion cycles), such as the Rocketdyne Space Shuttle Main Engine (developed from 1972–1980, see Chapter 7.8), the Russian Energomash RD-120 LPRE (developed from 1976–1983), or the Russian CADB RD-0120 (developed from 1967–1983) upper-stage engine.[25,26] They can be seen in Chapters 8.4 and 8.5. The number of LPREs that have flown with booster pumps as a part of the LPRE is relatively small, estimated at less than 10. Booster pumps can be seen in the flow diagrams of these engines: SSME in Fig. 7.8-19, RD-170 in Fig. 8.4-8, and the RD-0120 in Fig. 8.5-19. By providing a small pressure rise with the booster pump, it was possible to further reduce propellant tank pressure and the inert mass of propellant tanks and piping. Booster pumps can also allow an increase of the speed of the main pumps, thereby reducing the size of the main pump impellers, the size of the main TP assembly, and the inert mass of the main turbopump. Booster pumps add to the complexity and costs of the engine, but they provide a suction performance margin for safe engine operation and allow optimization of the pump pressure increases to minimize the total vehicle and engine masses. The turbines for the booster pumps have been driven 1) by warm gas from a GG or preburner, 2) by evaporated hydrogen from the cooling jacket, or 3) since about 1970 by a liquid turbine using high-pressure liquid propellant tapped off from a main propellant pump. Because the inlet pressures of a booster pump are usually low (typically 25 psi) and the pressure rise is modest (3–10% of the total pump pressure rise), the walls of these booster pumps are usually thin, and the booster pump mass is low. They can be large in diameter and size, sometimes larger than the main TP, but their masses and power are usually much lower.

In 1944 and 1945 your author was the test engineer for the booster pump on an Aerojet experimental engine called the Aerotojet. This LPRE is shown schematically in Fig. 7.7-8 and described in Chapter 7.7. This engine had a successful small (mixed-flow centrifugal impeller) nitric-acid booster pump. As far as the author knows, it was the first booster pump on any experimental LPRE (1944). The project had some technical problems (unrelated to the booster pump) and was canceled after component testing.

The Soviets are the only ones to have built and flown a lightweight sophisticated two-stage booster TP with two concentric shafts (each with an impeller and its own turbine) running at different speeds, similar in principle to some two-shaft jet engines.[16] This design is used on the oxygen booster pump of the RD-0120 engine, and it is shown in Fig. 8.5-20. It allows a further reduction in inert propellant tank mass and a better margin against cavitation, when compared to a single-stage, single-shaft booster TP. Of course it is a more complex and more expensive pump.

Gear Cases for Shaft Speed Reduction

Many of the early turbopumps for large LPREs used a gearbox for one of two reasons.[27] First the turbine and the pump(s) could be geared to run at their respective optimum speed for maximum efficiency; this uses a minimum amount of gas-generator propellant. However the TP becomes a more complex assembly with more parts and with provisions for lubrication and cooling of the gears and case bearings.[27] Examples are the TP gear cases in the General Electric Hermes LPRE (1946) shown in Fig. 4.4-9 and mentioned again in Chapter 7.3, the ARC Agena engine (1957) in Fig. 7.11-10, and the TP for the Pratt & Whitney RL 10 LPREs as in Fig. 7.10-3. Fig. 4.4-2 shows a Rocketdyne TP with a gear case, where the turbine is on a high-speed shaft and the two propellant pumps for LOX and RP-1 are on a lower-speed common shaft. This basic design was used, uprated, and modified to feed propellants in six different LPRE (not counting the different versions of each of these engines) over a period from 1953 to 2004. They were the TPs for the following engines: Navaho, Atlas, Jupiter, Thor, Delta, and Saturn I booster (H-1); these engines are discussed in Chapter 7.8.

The gears had to be lubricated and cooled to prevent gear failure, and initially a separate supply of lubricating oil was provided. There were some initial problems with foaming of the lubricant in the gear case. An improvement made in the United States around 1958 was to use the RP-1 (kerosene) fuel directly as a lubricant and coolant (initially with a small amount of oil additive), and this permitted the omission of the separate auxiliary oil pump and oil supply. In one early TP the General Electric Company used grease for lubricating gears and bearings, and they had a can of grease for supplying fresh lubricant to the gear case during engine operation.

Technology and Hardware **123**

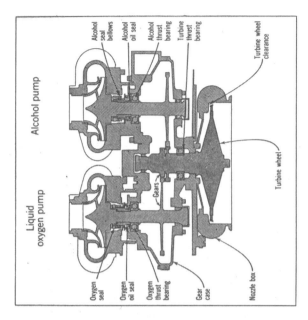

Fig. 4.4-9 Geared TP of the engine for the Hermes Missile. This TP is part of the engine developed by General Electric in the 1940s. All bearings and gears are lubricated by oil. Courtesy of General Electric Company, 1949. Copied with permission from Ref. 7.

The second reason for using gear cases was the driving of the auxiliaries, such as oil pumps at their optimum speed. Here the gears transmit only a minor portion of the full power of the turbine, and the gears as well as the gear case could be much smaller and lighter. On the Soviet TP for the RD-107 LPRE in Fig. 8.4-9, there is a small gear case with a small pump for liquid nitrogen (to pressurize the propellant tanks) and a pump for hydrogen peroxide (supply to gas generator). One version of the U.S. Altas TP had a gear case with a hydraulic pump (for control actuation) and an oil pump (for lubricating the gear case). The TP is very similar to the one shown in Fig. 4.4-2, where the small gears for the auxiliary pumps can be seen. In most of the more recent TP designs, there are no gear cases, and of course this further simplifies the engine.

Bearings and Seals

There has been amazing progress in bearing technology for TPs.[28-31] Some of the early TP shafts and gear cases started with conventional, grease-packed, premium quality ball and roller bearings. Prelubricated bearings, special materials in bearings smeared with grease, were not satisfactory at the power levels of LPREs. They were soon replaced with bearings that were cooled and lubricated by oil, which proved to be reliable. The TP with a gear transmission case then had a separate oil reservoir, an oil pump, and a cooler, where the oil was cooled in a heat exchanger with propellant. Fig. 4.4-9 shows a General Electric TP, where all of the bearings and all of the gears were oil lubricated. Later (late 1950s) propellant was used directly as a lubricant and coolant. In the TP shown in Fig. 4.4-2, oil was used initially for the gear case. Thereafter it was kerosene, to which a small amount of oil was added. Then it was proven that kerosene would work without the oil additive, and that simplified the engine and eliminated an oil pump, cooling device, and oil reservoir. Goddard lubricated and cooled his first TP (1934) with gasoline, and the first German TP used oil. All of the developers of hydrogen pumps have learned that liquid hydrogen can be a bearing coolant and lubricant; it is an excellent coolant, but a relatively poor lubricant. Liquid hydrogen was first used in pump bearings by Pratt & Whitney in the 1960s. Because the hydrogen is cold, any other liquid, such as water, would freeze in a bearing or valve, and therefore special procedures are needed to eliminate moisture in the system. During the start, when there is a poor supply of bearing lubricant available, the bearing runs almost dry, and the bearing materials have been selected to be able to withstand direct rolling bearing contact. In the 1980s the French developed a special dry lubricant material for coating of the bearing surfaces in order to ensure a good start and a good operation.[29]

The first TP ball bearings immersed in LOX and cooled and lubricated by LOX were developed by Goddard in the mid-1930s. Using liquid oxygen or other oxidizers as a lubricant is more risky because the bearing metal can be oxidized, if it gets hot enough to ignite a chemical reaction between the oxidizer and the metal. A bearing, seal, or impeller wear ring can also become locally hot enough to initiate the burning of the metal, if a foreign solid abrasive particle is caught in the clearance between the mating, moving surfaces between the rotating and stationary parts. With some bearing or seal materials a high force metal-to-metal contact can also start a metal fire with an oxidizer liquid. This author has seen a turbopump fail, burn up, and disappear in less than a couple of seconds once the metal started to burn and release heat. Metal ignition can usually be avoided by proper filtering of the lubricant or propellant, using proper clearances and tolerances, having clean parts (no grease or oil deposits in oxidizer passages, which can start a local chemical reaction), proper balancing of the rotating assembly, and good inspection techniques and fabrication procedures. Special bearing materials and the minimizing of direct bearing contacts seem to be important. With new materials and with careful design, the load-carrying capacity of ball and roller bearings has increased substantially, allowing smaller bearings.

Beginning in about 1991, work started on *hydrostatic bearings* using propellants as bearing fluids. They are sleeve bearings with precise tolerances, and they provide a very thin layer of lubricant at high pressure in a cylindrical bearing sleeve, good damping characteristics, and a rigid support. They eliminate friction, wear, or direct contact between rotating and stationary parts during operation, but can have momentary contacts during startup or shutdown. Initially this work was done at Rocketdyne (built bearing testers and experimental TPs) and then also by Pratt & Whitney. Recent experimental turbopumps from Pratt & Whitney (1998) used hydrostatic bearings for both radial and axial internal hydraulic loads; it was intended for a future engine of 50,000 lbf thrust (see Fig. 7.10-10). Rocketdyne tested an experimental pump with hydrostatic bearings in the 1990s. Some turbopumps had balancing pistons for compensating excessive axial loads, and some of them have been considered to be a form of axial hydrostatic bearing. As of January 2002, hydrostatic bearings had been tested in several experimental TPs, but as far as the author knows they had not yet been included in a TP that has flown.

In the 1970s and 1980s some designers advocated the use of magnetic bearings for turbopumps. It had the advantage of not needing a lubricant, providing an accurate centering of the shaft, no metal-to-metal contact, and no propellant bearing seals, leaks, or drains. It required a significant electronic control and a heavy separate power supply. This electromagnetic suspension of the shaft was built and tested, but was too complex to be adopted for a flying TP.

Positive Displacement Pumps

Piston pumps, gear pumps, and occasionally diaphragm pumps have been used to supply propellants under pressure to one or more thrust chambers (TCs). They have been used in applications with relatively low thrust or low propellant flow because conventional centrifugal pumps are notoriously inefficient at low flow, whereas positive displacement pumps can have a very high efficiency in this low flow regime. There are relatively few LPRE that have been developed and flown with positive displacement pumps. Four examples are given here.

For the feeding of pressurized propellants to the Soviet RD-1 LPRE and its subsequent improved versions of aircraft rocket engines, a gear pump assembly was used. It consisted of a nitric-acid gear pump and a kerosene gear pump on the same shaft. Rocket thrust levels were between 300–900 kg force or 660–1980 lbf. The power for these pumps was provided through a jack shaft and a coupling from the main aircraft engine.[6] With the gear pumps it is possible to accurately control the mixture ratio by the displacement volumes in the gear pumps, and there is no need for a separate mixture-ratio control. The gear pumps are shown in Fig. 8.2-6 and discussed in Chapter 8.2. It was flight tested in experimental versions of at least six different Soviet military aircraft.

In the late 1960s and the early 1970s gear pumps were used in the integrated secondary propulsion system of the Agena upper stage vehicle.[32] The Agena upper stage and its primary LPRE are discussed in Chapter 7.11. The fuel and the oxidizer gear pumps were driven by a single small electric motor, which obtained its electric power from the vehicle's electric system. A simplified schematic diagram is shown in Fig. 4.4-10. The two couplings between the motor and each of the pumps have a nonmagnetic metal barrier between the outer set of magnets, driven by the electric motor, and the inner set of magnets, which are connected to a gear pump. The barrier eliminated the need for a propellant shaft seal and prevents any seal leakage before, during, or after rocket operation. The propellants were high-density nitric acid (with 44% dissolved nitrogen oxides) and UDMH with 1% silicon oil, which caused the deposit of silica (silicon oxide) on the walls or the TC, reducing the heat transfer. The electric motor controller caused the pump to run only when a small thruster had to be fired. The two gears in each pump were supported in sleeve bearings lubricated and cooled by the propellant. The inlet and outlet

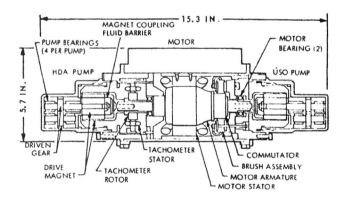

Fig. 4.4-10 Simplified cross section of motor-driven gear pumps for the Agena upper-stage secondary propulsion system. Courtesy of Lockheed Martin Corporation; copied with permission from Ref. 32.

Technology and Hardware 127

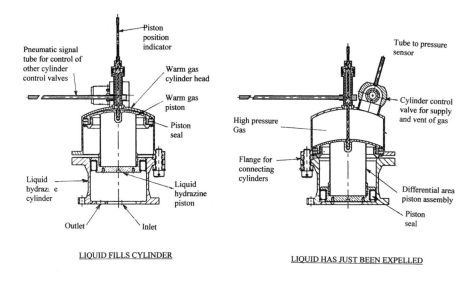

Fig. 4.4-11 Section through one of four small piston pumps flown in the Astrid experimental launch vehicle. Courtesy of Lawrence Livermore National Laboratory and provided by John C. Whitehead of that Laboratory.

of the two pumps are not shown in the figure. The motor shaft was supported by prelubricated ball bearings. This gear pump feed system for the Agena auxiliary thrusters was developed in the late 1960s by the Lockheed Missiles and Space Company, which today is part of Lockheed Martin Corporation.

A reciprocating piston pump is a different type of small flow positive displacement pump.[33] One of these piston pumps, developed at the Lawrence Livermore National Laboratory, is shown in Fig. 4.4-11. The pump is driven by a reciprocating warm-gas piston, and the piston motion is controlled by appropriate sequencing valves, but they are not included in the figure. The warm gas for the larger diameter driving piston was supplied with catalytic decomposition products from a small flow of hydrazine. Four of these small piston pumps were integrated into a single assembly, and they provided hydrazine to four 50-lbf catalytic thrusters, which propelled the experimental Astrid launch vehicle in 1994. The use of four pumps operating in sequence up to 20 cycles per second reduced the pump discharge pressure fluctuations inherent to a piston pump.

A related type of piston pump scheme, again driven by warm gas from the decomposition of hydrazine, was used in the British Chevaline submarine-launched missile. The LPRE had bipropellant thrust chambers using NTO and hydrazine propellants, and it was the auxiliary propulsion system for the flight control of the missile.* Each of two pump assemblies had an oxidizer piston

---

*Personal communications, M. Marvin and F. Boorady, ARC, Niagara Falls, NY, 2002–2004.

pump, a fuel piston pump, and a reciprocating gas-powered driving piston, all linked together. Two pumps were used in order to obtain a relatively steady pump discharge pressure, and each of the outlet pipes had a pulse attenuator. Also a bypass valve (allowing flow from the pump high-pressure discharge to the pump inlet) was used to recirculate propellant prior to TC start and to further reduce peak pressure. The thrusters operated initially for 4s, and the propellant valves and the GG valve were closed until the next demand for a thrust pulse. The chevaline missile has been decommissioned.

Start Turbines

In the Soviet Union some of the LPRE design bureaus had developed turbopumps with two turbines. One turbine is designed with relatively thin vanes or turbine blades, which are aerodynamically efficient, and they work throughout the engine operation with warm gas provided by a gas generator or preburner. The other turbine is a start turbine, and it is mounted on the same TP shaft. It is driven by the warm gas from a solid-propellant grain for the first few seconds of operation. This provides a rapid spin-up of the turbopump assembly and a fast engine start. Its warm gas has different gas properties than that of the GG, and it is usually a fuel-rich gas. The start turbines usually have thicker, less efficient blades or vanes, and they are more robust than the vanes of the other TP. The solid-propellant grain often contains some debris, such as pieces of rubber-type thermal insulation, small pieces of metal from the igniter and/or its wiring, and pieces of plastic grain support parts, and the impact of these solid pieces could damage a less rugged set of turbine blades. Some solid propellants have soot in their exhaust, and there have been instances when carbon deposits have formed in the turbine nozzles of vanes. If a single turbine would be used, the transition from a fuel rich gas (solid propellant) to the oxidizer rich gas (from a Soviet gas generator) could have a brief period, where the gas temperature would be high enough to melt or damage turbine components. After the solid-propellant grain has been consumed in the first few seconds of operation, this start turbine does not provide any further power, but continues to rotate during powered flight. Such a startup turbine can be seen in a TP developed by the Kuznetsov Experimental Design Bureau in Figs. 8.9-1, 8.9-2, and 8.9-3. This design bureau (DB) built about 10 different LPREs with this type of TP during the 1960s and 1970s, and the Yushnoye Design Bureau as well as the Korolev DB also developed engines that had a TP with an additional separate start turbine. The start turbine adds about 13–20% more inert mass to the TP. This author is not aware of any recent Soviet or Russian TP with an integrated start turbine since about 1975.

References

[1]Strangeland, M. L., "Turbopumps for Liquid Rocket Engines," *Threshold: Rocketdyne's engineering journal of power technology*, No. 3, Summer 1988, pp. 34–42.

[2]Ross, C. C., "Principles of Rocket Turbopump Design," *Journal of the American Rocket Society*, No. 84, March 1951.

[3]"Turbopump Systems for Liquid Rocket Engines," NASA Monograph SP–8107, Aug. 1974.

[4]Sutton, G. P., and Biblarz, O., *Rocket Propulsion Elements*, 7th ed., Wiley, New York, 2001, Chap. 10, pp. 362–392.

[5]Gachuna, G. G. (ed.), *Construction and Design of Liquid Rocket Engines*, Mashinostroenie Press, Moscow, 1989 (in Russian).

[6]Moshkin, Ye. K., *Development of Russian Engine Technology*, Mashinostroyeniye, Moscow, 1973; also NASA Translation Rept. TT F 15,408, May 1974 (in English).

[7]Sutton, G. P., *Rocket Propulsion Elements*, 2nd ed., Wiley, New York, 1956.

[8]*Liquid Rocket Engine Turbines*, NASA Space Vehicle Design Criteria (Chemical Propulsion), NASA SP-8110, Jan. 1974.

[9]Liang, H., and Chen, Z., "Characteristic Analysis of Flow Induced Vibrations on Inducers of an Oxygen Pump," *Journal of Propulsion and Power*, Vol. 18, No. 2, March 2002, pp. 289–294.

[10]Ek, M. C., "Solving Synchronous Whirl in the High Pressure Hydrogen Turbomachinery of the Space Shuttle Main Engine," *Journal of Spacecraft and Rockets*, Vol.17, No. 3, May-June 1980, pp. 208–218.

[11]Lalanne, M., and Ferraris, G., *Rotordynamics Predictions in Engineering*, Wiley, New York, 1998.

[12]Chopinet, J. N., and Huck, A., "Thermal and Thermomechanical Studies of the Vulcain Liquid Hydrogen Turbopump," *Revue Francaise de Mécanique*, No. 4, 1989, pp. 425–431. (in French).

[13]Barske, U. M., Saunders, D. J., and Saunt, C. G., "Development of the Turbopump for the Beta I Rocket Motor," Royal Ordnance Establishment, RPD-9, England, 1951.

[14]Kamijo, K., Sogame, E., and Okayasu, A., "Development of Liquid Oxygen and Hydrogen Turbopumps for the LE-5 Rocket Engine," *Journal of Spacecraft and Rockets*, Vol. 19, No. 3, 1982, pp. 226–231.

[15]Dimitrenko, A. I., Ivanov, A. V., Kravchenko, A. G., and Minick, A. B., "Advanced Liquid Oxygen Turbopump Design and Development," AIAA Paper 2000–3878, 2000.

[16]Brennan, C. E., *Hydrodynamics of Pumps*, Concepts ETI, Inc., and Oxford Univ. Press, New York and Oxford, 1994.

[17]Kassarik, I. J, Krutzsch, W. C., Frazer, W. H., and Messina, J. P., *Pump Handbook*, McGraw-Hill, 1976.

[18]*Liquid Rocket Engine Centrifugal Flow Turbopumps*, NASA Monograph SP 8109, Dec. 1973.

[19]Guichard, D., and Du Tetre, A., "Powder Metallurgy Applied to Impellers for Vinci Turbopump," *International Symposium for Space Transportation of the XXI Century*, [CD-ROM] May 2003.

[20]Huppert, M. C., and Rothe, K., "Axial Pumps for Propulsion Systems," *Fluid Mechanics, Acoustics and Design of Fluid Machinery*, Vol. 2, NASA SP-304, 1974, pp. 623–654.

[21]Scheer, D. D., *Liquid Rocket Engine Axial-Flow Turbopumps*, NASA Monograph 8125, April 1978.

[22]*Liquid Propellant Rocket Engine Turbopump Inducers*, NASA Monograph 8152, May 1971.

[23]Stripling, L. B., and Acosta, A. J., "Cavitation in Turbopumps-Part 1," *Journal of Basic Engineering*, Vol. 84, Sept. 1962, pp. 326–338.

[24]Stripling, L. B., "Cavitation in Turbopumps—Part 2," *Journal of Basic Engineering*, Vol. 84, Sept. 1962, pp. 339–350.

[25]Dimitrenko, A. I., Ivanov, A. V., Kravchenko, A. G., Mishin, A., and Minick, A. B., "Development Testing of an Advanced Liquid Oxygen Turbopump," AIAA Paper 2000–3852, July 2000.

[26]Demyanenko, Yu. Y., "Experience of Developing Liquid Propulsion Rocket Assembly Feedsystems Using Boost Turbopump Units," AIAA Paper 2003–5072, 2003.

[27]"Liquid Rocket Engine Turbopump Gears," NASA-SP-8100, March 1974.

[28]*Liquid Rocket Engine Turbopump Bearings*, NASA-SP-8048, March 1971.

[29]Marchal, N., Caisso, P., Alliott, P., and Souchier, A., "Dry Lubrication Applied to the HM–7 Engine Hydrogen Bearing," *Journal of Propulsion and Power*, Vol. 8, No. 5, Sept.–Oct. 1992, pp. 968–970.

[30]Bursey, R. W., Jr. Chin, H. A., Tennant, M. L., Olinger, J. B., Price, J. L., Moore, L. C., Thom, R. L., Moore, J. D., and Marty, D. E., "Advanced Hybrid Rolling Element Bearing for the Space Shuttle Main Engine High Pressure Alternate Turbopump," AIAA Paper 96–3101, 1996.

[31]Burcham, R. E. "Liquid Rocket Engines Turbopump Rotating-Shaft Seals," NASA Monograph SP-8121, Feb. 1978.

[32]DeBrock, S. C., and Rudey, C. J., "Agena Primary and Intergrated Secondary Propulsion Systems," AIAA Paper 73–1212, 1973.

[33]Whitehead, J. C., and Pittinger, L. C., "Design and Flight Testing of a Reciprocating Pump-Fed Rocket," (Astrid) AIAA Paper 94–3031, 1994.

4.5 Gas Generators, Preburners, and Chemical Tank Pressurization

These are combustion devices used in LPREs for creating "warm gases" (typically between 700 and about 1500°F or 371 and 815°C), which are used as a working fluid for driving the turbines of turbopumps. However chemically generated gases from GGs have also been used for pressurizing propellant tanks, typically at 250–500°F or 121–260°C. The term warm gas has been used for these temperature ranges, and this distinguishes them from the "hot gas" of the combustion in the thrust chamber, typically between 4400–6000°F or 2004–3315°C. These warm-gas combustion devices create gas and not thrust.

Gas Generators and Preburners

GGs or Preburners (PBs) of LPREs have usually used liquid propellant, but in a few engines a solid-propellant GG has been used. liquid propellant GGs create warm gases, which have either a fuel-rich or oxidizer-rich mixture ratio. Monopropellant GGs generate warm gas by catalysis. GG with solid propellant form reaction products, which are usually fuel rich. The warm-gas temperature is low enough to allow an uncooled turbine and an uncooled GG combustion chamber. They are smaller than the primary combustion device, namely, the thrust chamber, and they consume only about 2 to 7% of the propellant flow.[1,2] The exact percentage depends on the engine parameters. They usually use the same propellants as the main TC of the LPRE, and they usually obtain these propellants from the same TP, which also supplies the TC(s). However in some LPREs the propellants for the GG are supplied from a separate pressurized inert gas system with small separate propellant tanks, and this is discussed again in Chapter 4.8.

A GG consists of a combustion chamber (which is typically spherical or cylindrical in shape), an injector, which is often similar in concept to the injector for the thrust chamber, and a pipe or connection to the turbine inlet. There is no nozzle integral with the GG (or the PB), but there are supersonic nozzles at the turbine inlet, and these are usually a part of the TP assembly. The injectors and controls are similar to those used in TCs, and the same design approaches have been used ever since the first GG was developed. The ignition methods are basically also the same.

Gas generators are used with LPREs that run on a GG engine cycle (which was discussed and depicted in Chapter 4.2), their turbine exhausts are discharged overboard, and their chamber pressures have typical values beween 500 and 1800 psia. The operating pressure of the GG is typically the same or close to the chamber pressure of the main TC, and the flow through a GG is typically 2 to 6% of the total propellant flow. The first set of GGs were developed in several countries between 1933–1949, the same basic concepts are still used today, and they are described later in this chapter.

Preburners are defined as combustion devices used with engines running with a staged combustion cycle, which was discussed in Chapter 4.2. They are quite similar to gas-generator cycles, except the turbine exhaust gases go into the main TC, and the combustion pressure in the PB is higher. The chamber pressures of the main TC are typically between 1000 and 3700 psia, but some have operated at 4200 psia or more. The chamber pressures for PBs are 20 to 50% higher than these TC chamber pressures (to allow for pressure drops in the turbine, piping, and injector), and they are typically between 1300 and 5800 psi. In comparison the chamber pressures for GG chambers are typically between 150 and 1900 psi. The warm-gas temperatures are usually low enough to allow uncooled steel or nickel alloys to be used for the GG or PB chamber and the turbine hardware. The flow through a PB is much larger than the flow through a comparable GG: in many cases essentially all of the flow of one of the propellants and a small portion of the other propellant go through the PB. As discussed in Chapter 4.2, the LPRE with a staged combustion engine cycle and with a PB give 1.5 to 7% more performance and usually have a higher inert mass than an engine with a gas-generator cycle. It is theoretically possible to reduce this loss slightly by going to higher GG gas temperatures and to cooling the GG combustor and the turbine blades. This has been investigated in the 1960s; it becomes complex and expensive and was abandoned.

A variety of different GG and preburner designs (with different combustion chamber shapes, combustion volumes, injectors, or internal baffles) have been successful.[1,2] They have been designed with and without integrally mounted propellant valves, with propellants tapped off the main propellant pumps, such as in the Titan in Fig. 7.7-10, or alternately propellants provided by a separate small gas pressurized start tank feed system as in one of the versions of the Thor LPRE. For nonhypergolic propellants the GG has to have an igniter. Different ignition methods have been successful (hypergolic propellants, slugs of starting fluid, spark plugs, or solid-propellant cartridge igniters). An example of a GG with a solid-propellant start cartridge is the Rocketdyne H-1 LPRE, and this GG is shown later in this chapter. It has provisions to mount a solid-propellant start cartridge; it has enough energy to not only ignite the nonhypergolic propellant in the GG, but also to provide the initial rotation of the turbopump.

In the United States and several other countries, the GGs and PBs are usually operated at a fuel-rich mixture ratio. There are only a few exceptions, such as Goddard's GGs, which were oxidizer rich. In the Soviet Union and more recently in Russia, the vast majority of GGs and PBs are usually operated at an oxidizer-rich mixture ratio. The exceptions are LOX/LH$_2$ LPREs and a few others, where the gases are fuel rich. There are some reasons for the choice of fuel-rich or oxidizer-rich gases, when driving turbines, but the rationales are usually not very strong, and a new engine could be operated with either one of these gas mixtures.

Engines with a single PB or two PBs have been developed. Two preburners (with fuel-rich gas) were used in the U.S. space shuttle main engine (SSME) with LOX/LH$_2$ beginning in 1972. The first Soviet preburner used storable propellants, had an oxidizer-rich mixture ratio, and was developed and ground tested by the Soviets in 1958. Most Soviet-designed LPREs with a staged combustion cycle have a single preburner. However the RD-170 (LOX/kerosene) has two identical preburners feeding a single large turbine; this allows a more symmetrical thermal loading, a more even pressure distribution in the turbine inlet manifold, and smaller preburner assemblies. The U.S. shuttle main engine (LOX/LH$_2$, 1972) has also two preburners, but for a different reason, namely, to drive two different TPs without needing hot-gas transfer piping. They can be seen mounted on the turbopumps in Fig. 7.8-19.

Gas Generators for Tank Pressurization

Most of the early LPREs and many of the current LPREs use *pressurized feed systems* with inert gas, which is stored at high pressure, and this gas expels the propellants from their propellant tanks. Figure 4.2-2 shows a typical LPRE with such an inert gas feed system. In *chemical gas pressurization* warm gases are created by a combustion reaction in a liquid propellant GG or alternatively in a solid-propellant GG. They are attractive because they usually have a lower empty inert mass than a pressurized feed system with an inert stored gas. However they are more complex. Propellant tank pressures are the same for both of these pressurization systems, typically between 150 and 1200 psi.

There are a number of different schemes for chemical pressurization systems at high tank pressure. An example is the Bomarc LPRE developed by Aerojet in the 1950s, and it is shown in Fig. 7.7-16. It has two small GGs, one GG with fuel-rich gas for pressurizing the large fuel tank, and one GG with oxidizer-rich gas for pressurizing the large oxidizer tank. Two small propellant tanks, pressurized in turn by a small amount of inert gas, supply the propellants to the two GGs. A similar scheme with two small GG was used by the Soviet RD-253 LPRE for tank pressurization, again one fuel-rich GG and an oxidizer-rich GG. These GGs can be seen in the flow sheet of this LPRE in Fig. 4.2-8. The French use a single tripropellant GG (oxidizer, fuel, and water), and its gas pressurized both the main fuel tank and the main oxidizer tank in the Veronique, Coralie, Diamant A, and Diamant B. The GGs of these French engines have four small spherical tanks, one for the GG fuel, one for the GG oxidizer, one for water (it dilutes and cools the gas), and one for the inert pressurizing gas. It is described in Chapter 10 and is shown in Figs. 10-2 and 10-3 and has been used until the 1970s. Apparently the water dilution makes the gas reasonably compatible with both the oxidizer and the fuel propellants, and the French have not reported any significant chemical reaction in the propellant tanks. A third example is the solid-propellant GG for the Bullpup air-launched missile. This LPRE was developed by RMI in the 1950s and can be seen in Fig. 7.2-6. The fuel-rich gases from the solid-propellant GG pressurize both the fuel tank and the oxidizer tank. There was a chemical reaction or fire in the oxidizer tank, but the duration was so short that no noticeable damage was found. The French have used a slow-burning solid propellant in the Emeraude vehicle, but the gas was cooled and diluted with water. It is mentioned in Chapter 10.

For pump-fed LPREs the required propellant tank pressures are much lower (5–55 psi) than for a LPRE with a pressurized-gas feed system. Therefore the pressures of the warm gas is also much lower. There are a variety of chemical pressurization systems that have been used with pump-fed engines, and only a few selected types are presented here:

1) Some of the warm gas from the GG (or the PB), created for driving a turbine, is tapped off (usually at the discharge side of the turbine), and this small flow is reduced in pressure by an orifice or a warm-gas pressure regulator and cooled, usually in a heat exchanger, with cold propellant. If the GG gas were fuel rich, then this tapped-off cooled gas has usually been used for the pressurization of the fuel tank. This scheme is used in the Titan LPREs developed by Aerojet and can be seen in Fig. 7.7-19. Here a fuel-rich GG gas is tapped off the turbine discharge manifold, sent through a heat exchanger, where it is cooled by fresh fuel, and the flow of cooled gas is controlled by a sonic venturi. If the GG produces oxidizer-rich gas, then some of that gas can be tapped off, cooled, and reduced in pressure and used to pressurize the oxidizer tank. Some LPREs have separate GGs for pressurizing the two propellant tanks. An example is the RD-253 with two small GGs for tank pressurization, one that gives a fuel-rich gas and one that gives an oxidizer-rich gas.

2) Some of the Soviet flight vehicles and submarine-launched missiles have two LPREs. One is the main engine, propelling the vehicle, and the other is for the four vernier TCs, for attitude control, and they also provide 10 to 20% of the total thrust. The main LPRE has an oxidizer-rich GG (or PB), which also provides the warm gas for pressurizing the oxidizer tank, and the vernier LPRE has a fuel-rich GG, which also has a tap-off for pressurizing the fuel tank. This system is shown in Fig. 8.5-13 and 8.6-10. Each of the two gas flows to the propellant tanks passes through a heat exchanger, where it is cooled and a pressure regulator in order to maintain a predetermined tank pressure.

3) Direct injection of a small amount of a hypergolic propellant, tapped off the pump discharge, into the tank of the other propellant can allow a relatively simple tank pressurization scheme. Several organizations have failed to make it work reliably. However the Soviets have achieved this in the R 36M (SS-18, Mod. 1–3) ICBM in both the first and second stage NTO/UDMH propellant tanks and it was first flown more than 25 years ago.*

4) Monopropellant hydrazine has been decomposed by a catalyst, and the resulting gas has been used to pressurize hydrazine tanks.

5) With cryogenic propellants there is no need to tap gas off from the GG or the PB. It is simpler to just heat a small fraction of the cryogenic propellant (in

---

*O. Bukharin et al, *Russian Strategic Nuclear Forces*, Pavel Podvig, editor, Center for Arms Control, Energy and Environmental Studies at Moscow Institute of Physics and Technology, 2001. English Translation issued by MIT Press Cambridge, MA, and author only learned about this recently, personal communication, J. Morehart, The Aerospace Corp., 2005.

a heat exchanger in the turbine exhaust pipe or in a cooling jacket), evaporate the propellant, and use it to pressurize the tank. This scheme is used with LOX/LH$_2$ engines.

There have been problems with chemical tank pressurization in the past. Chemically generated gas, even if diluted with water, is not really completely neutral, and it reacts to some degree with the liquid fuel or the liquid oxidizer in the respective propellant tanks, or sometimes with both propellants. The warm gas from the GG, augmented by the heat of reaction of the gas with the propellant, causes the top layer of the propellant in the tank to be heated. When this top propellant layer is diluted by dissolving or absorbing some of the gas and is also heated, then the propellant density and viscosity will be changed, and the vapor pressure will increase. So when the propellants are nearly exhausted, this warmed propellant layer will enter into the feed pipes, and the changed propellant properties can cause problems with the injection, changes in the mixture ratio, changes in the thrust level, and possible cavitation in the pump impellers. The Soviets experienced these problems with chemical tank pressurization. They solved the problems by better cooling (lowering the propellant temperature) and reducing the thrust or propellant flow for a short time prior to cutoff, thus reducing the liquid flow velocity and with it reduce likely occurrence of cavitation.

If there are side accelerations during the flight, a vigorous sloshing of the propellants in the tanks can occur. The liquid will splash and suddenly cool the pressurizing warm gas in the tank causing an abrupt change in the tank temperature and the tank pressure. This phenomenon occurred in the development of the Bomarc missile and was remedied by changes in the tank's baffles.

History

The first GGs were developed in the three countries that were the most active in the early years of this LPRE history. The catalytic decomposition of 80% hydrogen peroxide was an early and simple solution for GGs. It was pioneered by the Hellmuth Walter Company (Chapter 9.3) in Germany during the 1930s and has been used thereafter as a GG for the LPREs of German aircraft and for the German V-2 beginning in about 1939. Chapter 9.3 describes several applications of this monopropellant hydrogen-peroxide GG. Figure 4.5-1 shows the GG for the LPRE of the German V-2 missile. The elliptical aluminum tank for the hydrogen peroxide is on top, the GG is at the bottom left, and the cylindrical liquid catalyst tank is next to the GG. The catalyst was an aqueous solution of potassium permanganate, which was also injected into the chamber of the GG. The plumbing and control of this GG are shown in the schematic diagram of the V-2 engine in Fig. 9.2-3. This GG has the unusual feature of operating at two different flow levels as explained in Chapter 9.2. The German rocket people liked the hydrogen-peroxide monopropellant decomposition because they had some prior experience with it in torpedo drives and because it is limited in the maximum possible gas temperature. This means there will

Fig. 4.5-1 This hydrogen-peroxide monopropellant gas generator of the LPRE of the German V-2 missile uses a liquid potassium permanganate solution as a catalyst (around 1937). Copied with permission from the second edition of Ref. 1.

be no burnouts of the GG chamber. Table 4.1-2 gives the maximum gas temperatures for the decomposition gas of hydrogen peroxide of different concentrations. These temperatures are low enough to allow the use of aluminum in the GG and the turbine parts. Most of the monopropellant hydrogen-peroxide GGs and LPREs after about 1948 used 90% hydrogen peroxide. The gas decomposition temperature of 90% will be about 1350°F or 732°C, which has been handled in steel-type materials. The Germans had extensive development experience, mass production experience, and experimental results of different catalysts with 80% peroxide. They used silver screens, liquid potassium or sodium permanganate solutions, or metal impregnated pellets, such as with cobalt, and several different catalyst bed designs in different engines. As described in Chapter 9.3, the Germans have used all three types of catalysis in different LPRE applications.

After World War II this GG concept of using the decomposition of hydrogen peroxide was quickly picked up by the Allies. This monopropellant GG design was used on the U.S. Redstone missile engine, the aircraft rocket engines for the X-1 and X-15 research aircraft (after 1950), and also in early LPREs of the Soviet Union (RD-100, RD-101, RD-103) and the United Kingdom (Britain). The Soviet Union had already independently developed this same monopropellant hydrogen-peroxide GG concept and used it on a few early small experimental aircraft engines before 1945. This was then not known outside of the Soviet Union. Figure 4.5-2 shows the GG for the U.S. Redstone engine using impregnated pellets as the catalyst, which was based in part of the German hydrogen-peroxide experience. Later reliable bipropellant GGs were developed, and the application of a third separate hydrogen-peroxide propellant for the GG in bipropellant rocket engines became less attractive. Using a bipropellant GG with the same propellants as the main engine eliminated the

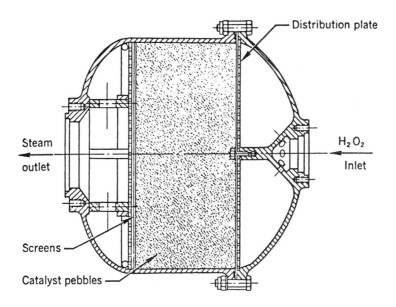

Fig. 4.5-2 Cross section of the hydrogen-peroxide monopropellant gas generator for the LPRE of the U.S. Redstone missile using a solid catalyst impregnated on ceramic pebbles. Courtesy of The Boeing Company, Rocketdyne Propulsion and Power. Copied with permission from third edition of Ref. 1

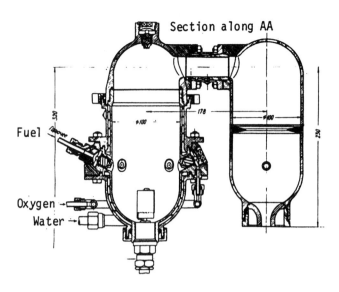

Fig. 4.5-3 One of the first gas generators built and tested by the Soviet Union around 1936 had two chambers. Courtesy Gas Dynamics Laboratory, St. Petersburg, Russia. Adapted from Ref. 3.

hydrogen-peroxide tank and the catalyst system, which simplified the engine system and allowed a simplification of the logistics. Therefore the use of monopropellant hydrogen-peroxide GGs was gradually phased out.

The first *Soviet bipropellant GG development* started in 1933, and testing took place between 1935 and 1936. Their GG-1 is shown in Fig. 4.5-3, and it had two separate chambers. Three propellants were used: nitric acid, gasoline, and water.[3] The water was added as a diluting and cooling agent to reduce the gas temperature. They used six simple spray injector elements around the middle of the chamber; three were for fuel, and three were for nitric acid, and the water was injected at the top. They started with a unit that had uncooled walls, but it often burned out. Therefore the combustion chamber was regeneratively cooled with the water, which was then injected into an uncooled mixing chamber with a central sharp-edged turbulence ring. Because of cold winter days in Russia, they changed to a type of antifreeze (75% water and up to 25% ethyl alcohol) instead of 100% water. This GG produced 40 to 70 liters/s with a gas at pressures of 20 to 25 atm at a feed pressure of about 30 atm and with gas temperatures of 450 to 580°C; it weighed 60 lb. They built and tested a similar two-chamber GG with an L-shaped configuration (the mixing chamber was at right angles to the combustion chamber) with an output of 100 liters/s. There was no information about these early Soviet gas generators being used with a turbopump or an engine. GG development work was apparently stopped in the Soviet Union for several years, as explained in Chapter 8.2. In the 1940s work on GG was resumed, but with other propellants, namely, LOX/ kerosene and extra water; it was used on one version of an aircraft LPRE. On the early GGs the Soviets used some form of liquid cooling with a cooling jacket; even though the final gas temperature of the GG was low enough to allow the use of uncooled chambers, the cooling provided an extra margin for poorly mixed gas (with hot streaks) or for inadvertent operation at a different mixture ratio. Later they managed to eliminate the water and used extra fuel or extra oxidizer for cooling the gas.

The RNII (Russian Science and Research Institute; see Chapter 8.1) also developed and used a monopropellant hydrogen-peroxide gas generator for generating the gas to drive turbopumps.[3] Several Soviet aircraft-type LPREs, developed by the team with Leonid S. Dushkin as their leader, in the 1940s used these hydrogen-peroxide monopropellant GGs with a solid catalyst that was developed in the USSR. This GG and its catalyst were developed independently of the German GG effort.

Figure 4.5-4 shows the bipropellant GG of the large Soviet RD-111 LOX/ kerosene LPRE developed by Glushko's team, which later became NPO Energomash. Its GG injector elements were similar to the injector elements used on the main combustion chamber of the RD-111, which were spray injection elements for the fuel and separate spray injection elements for the oxidizer. It still had regenerative cooling and was developed in 1959 to 1961; it was used on the first-stage engine of the R-9A ballistic missile.[4,*]

The Soviet Union developed its first experimental preburner at RNII and tested it in a ground-test engine with the first Soviet staged combustion cycle. The engine was first operated in 1958. No details are available on this engine or its PB. The S1-5400 LPRE was the first production engine with a staged combustion cycle using a preburner, and it was developed in the 1957–1959 time period.[4,*] This historic engine was used on an upper stage of the Molniya SLV, was developed by the Korolev Design Bureau, and it first flew in 1960; this was the first flight with a PB and with this novel engine cycle. The engine is shown in Fig. 8.11-1, and it is discussed further in Chapter 8.11. The Soviets operated preburners with LPREs featuring a staged combustion engine cycle using LOX/kerosene and NTO/UDMH and all at an oxidizer-rich mixture ratio. With their LOX/LH$_2$ engines, the Soviet preburners ran with a fuel-rich mixture. An estimated 35 different engines with staged combustion cycles and with preburners were developed and tested in the Soviet Union, and about 30 of them were flown, far more than all of the other countries together.

R. H. Goddard developed and operated the first U.S. bipropellant GG in a ground test of a LPRE in 1938. In his initial development (1936, 1937) he used three propellants: LOX, gasoline or kerosene, and water.[5] The water was added to cool the hot reaction gases to the desired turbine inlet temperature (210–500°F). His first U.S. bipropellant GG is shown in Fig. 4.5-5. The cylindrical chamber was approximately 2 in. in diameter and approximately 10.5 in. long and operated at around 150 to 165 psi, which was actually lower than the chamber pressure of its LPRE. Goddard chose an oxidizer-rich propellant

---

*Personal communications, J. Morehart, The Aerospace Corporation, 2000–2004.

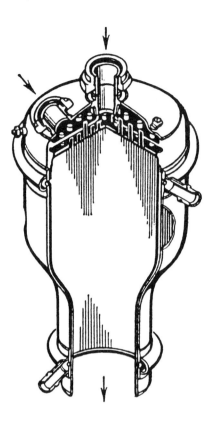

Fig. 4.5-4 This gas generator for the Soviet RD-111 LPRE was developed in the 1950s. It had regenerative cooling with a double-walled chamber and coaxial injection elements in the injector. Courtesy of NPO Energomash; from Technology Special Report, USSR Rocket Engines.

mixture for his bipropellant GG, probably because this helped with his start sequence, and a leak of this gas would not cause a fire with the air. He injected the oxygen at a rate of about 0.5 lbm/s tangentially into the GG combustion chamber and the fuel in an axial direction; the mixture ratio of oxidizer to fuel flows was estimated to be between 100 and 117. The warm gas drove his two turbines in parallel. In one test the resulting gas temperature was 210°F, but in some tests it was lower, and in some tests it was higher. In the three-propellant version of this GG, the water was also injected tangentially at the top of the chamber. In 1938 he undertook a number of engine ground tests and vehicle tie-down (no flight) tests of the new engine with a new GG and two new enlarged TPs. The first U.S. developed GG flew as a part of his propulsion system, in a sounding rocket in 1939 and again in 1941.[5] Goddard designed what he considered to be an advanced, complete engine with three propellants being fed into his GG, and he applied for a patent for this feed system in 1943; in this LPRE the water was used also to cool the thrust chamber. He also

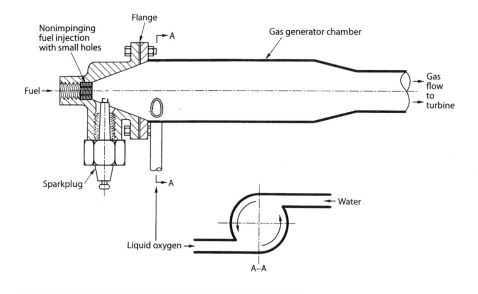

Fig. 4.5-5 Sketch of the first tripropellant gas generator by the American Robert Goddard, developed around 1936–1938 and first flown in 1939. Drawn by author from information in Ref. 5.

applied for a patent for an engine with a GG that only was supplied with fuel and oxidizer, and the TC cooling was done with fuel.[5]

The first large U.S. LPREs with a large bipropellant GG was the G-38 engine (three TCs at 415,000 lbf total vacuum thrust using LOX/kerosene) of the uprated Navaho missile; it was qualified, but it never flew.* It had a relatively large GG chamber volume that was needed for good mixing and uniform gas temperature. Most of the large U.S. LPREs developed since that time have used a bipropellant GG, and they have operated mostly with LOX/kerosene, LOX/LH$_2$, NTO/Aerozine 50, and one engine with LOX/ammonia. Several of the LOX/kerosene U.S. gas generators developed between 1953 and 1965 are shown in Fig. 4.5-6 for the Atlas booster engine, Fig. 4.5-7 for the H-1 booster engine of the Saturn I SLV, and Fig. 4.5-8 for the very large F-1 engine (1.5 million lb trust). These engines are described and shown in Chapter 7.8. The three figures show uncooled chambers, valves closely positioned to the GG injector (some poppets are at the face of the injector), an ignition device, and turbulence baffles for better mixing. One has provisions for mounting a solid-propellant starter cartridge for ignition and initial spin-up of the turbine. The GG is usually bolted to and supported by the turbine inlet of the turbopump assembly. In the F-1 LPRE a tank-head start is used (see Chapter 4.8 on Starting and Ignition), and the GG comes up to full flow relatively slowly over several seconds.*

---

*Personal communications, V. Wheelock, The Boeing Company, Rocketdyne Propulsion and Power, 1999–2004.

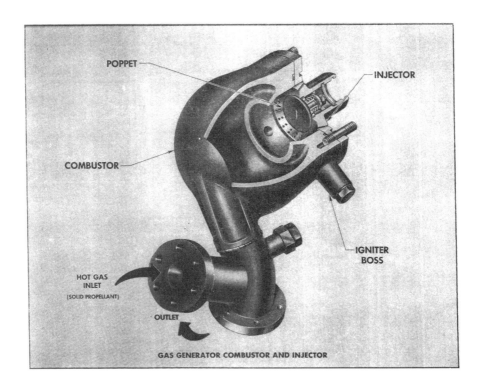

Fig. 4.5-6 Gas generator for the MA-3 booster LPRE for the Atlas flight vehicle. The injector has a single kerosene poppet valve, which injects a conical fuel spray and multiple holes for the injection of oxygen. Courtesy of The Boeing Company, Rocketdyne Propulsion and Power.

The GG for the upper-stage J-2 engine (LOX/LH_2 gas at 1200°F and 700 psi) was restartable and had a large poppet valve at the injector face. This GG is not shown here. The gas from the GG drove two turbines (for a fuel TP and a LOX TP) in series, and this gas was then aspirated into the diverging section of the TC's nozzle and provided some dump cooling of the nozzle-exit segment.*

The first U.S. preburner design was for the SSME, and two fuel-rich LOX/LH_2 PBs were installed right next to the inlet side of the two turbines in each engine. Figure 7.8-19 shows the installation of these two preburners right on the respective turbopumps. Figure 4.5-9 shows the larger of the two PBs, and it supplies power to the high-pressure LH_2 fuel turbopump.* It has coaxial spray injection elements and baffles, which are similar in concept to those in the SSME TC. A centrally located small internal chamber with a small flow of propellants and a spark plug serves to provide the igniter flame, which ignites

---

*Personal communications, V. Wheelock and others, The Boeing Company, Rocketdyne Propulsion and Power, 2004.

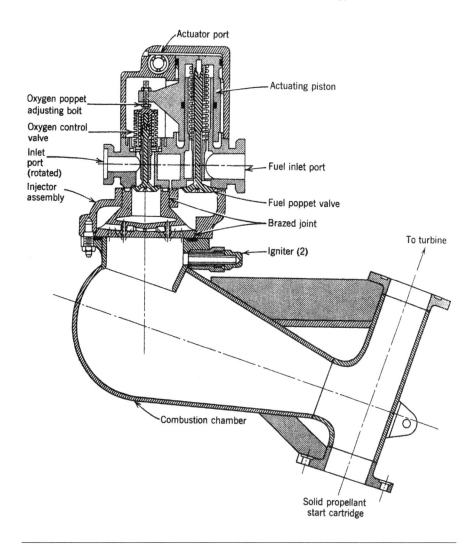

Fig. 4.5-7 Cross section of the uncooled bipropellant gas generator with its dual valve for the H-1 engine used on the Saturn I and IB SLVs. Courtesy of The Boeing Company, Rocketdyne Propulsion and Power: Copied with permission from fourth edition of Ref. 1.

the main flow of PB propellants. The two PBs can be identified in the SSME power head shown in Fig. 7.8-19.

Figure 4.5-10 shows the preburner for the Soviet RD-0120 LOX/LH$_2$ LPRE developed by the Chemical Automatics Design Bureau (CADB) in Voronesh, Russia. It used a multitude of coaxial spray elements in its injector, but only two are drawn in this figure. It uses a perforated curved sheet to equalize the flow, a fuel-cooled thin sleeve in the chamber, and an igniter, which is centrally

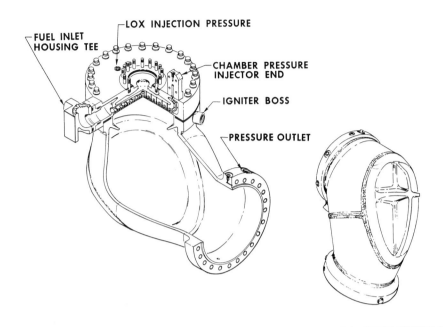

Fig. 4.5-8 Gas generator for the F-1 LPRE has a cruciform mixing baffle in the chamber. Courtesy of The Boeing Company, Rocketdyne Propulsion and Power.

located in the injector. The cooling fuel is dumped into the chamber. The large inlet on the right is for fuel, and the LOX inlet is on top.

Gas generators and preburners were also developed or operated by other nations than the three that were discusssed. The basic design concepts of their GGs or PBs were the same, but some of the details were different. The applications of their LPREs were for ballistic missiles or space launch vehicles, and the timing was not the same in each country. This includes Japan, China, France, India, Britain, North Korea, Iraq, and Syria. The work on large engines and thus on GGs and PBs occurred much later in these nations than in the three original countries, and this timing can be determined from the engine development dates in the chapters of each country.

Technology Issues

These technical issues required an effort during the development of a number of these GGs and PBs. For example the temperature distribution within the cross section of the GG or PB flow has to be even in temperature, and there should be no streaks of high temperature at any of the operating and throttling conditions of the engine. This means an essentially identical mixture ratio reaching every one of the turbine rotor and stator blades. This is particularly important in GGs or PBs, which are mounted direcly next to the turbine inlet, where there is not enough intermediate volume to achieve good

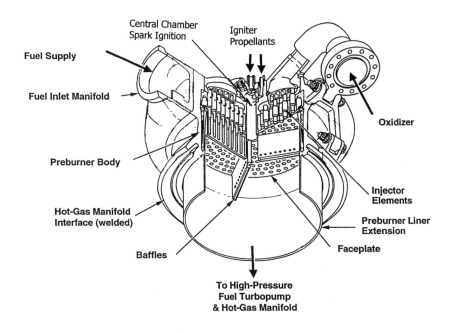

Fig. 4.5-9 This preburner, which is one of two for the SSME, has coaxial injection elements and provides hydrogen-rich gas for the high-pressure fuel turbopump. Courtesy and picture from The Boeing Company, Rocketdyne Propulsion and Power.

turbulent gas mixing. In several GGs it was difficult to achieve this uniform gas temperature.

The mixture ratio of a GG or a PB is either fuel rich or oxidizer rich, so that the gas temperature will be low enough to allow the use of uncooled metal alloy GG chambers and turbine assemblies. The GG designers in the early decades understood that a fuel-rich gas mixture can have major changes in gas temperature (and thus in the power delivered) with a relatively small change in the flow of the oxidizer. The reverse applies to an oxidizer-rich GG. For example a small increase (say 3%) in the flow of the propellant, which has the relatively very small flow, can result in a significant increase of the gas temperature (perhaps a change of more than 130°F). Therefore the GG's operating outlet gas temperature has to have a reasonable margin over the potential failure temperature of the GG wall material, and the flow of propellants to the GG has to be accurately controlled or calibrated, so that temperature excursions are limited and that start transients avoid those regions of mixture ratio that gave gas temperatures, higher than the allowable wall temperature.

Although early GGs used regenerative cooling of the GG combustion chamber, later designs used uncooled hardware. When it was not possible to remedy burnouts in uncooled chambers during off-design conditions, some form

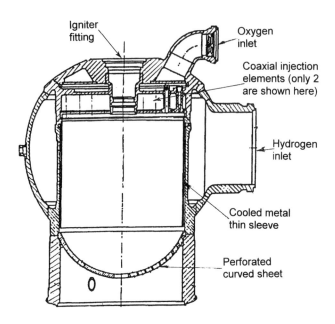

Fig. 4.5-10 Preburner for the Soviet RD-0120 has a cooled cylindrical thin sleeve. The perforated curved sheet improves the mixing and helps to get a more even temperature distribution of the gases. Courtesy of CADB; from Technology Special Report.

of cooling (film cooling, partial cooling jacket, internal refractory metal shield, thermal insulation) was again designed into recent GGs and PBs.

References

[1]Sutton, G. P., and Biblarz, O., *Rocket Propulsion Elements*, 7th ed., Wiley, Hoboken, NJ, pp. 193, 383, 384.

[2]Huzel, D. K., and Huang, D. H., *Modern Engineering for Design of Liquid-Propellant Rocket Engines*, Progress in Astronautics and Aeronautics Series, Vol. 147, AIAA, Washington, DC 1992, Secs. 4.6 and 5.5.

[3]Moshkin, Ye. K., *Development in Russian Rocket Technology* Machinostroyeniye, Moscow, 1973 (in Russian) also NASA TT F-15, 408, March 1974 (in English).

[4]Glushko, V. P., *Rocket Engines of the Gas Dynamics Laboratory-Experimental Design Bureau*, Novosti Press, Moscow, 1975 (in Russian) also NASA TT F 16847, Feb. 1976 (in English).

[5]Goddard, E. C., and Pendray, G. E., *The Papers of Robert H. Goddard*, Vol. 1, 2, and 3, McGraw–Hill, New York, 1970.

4.6 Small Thrusters for Attitude Control and Trajectory Corrections

Introduction and Applications

Work on small thrusters and their reaction control systems did not really start until the late 1950s in the United States. At that time it was understood how the flight paths of launch vehicles and satellites had to be controlled and how the angular orientation of the flying vehicles also had to be controlled. Rocket engines with multiple small thrusters (typically at thrust levels between 0.2 and 1000 lbf thrust each) appeared to be a suitable means for this purpose. However, the requirements of these small thrusters came as the result of earnest studies (of various space missions) performed by experts in several countries between 1940 and 1950. There were several options for controlling the flight path and the vehicle attitude, as discussed in Chapter 4.9. One of the options was to use multiple small TCs or thrusters to impart torques for vehicle rotation (in pitch, yaw, and roll) with one or more pairs of thrusters and to impart small translatory velocity changes in more than one direction to the vehicle. This then leads to engine systems with multiple small thrusters. Depending on the mission and the degree of redundancy, such an engine system had between two and 24 small thrusters. Subsequent studies showed that a pulsing operating mode of these trusters with short burn times and many restarts would be effective. Furthermore engines systems with small thrusters allowed complete rotational and minor translational maneuvers, while the main LPRE was not operating.

Although large thrust chambers are used basically for accelerating a flight vehicle, the small thrusters provide for the important tasks of flight-path corrections, minor orbit changes, maneuvers, and rotation of the vehicles flying in space. Multiple small thrusters are used in satellites, stages of launch vehicles, or space stations, and their applictions have been described in Chapter 2.3, in the Subsection Reaction Control and Minor Maneuvers. Typical applications are listed in the Subsection Spacecraft or Satellite Propulsion for Major Space Maneuvers in Chapter 2.3. Some of the contents of this chapter have been summarized in an abbreviated version in a technical paper while this manuscript was in its early preparation.[1] Small thrusters have been used for the following tasks: 1) attitude control for satellites, launch-vehicle stages, space stations, missiles (pitch, yaw, and roll), roll control for a LPRE with a single gimbal-mounted large main TC, Pointing/orienting antennas, solar cells, telescopes, mirrors, windows, and alignment for reentry; 2) flight-path or orbit changes or corrections, including station keeping; orbit insertion and/or deorbit maneuver; changing orbits; and divert maneuver of terminal interceptor stages; 3) correct the misalignment of principal (larger) thrust chamber(s); 4) warhead velocity changes (postboost control propulsion systems) for targeting; 5) propulsion for small flight vehicles, such as target drones or small missiles; 6) settling of liquid propellants in tanks, prior to gravity-free start in space of main engine; 7) flywheel desaturation; explained subsequently, and 8) space rendezvous maneuvers. This listing is not in any particular order and is not complete. The major applications of small thrusters have been for attitude control and orbit maneuvers.

In some spaceflight missions the small thrusters are used to *correct for thrust misalignment of a main engine* by counteracting misalignment torques in pitch or yaw. The minimum thrust level of the small thrusters depends on knowing the vehicle's instantaneous mass, the direction and magnitude of the thrust of the main TC, and the location of the center of the vehicle mass. The achievable accuracy, with which thrust vector can be aligned with the center of mass of the vehicle, has improved in the last 60 years, and therefore the amount of energy needed (thrust size) for a correction of the thrust misalignment has diminished, primarily because of better tooling, measuring, or inspection techniques. The missdistance between the thrust vector and the center of gravity has diminished by an order of magnitude since the first launches of vehicles propelled by LPREs.

One of the early applications of small thrusters for attitude control in unmanned spacecraft was in the *unwinding or desaturation of flywheels*.[1,*,†] These reaction wheels or flywheels act as a momentum storage device, and they allow highly accurate angular positioning of the spacecraft to less than 0.01 deg and low vehicle angular turning rates of less than 10^{-5} deg/s with relatively little expenditure of energy. This is more precise than the angular position control of a pair of small liquid propellant thrusters operating for a very short pulse. The vehicle is rotated (change in angular momentum) by the gyroscopic effect of accelerating (or decelerating) the spinning flywheel by an electric motor. A good percentage of all spacecraft were equipped with flywheels for very precise attitude control. A small increase in the rotating speed of a fly wheel will cause a small torque on the vehicle. Repeated torques can be applied by increasing this speed in steps until it reaches a maximum speed imposed by the maximum allowable stress in the wheel material. The wheel then becomes saturated, and no further speed increases can then be safely obtained. The "desaturation" of the flywheel is a process of slowing down the wheel, which by itself would impose a torque on the vehicle while at the same time operating a pair of small thrusters, which apply a countertorque in the opposite direction. If done properly, this prevents a net change in the vehicle's angular position. This scheme of loading or unloading flywheels with a pair of small thrusters was used beginning in about 1958, and it has been used frequently in satellites.

For some space flight missions with a long life and many maneuvers, the small TCs will need to be fired or "pulsed" many times. Typically thousands or ten thousands of pulses had to be made, and in some qualification tests a small LPRE might be required to demonstrate a life of between 100,000 and 500,000 pulses without degradation. Large numbers of pulses of small thrusters were first demonstrated in the 1960s. A rugged valve design is critical in surviving so many restarts without servicing or adjustment.

---

*Personal communications, O. M. Morgan and R. C. Stechman, Aerojet, Redmond Center, 2002, 2003.
†Personal communications, R. Sackheim, 1993–1995, and G. Dressler, TRW, 2001–2003.

The same set of small thrusters can be used for more than one of these tasks (listed in the opening Section) with a particular vehicle or spacecraft. If accelerations can be small, an electric propulsion system (with very low thrust) can today perform several of these tasks in some missions, but with higher specific impulse and with a supporting, often heavy electrical power supply. Several of the newer vehicle flight control systems have recently been designed to use both an electric and a chemical liquid propellant propulsion system, because some of the tasks cannot be performed by electrical propulsion systems alone.*

Requirements

All LPREs are developed because of a requirement for a specific propulsion system that was issued by a vehicle contractor and/or the government.[2] In addition to the usual LPRE requirements (such as the nominal, maximum, and minimum values of the thrust levels, specific impulse, propellant properties, reliability, mixture ratio, total firing duration, cost, or schedule), the small thrusters had additional requirements, that were gradually instituted over the period between 1960 and 1980. This includes duty cycle (cumulative firing time as a percent of total elapsed time), the number of thrusters, their location and orientation within the vehicle (firing direction), pulse durations (shortest, longest, and cumulative), cumulative number of pulses or restarts, allowable degradation of performance with pulse width or with use, ability to withstand vibrations and noise from the main engines, permissible thermal radiation to adjacent components, and minimum lifetime in space. These requirements were more flexible and not as numerous in the early decades of small thrusters (1960–1980), but were gradually refined as operating experience was gained in different applications. For example, in the 1950s it was learned that the average specific impulse decreased sharply when pulse width or firing durations became very small.

A comparison of different vehicle flight control systems in the 1960s indicated that a pulsing operation of the small thrusters was compatible with digital controllers and offered a better scheme than throttling of small engines. Typically a single pulse of thrust had a minimum firing duration of perhaps 0.30 s in the 1960s and only 0.015 s beginning in about 1985. The pause between pulses went from perhaps 0.15 s to about 0.01 s for the smaller sizes of thrusters. The time needed for opening and closing a small propellant valve was also reduced over the last 40 years by perhaps a factor of 10 with some types of small propellant valves. These improvements were made by better designs of injectors and valves. The vehicles command and control system would then select the number of pulses needed for specific thrusters in order to accomplish a desired maneuver or rotation.

---

*Personal communications, O. M. Morgan and R. C. Stechman, Aerojet, Redmond Center, 2002, 2003.

The thrust magnitude has always been specified, and it has varied widely, depending on the application. For small thrusters the low limit with liquid propellant has been about 0.1 lbf or 0.5 N (and as low as perhaps 0.001 lbf for gaseous propellants), and the upper limit is arbitrarily set at around 1000 lbf or 4500 N in a vacuum. The selected thrust values depend on the mission, the mass and moments of inertia of the flight vehicles, the moment arm between the thrust direction and the center of mass of the vehicle, and the angular rate or time period in which an angular movement has to be achieved. The thrusts with small satellites were typically between 0.1 and 50 lbf (0.5–22 N), and they were much bigger for large vehicles. For example the largest thrust of the multiple thrusters of the heavy Space Shuttle Orbiter is nominally 870 lbf. Some of the vernier thrusters in large Soviet vehicles were larger (1500–5000 lbf or 7200–22,500 N), and one was even 8500 lbf. These larger Soviet TCs were operated continuously and were not used for pulsing, rendezvous maneuvers, or stationkeeping. However the Soviet engineers used these thrusters for augmenting the thrust of the main LPRE of the vehicle and for attitude control during the main engine's operation. The did not use these larger TCs for rendezvous maneuvers, orbital corrections, or stationkeeping.

The number of small thrusters in a particular system is not mysterious or arbitrary. It depends on the flight mission and the kind of maneuvers that have to be performed.[2] To apply a pure torque to any flying vehicle in pitch, yaw, and roll (in each in two different rotational directions), it requires a minimum of 12 thrusters, namely, two pairs of two thrusters for rotation around each one of the three axes. This is shown in Fig. 4.6-1. It requires a minimum of six thrusters for pure translation maneuvers up and down any of the three axes, and their thrust vectors would have to go through the center of mass of the vehicle, which will often shift its position as propellant is consumed. With the arrangement of four thruster modules, each with four thrusters, as seen in the figure, it requires two selected thrusters to operate simultaneously to achieve translation maneuvers. So it takes 16 thrusters altogether for complete translation and rotation maneuvers without moving/rotating the vehicle. By operating thrusters pairs x and x simultaneously, it is possible to achieve a rotation maneuver. By firing x and the opposite x' together, a translation maneuver will be achieved. It uses some of the same thrusters, but in a different combination. It is possible to reduce the number of thrusters below 16 by judicious placement of the trusters on the vehicle, but it then will no longer be possible to achieve pure torques about three axes, but only torques that are combined with some translation forces. The number of thrusters for ACSs can be higher than the 16 shown in Fig. 4.6-1, if redundant thrusters are provided. These extra spares are to be used in case one of the active thrusters malfunctions. It requires only four small fixed thrusters (or two small hinged thrusters) to provide roll control in a vehicle with a single gimbaled main TC, which provides the pitch and yaw control. Another example is to use the attitude thrusters to turn the vehicle, so that the thrust vector of the main propulsion system is inclined to the flight path; this is how a turn in spaceflight direction can be obtained.

Technology and Hardware 151

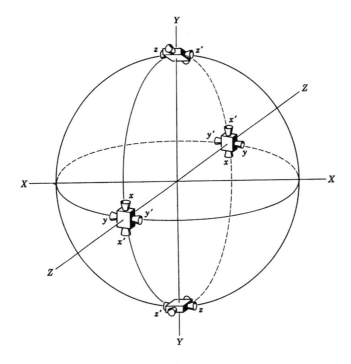

Fig. 4.6-1 Diagram showing placement of 16 thrusters. Copied with permission from Fig. 4-13 of Ref. 2.

Feed Systems

Feed systems were categorized in Fig. 4.2-1. The same feed system supplies anywhere from two to 24 small thrusters. The cold inert gas pressurization feed system has been used almost exclusively for expelling the propellants from their flight tanks and feeding them to the appropriate small thruster(s).[2] The pressurized feed system is the oldest one and was first used by Goddard in 1923, as described in Chapter 5.2. This type of feed system has been used for eight decades and is still used often today, particularly in satellites. Both a feed system with constant chamber pressure (using a gas pressure regulator) and a blowdown system (without regulator, without high-pressure gas storage tank) with constantly decreasing thrust have been used. These two feed systems were discussed in Chapter 4.2. Both monopropellant and bipropellant systems have used cold-gas pressurization. Figure 4.2-2 is representative of a bipropellant system with multiple thrusters and a pressurized-gas feed system with a regulator. It shows examples of provisions for filling the propellant and gas tanks, venting the feed lines, draining propellants, and using calibration orifices.[2]

There were some exceptions to the pressurized feed systems. The Soviets have used a pump feed system typically for a set of four hinged vernier TCs arranged around a single larger main TC. Most of these vernier TC are quite

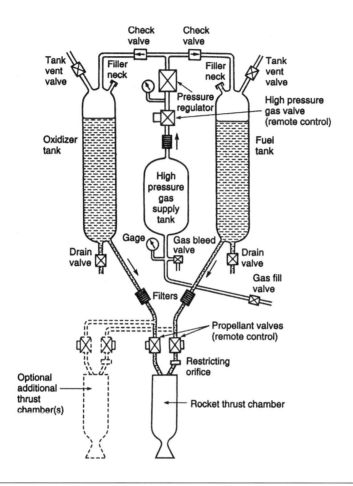

Fig. 4.6-2 Typical pressurized-cold-gas feed system with a pressure regulator designed for reuse. Adapted from Fig. 1-3 of Ref. 2.

large in thrust, but a few of the vernier thrusters were small enough to be considered as small thrusters. The Soviet vernier thrust chambers and their turbopumps had been mentioned before and are discussed in Chapters 8.5 and 8.6.

When the small thrusters operate continuously, while the main high thrust LPRE is running, then it is possible to feed the small thrusters by tapping off some of the propellants supplied by the turbopumps of the main engine. There is then no need for a separate pressurized-gas feed system for the small thrusters. An example is an early version of the gimbaled Thor LPRE, which had two hinged auxiliary small thrusters at 1000 lbf each for roll control only. All three TCs ran with LOX/kerosene, and they ran continuously for the same duration without restart. The firing of the two small hinged roll control thrusters and the larger Thor engine is shown in Fig. 4.6-3. The two small thrust chambers had a dual-wall construction, were regeneratively cooled, and were supplied from the same turbopump, which also provided the propellants to the main engine. One is shown in Fig. 4.6-4. The same small thrusters were also used for roll control with an early version of the Atlas sustainer engine. The design analysis of these small thrusters was done by your author in the late 1950s.

For those LPRE applications, that required engine start and/or restart in gravity-free space, it has been necessary to provide some form of positive expulsion to ensure a good propellant supply and prevent pressurizing gas to get into the TCs.[2] Many use flexible diaphragms, flexible bladders, pistons, or bellows to separate the gas from the propellant in the propellant tanks. If these tanks are properly loaded with propellant and vented, then a bubble-free

Fig. 4.6-3 The MB-3 LPRE of the Thor Missile is being test fired together with two small hinge-mounted roll control thrusters. Courtesy of The Boeing Company, Rocketdyne Propulsion and Power.

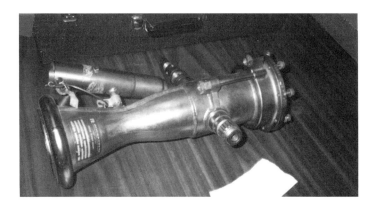

Fig. 4.6-4 This roll control thruster of 1000 lbf thrust, seen firing in Fig. 4.6-3, had regenerative cooling, two hollow hinge bearings, and a cartridge for the hypergolic start liquid fuel. Courtesy of the U.S. Air Force; and The Boeing Company Rocketdyne Propulsion and power; author's photo.

supply to the thrusters will ensure good starts and restarts at altitude as well as at sea level. Figure 4.2-3 shows different types of positive expulsion devices, and Table 6.5 in Ref. 2 gives a comparison between six different types of positive expulsion devices.* All have been used in a flying restartable LPRE with small thrusters. Most of these expulsion devices were known in the 1960s, but they were refined and further improved in the 1970s and later.

The early LPREs with small thruster did not have adequate filters, and there were failures as a result of valves not closing, when solid particles were on the valve seat, or as a result of metal shavings, debris, or sand plugging up small injection holes. Good filters and multiple filters are now included regularly with every feed system, particularly with LPREs that have small thrusters with small injection holes or with turbopumps.

Injectors

The injector designs used in small bipropellant thrusters are similar to those used in large thrust chambers as described in Chapter 4.3. The earliest injection scheme of Goddard, who did his original experimental work with small thrusters, was the shower head or nonimpinging type. Although it was stable, it gave poor performance and is no longer used today. The most common injection scheme in the United States has been some of the impinging stream

---

*Personal communications, M. Marvin and F. Boorady, regarding, ARC expulsion products and history, ARC, Niagara Falls, NY, 2001–2004.

injection elements such as doublets or triplets shown in Fig. 4.3-13; they often had extra film-cooling holes on the periphery of the injector.*,†,‡ For very small thrusters of 5 lbf thrust or less, it is typically only a single doublet or triplet injection element.

In the Soviet Union the most common injection element are spray-type injection elements, such as single-propellant sprays with fuel and with oxidizer sprays arranged in an orderly pattern. Later the Soviets used coaxial double-tube swirling bipropellant injection spray elements in many of their small thrusters, because of good performance. Figure 4.3-14 showed several types of them. Figure 4.6-5 shows such a bipropellant single central injection element designed at Nauchno-Proizvidstvennoye Obedinenie Mashinostroeniya (NII Mash), a Soviet organization that has specialized in small trusters.§ This company's stable of small thrusters is described in detail in Chapter 8.7. In this figure the prerotation of the propellant flows is achieved by spiral vane inserts. A conical UDMH fuel sheet emerges from the lip of the central hole. The conical sheet of the NTO oxidizer starts at the lip of the outer cylinder, and this top cone impinges with the fuel cone. Both of these thin conical sheets are atomized into small droplets. Many of the oxidizer droplets reach the outer wall, where they provide film cooling. This oxidizer film cooling is different from the U.S. practice, where some of the fuel is used for film cooling. In addition, this coaxial injector has a series of axially directed oxidizer injection holes, and the propellant jets from these holes break up the conical sprays or the conical flow of burning atomized propellant droplets. This allows the hot gas and vapors to flow between zones I and II in Fig. 4.6-5 and equalizes the pressure between these two zones. According to the chief designer of NII Mash, this feature ensures stable combustion operation.§

A pintle injector can be described as a spray-type injector with a central pin or pintle as shown in Fig. 4.3-14. A number of them have been used in small thrusters, and some are described in Chapter 7.9. A simple pintle provides a simple spray of one propellant without prerotation. A double valve can be built into the pintle injection element, which can achieve face shutoff.¶ This coaxial bipropellant version of the pintle provides intersecting sprays of both the fuel and the oxidizer.

---

*Personal communications, R. Carl Stechman, chief engineer, regarding injectors Aerojet, Redmond Center, WA, 2002.
†Personal communications, M. Marvin and F. Boorady, regarding ARC products and history, ARC, Niagara Falls, NY, 2001–2004.
‡Personal communications, Gordon Dressler, Northrop Grumman, 2001–2003.
§Personal communications, Y. G. Larin, chief designer, NII Mash (R & D Institute of Mechanical Engineering), Nizhnaya Salda, Russia, 2002.
¶Personal communications, Gordon Dressler, chief engineer, Propulsion Products Center, Northrop Grumman Corp., Redondo Beach, CA, 2001–2003.

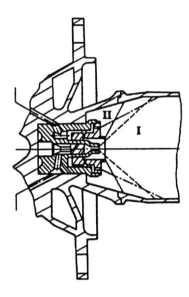

Fig. 4.6-5 Cross section of injector developed in the Soviet Union for a small thruster. It uses a single central coaxial injector element and is film cooled by the oxidizer. Courtesy of NII Mash, Nizhnaya Salda, Russia (copied from data provided through personal communications, Y.G. Larin, Chief designer, NII Mash, Nizhnaya Salda, Russia, 2002).

Reliability

Adequate reliability means that the small thrusters and their feed systems will operate flawlessly every time they are needed, without degradation in performance, under all likely operating conditions, and without any malfunction. Good reliability can be achieved, if the design has been well conceived, if extensive tests are used to prove this reliability, and if properly conceived and controlled manufacturing operations, inspections, and installations are performed. To achieve this high reliability, the qualification tests were made more difficult and more extensive.

Initially in the 1930s the early investigators did not do a thorough job of fabrication, had little documentation, and performed very few tests prior to flight, and the reliability of their LPREs and their thrusters was very poor. Only one out of two or three flight attempts had satisfactory propulsion performance. Slowly, decade by decade, the fabrication became more systematic, and more engine tests were added, including endurance tests, overstress tests, life tests, or operation at off-design conditions in mixture ratio and chamber pressure. It is not just the reliability of the thruster, but also the valves, instruments, or feed system components, that need to be proven. Demonstrations of operating thrusters for 100,000 restarts without malfunction and later for 500,000 restarts were impressive. Some valves have been cycled for more than a million open and close operations in a vacuum test facility. As a result, of all of these tests and many design improvements the reliability has improved remarkably and is now very close to 100%. Rigorous inspections and at least one hot-fire proof test are now done on each production thruster prior to delivery by most of the manufacturers of small thrusters. When there is a piloted flight, then the reliability requirements and demonstrations usually become more stringent. Although these tests help to gain a high confidence in the engine, they do not guarantee a 100% reliable flight operation. In contrast, a solid-propellant rocket motor is more difficult to inspect and cannot be hot-fire tested prior to use.

When reliability was believed to be not quite good enough, redundancies were designed into the engine. In the 1940s to 1970 it was quite common to fully duplicate a particular rocket engine with an identical spare rocket engine, particularly with manned vehicles. This happened with the attitude control system for the Mercury reentry capsule, which is shown later in this chapter and for the Apollo command module; both had a complete spare engine with multiple small thrusters, a separate feed system, and separate propellant tanks. This full engine redundancy was also used in early Soviet manned spacecraft and in the development of an unmanned lunar lander, which did not fly. Sometimes it is prudent to not duplicate a complete engine, but only parts of an engine. For example the Space Shuttle Orbiter attitude control system had spare thrust chambers with their own propellant valves, and these could be put into operation, if the primary thruster failed to perform properly. It still is a routine practice to put two check valves in series; if one leaked, the other would hopefully prevent backflow.

Some rocket engineers believe that the duplication of LPREs or of key engine components can actually decrease the overall reliability. This is because additional features and components have to be introduced. This includes sensors, which indicate that the engine has malfunctioned (decreasing thrust, burnout, failure of valve to open fully, propellant leak, etc.), and extra controls to turn off malfunctioning components and switch to the backup components. A failure of these extra features can be just as serious as the original malfunction. There have been some designers who believe the best approach is to make the LPRE as reliable as possible without redundancies.

Designs of Small Thrusters

Because pressurized feed systems have been well understood for decades, the principal differences in these LPREs were primarily in the type of propellant (e.g., storable monopropellant vs several storable bipropellants) and in the design and materials of the small thrust chambers or thrusters. Several of these will now be discussed briefly. In the 1940s and 1950s it was learned that for small thrust sizes (approximately 500 lbf or less thrust) regenerative cooling becomes marginal and needs to be supplemented (by film cooling) or replaced by another cooling method as described next. The reasons for this limitation of regenerative cooling were explained in Chapter 4.3

Historically some of the first ACSs used *cold gas* (such as compressed air or nitrogen) stored at high pressure and exhausted through simple valves and nozzles.[3,*] Even though cold gas is not a liquid, it has been customarily included in the discussion of liquid propellants. The gaseous propellant systems are simple, reliable, low in cost, and do not involve any high combustion temperatures or cold cryogenic temperature. In the United States and the Soviet Union cold-gas systems were used starting in the 1950s and continued sporadically in the United States until about 1980. They have low performance and sometimes were convenient to use when there was not enough time or resources to go to a better LPRE. The specific impulse for nitrogen or air is around 65 to 75 s, depending on the specific system parameters. Other gases, such as argon, ammonia, or freon, have been investigated for some systems, but their performance is not much better than air or nitrogen. Cold hydrogen gas has the highest performance (I_s = 280 s) of any cold-gas propellant, but it was never used because the hydrogen high-pressure storage tanks were much larger in volume and much too heavy. Occasionally helium (I_s = 180 s) has been used. Although its tanks are also relatively large and heavy, it is inert with all propellants and not explosive. With fast valves and cold gas, it has been possible to obtain a very small pulse width of perhaps 0.01 s. Cold gas was used with early simple satellites (see Chapter 7.9) and also for roll control of the last stage of the initial version of the air-launched orbiting Pegasus space launch vehicle. Cold-gas jets for ACSs have been abandoned for production vehicles because of low performance and heavy high-pressure gas tanks. They were replaced with propellants giving a higher performance. Occasionally a cold-gas system is still used because it is simple, reliable, and low in cost. Heated gas has a better performance than cold gas. Examples are electrically heated stored gas (see Chapter 7.9) or decomposed hydrazine monopropellant, which is stored after it has been gasified. Fakel (Chapter 8.12) has a heated system for very small pulses.

Monopropellant hydrogen peroxide thrusters were the logical improvement over cold-gas thrusters.[†,4-6] They had a rich technical heritage in several German World War II hydrogen-peroxide developments (1930s to 1945), particularly those of Hellmuth Walter (see Chapter 9.3). Historically they were the first ACS using liquid propellant, and they were used frequently in the United States beginning about 1948 and continuing intermittently until the early 1970s. While the Germans used 80% hydrogen peroxide, the United States, Britain and others usually used 90%. Thrusters ranging in thrust between 0.2

---

*Personal communications, Gordon Dressler, chief engineer, Propulsion Products Center, Northrop Grumman Corp., Redondo Beach, CA, 2000–2004.
†Personal communications, M. Marvin and F. Boorady, regarding ARC products and history, ARC, Liquid Operation, Niagara Falls, NY, 2001–2004.

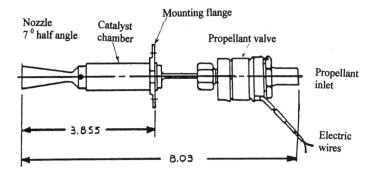

Fig. 4.6-6 Small monopropellant hydrogen-peroxide thruster flown in the mid-1960s. Developed and built by Walter Kidde and Company in the 1950s. Copied from Kidde catalog, circa 1960.

and 500 lbf were developed, but the most popular sizes were 1, 5, and 20 lbf. Figure 4.6-6 shows a typical hydrogen peroxide monopropellant thruster. Its nominal thrust was at 2.0 lbf, but could be operated between 1.1 and 2.5 lbf. The nominal steady-state specific impulse was 157 s at 135-psia chamber pressure. Its LPRE had multiple thrusters and a blowdown-type high-pressure nitrogen gas feed system. It has flown in the Early Bird and Syncom satellites. Thousands of thrusters using hydrogen-peroxide as a monopropellant were delivered in the United States alone between 1948 and 1972. Specific impulses were typically between 145 and 158 s, depending on the specific engine parameters. The key suppliers are listed in Table 6-2. All of the developers of hydrogen-peroxide thrusters had to solve some problems with the silver catalyst bed. This storable monopropellant is decomposed by the catalyst into steam and hot gaseous oxygen using silver screens or a pebble bed impregnated with a metal catalyst. The gas temperature is low enough (listed in Table 4.1-2) to use uncooled steel construction, and the system was relatively very simple.

The very first U.S. manned spaceflight of the historic Project Mercury on 5 May, 1961 had eight 24-lbf thrusters for pitch and yaw maneuvers, four 1-lb thrusters for fine control, and four 1- to 6-lbf thrusters for roll control, all using 90% hydrogen peroxide. Figure 7.11-2 shows a diagram of this engine system. This LPRE was built and qualified in the late 1950s by Bell Aircraft, a predecessor of ARC. Several redundancies can be noted. There were two separate feed systems with their own sausage-shaped propellant tanks, spherical high-pressure gas tanks, and valves. One system was automatically controlled, and the other could be operated manually as a backup. The pitch and yaw thrusters were also redundant. Only the four roll control thrusters and the four small thrusters for fine control of pitch and yaw were common to both feed systems. The X-15 research aircraft had other hydrogen-peroxide thrusters for an ACS, and it is part of Fig. 2-6. In the Soviet Union the first manned flight module with Yuri Gargarin in 1957 also had a monopropellant hydrogen-peroxide ACS. The Soviet Union actually worked with this monopropellant in gas-generator applications before they learned about the German experience during the world war. Because of the low performance and difficulties with slow self-decomposition during long-term storage, engines with this hydrogen-peroxide monopropellant were gradually replaced beginning in about 1960 by LPREs using hydrazine monopropellant and then by bipropellants. They are discussed next.

In *monopropellant hydrazine thrusters* the resulting gas temperature varies from about 1200 to 2500°F, depending on the amount of ammonia that is decomposed in a particular catalyst bed.[2] The hydrazine is catalytically decomposed into hydrogen, nitrogen, and ammonia. These thrusters were pioneered in the United States by TRW starting in about 1960 (today part of Northrop–Grumman, See Chapter 7.9), by the Rocket Research Corporation beginning development about 1963 (today part of Aerojet, see Chapter 7.7), and by Hamilton-Standard Division of United Technologies starting in about

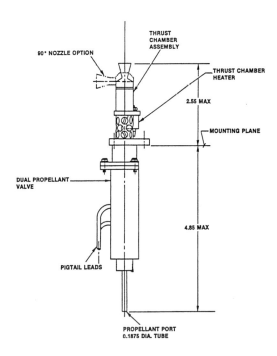

Fig. 4.6-7 Sketch of a hydrazine monopropellant thruster, which first flew in satellites around 1970. It had 0.5 lb (0.11 N) thrust, weighed 0.41 lbm, and minimum pulse duration was 0.015 s. Courtesy Hamilton Standard, United Technologies Corp.

1964 (today this product line is part of Atlantic Research Corporation, see Chapter 7.11) [7-12,*,†,‡] Today there are about 10 active organizations in the world who can supply qualified hydrazine monopropellant thrusters; they and others are listed in Table 6-2. There were other companies in this business, but they have dropped out. A variety of qualified thrusters in sizes between 0.1 and 200 lbf thrust, with different nozzle area ratios and chamber pressures, have been developed by these organizations, and standard models are readily available.[7-12] One unit had 600 lbf thrust, and others used different materials. The first U.S. flight with hydrazine monopropellant in 1960 was performed without a catalyst, which was then not ready, by using slug starts with NTO as a start propellant.[†] This has already been discussed in Chapter 4.1. An example is a 1970 thruster shown in Fig. 4.6-7. Today the hydrazine monopropellant thrusters are still popular, particularly with spacecraft that requires a simple system (simpler than the higher performing bipropellants), a clean exhaust gas, and demonstrated reliability.

Each small thruster contains its own catalyst bed to decompose the hydrazine and includes closely coupled propellant valves for fast start and stop. A thruster is seen in Fig. 4.6-8. Monopropellant hydrazine gives a good performance (I_s of 225 to 245 s vac is typical), allows a simpler system than a bipropellant, and often has been preferred for applications where the total impulse is modest. One of its disadvantages is its high freezing point (about 1°C or 34°F). Projects to find chemical additives that would lower the freezing point were not successful. The solution has been to apply electrical heaters to all parts of the LPRE including the tanks, valves, and the catalyst bed. A good pellet-type catalyst, known as Shell 405, was developed between 1957 and 1960 by the Cal Tech's Jet Propulsion Laboratory (in Pasadena, California) together with Shell Chemical Company. Shell then built a plant to produce it. When business dropped off, because other countries developed their own catalyst and domestic demand was reduced, Shell stopped making this catalyst a few years ago, and Aerojet now produces it.[§]

Initial catalyst bed designs were unsatisfactory and gave poor performance, incomplete catalysis, excessive overpressures during start, ignition delays, and short life. It took considerable development effort before the catalyst bed design was satisfactory, and short repeatable pulses could be achieved for more than 100,000 pulse cycles without significant performance degradation.[9-12] As shown in the lower part of Fig. 4.6-8, the distribution of the liquid

---

*Personal communications, O. M. Morgan, on monopropellant products and history, Aerojet Redmond Center, 2002, 2003.
†Personal communications, Carl Stechman, on the Hamilton Standard hydrazine monopropellant line, while he was at Marquardt, 2001.
‡Personal communications, Bob Sackheim, 1993–1995, and Gordon Dressler, 2001–2003, Propulsion Products Center, Northrop Grumman Corp., Redondo Beach, CA.
§Personal communications, O. M. Morgan, on hydrazine monopropellant thrusters and history, Aerojet, Redmond Center, Redmond, WA, 2002, 2003.

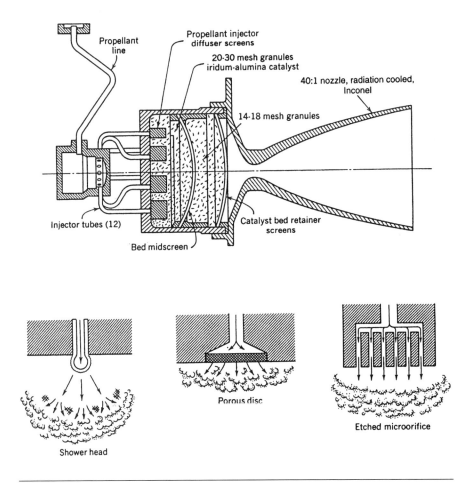

Fig. 4.6-8 Each hydrazine monopropellant thruster has its own bed of catalyst impregnated pebbles, usually held by a screen. Four different injection and distribution designs, which were investigated, are shown. Copied with permission from Ref. 2

propellant flow into the catalyst bed was one of the variables that had to be investigated to achieve long life. Other problems that were overcome were the attrition or physical loss of the catalyst material through the motion and abrasion of the pellets rubbing against each other and the decrease of catalyst activity caused by the poisoning of the catalyst by the impurities present in the hydrazine. The production process of hydrazine had to be modified to remove undesirable impurities such as minute quantities of aniline, a chemical that was used in the manufacturing process.

Figure 4.6-9 shows a small hydrazine monopropellant thruster developed by Bölkow in Germany, a predecessor of European Aeronautics Defence and Space (EADS).[13] To protect the electrical insulation in electromagnetic valves

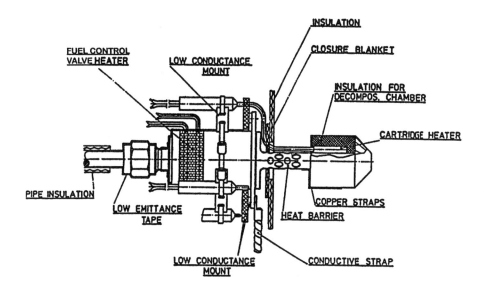

Fig. 4.6-9 This small German-designed hydrazine monopropellant thruster has a rated thrust of 0.5 N (0.11 lbf), electric heaters, and thermal insulation/barriers; around 1974–1976. Courtesy of MBB/EADS; copied from p. 309 of Ref. 13.

from becoming too hot, small thrusters use various means for thermal insulation, and thermal isolation, and several means are shown in the figure. To keep adjacent components from receiving excessive radiation heat from the hot glowing TC walls, an external thermal insulation is often provided around the TC. This thermal insulation is effective in reducing thermal radiation to the outside, but the insulation causes the TC wall temperature to rise.

A diagram of a complete engine system for the Radio Astonomer Explorer II Satellite is shown in Fig. 4.6-10. This LPRE was used for midcourse corrections and for lunar orbit retrofire to a lunar orbit.[9] It was designed by NASA's Goddard Space Flight Center and used only two hydrazine monopropellant thrusters (at 5 lbf thrust each) from Hamilton Standard of United Technologies. The four propellant tanks were pressurized by gaseous nitrogen using a blowdown-type system, and they were located on four outrigger arms. The spacecraft was rotated, and the centrifugal forces caused positive tank expulsion. Once the spacecraft was in lunar orbit, the tanks and piping were separated from the satellite and jettisoned, in order to allow unhindered observation without radio interference. In the figure the propellant valves at the thrusters were redundant. It flew in June 1973.

In the USSR hydrazine monopropellant thrusters were also developed independently, probably at about the same time (1960s) as in the United States. The Soviets developed their own catalyst. In some of their thrusters, they had a cobalt-based catalyst, which required preheating, and later they developed

164 History of Liquid Propellant Rocket Engines

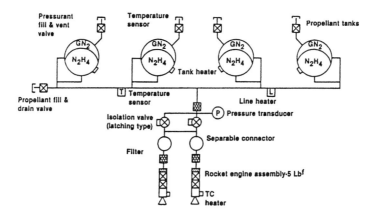

Fig. 4.6-10 Simplified diagram of a multithruster hydrazine monopropellant engine used on the Radio Astronomy Explorer satellite flown in 1973. From 1960 catalog of Hamilton Standard Division, United Technologies Corporation, today this technology is owned by ARC, from Ref. 13.

a catalyst with iridium, and it worked without preheating. Several USSR hydrazine thrusters were developed by the predecessor of the Khimmash Design Bureau located close to Moscow, and they were flown in USSR spacecraft or satellites. Some data on a couple of these USSR hydrazine thrusters are given in Chapter 8.6. Other countries (including Britain, China, France, India, Japan, Israel, and Germany) have also developed their own hydrazine monopropellant LPREs in the 1970s and 1980s. Many bought the Shell 405 catalyst, and some, like the Isrealis or the Chinese, developed their own different catalyst.[14] Most of these countries undertook the development after they read the literature and saw, copied, or licensed proven designs. For example, the Japanese obtained a license to copy and build the TRW design, before they built and flew a slightly modified version.*

Bipropellant Thrusters

The low-thrust *bipropellant thrusters* were a logical improvement because they gave higher altitude specific impulse (initially between 260 and 290 s and later over 320 s). They are discussed in Chapters 6.0, 7.2, 7.7, 7.8, 7.9, 7.11, 8.6, 8.7, 8.10, 11, and 13. They were usually preferred over monopropellants, when the total impulse of the RCS was relatively large and high performance was important for the mission. The initial fuel usually was hydrazine, and it was used with nitric acid or NTO oxidizers on the first few radiation-cooled TCs. Because of the high freezing point, the many electric heaters in the system, and an explosion of hydrazine in a cooling jacket, this fuel was

---

*Personal communication, Kenji Kishimoto, formerly MHI, Japan, 2004.

replaced in about 1963 with a mix of 50% hydrazine and 50% UDMH, which was called Aerozine 50 and had a lower freezing point and still a good performance. Later it was replaced by MMH in the United States. In the Soviet Union hydrazine was the initial fuel, and it was replaced with UDMH. A good number of small bipropellant TCs with storable propellants have been developed, built, and flown by all of the key countries in the LPRE business.[4,7,15–18,*,†,‡,§,¶] About 13 organizations in the world can provide or have in the past provided qualified bipropellant thrusters using storable propellants.

Bipropellant thrusters, which are pulsed or throttled, cannot be safely cooled by regenerative fuel cooling alone, as already mentioned. For small chamber sizes the heat from the combustion, which is transferred to the walls of the chamber and nozzle, exceeds the capacity of the coolant to absorb this heat without excessive boiling. Furthermore the heat in the hot wall will soak into the stagnant coolant during pauses between pulses, and the coolant might get hot enough to evaporate (form bubbles), decompose, or explode. Regenerative cooling can work in small thrusters, if there is no restart, no throttling, low chamber pressure, and if the heat transfer is augmented by supplementary film cooling. Therefore for pulsing operations other cooling methods, such as radiation cooling or ablative cooling, had to be developed in the late 1950s and 1960s; they are explained in the following.

The initial bipropellant thrusters for pulsing operations in the late 1950s used *ablative materials* in the chamber and nozzle.[2,§,¶] Ablative materials are made of fiber-reinforced plastics using fibers made from glass, silica, aramids such as Kevlar, or carbon, surrounded by a matrix of epoxy or phenolic resins. These ablative materials were originally conceived for nozzles of solid-propellant rocket motors. They absorb heat by a melting, decomposition, and charring processes. This cooling scheme is shown schematically in Fig. 4.6-11. The ablation process is a combination of heat absorption by melting, decomposition in depth, and evaporation of the matrix material, augmented by the film cooling of the decomposed gases. A porous carbonaceous insulating material forms under the surface of the ablative liner and preserves the nozzle contour and throat diameter. Often fuel film cooling is used to augment the heat absorption of the ablative materials. More recently a set of oriented carbon fibers in a matrix of carbon has been investigated for the walls and nozzles of small thrusters. An

---

*Personal communications on bipropellant thrusters, R. C. Stechman, Aerojet, Redmond Center, WA, 2002–2003.
†Personal communications on ARC products and history, M. Marvin and F. Boorady, ARC, Niagara Falls, NY, 2001–2004.
‡Personal communications, Y. G. Larin, chief designer, NII Mash (R & D Institute of Mechanical Engineering), Nizhnaya Saldar, Russia, 2002.
§Personal communications, R. Sackheim, 1993–1995, and G. Dressler, 2001–2003, Rocket Propulsion Center, Northrop Grumman Corporation, Redondo Beach, CA.
¶Personal communications, E. Shuster and V. Wheelock, The Boeing Company, Rocketdyne Propulsion and Power, Canoga Park, CA, 2001–2004.

ablative lined TC is shown in Fig. 4.6-12. The thrust chamber and nozzle use a relatively thick ablative liner fastened to a cylindrical metal shell. This thruster has a rated thrust of 93 lbf, is one of eight used to position the Apollo command module prior to and during its reentry, and its design was started in 1964. NTO and Aerozine 50 are the propellants. These thrusters are located close to the reentry heat shield of the capsule, and the larger diameter ablative material at the nozzle exit is needed to protect the thruster from the very high reentry heat and temperatures. The ablative liner is a layup of plastic impregnated cloth fibers, which are oriented directionally. During development, it was learned that the orientation of the fibers was very important in the wear or erosion of the ablative. Other ablative thruster designs had a ceramic sleeve in the chamber and a ceramic or graphite insert in the nozzle throat.

In the United States these small ablative TCs were developed by Rocketdyne (part of The Boeing Company) and TRW (part of Northrop-Grumman) and are discussed in Chapters 7.8 and 7.9. The U.S. Gemini manned spacecraft had two NTO/MMH bipropellant LPREs, each with its own gas pressure feed system and several ablative TCs. The Gemini Orbit Attitude Control and Maneuvering system (SE-7) had eight 25-lb ablative thrusters, two 85-lb thrusters and six 100-lb thrusters.* The Gemini Reentry Control System (SE-6) had two redundant sets of eight 25-lb ablative thrusters. These ablative thrusters for the Gemini Program are discussed in Chapter 7.8. Small ablative TCs were used in several other spacecraft in the United States, such as the ACS for the Apollo command module and the settling of liquid propellants or the attitude control in the Saturn S-IV upper stage.*,† The USSR has also developed some small ablative thrusters, perhaps using glass fibers, but the author has not been able to obtain any information on them.

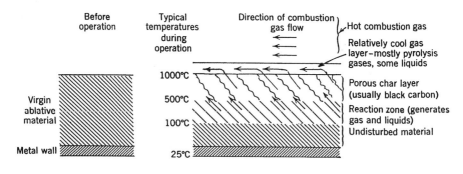

Fig. 4.6-11 Zones in an ablative material during rocket operation. Copied with permission from Ref. 2.

---

*Personal communications, E. Shuster and V. Wheelock, The Boeing Company Rocketdyne Propulsion and Power, 2001–2004.
†Personal communications, R. Sackheim, 1993–1995, and G. Dressler, 2001–2003, TRW.

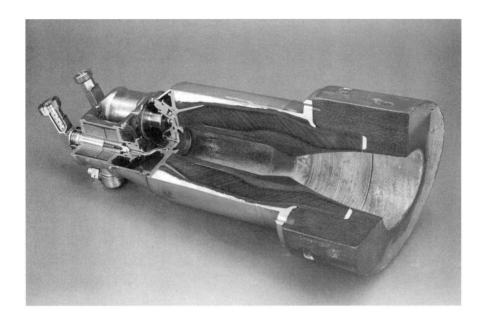

Fig. 4.6-12 Bipropellant ablative lined thruster (with a thrust of 93 lbf) of the ACS of the Apollo command module shows discoloration of the ablative material from the firing operation.* Courtesy of The Boeing Company, Rocketdyne Propulsion and Power.

Ablatives for small thrusters fell out of favor because they were heavy, the erosion products of ablative materials caused dirty exhausts, which would contaminate sensitive spacecraft surfaces, and sometimes the material would ablate unevenly and cause thrust misalignment. They were gradually replaced by refractory metal radiation-cooled thrusters, which by then had been fully developed.

A *radiation-cooled thruster* needs a high-temperature material for the thrust chamber and nozzle walls.[15-19] At a steady-state operation it glows white hot and radiates enough heat to the surroundings to maintain thermal equilibrium. The material has to be strong enough at the high temperatures to contain the pressures of the combustion gases. Figure 4.6-13 shows the melting points of several metals that have been investigated for radiation-cooled applications.[1] Also shown are the combustion temperature regimes and limits of monopropellant gas products and bipropellant gas products. Ideally this gas temperature should be lower than the melting point of the wall material. Stainless steel has been used, but it requires considerable supplementary film cooling, which reduces the performance, but keeps the wall temperature below its melting point. Niobium, also still called columbium by many people

---

*Personal communications, E. Shuster and V. Wheelock, The Boeing Company Rocketdyne Propulsion and Power, 2001–2004.

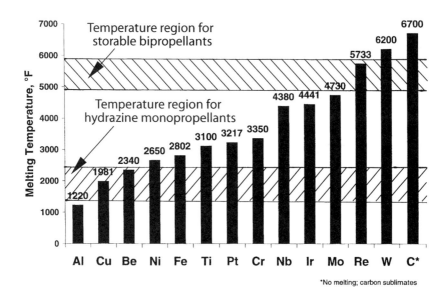

Fig. 4.6-13 Melting points of selected high-temperature materials and two regions of gas temperatures. Copied with permission from Ref. 1.

in the industry, has been a very successful material, and it requires less film cooling than the stainless steel; its performance is slightly better. Rhenium thrusters have flown, require even less film cooling than niobium, give slightly more performance, but rhenium is very expensive. Attempts were made to build thrusters from molybdenum, vanadium, tungsten, or titanium, and some were actually built and tested, but were abandoned for various reasons. Alloys of molybdenum and nickel were developed in Germany and made into successful small radiation-cooled thrusters. Carbon has a high enough sublimation temperature so that it would operate without any cooling, but it erodes and oxidizes readily at the high temperature, even when only a very small amount of oxygen or hydroxyl is present in the hot gas. Investigations for overcoming this oxidation have been undertaken for at least a couple of decades and are continuing.

The first small experimental radiation cooled thrusters in the United States were probably developed in 1958 and 1959 by Marquardt Corporation (today part of Aerojet General Corporation, see Chapter 7.7).* This first experimental TC had 25 lbf thrust, used NTO and hydrazine as propellants, molybdenum as the wall material for the chamber and nozzle, with an inside coating of molybdenum disilicide for oxidation protection. This TC design was intended for the Advent satellite and went through qualification tests, but it never flew. Soon

---

*Personal communications, R. C. Stechman, chief engineer, formerly with Marquardt, now Aerojet, Redmond Center, WA, 2002, 2003.

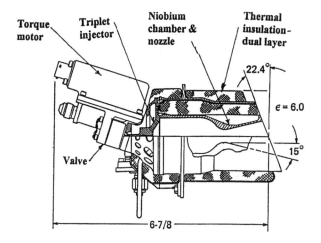

Fig. 4.6-14 This radiation-cooled, insulated niobium bipropellant thruster (23-lbf nominal thrust) is one of several used in the Minuteman postboost control system. Courtesy of ARC and U.S. Air Force (from personal communications, M. Marvin and F. Boorady, 2001–2004).

thereafter the molybdenum was replaced by niobium, which is also often called by its former name columbium, which has better properties. A radiation-cooled thruster developed by ARC is shown in Fig. 4.6-14. This is one of the pitch and yaw control thruster for the Minuteman III postboost control propulsion system with 23 lbf nominal thrust using NTO/MMH.* The nozzle is "scarfed" or cut off at an angle to fit the contour of the vehicle's skin. Operation can be in a pulse mode with the minimum impulse bit of 0.68 lb-s. The chamber pressure is 125 psia, and propellants are NTO/MMH. The external wrapped ablative insulation was intended to limit the maximum temperature of the outside of the TC to 400°F and thus protects adjacent vehicle hardware from excessive heat radiation. However it causes a significant rise in the temperature of the film cooled niobium TC walls. The valves are driven by a torque motor. Over 8000 of this type of thruster had been built up by the end of 2002. There have been no failures in the 175 Minuteman III launches (up to the year 2000), and some of these had been installed in the missile without maintenance for over 15 years.

Refractory metals (molybdenum, niobium, tungsten, vanadium, etc.) oxidize rapidly in rocket combustion gases that have only a small amount of free oxygen or hydroxyl; therefore, an oxidation-resistant coating was needed. After some trial and error the protective coating was obtained by dipping the niobium thrusters into a silicon solution with some chromium and titanium.

---

*Personal communications, R. Carl Stechman, chief engineer, Aerojet, Redmond Center, WA, 2002, 2003.

After baking this forms a Nb_5Si_3 coating. When exposed to hot combustion gas, the silicon is oxidized to SiO_2, and this then provides the protective layer.[†]

Radiation-cooled coated TCs have been used extensively in small LPREs, and several nations now manufacture them. Most have used niobium walls, but the Germans have also used a molybdenum-nicket alloy. Bipropellant thrusters were developed by the companies listed in Table 6-2.[13,15,18,20,*,†,‡,§]

Different design bureaus of the Soviet Union also developed small bipropellant trusters for flight control in the late 1950s and early 1960s. Nearly all used NTO and UDMH propellants. NII Mash, which is discussed in Chapter 8.7, is one of the few organizations that only specializes in small bipropellant thrusters.[‡] Their first-generation thrusters had stainless-steel walls in the chamber and the nozzle and were film cooled. The vacuum specific impulse values were between 240 and 254 s. The second generation of small thrusters used niobium with a silicate coating for the walls, required less film cooling, and allowed higher wall temperatures. With a higher nozzle area ratio (100 instead of 53) the specific impulses were between 290 and 315 s. Up to 2002 NII Mash had developed and qualified 30 different small thrusters, some with different propellants, and their products have flown in more than 800 space flights. The next two organizations concentrated on large LPREs, and small thrusters was not their principal product line. The second Soviet organization is the predecessor organization of the KB Khimmash or the Chemical Machinery Design Bureau located at Korolev in the Moscow region. Some of their small bipropellant and monopropellant thrusters are discussed in Chapter 8.6. The third USSR organization was Nauchno-Proizvidstvennoye Obedinenie (NPO) Yuzhnoye in Dnieprpetrovsk, a city that is today in the Ukraine. Its small thrusters were used primarily in the large guided missiles and space launch vehicles, which were developed in the same plant complex. The fourth organization is the Opitnoye Konstructorskoye Buro (OKB) Tumanskiy (Experimental Design Bureau Tumanskiy) in Tushino, which used to build these small thrusters, but has been phasing out of this business. All of the Soviet LPRE organizations are shown in Table 8.3-1.

In the last 12 years or so, a *rhenium* metal has been developed as a wrought and machinable material. It was tested successfully in small experimental TCs, and at least one thruster with a rhenium chamber and throat section has been developed, qualified and flown. Because rhenium melts at a higher temperature than niobium, it requires less film cooling, and the engine performance is

---

*Personal communications, R. Carl Stechman, chief engineer, Aerojet, Redmond Center, Redmond, WA, 2002, 2003.

†Personal communications, Mayne Marvin, regarding ARC LPRE products and history, 2001–2004.

‡Personal communications, Y. G. Larin, chief designer, NII Mash (R & D Institute of Mechanical Engineering), Nizhnaya Salda, Russia, 2002.

§Personal communications, R. Sackheim, 1993–1995, and G. Dressler, 2001–2003, Rocket Propulsion Center, Northrop Grumman Corp., Redondo Beach, CA.

somewhat better. However this material oxidizes readily at elevated temperatures, and a protective coating, usually iridium, has to be applied on the inside of the hot TC wall. Aerojet, formerly Marquardt, has developed a thruster with a rhenium metal wall in the chamber and nozzle throat regions, which is the high-temperature region, and with lower cost and less heat-tolerant materials in the nozzle extension.* Figure 4.6-15 shows such a thruster, and it has flown successfully. The coated rhenium chamber/throat wall section can be safely heated to a little more than 4000°F, the coated niobium section to about 3500°F, and the titanium nozzle-exit segment to about 1700°F. Titanium has a relatively low specific gravity compared to the two other materials. Therefore it has been possible to save some weight. Because rhenium is very heavy and expensive, a carbon nozzle with rhenium and/or iridium coatings, can offer a lighter-weight solution and is being investigated.

Carbon (or graphite) chamber or nozzle materials can withstand very high gas temperatures (over 5000°R is shown in Fig. 4.6-13), and they have been used extensively in solid-propellant rocket motors. There are several forms of the carbon that have been tried or used in small TCs with storable bipropellants. In order of increasing heat resistance, they are molded graphite of several grades, pyrolytic graphite, and carbon fibers in a carbon matrix. A few of these have also been used in experimental liquid propellant thrusters.[2] For example, a pyrolytic graphite nozzle (a special anisotropic form of carbon) was used successfully in an experimental TC with a fluorine oxidizer. Because these propellants do not include oxygen, there is no loss of carbon as a result of oxidation. Many carbon materials, including materials with carbon fibers, are porous, they can allow gas to flow through it, and they need to be sealed. If the propellant combustion products contain small amounts of free oxygen or hydroxyl or monatomic oxygen, the carbon will oxidize readily and erode. The carbon chambers and nozzles have been successful, if the mixture ratio is rich, the walls are protected by a fuel film coolant, or the hot gases have only a trace of oxidizing constituents. If a protective coating (such as ceramic or rhenium) is applied to the inner wall, the oxidation of the carbon can be diminished. Experimental TCs with these features are being investigated. The higher wall temperature offers a further improvement in performance.

Regenerative cooling has been applied in a few small bipropellant thrusters. Examples are shown in Figs. 4.6-4 and 8.7-3, which is a Russian 12 N (2.7 lbf) thruster. When there is no restart, then this cooling method might be suitable. When the thruster is pulsed, then heat soak-back from the hot wall can overheat the propellant trapped in the cooling jacket (between pulses) causing it to vaporize or perhaps decompose. Gas bubbles can form and cause problems in restart. If the frequency of pulsing and the time interval between pulses are limited, it can be possible to operate the small thruster, which is regeneratively cooled. Figure 4.6-16 shows a bipropellant thruster of 400 N

---

*Personal communications, R. Carl Stechman, chief engineer, Aerojet, Redmond Center, Redmond, WA, 2002, 2003.

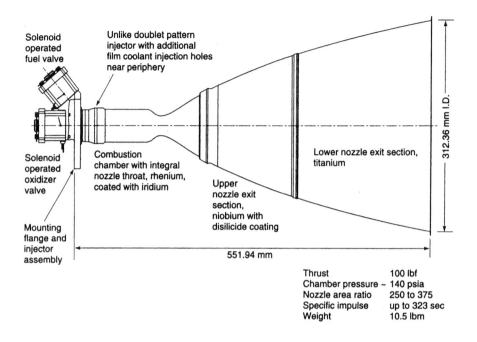

Fig. 4.6-15 This radiation-cooled bipropellant thruster with three different metals (rhenium, niobium, titanium) can deliver up to 323 s (vac) of specific impulse using NTO/MMH. Courtesy of Aerojet; originally developed at Marquardt, now part of Aerojet. Copied with permission from Ref. 2.

(93 lbf) thrust with regenerative cooling; it was developed by the Bölkow Design Bureau in Germany.[13]

There have been thousands of flying liquid propellant rocket engines with multiple small thrusters. As shown by the example in Fig. 4.2-9, they all have a pressurized feed system, multiple thrusters, and pressurized tanks. A special category is a *combination system*, which uses one or more bipropellant thrusters (for major maneuvers) and four or more simpler monopropellant thrusters (for attitude control or minor maneuvers). An example is shown in the simplified schematic diagram of Fig. 4.6-17 for the Advanced X-Ray Astrophysics Facility with imaging capability (AXAF-I). It has four bipropellant thrusters (NTO/hydrazine) of 105 lbf thrust for the apogee (orbit insertion) maneuver, which consumed the largest share of the propellants, and eight monopropellant hydrazine thrusters of 20 lbf thrust each for the reaction control. In addition, there are eight small monopropellant thrusters at 0.1 lbf thrust each for the desaturation of the flywheels. They are identified in the figure as the momentum unloading propulsion system (MUOS).

Another concept for cooling these small TCs, called *interregen* (internal regenerative), was developed by Rocketdyne in the late 1960s. As shown in Fig. 4.6-18, it relies on a combination of convective heat transfer from the

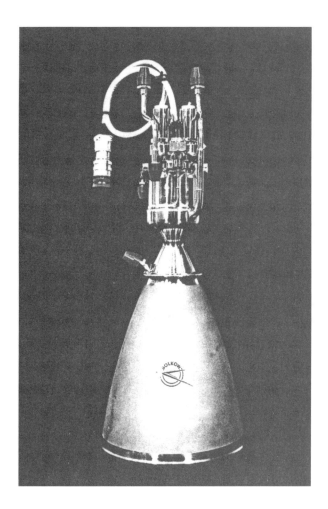

Fig. 4.6-16 German 400-N bipropellant thruster with partial regenerative cooling developed by the Bölkow design bureaus in the early 1960s. Courtesy of Bölkow/EADS; copied from Ref. 13, p. 124.

hot, fast moving gases to the beryllium metal in the nozzle throat region, where the heat transfer and wall temperatures are at their highest.* The heat absorbed at the nozzle is then conducted in the beryllium from the hot throat region to the cooler chamber regions, where this heat is transferred from the chamber wall into the cooler film coolant layer flowing along the inner wall. In this boundary layer this heat causes evaporation and decomposition of the film coolant. The nozzle-exit section is made of refractory metal, is radiation cooled, and insulated to protect adjacient components from excessive

---

*Personal communications, E. Shuster and V. Wheelock, The Boeing Company, Rocketdyne Propulsion and Power, Canoga Park, CA, 2001–2004.

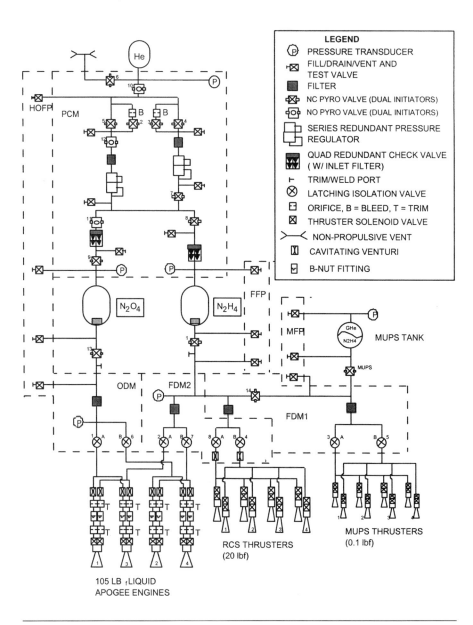

Fig. 4.6-17 Simplified flow diagram of the AXAF-I propulsion system with both bipropellant and monopropellant thrusters. Courtesy of Northrup Grumman (copied with permission through personal communications, Gordon Dressler, Propulsion Products Center, Northrop Grumman Corp., Redondo Beach, CA, 2001–2003).

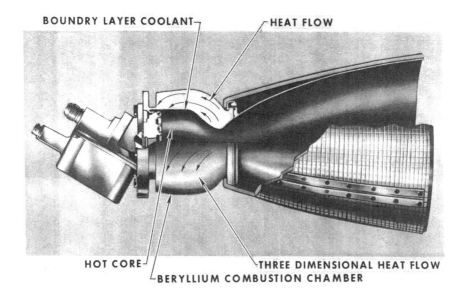

Fig. 4.6-18 Representation of the heat-transfer flow in an interregen bipropellant thruster. Courtesy of the Boeing Company, Rocketdyne Propulsion and Power.*

radiation heat. It uses a relatively thick wall of *beryllium* metal for the chamber/nozzle material. Table 4.6-1 shows this lightweight, low-density metal has good conductivity and a reasonably high melting point compared to other metals.

Beryllium is easy and safe to handle as a solid metal, and the author has personally worked with this material. However fine particles of beryllium or beryllium oxide are very toxic, and the machining process is usually performed in an enclosure, where the air is carefully filtered. As far as this author knows, there has to date not been any incident of inadvertently creating such small particles in an accident or exposing people to them. Some of the thrusters for the Minuteman and Peacekeeper postboost control systems have beryllium chambers (designed in 1966) similar to the ones shown in Fig. 7.8-27.

The Germans also developed several small thrusters using this interregen cooling method in the early 1990s.[13] They used a platinum alloy containing some rhodium as the heat-absorbing and heat-conducting material. One is shown in Fig. 4.6-19; it uses mixed oxides of nitrogen (NTO with some NO) as oxidizer and MMH as fuel and is rated at 400 N or about 93 lbf thrust. As seen in Table 4.6-1, pure platinum has a very good thermal conductivity, but it is very heavy. This German thruster has flown in European, Israeli, and Arab

---

*Personal communications, E. Shuster and V. Wheelock, The Boeing Company, Rocketdyne Propulsion and Power, Canoga Park, CA, 2001–2004.

Table 4.6-1 Unique Beryllium and Platinum Properties*

Property	Beryllium	2219 Aluminum	321 Stainless	Platinum
Melting point, °F	2345	1220	2800	3217
Specific gravity	1.85	2.70	8.00	21.45
Thermal conductivity, Btu/ft-s-°F	0.03	0.035	0.003	0.014
Pounds for equal heat absorption	1	2	4	1.4

satellites. Figure 9-21 shows the performance increases of this size German thruster over a 30-year period; the improvements were caused by increasing chamber pressure, increasing nozzle area ratio, and using less film cooling when going from a molybdenum alloy to a platinum alloy. More information is given in Chapter 9.5.

Because many SLVs have stages with liquid hydrogen and liquid oxygen, it was logistically and economically advantageous to want the small thrusters in these stages to have the same propellant. The development of small thrusters with LOX/LH$_2$ has been tried by several teams, including one at Aerojet, is technically difficult and, as far as the author knows, has not been successful to date. When a small flow of cryogenic liquid propellant absorbs heat from the hardware and/or the surroundings, the liquid is partly or fully evaporated. Thus the density, mass flow, bubble content, mixture ratio, combustion temperature, and thrust can vary greatly, particularly when pulsing thrusters are fired at random intervals. This makes it almost impossible to repeat or predict the performance of a small TC using LOX/LH$_2$. Small TCs with gaseous oxygen and gaseous hydrogen have been developed, and some have flown. An example is a 200-N thruster for the Buran vehicle from NII Mash.[†]

Combustion instability has occurred in a few small thrusters. Compared to large TCs, the combustion vibrations are not as intensive, do not often disable the thruster, and do not occur very often. For example, instability occurred in the 870-lbf thruster used on the Space Shuttle Orbiter, and it could be artificially initiated by injecting gas bubbles into the propellant lines. The injector of this thruster is shown in Fig. 4.10-5 and is discussed in Chapter 4.10.

Special Applications

The *postboost control systems of ballistic missiles* in the United States have a LPRE with a pressurized-gas feed system and multiple thrusters. Its task is to

---

*Table obtained through personal communication, V. Wheelock, The Boeing Company, Rocketdyne Propulsion and Power, 2003.
[†]Personal communications, Y.G. Larin, chief designer, NII Mash (R & D Institute of Mechanical Engineering), Nizhnaya Salda, Russia, 2002.

Thrust	400N/90 lbf
Oxidizer	mixed oxides of nitrogen
Fuel	MMH
Specific impulse	318 s
Mass	3.2 kg or 7.5 lb
Length	531mm or 20.91inch
Chamber pressure	142 psia
Nozzle exit area ratio	220:1.0

Fig. 4.6-19 This bi-propellant thruster has a platinum alloy for the chamber and nozzle throat region; it uses radiation cooling and some interregen cooling.

apply an accurate terminal velocity (in magnitude and in direction) to each of several war heads in the terminal stage of such a missile.[2] It has several small attitude control thrusters at right angles to the vehicle axis and one larger axial TC, which has its exhaust gas pointed in the aft direction. Figure 4.6-20 shows a simplified and generalized schematic diagram of a gas-pressurized restartable bipropellant rocket engine system used in the fourth stage of the Peacekeeper ballistic missile. The larger gimbal-mounted TC in the center is for providing translation or acceleration maneuvers, and the eight small thrusters, some with scarfed nozzle exits to fit the external contour of the vehicle, are for attitude control. The large TC uses an ablative liner, and the small thrusters use an interregen cooling approach with beryllium metal as just described. The pressurized feed system and the propellant tanks are drawn outside the vehicle's skin envelope, but they really are inside the vehicle. Chapter 7.8 has further discussion. The postboost control system used for the Minuteman III ballistic missile was developed by ARC, is shown in Fig. 7.11-3, and one of its small thrusters is in Fig. 7.11-4.

Another special version of this small LPRE category is intended for the final or terminal maneuver stage of an antiaircraft or antimissile vehicle system. Here several small pulsing thrusters are needed for the attitude control and at least one larger thrust chamber for moving or diverting the terminal vehicle

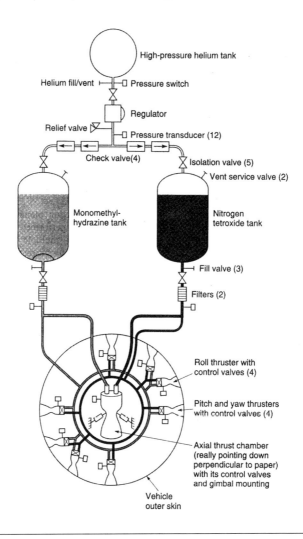

Fig. 4.6-20 Simplified diagram of a postboost control system with eight interregen small thrusters for attitude control and a single larger ablative gimbal mounted axial thruster. Courtesy of the Boeing Company, Rocketdyne Propulsion and Power; copied with permission from Ref. 2.

sideways in a direction normal to the flight path. The larger thruster often has been called a *divert thruster*, and the vehicle is often called a kinetic energy weapon.[20] These terminal maneuver LPREs typically operate at higher thrust and higher chamber pressure (500–800 psi) than those used for the more common ACS applications (50–200 psi). The higher pressure helps to makes the TCs smaller and easier to package, increases the performance, but is much heavier, and has a higher heat transfer. Several such experimental LPREs were developed in the United States and flown in a hover test facility in the 1980s and 1990s. The author does not have information about further developments or applications at the time of this writing.

Development and Manufacture of Small Thrusters

Table 6-2 lists all of the current and past developers and manufacturers of small thrusters known to the author. This table is sorted by the type of thruster. Some of the companies make more than one type and are therefore listed more than once. Thirteen organizations are listed for monopropellant thrusters and 16 for the radiation-cooled and interregen types, and they are in 10 different countries. The United States and Russia are the only countries with several organizations in this business. Some countries use privately owned companies and have this work done in government-owned or government closely controlled organizations.

For most of the organizations listed in this table, the small thrusters are a minor product line, and the main business is in other areas. Only a few organizations build primarily small thrusters, such as, for example, NII Mash (Chapter 8.7). A key asset of these organizations is the know-how in developing, testing, inspections, installations, and supporting the flights of their small thruster products. Each company seems to have its own way of doing some of these activities.

A good number of small thruster designs and materials are very similar or sometimes identical in several of the organizations or in several countries. Once a particular feature or material has been proven to be practical, another organization is most likely going to develop something very similar, to license the idea, or to outright copy it. For example, the first thruster made from niobium (also called columbium) is believed to have been developed in the United States around 1959. The same material or a niobium with slightly different minor ingredients and with a silicate coating is now used for radiation-cooled thrusters in most countries.

There are at least two ways in which a LPRE with small TCs and its pressurized gas feed system have been integrated with a vehicle. The vehicle design and manufacturing contractor designs and builds the tanks, piping, and pressurization system and buys only the TCs and their valves from a rocket propulsion contractor. This method is easier to integrate with the vehicle. The other way is for the rocket propulsion contractor to be responsible for the complete LPRE system with propellant tanks, controls, gas-pressurization system, heaters, and even parts of the vehicle structure. This is often easier to install into the vehicle and to check out as a self-contained assembly. The second approach was practiced in the United States in the 1960s and 1970s. Since then, there have been very few opportunities for a LPRE organizations to get new jobs for a complete engine system with small thrusters.

Concluding Comments

A large variety of small thrusters have been used successfully. Historically the earliest attitude control thrusters used cold gas as propellants. Then came the monopropellant hydrogen peroxide thrusters, which are now obsolete. Thrusters using hydrazine as a monopropellant and thrusters with storable bipropellants started to fly around 1960 and improved versions are being built and flown today. Ablative thrusters were an initial solution to pulsed bipropellant operation, but they fell out of favor. Partly regenerative cooled thrusters have been developed for special applications. The history of small thrusters is one of improvement with better materials, better designs, achieving higher performance, lower inert mass, and fabulous reliability. The selection of the type of thruster, its parameters, propellants, or feed system depend on the specific application and the prior experience and flight record. Recently most small thrusters have been developed and produced in the country or group of countries, where the flight vehicles are also built. Today there are a lot more LPREs with small thrusters than larger LPREs.

References

[1]Sutton, G. P., "History of small Liquid Propellant Thrusters," *52nd JANNAF Propulsion Meeting* Chemical Propulsion Information Agency, 2004.

[2]Sutton, G. P., and Biblarz, O., *Rocket Propulsion Elements*, 7th ed., Wiley, Hoboken, NJ, 2001.

[3]Mouritzen, G., "Cold Gas Rocket Propulsion," *Journal of the Astronautical Sciences*, Vol. 19, No. 1, July–Aug. 1971, pp. 50–70.

[4]Reaction Control Experiences, informal illustrated propulsion product publication, Bell Aerosystems Co., 1989. (Obtained through personal communication with M. Marvin, ARC Liquid Operation, Niagara Falls, NY, 2002.)

[5]Gribbon, E., Driscoll, R., and Marvin, M., "Demonstration Test of a 15 lbf Hydrogen Peroxide (90%) Thruster," AIAA Paper 2000-3249, 2000.

[6]Bloom, R., Davis, N. S., and Levine, S. D., "Hydrogen Peroxide as a Propellant," *Journal of the American Rocket Society*, Vol. 80, March 1950.

[7]Sansevero, V. J., Jr., Garfinkel, F., and Archer, S. F., "On-Orbit Performance of the Hydrazine Reaction Control Sub-System for the Communications Technology Satellite," AIAA Paper 78-1061, July 1978.

[8]Tolentino, A., and Grotter, R. J., "Advanced 5 lb Engine Demonstration," AF-Rocket Propulsion Laboratory Rep., March 1984.

[9]*Hydrazine Propulsion Experience*, Hamilton Standard Div., United Technologies Corp., Company Product Brochure SSS-89-01, data available from ARC, Liquid Operation, Niagara Falls, NY, 1989.

[10]Marcus, M., "Historical Evolution of the Hydrazine Thruster Design," AIAA Paper 90-1838, July 1990.

[11]Polythress, C. A., and Conkey, G. H., III, "Development of a Long Life 125 lbf Hydrazine Thruster," AIAA Paper 80-1170, 1980.

[12]Saenz, A., Jr., "Life Performance Evaluation of 22 N (5 lbf) Hydrazine Attitude Control Engines," Air Force Rocket Propulsion Laboratory, USAF/RPL, Technical Rep., Nov. 1980.

[13]Hopmann, H., *Thrust for Space Flight, Developments of Rocket Engines in Germany*, Stedinger Verlag, Lemwerder, Germany, 1999 (in German).

[14]Adler, D., Dubrov, E., and Mannheimer-Tinmat, Y., "The Performance of a Hydrazine Engine with an Improved Catalyst," *Acta Astronautica*, Vol. 2, pp. 613–625.

[15]Zhu, L., and Zhang, M., "Research on N_2O_4/MMH Thrusters," *Journal of Propulsion Technology*, Aug. 1999, pp. 20–32 (in Chinese).

[16]Coutrot, A., "Storable Liquid Propellant Thrusters for Space Applications," International Astronautical Federation, Paper 96-S108, Oct. 1997.

[17]Rath, M., Schmitz, D. H., and Stenborg, M., "Development of a 400-N Thruster for ESA's Atmospheric Reentry Demonstrator," AIAA Paper 96-2866, 1996.

[18]Stechman, R. C., "Advanced Thrust Chamber Materials for Earth Storable Bipropellant Rocket Engines," *Acta Astronautica*, Vol. 29, No. 2, 1993, pp. 109–115.

[19]Stechman, R. C., and Sumpter, D., "Development History of the Apollo Reaction Control System Rocket Engine," AIAA Paper 89-2389, 1989.

[20]Ruttle, D., and Fitzsimmons, M., "Development of a Miniature 35-lbf Fast Response Bipropellant Divert Thruster," AIAA Paper 93-2585, 1993.

4.7 Controls, Valves, and Interconnecting Components

What has to be controlled in a LPRE? The first four items listed next were controlled in all types of LPREs beginning in the 1920s:

1) Pressurization of the propellant tanks is the first item.

2) The starting procedure ensures a safe transition to full thrust for one or more thrust chambers, and it included the control of the ignition with non-hypergolic propellants.

3) Attain the desired thrust or thrust-time profile and the intended performance (specific impulse) by controlling the propellants' flow magnitude, mixture ratio, and pressure drops.

4) Shutdown or safely terminate the firing operation.

Furthermore there can be additional controlled features in some engines, and they were introduced in several countries between 1930 and 1980.

5) Precision automatic thrust control usually senses chamber pressure and regulates the propellant flow to the thrust chamber and/or the mixture ratio of the flow to the gas generator or the preburner.

6) Mixture-ratio control, also called propellant utilization control, minimizes the amount of residual propellant.

7) Condition monitoring (also called safety controls or health monitoring) will prevent engine damage by sensing out-of-tolerance parameters, such as GG overtemperature, chamber pressure, turbopump overspeed, or pneumatic pressure and initiate a remedial action and in some cases a safe shutdown.

8) The controls for thrust vectoring or applying side forces controls for attaining the desired vechicle flight maneuvers are usually separate from the engine controls and are treated in Chapter 4.9.

9) For large flight vehicles with multiple large LPREs, it is possible to provide differential throttling for TVC and also include an engine-out capability and complete the mission with fewer large engines. The malfunction of an engine has to be sensed and evaluated before the engine is shut down.

10) For an engine with redundant components, such as a spare thruster or a redundant second pressure regulator, the controller needs to sense and identify a component failure, shut down the flawed component, and immediately activate the spare or reserve component.

11) Control a restart during flight.

Controls

The controls have been called the brains of a LPRE, and the valves and igniters can be thought of being the muscles, which initiate the events in the operation of a LPRE. The electrical system or the pneumatic system provides the power to actuate the muscles or the valves. There is a large variety of controls and valves that have been used with LPREs, and their selection will depend on the engine requirements and preliminary engine designs.[1]

The very early starting procedure and shutdown method of LPREs had mechanical controls (e.g., pushing on a valve stem with a long rod or pulling strings) and was manually controlled. As mentioned subsequently in Chapter 5, the early pioneers, amateurs, and early LPRE experimenters used to make their own controls and valves because suitable commercial controls and lightweight valves were then not available. Some of the amateur and pioneers, like Goddard,[2] had to use three people to pull different strings or push different buttons upon spoken command, which in turn were based on visual observations of the flame or the reading of pressure gauges. The flame gave clues about the ignition and the functioning of the start procedure, and the test conductor would issue verbal commands to pull certain strings. Test personnel pulled on strings from their crude control station to open or close valves or to start an igniter. Goddard's work is described in Chapter 5.2. It took sometimes 6 to 12 s to start a simple LPRE with a gas pressure feed system. The history of the progress in shortening this start delay has been remarkable.

Because this start method took a relatively long time, Goddard invented some simple automation, so that some events would be automatically followed by another event without verbal commands to his test operators. Improvements using simple mechanical controls were made by Goddard to shorten his start sequence and to get away from relying on visual observation or verbal commands.[2] For example, he used a falling weight, triggered by a mechanical detent of a prior event, to pull a string over a pulley in order to initiate the next event. He switched to a timer control in about 1927 or 1928. With a *timer control* the various start events were commanded in a time sequence by using a clock with switches. Pratt & Whitney used a precision timer to control the earliest version of the RL 10 LPRE development in 1959 and found it to be unsatisfactory as described in Chapter 7.10. Beginning in about 1930, Goddard started to use pneumatic valve actuation and later some pneumatic controls, as discussed next.

The *step-ladder control* for the start was a significant innovation in about the late 1940s and the 1950s, but earlier (1930s) in Germany.[1,3] Here the next event of the operating sequence is commanded automatically only after there is confirmation that the previous event had in fact taken place. The validation confirming that a step had been completed was obtained from such measurements as pressure, valve position, heat from an igniter, or the reading of a flow meter signal. It was a safer start procedure than the use of a simple timer. The control systems were extended incrementally to allow for other additional functions. For example, features were included to maintain the thrust or propellant flow at a predetermined value, allow restart (late 1930s), and avoid unsafe regions of mixture ratio and chamber pressure during the start or thrust buildup (1950s). These dates are for a group of LPREs known to the author, and the years cited might not fit every LPRE.

LPREs have been controlled to achieve the desired thrust and the desired specific impulse.[3,4] This has been accomplished by proper calibration of the liquid flow through the liquid's path within the engine. The early pioneers did not have automatic thrust controls, but they calibrated the engine to predetermined propellant flows and hydraulic pressure drops.[1,3] The early calibrations were performed by flowing cold water through a pressurized feed system (from the propellant tanks through the piping, valves, cooling jacket, and injector), while measuring flows and pressure drops. A restrictor would be inserted into the nozzle to simulate the chamber pressure. From these flow data and from the propellant densities, one could then calculate the size and type of orifice that needed to be inserted into the propellant line in order to achieve the desired propellant flow. The sum of the oxidizer and the fuel flows would determine the chamber pressure and thrust level, and the ratio of the oxidizer to the fuel flow would determine the mixture ratio, which was preset to give the desired specific impulse. The pioneer Goddard did not put calibration orifices into the propellant lines of his early LPREs. Instead he used small petcocks or small needle valves, and he adjusted their valve position to obtain the necessary pressure drops for achieving the intended propellant flows.[2] These small valves can be seen in some of his early TC installation or test setups in Fig. 5-6 and 5-8. Some years later Goddard did use calibration orifices instead of adjustable valves. This basic calibration procedure is still in use today.

Pneumatic controls were introduced first in the 1930s. A compressed gas provided the power for opening the main propellant valves during start or closing them for shutdown. Inert gas pressure was used to activate most of the larger valves.[1,3] It allowed increases in the valve actuating forces, and the use of electromagnetic pilot valves really provided a form of power boost. High-pressure gas was already stored and used in most LPREs for pressurizing tanks or for purging propellant out of the piping or cooling jacket, and the use of this gas for actuating valves was a logical extension. Pneumatic controls were adapted for step-ladder control logic, and some of the sensors were based on pneumatic effects, such as pneumatic pressure. Most of the larger pump-fed LPREs designed since about 1940 have used pneumatic controls. For faster actuation and a stronger more precise actuating force, hydraulic actuators have also been used for a large valve or a thrust-vector-control cylinder. In some LPREs a separate hydraulic pump and an oil circuit were provided. In many cases it has been one of the pressurized storable liquid propellants that has been used as a hydraulic actuating fluid.

One form of flow control (and with it the control of both propellant flows and thus the mixture ratio) has been the use of *venturis*.[3,5] It is basically a bow-tie-shaped segment of a pipe with a high velocity in the narrowest section, often called the throat. One form of this venturi flow control was invented by R. H. Goddard, the early American rocket pioneer, and he received a patent on a

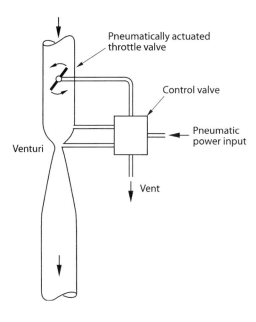

Fig. 4.7-1 Simple sketch of Goddard's patented concept of using a venturi flow section and a throttling valve to maintain a constant flow of propellant.

constant-propellant flow control device.[2] Its principle is shown in Fig. 4.7-1. He used the venturi as a flowmeter, a small control valve, and a throttling valve upstream of the venture section. The pressure difference between the upstream or venturi inlet pressure and the venturi throat pressure is supplied to a control valve, which Goddard designed. This control valve transformed these inputs into a pressure signal to actuate or move the throttle valve. It is basically a loop control system. This patent was filed by Esther Goddard after the death of her husband, but based on his writings and records U.S. patent 2,627,876 was granted in 1953. This method of flow control is still used.

As the flow velocity is increased, it will reach the velocity of sound at the throat of the venturi. A pressure wave or a pressure signal travels at the velocity of sound and cannot be transmitted through a sonic venturi. The flow will be choked and fixed in magnitude. The venturi is essentially a constant flow device; it requires a reasonable pressure drop to function and usually prevents pressure pulses from propagating in the liquid. Venturis have been used extensively in many LPREs. For example two venturis control the mixture ratio and the flow of propellants to the gas generator of the LPRE of the Titan flight vehicle as seen in Fig. 7.7-19. The lunar landing engine for the Apollo uses venturis inside the control valves to control the flow of propellants to the TC, and these valves can be seen in Fig. 7.9-7. The throat areas of these venturis could be varied to allow variable thrust, yet maintain the same mixture ratio. The low pressure in this throat can be below the vapor pressure of the liquid propellant, and bubbles can form there, which is cavitation. "Cavitating venturis" are used in LPREs.

Electronic computers allowed a major leap forward in engine control.[1,3] They were introduced in the 1960s, and the initial versions were relatively simple and had few functions. They were applied mostly to large- and medium-sized LPREs. More sophisticated multipurpose electronic controls were designed beginning in the 1970s. The size, weight, and heat rejection of the computers have diminished as electronic miniaturization progressed over the years. Today the controls are sophisticated and versatile and can include functions that are not really related to the control of the engine. Some computers allow very rapid sequencing of events, interpret measurements (received from sensors) of a large variety of engine parameters; command actions in response to sensor inputs; prompt recording of all events resulting in permanent records for later inspection or use; analyze, correct, process, and display data; instantaneously compare actual key parameters with those obtained by analysis (for example, the chamber pressure vs time curve during start); perform automatic decision making by the computer; self-test various parts of the control system; and automatically adjust or perform corrective action (or safe shutdown), if the actual values exceeded their limits.[6] For example, during start computers, connected to throttling valves, can avoid certain operating regions of chamber pressure and mixture ratio, where combustion instabilities were likely to occur. It also permits relatively sophisticated safety features. The computer can also be so designed that it will be possible to change some of the key engine parameters, sequences, and logic after development, if tests prove them to be desirable and without having to change engine hardware. There are no moving mechanical parts. Often the computer can be combined with the electrical system of the engine. There really is not much to see, except black boxes. Fig. 4.7-2 shows an external view of the space shuttle's main engine controller box, which was designed in the early 1970s.[6] The same controller today would be substantially smaller and consume less energy. Essentially all large LPRE designed since about 1972 used computers for part or all of the engine control system.

Many LPREs have their own computer controller, but in some engines the control functions are split between the vehicle's computer and the engine's computer. For simple pressurized feed systems and multiple thrusters the control of the engine is sometimes included in the vehicle's controller.

There are two types of automatic controls that are common in larger-size LPREs.[1,3] They are the automatic control of thrust and the automatic control of propellant mixture ratio. As mentioned in Chapter 5.2, Goddard knew about them[2] and actually tested engines with one or both of these in the late 1920s and early 1930s. One way for automatic thrust control to be achieved in a closed-loop control system (with feedback) is by measuring deviations from the intended chamber pressure (which is essentially proportional to thrust) and regulating the propellant flows through throttle valves to compensate for the deviation. One thrust control scheme is shown in Fig. 4.7-3 with a throttle valve in the oxidizer line that supplies a small flow of oxidizer to a fuel-rich gas

Fig. 4.7-2 The computer-controller of the space shuttle main engine is contained in a pressure tight box with cable connectors. The spikes sticking out of the side walls are for better heat rejection. Courtesy of NASA and The Boeing Company, Rocketdyne Propulsion and Power. Copied from Rocketdyne photo.

generator. In a fuel-rich GG or preburner the fuel flow is much higher than the oxidizer flow, and it is usually sufficient to control the flow of the oxidizer only. Small variations in this oxidizer flow will cause significant changes in the gas temperature and thus in the power of the turbine. Some controllers contain an algorithm that can correct for the changes in the ambient propellant temperature, which affects the propellant density and the viscosity. In pump-fed engines the thrust control is achieved by regulating the warm gas flow (in magnitude and/or temperature) by throttling (one or both) the flows to the gas generator or the preburner. Most of the pump-fed large LPREs have had a thrust control, which is either automatic or precalibrated.

The purpose of an *automatic mixture-ratio control*, sometimes called a *propellant utilization control*, is to minimize the amount of residual, unburned propellants left over after thrust termination in one or both of the propellant tanks and also in the pipes, cooling jacket, or valves of the engine and vehicle.[1,3] In effect, any residual propellant increases the inert mass of the vehicle at the end of rocket operation. This extra unusable propellant mass (typically 1 to 3% of the total propellant in that tank) can cause a significant flight performance loss. Some work on an active mixture-ratio control was undertaken in the 1940s and 1950s, and the control system has been improved since that time. Many of the larger pump-fed engines, but not all of them, have used a mixture-ratio control. Here the amounts of remaining propellants residing in the propellant tanks (at any one time) are measured in real time by one or

188 History of Liquid Propellant Rocket Engines

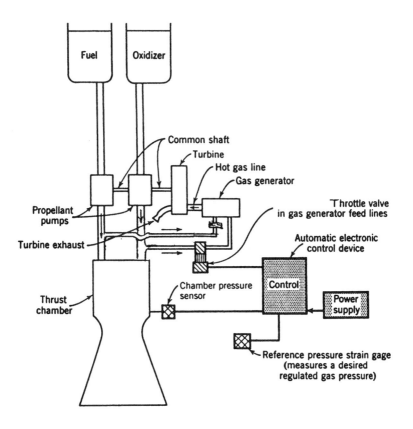

Fig. 4.7-3 Simplified schematic diagram of one type of servomechanism thrust control or pressure control of a LPRE with a TP feed system and a fuel-rich gas generator. Copied with permission from Ref. 3.

more sensors (allowing a determination of the instantaneous mixture ratio), and the automatic throttling of one of the propellant flows changes the operating mixture in the TCs in such a way that simultaneous emptying of both propellant tanks can be achieved and the maximum practical amount of the available propellants can be utilized for producing thrust. There have been a number of different schemes for measuring the propellant remaining in the tank during the firing including various level gauges and several different flowmeters.[7] Even with mixture-ratio control systems, there will always be some residual propellants (typically less than 1.0%), that cannot be utilized, such as the film of propellant liquid that wets the wall surfaces of pipes or tanks and the liquid that is trapped in pockets of valves, sensors, pipes, cooling jackets, or tube fittings. Figure 4.7-4 shows a diagram of one such a mixture-ratio control system with a throttling valve in each of the two propellant lines. Systems with just one of the propellant flows being throttled have been found to be feasible. A mixture-ratio control is usually provided if the total impulse is large, calibrations are not accurate enough, or the tolerances on the controls are large. Today more accurate flowmeters allow a more precise

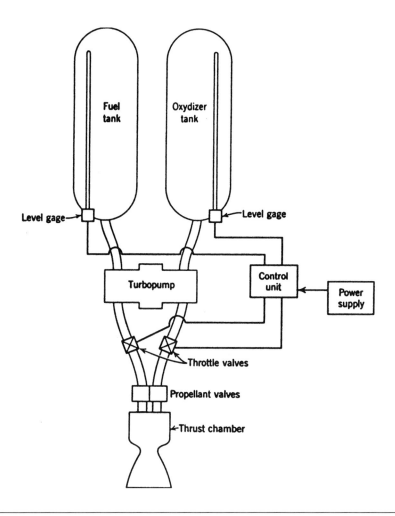

Fig. 4.7-4 Simplified schematic diagram of one type of propellant utilization or mixture-ratio control system. Copied with permission from fourth edition of Ref. 3.

engine calibration. In some engines, such as an advanced version of the Japanese LE-5, the mixture-ratio control system was actually deleted because this eliminated a set of relatively complex controls, sensors, and throttle valves and because tests showed that the flight performance improvement over a properly calibrated simpler system was relatively small.

Condition Monitoring System

For about 30 years there have been investigations of what has been called a *failure detection and correction system*, an *engine health monitoring system*, and more recently called an *engine condition monitoring system*. It has been considered to be a type of engine control. It senses periodically a variety of engine parameters.[1,8,9] Fast-changing quantities like chamber pressure or local accelerations can be monitored 1000 times per second, whereas others such as certain temperatures, which change more slowly, would be monitored one to five times per second. It then compares the measured values with expected values, which are derived from an analysis of a numerical model of the engine system and/or from data of prior tests.[10] If the monitoring system senses that a particular parameter is outside the normal expected range of values, it then analyzes the likely causes of the unusual operating condition and its likely consequences on the continued operation, and if serious enough the monitoring system commands an action to remedy the problem. The computer's decision of remedial actions can include a safe, but premature shutdown of the affected engine, continuing to operate under the present compromised condition, or operate at reduced thrust, provided the engine is designed to be throttled. The decision and action have to be done quickly before possible damage to the engine and the vehicle can occur.

Making a condition monitoring system work properly and quickly requires a good understanding of the engine's behavior under all likely operating conditions, a good mathematical simulation of the engine and its key parameters, and some prior test experience to validate these, before a logical selection of possible failure modes and remedies can be constructed. There has been good progress in simulating the engine performance and parameters.[10] Because there might be a malfunction of the sensor or data input, some control algorithms use and compare the input from several sensors before coming to the conclusion that the engine is misbehaving.

These engine condition monitoring systems have been very useful in saving expensive engine hardware in engine ground tests, by shutting the engine down quickly and safely when potentially serious malfunctions were sensed. This has happened during development programs and occasionally also during a qualification program, and the savings in hardware, cost, and time have been substantial. During a flight, the shutdown of the main engine of a particular stage is usually not an available option because it usually aborts the flight mission. Two other remedies have been options for a flying engine operating outside of its normal operating conditions. If the consequences are believed to be minor and will not disable the flight, one option is to continue to operate at the off-design condition, provided the engine will tolerate it. The other remedy is to continue to run, but at a reduced chamber pressure and lower thrust level, provided the engine can be throttled and the mission can be completed at the lower thrust level. Such a change in thrust causes a change in the flight acceleration and can only be allowed, if the vehicle's flight controller is able to make the necessary adjustment of the flight path.

Controls for Gas-Pressurized Feed Systems

As stated before, LPRE with gas-pressurized feed systems are simpler, have fewer sub-assemblies, and have no TPs or GGs. Their controls are also simpler than those needed for a pump-fed engine and are often mechanical and not electronic. The pressurization of the propellant tanks by an inert gas was initially controlled by maintaining a constant feed pressure (and thus essentially a constant chamber pressure) by using a pressure regulator.[1,3] Goddard designed and built several gas pressure regulators for his LPREs, and some are depicted in Ref. 2. This scheme has survived to this day as a simple and reliable means of pressurization control. The other common scheme, namely, the blowdown system, came about 20 years later. These engines do not have a high-pressure gas tank, and the pressurizing gas is stored directly in the propellant tanks, which are only filled part way with propellants. The gas is stored under pressure above the liquid level. This is a lower pressure than the pressure in a separate gas storage tank with a regulator. This system does not have an active regulated tank pressure control. It saves some inert mass (gas tank, regulator, valves), and it too is still used often. During operation, as the propellant is consumed, the gas in the propellant tank expands, and the gas pressure and temperature drop steadily. The TCs start at their highest thrust, and the thrust deceases steadily during the blowdown operation. The final thrust just before propellant exhaustion is reduced by a factor of two to five.

More sophisticated controls were needed for a warm gas pressurization system, when gas is obtained from a GG or a PB, as discussed in Chapter 4.5. Many of these systems use a warm gas pressure regulator (made of heat-resistant materials) or orifices for gas flow control. In another pressurization system, peculiar to a LPRE with a cryogenic propellant, a small flow of the propellant is evaporated and heated, and this small flow can be controlled by a pressure regulator. The control of the heat input to the cryogenic propellant becomes more difficult, if the LPRE is throttled. The pressurization alternatives described in this paragraph were explored in the 1940s and 1950s and have been used in flight vehicles since that time.

Engine-Out Control

When there are four or more main engines in a cluster propelling a particular large vehicle stage, then there is an opportunity to provide an *engine-out control option*, namely, to shut down one of the engines, where a condition monitoring system has detected a serious out-of-tolerance condition. The flight can then be completed on the remaining engines.[1] If the affected engine is an outboard engine, then the opposite outboard engine also has to be shut off to prevent sudden turning of the unbalanced vehicle. This engine-out scheme can only be considered, if the diminished total thrust will still go through the center of gravity of the vehicle, if the thrust is sufficient to fly and complete the intended mission, and if the controller is capable of controlling such an engine-out situation. This will change the flight path and require an automatic reprogramming of the guidance and vehicle flight control system and require the remaining engines to be able to operate for a longer period of time. Such a shutdown of an engine in a cluster happened during one flight of the Saturn I vehicle, when a J-2 engine malfunctioned. It was the center engine in a cluster of five J-2 LPREs, and the ascent was completed safely with the four remaining engines.

Valves

Many different types of valves are or have been used in LPREs, such as propellant valves, vent valves, various gas valves for warm and cold gases, pressure-regulating valves, pilot valves, check valves, control valves, throttle valves, fill valves, drain valves, safety valves, emergency shutoff valves, and others.[1-3] Of course, most LPREs do not have all of these valves. Early tests were often made with commercially available valves, but most of those did not do a good job, and they were too heavy. The investigators learned how to make better, lighter, and faster valves themselves. Valves and regulators developed by Goddard between 1923 and 1945 are depicted in Ref. 2, and some are mentioned in Chapter 5.2.

During the first 20 years, significant improvements were made to valves with better seals (both at the seat and the stem of the valves), longer life, lighter weight, faster actuation, lower pressure drop, valve position feedback, with or without a pressurizing fluid activated by a pilot valve, reusable or one-time-use only operation, improved ease of testing or servicing, better materials compatible with the propellants, or explosively activated. All variations seem to have been used, such as normally open, normally closed, three-way or four-way valves, fixed or variable pressure drop across the valve, high or low pressure, etc. The valves have been powered electrically (solenoid or electric motor), pneumatically by high-pressure gas, or hydraulically (by propellant or hydraulic oil). To this day, most LPRE organizations still develop many of their own valves. However beginning around 1950 valve manufacturers started to build lightweight valves and control components specifically for the LPRE industry. Companies, such as Moog and Perkin-Elmer, got into the business in the United States of supplying these special valves, particularly for small thrusters.

The variety of valves used with LPRE is very large.[1,3] The main oxidizer and fuel propellant valves ahead of each TC, for example, were coupled together mechanically (to ensure simultaneous operation) in many of the early LPREs, and a single actuator moved the valve stems with curved contour in both the fuel valve and the oxidizer valve. An example is shown in Fig. 7.7-10. It is the main propellant valve for the engine that propelled the first manned U.S. airplane. Today with better valve controls these two propellant valves are usually physically separated and actuated individually.

The simplest valve used in LPREs is a *burst diaphragm*. It is designed to burst at a predetermined burst pressure and to start the flow of propellant or pressurized gas. The diaphragm usually is a disk with small grooves (often in the shape of a cross) on its surface; under pressure it breaks at these grooves, and the four segments of the broken diaphragm pieces peel back to admit the flow of propellant. It provides a positive seal, can be opened by pressure or explosives, cannot be closed, and can only be used once. As far as it is known, burst diaphragms were first used by the Germans in the LPREs for some of their antiaircraft experimental missiles during the early 1940s. A diagram of the Wasserfall missile with five diaphragms is shown on Fig. 9-11 and an

Technology and Hardware   193

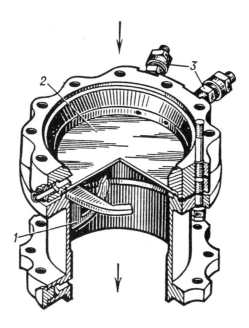

Fig. 4.7-5 Explosively actuated burst diaphragm for admitting nitric acid to the TC of the RD-216 LPRE; 1, finger for retaining the burst or sectioned diaphragm; 2, thin metal diaphragm; and 3, explosive charges. Courtesy of NPO Energomash (from personal communications, J. Morehart, Aerospace Corporation, 1990).

explosively actuated burst diaphragm in Fig. 4.7-5.* This metal diaphragm of the Soviet Energomash RD-219 has provisions to catch the folded-back diaphragm after it has burst open and prevents a large piece of the diaphragm from flowing downstream.

It is not possible to list or describe here all of the interesting valve concepts used on LPREs. So only a few examples of a few valves for early large LPREs will be shown here. For example, the high-pressure LOX valve for the space shuttle main engine in Fig. 4.7-6 was designed in 1971.[3] This ball valve operates at high pressure and has a low-pressure drop at full flow. The seal is a machined plastic ring, spring loaded by a bellows against the inlet side of the rotating ball. Two cams of the shaft lift the seal a short distance off the ball within the first few degrees of ball rotation. The ball is supported by two bearings and is rotated by a hydraulic actuator (not shown), which is connected through a thermal insulating coupling. The butterfly valve of Fig. 4.7-7 was originally developed by Rocketdyne in the early 1950s for the Thor IRBM engine and licensed to Rolls-Royce in England for their RZ.2 engine. This drawing was made by Rolls Royce. The valve has a pneumatic actuator, which turns a rotating shaft through a crank

---

*Personal communications, J. Morehart, The Aerospace Corporation, 2002.

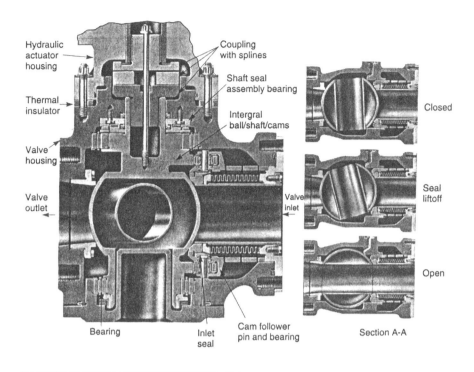

Fig. 4.7-6 High-pressure (over 3500 psi) oxygen main valve of the space shuttle main engine. Courtesy of the Boeing Company, Rocketdyne Propulsion and Power; copied from Ref. 3.

or rocker arm. This kind of butterfly valve was used often in the 1950s and 1960s because it does not take up much length in the high-pressure feed line. For cryogenic propellants (LOX) an electric heater prevents freezing of moisture at the joints of moving parts. The next three valves were all designed in the Soviet Union during the 1950s. The explosion-activated valve of the RD-216 LPRE Fig. 4.7-8 has a tapered pintle, which is driven into a conical seat. It has a burst diaphragm to relieve the pressure surge caused by the sudden valve closing, and it can only be operated once. In the unique liquid propellant throttle valve of Fig. 4.7-9, the flow is through slots of variable width in a shallow cone. It does not seem to have provisions for a positive seal. This valve was used in the Energomash RD-111 engine and was developed in the late 1950s. The liquid propellant pressure-reducing valve of Fig. 4.7-10 also belongs to the RD-111. These USSR engines are included in Chapter 8.4.

In the 1960s the valves for small pulsed thrusters were usually designed and built by the LPRE company because they had to be integrated with the injector and located close to the injector face. A closely coupled valve built into the injector minimizes the dribble volume of propellant between the valve seat and the injection holes or spray elements, so as to allow short thrust pulses and quick (milliseconds) restarts. The electromagnets used for repeated valve

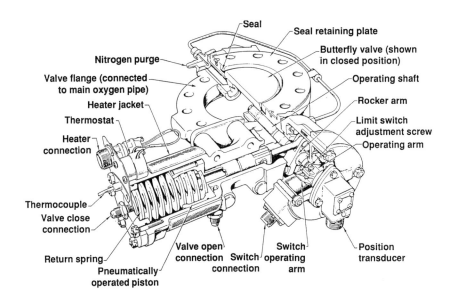

Fig. 4.7-7 Butterfly-type valve for LOX used in the early U.S. Thor LPRE and also in the British Blue Streak LPRE. Courtesy of the Boeing Company, Rocketdyne Propulsion and Power; copied with permission from third edition of Ref. 3.

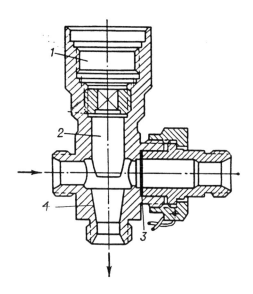

Fig. 4.7-8 Explosive shutoff valve of the nitric-acid flow to the gas generator of the RD-216 LPRE: 1, chamber for explosive and squib; 2, valve pintle; 3, free (unsupported) burst diaphragm for draining the liquid to minimize the hydraulic shock; and 4, tapered valve seat in housing. Courtesy of NPO Energomash (from personal communications, J. Morehart, Aerospace Corporation, 1990).

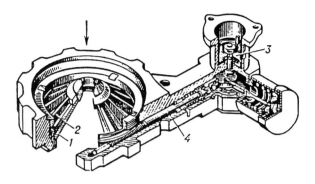

Fig. 4.7-9 Throttle valve of the RD-111 LPRE for the flow from the kerosene tank has a rotating conical set of slots: 1, cone-shaped movable grating with gear teeth; 2, fixed grating, part of housing; 3, half-coupling connector to electric actuator: and 4, threaded control rod for rotating the movable grating. Courtesy of NPO Energomash (from personal communications, J. Morehart, Aerospace Corporation, 1990).

actuation had to be designed for minimal heat absorption (to keep the insulation from decomposing), low electric power, fast actuation, light weight, and small size. Some of the early propellant valves were placed some distance away from its small thruster. Hamilton Standard's hydrazine monopropellant catalyst bed with a nozzle was physically separated from its electromagnetic valve, and the two pieces were connected only by a small propellant tube. This separation reduced the heat conduction from the hot catalyst bed to the electromagnet, which had temperature-sensitive insulation. This separation is shown for a hydrogen-peroxide monopropellant thruster in Fig. 4.6-6 and for a hydrazine monopropellant thruster in Fig. 4.6-8.

Figure 4.7-11 shows two electromagnets used in the valve designs of small thrusters developed by NII Mash (Russia, see Chapter 8.7).* The seat on the left valve is a soft Teflon®-type insert, which deflects under the spring load. The valve on the right uses a Bellville-type hollow spring. Both of these valves have been extensively tested with NTO and also with UDMH over many thousands of operating cycles. All of the electrical actuators had to be kept cool enough to prevent damage to the electrical insulation. In the 1960s a typical opening time for a propellant valve of a small thruster was about 30 ms. In the ensuing 25 years this was reduced to between 2 and 4 ms.

Interconnect Components

Interconnect components are tubes, ducts, pipes, and flexible joints for propellants and for warm gas.[1] It includes rotary hinge joints for hinged TC and flexible

---

*Personal communications, Y. G. Larin, Chief designer, NII Mash (R & D Institute of Mechanical Engineering), Nizhnaya Salda, Russia, 2002.

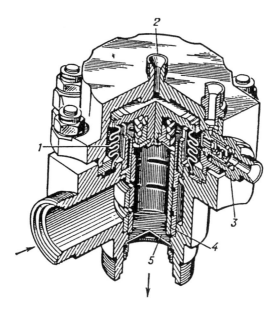

Fig. 4.7-10 Pressure-reducing valve for the LOX flow to the gas generator of the RD-111 LPRE with burst diaphragm: 1, bellows for separating propellant from the air; 2, inlet for control air feed; 3, relief or blowoff mechanism; 4, sliding cylinder valve sleeve; and 5, burst diaphragm allowing free flow to GG. Courtesy of NPO Energomash (from personal communications, J. Morehart, Aerospace Corporation, 1990).

bellows for transferring high-pressure propellants to gimbal-mounted TCs. These components are not as glamorous as TCs or GGs, but they are very essential to the proper functioning of a LPRE. They also include flexible wire-braided hoses, pipe fittings, flanges, plumbing connections, and small tubing for drains from TP seals, sensors, or measuring instruments. These interconnect components must be designed for thermal expansion or shrinkage, high and low internal pressures, high acceleration, cryogenic fluids, or warm gas. The early engine used standard joints, such as tube fittings, flanges, pipe-threaded tees, or threaded pipe joints. Since the 1960s, an all-welded construction (without flanges or fittings) has been used for joining components and engine parts of LPREs with high internal pressures. This welded construction makes it impossible to inspect certain inner parts. Flexible joints connect a moving hinged or gimbal-mounted TC with the nonmoving parts of the engine or the vehicle. They have been used since about 1945. Flexible pipes are also needed for thermal expansion/shrinkage of piping. Flexible joints with bellows usually have a structure (internally or externally) to prevent the bellows from stretching. An example of a flexible joint with an internal bellows and external universal-joint-type structure is shown in Fig. 4.7-12, it was designed around 1970.

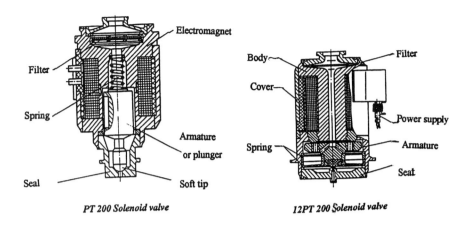

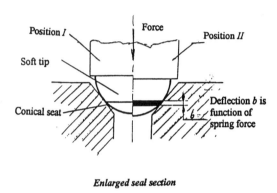

Fig. 4.7-11 Solenoid valves of two small thrusters developed in Russia. Courtesy of NII Mash (from personal communications, Y. G. Larin, Chief designer, NII Mash, Nizhnaya Salda, Russia, 2002).

Historically, interconnect components, most of which really are pressure vessels, have generally become heavier because the chamber pressure and the engine internal pressures have increased between 1930 and 1975. This tendency has been ameliorated by using better and stronger wall materials, which have become available. A good discussion of interconnect components can be found in Chapter 9 of Ref. 1.

Internal Cleaning of Components

Explosions of "unclean" valves, interconnect components, pumps, or tanks have happened, when the oxidizer, such as LOX, NTO, or hydrogen peroxide first enters. All such parts have to be thoroughly cleaned to eliminate residues or mere traces of grease, lubricating oils, organic shavings, preservation fluids,

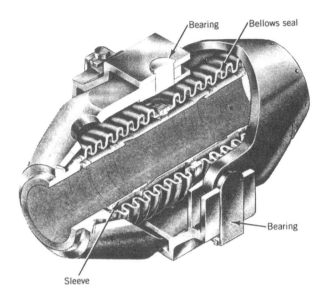

Fig. 4.7-12 Typical flexible joint with external gimbal rings for high-pressure hot turbine exhaust gas. Courtesy of the Boeing Company, Rocketdyne Propulsion and Power; copied with permission from Ref. 3.

fibers from clothes, cigarette ashes, or small pieces of wood or paper. A potential chemical reaction with the remains of an organic material has to be avoided. The early investigators learned quickly about having a clean oxidizer flow system and not having even a trace of contamination. It took a few years before an adequate procedure was worked out for routinely manually removing any organic material, cleaning and passivating of internal surfaces (often with an acid), flushing out with clean water, and inspecting properly. Such a procedure has been rigidly enforced in the fabrication, assembly, and installation of LPRE components since the 1940s.

There are many tales about unclean parts causing fires, problems, or even explosions. Wernher von Braun told the author the story of an oxidizer valve for an early version of the V-2 LPRE having unexplained internal fires and explosions. The valve parts, the valve, and all of the connecting piping had been thoroughly cleaned. The valve sleeve seal was a ring of square cross section made of specially treated and dried leather, and it was compatible with the LOX. O-ring seals were not known at that time. After extensive investigations the problem was traced to a particular assembly worker, who had always assembled this valve. This man had a private little can of special unapproved grease, which he applied to the leather, so that it would be easier to assemble. Even a small amount of this grease reacted with the LOX, particularly when there was a sudden rise in pressure. In the late 1930s and the early 1940s special greases were developed in Germany, which were more resistant to oxidation and were approved as lubricants for specific oxidizer valves and for application to dynamic and static seals. Similar greases/lubricants were developed in other countries, but at a later date.

References

[1]Huzel, D. K., and Huang, D. H., *Modern Engineering Design of Liquid Propellant Rocket Engines*, Progress in Astronautics and Aeronautics, Vol. 147, AIAA, Washington, D. C., 1992, Chaps. 7 and 9, pp. 219–283.

[2]Goddard, E. C., and Pendray, G. E., *The Papers of Robert H. Goddard*, Vol. 1, 2, and 3, McGraw–Hill, New York, 1970.

[3]Sutton, G. P., and Biblarz, O., *Rocket Propulsion Elements*, 7th ed., Wiley, Hoboken, NJ, 2001, Secs. 6.9 and 10.5, pp. 296–411.

[4]Petrov, B. N., Portnov-Sokolov, Yu. P., and Andrienko, A. Ju., "Control Aspects of Efficient Rocket Propulsion Systems," *Acta Astronautica*, Vol. 4, 1977, pp. 1127–1136.

[5]Zhu, S., "An Experimental Study of the Cavitating Venturi of Liquid Hydrogen," *Proceedings*, of the 16th International Symposium of Space Technology and Sciences, Vol. 1, AGNE Publishing, Tokyo, 1988, pp. 827–829.

[6]Mattox, R. M., and White, J. B., "Space Shuttle Main Engine Controller," *NASA Technical Paper 1932*, 1981.

[7]Demarais, J-C., and Deam, A., "An Ultra-Sonic Level Meter for the Measurement of Propellant Levels in Ariane First and Second Stage Tanks," *La Recherche Aerospatiale*, Jan.–Feb. 1980, pp. 9–22 (in French).

[8]Ray, A., Dai, X., Wu, M-K., Carpino, M., and Lorenzo, C. F., "Damage-Mitigating Control of a Reusable Rocket Engine," *Journal of Propulsion and Power*, Vol. 10, No. 2, March–April 1994, pp. 225–234.

[9]Zhang, H., Wu, J., Huang, M., Zhu, H., and Chen, Q., "Liquid Propellant Rocket Engine Health Monitoring Techniques," *Journal of Propulsion and Power*, Vol. 14, No. 5, Sept.–Oct. 1988, pp. 651–663.

[10]Bury, S., Dubedout, A., and De Boisherand, H., "Numerical and Analogue Models in Developing the Vulcain Engine," *Proceedings* of the 15th International Symposium of Space Technology and Sciences, Vol. I, AGNE Publishing, Tokyo, 1986, pp. 321–332.

4.8 Starting and Ignition

All new LPRE development groups had to learn the technique for starting the combustion or the firing operation of a TC (and also of a gas generator or preburner) and for the buildup of flow or thrust up to rated conditions. It has been common for most of the early LPREs and many new engines to experience starting problems, which can include explosions and misstarts. The starting procedure is different for different feed systems, different thrust levels, cryogenic vs storable propellants, hypergolic or nonhypergolic propellants, and for a single-start LPRE vs restartable or reusable LPREs. There are probably no two families of large LPREs with exactly the same start sequence. A safe start is critical, and a quick safe start can sometimes be rather tricky to achieve.

Starting a LPRE with a Pressurized-Gas Feed System

In every country the first rocket engines used a pressurized feed system. Starting such a system is influenced by several parameters, and all had at some time been investigated. The aim is to achieve a safe, predictable, and reliable transition to full thrust. The starting process depends on the particular propellants and their properties, the magnitude of the thrust, the number and location of thrust chambers, the particular pressurized feed system, the length of the feed lines, the timing of the two propellants to reach the combustion chamber, or the specific injector design. The basic starting process for LPREs with a gas-pressurized feed system has really not changed much in the last 80 years. Over the years various safety features were added, and the components (valves, regulators, controls, igniters, etc) have been improved. Starting reliability for qualified engines has been at 99.99% for several decades.

When the propellant valves are opened, there is already the full regulated gas pressure in the propellant tanks (which is 20 to 50% higher than the chamber pressure), but the initial pressure in the chamber is the ambient atmospheric pressure, which is very low. This high pressure differential causes the initial propellant flows to be higher than the rated flows. With the sudden high flow some pressure surges can occur in the liquid lines. If one of the propellants at high flow reaches the chamber well before the other, it might form a propellant puddle in the chamber, which can cause problems when mixed with the other propellant or when a restart is performed. The thrust buildup must avoid transient regions of chamber pressure and mixture ratio, where combustion instability is likely to occur. Gas bubbles in the initial propellant supply have in the past given starting problems. They can cause sudden changes in mixture ratio, which can change the chamber pressure, thrust, or combustion temperature or can initiate unstable combustion. Therefore, beginning in the late 1930s a procedure for eliminating trapped gas in the propellant flow system up to the propellant valves has been instituted for several LPREs by using venting of trapped gas and bleeding prior to start, sometimes in more than one location of the propellant flow system.

The start technique must also avoid an initial accumulation of unburned or unignited propellant in the chamber because this can cause an initial momentary overpressure or "hard start" (really an explosion), which can damage and sometimes destroy the chamber and also disrupt the propellant flow.[1,*] This is usually not an issue in small thrusters, because the amount of accumulated propellant in a small chamber cannot be very large.

In injectors that are typically larger than 2 or 3 in. diam or that have more than two or three injection elements, it is essentially impossible to have each propellant reach each injection orifice at exactly the same time. In regions where the oxidizer would be injected first, a local oxidizer-rich mixture would burn, and in locations where the fuel preceded the oxidizer an initial fuel-rich mixture would prevail. Some of these mixtures or some regions in the combustion chamber would temporarily have a combustion temperature higher than the burning temperature at the rated mixture ratio and chamber pressure, and this would cause temporary local overheating problems of the chamber or nozzle wall. When starting gas generators, this extra temperature spike has damaged turbine blades. The idea of a definite lead of one of the propellants, say, the fuel, would ensure a fuel-rich mixture at all injection element locations, once some of the oxidizer got to each of these injection elements.[2] This idea of a propellant lead was understood by Goddard and other early investigators.[3] A small intentional lead is therefore often included into the start of TCs. This propellant lead concept was not trouble free because some of the first propellant would also flow backwards through the injection holes of the second propellant into the distribution manifolds of the injector; so when the second propellant arrived in its manifold, it formed combustible or explosive mixtures inside the manifold.[†] Plugging of the injection holes was tried as a remedy, but it was abandoned. So in the 1940s a purge of inert gas was introduced in some large LPREs to precede the entry of the second propellant and prevent the other propellant from entering the manifold. Such purges have been used in several large LPREs. This development of a good starting technique used to require a lot of testing and good instrumentation. Today engine developers rely on earlier successful starting procedure with the same propellants and a similar TC and on transient analyses, and these have reduced the amount of testing required.

There have been several ways for reducing pressure surges and the potential initial accumulation of unburnt propellants in the chamber. Many large LPREs have an initial low (reduced) flow of propellants in the start procedure.[†] This is done in the German V-2 LPRE by adding bypass propellant valves that provide a period of initial low flow of propellants. Alternatively the opening rate of the main propellant valves can be slowed down, which really is a throttling of the initial flow. This reduces the initial flow of propellant, the surge

---

*Personal communications, F. Boorady, ARC, Liquid Operation, Niagara Falls, NY.
†Personal communications, V. Wheelock, E. Shuster, 2002–2004, M. Ek, 1970, The Boeing Company, Rocketdyne Propulsion and Power.

pressures in the piping, and also reduces the amount of possible accumulation of unburnt propellants in the chamber, but delays the time to reach full flow and full power.[4,5] A simple approach was to include a variable valve opening area in slow opening valves, as is shown in Fig. 4.8-1. In the United States this was first used by Navy researchers at the Naval Research Station in Annapolis in the early 1940s. Another approach was to design the volume of the feed pipes so that the two propellants arrive at the desired time interval. For example, a small extra volume was added to one of the propellant feed lines of the Rocketdyne lunar ascent engine in order to increase the time needed to fill that feed line and have both propellants reach the chamber at essentially the same time. The use of orifices or cavitating venturis in the feed lines can limit the maximum flow to the injector and chamber. This feature was used, for example in the Apollo lunar landing engine, and this engine's flow diagram is shown in Fig. 7.9-7, where the flow control valves have a built-in venturi with a variable throat area.

The startup procedure is the simplest for small thrusters with a pressurized-gas feed system, hypergolic storable propellants, and valves placed on top of the injector or close to the injector. The feed system would in concept be similar to the one in Fig. 4.2-9. For pressurized feed systems with a gas pressure regulator, the starting procedure is initiated by opening a valve or explosive diaphragm to allow gas flow through the regulator. The gas at regulated pressure then flows

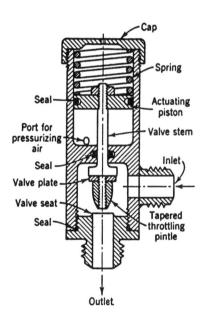

Fig. 4.8-1 This old sketch shows the concept of the tapered valve shank on the valve stem, which provided some throttling during a slow valve opening.

into the propellant tanks. The propellant valves at selected thrusters are then opened, admitting propellant flow to their combustion chambers, where hypergolic ignition starts the burning. The magnitude of the propellant flow is determined by the flow resistance (pressure drops through the pipe lines, valves, cooling jacket, and injector) and adjusted (usually by an orifice) to the desired flow rate.[4,5] When the propellant valves are commanded to close, the combustion stops very quickly because there is very little dribble volume, which is the volume between the propellant valve and the injector face in small thrusters.

The starting procedures for the larger LPREs with pressurized-gas feed systems have some additional features or complexities. For large TCs the flows are much larger, and the start surges are larger and more difficult to control because the tanks are bigger and taller and the pipe lines are bigger and longer. It takes a longer time for propellants to reach all of the injection elements of a larger injector. Usually there is an intentional, definite, but short time lead of one of the propellants. There can be an initial low flow intermediate phase in the start procedure. For nonhypergolic propellants an igniter is needed, and this requires some additional steps in the starting sequence, additional hardware, and usually additional electric circuitry and sensors to confirm that ignition has occurred. If the igniter uses solid propellants, the system is less complex than if it uses liquid propellants, which require additional valves, spark plugs or other igniters, and more calibrations and controls. If there should be a restart, then some purging or removing of the trapped propellants (downstream of the propellant valves) remaining in the engine might be needed. Again this requires additional steps in the shutdown procedure, more valves, and electric circuitry.

Starting a LPRE with a Turbopump Feed System

For a LPRE with a TP, there are some additional steps in the starting procedure.[4,5] Of course, an engine with a gas-generator cycle or a staged-combustion cycle has an additional combustion device that also needs to be started (it usually has its own valves, igniter, and controls), and enough energy has to be supplied to the turbine so that the TP is brought up to full speed. During a normal start of a LPRE with a TP feed system, the initial propellant flow can be low and can be gradually increased, which is contrary to the large initial propellant surge at the start of a LPRE with a pressure feed system.

As the pumps increase their rotational speed, the flow will go from a very low flow to full-rated flow, and from near zero to the rated pump pressure rise; they go through off-design conditions that are not necessarily efficient, and sometimes the pump can experience a stall. A *pump stall* occurs when the flow breaks away from the pump impeller blade at high angles of incidence and when cavitation also can be experienced locally. A stall is accompanied by energy losses and changes in the flow or discharge pressure.[5] A stall is more likely to happen with inducer impellers and axial-flow pumps, than with centrifugal pump impellers. These stall conditions must be avoided during the start transient. They can be momentary stalls during the start transient or of longer duration during operation. In the first flight of a new German Walter 109–509, the first model of a series of aircraft rocket engines, a pump stall with cavitation, caused a significant reduction in thrust, and the intended high flight speed was not attained. This incident and the remedy to fix the stall problem are given in Chapter 9.3.

There are three common ways for starting a LPRE with a TP; a "tank head" start and two types of "spin-up" start.[5] This tank head is the elevation of the liquid level above the pump inlet plus the head equivalent of the gas pressure in the propellant tank. It is a pressure, and it is expressed in meters or feet as the height of an equivalent column of liquid. With the low tank head as the initial pressure, the gas generator or preburner will have a low initial flow of warm gas, and it will take several seconds for the warm gas to spin the turbine and boot strap it up to full speed. This type of tank head start is used on the Rocketdyne F-1 engine shown in Fig. 7.8-10 and the MA-3 Atlas engine shown in Fig. 7.8-5 and the SSME shown in Figs. 7.8-17 and 7.8-18. The starting procedure for the SSME using a tank head start is described on page 401 of the seventh edition of Ref. 4. Although a tank head start can last 3 to 5 s for a large LPRE, the amount of propellant used is relatively small because most of this start period is at very low flow.

The concerns expressed in the preceding section about starting a LPRE with a pressurized-gas feed system also apply to a TP feed system, such as the lead of one propellant reaching the chamber ahead of the other propellant. There also is the concern about some of the lead propellant getting into the injector manifold of the other propellant and forming an explosive mixture inside this manifold. Purges have been one remedy. An alternate approach was used by the Germans in the TC of the V-2 in the 1930s. Paper cups were placed over the oxidizer injection elements to keep fuel from entering the oxygen holes.*

All of the considerations discussed here have caused considerable complexity in the engine system. Goddard's first pump-fed engine, with three propellants and provisions for future throttling, had more than 20 valves of various types.[3] The German V-2 engine with two thrust levels of operation and a separate propellant for the GG had about 34 valves as shown in Fig. 9-3. Clever designs have simplified the engine and the start system. For example, in the Aerojet Titan LPREs there are no separate valves for supplying propellant to the gas generator. It uses simple fixed venturis to control the GG propellant flows and mixture ratio and a check valve to prevent backflow. The Rocketdyne H-1 LPRE, whose start sequence is described later in this chapter, was relatively simple and had about 10 valves including the safety valves on the oxidizer tank.

Some form of external power can be applied temporarily during the start to accelerate the spin up of the TP to full power. This is the other way of starting a pump-fed LPRE. For large LPREs this will shorten the time for the start transient

---

*Personal communications, Walter Riedel, formerly German Army Experimental Station, Peenemünde, Germany, 1949.

206 History of Liquid Propellant Rocket Engines

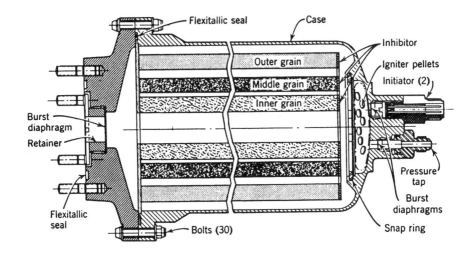

Fig. 4.8-2 Sealed solid-propellant start cartridge for gas generator of the H-1 LPREs used in the booster of the Saturn I SLV. Courtesy of The Boeing Company, Rocketdyne Propulsion and Power; copied with permission from Ref. 4.

by a factor of two or three. The earliest form of this external power was compressed air from an external source (such as a ground-based compressor or high-pressure storage tank), and this air was put into the turbine inlet manifold. Goddard used compressed air to start the tests for his first TPs (1938), and other LPREs have used these in-ground tests as a starting method in the 1940s and sporadically thereafter. The Rocketdyne J-2 LOX/LH$_2$ engine, described in Chapter 7.8, carries a tank of high-pressure hydrogen gas for starting and restarting the rotation of its two turbines. This spherical tank can be seen as a part of the engine in Fig. 7.8-13 and 7.8-14.

The most common form of power boost has been a *solid-propellant cartridge*, which not only supplies the initial power in the form of warm gas to the turbine, but also ignites the nonhypergolic propellants in the GG or the preburner.[4,5] An example of such a solid-propellant cartridge used in the Rocketdyne H-1 LPRE is shown in Fig. 4.8-2. It has a dual redundant set of squibs or initiators, a secondary charge of solid-propellant pellets, and multiple concentric cylindrical grains of double-base solid propellant.

The start sequence of the Rocketdyne H-1 engine is shown in Fig. 4.8-3 and described here in simplified from and taken from Ref.4.

1) *Prior to operation*, these steps occur:

   a) Inspect engine.

   b) Fill propellant tanks and gas tank in vehicle; validate or adjust regulated pressure; admit low pressure gas to fuel tank.

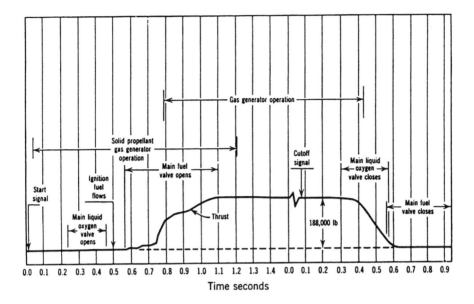

Fig. 4.8-3 Thrust–time relationship during start and shutdown of one version of the H-1 engine. Courtesy of the Boeing Company, Rocketdyne Propulsion and Power; copied with publisher's permission from Ref. 4.

c) Close tank vent valves on tanks; pressurize fuel tank with low-pressure gas; LOX tank pressurizes itself by evaporation.

d) Chill down oxygen pump and hardware by sporadic bleeding of LOX through passages into TC.

2) At the *start*, these steps occur:

e) Electric start signal ignites the solid-propellant spinner cartridge creating hot gas in the GG; the turbine starts to spin, and the pumps accelerate.

f) Pressure from the fuel pump discharge automatically opens the main propellant valves in the proper sequence and at the proper opening rates by means of the sequence valve.

g) Liquid oxygen together with a hypergolic start fuel (stored in small tank, confined by burst diaphragms, pushed out by fuel pressure) ignites in the combustion chamber. The fuel, which enters the chamber automatically a very short time thereafter, then burns with the LOX, and thrust is built up quickly.

h) As the pump discharge pressures rise, the LOX and fuel reach the gas generator, where they are ignited by the burning solid propellant. The resulting warm gas continues to supply power to the turbine.

i) A small flow of LOX is bled from the pump discharge, evaporated in a heat exchanger, and is used to pressurize the oxygen tank at a low pressure.

j) A small flow of fuel is blended with a lubrication additive and fed to the gear box, where it lubricates and cools the turbopump gears and bearings.

3) At *shutdown* these steps occur:

k) An electrical signal activated the close control of the main LOX valve causing it to fill the piston actuator of the main LOX valve.

l) Main LOX valve closes, causing it to starve the GG of oxygen, and therefore the GG stops to create gas. The pumps then slow down, and pump discharge pressures decrease. The decay of fuel pressure causes the main fuel valve to close.

The reader should refer to the schematic flow diagram of this engine in Fig. 4.2-7 to identify the components mentioned in the start procedure. This engine uses LOX/kerosene as propellants and reaches full thrust (205,000 lbf) in about 1 s.

The start sequence of the SSME engine, with a tank-head start, but a staged combustion cycle, is more complex and is given on page 401 in the seventh edition of Ref. 4. The start sequences for the Japanese LE-5 and LE-7 LPREs are shown in Fig. 11-5 and 11-10, and these are fairly complex engines operating with LOX/LH$_2$.

Another source of temporary power is a separate liquid propellant GG with its own small propellant tanks and a separate pressurized-gas expulsion system. This has been the third method for starting a pump-fed LPRE. A simplified schematic drawing of this type of engine is shown in Fig. 4.8-4 with a bipropellant GG. The first implementation was actually with the engine of the German V-2 missile using a pressure-fed monopropellant gas generator, where pressure-fed hydrogen peroxide was decomposed by a catalyst. This gas generator can be seen in Fig. 4.5-1 and in the engine flow sheet of Fig. 9-3. This missile was tested in about 1940, and it was flown in about 1942. In the United States this scheme was first flown in 1956 in the Rocketdyne LPRE for the Navaho missile, which carried onboard a set of small bipropellant start tanks. A variation of this method is in the Jupiter LPRE, where the two propellant start tanks and a high-pressure gas tank were on the launching platform on the ground and not on the engine or in the flight vehicle. The ground start system supplied propellants to the GG (through disconnect fittings and check valves) until the pump discharge pressures exceeded the start tank feed pressure. This ground-based start system reduced the inert hardware mass of the engine and reduced the amount of propellant onboard. Both of these engines can be found in Chapter 7.8. In LPREs with GG engine cycles, the nominal flow to the turbine is relatively small (2–5% of total propellant flow), but the pressure drop across the turbine blades is high (typically 300–1200 psi). The initial power boost during start by either a solid-propellant cartridge

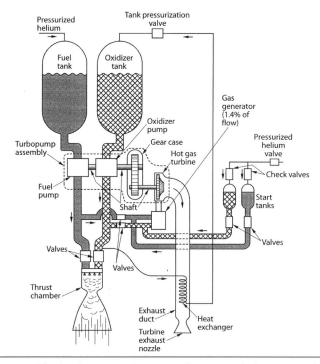

Fig. 4.8-4 Simplified schematic diagram of a pump-fed LPRE with a gas-pressurized liquid propellant start system for the initial supply of bipropellants to the gas generator. Copied with permission from seventh edition of Ref. 4.

or a set of liquid propellant gas generators with start tanks has been effective in shortening the engine start sequence. In LPREs with an expander cycle or a staged combustion cycle, the mass flow through the turbine is relatively very large (most of the fuel), and the pressure drop through the turbine is relatively low, typically 20 to 200 psi.

When a cryogenic propellant first enters an engine, it is partly on fully gasified by absorbing heat energy from the engine hardware, namely, the pipes, pumps, cooling jacket, and valves. This initial gas/liquid mixture contains bubbles or pockets of evaporated gas (or maybe all gas) and has a lower density than the cryogenic liquid. With the lower density the pump speed can rise, and there have been cases where the pump speed exceeded its safe maximum value, and this has caused pump overspeeding and pump failure. It is difficult to predict the reduction in the mass flow rate and the changes in the mixture ratio. It is possible to find a period during the start sequence where there is a short temporary oxidizer-rich mixture ratio, which has a combustion temperature higher than the rated value, and this can cause problems, such as damage to turbine blades or a burn-through of the inner wall of the chamber or nozzle. Therefore, it has been customary to bleed some quantity of the cryogenic propellant through the engine, in order to cool or chill down the hardware to cryogenic temperatures.[6,*] This helps in getting liquid propellant to the combustion chamber and not gasified propellant. This bleed procedure was first developed with cryogenic LOX by the Germans in the engine of the V-2 guided missile in the late 1930s. It has been used with cryogenic propellant ever since that time.

In the Soviet Union some of the LPRE design bureaus had turbopumps with two turbines, mostly with engines that ran on a staged combustion cycle. One turbine is designed with relatively thin vanes or turbine/stator blades, which are aerodynamically efficient, and they work throughout the engine operation. The other is a *start turbine*, and it is mounted on the same shaft. It is driven by the warm gas from a solid-propellant grain, but only for the first few seconds of operation. This provides a rapid spin up of the turbopump assembly and a fast engine start. This turbopump with an additional start turbine is discussed at the end of Chapter 4.4 and also in Chapter 8.10. After the solid-propellant grain has been consumed in a few seconds, this start turbine does not provide any further power, but continues to rotate. Such a startup turbine can be seen in a TP developed by the Kuznetsov Experimental Design Bureau in Fig. 8.9-3. This author is not aware of any Soviet or Russian TP with an integrated start turbine developed since about 1980.

If a particular engine in a cluster of engines fails to ignite, although the adjacent engine(s) starts normally, then an explosion of the premixed unignited propellants is likely to occur. Pratt & Whitney experienced such a destructive failure during the engine development of the Centaur, which is propelled by two RL10 LPREs. During a ground test, one of the two adjacent engines failed to ignite, while the other started properly and produced a hot radiating exhaust plume. The explosion of the unignited propellants was triggered by the hot flame and demolished the engine. This accident is mentioned again in Chapter 7.10. To avoid this problem, all engines in a cluster have instruments that sense the operation of an igniter and/or the initial combustion, and if ignition should fail to occur in one of the engines then all of the engines in a booster vehicle would be shut down. This has saved some hardware in the ground testing of clusters.

The *starting of large LPRE clusters* (three to 30 engines in the same stage) usually proceeds in a special sequence and has some unusual requirements. Because of structural problems of the vehicle and the hydraulic behavior (water hammer) of long propellant feed lines, these engines are usually never started simultaneously.[†] In the cluster of five F-1 LPRE of the Saturn V SLV, the central engine 5 is started first, then a pair of outboard engines 1 and 3 are started about 0.2 s later, and the remaining outboard engines 2 and 4 are started 0.6 s later than 1 and 3. Also there is a time interval of 0.120 s between the starting of the three main engines in the Space Shuttle Orbiter.[†]

---

*Personal communications, W. Riedel, formerly German Army Experimental Station, Peenemünde, Germany, 1949

†Personal communications, V. Wheelock, 2002–2004, The Boeing Company, Rocketdyne Propulsion and Power.

Ignition

Nonhypergolic (not self-igniting) propellants require a source of energy to ignite. Igniters are needed not only for the TC(s), but also for the gas generator or preburner. The igniter flame must impinge with and ignite the initial flow of propellants before an accumulation of unburned propellant can occur in the chamber.[4,5] For large diameter injectors with baffles, an ignition source must be provided for every compartment of the injector. A variety of different igniters have been developed and used in flying LPREs.[4] Four common types of ignition are given in Table 4.8-1.

The very early TCs used wadded rags or cotton fibers in the chamber (or located just outside of the nozzle), and they were manually ignited, typically with a small blowtorch. In some early TCs the burning rags ignited a series of match heads placed in an antichamber, and one can be seen in the TC of Goddards first flight vehicle in Fig. 5-8. In some of these amateur and Goddard tests, a brave individual would walk to the engine (whose tanks had already been pressurized) and hand carry a small blowtorch or matches to set fire to this wadded cloth starter material; then this person quickly ran to a shelter before the match heads ignited and before propellant valves were opened.

Since the 1930s, this ignition energy came often from a solid-propellant cartridge. Goddard, the American rocket pioneer, used two kinds of what he called "flare-type" igniters.[3] He already had experience with igniters for solid-propellant motors and made his own powders from ingredients or used available solid propellant. One of his technicians would assemble a squib and the

Table 4.8-1 Common Types of Ignition Mechanisms

Type	Comments	No. of starts
1) Gasoline-soaked cotton cloth or rags burning with air	Used by Goddard and amateurs in 1920s; flown in 1926; now abandoned	Single
2) Pyrotechnic or solid propellant	Goddard used in 1920s and 1930s with TCs; used with GG beginning about 1950	Single
Same but with rotating wheel and four solid cartridges	TC of German V–2 LPRE	Single
3) Electric power, spark plug, hot-wire glow plug,	Ignition prechamber in large LPREs, with small flow of both propellants; LOX/RP-1, or LOX/LH$_2$; used in space shuttle main engine	Multiple
4) Initial slug of hypergolic fuel	Aluminum triethyl will ignite with LOX or air, used in F-1, RD-170, H-1	Multiple

solid-propellant cartridge, usually in a cardboard tube. One type was inserted through the nozzle and held there by a string fastened to the forward end or the injector face of the chamber. He experienced ignition failures when the string burned and the igniter fell out of the chamber, which was mounted vertically. Thereafter he used a metal wire device to hold the igniter within the chamber. The other type had a similar solid-propellant cartridge, but in a metal chamber, and it screwed into the top of the chamber; he first tried this in the 1920s. He knew that a small primer charge of sensitive solid propellant would do a good job of igniting the main solid-propellant charge of the igniter.

A typical early U.S. pyrotechnic igniter used in the early 1940s for TCs of 300 to 1000 lbf thrust is shown in Fig. 4.8-5. It was attached to a central hole in the injector, burned for about 3 or 4 s, had dual squibs, and was tricky to assemble. This igniter was not the best design. Squibs, which are capsules filled with sensitive primer propellant and an electric heat source, such as a heated wire or an exploding wire, initiate the combustion, and the resulting flame ignites a solid-propellant charge. Some of the squibs have a metallic enclosure or cover, which is broken up by the ignition and ejected from the cartridge. In some tests the metal squib covers or pieces of wire are ejected by the combustion and have caused damage to turbine blades or to the inner TC wall, particularly in the nozzle throat region. A modern solid-propellant igniter for a gas generator is shown in Fig. 4.8-6. It has its own electrical conditioning and safety components (which avoids induced currents from external wires) and redundant dual-ignition squibs, which have a plastic enclosure. The wires are bonded to the wall of the igniter housing. About 10 to 20 years later the use of hot wires and exploding wires (by discharging a capacitor) became a common method for igniting a squib or primer with its own solid-propellant charge, which in turn ignited the main pyrotechnic cartridge.

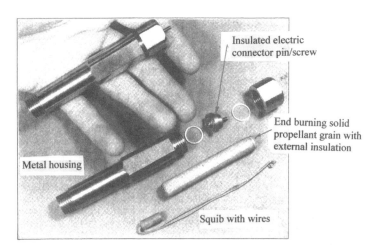

Fig. 4.8-5 Early U.S. pyrotechnic igniter for small experimental thrust chambers. Copied with permission from first edition of Ref. 4.

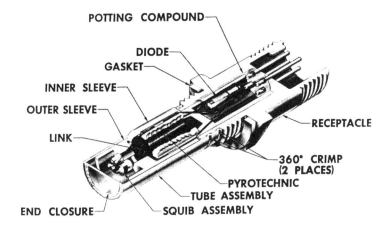

Fig. 4.8-6 Igniter for a gas generator with a pyrotechnic grain (solid propellant) and a built-in fusible link. Copied with permission from Ref. 5.

Goddard also developed a spark-plug igniter for repeated engine starts.[3] The spark-plug cavity was fed with a small flow of gaseous oxygen (supplied through a small line from the top of the oxygen tank) and a small initial flow of gasoline, which shut itself off when the pressure rose. The spark plug for early TC from Goddard and also early Soviet TCs was initially mounted on the side of the chamber, but later moved to the injector region. Goddard was granted two patents on igniters (number 1,879,186 in 1932 and 2,090,039 in 1937).[3]

Some restartable cryogenic LOX/LH$_2$ LPREs still rely on spark-plug ignition today, and many of them use gaseous oxygen and a gaseous hydrogen in a small ignition chamber attached to the injector. The igniter assembly has a set of small igniter propellant valves. The spark-plug assembly is integrated with its associated electrical equipment and is shown in Fig. 4.8-7. The Rocketdyne augmented spark igniter for the space shuttle main engine is shown in Fig. 4.8-7. It has dual redundant spark plugs, which have their own electrical conditioning equipment in the same assembly, including a high-voltage transformer.[5,*] The same augmented spark igniter assembly is used in three places in the SSME, the main TC, and two preburners.* It is installed in the middle of the respective injector and supplied with evaporated gaseous oxygen and gaseous hydrogen.

There have been other types of igniters. For instance, laser energy was applied successfully to ignite and reignite the vapors of cryogenic propellants in experimental TCs. The laser beam entered the ignition chamber through a glass window. Electrically heated wires, exploding wires (by discharging a condenser and vaporizing the wire), or glow plugs have worked in squibs

---

*Personal communications, V. Wheelock, 2002–2004, The Boeing Company, Rocketdyne Propulsion and Power.

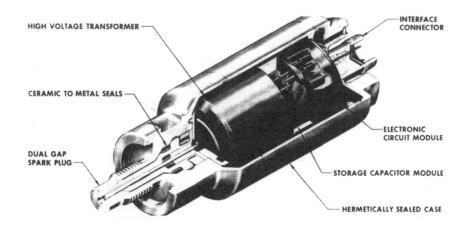

Fig. 4.8-7 Spark plug igniter for a thrust chamber with the electrical excitation equipment. Copied with permission from Ref. 5.

and primers of pyrotechnic devices. They are no longer very popular today, and as far as it is known they have not been used recently in production LPREs.

A shared ignition system has been developed for igniting and starting multiple thrust chambers or multiple engines at essentially the same time. This has been accomplished successfully in a set of 20 small thrust chambers also

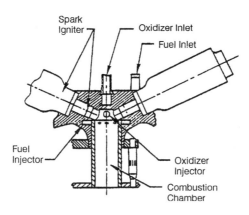

Fig. 4.8-8 Simplified sketch of an augmented spark igniter with two redundant spark plug assemblies and a small ignition chamber. It is located at the center of an injector, which is only shown in part. Courtesy of The Boeing Company, Rocketdyne Propulsion and Power; copied with permission from Ref. 5.

called thrust cells of an aerospike LPRE.[5,*] Called the *combustion-wave ignition* by Rocketdyne, it basically features a single common ignition chamber operating at a fuel-rich mixture of gaseous oxygen and gaseous hydrogen, dual spark igniters, and a series of warm gas pipes from the ignition chamber to the injectors of each of the 20 TCs. The warm gas pipe is placed into the center of each TC injector and emits a stream of warm gas, which then ignites the propellants supplied to each of the TCs. After startup the oxygen flow to the common ignition chamber is turned off, but the gaseous fuel, which is tapped off the cooling jacket, continues to flow through the combustion-wave chamber and the warm gas distribution piping during engine operation.

The detection of the proper functioning of the igniter and of achieving the burning of the initial liquid propellants creates an important signal because a failure to ignite can cause explosions of the mixed, but unignited propellant.[5] This signal is needed as a check before the remaining sequences of the start procedure can be implemented. In the V-2 LPRE (1942) the operator or the commander of the V-2 battery had to observe the exhaust and make a judgment of the appearance and color of the igniter flame and the initial combustion exhaust plume coming out of the nozzle before commanding full power. This method was subjective and not reliable. Since that time, various improved means have been successfully employed for sensing good ignition and initial combustion, such as optical instruments in the visual and also in the infrared spectrum, melting of wires strung across the nozzle exit, thermocouples, or changes in chamber pressure.[5]

The ignition system can be eliminated, if a *hypergolic or self-igniting fuel* is used,[4,5] as was already mentioned in Chapter 4.1. The Germans were first to fly a LPRE with the hypergolic ignition in an aircraft rocket engine in 1939 or 1940 using hydrogen peroxide and C-Stoff, which is methyl alcohol, hydrazine hydrate, and water. They were also the first to launch experimental tactical missiles with nitric acid as the oxidizer and several types of organic hypergolic fuels. The propellant combinations of NTO or nitric acid with N_2H_4 (hydrazine), MMH, UDMH, or their mixtures, and several other fuels are hypergolic. Examples are the propellant combinations for the LPREs of the Titan II, namely, NTO/Aerozine 50, and for the Energomash RD-216 (nitric acid/UDMH) used on the R-14 ICBM (also used on the first stage of Kosmos SLV), and the CADB RD-0233 LPRE (NTO/UDMH) used in the R-18 ballistic missile. Most reaction control LPREs with multiple small thrusters use a hypergolic storable propellant combination without a separate ignition system.

Hypergolic fuel ignition was first tested in Germany, and it was a xylidine fuel mixture with hydrogen peroxide as the oxidizer; this self-ignition was discovered around 1935 or 1936. Aniline could self-ignite with nitric acid, as was first discovered in the United States in 1940. However this aniline was known in Germany several years earlier. In 1933 the Soviet early investigators had a

---

*Personal communications, V. Wheelock, 2004, The Boeing Company, Rocketdyne Propulsion and Power.

fuel mixture of phosphorus, carbon disulphite, and turpentine, which was hypergolic with nitric acid. It probably was the first time a hypergolic start was accomplished with an experimental TC. The Soviet Gas Dynamics Laboratory then developed several fuels that would ignite with nitric acid. None of these early hypergolic fuels are used today. As discussed in Chapter 4.1, they were replaced by hypergolic fuels that had very short ignition delays, even at very low ambient temperatures.

Soon it was found that a small initial quantity or a "slug" of hypergolic fuel at the start was sufficient to ignite the nonhypergolic fuel, such as kerosene.[4,5] At first this slug was carried in a small separate tank, but for a one-time start only. It later was carried in the fuel line, just upstream of the injector, but separated by burst diaphragms from the regular fuel and from the injector.[5] One-time slug starts using aluminum diethyl (with additives) are used on a number of LOX/kerosene LPREs such as the U.S. Thor and Atlas LPREs or the Soviet RD-170 LPRE. Multiple slug starts were used on the Able 4A satellite in 1959 being propelled by a TRW (today Northrop Grumman) multiple-thruster hydrazine monopropellant engine capable of six start slugs of a small amount of NTO as described in Chapter 7.9. The initial burning of the hydrazine with the NTO was hot enough for the hydrazine to continue operation by thermal decomposition.

The analysis of the starting operations and the ignition processes began in the 1950s, but the early attempts did not yield good results. The analytical methods have been refined and expanded since that time. Since about 1970, the analytical simulation of the events in the starting transient operation of LPREs, given a better understanding of the process, has simplified the development and reduced the amount of testing. Since about 1985, it has helped in the design. Good simulation has been more difficult with cryogenic propellants because it is more difficult to predict the propellant properties of a gas/liquid mixture.

Restarting an Engine in Flight

Restarting during flight was required in the 1940s and 1950s on LPREs for aircraft propulsion, which is discussed in Chapters 7.7, 7.8, 8.2, and 9.3. Some aircraft engines had face shutoff valves, such as the Walter 509-109, and some used purging of the propellant system downstream of the main valves prior to restart. There are a series of additional requirements, if the LPRE has to be restarted. Beginning in about 1960, restarts were needed in some of the space flight LPREs. Many restarts were needed for the small thrusters of spacecraft ACSs, also for certain maneuvering spacecraft LPREs, and in some upper-stage LPREs. The placement of the propellant valves close to the chamber or on the injector face (pintle), positive expulsion tanks, and hypergolic ignition were found to be important in the restart of small thrusters.[4] In most cases a set of positive expulsion tanks (see Fig. 7.11-11 and 7.11-12 of ARC and Fig. 4.2-3) or a propellant management device will prevent the entry of gas into a propellant feed line during zero-gravity flight. Positive expulsion devices are discussed in Chapter 4.2.

Restarts were also needed in orbital maneuver engines (typically 100–3000 lbf thrust) and in larger upper-stage engines, such as the Pratt & Whitney RL10 (15,000 lbf up to 24,000 lbf) or the Rocketdyne J-2 at 230,000 lbf thrust. Positive expulsion tanks for the large volume of propellants are usually too heavy. During the unpowered flight in zero-gravity space, the propellants can float or wander around inside their tank, and the outlet pipes of these tanks can become uncovered, thus admitting some pressurizing gas into the propellant feed pipes. The aspirating of inert pressurizing gas into the liquid feed lines prior to and during restart must be avoided because gas bubbles in the propellant flow can cause erratic/lower thrust or combustion problems. For some large LPREs it also might be necessary to purge or flush the propellant pipes after shutdown and/or before restart. The solution has been to use one or two small thrusters or alternatively a couple of small solid-propellant rocket motors to induce a small acceleration on the vehicle. This is really a form of artificial gravity and it orients the propellants in the tank, so that the tank outlet pipe will be covered with liquid just prior to the start or restart of the large engine in the gravity-free environment of space.

In a *reusable LPRE* such as the SSME, there is an opportunity, between flights or after a predetermined number of flights, to perform inspections, maintenance, and possibly also some repair on the engine, before it is started again for the next flight. Such inspections are particularly important for specific components that may have experienced potential overstress or fatigue damage. This means turbine buckets that may have cracks from temperature cycling and vibrations or the inner walls of thrust chambers, which may have experienced high thermal stresses that exceeded the yield stress and may have formed crack upon cool down.

References

[1]Boorady, F. A., and Douglass, D. A., "Agena Gemini Rocket Engine—Hard Start Problem Resolved During Project Sure Fire," Paper Sept. 1966.

[2]Casillas, A. R., Eninger, J., Josephs, G., Kenney, J., and Trinidad, M., "Control of the Propellant Lead/Lag to the LEA in the AXAF Propulsion System," AIAA Paper 98-3204, July 1998.

[3]Goddard, E. C., and Pendray, G. E., *The Papers of Robert H. Goddard*, Vols. I, II, and III, McGraw–Hill, New York, 1970.

[4]Sutton, G. P., *Rocket Propulsion Elements*, 3rd ed., Wiley, New York, 1963.

[5]Huzel, D. K., and Huang, D. H., *Design of Liquid Propellant Rocket Engines*, Progress in Astronautics and Aeronautics, Vol. 147, AIAA, Washington, D. C., 1992, pp. 120–127.

[6]Davisson, J., and McHarris, G. A., "S-IVB Restart Chill-Down Experience," *Journal of Spacecraft and Rockets*, Vol. 8, Feb. 1971, pp. 99–104.

4.9 Steering or Flight Trajectory Control

To steer a rocket-propelled vehicle and maintain its intended flight trajectory, it is necessary to apply torques to turn the vehicle (in pitch, yaw, or roll) and side forces to the vehicle to change the direction of its flight velocity vector.[1] The steering control forces of a spacecraft also counteract the disturbing forces (e.g., wind, uneven Earth gravity, or moon attraction) and provide the means for correcting or changing flight direction. There are several common schemes for accomplishing this *thrust vector control* (TVC).[1-3] Four of these schemes used in LPREs are shown in Fig. 4.9-1. All have been used on engines for production vehicles at some time during the history of LPREs. The United States and the Soviet Union seem to be the only countries that have a history of having used all four of these TVC design concepts. The other countries have flown three or less. Each of the four schemes will be discussed in this chapter. There are other TVC concepts that have been used on solid-propellant rocket motors, but they are not included here.

Jet vanes are historically the first and the oldest method for the TVC of larger LPREs.[1] A jet vane is an aerodynamic lifting body immersed in the nozzle exhaust jet as shown in concept by the first sketch of Fig. 4.9-1. They were originally invented, built, and flown by Robert H. Goddard in the United States in 1937 and early 1938 as seen in Fig. 5-16. He used four jet vanes of a different shape and with a different mounting than subsequently developed by others. His jet vanes provide pitch and yaw control only.[4] Goddard's jet vanes were made of steel and were moved into the side of the nozzle exhaust jet, when a correction was needed. Goddard's vanes often burned up. The first successful development and production of jet vanes was accomplished by the

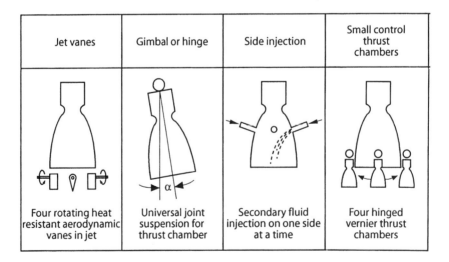

Fig. 4.9-1 Simplified sketches of four commonly used concepts for steering flight vehicles with LPREs. Adapted and copied in part from Ref. 1.

Germans. They used four rotating graphite jet vane in the exhaust of the LPRE of the V-2 missile (design about 1939). The concept was similar to the sketch of Fig. 4.9-1, and the vanes provided pitch, yaw, and roll moments to the vehicle. In a pitch maneuver the nose of the vehicle goes up or down, for yaw the nose turns right or left, and for roll the vehicle rotates about its longitudinal axis. The jet vanes for the engine of the German V-2 missile are shown in Fig. 9-5. In the late 1940s and early 1950s modified versions of the German jet vanes and their control systems were built and flown in the Soviet Union, China, and the United States. The U.S. Redstone missile (1950) and the Soviet RD-100 (R-1 missile, 1948) and the RD-101 (R-2 ballistic missile, 1950) used graphite jet vanes for TVC. The R-2 was copied by China in the 1950s. Even with the best available graphite or carbon materials, the jet vanes oxidized and eroded mostly at the leading edge of the vane, where the temperature and heat transfer were the highest, and they would recede at a rate of about 1 in. every 15 to 60 s (depending on the material), thus limiting the time for the vane's operation. Because of this limited duration, a 1 to 3% loss in thrust caused by vane drag and occasional uneven erosion of the vanes, they are no longer used today.

The next TVC development in the 1940s and early 1950s (after jet vanes) was large fixed main TCs or LPREs and a separate system of multiple small thrusters, which were pointed in different directions. For example, the U.S. Viking Space exploration vehicle and also the Vanguard launch vehicle had one large LPRE in each of its stages and a separate reaction control system (RCS) with multiple pulsing low-thrust small thrusters. This was also used on the Scout SLV, which used solid-propellant rocket motors and had a liquid propellant RCS with multiple thrusters in each of its four stages. For each specific maneuver, one, two, or more selected thrusters operated (often in a pulse mode) for a predetermined time period. These reaction control engines had their own separate pressurized feed systems and provided for maneuvers in pitch, yaw, or roll for the stage. It requires a minimum of 6 thrusters, but more commonly 8 or 12 thrusters to provide the control of the flight path. One method is shown schematically in Fig. 4.9-2 with eight small thrusters oriented for accomplishing the flight control. Two of the four roll control thrusters of this RCS are fired together (in opposing directions) to provide rotation (clockwise or counterclockwise), and they can provide a pure torque in roll. When one of the higher thrust pitch or yaw thrusters is fired, it provides both a torque and a small side force. Figure 4.6-1 explains how pure pitch and yaw torques can be imposed on a flying vehicle without departing from the flight path. A separate RCS had the advantage that maneuvers with the small thrusters could be performed before, during, and after the main engine operation.

As discussed in Chapter 4.6, this type of TVC using an RCS with multiple thrusters has been used effectively in satellites, spacecraft, and space stations, but of course without a large main LPRE. Depending on the mission and the degree of redundancy, anywhere from two to 24 thrusters are typically used for different applications.[1,2] However the Space Shuttle Orbiter has a total of 32 small thrusters including the spares or reserve thrusters.

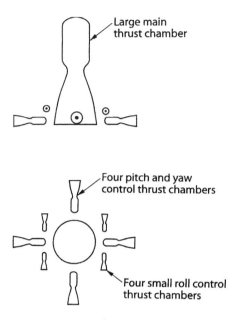

Fig. 4.9-2 Simple concept sketch of several fixed (nonmovable) small thrusters around a fixed main larger thrust chamber.

The second sketch in Fig. 4.9-1 is a simplified representation of a *hinged or gimbaled thrust chamber*. This was an American development and a major step forward in the technology of LPREs. A hinge allows rotation of the thrust vector about one axis only, like a door hinge. It requires a minimum of four hinged TCs to obtain pitch, yaw, and roll control. A gimbal is similar to a universal joint and allows rotation of the thrust vector about two perpendicular axes. A single gimbaled TC by itself can provide only pitch and yaw control, but two or more gimbaled TCs also allow roll control. For maneuvers during that flight portion, when the principal liquid- or solid-propellant rocket propulsion systems are not operating, another means for flight control has to be provided, such as the multiple small thrusters just described and in Chapter 4.6.

The first flight with a *hinge-mounted thrust chamber* occurred in late 1947. It was an advanced and uprated version of the Reaction Motors, Inc., historic 6000C engine (at 8000 lbf thrust) with four hinge-mounted TCs imparting control torques to the experimental small-scale model of a long-range missile. This is described in Chapter 7.2. Some vehicles have more than four hinged main TCs. The Soviet Proton SLV had six Energomash hinged RD-253 engines. This engine is shown in Fig. 8.4-25.

The first *gimbal-mounted* LPRE to fly was developed by RMI for the Viking sounding rocket. The engine had about 20,000 lbf thrust and had a turbopump. Design was in 1946, and the first flight was in 1949. This Viking LPRE is briefly described in Chapter 7.2. For large LPREs with pressurized feed systems, only the TC(s) are gimbal or hinge mounted. The preferred method of

TVC in the United States, Britain, Japan, and France has been to gimbal a complete engine, including the turbopump and its gas generator. Early gimbal-mounted large LPREs were developed in the United States for the Navaho G-38 (design in 1954), the Atlas missile (design in 1955; see Fig. 7.8-4), or the Titan missiles (1958; see Fig. 7.7-17). A typical gimbal bearing for supporting a large LPRE, the Rocketdyne SSME, is shown in Fig. 4.9-3. It is more compact than the gimbal rings used in the Viking, has two spherical bearing surfaces, and provisions for an accurate thrust vector alignment with the vehicle's center of gravity. It is mounted between the injector and the vehicle structure. The actual angular gimbal movements are usually just a few degrees, and they does not usually exceed 10 deg. Gimbal TVC has since been adopted for large engines in France, Japan, United Kingdom, Soviet Union, India, and China. The Soviet Union did not have a flying gimbal-mounted main engine until the late 1960s. However they did have hinge-mounted smaller engines in the 1950s. For applications where there are restrictions on the nozzle-exit skirt motions or the length of the TC (because of limited space in the engine compartment), dual gimbal rings around the nozzle throat region of the TC have been preferred over a single gimbal bearing assembly at the injector end of a TC. This was done for the axial TC of the Peacekeeper postboost control LPRE and the Apollo lunar landing engine shown in Fig. 7.9-7.

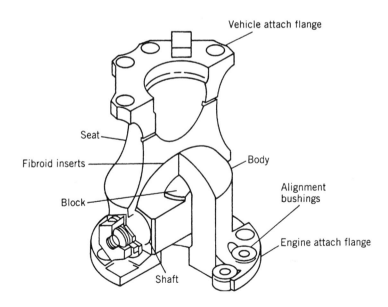

Fig. 4.9-3 Gimbal bearing of the space shuttle main engine has two spherical bearing surfaces and provisions for accurate thrust vector alignment during assembly. Courtesy of The Boeing Company, Rocketdyne Propulsion and Power; copied with permission from Ref. 1.

It requires one *actuator* to move or rotate a hinged TC and two actuators at right angles to each other to rotate a gimbal-mounted TC as shown in Fig. 4.9-4. The force provided by the actuator has to overcome the inertia of the rotating TC, the inlet pressure to the engine acting over the area of the inlet feed pipes, the spring forces of flexible joints in the pipes, the friction in the gimbal or hinge bearings, and the forces imposed by a misaligned TC.[2] In the late 1960s it was a surprise to experience broken actuators because of unexpected high nozzle side forces during the start. If an overexpanded nozzle

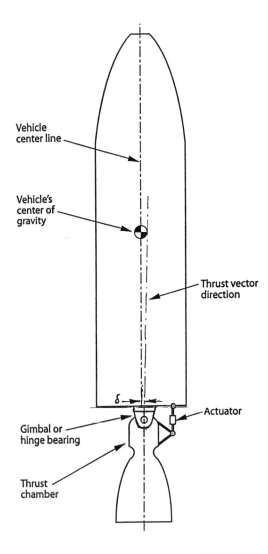

Fig. 4.9-4 Simple diagram of hinged LPRE with actuator, showing thrust alignment with center of gravity of vehicle. Adapted from figures in Refs. 1 and 2.

(with a large nozzle area ratio) is operated or tested at low altitude and if the exhaust jet separates unsymmetrically from the divergent nozzle walls, then significant side forces can be imposed on the nozzle. This nozzle flow separation phenomenon was explained in Chapter 4.3. Typically an actuator has to withstand a side force equal to about one-half to one-fourth of the nominal thrust of the TC.

Three kinds of actuators have been used.[1] Those that are driven by hydraulic oil have been common, give the highest forces per unit actuator mass, and allow high precision in the angular motion and position of the TC. Either hydraulic oil, with a separate oil pump or a pressurized oil reservoir, or high-pressure fuel (tapped off the fuel pump discharge) have been used effectively. Actuators using pneumatic pressure are also common; high-pressure gases are readily available in many of the larger engines, but they are not as accurate. The third type uses electromechanical forces; they are accurate, but usually limited to small forces or small TCs.

The responsibility for the design of actuators and their controls is usually assumed by the vehicle design organization, but with inputs from the engine designers. Only occasionally will a LPRE contractor have this primary responsibility, which has to be coordinated with the vehicle contractors. The control of the actuator and the angular movement of the TC use a closed loop servomechanism. For this control to work, the actuator has to have built in sensors or transducers to determine its position and the rate of the rotational movement. One version of servomechanism with feedback is shown schematically in Fig. 4.9-5.

The required actuator force for correcting thrust misalignment can be smaller if the TC is properly aligned with its vehicle.[1-3] In an ideal alignment the thrust

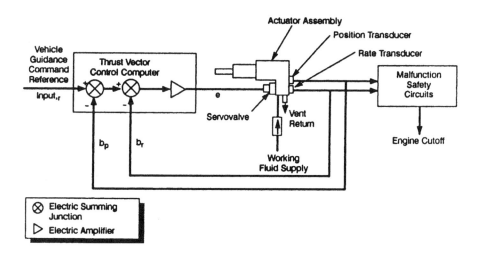

Fig. 4.9-5 Simplified closed-loop control diagram of the servomechanism for a hinged TC. Copied with permission from Ref. 2.

vector line coincides with the vertical axis of the vehicle, and the thrust vector goes exactly through the center of gravity of the vehicle as shown in Fig. 4.9-4. The thrust vector is usually determined as the centerline of the diverging nozzle segment, and on an actual nozzle it is established as a line between the center point of the nozzle throat circle and the center point of the nozzle-exit circle. An adjustment for the angle and the side distance or offset (identified as δ in the drawing) between the center of gravity and the thrust vector is usually made, and provisions for such an adjustment are seen in the gimbal assembly of Fig. 4.9-3. It has large holes in the mounting flange to allow a miss-distance correction, and thin metal shims are placed under the flange to change the angular alignment. In the 1940s a good alignment adjustment was within 1 in. of miss distance between the thrust vector and the vehicle's center of gravity and less than one degree of missed angle for a vehicle of the size of the German V-2 misssile. Today with laser optical equipment the angular and displacement alignment can be within about +/– 0.5 deg and +/– 0.25 in. for a larger vehicle and about half of these values for smaller tactical vehicles. Both are indicated in Fig. 4.9-4.

The first gimbal suspension of a thrust chamber was invented, built, and flight tested by Robert H. Goddard in 1937. His physical mechanism was different than what is used today, but the basic concept is the same.[4] He gimbaled not only the thrust chamber, but the whole tail end of his sounding rocket vehicle to impart pitch and yaw moments as described in Chapter 5.2. He called it tail motion control. It did not include roll control. The mechanism and its double bellows assemblies are shown in Fig. 5-17. His flight tests showed this TVC component (commanded by signals from his gyroscopes) to be truly effective in stabilizing the vertical flight of his test vehicles. The bellows were his actuators, and they also carried the full thrust loads. Today we prefer to use the actuators to carry relatively small side loads and not the full thrust, which is carried by a gimbal bearing.

For single gimbaled or fixed (not movable) rocket engines with turbopumps, it is also possible to obtain a small roll torque by using the momentum of the turbine exhaust gas.[1,2] A swivel joint or hinge device is placed at the end of the turbine exhaust gas pipe, just upstream of the movable turbine gas exhaust nozzle. In some engines the turbine exhaust gas flow is divided into two streams with two hinged nozzles on opposite sides of the main TC; this allows the application of pure roll torque to the vehicle. It was first used in the United States on the Vanguard booster-stage engine developed by General Electric and its first flight was in 1957. A turbine exhaust roll nozzle was provided in the Jupiter engine (1955 design), which is shown without this nozzle in Fig. 7.8-6; it flew also in 1957. Two hinged turbine exhaust nozzles are currently used by the French in their Vulcain second-stage LOX/LH$_2$ engine in a version of the Ariane SLV. This engine is shown in Fig. 10-13. The French HM-7 upper-stage LOX/LH$_2$ rocket engine uses hydrogen gas from the exit of the cooling jacket for this roll control. The Energomash RD-119 engine has a more sophisticated turbine exhaust system. The exhaust flow is divided into eight branch flows, two for pitch control, two for yaw control, and four smaller flows for roll control. There are special hot-gas valves in the hot-gas exhaust streams to regulate the magnitude of the flows and thus the magnitude of the corrections. A picture and a flow diagram of this RD-119 LPRE are shown in Fig. 8.4-13 and 8.4-14. Because the turbine flow represents only a relatively small portion of the total propellant flow, the magnitude of the corrections from the roll nozzles is limited. In the Soviet Lyulka D-57 engine, a small flow of preburner warm gas is bled off from the high pressure main turbine discharge, sent through another turbine (for driving the booster pumps) and then fed into four roll control nozzles. A flow diagram of this engine is shown in Fig. 8.8-2.

The alternative to the gimbal suspension was to use a fixed (not moving) main TC and two or four *auxiliary hinged TCs* (also known as *vernier TCs*) at substantially lower thrust than the main TC. Using four movable (hinged) vernier TC with one or more fixed (not movable) main TCs was a favorite method employed by the Soviets in their early vehicles, such as the R-7 missile or the Soyuz space launch vehicle. Typical vernier thrusts were between 2000 and 8000 lbf, and this was large enough so that four hinged vernier TCs could not only provide pitch, yaw, and roll control, but also provide some of the axial thrust to accelerate the vehicle along its flight path. This was first accomplished on the Soviet RD-107 and -108 LPREs, which were developed in the 1950s, and the 12 vernier TCs and the five main engines, each with four fixed TCs, can be seen in Figs. 8.4-7, 8.4-8, and 8.4-11. The propellant for these vernier TCs came from the TP of the main engine, and the vernier TCs could only operate while the main TCs were operating. Since about 1963, the four Soviet verniers had their own separate turbopump feed system as exemplified by the RD-0212 LPRE. Both the vernier and the main engine started at the same time, but the vernier engine could also operate before the main engine start and/or after the main engine was shut down. The Soviet Union had these four hinged vernier TCs with their own TP also in a number of other upper-stage engines, some booster engines, and their submarine-launched missile engines. An upper-stage engine, identified as the RD-0235, was really made of two separate engines, both with a pump feed, one for the main fixed TCs and one for the four hinged TCs. It is shown in Fig. 8.5-15. and the four TCs and the small TP for the vernier engine are visible in this figure. The forces needed to rotate the verniers were substantially less than those for rotating a large LPRE on a gimbal or hinge mount. In some of the engines, the vernier TCs could rotate through +/− 45 deg, and a satisfactory rotary transfer sleeve joint was developed for feeding propellants under pressure to the rotating hinged TCs. A key difference between the U.S. auxiliary TCs and the Soviet vernier TCs is that the Soviet verniers provided about 10 to 20% of the axial thrust. (The exact value depended on the specific LPRE.) These Soviet vernier engines thus augmented the axial thrust of the main engines. The vernier thrust levels are much higher than those used for spacecraft ACS in the Soviet Union or the United States.

In the United States these hinged auxiliary TCs were much lower in thrust, typically between 1.0 and 4% of the total thrust of the LPRE. The first large LPRE had a gimbal-mounted main TC and two hinge-mounted smaller roll thrusters. The main engine was gimbaled and not fixed, as in the early USSR large LPREs. In the United States this was first used in 1955 on one version of the Thor missile engine (135,000 lbf), where two small hinged 1000-lb thrust chambers were employed for roll control. Figure 4.6-2 shows this engine and the two hinge-mounted roll control TCs during a ground-firing test. The bright flame is typical of the LOX/kerosene propellants. The propellants for the small thrusters were supplied from the turbopump of the main engine. Two similar roll control thrusters were also used in the early versions of the sustainer LPRE for the Altas vehicle. The Atlas and the Thor engines are shown and discussed in Chapter 7.8, and later versions of these engines did not use two auxiliary roll thrusters. The small auxiliary thrusters used on U.S. upper stages typically used a gas-pressurized feed system and not a separate turbopump feed system as in the Soviet Union.

With multiple large TCs (such as two gimbal-mounted TCs or four hinged TCs) it is no longer necessary to provide auxiliary TCs. All flight correction can be achieved with relatively small angular motions of the nozzles of the large movable TC clusters. An example of four hinged TCs is the Soviet RD-170 shown in Fig. 7.2-16 and the British Gamma 8 LPRE shown in Fig. 12-14. An example of two gimbaled TC is the RD-180 shown in Fig. 7.10-11, the Titan booster engine in Fig. 7.7-18, and the British Blue Streak engine shown in Fig. 12-16.

In the 1950s and 1960s there was a great deal of interest to use *side injection of a liquid or a gas as a means for thrust vector control*. This fluid was injected through several small holes at one of four locations in the diverging section of the nozzle as shown in concept in Fig. 4.9-1; there are no gimbals, hinges, pipe joints, or actuators.[1,2] The U.S. Lance missile engine, shown in Fig. 7.8-25, designed in 1964, uses liquid fuel side injection for pitch and yaw control, in part because the space available for the engine in the missile would not allow enough room for moving a gimbaled TC. It is probably the only known flying LPRE with this liquid side-injection TVC scheme. The Soviets also used side injection, but with a moderately hot gas and not a liquid. Gas tapped off the gas generator or preburner of the main engine was used for the side injection, and the side injection was commanded by the vehicles flight control system. This scheme was used on the RD-857, an experimental engine, and RD-862 engines, which powered the second stage of the RS-16 ballistic missile. Both engines were developed by the Yuzhnoye Design Bureau in the 1960s and are described in Chapter 8.10. From a vehicle's performance point, the side-injection schemes did not use the propellants as effectively as an engine with a gimbal-mounted TC. Apparently there have not been any other flying engines with side injection for TVC.

An alternate method of flight control that can be implemented with four (or more) fixed large TCs is called *differential throttling*. This is shown schematically in Fig. 4.9-6. The thrust vectors are at a compound angle to the vehicle centerline and do not go through the center of gravity of the vehicle. The larger forces from the unthrottled TCs impose turning moments on the vehicle. It eliminates gimbals and other TVC systems, which usually require considerable

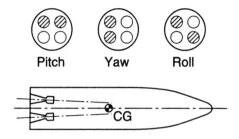

Fig. 4.9-6 Differential throttling of four fixed TCs is an alternate approach to TVC. The shaded nozzle indicates a throttled condition or reduced thrust. Copied with permission from Ref. 1.

auxiliary power, more vehicle space, and equipment. It was used in the experimental linear aerospike engine, shown in Fig. 7.8-37, where selected banks of small thrust cells were throttled for control about three axes. This engine has not flown, but the throttling features and side forces were demonstrated in ground tests. However, this differential throttling scheme was used and flown on the first stage and second stage of the large Soviet N-1 space launch vehicle intended for a moon mission. The first stage had 30 individual Kuznetsov NK-15 fixed LPREs of which 24 engines were in a ring around the periphery of the vehicle's base. By throttling six adjacent selected engines, the vehicle could be controlled in pitch and yaw. For roll control there were four separate smaller LPREs.

For vehicles that fly mostly in the Earth's atmosphere, the best way to apply moments or torques to the vehicle and the best flight performance is usually obtained by steering with aerodynamic surfaces, such as rudders and movable fins. The aerodynamic surfaces were used on rocket-engine-propelled tactical missiles, which spent a major part of their flight within the atmosphere. Examples are the German X-4 missile shown in Chapter 9.4 and the booster stage of the Navaho, a boost-glide long range missile. A rocket TVC is not very effective while flying within the atmosphere, but it is necessary for steering in flight above the atmosphere

References

[1]Sutton, G. P., and Biblarz, O., "Thrust Vector Control", *Rocket Propulsion Elements*, 7th ed. Wiley, Hoboken, NJ, 2001, Chap. 16, pp. 608–623.

[2]Huzel, D. K., and Huang, D. H., *Modern Engineering Design of Liquid Propellant Rocket Engines*. Progress in Astronautics and Aeronautics, Vol. 147, AIAA, Washington, D.C., 1992, pp. 222–225.

[3]Prescott, B. H., "Nozzle Design," *Tactical Missile Propulsion*, edited by G. E. Jensen and D. W. Netzer, Progress in Astronautics and Aeronautics, Vol. 170, AIAA, Reston, VA, 1996, Chap. 6, pp. 177–185.

[4]Goddard, E. C., and Pendray, G. E., *The Papers of Robert H. Goddard*, Vol. 3, McGraw–Hill, New York, 1970.

4.10 Combustion and Vibrations

The combustion behavior of a LPRE is the one area of science that has been analyzed, studied, and experimentally investigated more than any other scientific area in this field. This has been an ongoing effort in several countries ever since combustion instabilities were first encountered approximately 70 years ago.[1-3] Many TCs and LPREs have experienced abnormal combustion behavior, which can cause a sudden failure of the TC or the LPRE. The incidences of combustion instabilities were a surprise to the early investigators and almost every LPRE program that experienced this combustion oscillation problem ran into cost overruns and schedule delays. The gas oscillations can cause large pressure surges, which induce strong vibrations, thrust changes, and an instantaneous major increase in the heat transfer to the TC walls. Failure can occur quickly because of high-amplitude vibrations (values of more than 300 g acceleration has been measured on the injector and support structure of some LPREs), or by increased heat transfer, which is usually well beyond the capability of the cooling jacket. These instabilities can occur not only randomly during steady-state operation, but also during the startup and shutdown transients or when changing the thrust level. There is so much energy in the oscillating gases that severe damage or failure of the TC can occur in a brief time period, much less than a second. The instability phenomena and their resulting potential damage to engine hardware are most severe in large-sized thrust chambers, and the consequences are less pronounced in small thrusters. All LPREs designed since about 1960 must be proven by tests to be free of such instabilities before they will be allowed to fly.

Investigations of the Combustion Behavior

For perhaps 10 to 15 years (after the first few occurrences) the understanding of these combustion phenomena was inadequate, and remedies were empirical. The understanding and curing of instabilities became a high-priority industry-wide effort in the United States and the Soviet Union in the late 1940s and 1950s and to a somewhat lesser degree also in Germany, and later in France, Japan, and China. Special expert committees were assembled in the United States and in the Soviet Union to provide guidance in finding solutions. In the 1950s and 1960s projects with troubled engines were encouraged to try all sorts of different remedies, projects on more profound analyses were initiated, and more research efforts of the combustion process were funded.

Some outstanding work has been done and is continuing on the analysis and experimental investigations of combustion processes and the nature of combustion instabilities in LPREs. More specifically this work covered various aspects of liquid propellant combustion, including research on liquid jet/spray formation, atomization into small droplets, vaporization, mixing, diffusion of gaseous species, burning mechanism, the expansion of the gases in the chamber, turbulence, influence of propellants, mixture ratio, chamber pressure, film-cooling effects, different injection patterns, or other parameters. Combustion problems in specific LPREs have been investigated, and a variety of remedies to fix combustion instability has been successful, but these remedies were limited to a specific propellants, specific types and sizes of TCs or LPREs.[1,2] The progress in gaining an understanding of these phenomena was most rapid before about 1975, and some work has continued ever since that time. Early analyses and theoretical approaches aimed at understanding combustion phenomena in LPREs were undertaken.[4-6] However, a clear set of generalized validated design rules for preventing combustion instabilities in new TCs has not as yet been identified. Also a good universal mathematical three-dimensional simulation of the complex nonlinear combustion process has not yet been developed.

In the 1940s there were no suitable instruments or techniques to accurately measure or diagnose some of these phenomena. New high-frequency pressure sensors, flowmeters, and other new instruments were developed. The author participated in 1944 in the development of a pressure pickup, which would faithfully measure pressure oscillation at 20,000 cycles per second without distortion. The analytical simulations (both linear and nonlinear), the mechanisms for absorbing energy, and the techniques for estimating and predicting vibrations for a particular TC were all pursued. The few references cited in this chapter refer only to a small fraction of the papers and books that have been written on combustion and instabilities in a LPRE.

Three types of vibrations can be encountered.[2,3] Figure 4.10-1 shows some simple representations of them. The first is a low-frequency "chugging" or interaction of the liquid propellant feed system, if not the whole vehicle structure, with the gas behavior in the combustion chamber. Typical frequencies are between 10 and perhaps 200 cycles per second and sometimes up to 400

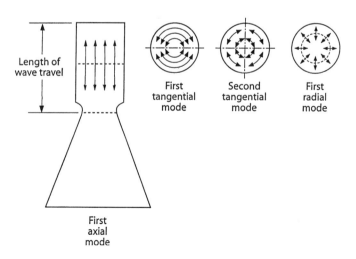

Fig. 4.10-1 Simplified sketches of three modes of instability Adapted from Ref. 3 and 7 with publishers permission.

cycles per second. It was probably first encountered in the early 1950s. This includes oscillations of propellants in long feed pipes from the propellant tanks to the combustion chamber, often called POGO instability.[7,8] The name was coined because of the resemblance to the oscillations of a POGO jumping stick. Remedies were developed and included modifications in the feed system, increasing the injection pressure drop, and for POGO instability the addition of damping accumulators in the pipe lines.

The second type of instability is characterized by intermediate frequency oscillations, called "buzzing", associated with mechanical vibrations and resonances of pieces of the engine structures, injector manifolds, pipes, and their interaction with gross combustion behavior, such as turbulence. Frequencies depend on the size and structural details and typically are between 400 and 2500 cycles per second. The frequency range given here for all three categories is overlapping and not rigorous. Changes in the chamber geometry, injector configuration, and in the structural resonance (by changing mass or stiffness) of the affected components were effective countermeasures. By about 1955 the understanding of the first two vibration types had advanced enough to diagnose most incidents and take effective remedial actions.

The last type of combustion vibration occurs at high frequency, typically 1000 to 30,000 cycles per second, and the harmonic vibrations can be even higher. It has been called "screaming" instability, and it was the most difficult to control. Because the vibration energy increases directly with the frequency, these high-frequency pressure oscillations and structural vibrations can be very powerful and destructive of the hardware. They have since been linked to the burning process itself, to pressure waves, combustion phenomena, and chamber acoustic resonances.[1-6] When instability did occur, particularly in the larger TCs, it would suddenly greatly increase the heat transfer, cause high-frequency large chamber pressure oscillations and often lead to the failure of the TC. Detection of this phenomenon was initially difficult, in part because the sensors were then not adequate and gave erroneous nonlinear readings. With marginal TC designs this instability would occur only in one test run out of perhaps 100 or 500 firing tests on the identical LPRE configuration. Therefore the only method for ensuring a stable design in these early days was to run hundreds of static tests on the same engine design without a single incident of damaging instability.

In the late 1950s three different types were identified, namely, axial or longitudinal, radial and tangential or circumferential resonance types of high-frequency gas oscillations as sketched in Fig. 4.10-1 (Refs. 2, 3, and 9). The arrows indicate the directions of the pressure waves moving alternatively in two directions. The axial oscillations are reflected from the injector face and the converging section of the nozzle and are similar to an organ pipe mode. The radial and tangential types were more troublesome in larger TC. Damping mechanisms, which absorb vibration energy, and physical barriers to prevent the buildup of vibration amplitude were introduced. The high-frequency gas vibration type is still not fully understood, but enough is now known to cure and to some degree predict and prevent its occurrence.

Rating Techniques

Several *rating techniques* were considered, but one developed between 1957 and 1967 had the best acceptance. Artificial disturbances are introduced into the combustion chamber, for example, by setting off several specific directional explosive charges, to induce large pressure surges in different modes, which in turn would trigger high-frequency vibrations.[1,2,10] Accurate high-frequency chamber pressure measurements can then determine if there is enough energy absorption for the magnitude of the pressure oscillations to diminish rapidly. The recovery time (in milliseconds) or the rate of decay between the artificial pressure surge and the resumption of steady relatively smooth combustion has became a rough measure of the inherent stability. By the mid-1960s this rating technique reduced the number of static firing tests needed to prove stability. Another method used to induce combustion instability was to squirt slugs of gas into the propellant feed lines, which formed bubbles, and these bubbles often caused the initiation of unstable burning. This method has been used also in small thrusters. Often a heavy-duty single-wall thrust chamber is used because this saves more expensive TC hardware and because an instability test can be run in less than a second.

Remedies

The traditional approaches were aimed at introducing damping to absorb vibration energy and on decoupling the periodic flow oscillations and the unsteady combustion responses. The highly turbulent combustion processes are complex and nonlinear, and it was usually not known in the early days which of the remedies listed next would be most effective in controlling a particular type of instability. From a practical point of view, several remedies have been satisfactory in eliminating the sudden occurrence of high-frequency destructive vibrations, but limited to specific LPREs and specific operating conditions. They included the following:

   1) Early solutions included changes to the resonant frequencies by changing *chamber geometry* (diameter or length, cylindrical vs conical chamber walls). These changes in geometry were used by early investigators and have been used occasionally since that time. This is expensive with a cooling jacket and successful only in some cases.

   2) Instabilities occurred more readily with certain *propellant combinations*, and a change of propellant has at times been very effective. For example, LOX/hydrocarbon is more likely to have combustion vibrations than LOX/alcohol. The German V-2 LPRE used alcohol as a fuel in part for this reason. The fuel used in the LPRE for the Bomarc air-launched missile was originally kerosene. When combustion instability was encountered, several remedies were tried, but did not work. The switch in fuel to UDMH and later a mixture of kerosene with UDMH was effective in eliminating the problem as discussed in Chapter 7.7.

3) If *gas bubbles* enter the combustion chamber, they can often trigger instability, particularly during the startup transient period. Therefore precautions were taken to properly drain propellant or vent trapped gas to eliminate gas bubbles in the propellant feed lines and to prevent tank pressurizing gas from entering the tank outlet during maneuvering flight operations. For LPREs with turbopumps, new procedures were developed to remove air or gas from the propellant lines and to properly prime the pump with liquid propellant prior to start.

4) The author learned early that an increase in the *injection velocity*, or really an increase in the pressure drop across the injection holes, was often a good remedy. In some injectors the pressure drop was as high as 50% of the rated chamber pressure. It required a higher pump discharge pressure and of course a more powerful and heavier TP. It worked often, but not always.

5) Vibrations occurred at times during *transient operations*, such as during thrust buildup or throttling and occasionally during shutdown or thrust changes. These seemed to occur at certain regions of the chamber pressure vs mixture-ratio spectrum. A change in start sequence, the transient flow, rate of pressure rise, or local mixture ratio was sometimes effective to eliminate these transient occurrences. A temporary brief hold at an intermediate thrust level during the start was also effective in some engines.

6) The Soviet engineers were the first to use cooled *metal baffles* as early as 1949, initially in the shape of a cross near the injector. In retrospect the baffle greatly reduced transverse gas vibrations. This was first achieved by Alexey M. Isayev early in one of his TC designs. Today his LPRE development organization is called the Khimmash Design Bureau. It is discussed in Chapter 8.6. In the United States cooled baffles were investigated starting in the mid-1950s, and they were introduced into large U.S. LPREs beginning in about 1958 (Refs. 1–3, 10, 11). They proved to be very effective in changing the resonant frequencies and mitigating transverse (radial or tangential) gas oscillations, which were perhaps the most destructive modes in larger TCs. A cooled baffle can be seen in Fig. 4.10-2 as used on the early version of the space shuttle main engine. Baffles were then retrofitted or designed into all large U.S. LPREs, such as the engines for the Thor, Atlas, Saturn, or Titan discussed in Chapters 7.8 and 7.7. In the development of the large Rockedyne F-1 LPRE, a set of complex and severe combustion instabilities was encountered (1959–1966). The final solution was an unusual baffle, which was selected after 14 different baffles had been built and tested.[11]. The selected configuration had cooled circular and radial baffles and is shown in Fig. 4.10-3. The extensive investigation suggested that the baffles protect the liquid propellant fans, which are formed by the impingement of individual injection streams, from the unsteady oscillations of the tangential and radial modes and that the major part of the combustion occurs downstream of the baffle tips. Roughly 3200 firing tests were conducted on the F-1 engine, and of these about 2000 tests were related to

Technology and Hardware 233

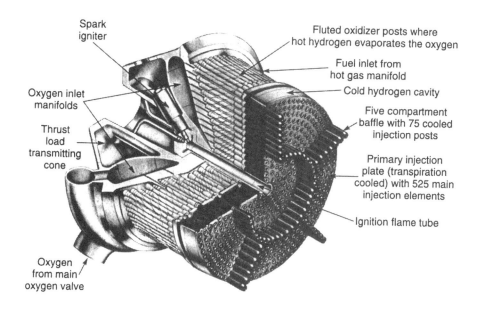

Fig. 4.10-2 Main injector assembly of the initial version of the space shuttle main engine showing a baffle with five outer compartments. Courtesy of The Boeing Company, Rocketdyne Propulsion and Power; copied with permission from Ref. 3.

the development of a stable injector configuration. In addition to the tests with different baffle geometries, some changes in the injection holes were also made. This major development effort had participation from different experts outside of Rocketdyne, was largely empirical, delayed the development program, and greatly increased its costs.

7) Another remedy discovered in the United States uses *acoustic resonance cavities* (to absorb gas vibration energy), was designed into LPREs beginning about 1963, and was reliable in preventing many incidences of high-frequency oscillations.[1,11,12] The concept is depicted in Fig. 4.10-4. These cavities are usually most effective when they are located at the boundary of a flat injector and the cylindrical chamber, where several types of acoustic waves seem to be anchored. They are designed or tuned for a specific vibration frequency, usually the estimated likely resonance frequency. These resonance frequencies can be predicted for different modes of acoustic oscillations. In the United States a number of production injectors, both for high and low thrust, were then redesigned to include resonance cavities. Some injector designs had both baffles and resonance cavities, such as the injector for the Apollo lunar ascent engine. Resonant cavities have been more effective than the baffles, and in some cases, such as on the space shuttle main engine, it was possible to later remove the baffles in the improved redesigned version. This remedy of resonance cavities is still used in some new modern TCs in the United States or Germany. The author

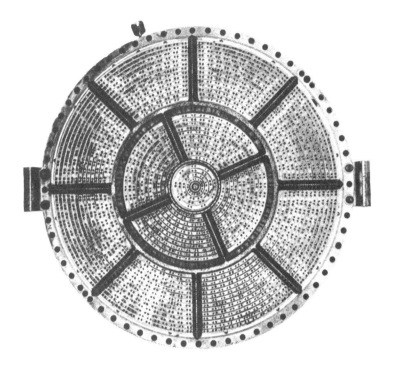

Fig. 4.10-3 Selected configuration of the baffles for the F-1 LPRE shows both radial and circular-cooled barriers. Internal diameter is about 38 in. Courtesy of The Boeing Company, Rocketdyne Propulsion and Power; copied with AIAA permission from Ref. 11.

could not find any evidence that the Soviets used resonance cavities as a remedy.

8) Some engineers believed that *long small-diameter injection holes* were more likely to allow stable combustion, than fewer, shorter, and larger holes. The long, thin hole has more friction loss and produces a laminar flow, which is hydraulically more stable than turbulent liquid flow, and this results into more uniform atomization into small droplets.* This has been the experience at TRW, which today is Northrop Grumman, as described in Chapter 7.9.

9) A *scale effect* became evident in the 1940s in the Soviet Union and in the 1950s in the United States. Some types of transverse vibrations were more likely to happen with larger chamber diameters. During an early LPRE development in the Soviet Union about 1946, when chief designer Alexey M. Isayev

---

*Personal communication, Gordon Dressler, chief engineer, Propulsion Products Center, Northrop Grumman, Redondo Beach, CA, 2003.

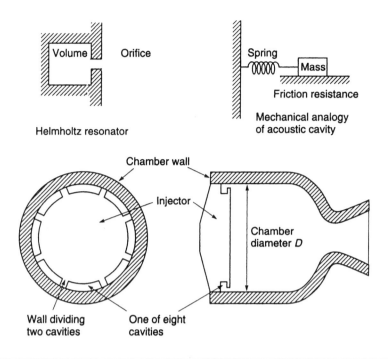

Fig. 4.10-4 Simplified diagram of acoustic energy absorber cavities at the periphery of an injector. Adapted from Ref. 3.

went from a single large TC to four smaller TCs of the same total thrust, he found it easier to cure a troublesome vibration. This is in part a reason for using four TCs in several early large Soviet LPREs. Small thrusters have experienced relatively few instabilities, and the severity is less, when compared to larger, higher thrust TCs. Occasionally instabilities occurred in a small thruster, but often the effect was not very destructive of the hardware. For example, in the 1970s combustion vibrations were observed in the 870-lbf thruster of the Space Shuttle Orbiter reaction control system.[13] It used the storable propellants of NTO/UDMH and was often used in a pulsing operation. When combustion gas vibrations did occur, they did not seem to cause rapid failure of the thruster, as usually happens with large TCs. The remedy was achieved by altering the injector face with drilled holes at its periphery as shown in Fig. 4.10-5, and these holes acted as quarter-wave cavity dampers. Many small thrusters had been developed and had flown successfully. These relatively infrequent high-frequency instabilities in small thrusters have led to the tentative conclusion that there is a scale effect. High-frequency instability was also observed in a 5-lbf thruster, also using NTO/UDMH.

10) Certain *injection elements,* such as certain sprays or injection orifices, were more stable and certain minor changes in their geometry sometimes

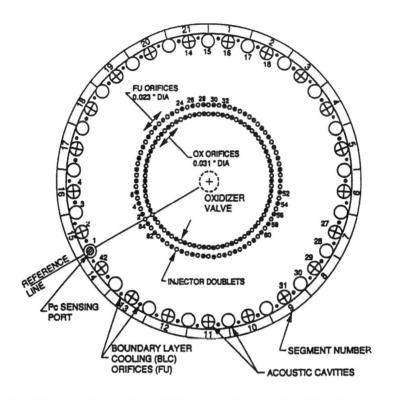

Fig. 4.10-5 Injector face of the 872-lbf thruster of the Space Shuttle Orbiter reaction control system. It has 82 unlike doublet injection elements, 41 fuel film-cooling holes, and 41 larger holes near the periphery for acoustic cavities. Courtesy of Aerojet; copied with permission from Ref. 4.

had a large influence.[1-3,7] The United States has had the most experience with drilled holes and jet impingement as injection elements, such as those shown in Fig. 4.3-13. For example, nonimpinging streams were believed to be more stable than impinging streams, but they required a larger chamber for achieving a high combustion efficiency. Small changes in hole sizes, number of holes, hole entry design, impingement angle, etc. were at times very effective to surpress instabilities. The Soviets had a long history of using spray nozzle elements in their injectors, both single-propellant spray elements and coaxial bipropellant spray elements. They learned how to change the detail dimensions of these elements to control the flow, swirl, and the location of the impingement of the two propellant sprays and thus the axial location of the maximum energy release. Several such injection elements are shown in Figs. 4.3-14 and 8.4-20. In several cases it was possible to eliminate instability by making certain changes in the geometry or the flow of individual spray elements, and this was done successfully without the use of baffles or resonance cavities.

11) If the *distribution and pattern of injection elements across the injector face* is changed, it has sometimes alleviated instability problems.[1,2] Originally designers attempted to have uniform flow densities and the same mixture ratios over the surface of the injector. Changing these parameters so that the flow density would not be uniform across the injector face has at times been effective. Different patterns of injection elements can be seen in Fig. 8.4-19, 8.4-22, and 8.5-8. Also by changing the impingement distance from the injector face (and thus the location of maximum energy release) has also been effective. This resulted in some complex injection patterns described later.

12) For the last 50 years it has been possible for designers to *calculate estimated values* of the likely *gas resonance and harmonic frequencies* in all three modes (axial, radial, or tangential) for a given thrust chamber and for its operating conditions.[1,2,6] This allows the designers to design the LPRE, its structure, and key components so that their natural resonance frequencies will not coincide with those of the combustion gas vibrations, thus avoiding uncontrolled vibration amplifications and potential overstressing of parts.

13) The use of a *flame holder*, which was regeneratively cooled, worked well at the General Electric Company in the 1950s. Test were conducted at the 10,000-lbf thrust level using LOX/alcohol and LOX/kerosene. Two examples of such flame holders are shown in Fig. 7.3-4. The original intent was to stabilize oscillations in the axial mode, but the flame-holder designs also seemed to influence the radial and tangential modes. This work was not well known outside of General Electric Company, and when this company went out of the LPRE business these holders seemed to have been forgotten. This remedy for combustion instability was recorded, and the records recently became available.

14) There was one remedy that worked well, and as far as this author knows was not practiced outside the Soviet Union. It is *temporary baffles* in the chamber, they can be seen in Fig. 8.5-9 (Ref. 14). They are made of felt-like material that is porous and combustible. They are glued to the chamber wall and are not part of the injector, as are U.S.-designed baffles. The temporary baffles work only during the start transient and the first few seconds of burning because they react with the combustion gases and are consumed.

One or more of these various remedies were successful in eliminating combustion instabilities during development of each new LPRE. None of the remedies just listed were 100% effective for different TC designs, transients, sizes, propellants, chamber pressures, or mixture ratios. It still is not always clear which of these approaches will be most effective and simplest for a particular TC that has experienced incidents of combustion instability

In the last few decades some TCs with certain injector designs (such as coaxial tubes with LOX/LH$_2$) operated stably without baffles or resonance cavities. Stability is obtained by using proven stable injector designs. The Japanese

LE-7A engine omitted the resonance cavities of the earlier LE-7 engine version. Many large and small TCs of the former Soviet Union have neither baffles nor cavities, but rely on multiple proven coaxial injection elements and the distribution of these elements over the face of the injector to achieve stable combustion. They have done this with several different propellant combinations including LOX/kerosene, NTO/UDMH and LOX/LH$_2$. This apparently was first done in the Soviet Union in the 1950s and 1960s.

Today there is an extensive experience background of stable operation with different injectors, different propellants, different operating conditions (chamber pressure, mixture ratio), different sizes or thrust levels, and alternate chamber geometry. Each organization uses its own experience base and analysis methods and has its own favorite injector designs. Today the incidence of combustion vibrations has been greatly reduced. The expensive, extensive engine static firings of the past (to get meaningful statistical data) have been replaced by analyses of vibration behavior and a few directed bomb tests for demonstrating stability and for rating the recovery time period. In the past 60 years it has been confirmed many times that the details of the injector design are crucial in ensuring stability and that some proof testing for instability is still desirable with each new TC.

References

[1] Yang, V., and Anderson, W. E. (eds.), *Liquid Rocket Engine Combustion Instability*, Progress in Astronautics and Aeronautics, Vol. 169, AIAA, Washington, D.C., 1995.

[2] Harrje, D. T., and Reardon, F. H. (eds.), "Liquid Propellant Rocket Combustion Instability," *NASA SP-194*, 1972.

[3] Sutton, G. P., and Biblarz, O., *Rocket Propulsion Elements*, 7th ed. Wiley, Hobaken, NJ, 2001, Chap. 9, pp. 342–361.

[4] Crocco, L., Grey, J., and Harrje, D. T., "Theory of Liquid Propellant Rocket Combustion Instability and Its Experimental Investigations," *ARS Journal*, Vol. 30, No. 1060, pp. 159–168.

[5] Culick, F. E. C., "High Frequency Oscillations in Liquid Rockets," *ARS Journal*, Vol. 1, No. 5, 1963, pp. 1097–1104.

[6] Sheveljuk, M. I., *Theoretical Basis for Liquid Rocket Design*, Oborongiz, Moscow, 1960, pp. 586–597 (in Russian).

[7] Oppenheim, B. W., and Rubin, S., "Advanced POGO Analysis for Liquid Rockets," *Journal of Spacecraft and Rockets*, Vol. 30, No. 3, 1993.

[8] About, G., Bouvert, P., Bonnal, C., David, N., and Lemoine, J. C., "A New Approach of POGO Phenomenon Three-Dimensional Studies on the Ariane-4 Launcher," *Acta Astronautica*, Vol. 15, Nos. 6 and 7, 1987, pp. 321–330.

[9] Huzel, D. K., and Huang, D. H., *Modern Engineering Design of Liquid Propellant Rocket Engines*, Progress in Astronautics and Aeronautics, Vol. 147, AIAA, Washington, D.C., 1992, Sec. 4.8, pp. 127–134.

[10] Reardon, F. H., "Combustion Stability Specification and Verification Procedure," Chemical Propulsion Information Agency *Publication 247*, Oct. 1973.

[11] Oefelein, J. C., and Yang, V., "Comprehensive Review of Liquid Propellant Combustion Instabilities in F-1 Engines," *Journal of Propulsion and Power*, Vol. 9, No. 5, 1993, pp. 657–677.

[12] Acker, T. L., and Mitchell, C. E., "Combustion Zone—Acoustic Cavity Interactions in Rocket Combustion," *Journal of Propulsion and Power*, Vol. 10, No. 2, March–April 1994, pp. 235–243.

[13]Hulbert, E. A., Sun, J. L., and Zhang, B., "Instability Phenomena in Earth Storable Bipropellant Rocket Engines," *Liquid Rocket Engine Combustion Instability*, edited by V. Yang and W. E. Anderson Progress in Astronautics and Aeronautics, Vol. 1 AIAA, Washington, D.C. 1995, Chap. 5.

[14]Rubinsky, V. R., "Combustion Instability in the RD-0110 Engine," *Liquid Rocket Engine Combustion Instability*, edited by V. Yang and W. E. Anderson, Progress in Astronautics and Aeronautics, Vol. 169, AIAA, Washington, D.C., 1995, Chap. 4, pp. 89–112.

CHAPTER 5

The Early Years, 1903 to the 1940s

First came the visionaries, who conceived and described the ideas of LPREs. Then came the pioneers and amateurs who actually built, tested, and flew the first crude LPREs. Then came organized groups who started research and development on this subject. The first LPRE efforts were in Russia, Germany, and the United States. A few years later there were key people working on LPREs in other countries such as Austria, Britain, France, Italy, Romania, or Switzerland.

In the early formative years of the development of LPRE, there were perhaps a dozen prominent individuals, who made significant early contributions to the state of the art. They were active between 1885 and 1940, and several of them might not have known about the specific achievements of some of the others. This section discussed three of the founders of the technology of LPREs, namely, the visionary K. E. Tsiolkowsky of Russia (theory and conceptual word description, first published in 1903), the pioneer, experimentalist, and visionary R. H. Goddard of the United States (designed, built, and tested first LPRE, 1921 to 1925 and first flight in 1926), and the visionary Hermann Oberth, originally of Romania and later a citizen of Germany (concept sketches and word descriptions in 1929 and actual firing tests in the early 1930s). There were other key people, such as F. A. Tsander (visionary and early LPREs) and V. P. Glushko in Russia (early LPRE tests, cooling and flights 1931 to 1936), or Eugen Sänger of Austria and later a citizen of Germany (designed, built, and test fired his first cooled thrust chamber 1933). They are mentioned later in Chapters 8.1, 8.4, and 5.4, respectively. The list of names of the early contributors and their contribution, which are mentioned here, is not comprehensive or complete.

Although they had some mechanics and assistants, the men of the 1920s and 1930s worked basically as individuals by themselves. This is in contrast to the team effort that has been necessary to build, test, and fly rocket engines since that time.

References for the next five chapters will be given at the end of Chapter 5.6.

5.1 Konstantin E. Tsiolkowsky

The first serious technical mention of a LPRE has been credited to the Russian teacher Konstantin Eduardovich Tsiolkowsky. He was born on 17 September

1857 in a Russian country village and died of cancer in September 1935. Even as a child, he had a remarkable curiosity about scientific things, an uncanny ability to learn by reading, and to set up simple experiments. At the age of 10 years (some historians say 9 or 11 years), scarlet fever left him with a permanent hearing deficiency. He could not hear his teachers and dropped out of school.[1-3] Thereafter most of his education came from books. Later in his life he devised a metal horn that he held to his ear and pointed into the direction of the sound. It allowed him to hear just enough to communicate. He was a self-made man, read avidly, and studied by himself. His parents sent him for three years to a technical school in Moscow, but he spent much of the time there reading technical or scientific library books of interest to him. He often lived on just bread and water, so that he would save enough to buy books that he could not find in the school libraries or to buy supplies for the technical experiments he performed in his room in Moscow.

Even though Tsiolkowsky never attended secondary school or graduated from a university, he easily passed in 1878 the exam for becoming a secondary school teacher. Most of his life he taught high school mathematics and physics. His first teaching job was in Borovsk, a small community 70 miles south of Moscow. Here he started thinking about space flight. He wrote several technical papers, which were not published until much later. His paper on the resistance of air was actually published in 1881. In 1892 he had the opportunity to get a better teaching job in Kaluga, a regional capital city 120 miles south of Moscow. He lived in Kaluga until he passed away. In his spare time he diligently pursued his analysis, writing (on various technical concepts and some science fiction), and his research. His life was hard, and at times there was not enough to eat. He could hardly afford the supplies needed for his research. For many years he was largely an unknown, far removed from the centers of technical excellence. His picture in Fig. 5-1 shows him at about age 65.

His technical interests centered on three areas:

1) He investigated airborne dirigibles with a metal skin and heated gas, but his ideas were never really supported.

2) He studied metal airplanes and aerodynamics; in 1897 he built an "aerodynamic pipe" or wind tunnel, the first in Russia. In 1900 he obtained a small grant from the Tsar's Academy of Sciences, and with it he used the wind tunnel to determined what is today the drag coefficients for a variety of simple aerodynamic shapes.

3) His avocation was space flight, and he wanted LPREs to accomplish it.[1,2,4]

He started his deliberations about space flight and rocket vehicles in about 1883. In the late 1890s he tried to publish his ideas, but several magazine editors rejected his manuscripts as being just fantasy. Finally, in 1903 the relatively obscure magazine *Nauchnoye Obozreniye (Scientific Review)* in St. Petersburg accepted his article "Exploitation of Cosmic Expanse via Reactive Equipment," which also has been translated as "Investigating Space

Fig. 5-1 Konstantin Eduardovich Tsiolkowsky, Russian visionary. Courtesy of American Astronautical Society, copied with permission from AAS History Series, Vol. 6.

with Reaction Devices." Because the article was long, it had to be published in two installments. The first part appeared in May 1903. Shortly thereafter the magazine was suspected of political activities and was shut down by the tsar's police before the second part could be published. This second part of Tsiolkowsky's paper was finally published in a revised form in 1911 (including most of part I) in the magazine *Vestnik Vozdukhoplavaniya* (*Heralds of Aeronautics*). It was republished again in 1912, 1914, and in an expanded edition in 1926.

In these classical papers this remarkable man described the use of rocket-propelled vehicles for investigating the upper layers of the atmosphere and, in the future, also for interplanetary flight.[4] He mathematically derived the laws of motion of a body with a changing mass. He was the first to show the equation for the Earth escape velocity, and he was the first to develop the equation of rocket motion, which has also been called the Tsiolkowsky equation. In their original form his equations were complex and difficult to follow. They define the velocity achieved by a vehicle that has been accelerated by rocket propulsion, while the vehicle mass decreases by the amount of propellant consumed. In today's idealized simple form it gives the flight velocity in a gravity-free space vacuum. He said that space flight was feasible and that LPREs were needed to make it a reality. At a later time Tsiolkowsky developed some approximate (but incomplete) methods for determining the drag and gravity forces, which are needed to determine an actual flight velocity and a flight trajectory.

He also wrote about an artificial space-ship like the moon, but brought arbitrarily close to our planet to a height just beyond the atmosphere.[1-4] Today we call it a satellite. He described weightlessness in space and rudders for steering. He superficially discussed interplanetary platforms as in-between stations on the way to the moon or the planets, and later he suggested gyroscopic stabilization.

Tsiolkowsky analyzed the energy needed for spaceflight, analyzed a number of different propellants, advocated bipropellants, and understood the concept of optimum mixture ratio. He concluded that liquid propellants had more energy than solid propellants, and therefore that liquids are preferable for spaceflight. He made only a few oversimplified conceptual sketches of a liquid propellant reaction device and relied mostly on rather general and sometimes somewhat disjointed verbal descriptions. He wrote about a pumped propellant feed system, the expansion and acceleration of the gases in the nozzle, and he believed that good mixing was necessary for high combustion efficiency.[4] These were all new ideas at the time.

In his conceptual sketches the LPREs had a very small combustion chamber (or no combustion chamber) and a very long, voluminous exhaust nozzle with a small cone angle and a trumpet shaped diverging exit.[1,2,4] In one of his sketches, Fig. 5-2, the nozzle-exit cone extended almost the full length of the vehicle. He wanted to burn the liquid propellants at high pressure, producing a lot of heat and very high thrust-chamber wall temperatures. His writings sug-

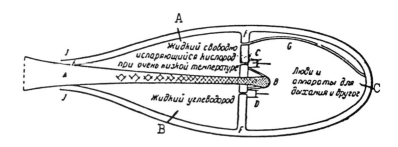

Fig. 5-2 One of Tsiolkowsky's imaginative sketches of his 1903 visualization of a future spaceship: A, container for freely evaporating LOX at a very low temperature; B, liquid hydrocarbon container; and C, relatively large crew compartment with oxygen supply for breathing and other apparatus. From Ref. 4.

gest a refractory material to line the chamber as well as the nozzle for heat protection. He also stated that the TC walls needed to be cooled. In one of the sketches, a part of the oxidizer flow is used as a coolant for the vehicle, presumably for reentry protection. His suggestion to use propellant liquids as a TC coolant was taken seriously by the Russians because they were one of the first to use cooling jackets and regenerative cooling in early Russian LPRE in the early 1930s.

Figure 5-2 indicated simple piston pumps, but it does not show the power source or control. He did not describe the propellant feed system, the design of the thrust chamber with its nozzle, or the cooling system. His propellant tanks were odd shaped, structurally heavy, and not of the right volume proportions. In spite of the inadequate detail, he was the first to seriously mention his concepts of a LPRE. His concepts, based solely on analysis and imagination, were remarkably close to what later was proven to be workable. He did not build, test, nor launch a LPRE. That was first accomplished by an American. Later in his life he had the satisfaction to learn that some of his ideas about LPREs were used by Russian early investigators (some had been his correspondents or disciples) and to hear about the first USSR launches of simple vehicles with LPREs.

Tsiolkowsky visualized the use of growing plants in spaceships for life support, the use of the sun's radiation to grow food and to supply thermal energy, the need of specialized spacesuits for people leaving the vehicle in transit, and the construction of large orbital human colonies.

In a 1929 paper Tsiolkowsky wrote about his concepts of multistaged vehicles. This idea was first described in the literature by Goddard in 1914 and in 1919 and by Oberth in 1923. However his Russian biographers contend that Tsiolkowsky conceived multistaged vehicles before Goddard or Oberth, but did not publish them at the time. There are other concepts, where there is controversy as to who did it first. When Oberth first published his version of a space vehicle, Tsiolkowsky wrote a letter to him saying that he had already described such a spaceship at a much earlier date.

Tsiolkowsky studied and suggested several liquid propellant combinations, an accomplishment, which is amazing for a visionary who never had any direct personal experience with them or the education to perform such analyses.[4] In his analyses liquid oxygen was the only oxidizer, but he proposed several fuels, namely, alcohol, hydrocarbon, methane, and liquid hydrogen. He selected liquid hydrogen and liquid oxygen as the best combination for spaceflight because they were more energetic, and this conclusion is still valid today.

Tsiolkowsky carried on an extensive correspondence, which became more voluminous after he became famous around 1919. He exchanged letters with several of the leaders in the Russian LPRE and spaceflight development, such as Valentin Glushko, the foremost Russian developer of large LPREs and Fridrikh Tsander, who was an early pioneer, space protagonist, and early LPRE developer.

Shortly after his death in 1935, Communist authorities broke into his home and confiscated and hauled away almost all of his laboratory equipment, books, sketches, correspondence, and writings. Later, when his home was made into a museum, some of these properties were returned. Most of his notes, papers, concepts, propulsion ideas, or theories were published 5 to 16 years after his demise.

Tsiolkowsy received recognition for his pioneering studies only late in life, mostly in his 60s and 70s. At that time the Soviet government wanted to publicize contributions to science, and Tsiolkowsky fit their purposes. In 1919 he was elected to the Socialist Academy (today the Russian Academy of Science), which was an honor. He was awarded the Order of the Red Banner of Labor by the Soviet government. He became well known in the Soviet Union, gave a nationwide speech on radio on May day, and became a world-renowned space scientist. With his fame also came a modest stipend and a pension, and he could then live a more comfortable life. Museums in Kaluga and Borovsk and two statues in his honor were built after his death. In 1954 the USSR government established the Tsiolkowsy Gold Medal, which is awarded every three years for interplanetary achievements. Today children in Russia learn about Tsiolkowsky in school.

5.2   Robert H. Goddard

This American physics professor, Robert Hutchings Goddard of Clark University in Worcester, Massachusetts, was a creative early researcher and the most important first developer of LPREs. He was the first to design, develop, build, static test and fire a small LPRE and the first to launch a flying vehicle with a LPRE.[5] He first designed a TC in 1921, developed and tested its key components (thrust chamber, propellant tanks, pressurization system, and propellant valves), and static fired LPREs during the years 1923 to 1925. He was the first to launch a sounding rocket with a simple LPRE and a simple pressurized feed system on 16 March, 1926 in Auburn, Massachusetts. It was the world's first launch with a LPRE. Thereafter he developed more advanced LPREs and flew many sounding rockets. His portrait is shown in Fig. 5-3.

There has been a claim by a man named Pedro Paulet, a young engineer from South America, of having invented, built, and tested a LPRE before 1900 (Ref.6). His story was published as a small article in a South American newspaper in 1920. It said that he had built and fired a bipropellant liquid propellant rocket thrust chamber in 1895 using nitrogen peroxide (it is not clear what this oxidizer really was) and gasoline as propellants with thrust up to 90 kg or almost 200 lbf. He used spark ignition, a conical nozzle of 52-deg angle, and intermittent propellant injection at 300 explosions per minute. He reportedly operated this engine for more than one hour. Tests were terminated in 1897. A reconstructed sketch of this Paulet rocket engine is shown in Fig. 5-4 as prepared by James Wyld, one of the early U.S. pioneers in LPREs and based on his interpretation of the newspaper article.[7] Paulet's concept description was short, incomplete, vague, and without proof. Today historians discount this claim because it was not published until 25 years after the event because there were no witnesses or substantiating documentation and no identification of the test location. It is unlikely that his oxidizer was then available in Peru and no one has since operated a thrust chamber with non-hypergolic propellants and an igniter at this pulsing frequency. It is also doubtful that a single individual could accomplish all this, when talented investigators in other countries took years to come up with such an advanced pulsing thrust chamber, with the claimed relatively high thrust level and firing duration, and a fancy ignition system. Nevertheless it is an interesting concept. There is a consensus among historians that Goddard was the first to fire and launch a LPRE.

Goddard was born in 1882. From childhood on, he was interested in physical sciences and in conducting experiments.[5] Throughout his life he was dedicated to his work, always curious, eager to learn, yet reticent and quiet. Besides his excellent work in LPRE, he was interested and pursued analysis or experiments on spaceflight, solid-propellant rocket motors, solar energy heating and propulsion, producing electrically charged particles (ion propulsion), balancing airplanes by means of gyroscopes, or generating hydrogen and oxygen on the moon. He had a good technical education and was strongly

Fig. 5-3 Robert Hutchings Goddard, rocket pioneer and visionary. Photo from AIAA.

influenced by his physics teachers. At the age of 41, he married Esther Christine Kisk in 1924, who was much younger. She not only was a good wife, but also his assistant, typist, and helper. For example, she operated the cameras recording his tests and flights and often pitched in as a member of the test or launching crew.

In 1913, at the age of 30 he was diagnosed with tuberculosis in both lungs and told he had not long to live.[5] Even though his doctor did not agree, he reasoned that slow, deliberate, deep breathing would bring fresh oxygen into his lungs and steam heat would keep the air dry in his room. His determination to rest, breathe deeply, and get well helped him to overcome the worst of the disease in about a year. During this convalescence, he had time to think and was allowed to write for one hour a day. He thus formulated the basic ideas, started to prepare patents on rockets, independently derived the principal theory, and wrote several important papers, but they were not published at this time

Goddard invented or discovered many of the features that are used in LPRE today, and several are enumerated in this chapter.[8] He launched sounding rockets with LPREs initially at Auburn, Massachusetts (1926–1930), and later at Roswell, New Mexico, during 1933 and 1938. His pioneering work was supported by the Smithsonian Institute, the Daniel and Florence Guggenheim

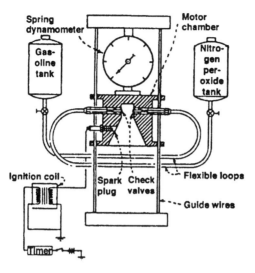

Fig. 5-4 Paulet's 1895 rocket engine as reconstructed from a brief newspaper article published in Lima, Peru in 1927. Copied with permission from Ref. 10, pp. 11–13.

Foundation, the Carnegie Institution of Washington, and his own Clark University. Charles Lindberg first visited him in 1929 and again later and he was instrumental in obtaining substantial Guggenheim grants for Goddard's rocket work.

His first analyses and deliberations about LPRE can be traced to 1908. In July 1914, shortly after his illness, he was granted two basic U. S. patents on solid-propellant and liquid propellant propulsion systems. One of the figures of his U.S. patent 1,103,503 established the basis for modern liquid propellant rocketry, is based on analysis and conceptual thinking, and was written before he did any building of LPRE hardware or any experimental work. It is copied here as Fig. 5-5 and describes the essential elements of a LPRE with a pump feed system. It was called "Liquid propellant rocket apparatus," and its explicit description includes these words:

> ... a combustion chamber (item140 in the figure) having a refractory lining 141 and a rearward extending tube 142 (nozzle), ... two tanks (144 and 143) containing liquid material, which when ignited will produce an exceedingly rapid combustion. This result may be attained, for instance, by filling the tanks with gasoline and liquid nitrous oxide (some years later he corrected it to mean nitrogen dioxide and then liquid oxygen). As this substance is a liquid only at low temperatures, it is necessary that this tank be filled immediately before the discharge of the apparatus. This tank is enclosed in another tank 145 and the space between is filled with a suitable non-conductor (thermal insulator) or a cellular vacuum casing. Force pumps 148 and 149 are used to feed the liquid materials through pipes 150 and 151 from the tanks to the combustion chamber. One form

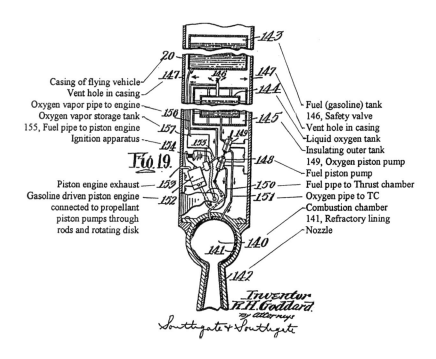

Fig. 5-5 Reproduction of Fig. 19 of Goddard's U.S. patent 1,103,503 issued 14 July, 1914 showing a LPRE with a pumped feed system.

of force pumps are piston pumps (which he later abandoned in favor of centrifugal pumps) operated by a single sliding rod to a crank pin on a rotating disk driven by a small gasoline engine 153. The force pumps should be so proportioned so that the proper mixture of gasoline and nitrous oxide will be at all times fed to the combustion chamber. The combustion is continuous and the propelling force is constant. . . .

Some of his colleagues urged him to publish some of his work. So he arranged for a paper, which he had previously submitted to the Smithsonian Institute, which was one of the sponsors of his research, to publish it.[9] Goddard paid for the cost of publication out of his research grant.[10] In 1919 his learned paper entitled "A Method of Reaching Extreme Altitudes" was published.[9] It discussed the theoretical possibility of flying to high altitudes by means of well-designed rocket vehicles and engines. It showed that high-exhaust jet velocities and reasonable payloads could be obtained. He included formulas for estimating payloads and sample calculations of possible vehicles. He mentioned that scientific measuring instruments could be sent to high altitudes. This publication had a good reception in technical circles, and it inspired enthusiast in the United States and in Europe. The press thought it offered a way to go to the moon. Although Goddard was not pleased about the moon publicity, which he considered unsubstantiated by his work, the press stories did give him unexpected fame. Goddard gradually gained an international reputation. However the 1919 paper gives no information about his concepts for LPREs or his ideas about the design of flight vehicles.

During World War I, Goddard volunteered his expertise and helped the U.S. Army to develop solid-propellant barrage rockets.[5,8] They worked well, but the Army did not put them into production because the war had ended in 1918. His early work with solid-propellant rocket motors (1914–1920) was abandoned in favor of liquid propellant engines because his theoretical analyses showed him that liquids would give more energy per unit propellant mass. His first detailed thrust-chamber (TC) design sketches were dated in 1921, and the first successful static firing tests took place in 1923 using liquid oxygen and ether as propellants.[8,11] He used high-pressure air or nitrogen to expel the propellants from their tanks, and he thus built and test fired the very first LPRE with the first gas-pressurized feed system. These early TCs were small (about 1.2 in. diam), and some of his TCs are shown in Fig. 5-6. This particular TC version had a ceramic sleeve in the chamber, a fuel hole and an oxidizer hole in the injector, and a long, slender nozzle with a nozzle half-angle of 4 deg. Static firing tests of different TCs continued periodically until about 1938. He adjusted the hydraulic flow resistance with rotary valves in the propellant feed lines in order to obtain the intended propellant flows and desired mixture ratio. He managed to get firing durations of more than a minute by using a fuel-rich propellant mixture and supplementary film cooling. Figure 5-7 shows an injector that had relatively large single nonimpinging injection holes (lined

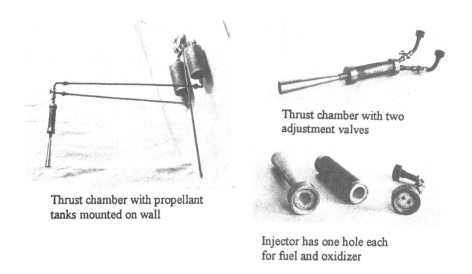

Thrust chamber with propellant tanks mounted on wall

Thrust chamber with two adjustment valves

Injector has one hole each for fuel and oxidizer

Fig. 5-6 Left figure shows one of Goddard's first test setups of a small TC and two welded propellant tanks. On the right is the assembled and disassembled TC. Copied from Ref. 8 with permission of McGraw–Hill Book Company.

Fig. 5-7 One of Goddard's early small injectors with relatively large-sized single fuel and oxidizer holes and many peripheral film-cooling holes. Copied from Ref. 8 with permission of McGraw–Hill Book Company.

with ceramic sleeves), one for the fuel and one for the oxidizer and a large number of small film-cooling holes near the periphery of the injector.[8] The film coolant would provide a relatively cool boundary layer to the chamber and nozzle walls and thus reduce the heat transfer and wall temperatures to acceptable levels. In some other injectors he used slots to inject the film coolant in a tangential direction. Over the next 15 years he tested different injection schemes, different chamber sizes, and different fuels (mostly gasoline, but also alcohol or kerosene).[8,11] One of his designs had a jacket around the chamber and the nozzle; oxygen was circulated through the jacket to evaporate it and to provide some cooling. He tested it, but did not pursue this design. He also tested other components, such as positive displacement pumps (and later centrifugal pumps), turbines, jet vanes, valves, pressure regulators, or igniters.

Judging from the volumes of several of his propellant tanks, Goddard often operated LOX/gasoline at a fuel-rich mixture ratio (the ratio between the oxidizer flow rate and the fuel mass flow rate) of perhaps 1.3 or 1.4, and he thus would have obtained specific impulses of about 150 to 190 s and nonoptimum lower combustion temperatures of perhaps 3600 to 4000°F. Although this is a poor performance compared to a modern equivalent LPRE, it is a most remarkable accomplishment that no one else had done before, and it was ground-breaking pioneering work. Today the optimum mixture ratio is known to be between 2.1 and 2.4, and the theoretical specific impulse (at 1000-psi chamber pressure expanding to atmospheric pressure) is about 245 s with a combustion temperature of about 6000°F. Goddard became aware of this and did change to higher mixture ratios in his later work.

Between 1926 and 1941 Goddard static fired about 95 different LPREs, attempted to vertically launch 49 different rocket vehicles propelled by a LPRE, and actually launched and flew a total of 33. A number of these did not achieve their intended flight path. They were all homemade sounding rocket vehicles with homemade LPREs, built and assembled in his own shop.[8,11]

Every one of these flight configurations had some new features, improvements, or design changes, and he never flew the same design twice. Although he had the help of some very talented mechanics and machinists, Goddard really was the designer, engineer, and manager of his projects, and he worked basically by himself. He was the guiding spirit, manager, and test engineer for his projects. He designed his own LPRE, his flight vehicles, and his rocket launchers and then had them built in his own shop by skilled craftsmen. He analyzed his own test data and kept his own records of his tests and design efforts, and he meticulously kept a diary.

Goddard's *historic first-flight* rocket vehicle with the first flying LPRE (1926) is shown schematically in Fig. 5-8. This diagram was actually made for the Science Museum in London just before the museum made its own copy of his first flying machine with a LPRE.[12] It was based on discussions with some of Goddard's technicians and is probably a better drawing than the sketches that Goddard made for his technicians.

Unfortunately these sketches do not seem to exist or are not readily available today. This first rocket vehicle rose 41 ft above the launch stand and flew 185 ft in about 2.5 s. It had the thrust chamber at the front of the vehicle and the long tanks at the aft end. He believed that this configuration was naturally stable and did not require a flight control system. The propellants were liquid oxygen and gasoline, which is called petrol in Britain, and this word is used in the call-out of the drawing. The propellant feed lines also served as the structure to tie the key components together. He used a crude simple cone as a heat shield to protect the tanks from being overheated by the rocket exhaust plume. The black powder igniter was in a tube on top of the TC. Ignition of the powder was achieved by a flame from some broken-off match heads inside a copper tube, which in turn was heated externally by some burning cotton. The cotton rags are not shown in this picture.

The two propellant tanks were both pressurized by gaseous oxygen, which was evaporated from the oxygen tank shown in Fig. 5-9 (Ref. 12). The needle valve at the bottom of the inner tank (but actuated by a spring loaded lever from the top of the tank) allowed some liquid oxygen to flow through an orifice down to the outer tank bottom, which is heated by an alcohol fire (not shown) from below. This liquid is evaporated into oxygen gas and used for pressurizing both propellant tanks. There is a check valve (identified as non-return valve in the drawing) in the pressurizing line to the fuel tank to prevent backflow of fuel to the oxygen tank and another check valve in the oxygen fill line. After the crew sees the flame of the igniter, they pull on a string to remove the roller-ended prop (on top of tank), which activates the needle valve and starts a small flow of oxygen to be evaporated. The cardboard sleeve around the inner brass tank is a thermal insulator. The mixture ratio was preset by two small needle valves near the TC. The lower part of the nozzle had burned off during the last part of the first flight.

Goddard was a reclusive individual, shy, and reluctant to give out technical information.[13] He did not disclose much detailed LPRE information that would have been useful to others in this field. In his 1927 autobiography, which was

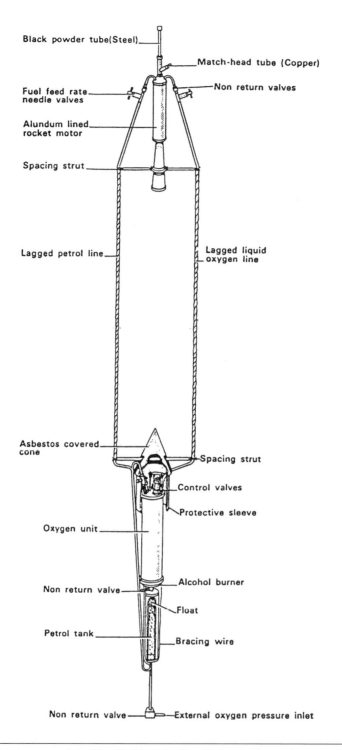

Fig. 5-8 Diagram of Goddard's historic first flying rocket vehicle with a LPRE, launched on 26 March, 1926 in Auburn, Massachussetts. Copied with permission from Ref. 12.

The Early Years, 1903 to the 1940s   255

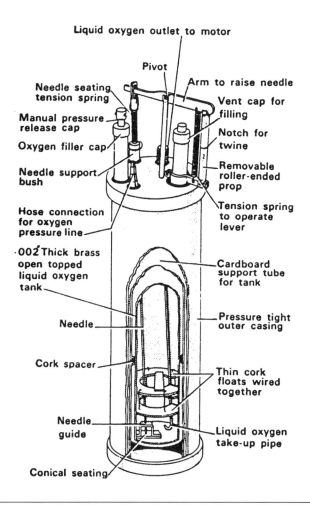

Fig. 5-9 Partially sectioned diagram of the liquid-oxygen tank used in the historic first LPRE flight. Copied with permission from Ref. 12.

released by his widow after his death, he mentions his 1926 first flight and other events, which are indirectly related to a LPRE, but the paper contained nothing about his LPREs or other detailed data.[10] In 1935 Goddard wrote a report to his sponsor, the Guggenheim Foundation entitled "Liquid Propellant Rocket Development," and it was published in 1936. Most of this report[14] is about his New Mexico operation, showing pictures of control stations, launch towers, the workshop, cameras, recording telescopes, and some photos of actual flights of sounding rockets. There is one short paragraph about the LPRE, and it gives the propellants, lifting force, size of TC, and duration, but no information on the materials, injection scheme, cooling, or development problems. This 1936 publication was quickly out of print, which limited the distribution. It was reprinted in 1949. The 1948 booklet of his liquid fuel research between 1929 and 1941 released after his death was more revealing,[11] but it too was short in data or details.

Goddard had really astonishing LPRE ideas and test results, but like all early engine developers he had failures in the tests, and some of the concepts or developments turned out to be unsuccessful. Most of the time Goddard understood the cause of each particular failure, and he tried to remedy reoccurrences with changes in design, operating procedures, and/or materials. Goddard's failures were typical of those encountered later by other pioneers and early LPRE organizations. They included complete or partial burnouts of chambers and/or nozzles, ignition malfunctions, inadequate welds, vendor delivery of liquid air instead of liquid oxygen, valve leaks and valve impairments, operator errors, seizing of pump shafts, burnout of turbine blades, defects in the tank pressurizing system, improper start sequence, etc. The same types of failures have been encountered by most of the other early pioneers and also by all developers of LPREs. Unfortunately these failures with their remedies (and the designs that were abandoned) were not always documented and never cataloged in a form suitable for helping other LPRE investigators in avoiding similar pitfalls.

Here are Goddard's more *significant contributions* to the state of the art of LPRE.[5,8,11,13,14]

1) Before he started his experimental work on rocket propulsion, Goddard proved experimentally in his university laboratory that combustion would take place in a vacuum and thrust could be generated in a vacuum.[8,10] He also proved experimentally that the oxidizer and fuel, when premixed, can form powerful explosives that burn very rapidly; however, with controlled mixing, as in an injector, he observed slower but steady burning. Both of these phenomena were subject to doubt among physicists of that time.

2) In 1923 he static tested or fired the very *first bipropellant thrust chambers* with live propellants.[8] Figures 5-6 and 5-7 show they were small in size. Like his later TCs they had a chamber with a cylindrical shape. Many of his early designs had burnout failures after 1 to 7 s of operation. After a number of trials, he learned how to design thrust chambers, so that they could be test fired for more than 25 s (1925) and then for a minute (1926) without burnout, by using an injector scheme with a central oxygen distributor and slots or orifices for tangentially injecting the fuel against the chamber wall. It was, what is called today, a type of film cooling, with the film coolant all being injected at the injector end of the chamber. Figure 5-7 shows one of his early injectors. It had a single oxidizer injection hole, a single parallel injection fuel hole (both using ceramic pipes) in a nonimpinging pattern, and a circular array of small holes for injecting extra fuel, which then forms a cooler protective fuel-rich gas boundary layer at the chamber and nozzle walls. Years later other investigators found that such a nonimpinging parallel stream injector does not give good combustion efficiency and required a large chamber volume for a reasonable performance. Actually in his first

tests of this injector, he used steam (formed by heating water) flowing through the small film-cooling holes, but in later design he used fuel for film cooling. Goddard also used tangential injection of film coolant in some injectors. He initially used more than half of the fuel flow for film cooling, but later he was able to reduce this to 15 or 20% of the fuel flow. For comparison modern TCs use less than 5%. Also his nonimpinging injector elements were later identified by others as having poor combustion efficiency. Therefore his early TCs had, what would be considered today, very poor performance. Nevertheless, he accomplished something no one else had done before; he could run his film-cooled TCs for more than a minute. He tried several different injector designs. He also tried, but later abandoned, a set of branched external pipes leading to individual injection spray nozzles. His early tests were at relatively low chamber pressures (less than 150 psia) and low thrust level (typically 40 to 100 lb). He later developed TCs with up to 10 in. diam and 3000 lbf thrust. Some of the early chambers had a ceramic sleeve insert for heat protection of the chamber wall, and one used a porous wall for injecting the film coolant, but these designs were not selected for application in his sounding rocket flights. Around 1938 or 1939 he learned about regenerative cooling (which was done earlier by others), but he continued to use his single-wall film-cooled thrust chambers for several more years. Goddard even received a patent for a spherical chamber with injection spray elements all around, which was expensive to fabricate. He also received patents for a rotating TC and for a TC with multiple nozzles. He did not pursue any of these three ideas. He also experimented with water-cooled nozzles. The resulting steam was injected into the thrust chamber, often as a film coolant, augmenting the propellant mass flow, but decreasing its performance (specific impulse). He abandoned water as a coolant for the nozzle in the late 1920s.

3) His TCs consistently used a simple *supersonic (bowtie-shaped) nozzle* as seen in Fig. 5-6. In the early years of his work, his nozzle-exit cones (with a cone half angle of 4 to 5 deg) were very long and relatively heavy. It seemed to have been the style because he and other rocket pioneers believed that a small angle reduced the divergence losses, but they seemed to ignore the friction or drag losses. Later (1940), after some optimization studies by others, he adopted shorter (lighter) nozzles with larger cone half-angles (12 to 18 deg), which is what was used in the United States for about 20 years (1938 to 1958) and is still used in a few applications today.

4) He developed and tested the *first propellant feed system using high-pressure gas* to expel the liquid propellants from their respective storage tanks into the thrust chamber (1923). In most of his static firing tests of TCs, he used compressed air or nitrogen for tank pressurization. As seen in Fig. 5-8, he pressurized both propellant tanks with vaporized oxygen in his first few flights, perhaps to save the inert weight of a heavy high-pressure gas tank, but he later abandoned this approach as being too risky because fuel vapor

and oxygen gas will form explosive mixtures. He also used a safety valve in the oxygen tank to prevent overpressurization, which can be caused by excessive evaporation. He subsequently tried carbon dioxide as a nonaggressive pressurizing gas in fuel tanks, but abandoned it, and he later settled on compressed inert nitrogen gas or compressed air for expelling propellants from both the fuel and oxidizer tanks. This type of feed system with propellant tanks, propellant valves, fill and drain provisions, check valves pressure regulator and safety valves is still used extensively today.

5) He developed and built a series of lightweight *propellant valves* specialized for use with LPRE. Suitable lightweight commercial valves were then not available. His first propellant valves were clever, lightweight, and were initially actuated by pulling strings from the test control station. He improved some of the valves by using pressurized gas (and in some versions also liquid propellant) for actuation beginning in the 1930s. He also developed his own safety valves, vent valves, and control valves. He also had solenoid operated valves in the early 1940s. Some of his valves had a controlled opening rate, thus ensuring low initial propellant flow. He developed some valves with position sensors and bellows-controlled opening programs. Many of the key features of several modern propellant valves find their heritage in Goddard's valve work.

6) He recognized the need for and developed several lightweight *gas pressure regulators* for LPREs for the use with pressurized feed systems. Such a regulator was needed to control the propellant tank pressure and thus maintain the thrust at nearly constant level. He used bellows and external springs for the design. The basic concepts of his regulators are still valid today.

7) Goddard was the first to build *lightweight propellant tanks* out of thin steel or aluminum or occasionally brass sheets. Early versions were cylindrical with flat or conical ends, and later versions had ellipsoidal rounded ends. From an early failure he learned to pressure test and leak test every tank and every pipe after installation. He was the first to introduce baffles into the propellant tank to minimize sloshing. He knew that sloshing caused unpredictable changes in the vehicle's center of gravity. One of his simple baffles can be seen in Fig. 5-10, which had to be installed before the tank was fully welded. Improved versions of this slosh-damping feature are found in many propellant tanks today. To prevent excessive ice formation (with cold cryogenic propellants), he was the first to apply thermal insulation (initially cardboard and a type of felt and later asbestos) to liquid-oxygen tanks, lines, and valves. He also used lightweight floats in his propellant tanks, partly to minimize evaporation and sloshing and partly to allow external measurements of tank liquid level, as described in item 11.

8) He started by using black powder for the *igniter* (1922–1926), which might have been a carryover from his prior work on solid-propellant rocket

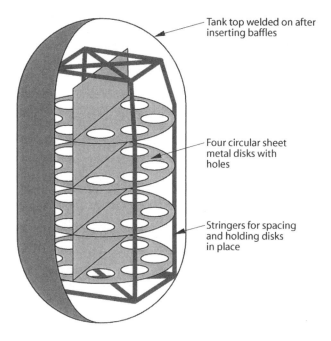

Fig. 5-10 Sketch of Goddard's antislosh baffles consisting of perforated sheet metal inside propellant tank.

motors. He next developed and invented a reliable pyrotechnic igniter with double-base solid propellant (1927–1928) and later a spark-plug igniter for repeated starts of the same TC (1937). He obtained patents on his igniters.

9) Goddard used *external high-strength steel wires* to reinforce his high-pressure gas steel or aluminum tanks, a clever design trick that is still used today in making lightweight tanks. Today high-strength synthetic fibers are sometimes used instead of the steel wire.

10) Only his first few LPRE-propelled sounding rocket vehicles had the thrust chamber at the vehicle nose because he believed that this provided flight stability when flying in an atmosphere. All of the other test vehicles *had the thrust chamber at the tail end*. With this configuration he provided flight stability (pitch and yaw moments to prevent tumbling) by using large aerodynamic fins at the vehicle's tail end. The use of tail fins was then not new and is still being used today in many missiles flying within the atmosphere. The aft location of the thrust chamber is still the standard for rocket-propelled vehicles today.

11) He made crude *measurements of propellant flow* and the propellant remaining in the tanks. He used floats (initially cork) on top of the liquid

surface of the propellant tank with a long wire sticking up, and the wire was visible in a vertical glass tube attached to the top of the tank. He also worked with venturi flowmeters. Later he invented and built an electrical level gauge by measuring electrical resistance. He then could calibrate or control the propellant flows with an orifice (or an adjustable needle valve). He conceived of an automatic adjustable propellant throttle valve, but it is not clear if this concept of flow control was actually implemented by him or flown. He understood the need for mixture-ratio control, which was way ahead of his time. Today this would be called propellant utilization system.

12) Goddard was the first to use *centrifugal pumps* for liquid propellants, the first to develop a *gas generator* (GG) for driving the turbine, and the first to design, test, and fly *turbopumps* (TPs). He recognized the need for feeding propellants by pumps, in order to reduce the propellant tank pressure, which in turn saved considerable tank weight (and considerable inert vehicle weight), and thereby increased the vehicle's propellant fraction and performance. His other initial motive was to use positive displacement pumps to achieve predicted propellant flows, a precise desired mixture ratio, and thus a nearly constant chamber pressure and constant thrust. So he initially tried and individually tested several types of positive displacement pumps, which are really constant-volume flow pumps, like vane pumps (1921), piston pumps (1923, with a piston inside a tube), gear pumps (1924), and rotary vane pumps. He did not achieve this goal, in part because of internal leakages of some of the pumps and the unpredictable evaporation of oxygen causing difficulty in maintaining a constant oxidizer density and thus a constant-propellant mass flow or mixture ratio; furthermore, the scale was too small to achieve precise flows. He also tried reciprocating bellows pumps (1927), actuated by pressurized gas, but he later switched to centrifugal pumps (1933–1934). He used a simple shaft with two ball bearings, a shrouded turbine on one end and a single centrifugal pump impeller on the other end. Figures 5-11 and 5-12 show some of his TP machinery. He also developed a special bearing cooled by the oxidizer that would work in a cold liquid centrifugal oxygen pump and received a patent on it. With liquid oxygen in the impellers or the bearings, the physical contact between the rotating assembly and the stationary parts had to be avoided because a prolonged contact or rubbing would cause the metal parts to become hot, and if hot enough the metal will burn with the oxygen, melt, and fail. For static component pump tests the power for driving these pumps often came from electric motors or air jets blowing on the blades of the turbine. He conceived of driving the fuel pump with a separate hot-gas turbines and the oxidizer pump with another gas turbine. Both turbines were partly immersed in the hot exhaust jet at the TC nozzle exit as shown schematically in Fig. 5-13. This became his first turbopumps to supply propellants to a TC. He developed and ground tested three or four different TPs and at least two different GGs. Ground TP tests were run in 1934 to 1936 with a TC. The hot TC exhaust gases stagnated locally (stagnation temperature of

Fig. 5-11 Technician holding two turbopumps. A conical diffuser extends outward from the pump discharge. The pump inlet shown with the elbow fitting is axially centered. Copied with permission from Ref. 11.

Fig. 5-12 Front and back view of a small centrifugal impeller. The holes are for equalizing the pressure on both sides of the impeller. Copied from Ref. 8 with permission of McGraw–Hill Book Company.

4000 to 5000°F) and often burned or melted the turbine blades; furthermore, the extra drag of the blades reduced the engine's thrust. He abandoned the scheme of using gas from the main nozzle of the TC. Later in his work (1938) he used a separate small combustion chamber with oxygen and hydrocarbon fuel, at an oxygen-rich mixture ratio. (About a dozen years later when the U.S. LPRE industry needed GGs for the large LPREs, the selected GG mixture ratio was fuel rich, in part because it was believed to be safer and did not involve high-temperature oxygen, a powerful oxidizer that might start fires of gaskets and metals. If it would have been known that Goddard successfully ran oxygen-rich GGs, it is likely that such a GG might have been considered more seriously. The Soviets used mostly

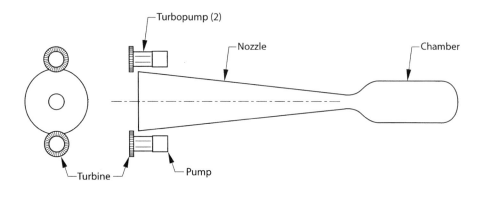

Fig. 5-13 Simplified diagram of the position of the two turbopumps at the nozzle exit of the TC for the initial hot-fire tests conducted about 1934 to 1936.

oxidizer-rich GGs and many oxidizer-rich preburners.) The resulting gas temperature of 190 to 300°C or 375 to 572°F at the turbine inlet was much lower than the gas in the main thrust chamber. This eliminated turbine blade burnout and also allowed an uncooled chamber for creating warm gas without the extra complication of a cooling jacket. Today it is called a *gas generator*. His earliest GG component tests (without turbine, 1932–1934) used three propellants, namely, liquid oxygen, hydrocarbon fuel, and water to dilute the gas and reach a gas temperature at which turbine blades would not be damaged. His initial gas-generator-type feed system therefore required three pumps, one each for oxidizer, fuel, and cooling water. He then went to a bipropellant GG operating at a very oxygen-rich mixture and thus at lower temperature. The hot gas from the GG was mostly hot oxygen gas and water vapor. A sketch of one of his gas generators is shown in Fig. 4.5-5 and a photo in Fig. 5-14. In this photo a screen-

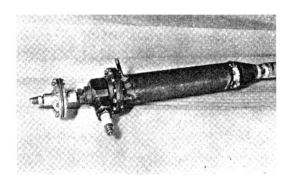

Fig. 5-14 Goddard built his first gas generator as a small cylindrical uncooled combustion chamber with tangential injection of LOX and water. Copied from Ref. 8 with permission from McGraw–Hill Book Company.

type filter is on the left, the fuel enters the uncooled cylindrical GG chamber through a small central spray element (not shown), and ignition was accomplished by a spark plug. He developed a simple circular gas inlet manifold with four nozzles for the GG gas to enter the turbine.

At least nine vehicle tie-down ground tests and two flight tests in 1939 and 1941 were run with a LPRE that had a turbopump feed system (with a gas generator and turbine-driven pumps).[8] He anticipated and predated turbopump and feed system features common in today's large LPRE, and he devised controls for starting and maintaining the thrust of such a LPRE. He also investigated a scheme where water was used in a partial TC cooling jacket (water is an excellent coolant), and the resulting steam was then used to drive the turbine. However he abandoned this scheme. He selected a feed system with excess oxygen in the GG to reduce the GG gas temperature. He obtained a patent for a restartable LPRE with turbopumps that used gas from a three-propellant GG (including water) and one patent with a bipropellant GG. He also developed an electromagnetic sensor to measure the shaft speed of the turbopumps.

13) In some of his LPREs, he used a small *ceramic-lined chamber for ignition* of the initial gaseous oxygen flow and a small flow of fuel. It was centrally located on the injector, and the ignition was started by a spark plug. This was the first U.S. TC with LOX/hydrocarbon propellants that was capable of restart. Today several cryogenic LPREs use spark-plug ignition and/or a pre-combustion chamber.

14) He made good *records of his tests and flights* and devised various techniques for recording data. He filmed the ascent of all of his flights, and he recorded time by simultaneously filming a clock. For static tests he used a film camera to record the readings of several pressure gauges, a timer, thrust meter, and indicator lights for solenoid-operated valves. Later he devised a magnetic speedometer and recorded the shaft speed of the turbopumps. He also developed a technique to measure propellant flow. He devised a technique for recording valve position during static tests and developed a recording barometer (for determining altitude) that was carried on several of his sounding rocket flight tests. Much of his simple recording techniques were later copied or independently duplicated by amateur rocket enthusiasts and by other early LPRE investigators.

15) Goddard understood how to stabilize the vehicle's flight with a gyroscope. He had done work with gyroscopes earlier in his life, and he developed and built small gyroscopes and designed them into his flight vehicle in order to stabilize the flight path. The gyroscope would measure angular deviation from the vertical flight path, and this would generate a signal to correct the flight path. This signal caused pneumatic valves to open or close, admitting gas to dual bellows, which activated or moved the vanes of the flight vehicle. He built vehicles with *air vanes* and engines with *jet*

vanes. His air vanes were segments of the vehicle's skin that were inserted or rotated into the airstream (causing more one-sided drag) to achieve pitch and yaw control. This is shown schematically in Fig. 5-15. When flying above the atmosphere, he would insert flat steel vanes into the nozzle exhaust gas flow to achieve pitch and yaw control. This is shown schematically in Fig. 5-16. He also conceived and tested a metal-bucket-type jet

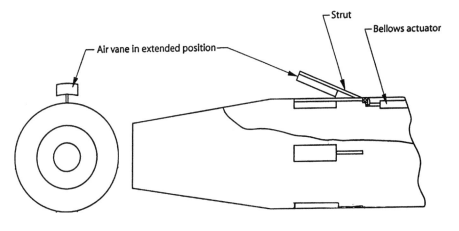

Fig. 5-15 Simplified diagram of two of the four air vanes used to achieve vertical flight control. When not extended, the air vanes are part of the vehicle's skin (first flown in 1937).

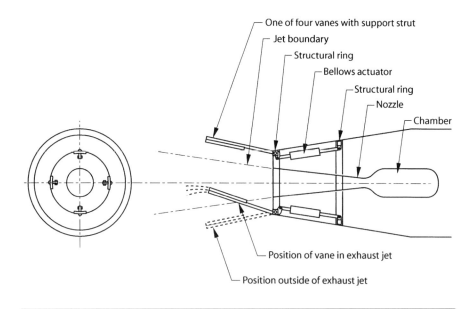

Fig. 5-16 Simplified diagram of two of the four metal jet vanes for pitch and yaw control (first flown in 1937).

vane, which did not work too well. He experienced frequent melting of these steel vanes because the melting point of the steel materials of the jet vanes was well below the stagnation temperature of the TC reaction gas. Goddard's flights with a gyroscope and vanes started in 1937. With these movable vanes his vehicles were able to fly a nearly vertical flight path, and his first flight with four jet vanes was in 1937. This solved his problems with his early flight-test vehicles, which often went off the vertical course and flew erratically, sometimes even horizontally. His steel vanes were the predecessors of the carbon jet vanes, which did not melt, but eroded gradually. Carbon jet vanes were first used operationally for thrust vector control on the German V-2 misssile around 1939, in the late 1940s also in several Soviet ballistic missiles, and in 1951 in the U.S. Redstone engine.

16) A few of his LPREs used a type of *gimbal* to support a swivelling thrust chamber. Goddard called it a "movable tail" because he tilted the whole tail end of his vehicle, including the TC, which was a part of the tail assembly. It was installed in a 1937 test vehicle. Although the references show photos of Goddard's tilting tail and its mechanism, they do not describe the function. The simplified diagram in Fig. 5-17 attempts to explain the concept. The TC is fastened rigidly to ring A, which can be tilted. The propellant feed lines to the TC are flexible high-pressure hoses, and they are not shown in the diagram. Each of the four push/pull rods are fastened to ring A and to

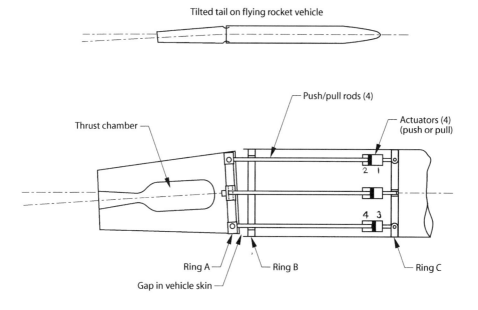

Fig. 5-17 Simplified sketch for Goddard's movable tail piece for steering or tilting a thrust chamber.

the center of four dual-bellows assemblies. For simplification a dual-piston assembly is drawn instead of the dual bellows. These bellows were in turn fastened to ring C. When the pressure in bellows number 1 and 4 is increased and at that same time the pressure in bellows 2 and 3 is decreased, then the rods will move and cause the nozzle to tilt downward. The small gas feed tubes for pressurizing or depressurizing the bellows and the control valves are not shown here. Although the assembly was suitable for tilting the tail by 6 deg, the side force on the vehicle was adequate when tilted by only 1 to 2.5 deg. Goddard said that this type of thrust vector control was more efficient because jet vanes in the exhaust plumes diminish the engine performance and often melt or oxidize. Today a rotary gimbal bearing is used to support the whole engine, and this linkage type of moving tail mechanism is no longer in use. Goddard's bellows had to accept the full thrust and axial vibration loads and had to be heavy, whereas the modern gimbal actuator only carries a smaller side load. However, the basic principle of a tiltable thrust chamber implemented by Goddard in 1937 has now advanced and been applied as a hinge (rotation about one axis) or as a gimbal (flexible joint with rotation about two perpendicular axes) in many large LPREs.

17) In the early 1940s Goddard designed and tested the first *U.S. variable thrust rocket engine*. Initial tests were with a gas-pressurized feed system, but he then also tested a variable thrust engine with a pump feed system and its gas generator. It had a small ignition chamber as a part of the TC to allow restart during flight. This engine was tested, but did not fly. However Curtiss–Wright Corporation (see Chapter 7.4) built such a variable-thrust LPRE based in part on Goddard's work.

18) Goddard flew the first rocket vehicle propelled with *multiple TCs*. When he had some problems with the larger 10-in.-diam TCs, he substituted a cluster of four smaller TCs of the same total thrust and flew them successfully in 1936. He found that this reduced the overall length of his vehicle and also saved some vehicle and engine structure and vehicle skin weight.

19) He pioneered the *controlling of the start and stop operations*, and he succeeded in making these transient durations shorter. As already mentioned in Chapter 4.7, in his early static tests and flights the valves and controls were actuated by his men pulling strings from the operating control station, which was some distance away from the test or launch facility. In some tests there were as many as 10 such strings, and it required three men to pull them in a sequence, which depended on observation, commands, or timing. For example, pulling a string can close the tank vent valves, start the tank pressurization, purge air from the fuel lines, start ignition, open each of the propellant valves (in a predetermined sequence), or open a vent valve in case of overpressurization. The start procedure was

not exactly reproducible, and it took 5 s or more. He soon substituted falling weights, which were tripped by a latch from a prior event, and using the weight to pull a string over pulleys to automatically initiate the next event. He then started to use pressurized gas (with a small pilot valve) to pneumatically open larger propellant valves; this is a form of power augmentation. He also used pressurized fuel for more accurate control of valve stem movement and actuation. He thus developed an early gas pressure control system and a semi-automatic, relatively fast, sequenced start control system. With the small TCs the start transient was shortened to less than 1 s.

20) He was the first to experience and remedy *oscillations in the feed system* to the gas generator and its propellant valves during his early GG tests in 1933 and 1934. He solved this problem by adding a damping device (dash pot) to the valve stem.

21) Even though most of his LPRE used gasoline and liquid oxygen as propellants, he investigated and tested *other propellants*. Alcohol, gasoline, and ether have already been mentioned. He did work with nitric acid and various organic fuels, including aniline, which is hypergolic or self-igniting with the nitric acid. He also designed and built a pump for nitromethane, a monopropellant; however, it never flew, and he abandoned this dangerous propellant. His analyses told him that liquid hydrogen with liquid oxygen would be superior in performance, particularly for spaceflight, but at that time liquid hydrogen was not available in sufficient quantities for him to use in experiments.

Goddard filed many *patents*, 48 during his lifetime, 35 more for which he had applied, but were issued after his death in 1945, and 131 filed by Mrs. Goddard as his executrix after his death, but based on his notes, sketches, films, and photographs.[8] The total number of Goddard patents is thus 214, and most deal with LPRE or sounding rocket vehicles. For example, he obtained a patent on a two-stage vehicle, film cooling, and a system for a restartable LPRE. The U.S. government bought the rights from his widow to use 200 of these patents for $1,000,000, and some of this money went to the Guggenheim Foundation that had supported much of Goddard's experimental work. His patents were publicly available, and German and Russian organizations acquired copies of many of them. The distribution of his patents within the United States to organizations, who might have benefited, did not seem to have been well implemented.

While still launching rockets at Roswell, New Mexico, he developed in 1941 a JATO unit for the Army Air Corps for 350 lbf thrust with LOX and gasoline.[5,8] In 1942 he and six of his assistants moved to Annapolis, Maryland. From 1942 to 1945 Goddard worked with the U.S. Navy Bureau of Aeronautics at Annapolis, Maryland, at the Navy's Engineering Experimental Station.[5] There he helped to develop a LPRE for JATO units for the Navy's PBY flying boat. It used LOX/gasoline propellants.[8] Goddard personally participated in the test flights in September 1942. These tests came a few months after GALCIT had demonstrated a JATO as described in Chapter 5.6. He also developed a variable-thrust JATO, which was probably the first U.S. variable-thrust LPRE. His control system for a restartable LPRE with a turbopump was fairly complex, and the concept was later refined and improved by a contractor, the Curtiss–Wright Corporation, as the XLR-25 rocket engine for the Bell X-2 research rocket airplane. He also became a part-time consultant to the Curtiss-Wright Corporation at Caldwell, New Jersey, during this period. To them he contributed information on LOX pumps, LOX/gasoline TCs of 3000 lbf thrust, nitric acid/aniline, TPs and TCs for about 350 lbf thrust, and a premixing chamber at the injector end of a TC. He did visit the General Electric Company's rocket test site at Malta in New York, but refused to work with this company. He was more cooperative with Reaction Motors, Inc., in New Jersey and visited there several times. He even joined the American Rocket Society, after refusing to become a member 12 years earlier. Goddard had signed a contract with Curtiss–Wright and was supposed to come to work for them in 1945, but he became ill with cancer and passed away before he could start the job. Most of his technicians and assistants accepted jobs with this company.

Goddard emphasized safety and trained his crew to work safely. He and his crew were careful and deserve credit for not having a disabling accident during the 24 years of working with hazardous energetic propellant materials. For instance, he used remote controls, a barricaded or reinforced control building, and in early tests he placed a heavy old steel boiler between the control station and the experimental area. Nevertheless, he took chances that today would be recognized as foolish and potentially dangerous. Examples are wooden structures for mounting his engine components in early static firing tests (wood will burn) and pressurization of ether or gasoline tanks with pressurized gaseous oxygen; this forms vapor mixtures above the liquid fuel level, and these vapors can readily explode by inadvertent heat, sparks, or impact. Fortunately Goddard recognized these risks and remedied both of them in later tests. Other early pioneers did not do as well. The French pioneer Robert Esnault-Pelterie lost three fingers in an explosion of a liquid monopropellant, and Max Valier (early Austrian/German pioneer) was killed in the explosion of a LPRE propelling his race car.

Goddard was very reluctant to disclose his concepts, designs, test data, or flight results to other people. Although he had correspondence with many people, including other noted rocket experts (e.g., Hermann Oberth of Romania/Germany, Robert Esnault-Pelterie of France or N. Rynin of Russia), he did not divulge very much useful information.[8] He was concerned about others using his concepts before they were fully proven, about disclosing information before it was patented, and perhaps also about others stealing his ideas. He did give some detailed information and advice to the U.S. Navy and to the company that hired him, namely, Curtiss–Wright Corporation. He turned down offers to work with General Electric (who had a very capable LPRE group at Schenectady, New York, at that time) and gave only very general information to inquiries from Aerojet, Rocketdyne, GALCIT, several government agencies, newsmen, and people from Europe. Professor von Kármán (an early U.S. space pioneer) at Cal Tech suggested a joint R&D effort, but was turned down by Goddard on the basis that unhindered independent development work would proceed more rapidly despite more limited resources.[13]

Although he wrote papers on other subjects, kept a diary and private records of his projects, he published very little about his work on LPREs. As was already mentioned, his 1919 publication[9] had essentially nothing about LPREs, and his 1936 publication on liquid rocket engines[14] was rather brief and devoid of useful detail or technical data. His collected works were more detailed, carefully documented, had more quotations, lots of useful photographs and sketches, and were published by his widow in 1970, which was 25 years after his death in 1945.[8] However they does not give much engine performance data, only a few sketches of his engine designs, and almost no information about his analytical LPRE accomplishments or how his engines were calibrated and inspected.

It is an ironic twist of history that Goddard's pioneering work in LPREs and sounding rocket vehicles had relatively little impact on U.S. LPRE development. The large U.S. LPREs, which were developed later by General Electric, Rocketdyne, and Aerojet, were designed and produced without the benefit of the work done by Goddard. For example, this author personally did the analysis and design for the first large U.S. thrust chamber (75,000 lbf thrust); it became part of the Redstone LPRE (which launched the first U.S. spaceflight). At that time (1947–1951) my fellow designers and I had not even heard of Goddard or any of his know-how or his unique contributions to the state of the art of LPREs. In retrospect it would have saved some time to have known at that time some details about Goddard's LPRE efforts This lack of Goddard's information might have been the fault of the author and his fellow designers for not noticing any of his patents. Instead Rocketdyne received a lot of help and data from the Germans and their V-2 LPRE information, which was very useful. In 1958 as president of the American Rocket Society (which merged later into the AIAA), your author finally learned about some of the work of this remarkable man and met his widow Mrs. Esther Goddard at an official visit at the dedication of a monument at his New Mexico launch site. It was only in 1970 (25 year after his death) when Goddard's papers were published (see referernces) that this author obtained a copy and became aware how far ahead of his time Dr. Goddard was in his work on LPRE.

By the time that his collected papers were published, most U.S. LPRE companies had already reinvented and developed their own LPRE designs, techniques, starting procedures, or test facilities, and some of these were very similar to what Goddard had done earlier. The release of his work in 1970 was anticlimactic. If Goddard could have been persuaded to release his information while he was still alive, it would have helped the companies to get started more quickly, and thrust chambers, pumps, lightweight tanks, and other hardware could have been available sooner. One can only speculate that this would have saved the U.S. government time and a lot of money.

For his historical and outstanding accomplishments Goddard has been honored. He received an honorary doctor of science degree from his own Clark University two months before he passed away. Unfortunately, most of the recognition and accolades were bestowed only posthumously. One of the major NASA Research Centers, at least five junior high schools, one high school, a couple of university professorships, and several streets were named after him. The U.S. Post Office issued a stamp in his honor. Several technical societies established annual Goddard awards and Goddard lectures. Exhibits of his work can be found in several museums including some foreign museums. Monuments in his memory were erected at his historic rocket launch sites. Many articles explaining and extolling Goddard's work have been written, and his biography has been recorded in articles and books.[5]

5.3 Hermann Oberth

Born in a German enclave in Romania, he worked there as a high school teacher. At an early age he became inspired by the fiction novels of Jules Verne, and he became interested in space and rocketry. Space exploration was originally a part-time hobby for him, but it became an all-encompassing personal research effort.[15,16] At the university of Heidelberg, where he was supposed to have studied medicine, he actually spent more time on physics, astronomy, and other subjects related to space and rockets. He submitted his work on space travel as a doctoral thesis, but it was rejected, in part because the faculty did not understand his subject. In 1917 he submitted a proposal to the German army to build a long-range missile using alcohol, water, and liquid air as propellants, but was turned down. His portrait is shown in Fig. 5-18.

After failed attempts to have the work of his rejected thesis published, a reputable Munich publisher produced his book in 1923 *Die Rakete zu den Planetenräumen (By Rocket into Planetary Space)* written in German.[17] Oberth might have paid for part of the publishing costs. This book became popular and was later revised, greatly enlarged, and republished in 1929 under a different title.[18]

The book *Wege zur Raumschiffahrt (Means for Space Travel)* is a serious technical, but speculative description of the scientific and technical problems of spaceflight and LPREs, as well as imaginative future space missions.[18] For example Oberth lists the equations for isentropic flow through nozzles, the relationship between flight velocity, vehicle and propellant masses, and the overcoming of gravitational attraction and air resistance (drag). It included preliminary, conceptual sketches of several single- and two-stage space vehicles

Fig. 5-18 Photo of Hermann Oberth, early visionary. Copied from Ref. 27.

and their LPREs. Figure 5-19, taken from his writings, shows two of his two-stage vehicle concepts with some detail about their LPREs. He planned the use of liquid hydrogen and liquid oxygen as propellants. The vehicle on the right is visualized as a manned spacecraft with a parachute for reentry. He wrote about life-support systems and explained how life in space will be managed. Its large TCs have a set of injection nozzle rings (with wing-shaped cross sections) and gaseous oxygen flows between these ring nozzles and is accelerated to high speed. The fuel is injected just upstream of each of these ring nozzles from manifolds indicated at the entrance of each of the seven injector-nozzle rings in the booster stage. The left drawing shows an alternate unmanned concept with one stage within the other. Alcohol and LOX are the intended propellants for the booster stage and LOX/LH$_2$ for the upper stage. The book described these vehicles in enough detail to make the reader believe that spaceflight was a real possibility. It is amazing that Oberth wrote these books with imagination, but without having done or seen any experimental work. Much of the hardware described in the book later turned out to be impractical.

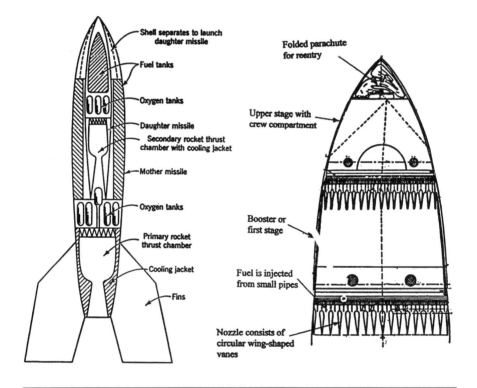

Fig. 5-19 These concept sketches of Oberth's two-stage launch vehicles were drawn before any experimental data were available to him. Unmanned version on left has second stage inside of the first stage. The manned crew compartment on right vehicle is surrounded by propellant and located on top of the combustion chambers. Copied from Ref. 18.

He visualized propellant pumps to be heavy-duty pressure vessels, where the liquid was displaced by high-pressure gas, which was then vented overboard so that the next charge of liquid would flow into the pressure vessel, also to be expelled by gas pressure. He needed at least two pumps to achieve a reasonably steady flow. The gas compressors were driven by gasoline engines. This type of pumping system was very heavy, inefficient, and was never actually considered for use with a LPRE.

Oberth was the first to propose the use of stars for navigation references and the use of aerodynamic drag (parachutes) to slow down a reentering spacecraft. He understood the benefits of multiple vehicle stages, and he made conceptual sketches of several of these. Oberth described three good spaceflight applications: 1) vertically ascending sounding rockets for atmospheric research and as a military observation post; 2) long-range rockets for Earth observation, fast mail transport, or military missiles; and 3) as a powerplant for airplanes, which could later be uprated and adapted to spaceflight. He understood the benefits of multiple vehicle stages and made conceptual sketches of them. Some of his predictions have indeed become true. He probably did have information that some of the items in his book had already been treated by others before. He corresponded with Goddard and Tsiolkowsky, but did not get a lot of help from them. Although Tsiolkowsky in Russia had worked on this subject for about 30 years, and although Goddard had patents issued to him and done work for almost 10 years, their specific investigations and the details were known only to a limited audience at that time.

Oberth became the technical advisor to a movie director in making the film *Die Frau im Mond (The Woman in the Moon)*. The movie contained scenes of multistaged space vehicles, which Oberth designed and configured, and sights of some dramatic launches and rocket flames, which he suggested. It became a very popular science-fiction film. He made sketches and did some experimental work on the multistaged spacecraft and engine design that he provided to the movie company. This film made him famous, but in ways he did not foresee. He also did some LPRE experimental work in support of his consulting contributions, but his LPRE work was limited to component tests on a very small scale, were not very successful, and were not widely known.

Parts of his books were criticized by the technical community; people like Goddard and Esnault-Pelterie of France disagreed with some of Oberth's technical reasoning and pointed out some flaws. However, some key people, such as Wernher von Braun (who later became the leader of the German V-2 effort and thereafter the U.S. Apollo space effort) and Eugen Sänger, were truly inspired by this book. It kindled the imagination of the public, the press, and governments and contributed to the start of amateur rocket groups. For roughly a decade Oberth probable was the world's intellectual leader of spaceflight. In 1936 he won a 100,000 French Francs literary prize and used the money to further his research.

Oberth built and tested small liquid propellant thrust chambers' in 1929 and 1930, after he published his books.[19] He believed that a conically shaped combustion chamber (with injection near the nozzle throat) would enhance the mixing. He called it Kegeldüse or cone burner, and one is shown in Fig. 5-20. It had single-injection holes for fuel and for the oxidizer, which were pointed at the apex of the cone. Compared to today's TCs, this cone burner had a combustion volume that was too small and the combustion efficiency must have been very low. It is unlikely that the liquids penetrated to the top of the cone;

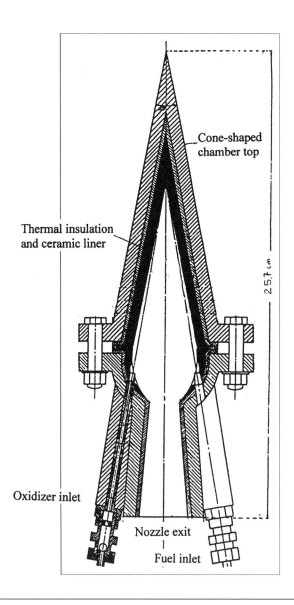

Fig. 5-20 Simplified diagram of one of Oberth's cone-shaped TCs with injection holes pointed forward and located in the converging portion of the nozzle. From Ref. 19.

probably a major portion of the propellant was pushed out of the throat as unburned liquids. It was not a useful contribution to the state of the art of TCs.

Oberth became a German citizen only in 1943. He was not offered a job in the German rocket effort during World War II. He did not seek much help from others, and he had occasionally difficulties in communicating with some of his assistants. In his later life he spent a few years as a professor at the technical universities of Vienna (Austria) and thereafter at Dresden (Germany), but his efforts on the university faculty were not remarkable. He spent three years in Alabama in the United States as a consultant to his friend von Braun, studying future propulsions systems, but the study results or his contributions have not been clearly identified. Your author met Oberth briefly in the 1950s in Germany, and he made a good impression. He seemed to be a serious man and somewhat reticent. His fame within the aerospace industry began to wane as the accomplishments of others became more widely known.

5.4 Other Pioneers

Many countries had one or more key people who were early participants, pioneers, enthusiasts, or promoters of rocket propulsion. The examples given next have been arbitrarily selected. There were other early contributors and pioneers.

Italy had General Arturo Crocco and his son Luigi Crocco.[20] The father worked on solid propellants, and the son Luigi designed in 1930 a liquid propellant thrust chamber, which is shown in Fig. 5-21. It was a very advanced design for its day, the first TC to feature regenerative cooling, with a fuel cooled jacket around the nozzle and an oxidizer cooled jacket around the combustion chamber. Propellants were benzine as a fuel and nitrogen tetroxide (NTO) as the oxidizer. This is the first application of NTO to a liquid propellant thrust chamber (TC) known to the author. It used stainless steel in the TC, a pressurized-gas feed system, and an elliptically shaped ceramic liner in the chamber (zirconia) to reduce wall temperatures. A torch was inserted into the TC for ignition, using gaseous oxygen and gaseous hydrogen; upon ignition the torch was immediately retracted. The thrust was only 1.25 kg at a nominal chamber pressure of 10 atms. There were several initial tests, but the project was stopped prematurely and never completed. This work was not reported until

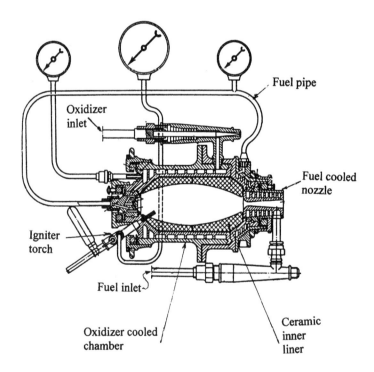

Fig. 5-21 Cross section of Luigi Crocco's experimental small LPRE partly tested in 1930 at University of Rome. Copied with permission from Ref. 20.

some time later. Luigi Crocco also was one of the first to try a mixed liquid monopropellant in 1930. He mixed nitroglycerine with methyl alcohol and somehow managed to test fire it in a TC. He measured fairly good performance, but experienced explosions with this dangerous mixture. Work was stopped for reasons unrelated to the LPRE project. Luigi Crocco later became a university professor and did some outstanding analysis on LPRE combustion stability.[21] The author was privileged to have had discussions with this man.

In Switzerland the early pioneer for propulsion was Josef Stemmer, who did some experimental work with LPREs. He also published what might have been the first simple educational "how-to-do-it" brochure on the subject.[22] It was used by amateurs.

Midshipman Robert C. Truax of the U.S. Naval Academy performed some work on LPREs between 1936 and 1938. He somehow found time to do rocket experiments in between his busy schedule as a student in this academy. His first design, Fig. 5.22, was operated on compressed air and gasoline and later on gaseous high-pressure oxygen and hydrocarbon fuel.[23] It had water film cooling at the throat and fuel regenerative cooling at the nozzle converging section. Some of the fabrication and testing was done at the Naval Engineering Experiment station across the river Severn from the Naval Academy. This author first met Truax in 1944 when he was a Navy officer assigned to the company, where the author was then working, cooperated with him as a fellow officer of the American Rocket Society, and has seen him several times since then. The U.S. Navy effort on LPREs was continued by several other Navy officers.

The Austrian Eugen Sänger (1906–1964) was an early enthusiast of spaceflight and a pioneer in regenerative cooling of TCs and knowledgeable about LPREs. In 1933 he published a book called *Raketenflugtechnik* (*Technology of Rockets*) in which he summarized his studies of the subject and also discussed the performance and operation of LPREs.[24,25] He independently conceived of the forced regenerative cooling of TCs. He started his work as a student at the Technical University of Vienna with laboratory experiments in 1932. His TC tests initially used water as a coolant (1933–1934), and later (1936) he used the propellant for regenerative cooling. Initial thrusts were between 10 and 20 kg (22 to 44 lbf), and later he had 100 kg or 220 lbf. In the early tests he used a ceramic liner, gaseous high-pressure oxygen, and diesel fuel, which was pumped by a manually driven three-piston hand pump. Later he used LOX. He had to relocate his test stand, which was originally in an empty building of the university, because loud noises from his tests caused extensive concerns by people in the surrounding area. His third-generation TC no longer used a ceramic liner, and the heat of the chamber was transferred through steel walls to the water in the cooling jacket. In 1934 he conceived of his first regenerative cooled TC; it had cooling coils wrapped around the smooth walls of the chamber and nozzle. In 1934 he published a short report on his tests and the feasibility of regenerative cooling in a special edition of the magazine *Flug* (*Flight*).[26] He ran a TC fully cooled by propellant in 1936. One of his cooled TCs is shown in Fig. 5-23.

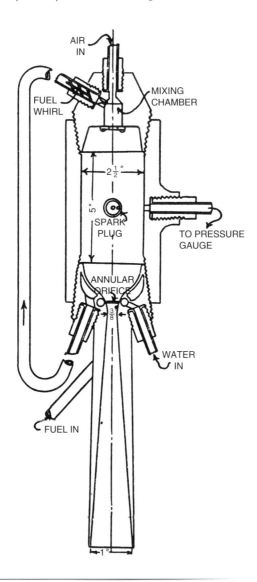

Fig. 5-22 Cross section of TC tested by Robert C. Truax at Annapolis, 1937, 1938. Copied with permission of AIAA from Ref. 23.

In some of Sänger's tests, he measured a water temperature rise in the cooling jacket of 480°C, and he was able to get steam out of it. In 1938, when Austria was annexed by Germany, Sänger became a German citizen and was able to continue his work with the German Research Institute for Aeronautics at Trauen near Lüneburg in Germany. There he designed and built a new test stand for his work. In 1940 he started to pump propellants, initially he used gear pumps with LOX and hydrocarbons, but like others he soon shifted to centrifugal pumps. He ran pump tests and larger TC tests (1000 kg) in the new

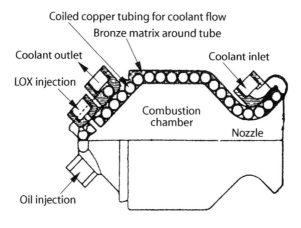

Fig. 5-23 Cooled TC designed by Eugen Sänger of Austria in 1936. Courtesy of AIAA; copied from the *Journal of the American Rocket Society* around 1937.

facility. He conceived of a unique complete pump-fed LPRE (novel at the time) using a steam turbine to drive three pumps: a LOX pump, a fuel pump, and a water pump. The water was used as a coolant, converted to steam for driving the turbine, condensed in a condenser with propellant as a coolant, pumped and reused as a coolant. He experimentally investigated this scheme with a closed-loop water circuit, but it never became a flight-worthy engine. In the 1940 he ingeniously invented a "slug start" by using a highly reactive hypergolic fuel (zinc-diethyl) at the start and by introducing the nonhypergolic fuel into the chamber only after a hot combustion had been started by the slug burning with the oxidizer. He simply filled the fuel line with this hypergolic fluid, which was than pushed into the TC by pressure from the fuel being pumped.

Sänger was a proponent of the winged boost-glide vehicle and performed several studies of it. He built a model of the 100,000-kg TC, which he would need for this vehicle. It had cooling tubes circumferentially wound around the chamber and nozzle. The German government was busy with the V-2 missile and really never gave Sänger adequate funding for his winged vehicle. Studies and some experimental work on boost-glide missiles were also done in the United States, the Soviet Union, and in other countries between 1946 and 1965. In 1945, immediately after the end of the World War II, the Soviet Union badly wanted to have Sänger's help with their own efforts on LPRE and boost-glide missiles. The Soviets organized a systematic hunt to catch Sänger in Germany; however, he had gone to France and was not interested in going to Russia. From a military point of view, the ballistic missile was considered to be superior to a boost-glide missile, and therefore this concept never became operational. Sänger also did some good analysis of the performance of different rocket propellants.

The Soviet Union had several important pioneers who contributed to the technology of LPREs. One is Fridrikh Tsander; he was a disciple of Tsiolkowsky, a space enthusiast and public speaker, and wrote about the subject and developed one of the first LPREs in the Soviet Union. His work is discussed and illustrated in Chapter 8.1 The most outstanding pioneer in large Soviet LPREs is Valentin Glushko. He developed some novel smaller TCs and LPREs early in his life and later became the leader of Energomash, the organization that developed most of the Soviet large LPREs. Many of his efforts, his photo, and some personal data can be found in Chapter 8.4. A few other Soviet pioneers are also mentioned in Chapters 8.1 to 8.11.

The outstanding early French visionary and pioneer was Robert Esnault-Pelterie, and some of his accomplishments are noted in Chapter 10.

5.5   Amateur Rocket Societies

The subject of spaceflight and rocketry was very intriguing, challenging, and adventurous to many people. They were enthusiasts and wanted to obtain information or participate in this new field. So in the 25-year period beginning in 1924, a series of voluntary amateur technical societies (estimated at more than 40) sprang up in many of the countries of the world.[27] The societies in the Soviet Union, United States, and Britain were probably the most visible on the international scene, and they are briefly discussed here.

These amateur rocket societies debated spaceflight issues, published articles and/or news bulletins, and held frequent meetings. Their biggest contributions were the popularizing of space travel and rocketry (they received lots of free publicity and press coverage) and the attracting, educating, and identifying of technical personnel who later became employed in the more serious endeavor to make rocketry happen. Furthermore, they peaked the interest of government and military officials, who later supported rocket flight vehicles, launch facilities, or rocket propulsion. Some of these societies also did practical, but limited experimental work on LPREs and launched simple sounding rockets with LPREs. Occasionally they did contribute a little to the state of the art of the technology of rocket propulsion, but mostly they duplicated work that had already been done by Goddard years before or by the then-secret government-sponsored groups in the Soviet Union or Germany. The results of these three sterling technical efforts were not available to the amateurs at that time and did not become public knowledge until much later.

The Soviet Union had the first amateur rocket society.[27] It was founded in 1924 and was called the *All-Union Society for the Study of Interplanetary Travel*. The impetus to the formation of this society probably came from a speech by Fridrikh Tsander before a Moscow group affiliated with amateur astronomy. Tsander was an early pioneer and a space enthusiast, who contributed greatly to create interest and build early LPREs. The Soviet amateur organization was initially focused in a group at the Military Science Division of the Soviet Air Force Academy in Moscow. The group very soon was disengaged from the military and became a true amateur organization called the *Society for Interplanetary Communication*.[27] This amateur organization attracted several top professors and researchers including Tsander. The visionary Konstantin Tsiolkowsky joined, but he was never very active as a member. The society soon had 200 members and a few years later over 1000. Similar amateur groups were set up in other cities of the Soviet Union.

For example, the group in Kiev organized an exhibition and lectures on spaceflight and rocketry in 1925. The Moscow group held an exhibition and meeting in 1927. Models of spaceships and associated mechanisms from different inventors and countries were on display. Although there were exhibits about the contributions of Hermann Oberth or Max Vallier of Germany, Robert Goddard of the United States, and others, these displays were put together by some of the society members and probably were not supplied or reviewed by the individual, whose name appeared at the top of each exhibit. In fact

Goddard or Oberth probably did not know about it. This exhibit attracted much attention in Moscow, but the information from the displays apparently did not reach other countries.

The Soviet government at that time encouraged scientific debate and discussion, and they supported and controlled these groups from Moscow. The society concentrated on spaceflight, but some of their papers, exhibits, and discussions related to LPREs. Visits and lectures by foreign experts were permitted. The acceptance by the technical community was enthusiastic. The Soviet Union was unique in this respect, for no other country at that time had intellectual technical debate on spaceflight or rocketry. Beginning in about 1930, after Joseph Stalin had come into power, this kind of freedom of technical discussion was no longer allowed, and these Russian amateur societies disappeared.

Der Verein für Raumschiffahrt (VfR or Society for Spaceship Travel) was founded in July 1927 by 10 enthusiastic members in Breslau, Germany.[27] By 1930 they had over 1000 members, including some of the best European experts. Their purposes were to publish and popularize rocketry and space travel and to perform practical experiments. The first credible technical publication, called *Die Rakete* (*The Rocket*), became the VfR magazine.

This German volunteer society built and ground tested LPREs beginning in 1930. Johannes Winkler, one of the founders of the VfR, ran a TC of his design with oxygen and a fuel consisting of 70% methyl alcohol and 30% water in the early 1930s.[28] This propellant combination gave more stable combustion and better cooling characteristics than other propellant combinations that were tried. A slightly modified mixture (25% water) was adopted by the German Army team then working on model TCs of what later became the V-2 LPRE. A number of the early TCs used a cone-shaped steel combustion chamber, similar to the one investigated by the pioneer Hermann Oberth, as shown in Fig. 5-20. Hermann Oberth joined the VfR and assisted with the design and launches of a few rockets.

The first flight of the VfR took place in central Germany with a design by Johannes Winkler, the first president of the Society.[28] It was a small vehicle, was about 2 ft long, and weighed 11 lb. The thrust chamber was on the top firing down between three equally spaced tubes, which were tanks for liquid oxygen, liquid methane, and pressurizing gas. Its first flight reached a height of only 10 ft, when it failed. Even though it did not complete its flight, it was nevertheless a first flight in Germany. The second flight in March 1931 near Dessau (100 km southwest of Berlin) was successful. The investigators and the VfR generally believed that this flight was the very first flight with liquid propellants; it was then not known in Germany that Goddard had flown a vehicle with a LPRE five years earlier in the United States.

The VfR acquired its own proving ground and launch facility, an abandoned military base at Berlin, and it was called Raketenflugplatz (Rocket Airport). One of the more notable early efforts was the MIRAK rocket series (Minimum Rakete or Minirocket). Most of them had a long aluminum tube, which was the fuel tank (and also acted as a flight stabilizer), a cone-shaped combustion chamber at its nose, but with the nozzle pointed aft. The oxygen tank surrounded the nozzle and some of the chamber. (It was expected to absorb some of the heat.) Carbon dioxide was used as the pressurizing gas in some tests. The first MIRAKs and some of the several subsequent versions failed. In July 1930 the first successful static firing test was made of a design conceived by Hermann Oberth, Wernher von Braun, and others. It also had a cone-shaped combustion chamber, a thrust of 15.4 lbf, used 13.2 lb of liquid oxygen, 2.2 lb of gasoline, and ran for 90 s.

The third-generation designs (called Repulsors) had fewer failures.[27] The first Repulsor flight happened in May 1931, reaching a height of 200 ft. During a two-year period, they conducted 270 static firings and 87 flights with simple finned vertical flight vehicles. (Some had parachutes that allowed recovery and reuse.) In late 1932 and in 1933 several of the society members, including Wernher von Braun, were offered positions with the German Army's secret development effort that ultimately led to the V-2 rocket.

By 1933 this VfR group had started work on a regenerative fuel-cooled thrust chamber in the 300- to 450-lb thrust range, several years before the United States tested one. However the work of the VfR was suspended because of financial problems and because the new government regime (Hitler came into power) supported their own efforts, which were secret at that time. Society operations stopped around 1934, but the VfR was revived after World War II.

The *American Interplanetary Society* (AIS) was founded in March 1930 in New York. It soon grew in membership and started to publish its own journal.[27] Goddard initially refused to join the Society. Like the Germans the AIS obtained the use of suitable land for static firing tests and the launching of their small sounding rockets. Several AIS members visited the VfR in Berlin and observed and learned from the German society's work. Therefore some of the early AIS flight vehicles and TCs resembled the German MIRAK rocket series. Figure 5-24 shows the first liquid propellant vehicle flown by this AIS.[29-31] The TC was uncooled and located near the nose because this allowed a drag-stabilized flight. The location of the two injection holes is at the nozzle entrance. The long impingement distance makes it unlikely that good impingement actually occurred. Ignition was initiated by burning cloth rags (placed below the TC), which were set on afire manually before launch. The fuel tank was only about two-thirds full. The ullage space above the fuel contained compressed nitrogen, which was fed through a tire valve prior to launch, and the fuel tank operated in a blowdown mode. The oxygen tank pressurized itself by heat absorption and evaporation. Three minutes were allowed for the oxygen pressure to become high enough for launch. The propellant valves were opened by long strings, which were pulled by the crew in the remote control station. In May 1933 this rocket made its first flight. The engine stopped running after a few seconds of powered flight. The oxygen tank had exploded, and the fins separated at an altitude of about 250 ft.

In 1934 the AIS changed its name to the *American Rocket Society* (ARS) because they wanted to get away from the word interplanetary, which the public

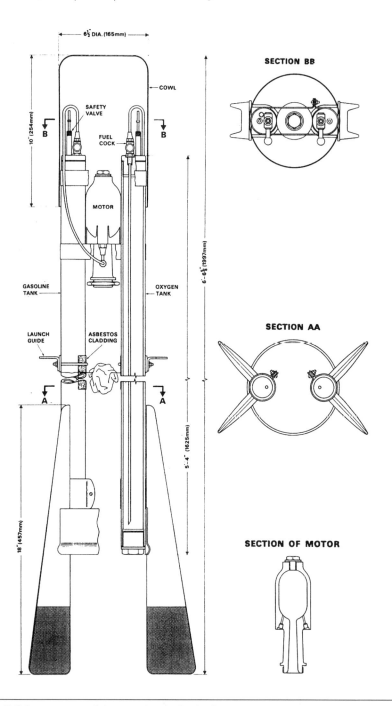

Fig. 5-24 Diagram of first rocket vehicle flown in 1933 by the American Interplanetary Society with a section through the first TC built in 1932. It produced 60 lb thrust for about 15 s. Courtesy of AIAA/AIS/BIS, copied with permission from Ref. 31.

and the press viewed with suspicion and considered a fantasy. Beginning in 1935 a series of tests was made using short and long thrust chambers and different injection schemes.[29] Figure 5-25 shows a version of a segmented uncooled thrust chamber of the ARS. It was easy to assemble in different ways and was a simple R&D tool for investigating different chamber volumes, lengths, and alternate nozzles. Equipment for ground tests and launches was simple, and often crude and safety protection for personnel was minimal. Figure 5-26 shows a static firing test of the ARS with a horizontally mounted TC on the society's portable horizontal firing test stand, which could measure thrust. The instrument panel was photographed by a movie camera for subsequent evaluation, and the high-pressure nitrogen gas bottle lay on the ground on the bottom of the photo. Figure 5-27 shows the ARS TC #4 designed by John Shesta, who later became one of the founders of Reaction Motors, Inc. Its four nozzles were pointed aft, and this TC was mounted on top of the fuel tank, followed by the LOX tank below it.[32,33]

Believing that thrust chambers could be cooled by fuel flow, an investigation of cooling methods was launched by the ARS. Society member James H. Wyld designed and successfully tested (first time in the United States in 1938) a fully regenerative fuel-cooled thrust chamber. His sketch was shown in Fig. 4.3-3. His TC was made out of machined and welded monel parts an alloy of copper and nickel. This was hailed as a major step forward in the technology because it allowed prolonged rocket operations without burnout failure. Wyld had built and tested earlier TCs with regenerative cooling with different chamber shapes and materials, but their tests ended in failures. Some people thought that Wyld 1938 TC was motivated by a picture and write-up of a cooled TC developed by Austrian Eugen Sänger around 1935. His work was published in Germany in 1936 and in the *Astronautics* magazine of the ARS in 1937. Figure 5-23 showed a picture of Sänger's approach using a spiral wound copper cooling tube embedded in a bronze wall. Water was the initial coolant

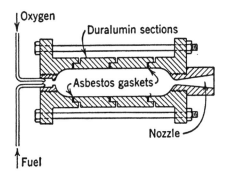

Fig. 5-25 Segmented experimental aluminum TC with a precombustion chamber, tested by the American Rocket Society in 1935. Courtesy of ARS/AIAA. from Ref. 29.

Fig. 5-26 Typical, crude, early test setup for an ARS static test firing of a small experimental LPRE with a bright flame. Most personnel was unprotected. The visual instruments are recorded by a movie camera. Nitrogen pressure cylinder is in the foreground. Copied with permission from Ref. 29.

for Sänger, but he also used fuel. It was then not known that the German and the Soviet Government teams had secretly accomplished this TC regenerative cooling a years or two earlier or that Tsiolkowsky had written about the use of fuel for cooling some 30 years before.

In 1941 the experimental work of the ARS was discontinued, in part because of the good and well-funded LPRE work being performed elsewhere. The Society concentrated on publishing refereed professional papers and holding technical meetings on the subjects of propulsion, space flight and rocket vehicles. Fifteen years after it started the ARS had about 15,000 members in about 20 local chapters. This author had the privilege to serve as a director and then as president of this society at a time when the first actual space flights occurred (1957/1958). In February 1963 the ARS merged with the US national aeronautical engineering society (NACA) to form the American Institute of Aeronautics and Astronautics (AIAA). Today this is a respected professional organization, that is still much concerned about LPREs, but this rocket propulsion technology is now only one of many fields of interest to this professional organization.

Besides the ARS there were other U.S. volunteer groups concerned with rocket propulsion.[34] The Cleveland Rocket Society was reported to have tested some simple vehicles driven by a LPRE in the early 1930s. There were amateur

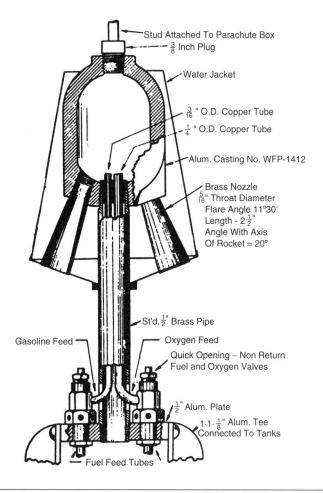

Fig. 5-27 Thrust chamber of ARS rocket #4 had four nozzles and was attached to the long cylindrical assembly of propellant tanks one on top of the other. Courtesy of AIAA/ARS; copied with permission from Ref. 32.

enthusiast groups reported in Boise, Idaho; two in California; in Urbana, Illinois; Detroit, Michigan Rocket Society; Chicago Rocket Society; Intermountain Rocket Society (Salt Lake City, Utah); and Cambridge, Massachusetts. Some of these groups engaged in testing and launches.

The British Interplanetary Society (BIS) was founded in October 1933 largely through the untiring efforts of one Phillip E. Cleator, a structural engineer. Initially the new society met in Liverpool, where its first office was established.[27] The *Journal of the BIS* started in 1934, and this journal is still today a leading respected professional publication. For example, the list of references in this book contains several papers published by this journal. In 1936 the BIS headquarters was moved to London, where most of its members resided. In the early years there were several experimental projects and launches undertaken by the BIS and also by other British amateur groups. During a launch demonstration by amateurs in Manchester in March 1937, an explosion caused some injuries to several spectators, and the police intervened. When the case came before a court, the prosecutor invoked the Explosives Act of 1875, which forbade all private experimentation with gunpowder, dynamite, certain other explosives, and rockets. The court ruled against any private rocket experimentation. There was no penalty put on the Manchester group, but the group had to promise not to use certain chemicals in propellants again. Although this Manchester amateur group was not part of the BIS, the court decision effectively shut down all further private tests and launches in the United Kingdom. So the BIS concentrated on its journal, technical meeting, planning of space projects, technical discussions, and even building inert models of spaceships and propulsion systems. During the war (1939–1945), the BIS suspended operations, but resumed its activities after the end of the war.

One testimony of the subject's popularity was that amateur societies were founded in other countries, including France, Bulgaria, Austria, Australia, Poland, Romania, India, Italy, Spain, and Switzerland. Later there were some others. Several of these groups of enthusiasts published interesting information about LPREs, and a few built and tested simple LPREs. Even today there are some amateur groups that build their own small LPREs and launch vehicles, then test them and fly them; they teach safety to the experimenters before they are allowed to undertake such amateur projects.

5.6 Early Team Efforts

The first organized team effort in the Soviet Union was in Tikhomirov's Laboratory, which was started in 1921. However their work on LPRE did not begin until about 1929, and this work is described in Chapter 8.1. In Germany a team under the technical leadership of Wernher von Braun started serious development of LPREs in 1933, and it is discussed in Chapter 9.1. The work of both of these two teams was secret at the time. These and the small teams listed next were informal, engaged in planned systematic R&D of LPREs, performed some experimental flights, but did not usually undertake any production effort.

The earliest U.S. team was GALCIT (abbreviation of Guggenheim Aeronautical Laboratory, California Institute of Technology) in Pasadena, California. This laboratory originally on the campus of the California Institute of Technology (Cal Tech) was perhaps the first academic group to undertake theoretical and experimental work in LPREs.[35,36] It was started in 1935 and operated until 1943, when most of its work was transferred to two other organizations as explained in the following. The chairman of the Aeronautical Engineering Department at the California Institute of Technology (Cal Tech) and the head of the GALCIT project was Professor Theodore von Kármán, a renowned aerodynamicist.[37] The GALCIT staff consisted of a few graduate students, some part-time faculty members, and a few employees. This laboratory performed laboratory tests of different propellants, designed and tested small TCs in their own off-campus test facility (beginning 1936), investigated different ignition techniques, and was the first in the United States to achieve hypergolic ignition using nitric acid and aniline as propellants (1940). Different propellants and thrust chambers with thrusts up to 1000 lb were investigated. In 1937 nitric acid was selected as a good potential storable oxidizer. In 1939 GALCIT improved this nitric-acid oxidizer by dissolving up to 30% nitrogen dioxide (a red colored gas that came out of solution and evaporated as reddish clouds), and this was henceforth known as red-fuming nitric acid or RFNA. The GALCIT people did not know that RFNA had been formulated in Germany several years earlier. It had greater density, slightly higher performance, and better ignition properties than nitric acid without the dissolved gas.

GALCIT published some historic analysis on rocket propulsion and was the first to use the concept of a thrust coefficient, which has been used in analysis ever since. They built and flew the first U.S. JATO on an A-20 aircraft in 1942 (Ref.38). Two JATO units were used on this first U.S. JATO flight with a relatively small airplane. The propellants were nitric acid and aniline, using a gas-pressurized feed system at a thrust of 1000 lbf each unit. A picture of this historical JATO engine in shown in Fig. 7.7-1. It was then not known that the Germans had flown JATO units about six years earlier; they successfully operated such LPREs on military airplanes for takeoff assistance, but with different propellants.

GALCIT was the predecessor of the Jet Propulsion Laboratory (JPL), which today is still administered by Cal Tech for NASA. Its first Director Frank

J. Malina was a key man at GALCIT and a former student of Professor von Kármán. Several projects that were initially conceived at GALCIT, such as a complete sounding rocket (WAC Corporal), were then pursued at JPL. The abbreviation WAC means without attitude control. Figure 2-1 shows the WAC Corporal in a launch tower. JPL has become a respected NASA laboratory, has done some work on LPREs in the 1940s, has continued at a low effort to investigate low thrust propulsion for unmanned space probes, and is today involved mostly with developing planetary probes, space exploration, and various other fields of space technology. The second organization spawned by GALCIT was the Aerojet Engineering Company in 1942. Some of the early Aerojet contracts, such as several JATO units, had their technical heritage in the work of GALCIT and JPL. Aerojet's first company chairman was Professor von Kármán and the first board of Aerojet directors included four GALCIT men. This company is discussed in detail in Chapter 7.7.

References for Chapters 5.1–5.6

[1]Parry, A., *Russia's Rockets and Missiles*, Doubleday, 1960, Chap. 6 and 7, pp. 81–110.

[2]Stoiko, M., *Soviet Rocketry, Past, Present and Future*, Holt, Rinehart, and Winston, New York, 1979, Chap. 1–3, pp.17–56.

[3]Sokolsky, V. N., "Some New Data on Early Work of the Soviet Scientist Pioneers in Rocket Engineering," *First Steps Toward Space*, edited by F.C. Durant, and G.S. James, American Astronautical Society History Series, Vol. 6, Univelt, San, Diego, CA, 1985, pp. 269–276.

[4]Yur'yev, B. N. (ed.), *Collected Works of K. E. Ziolkovsky*, Vols. 1-3, USSR Academy of Sciences, 1951; also NASA Technical Translation F-236, April 1965.

[5]Lehman, M., *This High Man, The Life of Robert H. Goddard*, Farrar, Strauss and Company, New York, 1963.

[6]Paulet, P. E., "Liquid Propellant Rockets," *El Commercio*, Lima, Peru, 1920.

[7]Wyld, J. H.,"The Liquid Propellant Rocket Motors," *Astronautics & Aeronautics*, No. 70, June 1947.

[8]Goddard, E. C., and Pendray, G. E., (eds.), *The Papers of Robert H. Goddard*, Vol. I, 1898–1924, Vol. II, 1925–1937, Vol. III, 1938–1945, McGraw–Hill, New York, 1970.

[9]Goddard, R. H., "A Method for Reaching Extreme Altitudes," *Smithsonian Miscellaneous Collections*, Vol. 71, No. 2, 1919; reprinted in Goddard, Robert H., *Rockets*, AIAA Reston, VA, 2002.

[10]Turchi, P. J. (ed.), "Autobiography of Robert H. Goddard," *Propulsion Techniques, Action and Reaction*, AIAA Reston, VA, 1998, pp. 17–23.

[11]Goddard, E. C., and Pendray, G. E., *Robert H. Goddard, Rocket Development, Liquid Fuel Rocket Research, 1929–1941*, Prentice–Hall, 1961.

[12]Ball, M., "Dr. Robert H. Goddard's First Successful Liquid Fueled Rocket as Launched at Auburn, MA on March 16, 1926," *Journal of the British Interplanetary Society*, Vol. 40, 1987, pp. 305–310.

[13]Marsh, P., "The Shy Eccentric Who Fathered the Rocket," *New Scientist*, Vol. 96, 7 Oct. 1982, pp. 26–32.

[14]Goddard, R. H., "Liquid Propellant Rocket Development," *Smithsonian Miscellaneous Collections*, Vol. 95, No. 3, publication 3381, 16 March, 1936; also reprinted in Goddard, Robert H., *Rockets*, AIAA, Reston, VA, 2002.

[15]Hartl, H., *Hermann Oberth*, Theodor Oppermann Verlag, Hannover, Germany, 1958 (in German).

[16]Oberth, H., "My Contribution to Astronautics," *First Steps Toward Space*, edited by F. C. Durant, and G. S. James, American Astronautical Society History Series, Vol. 6, Univelt, San Diego, CA, 1985, Chap. 13, pp. 129–140.

[17]Oberth, H., *Die Rakete zu den Planetenräumen (The Rocket for Planetary Space)*, R. Oldenbourg, Munich, 1923, (in German).

[18]Oberth, H., *Wege zur Raumschiffahrt (Means for Spaceflight)*, R. Oldenbourg, Munich, 1929, (in German).

[19]Hopmann, H., *Schubkraft für die Raumfahrt (Propulsive Force for Space Travel)*, Stedinger Verlag, Lemwerder, Germany, 1999, pp. 18–28 (in German).

[20]Crocco, L., "Early Italian Rocket and Propellant Research," Chapter 3, *First Steps Toward Space*, edited by F. C. Durant and G. S. James, American Astronautical Society History Series, Vol. 6, Univelt, San Diego, CA, 1985, pp. 33–48.

[21]Crocco, L., Grey, L., and Harrje, D., "Theory of Liquid Propellant Rocket Combustion Instabilities and Its Experimental Verification," *Journal of the American Rocket Society*, Vol. 30, No. 2, 1960, pp. 159–168.

[22]Waldis, A., "Origins of Astronautics in Switzerland," *History of Rocketry and Astronautics*, edited by R. C. Hall, American Astronautical Society History Series, Vol. 7, Part 2, Univelt, San Diego, CA, 1986, pp. 153–157.

[23]Truax, R. C., "Annapolis Rocket Motor Development, 1936-1938," *First Steps Toward Space*, edited by F. C. Durant and G. S. James, American Astronautical Society History Series, Vol. 6, Univelt, San Diego, CA, 1985, pp. 295–301.

[24]Sänger, E., *Raketen-Flugtechnik (Rocket Flight Technology)*, R. Oldenburg, Munich, 1933; reprinted Edwards Brothers, Inc., Ann Arbor, MI, 1945 (in German).

[25]Sänger-Brett, I., and Engel, R., "The Development of Regerative Cooled Liquid Rocket Engines in Austria and Germany," *First Steps Toward Space*, edited by F. C. Durant and G. S. James, American Astronautical Society History Series, Vol. 6, 1985, Univelt, San Diego, CA, pp. 217–246.

[26]Sänger, E., "Neue Ergebnisse der Rakenflugtechnik," ("New Results in Rocket Flight Technology"), *Flug*, 1934, (in German).

[27]Winter, F. H., *Prelude to the Space Age, The Rocket Societies: 1924 to 1940*. National Air and Space Museum, Smithsonian Inst., Washington, DC, 1983.

[28]Engel, R., "A Man of the First Hour—Johannes Winkler," American Astonautical Society History Series, edited by C. Hall, Vol. 10, Univelt, San Diego, CA 1990, Chap. 21.

[29]Pendray, G. E., "Early Rocket Developments of the American Rocket Society," *First Steps Toward Space*, edited by F.C. Durant and G.S. James, American Astronautical Society History Series, Vol. 6, Univelt, San Diego, CA, 1985, pp. 141–155.

[30]"The Flight of Experimental Rocket No. 2," *Astronautics*, No. 26, May 1933, pp. 1–11.

[31]Ball, M., and Turvey, P., "American Interplanetary Society's No. 2 Rocket; its Origin, Design, Construction and Launch," *Journal of the British Interplanetary Society*, Vol. 40, 1987, pp. 301–303.

[32]"Report on Rocket #4," *Astronautics*, No. 4, 1934, pp. 3, 4.

[33]"A concentric Tank Rocket," *Jet Propulsion*, Vol. 3–7, No. 26–38, 1933–37, pp. 2–6.

[34]"The Rocket Societies," *Astronautics*, No. 60, Dec. 1944, pp. 12, 13.

[35]Malina, F. J., "The U.S. Army Air Corps Jet Propulsion Research Project, GALCIT Project No. 1, 1939-1946: A Memoir," *History of Rocketry and Astronautics*, edited by C. R. Hall, American Astronautical Society History Series, Vol. 7, Univelt, San Diego, CA, 1986.

[36]Malina, F. J., "On the GALCIT Research Projects, 1936 to 1938," *First Steps Toward Space*, edited by F. C. Durant and G. S. James, American Astronautical Society History Series, Vol. 6, Univelt, San Diego, CA, 1985, pp. 113–127.

[37]von Kármán, T., and Edson, L., *The Wind and Beyond, Theodore von Kármán*, Little Brown and Co., Boston, 1967.

[38]Summerfield, M., Powell, W., and Crofut, E., "Development of a Liquid Jet Unit and its Operation on an A-20 Aiplane," California Inst. of Technology, *JPL Report No. 1–13*, Pasadena, CA, Sept. 1942.

Liquid Propellant Rocket-Engine Organizations Worldwide, 1932 to 2003

This chapter covers several aspects of LPREs from a global perspective. It deals with the reasons for countries wanting to enter into the LPRE business, the proliferation or spreading of LPRE work into more nations, the financial support of governments and their management of LPRE resources, the names of companies or development/manufacturing organizations on a worldwide basis and the first orbital launches of the first 10 of these countries. Chapters 7–13 describe the significant LPREs and their development organizations in more detail

Management and Funding

The government of each country is the *principal customer* for LPREs, either by funding programs directly under government contracts or indirectly through a flight vehicle prime contractor. The funding levels, funding continuity, national priority, and management skills of the concerned government offices have been crucial in the development of LPREs and of flight vehicles and their other major components. Only a very small portion of the LPRE work is funded commercially, such as engines for satellite launches of commercial communication systems or company-funded projects.

Government budgets are at times irregular and subject to political influence and national interests. For example, it has been estimated that the cutting of the budget (by eager German bureaucrats) for the V-2's early development delayed the first operational launch by perhaps two years, and that might have affected the outcome of the World War II. As a consequence, the design of the V-2 engine had to be frozen to meet tight production schedules, and this did not allow incorporating various worthy technical improvements in reliability, performance, or cost savings. Having a relatively steady funding and a fairly predictable budget has helped some countries do a better job of government budget planning, such as in the Soviet Union. This has helped to make more rapid progress in aerospace programs, which includes LPREs, compared to countries, where the budget has changed dramatically from year to year.

The development and production of LPREs is performed by private, profit-making organizations in the United States, Great Britain, or Germany and by government organizations or pseudogovernment organizations in Russia (formerly the Soviet Union), China, or India. In addition there are government organizations that define the programs, arrange for the government funding, control the projects to be undertaken, and monitor the work being done as well as the expenditures..

Unfortunately there have been instances of government project cancellations and underfunding of important work or unnecessary duplication. The French Coralie program was canceled after a few flights and after a good engine had been developed. Similarly an improved version of the U.S. Navaho missile was canceled, and the innovative Rocketdyne G-38 engine, which had been qualified to become its booster LPRE, became surplus. Chapters 8.1–8.11 contain examples of perhaps 15 major Soviet LPREs that were qualified; some of these engines had actually been installed in flight vehicles, but they did not fly. They were scrapped or put into storage, mostly because the intended vehicle program was canceled. These program cancellations might have been justified, but they caused poor utilization of the available LPRE resources.

There were also projects where a company's management promised more than they actually could deliver or where they bid a lower than realistic price, hoping to make it up in additional funds for add-on work and contract changes. This author believes (and this is hindsight) that much of the world's LPRE development could have been done with less money, better planning, fewer companies, and fewer and smaller government laboratories.

Because it often requires more time to develop a complex large rocket engine than to develop a flight vehicle, several major LPRE projects were actually started well before the vehicle was fully defined or funded. For example, the developments of the large engines for the Saturn V SLV were started before the vehicle development.

Proliferation of LPREs[1]

The number of organizations and countries able to develop and build LPREs has proliferated or multiplied. Originally in 1932 small informal groups existed only in three countries, namely, Germany, the Soviet Union and the United States. The objectives then were to fire the engine without burnout and to produce enough thrust for a long enough time to lift the vehicle. Ten years later in 1942 there was much more activity. The Soviet Union had two fledgling R&D organizations devoted to LPREs and had flown some sounding rockets and its first manned aircraft propelled by a LPRE. The Germans had three design/development organizations: the Heerversuchsanstalt, which was established in Peenemünde; Hellmuth Walter Kommanditgesellschaft in Kiel; and the Bavarian Motor Works or BMW. They had flown airplanes with rocket engines and experimental missiles, which were the predecessors of the V-2. In 1942 the Japanese had a small Army effort and fired a small TC, but this effort was not noticed and not continued. In the United States the first two LPRE companies (RMI and Aerojet) were in the early phases of JATO LPRE development, and GALCIT as well as Professor Goddard at the Naval Experimental Station had flown the first U.S. JATO units. Furthermore in 1942 there were active amateur organizations in several countries, including the United States, Soviet Union, Germany, France, Canada, the Netherlands, and Great Britain.

Forty years later, in 1982, there were well-organized experienced groups of LPRE specialists and experts on some aspect of LPREs in at least a dozen countries. There were more than six organizations each in the United States, Russia, and the cooperative European Space Community (France, Germany, Norway, Italy, Belgium, Netherlands, etc.) and a smaller numbers of active groups in the United Kingdom, Japan, China, and India. These countries had launched satellites and space probes with LPREs, and manned spaceflight programs had been undertaken in the Soviet Union and the United States. Each of the major countries had developed, built, and flown more than one type of LPRE. The objectives had changed dramatically in the 40 years; the aim now was to develop and build reliable, low cost, and high-performance LPREs on schedule for specific applications and to fly them. These groups were capable then and are capable today of designing and building a variety of different LPREs with confidence. Some of the historical accomplishments made in these countries will be described later.

In 1992 (10 years later) there were other countries in LPRE work. These countries had at least one small company, laboratory, group of enthusiasts and/or technical people that had been or were still involved in some phase of LPRE work (operation, production, component development, experiments, research, teaching, and investigations). They included amateurs, researchers, academicians, administrators, and rocket engineers. In some countries they developed (indigenously) or bought or otherwise acquired one or two foreign LPREs. The countries, which have or used to have a LPRE capability were Argentina, Brazil, Indonesia, Iraq, Iran, Israel, Kazakhstan, Libya, North Korea, Pakistan, Syria, Taiwan, Turkmenistan, and the Ukraine. There were in 1992 additional countries that had in the past been active or still were moderately interested at that time, but their efforts were limited: Austria, Australia, Bulgaria, Canada, Egypt, Romania, Serbia, Slovakia, Spain, or Switzerland. Altogether LPRE had been pursued in more than 30 different countries in the past 60 years.

There are many more world countries, that have not had an interest or a capability to develop space vehicles or ballistic missiles and thus no interest in LPRE. They typically consider this field to be an expensive endeavor, of low national priority, not directly useful or beneficial, or beyond their country's budgets. Also many of these countries do not have a background of high technology work or trained technical personnel. Some countries, which had been active in LPREs and other aerospace developments, have decided to leave this field. For example the United Kingdom or Britain used to have a formidable LPRE capability between 1940 and about 1972. For reasons that are explained in Chapter 12, the United Kingdom deliberately opted out of the launch vehicle and the LPRE business.

Rocket Engine Organizations

What companies or organizations developed and produced the LPREs? The answer to this question is given in two tables, one for organizations working on relatively complex and usually large LPREs and one for small simpler thrusters. Table 6-1 has a list of those that created and produced relatively large engines, all with a pump feed system.[1,2] It includes organizations that have ceased to exist or have been "disbanded" or companies that are today in some other business, but stopped their work in LPREs and have been "inactive" in this field. In Russia (formerly the Soviet Union) the mass production of LPREs is performed by separate production establishments and not by the organizations that do the development. A list of Soviet production organization is given in Table 8.3-2. For each entry in Table 6-1, there is a corresponding chapter in this book, where further details can be found. The order in which these companies are listed is by the chapter number of this book, and the order has no significance. There were a few other organizations with limited efforts in LPREs, and they are not listed.

Table 6-2 shows a list of developers and fabricators of small LPREs or small thrusters, all with a pressurized-gas feed system.[1-4] Some of the organizations in these two tables have gone out of the business, either voluntarily or by government choice. Tables 6-1 and 6-2 are not complete because there were others. For example, there are companies where the LPREs are merely a sideline to a larger, but different business. For example, Beal Aerospace has built and tested large TCs with hydrogen peroxide, and Microcosm has built and tested some ablative TCs.

It takes a few years for a new LPRE organization to become proficient in this business. Each had to learn the technology and understand the technical and business relationships with their customers. They had to repeat some experimental investigations, which had already been undertaken earlier by other LPRE organizations. They had to take the time to invest in test facilities, computers, computer programs, inspection equipment, fabrication facilies/machinery, and hiring and the training of their personnel. Most of these organizations worked on initial LPRE projects, which were not delivered and did not fly. Typically these were simple LPREs with inert gas-pressurized feed systems, such as JATO units or simple sounding rockets.

Motivation for Entering into the LPRE Business

What motivated these countries to be in the LPRE business? What induced them to allocate the resources needed to be successful in LPRE? There are several reasons, and in some countries they seem to change with time or with major aerospace programs. A motivation that was strong in the United States and some other countries is to advance the state of the art of the technology and be in the forefront of scientific discovery and novel fields of endeavor. When President Kennedy set a goal of going to the moon, it became a strong objective for the country and the U.S. space industry. Another primary moti-

Table 6-1 Development Organizations of Large Liquid Propellant Rocket Engines with Turbopump Feed Systems

Organization	Country	Work status	Chapter no.
Reaction Motors, Inc. (closed business)	USA	Disbanded	7.2
General Electric Company	USA	Group disbanded	7.3
Curtiss Wright Corporation	USA	Group disbanded	7.4
M. W. Kellogg Company	USA	Group disbanded	7.5
Aerojet General Corporation	USA	Active	7.7
The Boeing Co., Rocketdyne Propulsion and Power	USA	Active	7.8
Northrop Grumman Corp. Propulsion Products Center (formerly TRW)	USA	Inactive	7.9
Pratt & Whitney, Space Propulsion, UTC	USA	Active	7.10
ARC (was in large LPREs with TPs)	USA	Inactive	7.11
NPO Energomash	Russia	Active	8.4
Chemical Automatics Design Bureau	Russia	Active	8.5
KB Khimmash	Russia	Active	8.6
NPO Saturn, formerly OKB Lyulka	Russia	Inactive	8.8
NPO Samara, formerly OKB Kuznetsov	Russia	Inactive	8.9
NPO Energiya, formerly KB Korolev	Russia	Inactive	8.11
NPO Yushnoye	Ukraine	Active	8.10
Heeresversuchsanstalt Peenemünde	Germany	Disbanded	9.2
Hellmuth Walter Kommanditgesellschaft	Germany	Disbanded	9.3
BMW, Rocket Propulsion Operation	Germany	Operation disbanded	9.4
EADS, formerly Astrium, MBB, DASA	Germany	Active	9.5
EADS, formerly Snecma, SEP, SEPR	France	Active	10
LRBA, merged into Snecma 1971	France	Reorganized	10
Mitsubishi Heavy Industries	Japan	Active	11
Ishikawa Harima Heavy Industries	Japan	Active	11
Rolls Royce, Rocket Propulsion Operation	Britain	Group disbanded	12
China Academy of Launch Vehicle Technology	China	Active	13
Shanxi Liquid Rocket Engine Company	China	Active	13
Liquid Propulsion Systems Center, ISRO	India	Active	14

Table 6-2 Developers and Manufacturers of Small Thrusters[a]

Organization and location	Status of product	Chapter
Hydrogen-peroxide monopropellant (no longer produced)		
Walter Kidde and Company, Belleview, NJ	Abandoned	7.6
Bell Aircraft (today ARC), Niagara Falls, NY	Abandoned	7.11
KB Khimmash, Korolev, Moscow Region, Russia	Abandoned	8.6
General Electric, Schenectady, NY	Abandoned	7.3
Hydrazine monopropellant		
Hamilton Standard (UTC), (today licenced to ARC)	Active	7.11
Aerojet formerly Rocket Research Corp, Redmond, WA	Active	7.7
Northrop Grumman (formerly TRW), Redondo Beach, CA	Active	7.9
China (manufacturer not identified)	Active	13
KB Khimmash, Korolev, Moscow Region, Russia	Active	8.6
OKB Fakel, Kaliningrad, Russia	Active	8.12
Israel (manufacturer not identified)	Active	—
Royal Ordnance, Summerfield, England (today part of ARC)	Active	7.11
Bölkow (today EADS), Ottobrunn, Germany	Active	7.5
Ishikawajima Harima Heavy Industries, Tokyo, Japan	Active	11
SEP (today EADS), Vernon, France	Active	10
Walter Kidde and Company, Belleview, NJ	Disbanded	7.6
Hughes Aircraft Company (Abandoned this business)	Disbanded	—
Regenerative cooled (partly), bipropellant		
Rocketdyne Propulsion, The Boeing Co, Canoga Park, CA	Active	7.8
EADS, Ottobrunn (formerly Bölkow), Germany	Active	9.5
Northrop Grumman (formerly TRW), Redondo Beach, CA	Active	7.9
NII Mashinostroyenia, Nishnaya Salda Russia	Active	8.7
Interregen, bipropellant		
Rocketdyne Propulsion, The Boeing Co, Canoga Park, CA	Active	7.8
EADS (MBB), Ottobrunn Germany	Active	9.5
Ablative thrusters (no longer produced), bipropellant		
Rocketdyne Propulsion, The Boeing Co, Canoga Park, CA	Abandoned	7.7

(Continued)

Table 6-2 Developers and Manufacturers of Small Thrusters[a] (continued)

Organization and location	Status of product	Chapter
Northrop Grumman (TRW), Redondo Beach, CA	Abandoned	7.9
KB Khimmash, Korolev, Moscow Region, Russia	Abandoned	8.6
Radiation cooled, bipropellant		
Aerojet (formerly Marquardt Corp.), Redmond, WA	Active	7.7
Rocketdyne Propulsion, The Boeing Co, Canoga Park, CA	Active	7.8
ARC Liquid Propulsion, Niagara Falls, NY	Active	7.11
Northrop Grumman, formerly TRW, Redondo Beach, CA	Active	7.9
EADS (formerly MBB and DASA), Germany	Active	9.5
Royal Ordnance, Summerfield, Britain (now part of ARC)	Active	12
NPO Yuzhnoye, Dnepropetrovsk, Ukraine	Unknown	8.2
KB Khimmash, Korolev, Moscow Region, Russia	Active	8.6
NII Mashinostroyenia, Nishnaya Salda, Russia	Active	8.7
China (company not identified)	Active	13
India (company not identified)	Active	14
Ishikawajima Harima Heavy Industries, Tokyo, Japan	Active	11
EADS (formerly SNECMA and SEP) Vernon, France	Active	10
OKB Tumansky, Tushino, Russia		
NPO Soyuz, Turayeva, Russia	Unknown	—
Reaction Motors Division, Thiokol Corporation, Pompton Plains, NJ	Disbanded	7.3

[a] Table has been copied with permission from Ref. 2 and adapted by author.

vation (strong in the 1950s to 1970s for the United States and the Soviet Union) was to build the propulsion for military weapons. The LPREs were the key to long-range ballistic missiles. These weapons also at one time motivated China, Britain, or France, when they perceived the ballistic missiles of other countries to be a potential threat to them and they searched for a defense or for the capability to strike back. When the Soviet Union decided that it needed an antiaircraft missile capability, it applied large resources to make it happen.

Motivation is also connected with how a country wants to be perceived by the world. A country might want to have the image of an advanced nation, one that is active in learned endeavors, raises its own standard of living, and knows about high-technology hardware, which includes spaceflight and thus also LPREs. It might want to be known as a participant in ushering in the space age.

Another rationale was that some countries did not want to depend on foreign sources for LPREs and other aerospace hardware. They want to build the high-technology hardware, which was believed to be critical to their military posture, in their own country and not depend on foreign sources, which could be cut off. With some exceptions, they want to do it themselves. For example India wanted to have the capability to observe Earth events from their own synchronous satellite and not depend on information from foreign observation satellites. China wanted to demonstrate their advanced state of the art in spaceflight vehicles by launching an astronaut into an Earth orbit.

First Satellite Launches[2,3]

One of the crowning achievements of rocket propulsion systems is to place a useful payload into an Earth orbit. The first such launch of the country is an event that has been noticed in the world. LPREs have enabled 10 different nations to launch satellites into orbits. Table 6-3 gives a list of these 10 nations together with the booster propulsion, which lifted the vehicles off the ground. It also gives the chapter number where more information can be found. The first satellite, called Sputnik 1, was launched in the Soviet Union on 4 October 1957. The launch vehicle was a modified R-7 ICBM using a RD-107 engine in each of four droppable boosters and one RD-108 sustainer engine in the vehicle center core. This particular combination of the first and second stages forms a relatively large launch vehicle and has been used in the Soviet Union and later in Russia to launch many upper stages and payloads into orbits ever since.

The second nation was the United States sending the Explorer 1 into orbit. It was launched by a Juno 1 launcher, a modification of the Redstone missile, using a booster engine of 78,000 lbf thrust. Your author designed, developed, and tested the TC of this engine. Thereafter the United States used larger launch vehicles (such as Thor, Atlas, or Titan, all starting as ballistic missiles), and they allowed much higher payloads.

Four countries, listed in the table, used solid-propellant rocket motors for the propulsion of the first stage in each of their first takeoffs. The Japanese soon replaced their small launch vehicle, which used solid-propellant motors, with more powerful, higher-thrust larger space launch vehicles, propelled by LPREs, giving substantially more payload. These newer Japanese SLVs use LPREs in all stages and have been improved and uprated since that time. The Indians have used their solid-propellant booster for many years, but they used LPREs in the second stage and in some SLVs also in the third stage. The launch by Great Britain seems to have been a one-time effort only, sending a small payload into a LEO, and the development of another British SLV will likely not be repeated. The launch vehicle from Israel had three stages, each was propelled by solid-propellant rocket motors, and the payload is limited. The vehicle is believed to have been based on Israel's Jericho 2 ballistic missile with essentially similar propulsion systems. Israel has launched four more similar vehicles up to 2003 with slight improvements and with some more payload.

Table 6-3 First Satellite Launches by Different Countries and their Booster Rocket Propulsion[a]

Country	Date	Spacecraft	Launcher	Booster propulsion	Chapter no.
Soviet Union	4 Oct. 1957	Sputnik 1	R-7 missile	RD-107/RD-108	8.4
United States	31 Jan. 1958	Explorer 1	Juno 1	Restone LPRE	7.8
France	26 Nov. 1965	Arterix	Diamant A	Vexin LPRE	10
Japan	11 Feb. 1970	Ohsumi	Lambda-45	Solid-prop. motor	11
PRC[b]	24 April 1970	DFH-1[c]	Long March 1	YF-2 LPRE	13
Great Britain	28 Oct. 1971	Prospero	Black Arrow	Gamma 8 LPRE	12
ESA[d]	24 Dec. 1979	CAT	Ariane	Viking LPRE	10
India	18 July 1980	Rohini 1	SLV-3	Solid-prop. motor	14
Israel	19 Sept. 1988	Ofeq-1	Shavit	Solid-prop. motor	—
Brazil	23 Oct. 2004	—	—	Solid-prop. motor	—

[a]This table has been copied with AIAA permission from Ref. 3. It includes author's modifications.
[b]People's Republic of China.
[c]DFH is for Dong Fang Hong or "East is Red".
[d]European Space Agency.

The Israeli SLV had hydrazine monopropellant small ACS thrusters. Brazil, the 10th nation to launch a satellite, also has an all-solid-propellant set of stages, and the payload is limited by the thrust of the first stage. France has stopped sending up satellites of its own and has put all of its LPRE work in support of the Ariane, the European multinational SLV program. The newer SLVs in India have continued to use much larger solid-propellant rocket motors in the first stage, in part because they seem to have the capability to develop and build them. A set of LPREs in the first stage would give them higher payloads for the same takeoff mass.

References

[1]*Space Directory*, Jane's Information Group, Inc., Coulsdon, Surrey, England, 1996–1997, 2000–2001.

[2]Isakovitz, S. J., Hopkins, J. B., and Hopkins, J. P., *International Reference Guide to Space Launch Systems*, 4th ed., AIAA, Reston, VA, 2004.

[3]Gruntman, M., *Blazing the Trail, the Early History of Spacecraft and Rocketry*, AIAA, Reston, VA, 2004.

[4]Sutton, G. P., "History of Small Liquid Propellant Thrusters," Chemical Propulsion Information Agency, Columbia, MD, 2004.

Liquid Propellant Rocket Engines in the United States (Summary)

This chapter summarizes many of the key technical innovations, industry attributes, and historical events of LPRES in the United States. More different LPREs were developed, produced, and/or flown in the United States than any other country, except the Soviet Union, which today is Russia. As mentioned earlier, an estimated 300 different LPREs have been developed, demonstrated, and/or tested in this country. However only some of these LPREs were put into production. In the United States the company that develops the LPRE usually is awarded the production contract. There are separate chapters for each of the key U.S. companies, which are or were active in this LPRE field. An attempt is made here to summarize the major U.S. innovative accomplishments and historic events in the following numbered paragraphs.

1) The American Robert H. Goddard was the most important hands-on early pioneer in this field, the first to build and test a LPRE, the first to launch a vehicle with a LPRE, the inventor of jet vanes, gimbals, or gyroscopic flight control. His technical and historic contributions are almost legendary. They are described in Chapter 5.2.

2) The demands for new technology and production were very high during the Cold War with the Soviet Union during the 1950 to 1970 period, and the U.S. LPRE industry successfully met these demands and did its share to put missiles and military spacecraft into the arsenal. The period of the 1960s and the 1970s saw U.S. employment in LPRE work at its highest levels.

3) In 1965 the United States launched the highest thrust engine at that time, namely, a cluster of five F-1 LPREs at 1.5 million pounds thrust (sea level) each, and it was flown in the Saturn V SLV. This record stood until 1985 when the Soviets flew the RD-170 engine that had just a little more thrust, and they clustered five of those together in the Soviet R-7 ICBM.

4) There were a good number of inventions, innovations, or first implementations of technology pertaining to LPREs, and they can be credited to organizations in the United States. This includes the first feed system with pressurized cold gas, first flight of an engine with liquid hydrogen as a fuel, the first engine with an expander engine cycle, the precise bell-shaped nozzle contour, the first booster pump, and the first applications of gimbals to TCs for TVC. Furthermore, one can include the development of the tubular thrust chambers for regenerative cooling, the first use of ablatives on LPREs, development of special materials for hot TC walls, flat-plate machined injectors made of forgings, prepackaged LPREs with storable propellants, certain clever valve designs, certain injection patterns, including platelet injectors or pintle injectors, the initial pneumatic and electronic engine controls, pressurizing propellant tanks with warm gas from a GGs, liquid side injection for TVC, a catalyst for decomposing hydrazine, and very fast small propellant valves for small thrusters. Experimental LPREs were developed for several aerospike nozzles (up to 250,000 lbf), a linear aerospike engine, two early TCs with expansion/deflection nozzles (up to 50,000 lbf), and early experimental engines with a topping engine cycle (TC combustion gas to drive TP).

5) There are a number of clever innovations that were conceived and implemented in the United States, but very similar or identical innovations were actually accomplished at an earlier date in Germany and/or the Soviet Union. The LPRE work in these two countries was secret at the time and not known to U.S. LPRE personnel. Examples include engines with a staged combustion cycle, the first JATO units, using turbine exhaust gas for TVC, the first nozzle exit with circular arcs for its curved shape, early reinforced concrete test facilities, the highest-thrust large engine, LPREs specifically for aircraft installation, the first airplane flights with LPREs, radiation-cooled niobium thrust chambers, or TVC by auxiliary or vernier thrust chambers supplied by the main engine feed system. It also includes large LPREs with the highest chamber pressures, hydrazine monopropellant LPREs and their catalysts, and pump-fed experimental LPREs with certain high-energy propellants. It was only many years later that the United States learned about these earlier accomplishments in Germany and the Soviet Union.

6) The RL 10 is the first LOX/LH$_2$ engine (1963) that has flown. One version, the RL10B2 has the highest known specific impulse, namely, 467s. In about 1990 an upgraded version of this engine has flown with the highest nozzle area ratio for a pumped LPRE of 290, and this also was the first application of an extendible nozzle-exit segment in a LPRE (see Pratt & Whitney, Chapter 7.11).

7) The United States had in 1965 several small radiation-cooled niobium-walled thrusters that could demonstrate over 100,000 cycles or restarts.

8) Several of the propellants probably originated in the United States. This includes inhibited red-fuming nitric acid, Aerozine 50, mixed oxides of nitrogen,

hydroxylammonium nitrate, and ultrapure hydrazine that would not contaminate its catalysts.

9) The United States was a leader in the application of computers and software to LPREs. The RS-68 was the first LPRE to be designed entirely by computers. This covers not only design or analysis programs, but also computer-aided manufacturing, test operations, engine controllers, inventory control, cost and schedule control, or project management.

10) The technology of rocket propulsion exhaust plumes was initiated and is continuing to be developed in the United States. This includes plume observations, measurements, analysis, and prediction of plume behavior and properties.

11) The Nike–Ajax antiaircraft missile used a LPRE for the upper stage, and it was the first to be mass produced (16,000 missiles) and deployed. The delivery of about 50,000 pressure-fed engines for the Bullpup air-launched missile, delivered between 1960 and 1967, is the largest known production quantity of any LPRE in the world.

12) The United States has had extensive interactions about LPREs with international partners. This was encouraged by the U.S. government. Licenses for using specific U.S. know-how or technology and building certain LPREs were granted to organizations in Japan and the United Kingdom. Personnel from U.S. LPRE companies worked at organizations in France. The U.S. military stationed missiles with LPREs in Italy, United Kingdom, Turkey, and Israel and even trained foreign troops in their operation. In turn the U.S. licensed technology of reinforced carbon and carbon/ceramic nozzle materials from France (ceramics and extendible carbon nozzle) and acquired LPREs (NK-33 and RD-180) from Russia.

13) Funding for LPRE work has at times been irregular and not always well planned in the United States. Several major LPRE programs were cancelled after spending millions of dollars on them; this includes for example the M-1 LOX/LH_2 LPRE, Titan I, and the RS-2200 aerospike, several JATO units, and large uncooled thrust chambers,. Usually these cancellations were not caused by poor performance or problems with a particular LPRE, but by canceling of the vehicle programs, for which the engine was originally intended, or by cuts in the government's budgets.

14) The government's policy on LPRE test facilities has changed. Very early test facilities (Goddard, amateurs, or early RMI) were financed by the organization itself or by grants from private foundations (1920s to 1940). Thereafter larger test facilities for LPREs have been funded by the U.S. government, but built at the plant of the engine developer. This includes the first large test stand at General Electric Company, and it was supported by the U.S. Army Ordnance

(1940s), the extensive facilities of Aerojet at their Sacramento, California plant, the Rocketdyne Santa Susanna test operation, or the test facilities at ARC in Niagara Falls (all approximately 1940–1970). An approach that started in the 1950s was the building of large test facilities at government laboratories, and this required the contractors to bring their LPRE hardware to the government facilities for tests. It includes for example the altitude test facilities at the Arnold Engineering Center in Tullahoma, Tennessee, the large test facilities at NASA Marshall Space Flight Center, or the Air Force Rocket Propulsion Laboratory at Edwards, California. This saved government facility funds by not building similar facilities at contractors' plants, but makes the test operation for the contractors more difficult and expensive. It also helps to justify government laboratories. In 2005 all of the test facilities are not as busy as they used to be.

7.1 Liquid Propellant Rocket Engine Developers and Manufacturers in the United States

This chapter describes the organizations of the US LPRE industry and gives a listing of the principal companies, universities, and government organizations, that make up this industry.[1,2] Each of the following items will again be discussed in one or more subsequent chapters. For further references the reader is referred to these other chapters.

The U.S. LPRE industry has changed since the first company was founded in 1941 and the first government laboratory studied the subject. In many countries, such as Russia, China, or India, there are no private companies in this business; only government or pseudogovernment organizations to do this work. In the United States, Germany, Britain, or Italy, the government relies heavily on private industry. These commercial companies and their competitions with each other are essential to their continued viable existence of the United States and the U.S. form of governace. Fourteen U.S. companies have been engaged in the design, development, manufacture, testing, and flight support operations of some type of LPRE since 1941. Table 7.1-1 gives names and shows that there have been consolidations, acquisitions, mergers, and demises of companies in the LPRE business. As of mid-2004, five of the companies are still active: Aerojet, ARC, Northrop Grumman, Pratt & Whitney, and Rocketdyne.

There were other companies, but they are not listed in the table. They were component developers or suppliers to an engine manufacturer (and they also did some work on LPREs such as Moog), who developed one or two LRPEs in house, some in connection with their own vehicle development (such as Microcosm or Beal Aerospace), or were involved only in one or two LPRE projects of a narrow scope.

Employment in these LPRE companies peaked around 1955 to 1970, and the exact peak years were different in each company. An estimated peak of perhaps 140,000 people worldwide was engaged full time in the research, design, development, production, installation, maintenance, flight support, or testing of LPREs in this period. This does not count the propulsion personnel in flight-vehicle organizations or military operating units. This was during the Cold War between the Soviet Union and the United States and historically the busiest time in the LPRE industry. There were two other surges in LPRE activity, but they had much less impact. The decision of the United States to send men to the moon resulted in the development of several new LPREs. The increase in space launches for communication satellites, scientific research, and military purposes has caused some increase in the Liquid Propellant business. However this number of new satellites has decreased in the last decade and with it the number of LPREs. Because of the maturing of the technology, with diminishing research and development efforts the urgency of other government projects, and the increase in the life of reusable engines, the amount of funding available for LPRE work has decreased. So has the number of employees and the number of companies. For example, the peak total employment at Rocketdyne was

Table 7.1-1 U.S. Companies in the LPRE Business

Company	Typical LPRE work	Comments
Reaction Motors, Inc	Engines for experimental aircraft, Viking, prepackaged LPREs, TCs for ACS	Started December 1941; merged into Thiokol 1958; stopped operation 1971
General Electric Co.	LPREs for Hermes and Vanguard, U.S. V-2 launches, H_2O_2 monopropellant thrusters	Started 1942; LPRE dept. was dissolved in late 1960s
Hamilton Standard Div. United Technologies Corp.	Hydrazine monopropellant thrust chambers	Started 1943; bought by Marquardt in 2000
Marquard Corp., Van Nuys, CA moved to Redmond, WA	Small bipropellant LPRE	Started LPRE 1958; bought by Kaiser, later by Primex 2001.
Aerojet General Co., was Aerojet Engineering Corp. then part of General Corp.	All types of LPRE	Started in 1942; bought by General Tire Corp. 1944; acquired General Dynamics 2003; acquired and divested ARC Liquid (2004)
Rocketdyne Propulsion and Power, The Boeing Company, 1996	All types of LPRE, except monopropellant thrusters	Started 1946 as part of North American Aviation; merged with Rockwell International 1964; acquired by Boeing 1996
Pratt & Whitney, a United Technologies Company	Upper-stage LOX/LH_2 engines, Russian RD-180	Started 1957/58
Rocket Research Corp., in 1963; changed to Primex, then bought by General Dynamics	Hydrazine monopropellant and bipropellant TCs, ACS/maneuver systems	Started LPREs acquired by Olin Corp. 1985; GD bought Primex in 2001; bought by Aerojet 2003
Ampac ISP (2004) American Pacific Corp. formerly ARC, Atlantic Research Corp., Liquid Div.	Agena upper-stage, small bipropellant LPREs Positive expulsion tanks	Formerly Bell Aircraft Co., became div. of Sequa 1988; acquired Royal Ordnace (Britain) and Hamilton Standard line of N_2H_4 thrusters
Propulsion Products Center, Northrop Grumman Corp. formerly TRW	ACS LPRE, lunar lander, bipropellant and monopropellant thrusters	Started 1960 as part of Space Technology Laboratory, 2003 became TRW

(Continued)

Table 7.1-1 U.S. Companies in the LPRE Business (continued)

Company	Typical LPRE work	Comments
M. W. Kellogg Company	JATO units, R&D on LPRE	1945 until late 1950s; voluntarily stopped work on LPREs
Curtiss–Wright Corp.	LPRE for research aircraft	1943 until 1960s; voluntarily stopped work on LPREs
Walter Kidde and Co.	H_2O_2 monopropellant and N_2H_4 thrusters	1945–1960s stopped LPRE work
Hughes Aircraft Co.	Hydrazine monopropellant thrusters	Discontinued LPREs

about 20,500 people in 1964; it hit a low in 1971 but went up to about 6000 in 1985.* This includes some people who worked projects unrelated to LPREs. In June 2001 the number of Rocketdyne employees engaged exclusively in LPREs had shrunk to about 2800.* Aerojet's employment in LPREs personnel peaked in 1963 at about 10,000 and was about 1000 in the year 2000.

Universities and Government Laboratories

Although some very good R&D work related to LPREs has been done in the United States by certain universities, some private research organizations, and by government laboratories, their efforts are not mentioned here because they did not directly develop or qualify LPREs. The ultimate application of good research project results leads to a useful LPRE product, improvement, or service. This includes for example R&D work on topics of interest to LPRE people at Princeton University, Cornell University, Purdue University, Pennsylvania State University, University of Alabama, the Navy's Post-Graduate School, or the California Institute of Technology. Nongovernment laboratories doing work related to LPREs included the Batelle Memorial Institute and SRI (Stanford Research Institute). The graduate and undergraduate education of qualified technical LPRE personnel is perhaps the universities' most important contribution, and more than 25 U.S. universities have at one time taught or are still teaching courses concerned with LPREs. There have been some excellent research professors and some early relevant research projects that did trickle down to the industry and were helpful to companies in gaining a better understanding or doing a better job with LPREs. Unfortunately the linking of the

---

*Personal communications, V. Wheelock, The Boeing Company, Rocketdyne Propulston and Power, 2002, and C. Fischer, Aerojet Propulsion Company, 2001.

university research with practicing engineers in commercial companies has been tenuous and inconsistent. Much of what the industry needed was not well defined and not communicated to the universities. Conversely, very few engineers or designers were aware of the work conducted by universities. As a result, a good many university research programs were never known or understood by those people in industry, who actually made the design decisions. Also as the result of poor communications, a good number of university research projects in the last few decades were interesting to the professors, but of no use to the industry.

Government laboratories doing work on LPRE include the Rocket Propulsion Laboratory (now part of Philips Laboratory of the U.S. Air Force) at Edwards Air Force Base, Edwards, California, Arnold Engineering Test Center, Arnold Air Force Base, Tennessee; NASA Marshall Space Flight Center, Huntsville, Alabama; Jet Propulsion Laboratory, Pasadena, California; Stennis Space Center, Mississippi; White Sands Proving Grounds, New Mexico; and NASA John H. Glenn Research Center, Cleveland, Ohio. These and other government laboratories are indeed important in unbiased testing of LPREs, which were developed by companies, in defining, selecting, and monitoring work at U.S. and some non-U.S. contractors, doing some R&D, and in maintaining and providing propellants and testing facilities for LPREs, including hover test facilities, and simulated altitude test facilities. Test facilities for large LPRE are now so expensive that private companies can no longer afford them, and they have to rely on government facilities.

References

[1]Seifert, H. S., "Twenty Five Years of Rocket Development," *Jet Propulsion*, Vol. 25, No. 11, Nov. 1955.

[2]Doyle, S. E. (ed.), *History of Liquid Rocket Engine Development in the United States 1955–1980*, American Astronautical Society History Series, Vol. 13, Univelt, San Diego, CA, 1992.

7.2 Reaction Motors, Inc.

Reaction Motors, Inc. (RMI) was the first American company dedicated to developing and manufacturing LPREs.[1] It was founded by four amateur experimenters of the American Rocket Society and incorporated in December of 1941. The words *rocket motor* were used in those days for what today is designated as a *rocket engine*. One of the founders, James Wyld, was the first in the United States to develop full regenerative cooling for thrust chambers in 1938 as shown in Fig. 4.3-3. Robert H. Goddard, the first man to build and launch a LPRE, met from time to time with RMI people between 1942 and 1945 and consulted with them. This company contributed significantly to the history of American rocketry and was the first to develop a number of different LPRE types as mentioned next.[1-5] In 1958 RMI was acquired by and became a division of the Thiokol Corporation (a leading manufacturer of solid-propellant rocket motors). Thiokol wanted to balance its product line and participate in the LPRE work. In 1972 Reaction Motors was shut down by the Thiokol top management and ceased operations because of a lack of business.

The very first TC built by RMI was small (100 lbf with LOX/75% ethyl alcohol) and was used in part to learn, but also to demonstrate the rocket propulsion principles to potential customers. Soon they successfully tested a 1000-lbf TC. RMI's first contract efforts were in liquid propellant JATO rocket engines, and they developed a number of them with different thrust levels.[2,3] Their large JATO unit at 3000 lb thrust with regenerative cooling was successfully flight tested in 1944 in a Navy PBM-C3 flying boat, even achieving underwater ignition. However none of these JATO engines were adopted by the military services, in part because liquid oxygen could not be stored and proved to be impractical for operational use.

Aircraft Rocket Engines

The best known and perhaps the most historic of their engines was the RMI 6000-C4 aircraft rocket engine with four fuel-cooled thrust chambers shown in Fig. 7.2-1 and described in Table 7.2-1. It was designed for the Bell Aircraft manned research airplane X-1 sponsored by NACA, which is today NASA. This engine propelled this research aircraft on 14 October 1947 to a record speed of Mach 1.06, a truly historical event. The pilot Captain (later General) Charles "Chuck" Yeager became the first U.S. flyer to break the sonic barrier. Because this engine was painted black (to prevent unsightly rust), this engine had the nickname "Black Betsy."[2,4,5]

Each thrust chamber had a small igniter chamber in the center of the injector designed to allow multiple restarts. The igniter used a spark plug to ignite a small flow of fuel and gaseous oxygen that had been evaporated in coils around the fuel feed pipe. The igniter operated at a pressure of 50 to 75 psi, which is much lower than the chamber pressure at full flow. It used a system of pressure switches and relays to make the start control automatic. The concept of this igniter, the evaporation of LOX by winding a coil of tubing around

312 History of Liquid Propellant Rocket Engines

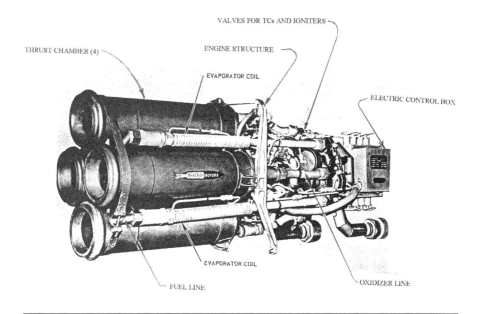

Fig. 7.2-1 RMI 6000C-4 LPRE for the X-1 research aircraft was historical. It had four TCs. Each could be turned on or off individually thus giving a stepwise change in thrust. Courtesy of Reaction Motors, Inc.; adapted from Ref. 5.

the main fuel pipe, and the controls were attributed to Lovell Lawrence, the company's first president and a former rocket amateur. Each of the four thrust chambers could be turned on or off, allowing four step changes in total thrust. Initial versions used a gas-pressurized feed system with relatively high pressures (280 to 370 psi) in the propellant tanks. The development of the initial

Table 7.2-1 Data for the RMI 6000-C Rocket Engine

Number of thrust chambers	4
Thrust for each thrust chamber	1500 lb
Total thrust (all four TCs)	6000 lb
Oxidizer	Liquid oxygen
Fuel	75% ethyl alcohol with 25% water
Chamber pressure	220 psia
Specific impulse (sea level)	209 s
Propellant supply, initial	Pressurized feed system
Propellant supply, later version	Turbopump feed with gas generator using 90% hydrogen peroxide and a catalyst

version of this engine with a gas-pressurized feed system was completed in about two years. Later versions (development started about 1947) used a small turbopump shown in Fig. 7.2-2 allowing much lighter propellant tanks and gas tanks.[2,5] This was the first U.S. turbopump to fly in a U.S. aircraft. It was a single-shaft TP with the turbine in the middle and overhung centrifugal pumps at each end. The turbine was driven by decomposed steam and hot oxygen gas from a separate hydrogen-peroxide monopropellant gas generator, with a silver-plated set of catalyst screens. With the GG, dual ignition chambers, restart, throttling, and pump drive the engine control system became quite complex.

The engine was improved and used to fly several later versions of the Bell X-1 research aircraft, the Douglas D558-2 Skyrocket research aircraft, several unmanned research lifting bodies, and as a dual engine (with total of eight TCs) for an interim power plant for the North American X-15 research airplane. An uprated version of this four barrel engine (at 435 psi chamber pressure and 7600 to 8400 lb total thrust) launched the small-scale model of what

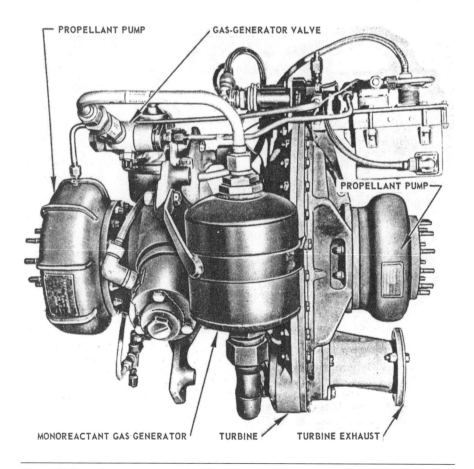

Fig. 7.2-2 Turbopump assembly for later versions of the RMI 6000C-4. Courtesy of Reaction Motors, Inc.; adapted from Ref. 5.

later became a U.S. long-range ballistic missile (Project MX-774).[1,2,4] This version did not have a restart capability, which simplified the design and controls. It was the first U.S. engine with hinged thrust chambers, which allowed flight-path control. There were engine-related problems in two of the only three flights in 1947 and 1948 at the White Sands Proving Grounds; in one the oxygen tank disintegrated (because of a pressure regulator failure and overpressure in the tank), and the vehicle broke up. In another flight a faulty electric system caused a premature cutoff of the engine.

The largest and most complex RMI LPRE was the throttled (down to 30% thrust), restartable (six times), *reusable engine for the North American Aviation X-15 research airplane*.[3,4,6] It was identified as the XLR-99 LPRE. Some of its nominal data are shown in Table 7.2-2 and a flow diagram in Fig. 7.2-3. At 50,000 lb thrust it was the largest engine ever delivered by RMI. The engine development program turned out to be difficult. It used liquid oxygen and ammonia as propellants, and it was the only flying engine with this propellant combination. It was based in part on an earlier experimental engine of about the same thrust and the same propellants.

The XLR-99 had the first large tubular TC built by RMI with a single pass of the fuel, as discussed in the next section of this chapter. The two-stage, two-chamber igniter became necessary to avoid an accumulation of unburned

Table 7.2-2 X-15 LPRE Data (XLR-99)[a]

Thrust at 18,700 ft altitude	50,000 lbf or 220 kN (100%)
Thrust in vacuum (100%)	50,870 lb or 224 kN (101.7%)
Throttle range	Down to 30% or about 15,000 lbf
Duration, at full thrust	180 s
Engine dry mass	910 lbm
Engine wet mass	1025 lbm
Specific impulse (@ 18,700 ft)	230 s at full thrust
Overall length and diameter	82 and 39.3 in. (208 and 100 cm)
Weight, dry and wet	910 and 1025 lb (412 and 465 kg)
Mixture ratio, oxidizer to fuel flow	1.25
Engine life before overhaul	1.0 h and/or 100 starts
Chamber pressure	600 psia (100%) and 225 (30%)
Gas-generator flow rate (hydrogen peroxide)	8.10 lbm/s (3.6 kg/s) at 100%
Turbine speed at full flow	12,700 rpm
Nozzle area expansion ratio	9.8
Nozzle design altitude	18,500 ft (5638 m)

[a]From Refs. 3, 4, and 6.

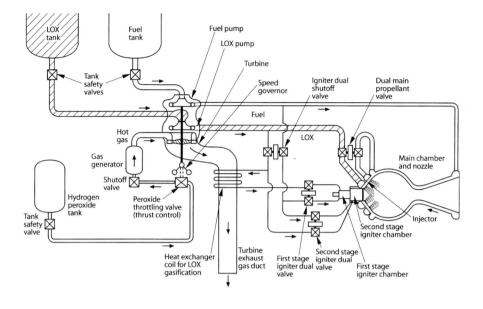

Fig. 7.2-3 Simplified flow sheet of the XLR-99 LPRE with 50,000 lbf thrust for the X-15 research aircraft. Redrawn by author with permission from Refs. 3,4, and 6.

propellant in the chamber, which resulted in momentary overpressures during the start or what was commonly called a "hard start." The two igniter chambers ran throughout the engine operation. The turbopump and the two igniter prechambers could operate in an idle mode allowing a checkout of over 90% of the key engine components, prior to turning on the full propellant flow. A centrifugal swinging-ball-type rotating governor throttled the flow to the GG and thus regulated the TP shaft speed and prevented a TP overspeeding. This governor was tied into the hydrogen-peroxide throttle valve, which indirectly controlled the gas supply to the turbine. It is the only TP-fed LPRE known to the author with such a mechanical shaft speed governor. The concept of the ball-type rotating speed governor might have come from Goddard, who at one time considered this idea.

The pilot's engine throttle lever (not shown in flow diagram) was electriclly connected to the hydrogen-peroxide throttle valve and also to the setting of the rotating ball speed governor. The position of the lever also initiated the start sequence and the cutoff sequence and gas purge of remaining propellant from the engine. Figure 7.2-3 shows that the TP was supplied with steam and hot oxygen gas from a hydrogen-peroxide gas generator with a solid-bed-type catalyst. Not shown in the figure are the valves for filling or draining propellants and the pressurizing helium gas, helium tank, helium purge lines and valves, helium lines to actuate the principal valves and pilot valves, flushing of propellants, pressure relief, instrument sensors, the mechanism for setting the shaft speed governor, or the redundant spark plugs and their electric power supply.

The delivery of the XLR-99 was more than a year late because of several R&D problems associated with the engine complexity and pilot safety.[6] Many of the problems described here were a result of learning about new features that had not been proven earlier and their interactions in a new engine For example, there was an explosion of the ammonia tank during a static engine test, while installed in the airplane on the runway. It was found to have been caused by a failure of the pressure regulator in the helium system, which put too much pressure into the fuel tank. There was no injury to the pilot, and the airplane was repaired. The specific impulse was below the nominal value in initial tests and particularly at the low thrust level. After trying several different injectors the performance came up to the nominal value of 230 s, but this was still below the company's expectation of 241 s. The injector had 18 heads or welded inserts, each with a nonimpinging set of injection holes and 66 separate film-cooling holes at the periphery. The inner wall of the tubes in the chamber and throat region of the TC were coated with a flame sprayed layer of rokite, a mixture of nichrome and zirconia. During some of the development tests and the early flight qualification tests, this coating spalled off locally or flaked off, probably because of thermal cycling or vibration or both. The local heat transfer to the exposed unprotected tube wall was very high, causing the ammonia inside the tube to boil, and the exposed local tube area would exceed the yield point of the metal and form cracks upon cooldown. The crack would become bigger and deeper with the next start of the engine, and the ammonia would leak into the chamber after a few starts. A program to improve the coating was undertaken at RMI in early 1961. A graded coating of sprayed nichrome and zirconia on a base of molybdenum primer coating turned out to be satisfactory.

Because a single igniter with spark plugs did not give a satisfactory start, a second ignition chamber was added, and both ignition chambers operated continuously throughout the engine's operation. An idling low-thrust stage could be achieved with the two igniter chambers providing the hot gas and with the turbopump running. During this idle stage, about 90% of the various valves and controls could be checked out automatically, and if these did not indicate safe operation the engine could be shut down.

There were instances of strong vibrations during the startup of the XLR-99 (Ref.6). Putting some damping into a modified engine structure did not help. Because good high-frequency measurements were not available in those days, RMI really did not know what happened. They did have acceleration sensors, but the measured values were questionable. During an incident, the accelerations were initially low, but could quickly build up to 100 g (g is the acceleration of gravity). Above 200 g the engine failed and was destroyed. The remedy was to shut off the engine, if their accelerometers indicated accerations above 120 g, thus forcing the airplane to make an emergency landing. This saved some hardware, but also caused the abortion of a couple of test flights. This high acceleration was more likely encountered at mixture ratios lower than the design ratio. Also the engine experienced a strong vibrations during shut down, and this was eventually remedied by changes in the injector.

The XLR-99 program started in 1956, but the completion of development was delayed. With this delay in schedule, an interim powerplant solution was found, which allowed a part of the aircraft flight-test program to proceed. Two proven RMI-6000 LPREs were quickly adapted and installed in the X-15 aircraft, and they worked well, but the low thrust (12,000 lbf) limited the flight maneuvers, altitude, and maximum speed. Beginning in about 1959, the X-15 made 29 flight tests over a period of about a year using these two interim engines. Because of the development problems and the delivery delays, NASA started a backup engine program at General Electric Company, but it was canceled, when most of the XLR-99 engine problems had been solved.

The first satisfactory XLR-99 engine ground-test firing was performed in 1959, the flight rating tests in 1960, and the first flight in 1961 (Ref.6). Even though the engine flew many times reliably and safely, it was not considered safe to operate it in certain modes, and the engine operating regime was restricted during flight.[6] For example, it could not fly in the idle mode (with igniters and TP runing) for more than 30 s. As already mentioned, the engine automatically would shut down if local engine accelerations exceeded 120 g.

The X-15 airplane, shown in Fig. 2-6, was called a flying set of propellant tanks.[6,*] These tanks and the monopropellant hydrogen-peroxide attitude control system can be identified in this figure. The monopropellant thruster locations in the wings and the nose of the aircraft are visible. With this LPRE the X-15 airplane exceeded all prior altitude and speed records. In spite of the handicaps of the engine operation, the flights were very successful in learning about supersonic aerodynamics. A total of 169 flights were made with the XLR-99, and eight different engines were used during these flight tests. Eight of the flights had some engine anomaly or nondestructive failure, which caused emergency landings. However it is to the credit of the designers that these emergencies did not cause any damage to the aircraft and did not put the pilot into jeopardy. With this XLR-99 engine the X-15 aircraft achieved a speed record of 6700 ft/s or a Mach number of 6.7 and a new altitude record of 67 miles.

Tubular Thrust Chamber

An important advance in the technology was the invention of a tubular thrust chamber.[2,3,*] This concept is explained in Chapter 4.3, and some of this information is repeated here. It was first conceived at RMI, first built by Aerojet, and first flown by Rocketdyne. Edward A. Neu, Jr., an engineer with RMI, supposedly was the first to conceive the idea in 1946. All of the references say

---

*Personal communications, P. F. Winternitz, Propellant Laboratory Manager, RMI (circa 1970), and F. H. Winter, Space History Division, National Air and Space Museum, Smithsonian Institution (circa 2002).

that RMI applied for a patent on the tubular TC in 1947. However, according to the U.S. Patent Office records, RMI applied for a patent on the tubular cooling jacket in Ed Neu's name only in 1950 (after RMI had tested one), and it took an unusually long time, 15 years, before this patent was granted in 1965 (U.S. Patent 3,190,070). The delay was in part caused by the patent examiner evaluation of claims, which more properly belonged to other companies. Two drawings from this patent are shown in Fig. 7.2-4. RMI's first small tubular TC was reported to have been built in 1949, but it was not tested until later, and a larger TC at about 500 lbf thrust was tested in 1952. The original TC started fabrication with constant diameter tubes. The multiple tubes were bent and shaped to fit the contour of the combustion chamber and nozzle. They were welded together and formed the combustion chamber as well as the cooling passages of the single-flow-pass cooling jacket. Ammonia entered the cooling jacket through a manifold at the nozzle exit, flowed through the tubes and was collected in a coolant exit manifold (just next to the injector), sent through the main fuel valve, and then supplied to the injector. After examining a number of tube materials, RMI selected nickel because it had good physical properties

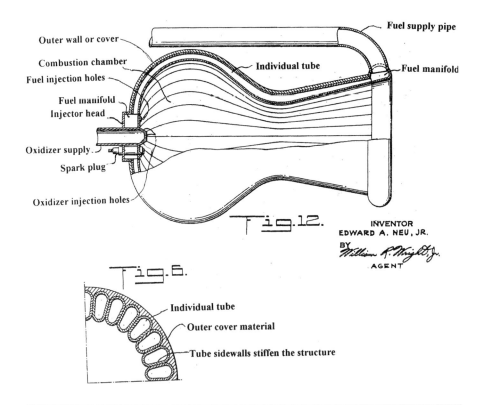

Fig. 7.2-4 Two drawings from U.S. Patent 3,190,070 describing a tubular thrust chamber construction (1965).

at elevated temperatures, was ductile, had a reasonable thermal conductivity, and had good elongation. It was based on the design and fabriction techniques of smaller tubular TCs developed at RMI and also on observations of tubular cooling jackets of their competitors General Electric and Rocketdyne. Considerable difficulties were encountered in building and repairing the welded nickel tubes in the XLR-99. Later engineers at another company learned that a small amount of sulphur, contained in the fuel or the welding flux, would cause embrittlement of the nickel, and this caused cracks and leaks to form during temperature cycling with multiple restarts. After many tube repairs RMI switched to stainless steel tubes for the TC of the XLR-99 LPRE, and it flew first in the X-15 aircraft in 1961. The stainless was not as good a heat conductor, but it could be fabricated and welded and would not readily form cracks.

Rocketdyne and Aerojet apparently conceived very similar ideas a year or two later, but their designs and fabrication techniques were different. Rocketdyne selected stainless steel, double tapered each tube (not constant diameter tubes), and used brazing (rather than welding) to seal the space between tubes and external reinforcing bands to hold the tubes together. Rocketdyne selected a design with alternate, but adjacent tubes for downflow and upflow, and used brazing to bundle and seal the tubes together. Figure 4.3-4c describes the concept. As Fig. 4.3-5 shows, the turnaround of the flow at the nozzle exit required a very small fuel manifold. The flow from the upflow tubes went directly into the injector. Rocketdyne designed the first tubular TCs in 1948, investigated various materials, alternate fabrication techniques, tube shaping, and joining techniques in 1949. The first large TC with tubular construction at 120,000 lbf thrust was tested in 1950 and was the first tubular TC to fly in the Navaho missile in 1956.

Aerojet was the first to build a small tubular TC in 1948, which was the earliest test TC. It was made with constant diameter aluminum tubes, which were welded together. Each tube had a U turn at the nozzle-exit area, so that the same tube could be used for downflow as well as upflow. Aerojet had some difficulties with the welding of thin aluminum tubes. The tubular concept was then not actively pursued at Aerojet for several years. The Titan-I thrust chamber (150,000 lbf thrust) was built with stainless-steel tubes brazed together in 1957 and flew in an engine in about 1959, as discussed in Chapter 7.7.

Viking

RMI designed the booster engine for the Viking upper-atmosphere sounding rocket. It was one of the early RMI projects; it began in 1946, and engine static firing tests started in 1947. It was rated at 20,000 lbf (9100 kg) thrust, used liquid oxygen and alcohol, and it was the first LPRE in history, where a thrust chamber was mounted on a gimbal.[2,3,5,7]

Gimbals have been applied to many LPREs since this time. The inner wall of the thrust chamber was made of pure nickel because of its superior themal conductivity and good elongation. Its 300-hp single-shaft turbopump ran at

10,000 rpm and had a single-stage turbine and centrifugal pumps for the propellants. It was powered by a hydrogen-peroxide gas generator with a catalyst. This type of GG was based on German technology. A total of 13 flights were launched, and some of these did not achieve the intended flight objectives. The RMI engine did not perform properly in at least two of these flights. There were modifications made to the engine to improve its reliability and to slightly raise its performance. The last two flights (1957) were successful; they did not carry atmospheric research instruments, but experimental equipment for trying some novel features of the upcoming Vanguard SLV program. Although the Viking flights provided some new useful atmospheric data, there was no follow-up to this sounding rocket program because the vehicle was too small and too expensive. Captured German V-2 vehicles had been used for the exploration of the atmosphere, and they were less costly.

Lark Missile

RMI built a dual-thrust-chamber sustainer engine for the Lark surface-to-air missile.[2,5] Development of the LPRE started in 1944 during the war. It is shown in Fig. 7.2-5. The Lark missile was boosted by solid-propellant rocket motors, used a LPRE in its second stage, and was flight tested at Cape Canaveral in 1950. About 500 Lark engines were built by RMI, and most were used in Navy experimental missile flights. The design is based on development

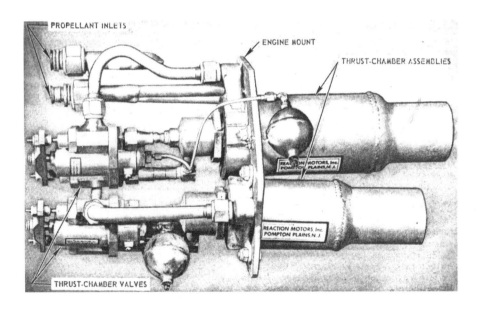

Fig. 7.2-5 RMI LPRE for the Lark missile had a 400-lbf thrust booster TC and a smaller 220-lbf thrust sustainer TC both with double-wall construction Courtesy of Reaction Motors, Inc.; copied from Ref. 5.

work done by the U.S. Navy at Annapolis at about 350 lbf total thrust. It used the hypergolic propellant combination of nitric acid and aniline (with some minor additive), which was selected after trying several different hypergolic fuels, such as aniline diluted with ethyl alcohol or isopropyl alcohol. The engine had a simple gas pressure feed system, and the two double-wall TCs were regenerative cooled with a spiral flow cooling jacket made of welded sheety metal. The smaller (220-lb) thrust chamber was the sustainer and operated continuously, and the larger engine (400 lb) provided the initial boost. When the missile reached a speed of approximately Mach 0.85, the larger engine was stopped, but it was restarted when the speed dropped below its lower limit or when the vehicle came close to its target. Typical powered flights lasted 4 min, but for trajectories when there was little use of the larger thrust chamber it could propel for a longer time. Initial hard-start problems (pressure peak caused by the very high initial flow of propellants in the chamber) were solved by using a tapered pintle on the valve stem of the propellant valves. Figure 4.8-1 shows a curved tapered pintle, it needed to be opened slowly, and this design idea was suggested by the Navy. This type of valve allowed a more gradual increase of flow, reducing the high initial propellant flow and minimized accumulation of poorly mixed unburnt propellant, and was soon used by other developers of thrust chambers. RMI did not win the follow-on contract. Rocketdyne was selected for the product improvement and additional production work on the Lark LPRE.

Prepackaged Storable LPREs Including Bullpup

RMI developed several storable propellant prepacked rocket propulsion systems.[3] In the 1940s solid-propellant motors had problems operating at ambient temperatures lower than about −40°F, but the liquid engines could easily meet the required minimum temperature of −65°F. Furthermore the liquid propellants were essentially smokeless, which the Navy wanted, whereas the solid propellants usually left clouds of smoke. For this reason the government initiated in the mid-1950s the development of the Liquid Propellant engine for the Sparrow III air-to-air missile. Here the storable propellants were permanently sealed in their tanks, and pressurization was accomplished by burning a small solid-propellant charge. There were no real valves per se, but only burst diaphragms. This was a very tough job because the prime contractor (Raytheon) insisted that the LPRE have the same envelope, interface, and thrust as the existing solid-propellant motor, and be fully interchangeable. RMI developed a 5000-lb thrust unit with 2.5 s duration using RFNA with 20% NTO and hydrazine with ammonium thiocyanate additive as a freezing point depressant. The concept was very similar to the German Taifun missile, which was developed toward the end of World War II. The Sparrow LPRE worked, but the initial ignition spike caused a momentary overpressure and a short high acceleration pulse that gave problems to the guidance system. RMI managed to reduce the pressure surge, but did not succeed in eliminating it altogether. The project was canceled in 1960, even though RMI had produced about 400 units because by that time a suitable solid-propellant version had been developed, and it was capable of being cooled to −65°F, and it did not have an ignition overpressure surge.

The next prepackaged LPRE was for the *Bullpup air-to-surface missile*. This was a highly successful program, which culminated in mass production. Work started in 1955. It had the largest production of any LPRE in the world, and approximately 50,000 units were delivered between 1960 and 1967.[3,5] Work started while the Sparrow effort was near its peak. As shown in Fig. 7.2-6, it used the same basic concept, storable propellants, but a different hypergolic fuel than the Sparrow (RFNA and mixed amine fuel, namely, 50.5% diethylenetriamine, 40.5% UDMH and 9% acetonitrile).[8] A solid-propellant (double-base) short-duration cartridge was used for tank pressurization. As seen in Table 7.2-3, there were two versions: a smaller one called LR58 and one with a much larger payload and a LPRE that was almost three times larger (LR62). It started with the ignition of a small powder cartridge moving a piston (the only moving part), which then sheared or broke a diaphragm and initiated the burning of the solid-propellant GG grain. Full thrust was achieved in about 0.1 s. Burst diaphragms confined the stored propellants. The thrust chamber was inside a propellant tank, had a shortened, inefficient nozzle, was regeneratively

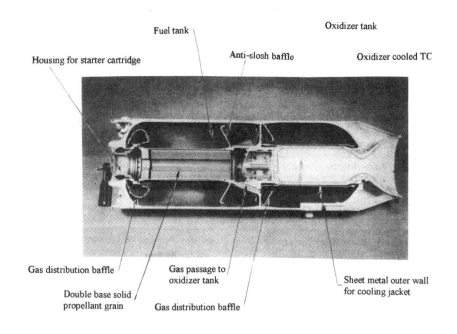

Fig. 7.2-6 Cutaway section of the Bullpup A (LR-58) with a central internal-burning propellant grain for pressurizing the two annular propellant tanks. More than 50,000 of these were built, the highest production quantity for a LPRE. Copied with AAS permission from Ref. 3.

Table 7.2-3 Bullpup LPRE Data[a]

Parameter	Bullpup A	Bullpup B
Engine designation	LR58	LR62
Diameter, in. (cm)	12.1 (30.7)	17.3 (43.9)
Length, in. (cm)	40.5 (102.7)	61.2 (155.4)
Weight, loaded, lb (kg)	203 (92)	563 (255.3)
Weight, dry, lb (kg)	92 (41.7)	205 (92.9)
Thrust, (lbf/kN)	12,000 / 52.8	30,000 / 132
Duration, s	1.9	2.3
Total impulse, lb-s/kN-s	22,800/101	69,000/307

[a]Copied from Ref. 3.

cooled, and had an additional protection of a zirconium oxide coating in the nozzle throat region. The Bullpup had a minimum storage life of 5 years and a safe storage temperature range between −80 and +160°F (−62 to +71°C). It became an operational military missile in 1959. There were thousands of test firings, both on a test stand and in flight. Qualification and preflight testing was tough and included drop tests from heights of 1.8 m and a few at 12 m (caused some tanks to split open), vibration exposure, impact from 45-mm tracer bullets, which did not cause ignition, and 30-min external fire exposure. Reliability of both the larger and the smaller version was rated at 0.9972. Military cutbacks and changing military requirements eventually caused the Bullpup to be taken out of service in the 1970s.

The development of Bullpup was not trouble free. For example, the plastic polymer material, which was used for sealing the fuel fill plug of the fuel tank, was plasticized and effectively dissolved by the fuel and caused leaks in the field. Another polymer was the solution. A chemical reaction of the hot fuel-rich gas (from the solid-propellant GG) with the oxidizer inside the oxidizer tank caused high-ullage gas temperatures. The gas inlet to the tank was redesigned. With the short operating duration the walls of the oxidizer tank did not become hot enough for the wall to become weak. Furthermore, sloshing in the tank caused sudden cooling of the warm gases in the ullage space above the liquid and sudden changes in gas temperature and pressure. Baffles in the tanks and modification of the solid propellant alleviated the problems.

RMI was also involved, but only for a relatively short time, in two other prepackaged LPREs for the Corvus and Condor air-to-surface missiles, which were to be launched against enemy surface ships.[3] The *Corvus* LPRE had a thrust of 1030 lbf (4580 N) for 177 s, used nitric acid and a mixed amine fuel, a small turbopump with a gas generator using the same hypergolic propellants, but at a low mixture ratio, and a solid-propellant starting cartridge. Work began in 1957, and the first test flight took place in 1959, and it was successful. The U.S. Navy canceled the program, after millions of dollars had been spent on the vehicle and its engine, mostly because another Navy missile program was then believed to lead to a better waepon.

The *Condor* program was started by the Navy in the mid-1960s using a high-energy, high-density oxidizer, namely chlorine trifluoride (ClF_3) and the fuel was a mixture of hypergolic amine compounds.[3] This oxidizer is very toxic and corrosive with most materials. Some of the Navy operating people were not enthusiastic about putting ClF_3 into any ship. The thrust was in the 3000- to 4000-lbf thrust range with deep throttling (15 to 1.0 range). It used thin collapsible stainless-steel bladders to confine the propellants and separate them from the pressurizing gas, which would otherwise dissolve in the liquid propellants. The technical problems of corrosion and deep throttling caused delays and overruns, which in turn brought on the canceling of the project in 1967.

Small Thrusters

RMI's reputation was tarnished as the result of the delivery delay and flight restrictions of the XLR-99 (X-15 research airplane) program and inadequate performance on other large engines. So RMI decided to concentrate on small thrusters.[3,5,8] In July 1964 RMI obtained a contract to develop small thrusters for the Surveyor spacecraft, which successfully accomplished an unmanned moon landing with a variable-thrust small TC. Propellants were hypergolic, namely, NTO with 10% NO (to reduce the freezing point) and MMH with 28% water. The water was added to reduce the combustion temperature, but it also reduced the specific impulse. The thruster's chamber and throat regions were regeneratively cooled, and the nozzle extension was radiation cooled and made of thin molybdenum. The trust could be varied continuously between 30 and 104 lbf, and the vacuum specific impulse was between 278 and 287 s, and these were technically difficult to achieve. The thruster passed the required 10,000 start cycle demonstration. The exterior of the thruster assembly (TC and valves) was gold plated because this would reduce the heat absorbed from solar radiation during long periods in space. Gold plating was later found to be unnecessary. The coolant became quite hot during operations at low thrust, and there was the concern about boiling or possible decomposing the coolant. These heat-transfer problems were solved by going to a ceramic nozzle throat insert and a change in the shape of the chamber. There were no thruster failures during the actual flights, and the flight reliability was 100%.

RMI developed a couple of other small thrusters, but they were not qualified and did not fly. Although RMI expected to be selected for other small thruster programs, this did not happen. Other developers had simpler thrusters with 100% radiation cooling (no throat insert and no regenerative cooling), used film cooling, a better wall material (niobium), and also had good performance.

Other RMI LPREs

RMI also developed a number of other unique experimental LPREs.[2-5] This included a project for helicopters to be powered by a LPRE. Small monopropellant hydrogen-peroxide 50-lb thrust chambers were built into each of the helicopter's blade tips, and the rotation of the blades caused a pumping action. This concept was successfully demonstrated, but it required a relatively large amount of propellant. There was no military interest and no follow-on contract.

RMI also tested experimental engines (5000 and 50,000 lb thrust) with a *topping engine cycle*, where hot gas for driving the turbine was bled off the main combustion chamber and theses gases were diluted with fuel and thus cooled, prior to driving the turbine.[3] This cycle has also been called a *combustion tap-off engine cycle*, and it is shown with other engine cycles in Fig. 4.2-5. There was no gas generator. A small portion of the hot combustion gas from the combustion chamber flowed through multiple holes into a manifold surrounding the injector assembly. Fuel from the cooling jacket was injected into each of the streams of tapped-off gas flowing to the manifold. The fuel mixed with the hot gas, thus cooling the gas. Each individual stream of hot gas and of cooling fuel was metered through an orifice. This was the earliest known LPRE with a topping engine cycle. Initial tests had low engine performance. There apparently was no suitable application. The development of the engine was not completed, and the program was stopped.

RMI also has done some interesting propellant research in their propellant laboratory headed by Paul F. Winternitz.[2,8,*] They investigated the burning of hydrogen peroxide with organic fuels and managed to lower the freezing point of aniline by adding methyl aniline. They tried to lower the vapor pressure of ammonia (used in the X-15 engine) by adding methyl amine and other chemicals, but this effort was not successful. For the Lark program they selected the propellant combination after examining several hypergolic fuels, including turpentine and furfural alcohol, and variations of nitric acid to find a suitable hypergolic propellant combination that was storable for a long time and still gave good ignition and reasonable performance. The first performance analysis of several high-energy propellants, such as with boranes, was performed in this propellant laboratory. A small TC with diborane and hydrazine was also tested. They tested also small experimental TCs with LOX gasoline and added water to reduce the combustion temperature, but it also drastically reduced the performance. They also used LOX with hydrazine hydrate (in lieu of gasoline), but there was no further effort with these propellants.

---

*Personal communications, P. F. Winternitz, Propellant Laboratory Manager, RMI (circa 1970), and F. H. Winter, Space History Division, National Air and Space Museum, Smithsonian Institution, 2002.

Concluding Comments

As the first U.S. LPRE company, RMI pioneered and innovated in this new field. They developed the first U.S. LPRE with four TCs, the first U.S. aircraft rocket engines (with throttling and restart), designed novel tubular TCs, built 50,000 Bullpup missile rocket engines (largest production of any LPRE in the world), the first to develop several engines for storable tactical maneuvering missiles with diaphragm positive expulsion, the first to try out a topping cycle engine, and the first to fly a throttling small thruster. However the company's reputation was tarnished by flight failures in their Viking booster engine, a few Bullpup flight failures, being late in developing the XLR-99 for the X-15 experimental aircraft and having to place restrictions on the operation of this engine in flight. They were late in several deliveries. The president and several key engineers left the company when it was acquired by Thiokol. With this blemished image and their high prices, Reaction Motors was not selected for another major U.S. LPRE program and the organization was disbanded for lack of work and lack of future prospects.

References

[1] Shesta, J., "Reaction Motors, Inc, the first Large Scale American Rocket Company: a Memoir", presented at the 12th History Symposium of the International Academy of Astronautics, Oct. 1978; Chapter 11, *History of Rocketry and Astronautics*, American Astronautical Society History Series, Vol. 10, 1990, pp 137–148

[2] Ordway, III, F. I., and Winter, F. H., "Reaction Motors, Inc.: A Corporate History, 1941–1958, Part I, Institutional Development", and Winter F. H., and Ordway, F. I., "Reaction Motors, Inc.: A Corporate History, 1941–1958, Part II, Research and Development Efforts", two papers presented at 16th History Symposium of the International Academy of Astronautics, Paris, 1982. Chapters 7 and 8, *History of Rockets and Astronautics*, American Astronautical Society History Series Vol.12, 1983, pp 75–100,101–127

[3] Ordway, F. I., "Reaction Motors Division of Thiokol Chemical Corporation, An Operational History, 1958–1972 (Part II)," and Winter, F. H., "Reaction Motors Division of Thiokol Chemical Corporation, A Project History, 1958–1972 (Part III)," both presented at the 17th History Symposium of the International Academy of Astronautics, Budapest, 1983; Chapters 11 and 12, *History of Rocketry and Astronautics*, American Astronautical Society History Series, Vol. 12, 1990, pp 137–165, 175–197.

[4] Winter, F. H., "Black Betsy, The 6000C-4 Rocket Engine,1945–1989 – Part 1, and Part 2", papers at the 24th History Symposium of the International Academy of Astronautics, Dresden, 1990; also Chapter 11, *History of Rocketry and Astronautics*, American Astronautical Society History Series, Vol 19, 1990, pp 237–257.

[5] Herrick, J., Chief Editor and Burgess, E., Associate Editor, *Rocket Encyclopedia, Illustrated*, Aeropublishers, Inc., Los Angeles, CA, 1959.

[6] Wiswell, R., "X-15 Propulsion System", AIAA paper 97-2682, 1987

[7] Rosen, M., *The Viking Story*, Harper & Brothers, New York, 1955.

[8] Clark, J. D., *Ignition*, Rutgers Univ. Press, Rutgers, NJ, 1972

7.3 General Electric Company

This company (abbreviated as GE) established a group to work on missiles and rocket propulsion systems in 1944 before the end of the war. The Army did not have a good missile and propulsion capability at that time, and the Army wanted to work with an experienced engineering organization. GE was an established hardware developer with experience in different products ranging from light bulbs to power systems and electrical machinery. In particular, their engineering and manufacturing background in steam turbines, chemicals, and combustion were considered to be relevant to their expected performance in LPREs. GE agreed to enter into the flight vehicle and LPRE business in early 1944, and actual work started in that year

Project Hermes

A flexible contract arrangement was signed in November 1944, called Project Hermes, whereby GE would work for the Army Ordnance Department on a variety of missile and propulsion jobs.[1] The Army wanted GE to work in several related areas, namely, missiles, the U.S. versions of German missiles, and several types of propulsion (ramjets, solid-propellant rocket motors, hybrid propulsion, and LPREs). The GE group responsible for the propulsion efforts was also called the "Malta Test Operation" after its test station in Malta, New York. As described next, GE became an excellent early developer of different types of LPREs for flight vehicles and an innovator of several engine features.[2] Unfortunately many of their engine projects were canceled (mostly for reasons not related to GE), and only some of their engines were actually flown. GE was active in the LPRE business for about 35 years.

When World War II ended, GE personnel went with the Army to Germany and assisted in collecting technical information and hardware from the German military missile effort.[2] The Army captured and transferred a good number of German V-2 missiles to the United States. Under Project Hermes GE, with the help from the German team of Wernher von Braun, inspected and launched captured V-2 missiles, initially from the Army's White Sands Proving Ground in New Mexico. It also included making modifications to the V-2 missiles, such as different payloads (including a ramjet) or lengthening the propellant tanks. During Project Hermes, more than 200 GE employees were stationed at Fort Bliss in Texas, where Wernher von Braun and his team of Germans were located.

The engineering and testing of the first two-stage missile in the United States, namely, a WAC Corporal missile (from GALCIT, mentioned in Chapter 5.6) on top of a modified V-2, was also largely done by GE people with the input and help of others. Figure 7.3-1 shows four of the vehicles from Project Hermes, and it includes a simple sketch of such a two-stage vehicle, which was called the *Bumper-WAC*. The purpose was to investigate stage separation techniques and upper-stage engine blast effects. Altogether eight bumper WAC vehicles were launched between 1948 and 1950; three of them were fully

328 History of Liquid Propellant Rocket Engines

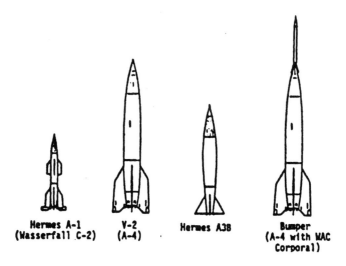

Fig. 7.3-1 Simple sketches of several vehicles developed under Project Hermes. The Bumper-WAC was the first U.S. two-stage flight vehicle with a modified German V-2 missile and a WAC Corporal sounding rocket for the second stage. Adapted with permission from Ref. 1.

successful, two were partially successful, and three were failures. One flight set an altitude record in 1949 for reaching 250 miles.

GE's work on LPREs started under Project Hermes in 1944 before the end of World War II. By August 1945 small experimental thrust chambers had been tested. The LPRE work was focused on LPREs for several Hermes Missiles, as explained next.[1-3]

The first of several experimental missiles designed, built, and tested by GE was the *Hermes A-1*. The vehicle configuration was patterned after the German experimental Wasserfall antiaircraft missile, and it can be seen in Fig. 9-10. The Hermes A-1 vehicle was considered to be a copy of this Wasserfall because it had the identical aerodynamic configuration, the exact same size, and the same basic guidance and control features.[1,2] However GE did not copy the Wasserfall LPRE, which had used storable propellant, had some clever novel features, and is shown schematically in Fig. 9-11. Instead they used a version of the GE-developed LOX/alcohol LPRE for the A-1 flight vehicle. This substitution was satisfactory because the engine was not too critical to the flight tests. The engine flow sheet of the Hermes A-1 vehicle, new at the time, is relatively conventional for today's large gas-pressurized LPRE feed systems and is shown in Fig. 7.3-2. The LPRE thrust was 13,500 lbf, it used LOX and diluted 75% ethyl alcohol as propellants, and had a double-walled cooled TC similar to the one shown in Fig. 4.3-1. The first flight was in May 1950 at the White Sands Proving Ground, and it was followed by more good flights within a year.

Liquid Propellant Rocket Engines in the United States (Summary) 329

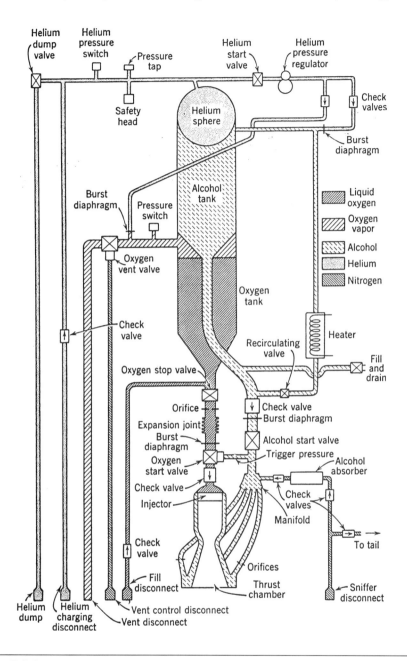

Fig. 7.3-2 This schematic flow diagram of GE's first flying LPRE in the Hermes A-1 test vehicle. It used a gas-pressurized feed system. Courtesy of General Electric Company; copied with permission from second edition of Ref. 4.

The project was canceled because home-grown antiaircraft missiles, such as Nike and Terrier, were then almost operational. The military in the Soviet Union was also very impressed with the Wasserfall missile, and they too built copies of it, tested, and launched them. As in the United States, there was no follow-up program for this copied Wasserfall missile in the Soviet Union.

A historic and *novel type of injector* was developed for the LPRE of the Hermes A-1.[1–4] It is shown in Fig. 7.3-3, and the concept is illustrated in Fig. 4.3-16b. It became the model for other, larger U.S. injector designs at other companies, such as Rocketdyne or Aerojet. The injection holes (with doublet impinging stream injection patters) were drilled in concentric multiple rings on the face of the injector. These rings were brazed into the forged injector body. Behind the rings were circular groove manifolds, one for fuel and the next one for oxidizer. The oxidizer was fed to their respective grooves through holes from a common dome, and the fuel was fed into alternate grooves from the cooling jacket through multiple radial passages. This author visited GE in

Fig. 7.3-3 Injectors for a 3-in. diam TC and the Vanguard TC with predrilled rings of self-impinging doublet-propellant streams. The rings were brazed into a forged injector body. The radial fuel feed holes can be seen in the lower photo. Courtesy of General Electric Company; copied from Ref. 2.

the late 1940s and was responsible for transferring this injector technology to Rocketdyne, where it was modified for injectors with much larger diameters.

Combustion Stability Concerns

In 1946 GE encountered its first instance of *combustion instability* during the early phases of the TC work on Project Hermes. GE used several approaches to understand and cure this combustion problem.[3] They conceived and built a TC with a long slit glass window to allow high-speed photography of the flame phenomena in the combustion chamber. From the data it was possible to determine the velocity of high-intensity combustion layers and to distinguish between stable and unstable burning. This apparatus furnished a means for investigating the combustion stability with different propellants and injectors. GE then systematically tried various chamber geometries and various injection patterns using discrete liquid streams of injected propellants. The like-on-like doublet injector scheme was found to be stable, and it was used on most of GE's injectors thereafter. In this scheme each of the many fuel streams impinges (close to the injector face) with another fuel stream, and pairs of oxidizer streams impinge upon each other, in order to atomize propellant streams into small droplets.

Another novel and clever approach to the combustion instability problems was the use of a flame holder.[3] It was successful in stabilizing the combustion process in several TCs. This technology was obtained from ramjets and jet engines, which at the time were also being developed at GE. Different flame-holder patterns were tested, some that were uncooled and others with regenerative cooling, and all were successful. Figure 7.3-4 shows such flame holders for a small diameter TC. This author could not find any information on why GE did not contine work on this simple promising device for avoiding combustion vibration and the damage it can cause.

Test Facilities

In support of GE, the U.S. Army funded a *static test firing facility* at Malta, near Ballston Spa in New York. It was the first large U.S. test stand built for static tests (firing vertically down and also horizontally) of relatively large LPRE and thrust chambers.[2] It was based on German test stands that had been seen by GE engineers in early 1945. It served as a model for large test facilities later built at Rocketdyne, Aerojet, Huntsville, or Edwards Air Force Base. The first major tests at the GE Malta Test Operation were with a thrust chamber of 16,000 lbf and thereafter with a complete engine about 1948 or 1949. The highest thrust in this facility was provided by a GE experimental TC at 150,000 lbf thrust several years later.

In 1950 von Braun's design team moved from Fort Bliss, El Paso, Texas, to the Army's Redstone Arsenal at Huntsville, Alabama, and a contingent of GE employees, who had worked with the team on Project Hermes in El Paso, also moved to Huntsville. In 1952 the Hermes contract was phased out, partly

Typical coarse screen flameholder

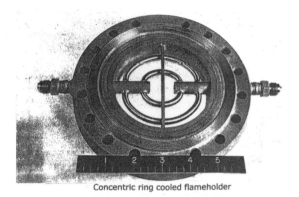
Concentric ring cooled flameholder

Fig. 7.3-4 Uncooled or a cooled flame holder, when placed just below the injector of a small thrust chambers, successfully eliminated combustion instability. Courtesy of General Electric Company; copied from Ref. 2.

because the Army had by this time developed an in-house capability at Redstone Arsenal.

LPREs for Flight Vehicles

The Hermes A-1 has already been mentioned. The Hermes 2 was propelled by a solid-propellant rocket motor and is not concerned with the subject of this book.

The *Hermes A-3 surface-to-surface missile* was planned to be a smaller, improved version of the German V-2, originally intended as a surface-to-surface missile with 150-mile range and a 1000-lb warhead. However, it turned out to be merely a test vehicle. It is also sketched in Fig. 7.3-1. Five of the six A-3 flights went well. It was driven by GE's first pump-fed engine. It originally was rated at 18,000 lbf sea-level thrust, was uprated to about 21,000 lbf thrust at sea level, used LOX/75% alcohol as propellants at a mixture ratio of 1.12 (oxidizer-to-fuel flow ratio), and a specific impulse of 200 s at sea level. This is a more fuel-rich mixture than the German V-2 (1.24), and the specific impulse at sea level is a little less than the V-2 (203 s). The TC had a flat injector and was gimbal mounted and provided pitch and yaw control during engine operation; these were both improvements over the V-2. It included the first geared U.S. turbopump, and the aluminum turbine was driven by decomposition gas of catalyzed hydrogen peroxide supplied from a pressurized hydrogen-peroxide tank. An engine flow diagram is shown in Fig. 7.3-5. The nitrogen pressurizing gas tank is toroid shaped in order to better use available vehicle space. A partial section view of the thrust chamber is in Fig. 4.3-1, and it was of a double-wall design with spiral fuel-cooling passages.

The design details of its novel turbopump with a gear case are quite sophisticated and are shown in Fig. 7.3-6. The gear case allows the turbine to run at a higher, more efficient shaft speed, and the two pumps run at slightly different, but lower speeds.[2–4] The aluminum turbine is at the lower left. Aluminum can be used (as in the V-2) because the gas temperature from the hydrogen-peroxide decomposition products is lower than the melting point of aluminum. The oxygen pump components are from left to right an aluminum pump housing, shrouded aluminum impeller, seal ring assembly, cover plate

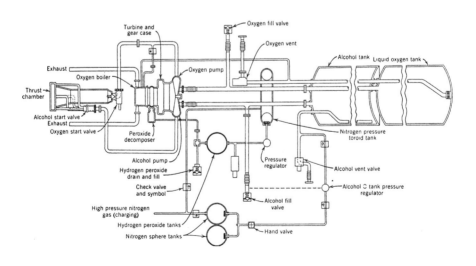

Fig. 7.3-5 Simplified schematic diagram of the LPRE of the Hermes A-3 flight vehicle. The turbine is driven by the decomposition products of hydrogen peroxide. Courtesy of General Electric Company; copied with permission from second edition of Ref. 4.

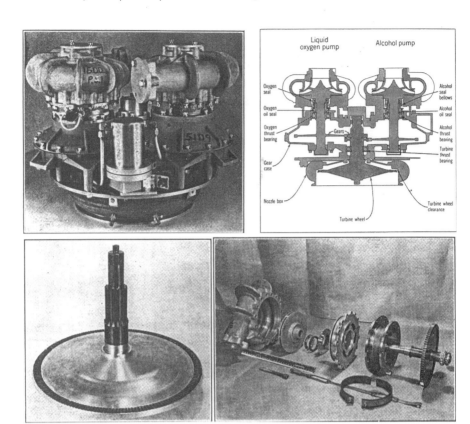

Fig. 7.3-6 Three views and one cross section of the turbopump assembly of the LPRE for the Hermes A-3 missile. Courtesy of General Electric Company; copied with permission from Ref. 4.

assembly, bearing housing, and gear. The shaft and the two-piece neck heater are in the foreground. The heater prevents the oil or grease from becoming stiff or frozen, when cooled by LOX. In improved versions of the TP, the company managed to change from oil lubrication of the gear case and the bearings to a pregreased design, which is a simple subsystem and has no oil reservoir and no oil pump. During the operation, additional grease was squeezed out of a canister and injected into the gear case, and this prevented overheating of the gears, bearings, or grease. This canister can be seen in the top-left view in front of the TP in Fig. 7.3-6. The vehicle and its guidance and control subsystem were also developed by GE. The work on the Hermes A-3 vehicle and its LPRE was stopped because the Army believed that a lower-cost Army defensive missile could be developed. The A-3 program was canceled.

As a follow-on project to the pump-fed LPRE for the Hermes A-3 program, an advanced experimental engine was developed. It had a thrust of 27,000 lbf, a higher chamber pressure, and was identified as the X-400 LPRE. It also used

LOX/RP-1 propellants, and it included features that were later helpful on the Vanguard engine, which is described next.

The *Hermes C1* was a single-stage booster designed for a missile carrying a 1500-lb payload for a distance of 500 miles. Its outline is sketched in Fig. 7.3-1. It was a more advanced missile than the V-2. For this launch vehicle GE designed, built, and component tested an engine with 80,000 lbf thrust.[3] It had a turbopump feed system with a hydrogen-peroxide catalytic gas generator. However this project was stopped and not completed, so that GE could concentrate on the Vanguard engine, which was in trouble at the time and which is described next.

GE developed the booster engine or first-stage engine for the three-stage *Vanguard* space launch vehicle, with the Glen Martin Company as a prime contractor. It first flew in October 1957, and in March 1958 it put the second U.S. satellite into orbit.[2,3,5] Its rocket engine might be the most successful GE LPRE because it was deployed in a space launch vehicle and launched 12 times. This 27,800-lb thrust LPRE had a turbopump feed system, a design specific impulse of 254 s at sea level, used liquid oxygen and kerosene propellants, and hydrogen peroxide for a catalyst gas generator. Ignition was accomplished by injecting an initial slug of triethyl aluminum, a highly reactive hypergolic start fuel. The TC was gimbal supported and could be rotated through +/–5 deg. The engine is shown in Fig. 7.3-7 and its simplified flow sheet in Fig. 7.3-8. It had some clever features new to U.S. LPRE designs at the time. The liquid-oxygen feed valve was located in the high-pressure portion of the feed line downstream of the oxygen pump; while filling the oxygen tank on the launch stand, the oxygen piping and the pump were automatically cooled to cryogenic temperature. This reduced the time delay for chill-down prior to launch and reduced the amount of liquid oxygen normally sent through the oxygen feed system. There were dual-turbine exhaust ducts with hinged nozzles at their lower ends for roll control during powered flight. The mechanism for this roll control is not shown in Fig. 7.3-8. This feature was first seen in the United States on this engine and has since been used on other engines, such as the engine for the Jupiter missile. To achieve adequate roll control during the critical period of stage separation, when there was little turbine exhaust gas produced or available for roll control, the remaining helium from the tank pressurization system was dumped into the two swivelled turbine exhaust ducts during and immediately after the shutdown. It was also one of the early U.S. large TP-fed engine with a gimbal suspension of the thrust chamber, allowing pitch and yaw control. The Vanguard TC used double-wall construction, and the fuel coolant enters through a manifold at the nozzle end, flows through the cooling jacket, and enters directly into the injector. In the 12 Vanguard flights, this engine failed once; it was during the second flight. The Vanguard program was not continued because it was relatively expensive for the small orbital payloads and because larger U.S. space launch vehicles with much more payload became available.

GE also developed an experimental tubular TC for this Vanguard engine application with a pass and a half-cooling path.[3] The inlet manifold is on the

336 History of Liquid Propellant Rocket Engines

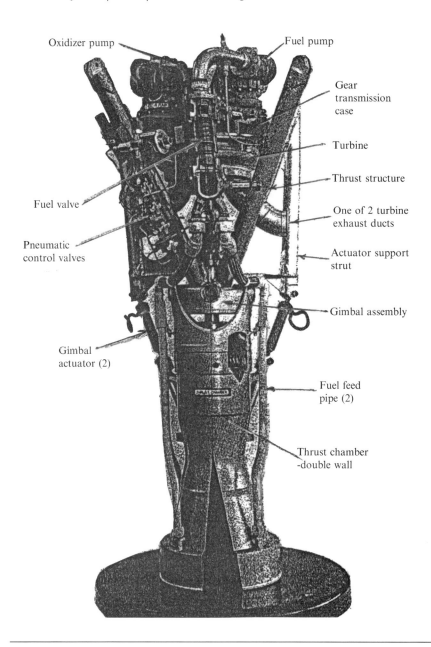

Fig. 7.3-7 View of GE Vanguard booster engine of 27,800 lbf thrust with gimbaled TC. Courtesy General Electric Company; copied with permission from Ref. 2.

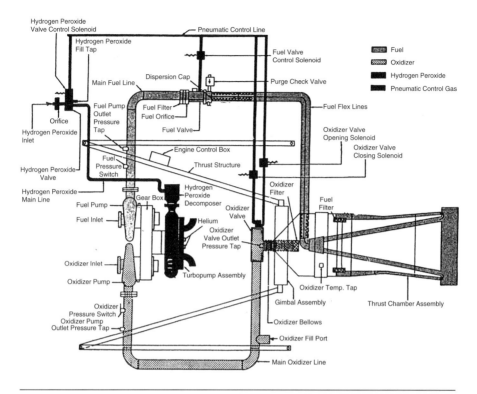

Fig. 7.3-8 Simplified schematic diagram of GE's Vanguard booster LPRE. Courtesy of General Electric Company; copied with permission from Ref. 2.

diverging section of the nozzle, the coolant flows down toward the nozzle exit in every second tube and returns in alternate tubes; the flow through the tubes at the nozzle throat region and the chamber region is unidirectional. This technology of using tubes in the cooling jacket was tested earlier at GE in TCs using aluminum and stainless-steel tubes. Furthermore GE had the advantage to learn about tubular TCs developed earlier by Rocketdyne, Aerojet, and RMI. This GE TC with multiple tubes in the cooling jacket was ground tested, but has not been flown.

The engine for the *Vega SLV's* upper stage was proposed by NASA in the late 1950s. GE got this job on the basis of a further uprating and improvement of the basic Vanguard engine.[3] This Vega stage was to be carried on top of an Atlas booster stage, and it was intended as an interim solution prior to the deployment of the Centaur LOX/LH$_2$ upper-stage engine, which then was being developed at Pratt & Whitney (discussed in Chapter 7.10). GE received a contract in 1958 for the 405H *Vega upper-stage engine* using LOX/RP-1 propellants.[3] This engine is shown in Fig. 7.3-9. It had a higher thrust than the Vanguard booster engine (33,840 lbf), a larger nozzle area ratio (25:1), an altitude specific impulse of 300 s, and a gimbaled TC. It allowed the fuel flow to be controlled by +/− 8% in order to fully utilize all of the propellant. The supply subsystem for hypergolic ignition fluid was altered to allow three altitude/vacuum starts.

This Vega engine had a large *TC with tubular construction*, and it was developed by GE. The Vega TC used stainless tubing with the inlet manifold and the outlet manifold of the cooling jacket being near the injector. This means that the fuel flow was down to the nozzle exit through every second tube and back through the alternate tubes. A cross section showing the double manifolds, the upper part of the tubes, and the injector is shown in Fig. 7.3-10. This figure also shows the injector's inlet holes to the annular grooves (behind the injector rings), the oxygen dome, and the fuel supply holes. GE's upper-stage Vega engine successfully passed preliminary flight rating tests. The first flight was scheduled for 1960, but in December 1959 NASA decided to drop the Atlas/Vega SLV in favaor of an Atlas/Agena SLV, which had been supported by the Department of Defense. So further work on the Vega stage and engine was stopped.

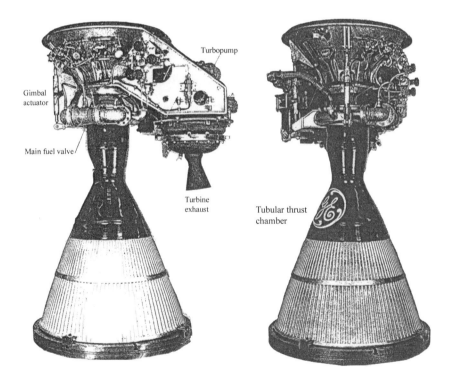

Fig. 7.3-9 Two views of the Vega upper-stage engine with turbopump mounted on the side of a tubular thrust chamber. Courtesy of General Electric Company; copied with permission from Ref. 2.

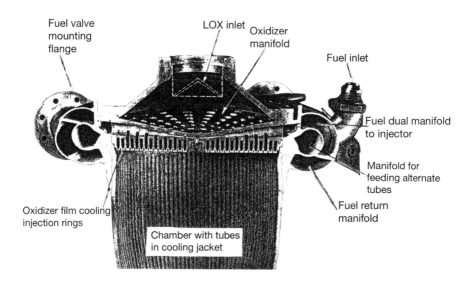

Fig. 7.3-10 Two views of the Vega engine injector with triple fuel manifolds. Courtesy of General Electric Company; copied with permission from Ref. 2.

GE also developed a *twin-engine system* identified as the 412A LPRE of 26,000 lb thrust each (52,000 lb total) intended for the X-15 manned research aircraft.[3] It was a backup engine to the XLR-99 being developed at RMI in the late 1950s; RMI's development was delayed by technical problems at the time. The 412A was a slightly modified version of the Vanguard engine. It had an engine-out capability because flights could be completed with only one of the two engines. Only one engine was built and tested. It demonstrated a 10 to 1.0 throttling capability, which was a remarkable achievement at that time for an engine with a turbopump feed system. When RMI's engine finally got over its troubles in 1959 and flew for the first time in 1961, the GE 412A engine program was terminated.

Plug Nozzles

GE was the first to design, analyze, and test an experimental *engine with a plug-nozzle configuration* in about 1963.[3] It was later also called aerospike nozzle, a name coined by Rocketdyne, and its basic concept was shown in Fig. 4.3-9 and discussed in Chapter 4.3. This was a radical departure from regular nozzles and chambers. Early tests were on a scale of 16,000 lbf thrust; it had an uncooled central plug or spike (with a curved contour) and six simulated small uncooled combustion chambers. It is shown in Fig. 7.3-12. Each simulated chamber provided warm decomposition gas from one of six hydrogen-peroxide catalyst packages as a substitute for a combustion chamber with a bipropellant injector. Operation of this test unit showed good performance at off-design pressure ratios and demonstrated the feasibility of thrust vector control by throttling of the propellant flow through selected decomposition beds.

The next version of the plug nozzle at 50,000 lb thrust used eight segments of tubular cooling jackets fastened together. Each tubular cooling jacket segment covered only a portion of the contoured plug and the combustion chamber walls. Each segment had its own bipropellant flat-face injector in the form of an arc with curved grooves and injection elements. The engine was tested by GE in the mid-1960s, and it worked well in the first test.[2],* This test firing can be seen in Fig. 7.3-12. They demonstrated that a segment type of engine with a plug nozzle could be developed, built, and assembled relatively inexpensively, that it gave good performance, and that it could be cooled. They also demonstrated that thrust vector control could be obtained by throttling selective segments. Thereafter the company designed larger thrust LPREs with a plug nozzle for a 1,000,000-lb TC. GE also built one of the injector sectors with a piece of a tubular chamber and nozzle and partially tested it. GE went out of the LPRE business before they could continue this plug-nozzle work. This nozzle concept was actively pursued and improved by Rocketdyne, and this effort is described in Chapter 7.8.

Other Related Efforts at General Electric

GE was active in research, analysis, and technical investigations related to LPREs. There were GE publications on two-phase flow and heat transfer in cryogenic liquids, design studies of dynamic and static seals for LPREs, cavitation noise analysis, a leakage testing handbook, stability investigations of thermally induced flow oscillations in cryogenic heat exchangers, zero-leakage connectors for launch vehicles, investigations of jet vane materials, or alternate propellants for the hypergolic ignition of the Wasserfall.*

GE was the first to develop a *self-renewing thermal insulation coating* for the inner wall of TCs. This was achieved by adding a small amount (about 1%) of a soluble silicon compound to the kerosene or alcohol fuel.* In the hot combustion with LOX, the silicon oil additive decomposed, and the silicon is oxidized into fine small solid particles of silicon dioxide, which deposited themselves in a fluffy insulation layer on the inner wall of the combustion chamber and the nozzle. This fluffy coating was self-regenerating; if some part or small piece of the insulation layer flaked off and/or was blown away, another new layer of silicon dioxide would form immediately. The coating reduced the heat transfer by about 30%, and a corresponding reduction in wall temperature was also obtained. Silicon oil was also used successfuly as an additive to UDMH by ARC in their Agena engine. Rocketdyne and Aerojet also

---

*Personal communications, A. Kubica, 2003, and R. Porter, 1950s, both formerly with GE Malta Test Operation, New York, based in part on summaries of several company research reports.

Liquid Propellant Rocket Engines in the United States (Summary)

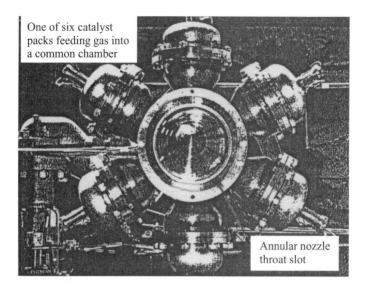

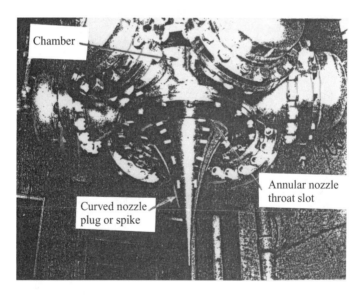

Fig. 7.3-11 Two views of the first plug nozzle with a curved full-length spike rated at 16,000 lbf thrust using monopropellant hydrogen peroxide decomposition product gases from six catalyst packs (1963). Courtesy of General Electric Company; copied from Ref. 2.

added about 1% silicon oil in experimental TCs and did obtain a significant reduction in heat transfer. However these two companies also had some unfavorable experiences with this additive and abandoned further work, as explained in Chapters 7.8 and 7.11. This method of forming a renewable insulating coating on the inner wall of TCs has not been used by GE in any flight of a LPRE. It was flown by ARC, but with a different propellant combination.

GE experimented with a novel tank *pressurization scheme for ammonia tanks* by direct injection of either nitric acid or liquid NTO (nitrogen tetroxide) into the liquid ammonia in the propellant tank.[2,3] Much of the work was at 1500-psig tank pressure and a 10% initial ullage volume, which means the tank was 90% filled with liquid propellant. Results showed smooth stable operation with a rapid initial pressure rise over a wide range of tank pressures,

Fig. 7.3-12 Enlarged film frame of the test firing of a plug nozzle TC at 50,000 lbf thrust with fuel-cooled chamber segments and a fuel-cooled truncated (cutoff) nozzle spike. Courtesy of General Electric Company from Ref. 2.

injection designs, and injection locations. The gas above the level of the propellant was between 500 and 900°F, but could easily become higher, if the NTO injection would not be properly controlled. The upper layers of the liquid would be heated, and this changed their physical properties, such as density or vapor pressure, which can cause changes in the mixture ratio or pump cavitation. This ammonia pressurization concept was not applied to an engine.

In 1948 with support from the U.S. Navy the *Operation Pushover* was conducted. In these tests two fully tanked and ready-to-launch V-2 missiles were pushed off their launch stands and toppled over onto a mock-up of a ship's deck to simulate a sea launch failure condition. Damage from the spilled propellants, explosions, and fires caused by this simulated accident was staggering and more severe than expected. The results of these tests were an important reason for the U.S. Navy's preference for solid-propellant motors rather than LPRE for propelling ship-launched missiles.

GE was the first LPRE organization to start work in 1945 on *boron-hydride fuels* under Project Hermes.[2,3,*] This was a part of the high-energy effort that was being pursued by the United States between 1945 and about 1960. These boron hydrides had been identified as a promising high-energy rocket propellants, rich in hydrogen content, with a specific impulse close to the theoretical value of over 350 s. The objectives were to develop processes and a pilot plant for manufacturing boron hydrides (with enough information to allow later scaling to a production facility) and to evaluate these fuels in rocket TCs. By 1947 GE had prepared small quantities of diborane in what was the first facility for making this ingredient in pure form. Combustion tests of gaseous diborane and gaseous oxygen started in 1948. In 1952 they were able to produce small quantities of both diborane (B_2H_6) and pentaborane (B_5H_9) in a new pilot plant. GE characterized these fuels, their physical properties, handling, storage, ignition, and combustion qualities. A number of TC firing tests of pentaborane with hydrogen peroxide followed. Firing tests with pentaborane and hydrazine were made because this propellant combination gives a high theoretical performance. These investigations proved that boron hydrides were manageable and could be stored for reasonably long periods. However the combustion efficiencies were very low, well below the theoretical values. This was because of the difficulties of burning boron compounds quickly and completely within the combustion chamber. These unfavorable results and the high toxicity of these boron fuels were the reasons for them not being pursued further. The Soviet Union built and tested a complete LPRE using pentaborane and hydrogen peroxide between 1960 and 1966, or about a decade later. This effort is described in Chapter 8.4. The Soviet engine development with these propellants was not continued.

GE also was active in *small thrusters* or *small thrust chambers*. They developed in the late 1940s at least three sizes of hydrogen-peroxide monopropellant

---

*Personal communications, A. Kubica, 2003, and R. Porter, 1950s, both formerly with GE Malta Test Operation, New York, included summaries of several company research reports.

thruster (5, 50, and 500 lbf thrust) using 90% hydrogen peroxide.[2,3] They were intended for use in early attitude control systems and for minor maneuvers of upper stages or satellites, as well as for gas generators, which GE used on several of their own engines. Figure 7.3-13 shows a sketch of a catalyst cartridge with multiple screens of silver wires coated with samarium nitrate. These screens were stacked up behind a perforated flow distribution plate. A spring-loaded support plate is situated below or downstream of the screens and holds them in place. GE claimed to have obtained excellent decomposition efficiency of over 95% and good performance of the flow through the catalyst bed, even when the flow was reduced to 15%. Some of these GE small monopropellant thrusters were supposedly installed in upper stages and/or satellites, but the author did not find any information to verify these flights or the performance of these thrusters. They used 90% hydrogen peroxide with catalyst beds in several gas generators, which have flown. Hydrogen-peroxide decomposition was also used in experimental plug nozzles and for bipropellant TCs, where the hot decomposition gases burned with a fuel giving a higher performance than the monopropellant by itself. Experiments were also conducted with higher concentrations of hydrogen peroxide (98.5%), but were not continued.

GE decided to voluntarily go out of the rocket propulsion business in approximately 1966. Several reasons contributed to this decision. The com-

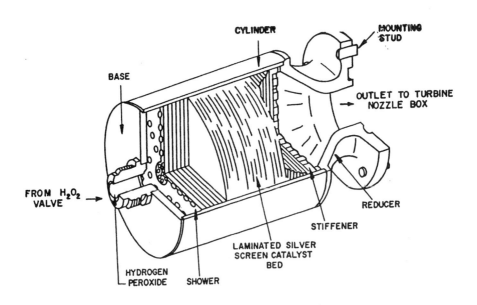

Fig. 7.3-13 Simplified partly sectioned view of the catalyst bed for a monopropellant hydrogen-peroxide gas generator with a perforated shower plate and perforated exit stiffener plate. Courtesy of General Electric Company; copied from Ref. 2.

pany, according to one former employee, was disenchanted by the rocket propulsion business, its frequent changes in government direction, untimely and frequent program cancellations, and the dim prospects for substantial future work. Another source hints that GE management did not allocate sufficient company funds to support in-house efforts leading to potential new LPRE projects. Supposedly the company funds were largely allocated to the GE jet engine operation, which was much larger in sales volume, had good prospects for major production contracts, and had more profit potential.* The extensive know-how, which GE had acquired through many years of LPRE experience, was apparently not transferred or sold to another LPRE organization.

References

[1]Braun, J. H., "The Legacy of Hermes," *History of Rocketry and Astronautics*, American Astronautical Society History Series, Vol. 10, Univelt, San Diego, CA, 1990, Chap. 5, pp. 135–142.

[2]*Liquid Bi-Propellant Rocket Propulsion Experience and Capabilities*, internal undated company document, Malta Test Operation, General Electric Company, circa 1968.

[3]Murphy, J. M., "Liquid Rocket Engine System Advances," American Astronautical Society History Series, Vol. 19, Univelt, San Diego, CA, 1990, Chap. 16, pp. 191–203.

[4]Sutton, G. P., and Biblarz, O., *Rocket Propulsion Elements*, 2nd and 7th eds. Wiley, Hoboken, NJ, 1956, 2001.

[5]Stehling, K., *Project Vanguard*, Doubleday and Co., Garden City, NY, 1961.

---

*Personal communications, A. Kubica, 2003, and C. Berman, both formerly with GE Malta Test Operation, New York.

7.4 Curtiss–Wright Corporation

This company was well known in the 1920s to 1940s for its aircraft piston engines and also its propellers. They foresaw the decline of the piston engine business and the propeller-driven airplane, and therefore the company decided to enter into three new businesses, namely, turbojets, solid-propellant rocket motors, and LPREs.* Because Curtiss–Wright (C-W) did not have any direct experience with LPREs, they approached Robert H. Goddard for help. At that time he was the only American with years of hands-on experience. He agreed to let C-W use his patents and to periodically consult for C-W starting in 1942. The LPRE activity started at C-W around 1942, and a new rocket engine department was set up under Charles Chillson, a capable engineer. This author and Chillson became acquainted and met on several occasions. Test facilities were built at the C-W propeller plant in Caldwell, New Jersey. Goddard also agreed to come to work full time for C-W in 1945, but he became ill and passed away before he could start this job. Goddard's technicians and assistants actually joined the company in 1945.

Initially C-W worked on small TCs and some JATO LPREs. They developed a JATO for the North American Aviation XB-45 bomber aircraft, which was flight tested in a modified B-45 aircraft. Originally it had three propellants: LOX, gasoline, and water, which was used for nozzle and film cooling. Because water would freeze in a cold climate, C-W switched to a LOX/alcohol propellant combination, where the alcohol fuel was diluted with water and it could meet the low temperature environment requirement of the Air Force. This engine had a gas-pressurized feed system. They also built and tested an experimental throttlable LPRE.[1]

In 1948 C-W obtained a contract for the XLR-25 rocket engine for the swept-wing research airplane X-2, which was intended to fly through the sound barrier and exceed Mach 1.0. The X-2 was carried to altitude underneath the wing of a mother airplane, and after it was released the rocket engine started to fire. This historic LPRE featured several innovations and helped to advance the state of the art of the technology.[1,2] This LPRE, shown in Fig. 7.4-1, used diluted (75%) ethyl alcohol and liquid oxygen as propellants had a maximum thrust of 15,000 lbf and could be throttled continuously to as low as 2500 lbf. Prior engines could not throttle continuously; for example, the RMI-6000 LPRE with four thrust chambers at 1500 lbf each could reduce the thrust by turning off individual TCs in steps of 25% of rated total thrust. The German Hellmuth Walter 109–509 LPRE had three steps of thrust change by turning off banks of injector elements in the same injector. This C-W engine had a smooth transition from minimum to maximum thrust. It had two thrust chambers capable of throttling, one at 10,000 lbf and one at 5000 lbf maximum thrust, and a turbopump feed system. The flow and the

---

*Personal communications, C. Chillson, Curtiss–Wright (circa 1950), and C. Ehresman, Purdue University, Indiana, 2001–2004.

Fig. 7.4-1 This XLR-25 LPRE was developed by Curtiss-Wright and flown on the NASA X-2 research aircraft. It had two TCS at 5000 and 10,000 lbf thrust and the engine could be throttled. Courtesy of Curtiss-Wright Corporation; copied with permission from Ref. 2.

chamber pressure could be reduced separately on either or both of the thrust chambers. It was a landmark aircraft engine for its day, and at that time it had the highest thrust for a LPRE as the sole powerplant for a manned aircraft. It was then not known that the Soviet Union had an earlier aircraft rocket engine with continuous throttling.

The thrust chambers had regenerative fuel cooling in the nozzle throat region only, and a single-wall uncooled chamber with extensive fuel film cooling.[2] The cooling scheme was patterned after Goddard's TC designs, and the combustion chamber had a relatively large volume (L^* = 90 in.). The first injector was a Goddard design, but it was soon replaced by a complicated C-W injector design with an internal fuel valve. The XLR-25 injector was unique as seen in Fig. 7.4-2. The oxygen was injected through relatively large holes in a direction parallel to the TC axis. The fuel was injected through three rows of

---

*Personal communications, C. Chillson, Curtiss–Wright (circa 1950), and C. Ehresman, Purdue University, Indiana, 2001–2004.

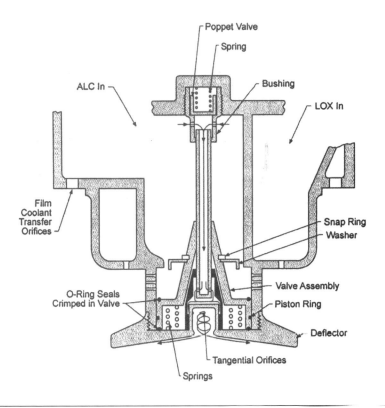

Fig. 7.4-2 Simplified section through the injector of the XLR-25 LPRE with built-in valves. The chamber wall is not shown. Courtesy of Curtiss-Wright Corporation; copied with permission from Ref. 2.

holes. About 60% of the fuel was used for film cooling at right angles to the oxidizer streams, radially outward, with some impingement. A deflector plate, which was fuel film cooled from underneath (through a swirling flow from a fuel cavity), helped to mix and divert the gas flow. At engine cutoff a dual valve, which was built into the injector, moved up by spring force and shut off the fuel flow to the three rows of radial fuel injection holes and also to the central whirling film-cooling fuel flow cavity. This feature prevented backflow of oxygen into the fuel manifold. For this same reason all of the 20 fuel film-cooling holes had small individual check valves, which are not shown in Fig. 7.4-2. It is the only TC known to the author with some valve on every fuel injection hole, including the film-cooling injection holes. For some of the flights, a nozzle extension with a higher nozzle area ratio, made of molybdenum, was attached to the TC in order to achieve a higher specific impulse. The engine performance was low, estimated at around 90% of theoretical specific impulse, because of the gas-generator cycle (2 to 3% loss), and the extensive film cooling (about 5 to 8% loss).[2] However, the combustion was stable over the throttling range.

The TP was advanced and unique because it had two stages of radial in-flow turbine rotors. It had a single shaft and the turbine was located in between the two pumps. The pump inlets were unobstructed and relatively large, which reduced the pump inlet velocities and avoided cavitation problems. As far as it is known to the author, this is the only radial inflow turbine that has flown in a TP of a LPRE. There were some others developed by other LPRE organizations, but they did not fly.

The engine controls and the piping became quite complex because of the requirements of throttling, altitude start, restart, ensured safety, and redundant ignition systems.* The TP was controlled in speed by bypassing some of the high-pressure fuel through a bypass control valve back to the pump inlet. The gas generator had two sets of propellant valves for high and low flow. The oxygen tank was pressurized by gasified oxygen heated by turbine exhaust gas in a heat exchanger. This method of pressurization had been done before in the V-2 LPRE. The fuel tank was pressurized by fuel-rich gas taken from the gas-generator (GG) discharge and reduced in pressure, a truly novel scheme at that time. Later other companies and several Soviet LPREs used this concept. The actuation of the principal valves was accomplished with high-pressure oxygen gas, which also was unusual and novel at this time. This engine therefore did not contain a heavy high-pressure inert gas tank, which was needed by most of the other contemporary engines.

The development, qualification, and engine demonstration efforts exceeded the anticipated costs and delivery schedule. This delayed the flight of the X-2 aircraft. After some flight-test problems unrelated to the engine (the first aircraft was lost), the first flight with the new XLR-25C engine occurred in 1955. It did not allow operation over the full intended throttling range. The technical problems, the design complexity, and the delay of the delivery of this engine tarnished the reputation of C-W, and thereafter C-W was not able to win a competition for a major new LPRE project.

The rocket engine group of C-W at Caldwell, New Jersey, performed several interesting R&D investigations and studies of advanced engines. Between 1962 and 1965 they experimentally investigated pyrolytic graphite stacked plates for a heat-sink nozzle design that would allow high heat-transfer rates in the throat region. This concept has subsequently been adopted for a good number of solid-propellant rocket motor nozzle throats, and some are still in production today. C-W used this pyrolytic wafer nozzle concept with experimental thrust chambers using liquid fluorine oxidizer. Because there is no free oxygen in the exhaust gas, the pyrolytic carbon throats did not experience oxidation and very little erosion. They successfully tested it in a liquid-fluorine/hydrazine rocket thrust chamber of 3750-lb thrust level. Test firings with NTO and amino-type fuels showed excessive erosion of the pyrolytic graphite, believed to have been caused by the small amount of free oxygen

---

*Personal communications, C. Chillson, Curtiss–Wright (circa 1950), and C. Ehresman, Purdue University, Indiana, 2001–2004.

present in the hot combustion gas products. In 1961 and 1962 they tested a small lithium-cooled rocket nozzle, but this idea was not used in a rocket engine. C-W went out of the Liquid Propellant rocket engine field in the late 1960s. Probably they were losing money and were not successful in getting major new LPRE jobs.

References

[1]Meyer, M., "Throttling Thrust Chamber Control," *Journal of the American Rocket Society*, June 1951.

[2]Ehresman, C. M., "The Goddard Connection in the Development of the Curtiss-Wright XLR25 on the Bell X-2 Research Aircraft," *AIAA Paper 2000–0803*, Jan. 2000.

7.5 M. W. Kellogg Company

This company was also an early entrant into this field. The main business of this company has been and still is the engineering and construction of complex chemical and petrochemical plants. The decision to compete in LPREs was made in about 1945. They assembled a competent group of personnel and started work in the LPRE field in 1946 at Jersey City, New Jersey. There they built a sophisticated LPRE test facility (with temperature conditioning of the engine and the propellants). Like several other U.S. LPRE companies new to the field, they did some small-scale initial testing and then developed several JATO LPREs.[1] One of these JATO units had a resemblance to the first Aerojet JATO (shown in Fig. 7.7-2) with three spherical tanks and an uncooled TC. One of their JATO TCs (1947) had an unusual design of the cooling jacket, and it is shown in Fig. 4.3-6. The outer wall was formed into spiral-shaped convolutions, which were then pressure brazed to the inner TC wall. This concept was a good idea for those regions of the cooling jacket, where the heat transfer was relatively low, such as in the nozzle-exit region. It was not a good idea for the nozzle-throat region, where the heat transfer and wall temperature were high, diminishing the material strength of the inner wall, which had to carry the imposed stresses and loads. As far as the author has been able to determine, none of the Kellogg JATO units were flight tested.

One of their early projects in 1946 was an important experimental evaluation of hydrazine, which was a new promising fuel at that time. They tested hydrazine (in the laboratory and in small TCs) with different oxidizers, which were liquid oxygen, two kinds of nitric acid, chlorine trifluoride, and hydrogen peroxide. At this time there was no catalyst for the self-decomposition of hydrazine, and so Kellogg used heat or thermal decomposition of hydrazine for generating gas in a gas generator. The test results were compared with theoretical performance values. The results were that hydrazine was a useful propellant with a high combustion efficiency, delivered a higher performance than any other storable fuel, and gave excellent hypergolic ignition with the storable oxidizers nitric acid, chlorine trifluoride, or hydrogen peroxide. It was a convenient GG monopropellant, and under certain conditions and with certain limitations it was also a good coolant.

The sophisticated JATO for the B-47 bomber with a turbopump feed system might have been Kellogg's most significant contribution to rocket history.[2] It is shown in Fig. 7.5-1. The engine used white fuming nitric acid (98%) and jet fuel, and this propellant combination requires an igniter. The TP was driven by bleeding compressed air out of the turbojet engines of the aircraft. There was to be one permanently installed JATO unit on each side of the B-47 bomber aircraft. The maximum thrust was 10,000 lbf for each of the two JATO units or a total takeoff assisted thrust of 20,000 lbf. Kellogg selected two TC at 5000 lbf each for each of the two JATO engines.

There were three unusual features.[2] The injector seen in Fig. 7.5-2 used a conical splash-plate design, roughly patterned after the German Enzian missile. This favored large injection holes, good mixing, and a good atomization

Fig. 7.5-1 One of two assisted takeoff LPREs of the M. W. Kellogg Company developed for the B-47 bomber aircraft. Each had two TCs at 5000 lbf thrust. Photo courtesy of Charles Ehresman; copied with AIAA permission from Ref. 2.

into small droplets. About half of the unlike doublet-propellant streams impacted on a central cone and the other half on an outer sleeve, which had holes for oxidizer film cooling. Ring valves in the oxidizer and the fuel manifolds of the injector, the second novelty, allowed face shutoff of propellants with very little dribble volume and also a gradual buildup of the flow. There were separate movable rings of different diameters for activating the oxidizer valves and the fuel valves. Each ring moves up, when opened by propellant pressure, or down when closed by two pneumatic piston actuators. Each ring is connected to a series of 32 small valve stems, which open or close a set of small valves with Teflon® seats directly at the inlet to each injection hole (see Fig. 7.5-3). There were problems with excessive flow of propellants in the chamber during the start, thus causing "hard starts." The fix was for one set of unlike doublets holes to receive propellant a short time before the others (they had different small valves), and this allowed a low initial flow and allowed a smooth start. The precision valves and rings make the injector assembly relatively complex, and there could be problems with the thermal deformation of the injector housing. No other injector with ring actuators and individual valves for each fuel and oxidizer injection hole is known to the author.

Liquid Propellant Rocket Engines in the United States (Summary)   353

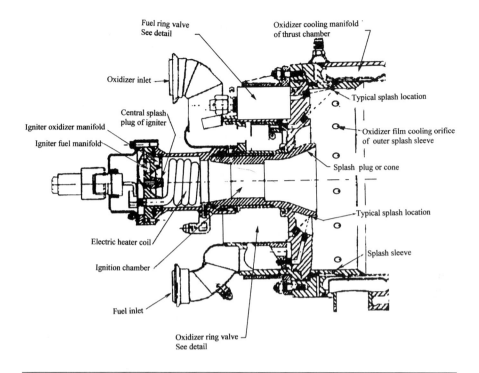

Fig. 7.5-2 Cross section of the splash-type injector with its ignition chamber. A detail of a ring-actuated valve is in the next figure. Modified from copies of figures from Ref. 2 with AIAA permission.

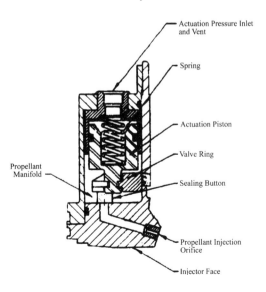

Fig. 7.5-3 One of several individual oxidizer injector valves, which are opened and closed by the movement of the actuating ring. Copied with permission from Ref. 2.

The injector also incorporated a central ignition cavity with three sets of unlike doublets impinging on a small central splash cone. An electric heater coil was used as the energy source for ignition. They can be seen in Fig. 7.5-2. An ignition sensor, namely, a pressure sensor in the igniter chamber, and a timer then gave a signal of satisfactory ignition of the initial igniter propellant flow. The thrust-chamber cooling jacket was of novel construction. The thin inner wall and the dimpled outer wall were resistance spot welded at the dimples, and this held the two walls together. The welded dimples can be seen in Fig. 7.5-1. The problems in perfecting this cooling jacket construction and allowing some thermal expansion of the inner wall were solved. The oxidizer was the coolant.

They also experienced combustion instability, which would quickly burn a hole in the TC. The problem was solved after trying out several changes in the injector design and much testing. When the Air Force customer changed the jet fuel from JP-3 to JP-4, the instability returned. A change in the wall contour and the removal of a shoulder on the outer splash ring seemed to remedy the instability.

The engine was to be started with the gas generator initially supplying warm gas to the turbine. There were considerable difficulties in obtaining a uniform GG gas mixture at 1250°F, particularly in the first few seconds. The requirement was changed, and the turbine rotation was started and initially driven by compressed air from the aircraft's de-icer system. A key lesson from prior experience was to prevent accumulation of unreacted mixed propellant in the chamber at shutdown because this propellant charge would explode at the restart of the LPRE. The ring valves were effective in preventing after-dribble or much flow after the valves were closed, and therefore the Kellogg ring valve design did not have explosions during shutdown or restart.

This Kellogg JATO LPRE was in competition with Aerojet, where another JATO LPRE for the B-47 was being developed for the same application and to the same requirements. Both companies developed satisfactory engines. The Kellogg injector design allowed excellent combustion efficiency and required a relatively very small chamber volume for essentially complete combustion. The Aerojet TC had a much larger combustion volume (by a factor of 2.7), and it too had hard-start problems, which Aerojet eventually solved. However Kellogg had more development problems and a major explosion of their engine during qualification tests in their new test facility in early 1951. Qualification tests were resumed after facility repair with the second engine. Shortly thereafter the Kellogg JATO contract was terminated, but Aerojet was allowed to proceed through flight testing with their version of the B-47 JATO. A description of Aerojet's twin LPRE is given in Chapter 7.7. Subsequently the requirement for this JATO unit was canceled by the Air Force, and neither of these two LPREs went into production.

Kellogg pioneered in the analysis of LPREs and their flow balance and pressure balances. They recognized early that the geometry of the injector and its pattern of holes had a dominant role in the engine performance and combustion behavior. They also determined that a smooth entry of propellant flow to an individual injection hole (by chamfering or a small inlet radius) could reduce the injector pressure drop by a factor of two or three.

The explosion in their New Jersey test facility and their failure to win a major competitive job contributed to their decision to abandon this field and close down their operations. Work on LPREs was stopped around 1954. Therefore Kellogg was in this business for only approximately nine years.

References

[1] Condello, V., and Kalitinsky, A., "Development of ATO Rocket," M. W. Kellogg Co., Special Projects Dep., Jersey City, NJ, May 1953.

[2] Ehresman, C. M., "The M. W. Kellogg Company's Liquid Propellant Rocket Venture," AIAA Paper 2000-3584, July 2002.

7.6 Walter Kidde & Company

Walter Kidde & Company was a multiproduct conglomerate organization and well known for its fire extinguishers. This company decided to enter into the small thruster field of the LPRE business. Kidde established a LPRE capability at their plant in Belleview, New Jersey, and was successful in developing a series of individual thrust chambers and also some complete engines using 90% hydrogen peroxide as a monopropellant. There were development problems in finding the right design for the catalyst package, but they were solved by Kidde's engineers. They used a stack of tightly fitted silver screens or silver-plated screens as catalysts for the oxidizer decomposition. This work was based in part on German technology. The specific impulse values for this monopropellant are inherently low, typically between 148 to 158 s at steady-state firing and as low as 90 s with very short pulses. However this performance was much better, and the engine and tanks were much lighter than an equivalent engine and tanks using stored pressurized inert gas. These monopropellant TCs were the key components of the reaction control systems during the 1950s and early 1960s. The exhaust products were environmentally benign and contained no liquid or solid species, which would deposit on sensitive vehicle surfaces.

In their 1962 catalog they offered a thruster with an integral valve in the following thrust sizes: 2 lbf (weight: 0.58 lbm), 5 lbf, 15 lb (weight: 1.75 lbm), 35 lbf, 300 lbf, and 500 lbf (Ref. 1). By reducing the chamber pressure, the thrusters could operate at lower thrust.

The 2-lb thruster, shown in Fig. 4.6-6, could be run at a 1.1-lbf thrust level. An example is the Kidde thruster shown in Fig. 7.6-1 developed around 1958. It is rated at 14 lbf thrust at an altitude of 200,000 ft, with 150 s of specific impulse and a chamber pressure of 277 psia. At sea level the thrust is only about 9 lbf. Although the rated duration was 10 min, the thruster could run as long as the propellant supply would last. The start time interval to reach 90% thrust was then about 30 to 45 ms with valves that were available 50 years ago. The first flight was in July 1960 as part of the attitude control system of the third stage of the Scout SLV. The nozzle is at right angle to the axis of the catalyst bed, which makes it easier to install in some vehicles. It has a conical nozzle exit with a relatively low half-angle of 7.8 deg because it was designed before the nozzle-exit angle was optimized and well before bell-shaped nozzles were known.

Complete LPREs using hydrogen peroxide as a monopropellant with gas-pressurized feed systems were also developed. For example 40 ship sets of a RCS for the Scout second stage were delivered.[1],[*] Each set had four propellant tanks (with flexible silicone bladders separating the propellant and the pressurizing gas), four thrusters at 35 lb thrust each, four more at 3 lbf each, and

---

[*]Personal communications, F. Winter, National Air and Space Museum, Smithsonian Institution, 2002, and A. Kubica, formerly with BECCO, Buffalo Electro-Chemical Co., a supplier of hydrogen peroxide to Kidde, 2001.

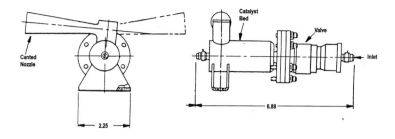

Fig. 7.6-1 Hydrogen-peroxide monopropellant thruster assembly with nozzle (at 90° to catalyst bed) and valve, rated at 14 lbf thrust at altitude. Courtesy of Walter Kidde and Company; copied from Kidde's Rocket Engine Manual, circa 1962.

four more at 1.0 lbf each for roll control. The nozzles of these 12 thrusters were at right angles to the axes of the catalyst beds because this took up less interior vehicle volume. A high-pressure nitrogen gas tank stored the gas for expelling the propellants from their tanks. A clever feature was used on the exhaust flow of excess pressurizing gas through the relief valve; this flow was sent through another catalyst bed, so that any propellant vapor (or liquid in case the unit was turned upside down) would be decomposed into harmless gases.

These Kidde monopropellant hydrogen-peroxide units were used successfully on a good number of space launch vehicles (such as the ACS for upper stages of the Scout SLV or the Titan I and II) and as ACS or orbit adjustment for spacecraft or satellites (such as Early Bird or Syncom II and III). Kidde reaction control systems were also used on some surface-to-surface misssiles, such as Little Joe II.[1] Several thousand thrust-chamber assemblies were delivered by Kidde.

Kidde anticipated that their line of hydrogen-peroxide thrusters would be replaced by the hydrazine monopropellant thrusters, which gave much higher performance (specific impulses of 220 to 245 s). A typical thruster is shown in Fig. 4.6-8. So Kidde developed some hydrazine monopropellant thrusters of their own in the 1960s and had some initial successes with them. Several were installed in satellites and were flown. There were some initial problems with the development of a satisfactory catalyst bed, but they were solved. They put small catalyst pellets confined by small screen baskets at each injection slotted tube as shown in one of the sketches at the bottom of Fig. 4.6-8. Larger sized catalyst pebbles are packed in the main bed held by a support plate with holes. One thruster model, which was flown in a Lockheed satellite, was very sophisticated and is shown in Fig. 7.6-2. It is gimbal mounted with a 6-deg vectoring capability. It had a thrust between 100 and 287 lbf, redundant valves, a large nozzle (88 to 1 expansion ratio) and a venturi flow control. Kidde developed some other hydrazine monopropellant thrusters, but they were not competitive in extended pulsing performance (over 100,000 cycles) or long life, and this became known in the industry. Thereafter the company's hydrazine thruster business dropped off. Kidde also tried unsuccessfully to develop a bipropellant thruster with

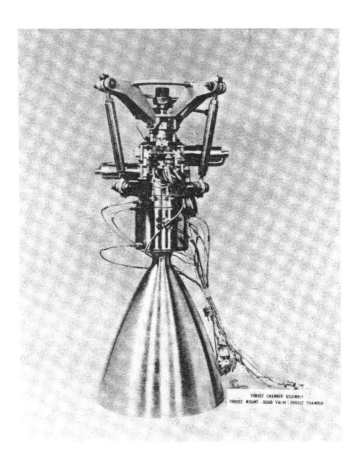

Fig. 7.6-2 Small hydrazine monopropellant thruster (287-lbf maximum thrust in vacuum) with its catalyst bed inside the chamber and mounted on a gimbal with two actuators. Courtesy of chemical Propulsion Information Agency, 1977.

concentrated hydrogen peroxide and several different fuels. These two failed attempts to develop a new product and the dwindling market in their hydrogen-peroxide monopropellant thruster business contributed to Kidde's decision to get out of the LPRE business in the late 1970s.

Reference

[1] "Available Liquid Propellant Rocket Thrusters and Systems," catalog, Walter Kidde & Co., circa 1962.

7.7 Aerojet Propulsion Company, a Subsidiary Unit of GenCorp, Inc.

This company started under its original name of Aerojet Engineering Company in 1942. It was the second new U.S. company dedicated to rocket propulsion, and it grew quickly.[1-3] The official company name today is as listed in the chapter heading, but it is usually simply called Aerojet. It was a direct outgrowth of the work done at the Guggenheim Aeronautical Laboratory of the California Institute of Technology (abbreviated as GALCIT) under the guidance of Professor Theodore von Kármán, a renowned aerodynamicist and head of the aeronautical engineering department at this institute. In 1942 GALCIT, after it had successfully flown the first U.S. JATO, expected the government to need JATO units in quantity. Because a university like the California Institute of Technology was never intended to produce military hardware, GALCIT approached several aeronautical companies asking them to complete the development and mass produce JATOs. Because none were interested, von Kármán decided to establish his own company, and this is how Aerojet came into being. Aerojet was thus founded in March of 1942, and its first board of directors included four men from GALCIT, and von Kármán was the first board chairman. This author joined Aerojet in spring of 1943 and was privileged to immediately work on two analytical projects directly with the famous von Kármán and to participate in several exciting projects described later in this chapter. In December 1944 the General Tire Company bought a controlling interest in the company and supplied badly needed cash funds for expansion. In 1953 the company was reorganized as Aerojet General Corporation and liquid rockets became a separate division of the company.

Aerojet's work in LPREs represented only about one-quarter of their overall business.[3] They were also engaged in solid-propellant rocket propulsion, nuclear propulsion, aerospace power, torpedoes, electronics, minisubmarines, and ordnance. In 2002 Aerojet acquired the Space Systems operation of General Dynamics at Redmond, Washington, which is well known for its small thrusters, and this product line is discussed near the end of this chapter. Aerojet has worked on a very large variety of propulsion schemes, but this chapter will discuss only some of their LPRE work.

Experiments in burning liquid methyl alcohol with gaseous oxygen started at the predecessor organization GALCIT in 1936.[4] It was followed by investigation of different propellants and thrust chambers with thrusts up to 1000 lb. In 1937 nitric acid was selected as the best potential storable oxidizer. In 1939 GALCIT improved this oxidizer by dissolving up to 25% nitrogen dioxide (a red colored gas, which came out of solution and evaporated as a red-colored cloud) in the nitric acid, and this was henceforth known as red-fuming nitric acid or RFNA. It had greater density, slightly higher performance, and better ignition properties than nitric acid without the dissolved gas. Because RFNA's vapor pressure is high, it is difficult to pump, and therefore highly concentrated nitric acid (known as white-fuming nitric acid or WFNA) is preferred for LPREs with pumped engine feed systems. All acids are toxic and corrosive propellants and the investigators soon learned to use acid-resistant materials, such as certain aluminum alloys or stainless steels (for tanks, pipes, or valves) and coroseal and a few other special plastics for seals and gaskets.* Teflon® and other flourocarbons were better, but were not available until several years later. The U.S. investigators probably did not know that RFNA was actually formulated and used in Germany about five years earlier.

In 1940 GALCIT learned about aniline being a self-igniting fuel from the Navy propulsion investigators in Annapolis and started its own laboratory tests. The first test firing with a TC occurred in 1942 with RFNA and aniline. The word "hypergolic" was coined in Germany, where other self-igniting fuels, such as xylidine, were discovered five years earlier. Self-igniting or hypergolic propellant combinations made a separate ignition system superfluous, simplified the engine, simplified the combustion processes, and allowed a much simpler system for multiple engine restarts. Much of the early work at Aerojet was done with this propellant combination of RFNA with aniline. They also worked with mixtures of aniline with other liquids, such as with 20% to 35% furfural alcohol, because this lowered the freezing point of aniline and made the fuel more appealing to military customers.

Jet-Assisted Takeoff Engines

Aerojet developed more JATO engines than any other U.S. company. Aerojet had the advantage of inheriting the technology and some of the personnel from GALCIT, who had designed (1941) and flown JATO rocket engines with an uncooled thrust chamber of 1000 lb thrust using RFNA and aniline.[2] The first successful application of a Liquid Propellant JATO with a manned U.S. aircraft, namely, a Douglas A-20A attack bomber, with two such JATO LPREs was made in mid-April of 1942 at Muroc dry lake in California using a GALCIT design with some assistance of Aerojet personnel. This unit is shown installed in one of the airplane nacelles in Fig. 7.7-1 with the nozzle exit near the bottom.

The first development of a LPRE of Aerojet was an improved version of the GALCIT-developed JATO that flew on the A-20A airplane.[2,3] Figure 7.7-2 shows this engine, which was identified as 25AL-1000. It was more compact and lighter. It had three spherical tanks, one for the RFNA, one for aniline fuel, and one for the high-pressure nitrogen pressurizing gas. The thick-walled TC was uncooled, and the simple injector, shown in Fig. 7.7-3, with four unlike doublets was basically copied from a GALCIT design. One version was recoverable and reusable; it included a recovery parachute and features for draining, purging, flushing, and cleaning the system. It was the first recoverable and reusable JATO. By 1944 some 64 of these were produced and delivered at a cost of $3450 each. Later additional deliveries were made.

These first Aerojet JATO units were flight tested in a Douglas A20 bomber aircraft, in a P-38 fighter aircraft, but unverified data about flights with other U.S. military aircraft were also reported.[2] The results showed an increased takeoff performance, such as shorter runway distance or clearing a 50-ft obstacle by a large margin. Furthermore, a JATO allowed an increase of the aircraft

Liquid Propellant Rocket Engines in the United States (Summary) 361

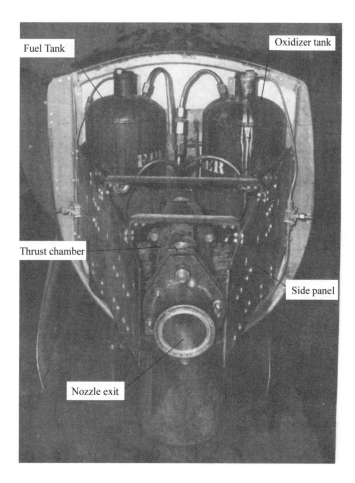

Fig. 7.7-1 First U.S. JATO developed by GALCIT installed in one of the two nacelles of the A-20 aircraft with the cover removed. Thrust was 1000 lbf. Copied with permission from Ref. 2

takeoff weight, which translates into more payload or more fuel. It would allow takeoff at a high-altitude airport with a full load, and this could not have been done without the JATO. However the JATO could not be stored with propellants in the tanks for more than 6 to 12 months because of acid corrosion and deterioration. Also leaks or acid spills in loading would cause corrosion problems in the aircraft. A few of the Air Force maintenance people got acid on their skin, and these skin burns heal very slowly.

The Navy wanted a larger JATO that could be dropped into the ocean. Another similar JATO (38ALDW-1500) with 1500 lb thrust, the first with a regenerative cooled thrust chamber, and with 38-s duration was successfully developed for the PB2Y-2 flying boat.[3] It also used a gas-pressurized feed system. Almost 100 of these were delivered to the U.S. Navy. Most of these JATO

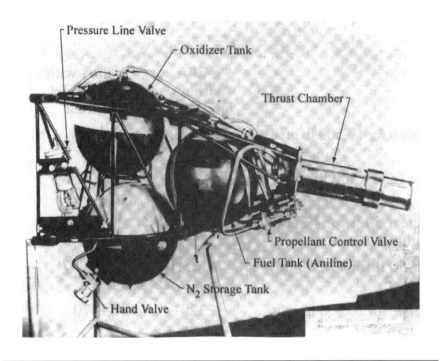

Fig. 7.7-2 First JATO produced by Aerojet with 1000 lbf thrust for 25 s. Courtesy of Aerojet; copied with permission from Ref. 2.

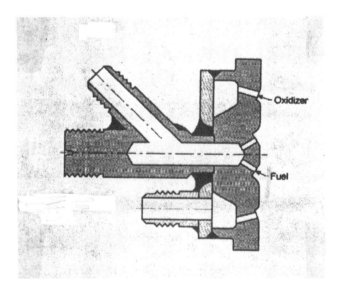

Fig. 7.7-3 Simple injector for first Aerojet JATO. It was developed by GALCIT. Courtesy of Aerojet; copied with permission from Ref. 2.

LPRE were later given to the Coast Guard, who used them occasional for difficult sea rescue missions and/or takeoff in rough seas.

Subsequently a job for a more advanced JATO unit (35AL-6000) was given to Aerojet for the PB2Y-2 flying boat, but with higher thrust and with a pumped feed system, which was permanently installed in the aircraft.[3] Figure 7.7-4 shows the pump package and its power sources, a small separate dedicated aircraft-type piston engine. It used concentrated nitric acid (not RFNA) and a mix of aniline and furfural alcohol as propellants and had three regenerative cooled thrust chambers. Flight tests included takeoff from the ocean in rough weather with the TC submerged in sea water at the start. The flight tests were successful, and as a result the unit was put into limited production. Several versions of this unit were developed (different durations, more thrust) and delivered for further Navy tests.

Next came a JATO for the Air Force B-29 bomber, which Aerojet identified as ALD-3000 (Ref. 3). It went through development, but was not put into production because the contract was canceled. It was followed by a contrat for the X60ALD-4000 for the B-45 medium-range bomber aircraft. It also used a pressurized-gas feed system, a regeneratively cooled TC, and was equipped with a parachute for air drops and recovery. The JATO was checked out, refilled, and reused. Flight tests with the B-45 were successful.

Fig. 7.7-4 Installation of propellant pumps driven by an aircraft piston engine with gear case for a JATO LPRE in a PB2Y-2 flying boat. Courtesy of Aerojet and Aerojet History Group; copied from Ref. 3.

With this success the Air Force decided to equip a portion of its B-29 fleet with JATOs. Aerojet was contracted to develop the LR13-AL-3 with a propellant that would not freeze in the cold Alaska climate, and the selected propellants were concentrated nitric acid and a mix of 65% xylidine (which the Air Force had stock piled) and 35% gasoline.[3] Figure 7.7-5 shows this JATO with three tanks and an inclined TC of 4000 lbf thrust. The aircraft had four of these droppable disposable nonrecoverable JATO units under the wings for a total takeoff thrust of 16,000 lbf and a duration of 40 s. The figure shows an early model with a parachute for recovery. The production version did not have a parachute. Because this was an expendable JATO unit, it was designed to be of relatively low cost using a pressurized feed system. For example, the regenerative cooled TC was replaced by a ceramic-lined uncooled TC. The first test firing took place six weeks after contract award. With this takeoff assistance the B-29 could carry enough fuel to reach any location in the Soviet Union and return to a friendly base; this was not possible without the JATOs. Approximately 150 units were delivered. The Air Force modified two B-29 bombers to accept the JATOs, and test flights were then conducted. However, there was no follow-on production contract.

The JATO for the B-47 bomber had a higher thrust than any prior Aerojet JATO and represents a historic achievement.[3,5] There were two permanently installed units on each aircraft, one on each side of the fuselage. They were mounted on doors, which were swung out of the fuselage into a firing position as shown during takeoff in Fig. 2-2. Figure 7.7-6 shows that each unit had two 5000-lb thrust chambers, giving a total thrust of 20,000 lb. The Air Force insisted on WFNA and jet fuel as propellants because the fuel could be taken directly from the aircraft tanks and WFNA was easier to pump than RFNA (less likely to cavitate). The nitric-acid tank was permanently installed in the fuselage. This propellant combination requires an igniter, and ignition of a small

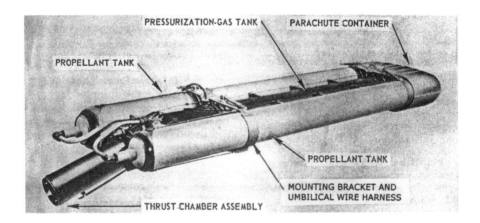

Fig. 7.7-5 One of four JATO units for the B-29 bomber. Courtesy of Aerojet; copied from Ref. 3 on p. 3 of this book.

flow of both propellants was done wih an electrical glow plug. This ignition system could be restarted. Therefore the equivalent of a 50-lb thruster was built as an igniter chamber into the injectors of each of the 5000-lb TCs. The initial TC shown in the figure had an acid cooling jacket, but a later version had a *ceramic-lined chamber* using a commercial silicon carbide called Niafrax (later other ceramics, such as a zirconium oxide were tried). Propellants were supplied by two *propellant pumps* with single directly coupled turbines, driven by warm *air bleed from the compressors of the B-47 aircraft's jet engine*. (This reduced the inert tank hardware weight.) It included some new safety provisions. Just before the preliminary flight qualification tests were to begin, this B-47 Aerojet JATO unit exploded during shutdown. This accident triggered a major project review, some redesigns or revisions, a delay in the development effort, and an indefinite postponing of the planned production. Further R&D provided these corrections of the problem: minimizing the amount of after-dribble of propellants, better purges of propellant lines, venting and changes in the control sequences. The control actually became rather complex.

The M. W. Kellogg Company had a contract for a competitive JATO for the B-47 bomber with the same rocket engine requirements; it was more efficient but also more complex than the Aerojet unit and had several significant development problems, which are described in Chapter 7.5. Kellogg's JATO did not

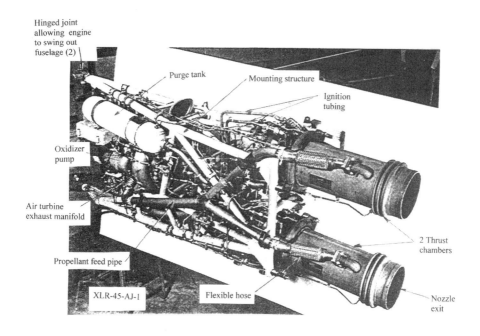

Fig. 7.7-6 One of two permanently installed JATO LPREs for takeoff of a modified B-47 bomber airplane. Courtesy of Aerojet; copied from Ref. 2 of Chapter 7.5 in this book.

fly. Aerojet's B-47 JATO was selected to undergo flight tests with a modified B 47 aircraft, and these flights had no significant rocket engine problems. In one of the flight tests, a safety feature malfunctioned and caused an involuntary shutdown of one of the two JATO units, but the restart feature allowed the pilot to quickly restart the JATO unit and to achieve a satisfactory takeoff. A spectacular view of the B-47 takeoff is shown in Fig. 2-2, which had just been mentioned. Subsequently, the requirement for this B-47 JATO was cancelled by the Air Force.

The last JATO unit (YLR63-AJ-1) to be developed by Aerojet (1948) was for the F-84 fighter aircraft built by Republic Aircraft.[3] The propellants were white-fuming nitric acid and jet fuel. The JATO is shown in Fig. 7.7-7 without its cowling, and it had to be carried in a small long pod or bulge underneath the aircraft, which was volume limited. Aerojet's hope of using proven components from the B-47 JATO were dashed because the pump package or the thrust chamber were too large to fit into the limited available volume. So a new pump package and a new ceramic-lined TC had to be developed. The turbine of the JATO was driven by bleeding compressed air from the compressor of the turbojet's engine. The pumps had gold sleeves cooled by acid in the oxidizer pump bearings and fuel-cooled silver sleeves in the fuel pump bearings. The igniter was really a separate small combustion chamber attached to the main injector, and the ignition energy came from an electric glow plug. The flight tests with the F-84 went well. Unfortunately, Westinghouse Electric, the contractor for the turbojet engine, was seriously behind schedule in deliveries, and more than 150 of the F-84 aircraft were parked at the aircraft manufacturer's plant waiting for jet engines. The Air Force canceled the Westinghouse engine contract and instead used a new General Electric jet engine, which did not need a JATO for takeoff. The production of the G-version of the aircraft, which had propellant tanks and the JATOs, was also canceled. Aerojet JATO production contract was stopped.

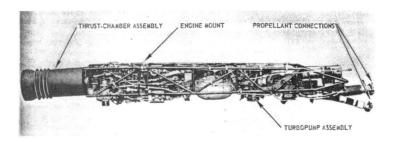

Fig. 7.7-7 JATO for mounting at the bottom of the fuselage of the F-84 fighter airplane, shown without the sheet metal cover. It was also used for superperformance maneuvers, and it could be jettisoned. Courtesy of Aerojet; copied from Ref. 6 on p. 107 of this book.

The U.S. military services were never really enchanted with Liquid Propellant JATO units, in part because the servicing and refurbishing of used units (with remnants of acid propellants) was hazardous and because it required new acid storage, transfer and supply equipment at military air bases. Many pilots believed that there was relatively little need for takeoff assistance. Also there was concern about combat-damaged airplanes containing concentrated nitric acid. None of the U.S. JATO units were ever operational. However, the Germans and the French used them in their air force operations. Solid-propellant JATOs, which had overcome their earlier problems with operation at extreme ambient temperatures, gradually replaced them. Aerojet also produced these solid-propellant JATO rocket motors in large quantities. Later, when runways were made longer and when turbojet engines with afterburners were developed, the requirement for JATO units became unnecessary.

Liquid Propellant Rocket Engines for Aircraft

During World War II, there was a concern that German fighter aircraft would outmaneuver the U.S. and British fighter airplanes, which were then escorting and defending allied bombers flying raids over Europe. The Allies had intelligence information that the Germans would soon put a new rocket-powered fighter airplane (Me 163) into service. The military command was looking for ways to counteract this threat. A new rocket engine would give a U.S. fighter airplane a superior flight performance, particularly at high altitudes. Therefore there was great military interest in LPREs for aircraft, both as a primary aircraft rocket engine or as an auxiliary engine to augment the aircraft's main engine.

In 1943 Aerojet conceived, designed, and proposed a new concept for an aircraft rocket engine, called *Aerotojet*.[3,*] In 1944 they received a contract to build and develop this engine. It was intended to propel a new (at the time secret) flying-wing bomber design of John Northrop, the founder of Northrop Aircraft Company. For this application RFNA (higher density and performance than WFNA) and a fuel consisting of aniline with 20% furfuryl alcohol were selected. This author was a young test engineer in the early phases of this unique project. This LPRE is shown schematically in Fig. 7.7-8. It had two stationary thrust chambers of 750 lb thrust each and two rotating thrust chambers of 200 to 300 lb each. The latter were mounted on a rotating hollow shaft, fired almost parallel to this shaft, but canted at 20 deg so as to produce a torque. The shaft (through a gear case) drove three centrifugal pumps, an RFNA main pump, an aniline pump, and a low-speed RFNA booster pump. The booster pump was necessary because RFNA's high vapor pressure would otherwise have caused pump cavitation. This was the first known application of a booster pump for preventing cavitation with a high vapor

---

*Author's personal recollections while employed at Aerojet, 1943–1946.

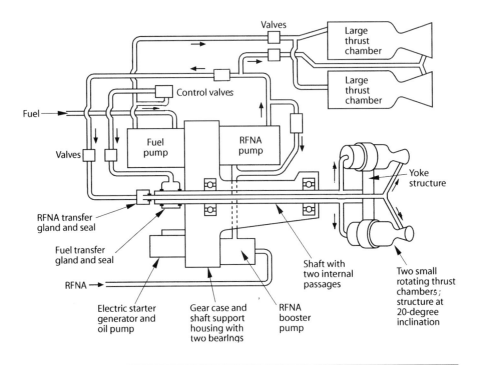

Fig. 7.7-8 Simplified schematic diagram of Aerotojet LPRE concept. Courtesy of Aerojet.

pressure propellant at the main oxidizer pump. It also had to be throttled with a large reduction in thrust; this was accomplished by reducing the flow (and thus the thrust) to the rotating chambers, which in turn reduced the pump speed. The rotation of the chamber and injector caused centrifugal forces to act on the fluids causing an additional pressure rise and a mismatch of the impingement of the fuel and oxidizer jet injection streams, which in turn caused incomplete combustion and the accumulation of unburned and poorly mixed propellants in the chamber. The explosions of these accumulated propellants broke experimental rotating thrust chambers on two occasions. In the author's opinion, this centrifugal mismatch effect was the most severe problem, but there were others.

Different bearing materials for the acid pump were unsuccessful, but gold sleeves or thick gold plating did work.* Aerojet henceforth used gold sleeves in most of its nitric acid pumps. The materials and design of the seals in the acid pumps, with rubbing, wear, and local heat generation, were never completely satisfactory; however, several years later this was solved. Also there was the problem of the shaft seal at the transfer gland, where the high-

---

*Author's personal recollections while employed at Aerojet 1943–1946.

pressure acid propellant was transferred into the rotating shaft; the seals at that time seemed to hold for at least one run. When this shaft seal did leak, the amount of leakage was small. There were also combustion problems at the low thrust and cooling problems with the gear case. Perhaps there were just too many new features in this early aircraft rocket engine, or perhaps the concept was too far ahead of its time. The delays in this engine development forced Northrop to go to an alternate propulsion scheme for their flying wing-shaped large aircraft, and the Aerotojet project was canceled. Much later it was learned that a similar project was investigated a few years earlier in the Soviet Union; components were built, but the engine was not; it too was not put into a vehicle. No other similar effort with rotating thrust chambers is known.

A small-scale *model of this Northrop flying-wing aircraft* was built starting in 1942 to test the aerodynamics of this novel wing design. It was known as Project MX-324. This small aircraft model had less than 20-ft wing span, and a picture is shown in Fig. 7.7-9 and during engine operation in Fig. 2-5. This flying-wing experimental aircraft was so small that the pilot was not able to sit, but had to lie down on his stomach with his head in the plastic nose. Aerojet built, tested, and flew a small LPRE for this model airplane, and your author was a part-time member of the team. It used RFNA and aniline as propellants, had a single small regeneratively cooled (with fuel) thrust chamber of about 200 lb thrust (aluminum with a copper nozzle), a pressurized feed system, and it was restartable in flight. The small aircraft was taken to altitude by a larger airplane, and the Aerojet rocket engine was started after the small experimental aircraft was released from its mother airplane. Figure 7.7-10 shows the Aerojet cooled TC and dual valve with a single common actuator. These dual valves were a favorite early method for achieving simultaneous opening and closing or the fuel valve and the oxidizer valve. It flew successfully for the first time in July 1943 and several times thereafter. It was the first piloted U.S. airplane propelled by only a LPRE.

Aerojet also developed *auxiliary rocket engines for three fighter airplanes* (which already had their main turbojet engine) to augment the flight performance.[3] These were often called *superperformance rocket engines* because they allowed a figher airplane to climb to high altitude in record time, and they enhanced the flight velocity of the fighter airplane, thus improving its maneuverability. One such LPRE was developed by Aerojet for the P-51 Mustang fighter aircraft. This application is different than the JATO mission, where aircraft are given an extra push during takeoff (with heavy loads), while on the runway. Work on Aerojet's auxiliary engine for the P-51 fighter aircraft was started in late 1944. It had 1300 lb thrust for 1 min, a pressurized feed system, a regeneratively cooled TC, and used RFNA with 80% aniline 20% furfural alcohol as propellants. Test flights were conducted with one of the P-51 aircraft, which was modified to accept the LPRE and its feed system. The war ended in 1945, and the project was canceled. The JATO LPRE (1948) installed in the experimental F-84, which was already mentioned, could also be used and was tested as an auxiliary or superperformance LPRE at altitude.

Fig. 7.7-9 This small airplane modeled a full-scale Northrop Flying Wing airplane. It was the first aircraft in the United States to be powered by a LPRE, and it flew first in July 1943. Courtesy of Northrop Grumman; photo from Sam Smyth.

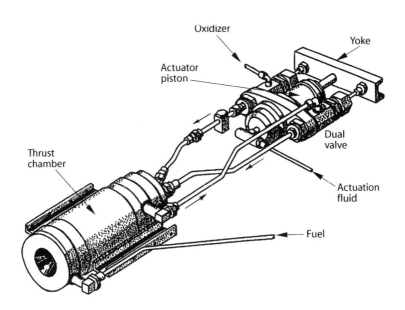

Fig. 7.7-10 TC and dual-propellant valve (with actuator in center) of the U.S. LPRE, which propelled the first U.S. piloted aircraft. The chamber diameter was about 5 in. Courtesy of Aerojet; adapted with permission from Ref. 2.

In 1949 Aerojet developed and delivered another auxiliary LPRE for the F-80 jet fighter aircraft; it had enough propellant for 4 min of powered flight. Satisfactory flight tests were made, but production was never started, probably because the extra weight of the LPRE and its propellants severely diminished the range and thus the military value of the F-80 aircraft. The last Aerojet auxiliary LPRE was for the aircraft F-86 fighter airplane during the Korean war to counteract the reported superiority of Soviet MIG fighter aircraft, which could fly at higher altitude and had good maneuverability. The LPRE for the F-86 was more complex than earlier units. A turbopump feed system (using WFNA and UDMH propellants) was modified for this application in 1953. Instead of air gas bleed from the compressor of the turbojet (this air was not really available at high altitude), it had its own gas generator with its own supply tanks using a small gas-pressurized feed system. It provided six restarts and a preset thrust between 3500 and 5000 lb. Again flight tests showed improved aircraft performance, but the Korean war ended, and the LPRE was not retrofitted into operational aircraft.

Just as it happened with the JATO units, many military U.S. pilots were never convinced that they needed superperformance rocket engines, particularly with the corrosive toxic nitric acid. It is regrettable that Aerojet spent two decades, much of its capability and talents on JATO and superperformance projects, that never were deployed with the military forces. However the French and the Germans JATO rocket engines in their air force.

Upper-Stage LPREs

A series of successful pressure-feed *storable LPREs for upper stages of launch vehicle* were also developed by Aerojet for the *second stage of several SLVs and missiles*. About 49 different versions and modifications were developed, and several hundreds were delivered and flown in a period of about 45 years. It started in 1944 with the sustainer LPRE for the WAC-Corporal, which was initially developed at the Jet Propulsion Laboratory (which in turn was based on work of GALCIT), but with a few components and a little assistance from Aerojet. The WAC Corporal is shown in Fig. 2-1, was first launched in 1945, and was the engine on which the Aerojet upper-stage LPREs were based. All had relatively low chamber pressures, and all used gas-pressure feed systems. In addition to propelling upper stages, as explained next, some of these engines were also delivered for the wingless flight models, various research test vehicles, and several versions of the successful Aerobee sounding rocket upper stages, which are explained later in this chapter.

The sustainer engine for the Nike-Ajax surface-to-air missile (2600 lb thrust, 21s, with nitric acid and aniline propellants, a pressurized-gas feed system, regenerative cooling) was first fired by Aerojet on a test stand in May 1946.[3] In 1953 an improved version of this engine became an integral part of the world's first operational antiaircraft missile, which can be seen in Fig. 2-8. The Aerojet-designed engine propelled the second stage, and a solid propellant booster was in the first stage. During the development period and also during the early production, Aerojet built more than 100 complete LPREs and perhaps 6000 TCs for the test and development program. However Aerojet lost the main production of this engine to the Bell Aircraft Company, where thousands of operational missiles were built, and this included the building of the upper-stage LPRE. These missiles were deployed with the U.S. Army in several places in the United States and later in several foreign countries. Altogether about 16,000 units were delivered. Nike-Ajax remained in the Army's operational inventory until 1958, when a gradual replacement began.

A greatly uprated version of 7500 lb (later 7800) thrust with WFNA and UDMH propelled the second stage of the Vanguard space launch vehicle. This Vanguard was conceived as a U.S. satellite launcher for the international Geophysical year (1958), and it was a three-stage vehicle. The first stage was propelled by a LOX/kerosene engine of General Electric shown in Figs. 7.3-9 and 7.3-10. The complete second stage including its engine was developed (1955–1961) and built by Aerojet. Its LPRE used NTO/UDMH, a set of formed stainless-steel tubes for a regenerative cooling jacket, and a gimbal suspension of the TC.[3] The pressurized-gas feed system had a solid-propellant GG inside the high-pressure gas tank in order to heat the gas and reduce the amount of gas needed, as well as the inert mass. Aerojet initially built and tested TCs with aluminum tubes and also TCs with stainless-steel tubes in the cooling jacket. There were some failures and problems with burn-through incidents, particularly near the end of the burn period. They selected the stainless-steel tubes, which were heavier, but proved to be more reliable. The first successful Vanguard satellite launch was in 1958.

The *Able upper stage and its Able LPRE* was originally designed to fit on top of a Thor first stage or Atlas SLV booster stage, and it was a successful series that existed for more than 40 years.[3] It was an outgrowth of the Vanguard second stage and its engine and a version of the Aerobee sounding rocket engine, which is discussed later in this chapter. The LPREs had regeneratively cooled aluminum thrust chambers and aluminum injectors. At the low chamber pressures of these TCs (90 to 155 psia), the heat transfer is relatively modest, and it is possible to use aluminum in the injector and the cooling tubes. An improved version, called Able-Star, had larger vehicle tanks, a longer duration (5 min), a higher nozzle area ratio (over 40), an uncooled nozzle-exit skirt, a restart capability, used IRFNA and UDMH propellants, and had an initial sea-level specific impulse of 276 s at 8000 lb thrust. Figure 7.7-11 shows an Able-Star LPRE. At least six different configurations of the Able and Able-Star series were developed with different TCs and tank sizes. A total of about 40 upper Able/Able-Star stages with their engines were delivered and flown between 1956 and 1980.

The *Transtage* was an outgrowth of the Able series and became the third stage for the Titan SLV. It had two gimbal-mounted upper-stage engines of about 8000 lbf thrust each and a large nozzle area ratio. It not only could provide pitch and yaw control, which the Able engines had provided, but also roll control with two gimbaled TCs. It had ablative liners for cooling and restart, a nozzle skirt made out of niobium, which was new at that time, an aluminum injector, and a pressurized feed system.[3] Most of the Transtage engines reached a steady-state altitude specific impulse of 320 s.

The experience with the Able-Star upper stage on the Thor missile led to the *Delta launch vehicle upper stage*, where variations of this second-stage engine were successful in propelling payloads into space orbits. One version is shown in Fig. 7.7-12.[3,6] Similar Aerojet engines were also used on the third stage of Titan between the 1960s and the 1990s. The Delta LPREs had a radiation-cooled nozzle-exit skirt, a thrust of about 9800 lbf (later uprated to 10,200 lbf), a specific impulse of 320 s (vac) at a chamber pressure of 125 psia using NTO/Aerozine 50, and different durations or tank volumes. All used pressurized-gas feed systems, with low chamber pressure, and a gimbal-mounted TC, with a large nozzle-exit area ratios of 65 to 1.0. Some provided also roll control with two cold-gas jets with hinged nozzles, and some were able to provide limited throttling during flight. Many used platelet injectors, which are described near the end of this section. Some of these upper-stage pressure-fed LPREs used aluminum tubing in the cooling jacket, some used ablative lined chambers and nozzle throats, and some used a fuel regenerative cooling jacket with an ablative nozzle-exit skirt. An ablative TC makes restarting in space a lot simpler and safer. Between 1960 and 1993

Fig. 7.7-11 Installation of Able-Star thrust chamber with instrumentation. Courtesy of Aerojet and Aerojet History Group; copied from Ref. 3.

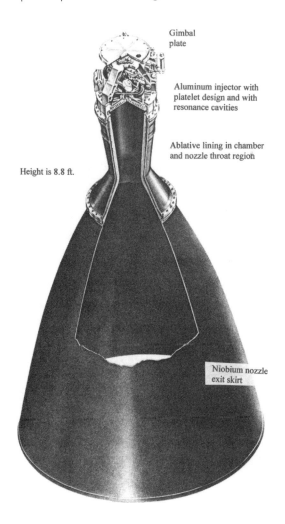

Fig. 7.7-12 This TC was used on the second stage of the Delta II SLV. It has an ablative liner in the chamber and nozzle-throat regions and a niobium nozzle extension to an area ratio of 65. Courtesy of Aerojet; copied from data sheet Ref. 6.

a total of about 247 Delta upper-stage engines were produced. There were 13 flight failures or malfunctions with these Delta LPREs, mostly in the first two decades. The overall reliability of these engines was estimated at 94%. Today the customers want 99.9% reliability in flight.

A version of the upper-stage Aerojet engines was developed as an orbital maneuver engine for the planned Japanese Hope X flight vehicle, and it was delivered to Ishikawa Harima Heavy Indstries in Japan.[3,6] The LPRE was rated at 3822 lbf thrust, had a radiation-cooled niobium nozzle-exit skirt with a nozzle-exit area ratio of 86 to 1.0, and a relatively high chamber pressure of 155 psia for a pressurized feed system. The thrust chamber was regeneratively cooled and featured a platelet injector. Eight complete upper stages with this engine, its gas pressurization system, and with roll control for this upper stage were built and delivered to Japan, and they flew successfully in the upper stage of the Japanese N-2 SLV.

The *space shuttle orbital maneuver engine system (OMS)* started in 1973 and is still an active program.[3,7] It is another in the successful series of pressure-fed upper-stage gimbaled engines using storable propellants (NTO/MMH). It had a critical mission, which included the final Apollo orbit injection, Earth orbit circularization, orbit transfer and deorbit maneuver. There are two TCs of 6000 lbf thrust each at a chamber pressure of 130 psia, one on each side of the aft end of the Space Shuttle Orbiter as shown in Fig. 2-11. The nominal vacuum specific impulse is 313 s. At a mixture ratio of 1.65, the volume of the fuel tank and the oxidizer tank are equal, and the same basic tank can be used for both. The TC is shown in Fig. 7.7-13. It uses a milled channel design for the cooling jacket, and this design allows a high heat-transfer rate as explained in Chapter 4.3. Acoustic resonance cavities are built into the injector to stabilize the combustion, and a radiation-cooled nozzle extension is between the nozzle area ratio stations of 6 and 55. Aerojet also developed a pump-fed version of the OMS engine, which would have saved some inert mass, but NASA did not want to put it into the space shuttle because NASA did not want to abandon a well-proven, properly functioning pressurized-gas feed system and the tie-in with the reaction control system would require extensive vehicle changes.

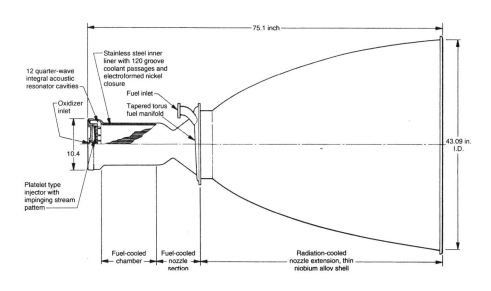

Fig. 7.7-13 Simplified half section of one of the two TCs of the orbiting maneuvering engine (OME) used on the Space Shuttle Orbiter. Courtesy of Aerojet; copied with permission from Ref. 7.

The historic *Apollo Service Module* was propelled by an uprated (considerably larger) version of these Aerojet upper-stage engines.[3,6] It had a critical mission, namely, the retroaction while approaching the moon, the circularization of the lunar orbit, rendezvous with the lunar ascent module, and the departure maneuver from its lunar orbit. Figure 7.7-14 shows the TC assembly with the propellant valves, but without the nozzle extension. Work on the LPRE started in 1973. It became a 100% reliable flight engine, and 50 thrust chamber assemblies were produced. With 20,000 lb of nominal thrust in a vacuum, it had enough propellant (nitrogen tetroxide and Aerozine 50) for a cumulative duration of 12.5 min, with firings ranging from ½-s to 7.3-min duration. For a chamber pressure of 100 psia, it had a relatively large pressure drop in the injector, valves, and pipes (65 psi). The thrust chamber and nozzle-throat region had a thick ablative liner construction (suitable for restart), the largest piece of ablative built up to that time (1974). The gimbal was mounted around the nozzle throat, it had an aluminum injector with baffles for combustion stability, and the fixed nozzle extension, with a nozzle area ratio of

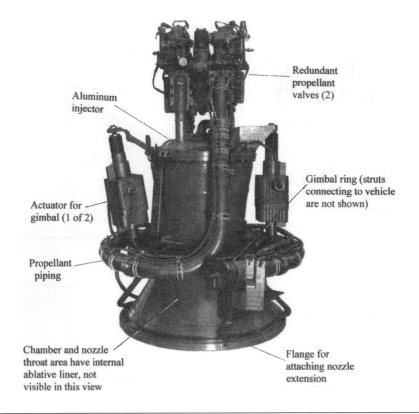

Fig. 7.7-14 View of thrust-chamber assembly for the Apollo service module, but without the large exhaust nozzle extension. Courtesy of Aerojet; copied with permission from Ref. 3.

62.5 to 1.0), was made of niobium. Its qualified total firing life was 33 min, a record for ablative TCs at that time. It had multiple vacuum restart features, redundant valves, and safety provisions for manned flight.

Beginning in about 1965 Aerojet developed what is known as their *platelet injector technology*. Platelets are thin metal sheets in which fluid paths or opening are stamped or etched in predetermined patterns.[7–9] The thin metal plates, with cutout patterns, are stacked in a predetermined sequence to form hydraulic passages and injection orifices. The stacked metal sheets are diffusion bonded together forming a rigid structure. Figure 7.7-15 shows the platelet injector concept in two simple injector configurations. With this technique it is possible to design injectors with a very large number of small holes per unit area of injector face, either impinging or nonimpinging injection hole elements; this promoted good mixing of propellants, uniform gas composition and uniform gas temperature across the chamber, a small chamber volume, and high combustion efficiency. It also allows a small manifold volume (fast start and low dribble mass) and permits a large variety of different injection patterns. A good number of the Aerojet injectors have used this platelet design and have obtained a high combustion efficiency. Examples are a number of the Delta TCs and the Apollo orbital maneuver engine. None of the other LPRE development organizations seem to have used this platelet concept in their injectors, probably because their own injection concepts gave also equally good performance.

Aerojet also applied this platelet concept to construct lightweight cooling jackets and tested several of these up to a thrust level of 40,000 lbf with LOX/LH$_2$. The tooling can become complex and expensive. There has not been any further work on using such a cooling jacket.

Bomarc

The *booster LPRE for the Bomarc area defense system* (ramjet-powered supersonic missile) was developed by Aerojet beginning in 1951. It is a historic LPRE because it was Aerojet's first major occurrence of combustion vibration, a problem that was solved, and because it had a novel tank pressurization scheme.[3] The TC had a ceramic liner and was one of the first to use a gimbaled TC in U.S. military operation. Figure 7.7-16 shows a reconstructed flow diagram of one version of the Bomarc booster LPRE. Originally it used WFNA and jet fuel propellants, which need an igniter. It was designed to deliver 35,000 lb of thrust for 4 s. The tank pressurization system was unique and for the first time gas came from two separate small gas generators (one was fuel-rich gas for the fuel tank and one was oxidizer rich), both supplied from a separate set of gas-pressurized small propellant tanks. This scheme saved some inert gas pressurization hardware weight. This type of chemical tank pressurization has been used on other subsequent LPREs.

The first test was a disaster. The LPRE failed because of a high-frequency combustion instability, which caused excessive heat transfer, erratic operation, and a burnout. These combustion vibrations are discussed in Chapter 4.10. Special expert teams were assembled, and several approaches to achieving

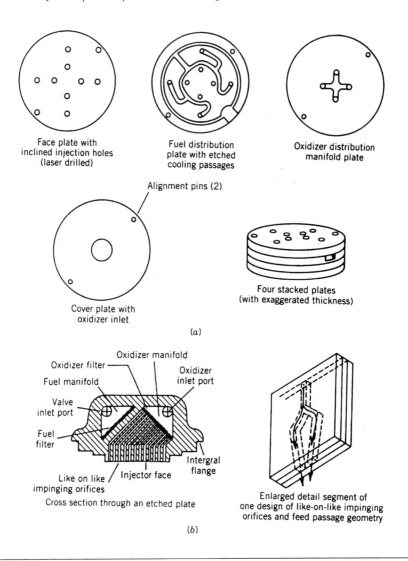

Fig. 7.7-15 Simplified sketches of two types of injectors using a patented bonded platelet construction; a) with four impinging unlike doublet liquid streams and b) like-on-like injector with 144 orifices. Courtesy of Aerojet; copied with permission from Ref. 7.

stable burning were arbitrarily tried. The injection of water was helpful, but not a complete solution. Finally a change in fuel from jet fuel to UDMH solved these vibration problems. This fuel was quickly changed to a mix of 60% JP-4 jet fuel and 40% UDMH, which was lower in cost and worked just about as well at low ambient propellant temperatures. In addition to the high-frequency combustion instability, a lower-frequency instability was also observed in several later tests. This lower-frequency "chugging" instability was apparently

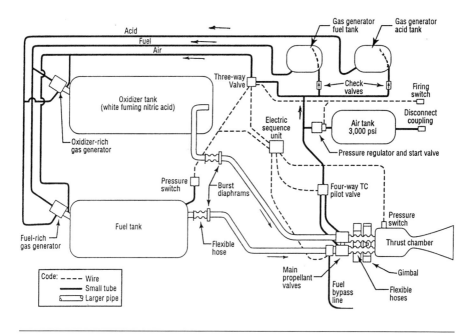

Fig. 7.7-16 Reconstructed flow diagram of one version of the Bomarc booster LPRE with a gimbal for the TC and two gas generators for pressurizing the main propellant tanks. Courtesy of Aerojet and Aerojet History Group; adapted from Ref. 3.

amplified by a resonance of the gimbal and its support structure between the thrust chamber and the oxidizer tank. The strengthening of these structures remedied the problem.

There were also problems with the unique hot-gas pressurization system. During flights with high side accelerations, the liquid propellant in the tank can slosh excessively, and in turn this can cause a sudden cooling of the "warm" ullage gas, which quickly leads to a lower tank pressure and thus to erratic thrust variations. Antisloshing baffles in the propellant tanks solved this concern. Qualification test of the engine were completed in about 1958. Bomarc was deployed by the Air Force and operated with the military services until 1972. The prime contractor, The Boeing Company, delivered 366 missiles.

Liquid Hydrogen

In 1945 Aerojet might have been the first U.S. rocket propulsion company to test gaseous hydrogen with oxygen in a small (45-lb) thrust chamber, which burned out. Several larger thrust chambers using these propellants were more successful in 1946. A hydrogen liquification plant was installed at Aerojet, and liquid-hydrogen firings (1948) and pump tests (1949) were conducted.[3,10] These were the first known LOX/LH$_2$ firing tests in the world. In the next decade studies of larger LOX/LH$_2$ LPREs were performed at Aerojet and at other companies.

In 1962 Aerojet was awarded a contract to develop the large *M-1 LPRE* intended to have originally 1.0-million-pound thrust, and later it was raised to 1.5-million-pounds thrust. It operated on liquid hydrogen and liquid oxygen, the propellant combination identified by early rocket propulsion pioneers as being the most practical for high specific impulse. This *M-1 engine* was designed for a chamber pressure of 1000 psia, a specific impulse of 430 s, a nozzle expansion area ratio of 40 to 1, and a gas-generator engine cycle. It used bundled, shaped tubes for a regenerative cooling jacket in the chamber and upper part of the nozzle or the throat region. The nozzle extension beyond an area ratio of 14 was cooled by turbine exhaust gas. This dump cooling feature had been designed into the Rocketdyne F-1 and J-2 engines several years before. The M-1 had separate TPs for the fuel and the oxidizer, again similar to the earlier J-2 LPRE. The hydrogen pump had nine axial pump stages, a spiral inducer, a separate dual-stage direct-drive turbine of about 75,000 hp. It was similar in concept to the J-2 pump with seven axial pump stages. The engine was big, about 311 in. or 7.898 m high, and the nozzle-exit diameter was 212 in. or 5.38 m. It would have been the largest liquid-hydrogen/liquid-oxygen rocket engine, if the program would not have been stopped. A few component tests were performed before the contract was canceled in August 1965. At that time NASA was supporting two large engine programs for a moon flight and was asked to stop one of them. The survivors for Apollo/Saturn program were the F-1 and J-2 LPREs, in part because they were several years further along in their work.

Titan LPREs

Aerojet's most successful set of large LPREs was that for the booster and sustainer stages of the several versions of the Titan vehicle.[3,6,11,12] Figure 7.7-17 shows the twin booster engine of the Titan I, and selected data for the four major Titan LPRE programs are given in Table 7.7-1. As of the year 2002, over 1500 individual LPREs have been delivered over a program period of 47 years. The booster engine has twin LPREs with regeneratively cooled gimbal-mounted thrust chambers, each with its own turbopump.

The single high-speed turbine was geared to the two lower-speed centrifugal propellant pumps and a lubricating oil pump. This TP configuration was chosen to allow high pump and turbine efficiencies and thus low gas-generator flow, which in turn allowed a slightly better specific impulse. The sustainer engine has a lower thrust and a single cooled gimbaled thrust chamber with an ablative liner at the exhaust nozzle-exit section. The sustainer engine's TP was smaller, but similar to the booster TP; however, the turbine and the fuel pump were on the same shaft, and only the oxidizer pump was driven through a gear reduction and ran at a lower shaft speed. The bearings and gears of both the booster and sustainer engine were lubricated and cooled by oil supplied from as small oil pump, which was driven by an auxiliary set of small gears off the gear case.

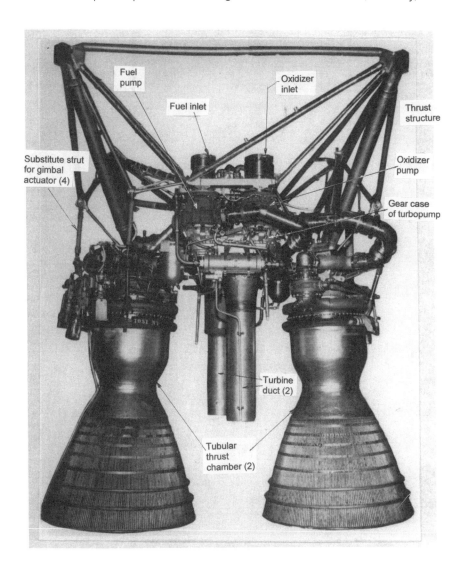

Fig. 7.7-17 Aerojet dual booster engine for the Titan I vehicle had two gimbal-mounted TCs and two geared TPs with heat exchangers in the turbine exhaust ducts. Courtesy of Aerojet; copied with AAS permission from Ref. 11.

One of the unusual features of these Titan TPs was the use of three bearings for each of the shafts, two roller bearings for radial loads and one ball bearing for axial loads. Other TPs have only two bearings on each shaft because it is usually difficult to properly align three bearings.

At the insistence of the Air Force, the propellant combination was changed from the cryogenic LOX/kerosene in Titan I to a noncryogenic or storable propellant combination, namely, nitrogen tetroxide with Aerozine 50 (a mix of

Table 7.7-1 Several Key Parameters of Titan LPREs

Vehicle	Application	Stage	Thrust, lbf (vac)	Thrust, lbf(SL)	Specific impulse, s (SL)	Specific impulse, Is, S (vac)	Propellants	Mixture ratio	Chamber pressure, psia	Nozzle area ratio
Titan I	Experimental	1	328,000	300,000	278	315/329	LOX/Kerosene	202	NA	8
	SLV	2	100,000				LOX/Kerosene	NA	NA	NA
Titan II	ICBM	1	473,800	430,000	258.8	285.2/309	NTO/Aerozine 50	1.930	795	8
		2	100,000				NTO/Aerozine 50	1.800	827	49.2
Titan III	SLV	1	527,000	460,000	258	296/318	NTO/Aerozine 50	1.925	823	12
		2	100,000				NTO/Aerozine 50	1.75	853	65
Titan IV	SLV	1	550,900	489,000	NA	303.5/316.2	NTO/Aerozine 50	1.91	854	16
		2	105,000				NTO/Aerozine 50	1.775	873	49.2

NA = not available to author

50% UDMH and 50% hydrazine originally conceived by Aerojet) for the Titan II. Later the Titan III and IV versions continued to use the same propellants. Aerojet selected Aerozene 50 as the fuel; it was a compromise between the performance, which is highest with pure hydrazine, and an acceptable freezing point, which is unacceptable with hydrazine only.

In 1960 work began on Titan II, which went into production. Figure 7.7-18 shows the twin booster engine of Titan II, Fig. 7.7-19 shows its flow diagram, and Fig. 7.7-20 shows the sustainer or second-stage engine. The turbine exhaust of the sustaine engine was sent through a hinge-mounted nozzle for roll control of the stage. Titan II was an ICBM, was installed in heavily armored underground silos and was operational with the U.S. Air Force between 1963 and 1987. When these Titan II missiles were decommissioned (beginning about 1984) from their service as a military deterrent, they became available as SLVs. There were some modifications to achieve this engine conversion.

There were other changes, besides the propellants, in going from Titan I to Titan II. The controls were simplified, the ignition system was eliminated (the new propellants were hypergolic), a solid-propellant cartridge was used to start the TP, and the injector was simplified. The oxidizer tank was pressurized by evaporated NTO, which was heated in a heat exchanger by turbine exhaust gas. The fuel tank was pressurized by a bleed from the fuel-rich GG gas from the turbine discharge and cooled with fuel in a heat exchanger. The cold helium gas pressurizing system with its very heavy tanks was no longer needed.

This is the one of the few large flying pump-fed production LPRE where both the oxidizer and the fuel were changed. One Titan I booster engine with a new fuel pump was later also ground tested with LOX/LH_2. This might have been the only large LPRE with this propellant, which used an impinging stream injector and not a set of coaxial injection elements, which are preferred in all of the other engines with this cryogenic propellant combination. Because a Titan I engine was operated also with NTO/aerozine 50, prior to the formal change to Titan II, it was one of the few known large LPRES that has been run with three different sets of propellants.

Unfortunately combustion instability was observed in a few engine tests. After the injectors were equipped with baffles, the combustion appeared to be stable. A method of testing for stability (by detonating of specified oriented explosive charges in or on the chamber as explained in Chapter 4.10) was successfully applied, as explained in Chapter 4.10. This method had been developed by a national effort as a rating technique, which was a response to the earlier stability problems with the Bomarc engine and several Rocketdyne engines.

A surprise in the first flight of Titan II was that the booster engine thrust and the chamber pressure oscillated by ± 150 psi at a relatively low frequency. This oscillation was subsequently identified as a POGO instability (pulsing or flow oscillations in the propellant feed lines to the engine) and remedied by installing spring-loaded surge chambers in the propellant feed lines. This is also discussed in Chapter 4.10. When the Titan II was used for launching the

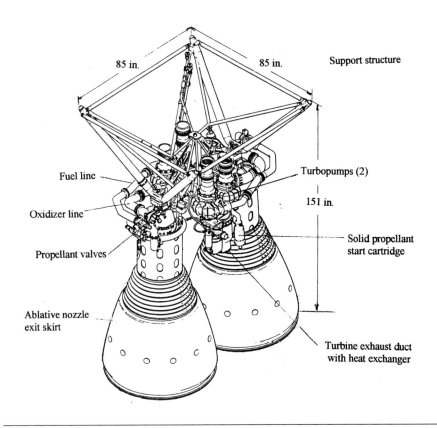

Fig. 7.7-18 Aerojet's Titan II booster dual engine LR87-AJ-5 with two TCs. It is 151 in. tall and has a 8:1 nozzle area ratio. Courtesy of Aerojet; copied from Ref. 6.

Gemini manned space capsule, the engines were modified by introducing some redundancies and additional safety features.

Titan III and IV are bigger, uprated SLVs specifically for heavier spaceflight payloads. Several changes were made in the engines to accommodate space launch requirements. The thrusts and the nozzle area ratios of the Titan III and IV engines were uprated progressively in several steps as seen in Table 7.7-1. Not all of the variations in Titan LPREs are contained in this table because there were intermediate versions This uprating caused the turbopumps, piping, and other parts to be heavier. Figure 7.7-21 shows an assembly line of TCs and TPs (with their exhaust heat exchangers) of the booster engine of Titan IV. In Titan II, III, and IV the turbine exhaust gas from the sustainer engine was used for roll control. The flow to the gas generator and to the propellant tanks (for pressurization) had a very simple set of controls, as shown in Fig. 7.7-19. There were only venturis to calibrate the flow and check valves to prevent back flow.

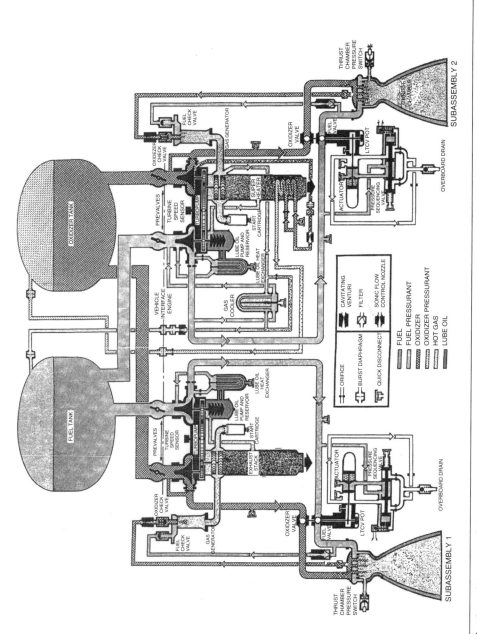

Fig. 7.7-19 Flow diagram of the Titan II dual booster LPRE. Courtesy of Aerojet; copied from Ref. 6.

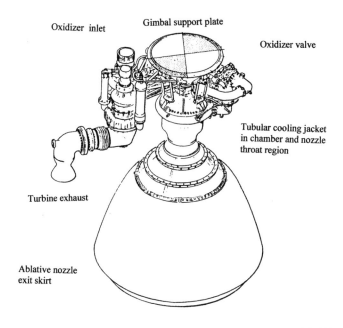

Fig. 7.7-20 Aerojet's Titan II sustainer engine LR91-AJ-5. Height is 110.6 in., and nozzle exit diameter is 64 in. Courtesy of Aerojet; copied with permission from Ref. 11.

Hot Separation

In the Titan II a "hot separation" was used to separate the Titan second stage during flight from the Titan expended first stage, which was then dropped off. In earlier SLVs and missiles the stage separation is initiated by shutting off the operation of the LPRE in the lower stage, and its thrust is allowed to decay to a very low value (essentially zero). The stages are then separated (usually with explosive bolts and by activating a separation mechanism, such as springs or small solid propellant rocket motors). The upper stage LPRE is started, when there is a reasonable distance (50 ft or more) between the stages, and the upper stage then accelerates and pulls away from the lower stage. All of this takes time (decay of lower-stage engine's thrust, stage separation, and startup of upper-stage engine), typically 5 or more seconds, depending on the specific vehicle. During this staging interval, there is very little forward thrust, and the Earth's gravity pull causes a slowdown or a decrease of the velocity of the upper stage, typically by an estimated 100 to 300 ft/s.

In a hot separation this time interval is shortened, and the velocity decrease caused by gravity is less. The first step is the command to shut down the first-stage engine. The second-stage engine is actually started when a predeter-

Fig. 7.7-21 Assembly line for the thrust chambers and turbopump assemblies of the Titan IV booster engines LR87-AJ-11. Courtesy of Aerojet; copied from company data sheet Ref. 6.

mined decay of the thrust of the first-stage engine is sensed. The initial hot exhaust gases from the TC of the second-stage engine are ducted overboard by flowing through symmetricall openings or blast ports in the Titan's interstage structure assembly. These blast ports can be seen in Fig. 7.7-22 of the Titan II. When the thrust of the second-stage LPRE exceeds the thrust of the first-stage engine, the explosive bolts are fired to release the two stages from each other, the two stages begin to separate, and the upper stage is propelled by its engine and pulls away. In the Titan II is a series of openings above the top of the upper oxidizer propellant tank of the lower stage, which allows the initial nozzle exhaust gases to escape. In other vehicles there can be doors in the skin of the interstage, and they open to let out the exhaust gas.

Hot separation depends on a reproducible shutdown thrust decay timing of the lower-stage engine and a reproducible rapid start transient of the upper-stage engine. The very hot flame of the upper-stage engine can cause some damage to the lower stage during this separation (such as local melting or deformation of the structure), but the lower stage is then no longer needed. Titan II and IV use a similar, but improved design for the hot separation of the first and second stages.

Hot separations have been used in several ballistic missiles and SLVs of Russia (formerly the Soviet Union) and also in the People's Republic of China. In the Soviet N-1 large moon launch vehicle, seen in Fig. 8.9-5, there is no vehicle skin around most of the interstage structure, but a series of open crossed columns, which can be seen in the picture; this allows the exhaust gas

Fig. 7.7-22 Launch of Titan II in 2003. It shows the rectangular and oval opening in the interstage structure between the first and second vehicle stages for letting the initial exhaust gases from the second-stage engine escape during hot stage separation. Courtesy of U.S. Air Force; personal communications, Lt. Col. D. Lileikis, U.S. Air Force, Titan Program Office.

to be dumped symmetrically. In the N-1 SLV the hot separation occurs not only between the first and second stages, but also between the second and third stages. Chapter 13 discusses the hot separation used in China's ballistic missiles and SLVs.

Small Thrusters

Aerojet has two different internal organizations working on small thrusters. The first is located at Sacramento, started probably around 1969, focused on small bipropellant thrusters, and is described briefly in the next paragraph. The second small-thruster organization was acquired from General Dynamics in 2002 and is located in Redmond, Washington. Its predecessor organizations started on bipropellant thrusters in 1958 and on monopropellant hydrazine thrusters in 1963. It and its heritage will also be described next. These small thrusters are a separate subcategory of LPREs, use storable propellants, and are usually supplied with integral propellant valves.

In their Sacramento, California, facility Aerojet has developed a number of small bipropellant thrusters for reaction control systems beginning approximately in the early 1970s.[3,13,14] The company, according to one of its employees, got into this business on a fluke. One of its chemists needed samples of actual rocket exhaust gas, and he arranged to build a small thrust chamber of about 5 to 8 lbf equivalent thrust for getting these samples. A customer, who might have known about this small chemistry thrust chamber, wanted a light weight version for his satellite reaction control system. So, with this background, a light weight small thruster with the storable propellants NTO/MMH and 5 lbf thrust was quickly developed.[3] One version is shown in Fig. 7.7-23, and it was flown in a satellite. After qualification 49 units were delivered. It has flown in the Milstar II satellite of Lockheed-Martin. An improved version of this 5-lbf thruster weighed 1.7 lbm, had an area ratio of 150, and a specific impulse of 310 s (vac). Later Aerojet developed a 25-lbf and a 100-lbf thruster. There were different models or versions of each, and in the late 1980s Aerojet was able to count 36 different variations of small thrusters (different area ratio, packages, valves, propellants, etc.) developed by the Sacramento group for small thrusters.[3] Some of these had platelet injectors.

At its Sacramento organization Aerojet also developed and tested several radiation-cooled rhenium thrusters protected from oxidation by an internal iridium coating. One such thruster was rated at 490 N or about 110 lbf, had a niobium nozzle extension with an area rtatio of 286, and a vacuum specific impulse of 321 s with NTO/MMH. It is not clear to this author if these rhenium thruster have been flown.

In the 1990s Aerojet obtained the job for the reaction control system (RCS) of the X-33 experimental reusable launch vehicle with Lockheed-Martin as the prime contractor.[6,*] Because the principal propulsion system of this research vehicle used a LOX/LH$_2$ linear aerospike, the customer wanted to use the same propellants in the RCS. As explained in Chapter 4.1, this is almost impossible with cryogenic liquid propellants at low thrust with irregular pulsing. The requirements then were for gaseous oxygen and gaseous hydrogen, but Aerojet was not able to make these work properly and predictably. So the requirements were changed to gaseous oxygen and stored high-pressure methane (with known properties), and this propellant combination proved to be feasible and more predictable under all operating conditions. The multiple

---

*Author's personal recollections, 1943–1946, and visits to Aerojet in 1985, 1987, and 2001.

Fig. 7.7-23 Small bipropellant thruster of 5 lbf thrust and valves, flown on the Millstar II satellite. Courtesy of Aerojet; copied from company data sheet Ref. 6.

thrusters were at a nominal thrust of 500 lbf each, and the methane gas was stored at 5000 psia in heavy tanks. Although there was some success in the development, this RCS system was never fully developed because the X-33 program was canceled for reasons unrelated to the RCS.

In August 2002, when the General Dynamics facility at Redmond, Washington, was acquired, Aerojet obtained a second organization doing work on small thrusters. It had a longer and illustrious history, more different small thrusters, and more production experience than the Sacramento group. The history and heritage of this part of General Dynamics has two principal roots.* The first is the *Rocket Research Company*, which started in 1963 to work with hydrazine monopropellant rocket engines and became a leader in this field, and the other is the *Marquardt Corporation*, which started in 1958 to work on small storable bipropellant rocket engines and became a major supplier in this business.* Their LPRE histories are discussed next. These small LPREs have found applications as ACSs, minor flight trajectory changes, apogee kick thrusters, stationkeeping, and other missions mentioned in Chapter 2.3. They

---

*Personal communications, J. Galbreath, D. Swink, C. Stechman, and O. Morgan, Aerojet Redmond Rocket Center and predecessor organizations, 1992–2004; C. Fisher, 2001.

sold mostly assemblies of thrust chambers with their control valves, but they also developed modules of several TCs in a single assembly or complete LPREs with pressurized feed systems.[15]

Rocket Research of Redmond, Washington, which started as an independent company in 1960, changed its name in 1977 to Rockor, Inc., and was acquired by Olin Corporation in 1985. In 1997 Olin spun off this rocket operation as a part of Primex Technologies, the successor of Rocket Research.*,† The other root organization, Marquardt Corporation of Van Nuys, California, was acquired by Kaiser and was called Kaiser–Marquardt starting in about 1990. In 1995 Kaiser–Marquardt purchased the product line of hydrazine monopropellant rocket engines that previously had been developed by Hamilton Standard, a division of United Technologies Corporation. At that time this product line was competitive with a similar product line in Primex. In June 2000 Primex acquired Kaiser–Marquardt and moved its operation into its own plant in Redmond, Washington. However, as a condition of the merger, mandated by the U.S. government, the Hamilton Standard product line had to be divested and was sold to ARC (discussed in Chapter 7.11). In January 2001 Primex itself was acquired by General Dynamics.* In turn this operation of General Dynamics was acquired by Aerojet in August 2002 and renamed Aerojet Redmond Rocket Center.

Rocket Research Company, one of the predecessor organizations, was founded in 1960, but became active in small rocket engines only in 1963.† In the early years there were critical R&D contracts to design, build, and test catalyst beds, understand scaling criteria, and evaluate the monopropellant hydrazine systems against potential bipropellant systems, such as, for example, the Miniteman III RCS. A major step forward in the history of these monopropellant thrusters was the development of a suitable catalyst for hydrazine decomposition. The work on the catalyst was actually performed by Cal Tech's Jet Propulsion Laboratory under NASA funding beginning in the late 1940s. It resulted in a good catalyst, and it was done in a joint effort with the Shell Development Company, who then produced the Shell 405 catalyst for many years. When Shell decided to go out of the business of making this catalyst, whose market had declined, the Aerojet center at Redmond, a major user of the catalyst, decided to produce and sell the catalyst beginning in 2003.

Figure 7.7-24 shows a typical hydrazine monopropellant thruster assembly developed by Rocket Research.[15] The original early designs of the catalyst bed were beset by technical problems, which caused poor performance, ignition delay, overpressures, short catalyst life, or progressive decreases of thrust.[7,‡] Remedies were concerned with the proper distribution of the liquid flow into the bed, as seen in Fig. 4.6-8 and avoiding catalytic attrition or physical loss

---

*Personal communications, J. Galbreath, D. Swink, C. Stechman, and O. Morgan, Aerojet Redmond Rocket Center and predecessor organizations, 1992–2004; C. Fisher, 2001.
† Author's visits to Aerojet Marquardt, Rocket Research, 1975–2001.
‡ Personal communications, J. Galbreath, D. Swink, and O. Morgan, of Aerojet Redmond Rocket Center and predecessor organizations, 1992–2004.

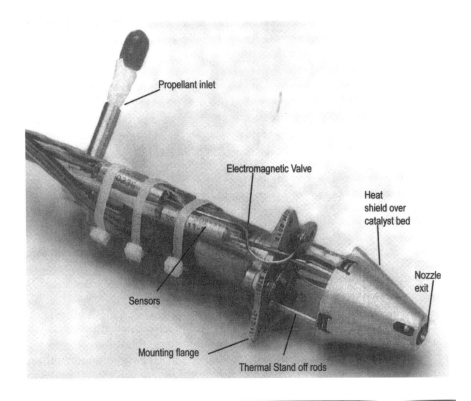

Figure 7.7-24 Monopropellant hydrazine thruster with electrical heaters, originally developed at Rocket Research. Courtesy of Aerojet; copied from Ref. 15.

of catalyst material from the motion, miniexplosions, and abrasion of the catalyst pellets. Catalyst activity did also decline because of the catalyst poisoning by impurities present in very small quantities in the hydrazine. It took Rocket Research some time to learn how to design catalysts and catalyst beds for good performance, long life, for short pulse width and multiple (up to 500,000) pulses. By using a thermal isolation structure between the valve and the catalyst bed, the valve electric insulation did not get hot enough to deteriorate. In the late 1960s the manufacturing process of hydrazine was changed to produce a product that was free of aniline and other impurities, which had been the cause of catalyst poisoning.

The first Rocket Research contracts for monopropellant hydrazine thrusters were for Transstage (25 lbf thrust, flown 1966) and thereafter for the ACSs of Titan.* Other contracts have since followed. Several sizes and types of monopropellant hydrazine thrusters were developed and qualified. Because of the high freezing point (34°F), it was found necessary to heat the thrusters, pip-

---

*Personal communications, J. Galbreath, D. Swink, and O. Morgan, of Aerojet Redmond Rocket Center and predecessor organizations, 1992–2004.

ing, tanks, and valves with electrical heaters. All LPREs with hydrazine propellant therefore have a number of electric heaters. Starting with a very cold catalyst bed showed erratic results, and in some designs the bed had to be heated to ensure acceptable initial performance. In the last few years the company has learned how to start a hydrazine monopropellant thruster at an ambient temperature of −30°C (−22°F).

In the late 1960s work was performed on radial flow catalyst beds and on throttling of the propellant flow and thrust. It soon led to the development of a 600-lbf rocket engine for the successful NASA Viking Mars Lander Project.[16] This engine is seen in Fig. 7.7-25. Three such engines were needed for a soft landing of the spacecraft on Mars. It had 18 nozzles because multiple exhaust jets allow a shorter assembly and would not kick up as much Martian dust as a larger single jet. It had a radial flow two-layer catalyst bed (shorter than an axial-flow bed), and it had to be sterilized (to prevent Earth bacteria from contaminating Mars). The propellant combination of NTO and MMH gave a relatively clean exhaust. Throttling controlled the rate of descent on Mars. This Mars lander, with a total thrust of 1,800 lbf, was the highest known thrust of a flying monopropellant engine.

The long life of these LPREs has been proven. For example the NASA Voyager Deep Space Probe was launched in 1977 with monopropellant hydrazine engines, and these were still working in space in 2002. The company has now a proven stable of catalytic monopropellants engines ranging in thrust from 0.1 lbf to over 300 lbf and a history of successful applications in approximately 60 different space launch vehicles and spacecraft.[15] As a customer, the author was involved in several of the 50-lbf thrusters and became acquainted with several of the key personnel. A few characteristics of four of their thrusters are shown in Table 7.7-2, and the data were dated about 2001. By 1999 Primex had delivered 10,000 such monopropellant thruster. The small thruster (0.2 lbf) had satisfactorily completed over 300,000 pulse cycles in ground tests. Two to five individual thrusters and their valves were often packaged into a subassembly or rocket engine module because this simplifies their vehicle installation or assembly, their checkout, and servicing. Examples are the ACSs for the Titan I and II and the Centaur upper stage of a space launch vehicle. Typical modules of two thrusters each can be seen in Fig. 7.7-26. The company also designed, developed, tested, and delivered complete LPREs using hydrazine monopropellant with a pressurized feed system and multiple thrusters. An example of a complete LPRE with hydrazine monopropellant is shown in Fig. 7.7-27. It features a simple blow-down feed system, pear-shaped tanks, and multiple thrusters. This engine was flown on the MSTI-3 satellite.

For the past eight years the company has investigated a relatively new synthetic propellant, namely, hydroxylammonia nitrate (HAN). This was mentioned in Chapter 4.1. HAN is environmentally more friendly, less toxic, has a lower freezing point, and might allow a somewhat higher monopropellant performance than the hydrazine currently being used. It can also be used as a bipropellant.[17,18] As of 2003, the investigation was continuing, and this propellant had not yet been installed in a flying vehicle.

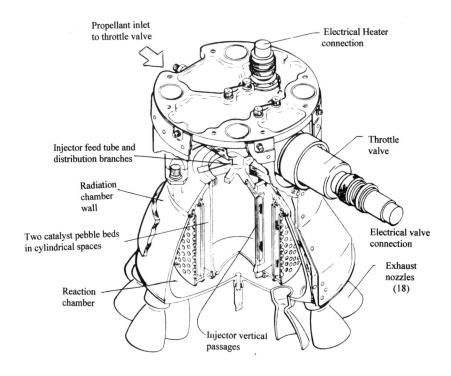

Fig. 7.7-25 Hydrazine monopropellant propulsion system for the Viking unmanned Mars landing mission with 18 nozzles and 600 lbf thrust developed at Rocket Research. Courtesy of Aerojet; copied with permission from Ref. 16.

Table 7.7-2 Characteristics of four Monopropellant Hydrazine TC/Valve Modules

Thrust, lbf nominal	Nozzle expansion ratio	Chamber pressure, psia	Specific impulse, s, (vac)	Max. number pulses tested	Cumulative firing time, h
0.20	100	340	220	745,000	40
1.2	74	175	227	6,000	4
7.2	40	84	225	12,300	No data
50	21.5	120	230	26,600	0.8

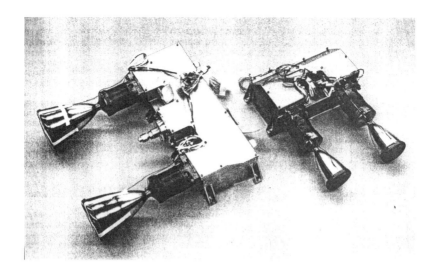

Fig. 7.7-26 Two modules, each with two monopropellant thrusters, which were used on Space Station *Freedom* for axial maneuvers and attitude control developed by Rocket Research. Courtesy of Aerojet; copied from Ref. 15.

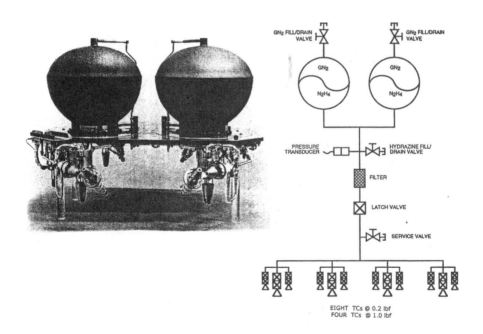

Fig. 7.7-27 View and diagram of a propulsion system with multiple small monopropellant thrusters used for the MSTI-3 satellite, first flown in 1994. Courtesy of Aerojet, adapted from Ref. 15.

Marquardt, the other predecessor of Primex, started in the LPRE business in 1958 by conducting a study of the future requirements for reaction control systems.[15,*] The report indicated a need for bipropellant (storable) small thrust chambers capable of operation in a gravity-free vacuum with many restarts. They then started in-house work on such a thrust chamber using nitrogen tetroxide and hydrazine. The long start delays of early thrust chambers were shortened by placing the propellant valves closer to the injector. In some small thrusters the nozzle axis is not in line, but at right angles to the chamber axis in order to accommodate space restrictions in the flight vehicle. Because there were then no suitable fast-acting commercial lightweight valves available, they initiated their own design of such valves. A few years later they mounted these valves directly on the titanium injectors so as to minimize the propellant volume between the valve seat and the injector face. This not only reduced the start delay to a practical minimum, but it also reduced the cutoff dribble or the total impulse during thrust decay and the potential postfire accumulation of mixed propellants in the chamber after cutoff or prior to restart.

Marquardt's first contract for a bipropellant rocket engine of 25 lb thrust was for the Advent satellite of General Electric.[15,*] This LPRE experienced some failures during development, but did pass the qualification tests. However for unrelated reasons the Advent satellite never flew. Marquardt experimented with a variety of liquid storable propellants. They selected NTO and MMH for their thrusters. However government requirements led Marquardt to also use Aerozine 50, a fuel mix of 50% hydrazine and 50% UDMH. In the last couple of decades, Marquardt learned how to use the same thruster with several different propellants, namely, hydrazine, UDMH, MMH, or a mixture of any of these.

The 100-lb thrust (R-4D) unit was originally developed for the Lunar Orbiter, which mapped the moon in 1964. It was a key historical product for the company. Modified versions of this thruster have been successfully employed in several applications, such as the Apollo Service Module (16 thrusters) or the Apollo Lunar Lander (16 thrusters). This R-4D has an 8-on-8 unlike doublet impinging stream injector with eight film-cooling holes and valves positioned very closely to the injector face. Two radiation-cooled versions of this thruster family with niobium (also called by old name columbium) chamber/nozzle walls and titanium injectors are shown in Fig. 7.7-28. The RD-4 has a nominal thrust of 110 lbf, a specific impulse of 312 s in a vacuum, and the R-1E has 25 lbf thrust, a specific impulse of 290 s at a nozzle area ratio of 100 to 1.0. The R-1E, which was flown in the space shuttle, has external thermal insulation, which reduces the radiation heat flux from the white glowing TC to surrounding heat-sensitive vehicle components; however, this insulation causes a significant rise in the chamber wall and nozzle wall temperatures. Table 7.7-3 gives data on a version of the RD-4 (110 lbf thrust), the R-1E (25 lbf), and also lists data of several other bipropellant thrusters of Marquardt.[15] All of the units

---

*Personal communications, C. Stechman, Aerojet Redmond Rocket Center and predecessor organization, 2001–2004.

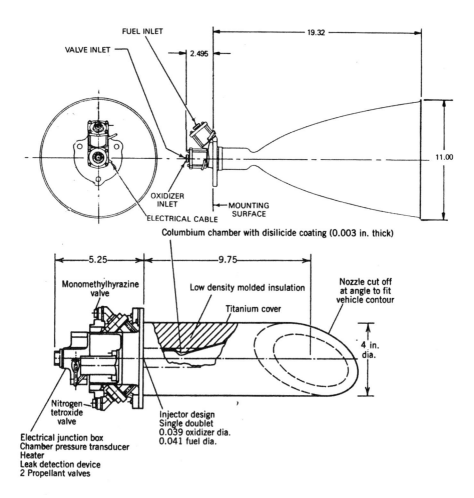

Fig. 7.7-28 Two examples of radiation-cooled thrusters with niobium chambers and nozzles. The chambers and nozzle throat regions will be glowing white hot. The lower sketch shows an insulated R-4 thruster with the nozzle cut off at an angle to fit the vehicle contour. Not drawn to same scale. Courtesy of Aerojet; copied with permission from Refs. 7 and 15.

listed show a mixture ratio (NTO to MMH flow) of 1.65 because at this value the volume of the NTO tank is equal to the volume of the fuel tank, and identical tanks can be used; this results in a cost savings. The optimum mixture ratio of 2.05 gives only 1 or 2 s more specific impulse. Usually these bipropellant TCs will operate satisfactorily over a mixture range from 1.0 to 2.4. Over the years Marquardt has developed a series of bipropellant thrusters ranging in thrust from about 1 lbf (5 N) to about 900 lbf (4000 N), with several materials, nozzle geometries, thermal insulation, durations, duty cycles, total

Table 7.7-3 Characteristics of Several Bipropellant Thrusters with Integral Valves

Thrust, lbf	Nozzle-exit-area ratio	Feed pressure, psia	Mixture ratio	Specific impulse, s (vac)	Minimum pulse width, s	Weight, lbm
2.2	150	220	165	285	0.007	1.38
5	100	220	165	289	0.013	1.48
110	100	360	165	305	0.6	8.8
25[a]	220	220	165	290	0.2	3.5
870[b]	Various	238	165	281	NA	22.6

[a] This vernier of the Space Shuttle Orbiter has been demonstrated for a cumulative life of 22.77 h and 300,000 pulses.
[b] This thruster of the Space Shuttle Orbiter has been demonstrated for a cumulative life of 15,319 s with the longest burn of 500 s.

number of pulses (over 100,000), and very short pulse widths (minimum cutoff total impulse of 0.01 lbf-s).[15,*]

Some of the smaller thrusters have been tested for over 100,000 cycles and one type of propellant valve for more than two million cycles in a vacuum environment.

Marquardt had significant development problems with the bipropellant engines, particularly at restart in space.* Burnout failures of the initial molybdenum chambers [and later niobium (or columbium) chambers with a disilicide coating, which had to be developed] occurred until the injector pattern was changed to include a small preignition chamber. At the low ambient temperatures of space, unexpected pressure rises (really explosions) occurred in the chamber during restart. Eventually this was traced to accumulations of unburned cold fuel mixed with some NTO vapor, which came into the chamber during the shutdown transient of the prior firing and from dribbled-out trapped propellant downstream of the propellant valves. The unburnt mix stuck to the chamber wall. This propellant layer exploded during restart, damaging the TC. The high-pressure spikes (4000 to 5000 psi) were eliminated by electrically heating the thruster to 70°F, when using MMH, and to 125°F, when using Aerozine 50. This heat evaporated the deposits at the low ambient pressures. There were also restart problems because after firing the titanium injector soaked up heat from the chamber and could reach 400 or 500°F before restart. This heat evaporated the initial inflow of NTO causing unsteady oxi-

---

*Personal communications, C. Stechman, Aerojet Redmond Rocket Center and predecessor organization, 2001–2004.

dizer flow with bubbles, very fuel-rich start mixtures, and hard starts. When the film cooling was added, the injectors did not get so hot, and this injector overheating problem, caused by heat soak-back, was solved. However film cooling reduced the performance slightly.

In 1973 and 1974 Marquardt built the LPREs for the reaction control of the Space Shuttle Orbiter.[19,20] Each shuttle vehicle had 28 reaction control thrusters (R-40) at 870 lbf thrust each and six vernier thrusters (R-1) at 25 lb thrust each. Many of these 34 thrusters were backup or spare units, and this redundancy enhanced the reliability. They are used for attitude control, some orbit maneuvers, orbit maintenance, rendezvous maneuvers, and some emergency maneuvers. Many have their nozzle cut off at an angle (also called scarfed nozzle) to fit the skin contour of the vehicle.

Marquardt pioneered in using different metals in the same radiation-cooled nozzle-exit section of high-nozzle-area-ratio thrusters, and this improved costs and reduced weight. Figure 4.6-15 shows a thruster with three different metals at an area ratio of 375 and a specific impulse of 322 to 324 s with MMH and 325 to 327 s with hydrazine.[21] A comparison of the melting points of these three different metals for TC walls is given in Fig. 4.6-13. By using titanium, which has a lower density than niobium or rhenium at the nozzle-exit region, which has much lower heat transfer, a weight savings can be realized. The rhenium material in the chamber and nozzle-throat region needs very little film cooling, and this allows a higher performance. By 2002 about nine of these three metal thrusters had been flying in space.

Other Liquid Propellant Rocket Engines

In the early 1990s Aerojet obtained the right to sell in the United States several of the Soviet rocket engines designed and originally developed by the Kuznetsov Design Bureau in the 1960s for the Soviet N-1 moon flight vehicle program, which was canceled. Two of these engines, the NK-33 and the NK-43 engines, were sold to *Kistler Aerospace Corporation* of Kirkland, Washington, for their unique Kistler SLV.[22] This vehicle has been designed to have three modified NK33 LPREs in its first stage and a modified NK43 engine in its second stage. At the time of the preparation of this manuscript, the Kistler SLV development was in progress. More information about these Kuznetsov engines is found in Chapter 8.9. Several dozen of these large LOX/kerosene engines were taken out of Russian storage and shipped to Aerojet. A firing of the NK33 at Aerojet in 1995 (339,400 lbf thrust, LOX/kerosene) is shown in Fig. 7.7-29. Aerojet developed a gimbal mount for these engines and also the electromechanical gimbal actuators. There is no coking (carbon deposit formation) in the cooling jacket or the turbine, which is supplied by an oxidizer-rich preburner. Aerojet also developed LPREs with small thrusters for the attitude control of the Kistler vehicle. One of the peculiar aspects of this deal with the Kuznetsov DB is the fact that the NK-33 and NK-39 have never flown. They were qualified and installed in a couple of Soviet N-1 vehicles, which were ready to fly, but the N-1 program's termination stopped further flights. However, these two engines

Fig. 7.7-29 Test firing of the Russian NK-33 LOX/kerosene LPRE in 1995 at Aerojet's facility in Sacramento, California. The power level was varied from 77 to 104%. The engine diameter is approximately 1.5 m. Courtesy of NPO Samara (formerly OKB Kuznetsov) and Aerojet.

are improved versions of earlier similar engines of the same thrust, which have flown. Aerojet also had negotiated an arrangement to obtain the Soviet D-57 LOX/LH$_2$ upper-stage engine (designed by the NPO Saturn/Lyulka Design Bureau in the 1960s and stored since about 1977; see Chapter 8.8) from the USSR for possible application to a potential U.S. space vehicle. However Aerojet did not find an application or a customer for it.

As a result of the Nike-Ajax upper-stage work, Aerojet was in a favorable position to become the supplier of sounding rockets. James van Allen, who discovered the magnetic radiation belts around the Earth known as the van Allen belts, was instrumental in persuading the U.S. Navy to support sounding rockets for upper-atmosphere research.

As a result, Aerojet began in early 1947 to develop the first of a series of complete sounding rockets, identified by the name *Aerobee*.[3] The first stage was a short-duration solid-propellant rocket motor, and the second stage was a modification of the Nike-Ajax LPRE. It has 2600 lbf thrust for about 45 s and used nitric acid and a mixture of 65% aniline and 35% furfural alcohol. A series of at least seven different sounding rockets were developed, some with larger diameter and more propellant and some with different thrust levels. Aerojet also built some all-solid-propellant versions of these sounding rockets. The Aerobee series was a successful production program. Production ended around 1979, and the last Aerobee flights were in the mid-1980s.

On several occasions Aerojet undertook joint programs with one of its competitors. For example, when Lockheed-Martin and ARC wanted to promote a modernized version of the Agena engine, they teamed with Aerojet in a program called *Agena 2000*. The objective was to develop and market an improved modernized version of the workhorse ARC Agena engine, which had been a successful upper-stage engine in the 1960s up to the 1980s. Aerojet and ARC did a joint redesign of this engine of 15,000 lbf thrust, and the testing was done at Aerojet in the 1990s.[23] Propellants were changed from nitric acid/UDMH to NTO/MMH. The specific impulse was expected to be 336 s (vac) using a high nozzle area ratio, which would be an improvement of about 12%. However no application of this improved Agena engine could be identified, in part because LOX/LH$_2$ upper-stage engines were then available and their performance was vastly superior. Aerojet also teamed up with Pratt & Whitney in joint studies of advanced engines.

Concluding Comments

Aerojet has had a distinguished history in both solid propellant rocket motors and liquid propellant rocket engines. Aerojet has been a successful and respected developer of a large variety of different LPREs. In a way the history of LPRES is mirrored in Aerojet's successive engines and their technical advances. Their stable of engines ranges from very low to very high thrusts, covers most of the common applications, and makes use of different propellants. Aerojet probably has provided their LPREs to more different customers than any other U.S. LPRE company.

They have developed, built, and flown LPREs for many different applications, such as sounding rockets, JATOs, aircraft superperformance, anti-aircraft missiles, first second and third stages of SLVs, first and second stages of ICBMs, various LPREs for the Apollo Moon mission and the space shuttle, a variety of small thrusters for ACS, space maneuvers, and a variety of propellants. With the prior decommissioning of the Titan II ICBM, and the expected phase-out of the Titan SLV and the space shuttle, the company's major LPRE business in large LPREs is coming to an end. There is currently no other major large LPRE program to take its place. The small thruster business, which seems to be continuing, various studies (in-house or together with another company) and some limited R&D component projects, seems to be the current work at Aerojet in LPREs. Their level of activity in large LPREs has greatly diminished.

References

[1]Winter, F. H., and James, G. S., "Highlights of Fifty Years of Aerojet, a Pioneering American Rocket Company, 1942-1992," *History of Rocketry and Astronautics*, American Astronautical Society History Series, Univelt, San Diego, CA, Vol. 22, 1998, Chap. 4, pp. 53–103.

[2]Ehresman, C. M., "The First Operational Liquid Rocket for Aircraft Developed and Produced in Quantity in the United States," AIAA Paper 2000-3280, July 2000.

[3]Umholtz, P., "Liquid Rocketry," *Aerojet—The Creative Company*, edited by B. L. Dorman, S. Feldbush, R. Gordon, C. Hawk, et al., Aerojet History Group, 1995, Chap. 3.

[4]Malina, F. J.,"The U. S. Army Jet Propulsion Research Project GALCIT Project No.1, 1939-1946: A Memoir," *History of Rocketry and Astronautics*, American Astronautics Society History Series, Vol. 7, Univelt, San Diego, CA 1986.

[5]Ehresman, C. M., "A Big Boost from Rocket Assist for the Boeing B-47 Bomber," AIAA Paper 98-3192, July 1998.

[6]"Spacecraft Propulsion for near Earth Astoroid Rendezvous," "Delta Second Stage Engine (AJ10-118K)," "Engines for Titan II, III and IV," "Divert and Attitude Control Systems," "AJ26-58 engine (modified Russian NK33A)," "Hope-X Orbital Maneuvering Engine," "Kistler K-1 Launch Vehicle," "Agena 2000," and "ALAS or Advanced Liquid Axial Stage," Aerojet Data Sheets or Product Pamphlets, Aerojet Corp., Sacramento, CA, 1990–2004.

[7]Sutton, G. P., and Biblarz, O., *Rocket Propulsion Elements*, 7th ed., Wiley, Hoboken, NJ, 2001.

[8]Mueggenburg, H. H., and Repas, G. A., "A Highly Durable Injector Face Plate Design Concept for O2/H2 Propellants," AIAA Paper 90-2181, July 1990.

[9]LaBotz, R. J., "Development of the Platelet Micro-Orifice Injector," *35th IAF Congress*, International Astronautical Federation, 1984.

[10]Osborn, G. H., Gordon, R., Coplen, H. L., and James, G. S., "Liquid Hydrogen Rocket Engine Development at Aerojet," *History of Rocketry and Astronautics*. edited by C. Hall, American Astronautical Society History Series, Vol. 7, Part 2, Univelt, San Diego, CA 1986, p. 279.

[11]Adams, L. J., "The Evolution of the Titan Rocket—Titan I to Titan II," *History of Rocketry and Astronautics*, edited by J. D. Hunley, American Astronautical Society History Series, Vol. 19 Univelt, San Diego, CA 1990, Chap. 7.

[12]Meland, L.C., and Thompson, F. C., "History of the Titan Liquid Rocket Engines," AIAA Paper 89-2389, June 1989.

[13]Rosenberg, S. D., and Schoenman, L., "New Generation of High Performance Engines for Space Propulsion," *Journal of Propulsion and Power*, Vol. 10, No. 1, Jan.-Feb. 1994, pp. 40–46.

[14]Schoenman, L., Rosenberg, S. D., and Jassowski, D. M., "Test Experience, 490 N High Performance [321 sec Specific Impulse] Engine," *Journal of Propulsion and Power*, Vol. 11, No. 2, Sept.–Oct. 1995.

[15]Marquardt Corp., Rocket Research, Primex Space Systems, and General Dynamics, Product Brochures and Technical Data, 1998–2003.

[16]Morrisey, D. C., "Historical Perspective: Viking Mars Lander Propulsion," *Journal of Propulsion and Power*, Vol. 8, No. 2, March – April 1992.

[17]Morgan, O. M., "Monopropellant Selection Criteria—Hydrazine and Other Options," AIAA Paper 99-2595, June 1999.

[18]Zube, D. M., Wucherer, E. J., and Reed, B., "Evaluation of HAN-Based Propulsion Blends," AIAA Paper 2003-4643, Summer 2003.

[19]Stechman, R. C., and Sumpter, D., "Development History of the Apollo Reaction Control System Rocket Engine," AIAA Paper 89-23, July 1989.

[20]Nandiga, M., Dorantes, A., and Stechman, C., "Design, Fabrication and Testing of Extended Life Space Shuttle Vernier RCS Engines," AIAA Paper 2003-4929, 2003.

[21]Krismer, D., Dorantes, A., Miller, S., Stechman, C., and Lu, F., "Qualification Testing of a High Performance Bipropellant Rocket Engine Using MON-3 and Hydrazine," AIAA Paper 2002-4775, 2002.

[22]Limerick, C., Andrews, J., Starke, R., Werling, R., Hansen, J., and Judd, C., "An Overview of the K-1 Reusable Propulsion System," AIAA Paper 97-3120, July 1997.

[23]Gribben, E., Driscoll, R., Marvin, M., Wiley, S., Anderson, L., and Fischer, T., "Design and Test Results on Agena 2000 Main Axial Engine," AIAA Paper 98-3362, July 1998.

7.8 The Boeing Company, Rocketdyne Propulsion and Power*

This company (henceforth abbreviated simply as Rocketdyne in this book) has been the largest commercial LPRE organization in the United States and the largest in the non-Soviet world. It has developed and produced LPREs of every kind, with the exception of small monopropellant engines. It was started in 1945 as a section of the Aerophysics Laboratory of the aircraft company North American Aviation (NAA).[1] This company was then well known as a designer and manufacturer of airplanes, and its main operation was located at an airport, which later became the Los Angeles International Airport. In 1967 NAA merged with Rockwell Corporation and was renamed North American Rockwell. In 1973 the name of the company became Rockwell International Corporation. In 1955 the liquid-rocket-engine work had grown so much that it became a separate division of the parent company, and at that time the name Rocketdyne was adopted. Sam K. Hoffmann became its leader, and he guided Rocketdyne through its busiest period until 1970, when he retired. Rocketdyne also was in the solid-rocket-motor business between 1959 and 1978, worked on chemical lasers, did some work in the electrical propulsion and nuclear propulsion, and was, and still is, active in the space power supply business. Together with some other divisions, Rocketdyne was sold to the Boeing Company in December 1996, when the name listed as the title of this chapter was adopted for the organization concerned with LPREs. As already mentioned Rocketdyne had about 20,500 employees at its busiest period (1964), and all except 600 were in the LPRE business. In 2003 the LPRE work had less than 2000 people.

In its 60-year history Rocketdyne had flown and/or put into production about 19 large LPREs (1500 to 1,500,000 lbf thrust) and 12 smaller LPREs (1.0 to 1500 lbf); several of these had between one and five major redesigned or upgraded models, and they were not counted as a separate LPRE. Up to 1 June 2001, Rocketdyne engines had boosted 1516 rocket space vehicles. In addition Rocketdyne developed and tested about 40 large experimental engines or thrust chambers and approximately 35 small ones, all aimed at demonstrating some particular feasibility, reliability, or advance in technology. Like several other U.S. LPRE companies, it started out with the testing of small thrust chambers (TCs) and JATO units.[1,2] In the last two decades this company has pioneered the use of new software and computer programs for engine design, engineering analysis, test data recording, analysis and display, computer-aided design, computer-aided manufacturing and engine controls. The author was fortunate to have been part of that organization beginning in 1946 until 1975.

---

*While this book was being edited, Pratt & Whitney, a United Technologies Company, offered to buy the Rocketdyne Propulsion organization from The Boeing Company. The new company would be called Pratt & Whitney-Rocketdyne. As of late July 2005 the negotiation for this acquisition had not been completed.

Early Efforts, 1945 to 1949

North American Aviation had the foresight to anticipate the future need for rocket-propelled guided missiles and initiated some in-house studies. In 1945, when wartime German missile developments became known, NAA decided to enter into this new field. They started the Technical Research Laboratory, which was quickly renamed the Aerophysics Laboratory and began to retrain some of their own people and to hire engineers with at least some experience in fields like gyroscopes, supersonic aerodynamics, rocket-propellant chemistry, or high-temperature materials. The first propulsion section was formed in early 1946 headed by Edward M. Redding. They also started sections on guidance and control, flight-vehicle design and testing, and supersonic aerodynamics. By 1945 the aircraft business had shrunk from a wartime high of 90,000 employees to 5000, and they kept the organization together by developing, testing and selling a small six-seat executive aircraft, on which they lost money. NAA was eager to work in the new missile field. The company bid and won a major contract to develop a surface-to-surface missile, known as Project MX-770. Phase I was a study of German missiles and LPREs, including the V-2, phase II included building and ground testing a copy of the V-2 LPRE and the development and launching of experimental small missiles, and phase III included an entirely new, uprated LPRE and the development of a long-range missile.

This author joined the new organization in early 1946 as a rocket propulsion engineer. In those days it was difficult to find experienced technical personnel, and the author's 3.3 years of rocket experience at Aerojet was not matched or exceeded by anyone else in the new organization at the time. The author became quickly involved in designing small thrust chambers (50, 200, and 300 lbf thrust), which were tested at a newly established and improvised company test facility. It was built in a fenced-off portion of the company's east parking lot in 1946.[1,2] The first initial test cell was crude, but was soon refined with better safety provision and better and more measurements. By the time the test facility closed in 1949, there were seven test cells; some were used for thrust chamber firing, gas-generator testing, pressure testing, and propellant testing. The laboratory pioneered in adapting available instruments and data recorders for measuring LPRE parameters and in calibrating these instruments.

Early tests were conducted at this east parking lot facility with the propellants used in the V-2 (LOX and 75% ethyl alcohol with 25% water).[2] Through the government the company obtained a 25AL-1000 and a 38AL-1500 JATO units, developed and delivered by Aerojet, and they were fired. It was a great learning experience. Simple water flow tests were conducted on the German V-2 gas generator (75% hydrogen-peroxide monopropellant), which was also run with actual propellants. Different injector designs were experimentally evaluated in a 1000-lb thrust chamber, including effects of changes in chamber pressure, propellant composition, and mixture ratio. Other propellants, including high-energy propellants, were also tested in small TCs, namely, nitric acid with hydrazine, fluorine with hydrazine, hydrogen peroxide as a mono- and bipropellant, and chlorine trifluoride with hydrazine.

In 1946 NAA started to look for a remote site near Los Angeles for building a new set of test stands, laboratories, and facilities suitable for larger LPREs, which were then expected to be needed in late 1949.[1] The selected location was in the Santa Susana mountains near Los Angeles, and in 1947 NAA was granted a county permit to build and use such a facility. The first large test stand was sized to take up to 1,000,000 lb of thrust firing vertically down or horizontally, and the design was based in part on earlier known test facilities in Germany and at General Electric (Malta, New York). The first water flow tests were run of a U.S. copy of the German V-2 thrust chamber in late 1949. Your author was the engineer on these tests.

NAA needed some experience in launching and flying guided missiles and to make and record in-flight measurements of the guidance and control systems and the rocket engine.[1] In 1947 the company designed and built 15 simple small guided vehicles, called NATIV (North American Test Instrumented Vehicle). The propulsion was a rocket engine from Aerojet, a modified version of the upper-stage WAC Corporal engine at 2600 lbf thrust using nitric acid and aniline. NAA developed the carbon jet vanes for the vehicle's flight control in pitch, yaw, and roll, which was the steering concept used in the V-2. Several different types of carbon vanes had been experimentally evaluated in the exhaust of the rocket engine at the east parking lot test facility. The first flight was at the Holloman Air Force Base in New Mexico in May 1948. Only four of these NATIV test vehicles were launched, and they provided good experience in launching and flight data recording and evaluation. By 1949 the concepts for guiding and controlling a long-range missile had advanced sufficiently so that the NATIV's guidance system tests were no longer relevant.

Large Liquid Propellant Rocket Engines, Early Activities

Rocketdyne developed more large LPREs than any other company (except for USSR's Energomash). The first effort was to *copy, build, and flow test three American copies of the German V-2 rocket engine*. Your author was responsible for the effort of copying the thrust chamber of the V-2. Figure 7.8-1 shows the U.S. engine copy next to an original German V-2 engine. The U.S. copy (on the left) is shorter and more compact. The turbine exhaust pipe is missing in this picture. It was not an exact copy because it used available U.S. materials (similar, but not identical to the German materials) and U.S. instead of German standard sheet thicknesses. A close look will show the U.S. copy had standard U.S. pipe sizes and fittings and U.S. standard electrical connectors or components. The Rocketdyne-made copy was water flow tested on the new test stand. Just as this U.S.-built copy of the V-2 TC was ready to be hot fired with propellants, it was pulled off the only existing high-thrust vertical Rocketdyne test stand in favor of the first U.S.-designed large (Redstone) thrust chamber, which then also was ready to test.

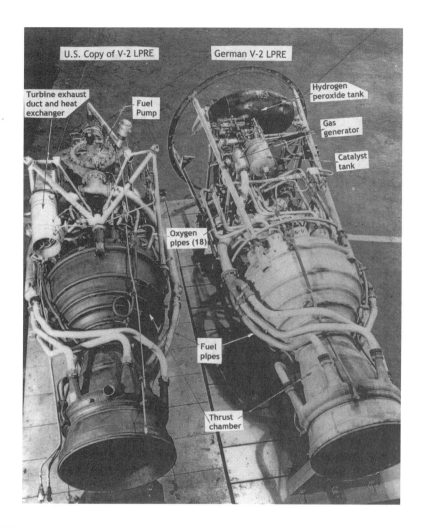

Fig. 7.8-1 Comparison of U.S. made copy (left) and the German V-2 rocket engine (right). One of these engines would need to be turned 180 deg on its axis to have the same components on top as the other engine. Courtesy of The Boeing Company, Rocketdyne Propulsion and Power, Rocketdyne photo.

Table 7.8-1 gives summary data for several of the large Rocketdyne booster and upper-stage rocket engines. Each will be discussed in this chapter. The table was meticulously put together by Vince Wheelock of Rocketdyne. For engines, which had several models in the same engine family, the data for the last or most recent version are listed in this table. The dates for the design year and the first flight give an idea of the historical sequence between engines.

The first indigenous large engine development effort was a pump-fed LPRE of 75,000 lb thrust, which soon became known as the engine for the Army's

Table 7.8-1 Performance Data for Selected Large Rocketdyne Production Engines/Engine Families

Missile/Launch-Vehicle Application	Redstone MRBM and SLV[b,c]	Navaho Cruise Missile	Atlas ICBM and SLV	Jupiter IRBM and SLV	Thor and Delta IRBM and SLV	Saturn I and IB SLV	Saturn V SLV	Saturn IB and V SLV	Space Shuttle SLV	Delta IV SLV
Engine family designation (data from last-in-family engine)	A-6 and A-7	G-26	B-2C, MA-1, MA-2, MA-3, MA-5, and MA-5A	S-3D	S-3E, MB-1, MB-3, RS-27, and RS-27A	H-1 A, B, C, and D	F-1	J-2	SSME Phase I and II Block I, IA, II, and IIA	RS-68
Initial family engine design year	1948	1950	1952	1953	1953	1958	1959	1960	1972	1997
Thrust (1000 lbf)										
Sea level	78	240	430 (B) 60 (S)	150	200	205	1,522	DNA[i]	374 (100%)	650
Vacuum	89	278	480 (B) 84 (S)	174	237	237	1,748	230	470 (100%)	745
Specific impulse, s										
Sea level	218	229	265 (B) 220 (S)	248	255	263	265	DNA	361 (100%)	365
Vacuum	249	265	295 (B) 309 (S)	288	302	295	305	425	452 (100%)	410
Oxidizer	LOX	LOX	LOX	LOX	LOX	LOX	LOX	LOX	LOX	LOX
Fuel	Alcohol (75%)	Alcohol (92.5%)	RP-1	RP-1	RP-1	RP-1	RP-1	LH_2	LH_2	LH_2

Mixture ratio (oxidizer/fuel)	1.354:1	1.375:1	2.25:1 (B) 2.27:1	2.4:1	2.24:1	2.23:1	2.27:1	5.5:1	6.03:1	6.0:1
Chamber pressure, (psia)	318	438	719 (B) 736(S)	527	700	700	982	717	2,747 (100%)	1,460 (100%)
Nozzle area ratio (exit/throat)	3.61	4.6:1	8:1 (B) 25:1 (S)	8:1	12:1	8:1	16:1	27.5:1	69:1	21.5:1
Nominal flight duration	121	65	170 (B) 368 (S)	180	265	150	165	390 (S-II) 580 (IVB)	520	400 max.
Dry mass, (Pounds Mass)	1,478	2,501	3,336 (B) 1,035 (S)	2,008	2,528 2,041 (D)	2,010 (C)	18,616	3,454	7,774	14,850
Engine cycle	GG	GG	GG	GG	GG	GG	GG	GG	SC	GG
Gimbal Angle, deg	None	None	8.5	7.5	8.5	10.5	6	7.5	11.5	10 -MPL 6 -FPL
Diameter/ width, in.	68	77	48 (B TC) 48 (S)	67	67	66	149	81	96	96
Length, in.	131	117	101 (B) 97 (S)	142	149	103	230	133	168	205
Operating temp. limits, F	−25 to +110	−20 to +110	−30 to +130	+40 to +130	0 to +130	0 to +130	−20 to +130	−65 to +140	−20 to +130	−20 to +140

(continued)

Table 7.8-1 Performance Data for Selected Large Rocketdyne Production Engines/Engine Families—(continued)

Missile/launch-Vehicle Application	Redstone MRBM and SLV	Navaho Cruise Missile	Atlas ICBM and SLV	Jupiter IRBM and SLV	Thor and Delta IRBM and SLV	Saturn I and IB SLV	Saturn V SLV	Saturn IB and V SLV	Space Shuttle SLV	Delta IV SLV
First flight date in family	08-20-1953	11-06-1956	06-11-1957	03-01-1957	01-25-1957	10-27-1961	11-09-1967	02-26-1966	04-12-1981	11-20-2002
Comments	---	---	---	---	---	---	---	Restart in space	Throttleable, power level 67% to 109%	Throttleable, power level 57 to 102%

Table prepared by Vince Wheelock of the Boeing Company, Rocketdyne Propulsion and Power, August 2003. Data from last in-family engine.

MRBM = medium range ballistic missile.
SLV = space launch vehicle.
ICBM = intercontinental ballistic missile.
IRBM = intermediate-range ballistic missile.
A&C = inboard engines.
B&D = outboard engines.
B = booster.
DNA = does not apply.
S = sustainer.
GG = gas generator engine cycle.
SC = staged combustion cycle.
MPL = minimum power level.
FPL = full power level.

Redstone ballistic missile. The missile was built by Chrysler, but some of the vehicle's engineering was done by the Germans, formerly of Peenemünde under the leadership of Wernher von Braun. The Rocketdyne-developed Redstone engine is shown in Fig. 7.8-2, and some parameters are listed in Table 7.8-1 (Refs. 3–8). Some people consider it to be an upgraded V-2 engine because it used the same propellants (liquid oxygen and 75% ethyl alcohol) and had some similar features. Like the V-2 engine, it had a heavy steel sheet, double-walled regenerative cooling jacket with supplementary film cooling, the turbopump had an aluminum turbine between the aluminum fuel and the oxidizer pumps, and jet vanes were used for thrust vector control during powered

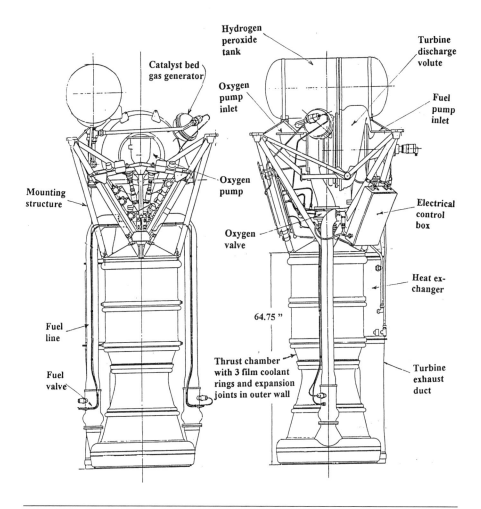

Fig. 7.8-2 Two sketches of Redstone rocket engine, the first large U.S. LPRE. Courtesy of The Boeing Company, Rocketdyne Propulsion and Power, Copied with permission from Ref. 3.

flight. Although the carbon jet vanes eroded severely during powered flight and caused a small loss (1 to 3%) in performance, they did work well, and historically they provided the first method of thrust vector control (TVC) in the United States as well as Soviet missiles as discussed in Chapter 4.9.

However, these were some of the significant improvements and differences with the V-2 LPRE.[3] The Redstone engine thrust and chamber pressure were higher, namely, 78,000 lbf vs 56,000 lbf and 317 vs 220 psia. For comparison some data and illustrations of the V-2 engine are in Chapter 9. Here are some of the other changes, when compared to the V-2: It had a new type of large diameter (flat surface) injector, a concept that was originated a few years earlier by the General Electric Company, as seen in Fig. 7.3-3. It was a stainless-steel forging with machined internal circular grooves at the face, radial internal fuel supply passages, and a bolted-on dome to distribute the oxidizer flow through axial holes to alternate circular grooves. Alternate rings had drilled holes for fuel injection and for oxidizer injection, and the flat rings were brazed into the face of the grooves. The flat surface of all of the rings formed the face of the injector as seen from the combustion chamber. The hole pattern in the rings had doublet self-impinging pairs of liquid jets, as depicted in Fig. 4.3-13. Extra film-cooling holes were placed at the periphery of the injector. The thrust chamber and nozzle entrance did have three sets of film-cooling holes in the wall (as in the V-2), but the locations were not the same, and the percent of fuel used for film cooling was much less. The chamber geometry was cylindrical and not pear shaped. The new thrust chamber gave better combustion efficiency and performance than the V-2 engine even though the relative chamber volume was less (L of 60 in. vs 90 in. for the V-2). The engine control system was more sophisticated. The gas generator used hydrogen peroxide with a solid catalyst bed, which is shown in Fig. 4.5-2, instead of the liquid permanganate solution of the V-2. There were two different versions of the Redstone engine, with a key difference in the fuel piping.

The first hot firing of the new large thrust chamber (with a pressurized test stand feed system) occurred in January 1950, which at 75,000 lb was then the highest rocket thrust in the United States.[2] Your author was the designer and the development engineer for this TC. It was part of the job to inspect the chamber and injector for discoloration or burned spots immediately after a test run. To do this, I had to stand on a short ladder with a flashlight in hand and squeeze through the tight 15.5-in.-diam nozzle into the chamber. There were fumes in the chamber from the denatured (poisoned) alcohol, and I quickly became drunk and got a big hangover from breathing these fumes all on company time. Needless to say, I used a fresh air supply on my next inspection.

The first static test of the complete Redstone engine took place in late 1950, and the first flight was in August 1953. It was part of the first U.S. ballistic missile to become operational, and the missiles were also deployed over sea in June 1958. This engine propelled a Redstone missile on 31 July 1958 at Johnston Island in the Pacific; it was carrying a live nuclear warhead to its first high-altitude detonation (at altitude of 47.7 miles). This engine also launched the first U.S. satellite (Explorer on 31 January 1958). A series of spaceflights of a modified Redstone missile with solid-propellant upper stages was launched as the Juno SLV. It had several flight failures, but they were not related to the large booster engine. The modified Redstone missile also launched two U.S. astronauts in their Mercury capsule on their first suborbital space flights in 1961.

The Army used the Redstone as a mobile missile, which means it could be launched from many different locations, but it required some special support equipment and more than a dozen special road vehicles. The Redstone missile needed a large (10-ton) launch erector/transport trailer with a tow truck, two heavy liquid-oxygen tank trucks with transfer pumps, one alcohol fuel tank truck with transfer pumps, one hydrogen-peroxide trailer, a warhead transporter, a launch control/computer vehicle, trucks for power generation, air compressors, accessories, troop transport, a small repair shop, and a fire engine for handling potential spills or fuel fires. Not all of these road vehicles relate to the LPRE. Similar support vehicles were used earlier by the Germans for their mobile V-2 missile and later by the Soviets for the Scud SRBM and by the Iraqi in their uprated Scud missiles during the Desert Storm war.

There are perhaps a dozen occasions in the history of LPREs when the same production large engine has been operated with different propellants.* To increase the flight performance or the payload, the Army ran this Redstone engine with "Hydyne," a mixture of UDMH and triethylene amine concocted at Rocketdyne, instead of the 75% alcohol fuel. The thrust was 10% higher, and the specific impulse went up almost as much. However because of the fuel's toxicity, this fuel was not used in any flights or the operational Redstone missiles.

Large Liquid Propellant Rocket Engines with Tubular Thrust Chambers

The second high-thrust engine developed by Rocketdyne was for the *Navaho G-26* intercontinental boost-glide cruise missile. The engine propelled the booster stage, which was pushing the upper stage driven by a supersonic ramjet.[5-10] This LPRE, shown in Fig. 7.8-3, had two fixed (not gimbaled) thrust chambers of 120,000-lb sea-level thrust each and two new turbopumps. It had the same propellants as the Redstone engine, namely, liquid oxygen and ethyl alcohol, but at 92.5% fuel concentration and not 75%, which caused a somewhat improved performance, and a conical exhaust nozzle exit. It was a large LPRE, the first to fly with tubular thrust chambers (see Chapter 4.3), the first with large geared turbopumps, which had a gear case and oil lubrication, and a single gas generator supplying both TPs with warm gas. The new TP is similar to the one shown in Fig. 4.4-2. The gear transmission case in the TP allowed the turbine to rotate much faster than the pumps, which allowed better efficiencies and consumed less GG propellant. The gas generator was a scaled-up version of the Redstone GG and used hydrogen-peroxide decomposition for cre-

---

*Personal communications, V. Wheelock, J. Halchak, H. Minami, and J. Lincoln, retirees of The Boeing Company, Rocketdyne Propulsion and Power, Rocketdyne, 1990–2004.

Fig. 7.8-3 View of Navaho G-26 dual-TC LPRE on test stand. The individual tubes of the cooling jacket are visible between the reinforcing bands. Courtesy of The Boeing Company, Rocketdyne Propulsion and Power; copied with permission from Ref. 3.

ating the warm gas. Both turbines were supplied from a single GG. This Navaho engine is the only known large flying LPRE, without any thrust-vector-control provisions; there were no gimbals, vernier TCs, or jet vanes. The upper stage was ramjet propelled, and the missile flew only within the atmosphere. Flight-path control was achieved by aerodynamic surfaces on the Navaho flight vehicle. This is a very effective method of thrust vector control, but it works only while flying through the Earth's atmosphere.

This engine program started in 1949, passed qualification tests, and was first flown in 1956. The rocket engine propelled the upper stage satisfactorily

11 times. When it first flew, it was then the highest thrust U.S. LPRE at 240,000 lbf at sea level.

Starting of the Navaho engine was accomplished by using the gas-pressurized onboard hydrogen-peroxide tank to initiate a low flow to the GGs, which in turn started the rotation of the turbopumps and by activating pyrotechnic igniters in the two TCs. During the early engine development tests, it required about 16 s from the start signal to reach 90% of rated thrust.[3] With full automation a modern engine of this size would take only perhaps 2 to 4 s. The long start time was caused in part by the tank pressurization being done after the start command was given and in part caused by the requirement for the launch or test operator to manually flip two switches. Like the V-2, the Navaho had a low-flow preliminary start stage and solid-propellant igniters in the TCs and the GG, which would create hot gas for up to 30 s. He pushed the start switch only after he had visually checked the instrument readouts for the initial pressures in the propellant and gas storage tanks and the voltage of the igniters. After he observed the flame of the preliminary stage burning for up to 5 s and approved it, the operator would then at his discretion manually push the full start switch, which caused the hydrogen-peroxide flow to increase, the fuel pressure and the oxidizer pressure to rise, and the thrust to climb. Before the engine development was completed, this start sequence became more automated and much shorter.

As said before, the Navaho G-26 TC was Rocketdyne's first large TC with tubular construction in the cooling jacket to fly, and at the time it was the highest thrust set of two TCs in the United States. Figure 4.3-4c shows a typical cross section of a cooling jacket using formed tubes. Figure 4.3-5 shows a TC with a tube-type cooling jacket similar to the Navaho TC, and Chapter 4.3 discusses this and other cooling-jacket constructions and their fabrication techniques. The Navaho TC used stainless-steel tubes and a double-pass flow scheme with downflow in each second tube and upflow in the in-between or alternate tubes. The relatively thin wall of the tube has a relatively small temperature gradient compared to the thick walls of the prior double-wall scheme, and this reduces the thermal stresses and the formation of cracks in the nozzle-throat region upon cooldown. This tubular cooling-jacket design has been used by Rocketdyne in all of its large subsequent LPREs with regenerative cooling, either for the full thrust chamber (as with Thor, Jupiter Atlas, H-1, or J-2 LPREs) or for a part of the cooling jacket (as with the SSME and the RS-68, described in this chapter). Tubular large TCs have been developed by RMI (Chapter 7.2), General Electric (Chapter 7.3), Aerojet (Chapter 7.7), Mitsubishi Heavy Industries in Japan (Chapter 11), SEP in France (Chapter 10), DASA/Astrium in Germany (Chapter 9), and Rolls Royce in Britain (Chapter 12).

An *enlarged Navaho booster fight vehicle* with three uprated engines at 135,000 lb thrust each (a total of 405,000 lbf) was the next major project. Its engine was designed in 1954, was fully developed, but the missile program was canceled before the engine could fly. It had three novel attributes: the TCs were hinged, the first at Rocketdyne, and the fuel was changed from alcohol to

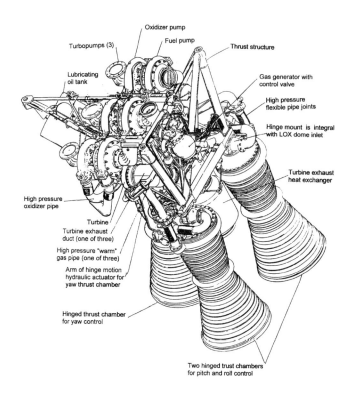

Fig. 7.8-4 View of the LPRE for the G-38 cruise missile. It consists of three rocket engines of 135,000 lbf thrust each. It was the first large Rocketdyne LPRE with bipropellant gas generators and hinge mounting of the thrust chambers. Courtesy of The Boeing Company, Rocketdyne Propulsion and Power; copied with permission from Rocketdyne line drawing.

RP-1 (kerosene), which resulted in a higher performance.[3,*] A view of this engine for the G-38 cruise missile is shown in Fig. 7.8-4. A cone-shaped nozzle was used for the three thrust chambers. The hinge mounting proved to be a better method of thrust vector control than jet vanes (no performance loss, no time limit from erosion). A gimbal mounting was used subsequently in most large U.S. LPREs. The turbine exhaust gases from the three TPs were combined sent through a relatively large heat exchanger, and discharged between the three TCs.

The preliminary design of the engines for the *Convair ATLAS ballistic missile* was started in 1952 and was unique.[3,5-8,11] The author's engineering group went through several iterations and discussions with Convair, the prime con-

---

*Personal communications, V. Wheelock, The Boeing Company, Rocketdyne Propulsion and Power, 2001–2004.

tractor, and the Air Force in arriving at the preliminary design. The two booster engines of 150,000 lb thrust each were mounted in a ring or doughnut-shaped structure at the missile's tail, and this structure was dropped from the flying vehicle after booster engine cutoff at about 150 s. The sustainer engine with 60,000 lbf (located in the center of the aft end of the vehicle) is also started at launch, but runs continuously for a total of about 310 s. Figure 7.8-5 shows a test of this three engine configuration. This one-and-one-half set of stages was selected in part because Convair, the vehicle developer, and Rocketdyne were not sure at that time if altitude ignition of a sustainer could be reliably achieved. Detail design started in late 1954. The total three-nozzle thrust of 360,000 lb at sea level (about 414,000 at high altitude) was increased in steps in the several subsequent modifications and in uprated versions of this engine until it reached an altitude three-nozzle thrust of 564,000 lbf as shown in Table 7.8-1. The altitude specific impulse reached 309 s (at 84,000 lbf thrust) for the sustainer engine. TVC during the booster flight was provided by the two

Fig. 7.8-5 Static firing test of the Atlas MA-5A LPRE. The nozzles of the two booster engine are flowing full, and their bright radiating plumes have sharp boundaries. The center sustainer engine is overexpanded with its 25:1 nozzle-exit-area ratio at the low altitude of the test facility. Its jet has separated from the nozzle wall. Steam clouds from the test facility water spray are being aspirated into the central jet, which is not visible, but has a smaller diameter than the nozzle exit. Courtesy of The Boeing Company, Rocketdyne Propulsion and Power; photo copied with permission from Refs. 3 and 12.

gimbal-mounted large booster engines. During the sustainer-only portion of the flight, the roll control was provided by two small regeneratively cooled small hinge-mounted TCs of about 1000 lbf thrust each. They received propellants tapped off the main sustainer engine's TP. A similar roll control scheme was used in the Thor missile, and it is shown in Fig. 4.6-3 and 4.6-4. The propellant tanks developed by Convair had very thin walls and had to be inflated by gas pressure at all times (during storage, launch preparations, or road transport) to prevent buckling or collapse. This also caused some extra steps in the vehicle transport, launch operating sequence, and in the engine installation.

The new *turbopump* for the booster stage engine, seen in Fig. 4.4-2, featured an alloy steel turbine, connected to the propellant pumps through an oil-lubricated gear transmission case.[9–12] The turbine ran at a higher shaft speed than the two pumps, which were on one common shaft. In prior engines a 90% monopropellant hydrogen-peroxide GG was used, and its maximum flame temperature was limited to about 1360°F. With a bipropellant GG the warm gas temperature can rise quickly (by 100 to 300°), if the mixture ratio shifts even only a percent or two. A high-temperature turbine material and an adequate margin are therefore necessary. In later versions of this TP, with other large LPREs some or all of the lubricating or cooling oil was replaced by kerosene fuel, and General Electric replaced the oil with grease in one of its gear cases (Chapter 7.3). The Atlas TP was an uprated version of the Navaho TP. The turbine was driven by a gas generator using the same propellants as the main chamber, but at a fuel-rich mixture resulting in a gas at about 1000 to 1350°F. The gear case allowed the pumps (about 5000 rpm) and turbine (about 30,000 rpm) to operate at different shaft speeds; the resulting higher efficiencies reduced the amount of propellant needed by the gas generator and raised the specific impulse by perhaps 1 or sometimes up to 3 s (Ref. 12). Early versions of this turbopump and the TPs of other early engines did not have inducer impellers, which provided a margin for avoiding cavitation and allowed a somewhat higher pump speed of the main TP. An inducer impeller is usually an axial-flow low-head pump, and an example is shown in Fig. 4.4-8. Very similar TPs with inducers were used on later versions of the Atlas, Thor, Jupiter, and on all of the versions of the H-1 and the RS-27 engines, and these engines are discussed later. Early versions had two auxiliary pumps attached to the gear case, an oil pump to cool and lubricate bearings and gears, and a hydraulic oil pump to supply pressurized oil to gimbal on hinge actuators and to some valve actuators. The gas generator, which supplied the TP with chemically heated gas, is shown in Fig. 4.5-6. In one version of the Atlas propulsion system, the two TPs for the booster engines were located next to each other and supplied with warm gas from a common single-bipropellant GG.

Although the initial Atlas engine designs had conical nozzle-exit cones, they were soon changed to a bell-shaped nozzle exit beginning in about 1958. This specially contoured nozzle exit was believed to have been an original contribution of Rocketdyne to the U.S. technology of propulsion. Figures 4.3-5 and

7.8-2 show large early conical nozzle exits of the Thor TC and the Navaho, but Figs. 7.7-18 and 7.8-13 show the large Titan TC and the J-2 engine with typical bell-shaped nozzle exits. The genesis and advantages of bell nozzles are explained in Chapter 4.3. In fact Rocketdyne had the first production thrust chamber with a tubular cooling jacket for the chamber and with a bell-shaped diverging nozzle. Rocketdyne had experimented with nickel, inconel, and stainless-steel tubes, developed the process for tapering the straight tubes, bending them into the contour of the combustion chamber and nozzle, pressing and forming the tube cross section into a nearly rectangular shape, and then brazing the tubes together in a special furnace. Different brazing compounds had been investigated.

The start features were different in the five models of the Atlas LPRE, each with a somewhat different vehicle stage.* In all models the two booster engines were dropped off. In the MA-1 model the starting was accomplished with pyrotechnic igniters in both TCs and both GGs, which were supplied through two onboard pressurized propellant start tanks. The MA-1 sustainer engine had pyrotechnic igniters in its main TC, its GG, and the two small vernier thrusters, which were used for roll control. In the MA-2 and MA-5 models slugs of a hypergolic start liquid fuel ignited the initial flow in all five thrust chambers, but they retained the pyrotechnic igniter in the three GGs. The Atlas MA-3 was equipped with solid-propellant start cartridges for igniting the initial flow to the three GGs and for starting the spin-up of the turbines; hypergolic start fluid ignition continued in all five TCs. In the last Atlas version, the MA-5A, the start tanks, and their pressurizing gas systems were on the ground, with two pyrotechnic igniters in each GG and hypergolic start fuel in the three main TCS. There were no small hinged vernier thrusters in this model. However the sustainer stage of the MA-5A had a separate hydrazine monopropellant multithruster system for roll control during the main engine's operation of the sustainer engine, and it was also used for other maneuvers.

The *Atlas missile was the first U.S. ICBM* and was operational in the U.S. Air Force between 1960 and 1965 (Refs. 3-8, 11,12). Several versions of this Atlas engine also served in propelling satellite launches for many military and space exploration payloads.[5-8] This included the Surveyer, Pioneer, or Intelsat satellites. A takeoff of an Atlas SLV is shown in Fig. 2-9. Different versions of the Atlas SLVs had various upper stages, including Agena with a LPRE from ARC and Centaur (LOX/LH$_2$) with two LPREs from Pratt & Whitney. The ATLAS engines drove the booster for the Mercury manned spaceflight program. It was an active rocket engine program for 46 years, in production intermittently between 1956 and 1996, and 571 engine sets (consisting of two boosters and one sustainer engine) had been delivered up to December 2002. There were 24 failures or malfunctions during flight out of 1713 flying engines, which were attributed to the propulsion system, but only some of these caused an

---

*Personal communications, V. Wheelock, The Boeing Company, Rocketdyne Propulsion and Power, 2001–2004.

abort of the mission. This gives an average overall engine reliability of about 0.97. Most of the flight failures of the engine occurred during developmental flights, and the most recent series, the MA-5A, had no engine flight failures.

Rocketdyne built the engines for the *Thor intermediate-range ballistic missile (IRBM)*, which was developed and produced by the Douglas Aircraft Company (later McDonnell Douglas and today Boeing) for the Air Force.[3, 13] Its engine was very similar to the engine for the *Jupiter IRBM*, which was developed in part by Wernher von Braun's team at Huntsville for the Army.[14] This Jupiter missile was produced by Chrysler Corporation, who has since left the missile business. Figure 7.8-6 shows these two engines together with the Redstone engine. Both the Thor and the Jupiter LPREs had the same gas generator engine cycle, same basic TC concept, and the same TP design configuration as the two Navaho engines, or the booster engines of the Atlas. As the thrust was increased in more advanced versions, the TP weight also increased. The butterfly valve shown in Fig. 4.7-7 was used here as a propellant valve.[12] The Jupiter engine design started in 1955 and Thor engine also in 1955. Their key design parameters are listed in Table 7.8-1. They used the same propellants (liquid oxygen and RP-1). The Thor and Jupiter engines used different engine mounting structures for transferring the thrust loads into the vehicle and somewhat different controls and piping. The thrust chambers of these two IRBM engines were almost identical. The early versions used a conical nozzle-exit section, as shown in Fig. 4.3-5, but a bell-shaped nozzle exit was used soon thereafter. An early MA-3 Thor engine version is shown in Fig. 4.6-3 while it and its two verniers are firing.

There were several increases of the thrust in successive versions of the Thor

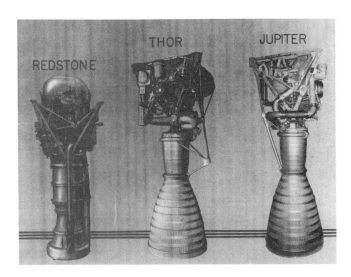

Fig. 7.8-6 Comparison of the LPREs for the Redstone, Thor, and Jupiter missiles. Courtesy of The Boeing Company, Rocketdyne Propulsion and Power photo copied of with permission from Ref. 3.

engine.[11,13,*] The MB-1 had 135, 000 lbf thrust and was used for initial flight tests. The first series of the MB-3 engines had conical TC nozzle exits with a half-angle of 15 deg, but in the late 1950s all versions were switched to a bell-shaped nozzle-exit contour, which gave a little more performance and/or allowed a shorter nozzle. The MB-3 had initially 150,000 lbf thrust, then 165,000, and finally 170,000 lbf. The early MB-3 versions of the engine were used in the IRBMs, which were deployed by the U.S. Air Force. The thrust increases allowed more propellants and/or more payload to be carried in the Thor or in the Delta SLV, the name given to the civilian SLV version of the Thor missile. The RS-27 engine was a further uprating in thrust and was specifically developed for an uprated Delta SLV of McDonnell Douglas Space Systems Company working for NASA. It came after two related engines had been developed, the experimental X-1 and the H-1, and both are described briefly next. The RS-27 incorporated improvements proven in these other engines, such as a simpler control, bigger TP and heavier TC and a simpler start system. Its thrust was 207,000 lbf at sea level with a nozzle-exit area ratio of 8 to 1.0. The last version identified as RS-27A had a larger nozzle area ratio (increased from 8.0 to 12) and thus a little better performance. It provided 237,000 lbf vacuum thrust and a specific impulse of 302 s (vac). Roll control of the Thor/Delta was by two hinged small thrust chambers of 1000 lbf each, similar to the Atlas sustainer or Thor engine described before. It is shown in Fig. 4.6-3. The small thrusters were provided with propellants directly from the turbopump of the Thor/Delta engine.

The Thor and Jupiter engine were single-start engines with many common or similar components; however, the start systems were different.[3,13,14,*] The Jupiter S-3D had ground liquid propellant start tanks, two redundant pyrotechnic igniters in the GG, and a larger single pyrotechnic igniter in the TC. The Thor rocket engine (MB-1) had onboard small start propellant tanks for the initial flow to the GG. Pyrotechnic ignites were used in the main TC and the two hinged vernier thrusters. The MA-3, RS-27, and RS-27A used a hypergolic start fluid in the main TC and the two vernier thrusters, but two pyrotechnic igniters in the GG. The turbopumps were of the same family and were similar to the one in Fig. 4.4-2. The Jupiter engine did not have two small vernier thrusters, but instead used the turbine exhaust gas with moveable nozzle, for roll control.

Jupiter flew first in early 1957, and after several launch failures the Thor flew successfully in September 1957.[13,14] The Jupiter flight-test program had 25 flights. Both were also used extensively as a space launch vehicle. The Thor missile was converted into the Delta launch vehicle with the same engine, and it successfully launched many different satellites over a period of many years.

Both of these 1500-mile intermediate-range ballistic missiles were deployed in the military services, and both had operational military units stationed in Europe. The Thor was operational between 1959 and 1964. For several years the Thor missiles were deployed in England, and the Jupiter missiles were in Turkey and Italy and were serviced by Italian troops.

---

*Personal communications, V. Wheelock, S. Claflin, and J. Halchak, The Boeing Company, Rocketdyne Propulsion and Power, and H. Minami, and J. Lincoln, retirees of Rocketdyne, 1980–2003.

Some of the development problems of the Thor, Atlas, and Jupiter engines were similar in nature.[11,13,14] Turbopumps gear cases had problems with adequate lubrication and axial loads in bearings. The remedies were better distribution of the lubricant, better bearings, and better bearing supports. During component testing, some of the turbine blades were found to have incipient cracks at the turbine roots. This was diagnosed as fatigue failures under high-frequency vibrations. The blades were strengthened by changing the blade profile, and the mass was changed by adding blade-tip shrouds. During engine development and TC development testing, a few incidents of unstable combustion occurred under certain conditions. This was believed to be caused by gas oscillations in tangential and radial modes, mostly in the combustion regions near the injector, as described in Chapter 4.10. The combustion instability was eliminated after using injector baffles, similar to the ones shown in Fig. 4.10-2 and after some changes in the injection elements and pattern.

The Atlas sustainer engine LOX pump burned and exploded during development testing. It was caused by an overhung inducer impeller rubbing against the pump housing, when the shaft experienced deflections caused by dynamic and static loads. The pump clearances were increased, and the housing was lined with a Kel-F fluorocarbon sleeve.

The *X-1 experimental LPRE* (1957–1959) was intended to investigate product improvements for IRBMs and more specifically evaluate several approaches to simplify the engine, increase reliability, and reduce inert mass.* It included a simpler hydraulic system, which used a pressure ladder start and shutdown for activating propellant valves and gimbal actuators, a lightweight beam engine mount, a novel gimbal design, simpler and fewer oxidizer and fuel flow components, a hypergolic ignition system, a solid-propellant cartridge for the gas-generator start and initial turbine power, a lubrication system that used additive enhanced fuel (RP-1) instead of oil, and a hinged roll control nozzle for the turbine exhaust, which replaced the two small hinged 1000-lbf thrust chambers. The X-1 started at 165,000 lbf thrust, but it was operated at 198,000 lbf thrust, and it was also used to investigate off-design operating conditions. The work on the X-1 engine was also useful in the design and development of the H-1 engine (for the Saturn V SLV), which is described later.

In 1957 Rocketdyne activated a production plant for LPREs at Neosho, Missouri, and it produced, calibrated, and proof tested rocket engines for the Atlas, Thor, and Jupiter programs.† The plant operated for 11 years and was closed when production quantities and forecasts for future deliveries decreased. A chart of Rocketdyne engine deliveries from both its Canoga Park

---

*Personal communications, V. Wheelock, and J. Halchak, The Boeing Company, Rocketdyne Propulsion and Power, and M. Yost, retiree of Rocketdyne, 1990–2004.

†Personal communications, V. Wheelock, The Boeing Company, Rocketdyne Propulsion and Power, and H. Minami, and J. Lincoln, retirees of Rocketdyne, 1990–2004.

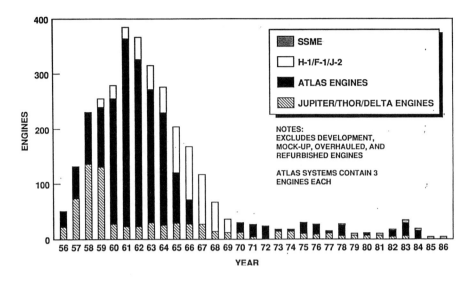

Fig. 7.8-7 Rocketdyne's history of deliveries of several large LPREs. Courtesy of The Boeing Company, Rocketdyne Propulsion and Power data obtained through from Ref. 3.

and Neosho production facilities is shown in Fig. 7.8-7. It shows that these plants were very busy for about 15 years.

Because the United States was at that time in an unofficial but serious international competition with the Soviet Union (called the Cold War), there was great urgency to have ICBMs and IRBMs in the military services and the Redstone, Thor, Atlas, and Jupiter missile programs received the highest national priorities. For example, engine production was started before flight testing or R&D was really completed. In retrospect some of these decisions and the steps to reduce the schedule did lead to some program inefficiencies.

In accordance with a U.S.–Britain government agreement, the Thor/Delta vehicle technology and the current version of the Thor MA-3 engine (at that time it still had a conical exhaust nozzle) were licensed in 1955 to Rolls-Royce.* This formed the basis for the British development of the Blue Streak missile engine, which is discussed in Chapter 12 and is shown in Fig. 12-16 and 12-17. In 1965 the technology of the Delta SLV and a more advanced MA-3 version of the Thor engine were licensed and transferred to Mitsubishi Heavy Industries, implementing a government agreement between the United States and Japan. It helped the Japanese, and they copied and produced their version of this engine. They flew it in the first stages of three different SLVs as discussed in Chapter 11.

---

*Personal communications, V. Wheelock, The Boeing Company, Rocketdyne Propulsion and Power, and H. Minami, and J. Lincoln, retirees of Rocketdyne, 1990–2004.

Liquid Propellant Rocket Engines for the Saturn/Apollo Program

Rocketdyne was selected to provide the large LPREs for the booster stages and second stages of the Saturn-I SLV and for all three ascent stages of the Saturn-V moon launch vehicle.[15–17] These engines propelled astronauts into orbits and the Apollo modules toward the moon. Late in the schedule Rocketdyne was asked to remedy ARC's injector problems on the lunar ascent engine for the lunar excursion module of Apollo. Thereafter Rocketdyne assembled this engine.[17] The company also provided small ablative thrusters for the attitude control system of the Apollo Command Module and for the S-IVB stage, but these two sets of small thrusters are discussed later in another section of this chapter.[18] *Saturn-I* was boosted by eight H-1 (LOX/RP-1) engines, each initially at 165,000 lbf thrust, but later versions were upgraded in steps to 205,000 lb thrust each. Its second stage, the S-IVB, was powered by a single J-2 (LOX/LH$_2$) engine. The *Saturn-V* SLV has five F-1 booster engines in the first stage (at 1.5 million lb thrust each) with a takeoff thrust of 7.5 million pounds, five J-2 oxygen/hydrogen engines at 230,000 lb (vac) each in the second stage, and one J-2 engine in the third stage (another S-IVB). These engines will be described next. This huge Saturn-V-Apollo space launch vehicle has six stages, three launch stages, one service module, one command module, and a two-piece lunar excursion module. It is a rocket engineer's dream because it had a total of 83 rocket propulsion nozzles; of these 66 were nozzles of LPREs. Figure 7.8-8 shows these stages and identifies all of the rocket propulsion systems.[10] This is the heaviest and most complex manned spaceflight vehicle assembly that has ever flown successfully and returned crews safely back to Earth. The Soviet Union developed an even larger and heavier N-1 launch vehicle for lunar exploration and for landing a man on the moon. After four unmanned failed launch attempts, the program was canceled. Chapter 8.9 gives more details.

The *H-1 engine* was a logical upgrade (higher thrust) and simplification of the Thor LPRE and used features proven in the experimental X-1 engine. Figure 7.8-9 shows a simplified line drawing of an H-1 engine, Fig. 4.2-6 in Chapter 4.2 shows a simple schematic flow diagram, and Table 7.8-1 shows some data.[3,7,8] In this engine the turbopump was gimbaled together with the TC and the flexible pipe lines were in the low pressure piping on the suction side of the pumps, thus eliminating the need for high pressure flexible joints. Thrust levels were increased progressively with newer models, starting at 165,000 lbf, then 188,000, then 200,000, and finally 205,000 lbf thrust each. All together 152 H-1 engines were flown in 19 missions of the Saturn-I.

The H-1 engine start sequence diagram is shown in Fig. 4.8-3, its gas generator in Fig. 4.5-7, and the solid-propellant starter cartridge used for igniting the GG and starting the spin of the TP is shown in Fig. 4.8-2. The main H-1 TC used a slug of hypergolic fuel for ignition of the initial propellant flow to the TC and two pyrotechnic igniters in the GG. The H-1 engine's heritage can be traced to Atlas booster, Thor, Jupiter, and X-1 engines and is an extension of earlier large Rocketdyne LPRE technology. It was the first flying U.S. LPRE, where the turbine exhaust gases were aspirated into the nozzle-exit jet of the TC. This gives about the same or slightly higher specific impulse and mini-

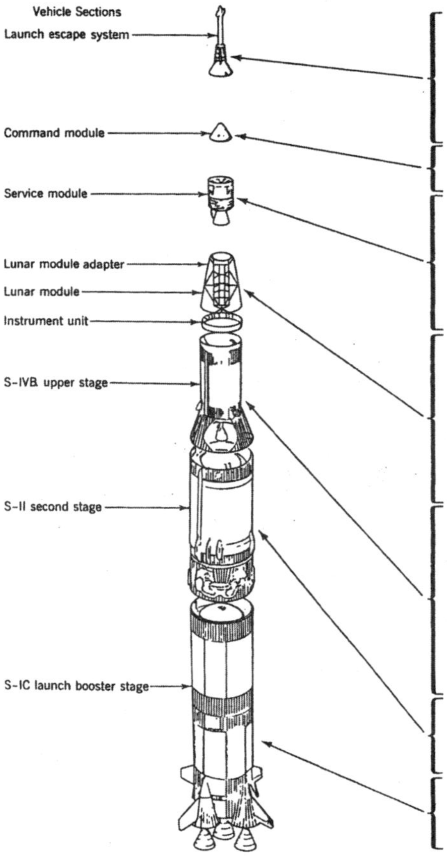

Vehicle Sections	Number and Model of Propulsion System	Propellants, Starts, Burn Time	Thrust (Lbs)	Usage or Missions
Launch escape system	1 4-Nozzle	Solid 1 burn, 8 sec	147,000	Payload emergency separation from vehicle during early flight
	1 PCM	Solid 1 burn, 0.5 sec	2,400	Pitch control during emergency separation
	1 TE-380	Solid 1 burn, <1 sec	31,500	Normal (no emergency) separation of entire tower at +180 sec or in emergency separation, after tower and spacecraft have been carried safe distance (1 mile) from vehicle, will pull tower from spacecraft
Command module	12 SE-8 (6 redundant)	NTO/MMH Multiburns (230 sec max.)	(93 each)	Reaction control of command module during reentry
Service module	16 R4D-1	NTO/50% UDMH and hydrazine Multiburns	(100 each)	Reaction control, propellant settling, lunar module docking adjustments, minor velocity corrections, and last separation of command and service modules upon reentry
Lunar module adapter	1 AJ10-137	NTO/50% UDMH and hydrazine Multiburns (12.5 min max.)	21,900	Payload emergency separation from vehicle *after* escape tower jettison; deceleration during earth-orbit missions; midcourse velocity correction during translunar coast; plane changes; spacecraft insertion into lunar orbit and transearth orbit
Lunar module	16 R4D-2	NTO/50% UDMH and hydrazine Multiburns	(100 each)	Reaction control system for lunar module and aid in rendezvous/docking with spacecraft after return from lunar surface (in ascent section of lunar module)
	1 8285/RS-18	NTO/50% UDMH and hydrazine 2 burns (5 min max.)	3,500	Ascent power from lunar surface to lunar orbit and rendezvous/docking with spacecraft
	1 TR 200		1050 → 10500 (throttlable)	Deceleration, midcourse correction, descent thrusting, and hover to lunar landing; also can return lunar module to lunar orbit if emergency arises
Instrument unit	6 TR-402	NTO/MMH Pulsefired (20 min max.)	(150 each)	Reaction control during 4-hr coast period after first J-2 burn and during translunar maneuvers
S-IVB upper stage	2 SE-7-1	NTO/MMH 2 burns (425 sec max.)	144	Propellant settling after first J-2 burn and during restart chilldown prior to second J-2 burn
	1 J-2	LOX/Liquid hydrogen 2 burns, (8.5 min max.)	230,000	Final vehicle boost into earth orbit (142 sec) and after 4-hr coast period, reignite (at apogee or earth orbit) to insert payload (360 sec) into translunar trajectory
	2 TX-280	Solid 1 burn (3.9 sec)	6,780	Propellant settling prior to J-2 first burn
S-II second stage	4 TE-29-1B	Solid 1 burn (1.5 sec)	140,000	Separation by *deceleration* of S-II stage: four retros mounted on aft interstage assembly facing *upward*
	5 J-2	LOX/Liquid hydrogen 1 burn (400 sec)	1,150,000	Boost second stage
S-IC launch booster stage	8 BSM	Solid 1 burn (0.67 sec)	704,000	Separation retro-action
	5 F-1	LOX/RP-1 1 burn (150 sec)	7,610,000	Boost first stage

Fig. 7.8-8 Propulsion systems for Apollo/Saturn V lunar landing and return mission. Courtesy of NASA; copied with permission and modified from fourth edition of Ref. 12, pp. 18,19.

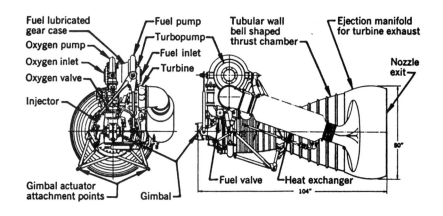

Fig. 7.8-9 Two simplified drawings of the H-1 LPRE used in the Saturn I SLV. Courtesy of The Boeing Company, Rocketdyne Propulsion and Power; copied with permission from fifth edition of Ref. 12, p. 146.

mizes turbine exhaust gas backflow problems at altitude, when compared to an engine with a separate duct and nozzle for the turbine exhaust. Like its predecessors, it uses a gas-generator cycle, LOX/RP-1 propellants, a tubular TC with a bell-shaped nozzle, and a very similar, but stronger and heavier geared TP. Problems in development were often similar in two or more of these engine versions, such as inadequate lubrication and cooling of the gear cases for the turbopumps, cracking of turbine blades during component tests, or the rubbing of an overhung inducer impeller against the inlet pump housing.[3,8,*] The gas generator is different from other GGs and is shown in Fig. 4.5-7.

In two of the flights, one of the eight H-1 engines' monitoring system detected an out-of-tolerance engine parameter, and this caused a premature, but safe shutdown of that engine.[†] The flight proceeded satisfactorily with the other seven engines, and the first stage flew an alternate flight path. This was an unintended demonstration of the engine-out capability of the engine cluster and the reprogrammed guidance and control system.

Combustion vibration problems in these and some other earlier engines plagued Rocketdyne for at least a decade. The *combustion instability* phenomena are complex and occurred in several other Rocketdyne engines. These vibrations are discussed in Chapter 4.10 of this book, and even today they are not fully understood.[12] The only way for solving these problems in the 1960s was to do a lot of hot-fire testing with several supposedly identical engines, induce artificial pressure pulse disturbances to the combustion, observe their combustion behavior, measure the time for recovery or return to normal com-

---

*Personal communications, V. Wheelock, H. Minami, J. Lincoln, and J. Halchak, The Boeing Company, Rocketdyne Propulsion and Power, 1990–2004.
†Personal communications, V. Wheelock, The Boeing Company, Rocketdyne Propulsion and Power, 2000–2004.

bustion, and statistically analyze the test results. Eventually this national effort led to a better understanding and the use of baffles on the injector to inhibit transverse acoustic resonances. All H-1, Thor, and Atlas booster engines had a five-compartment baffle similar to the one shown for the F-1 engine in Fig. 4.10-2. Later it was learned that acoustic resonance cavities at the edge of the injector were often more effective than baffles by themselves. Today there are some analysis tools to predict the frequencies of various instability modes, and therefore the cavities can be designed or tuned to dampen specific frequencies.

The large *F-1 LPRE* has the highest thrust (1,522,000 lb at sea level) of any U.S. engine and for more than a decade the highest thrust in the world. An early version of this engine is shown in Fig. 7.8-10, and some data are in Table 7.8-1. Engine detail design started in 1962.[3,7,8,15–19],* It was the second U.S. LPRE where the bottom nozzle-exit section (between area ratio of 10 and 16) is film cooled with warm (about 900 °F) turbine exhaust gas. This method has also been called "dump cooling" because the turbine exhaust gas is dumped into the nozzle-exit section through slots between the tubes. The manifold for introducing this warm gas can be seen in the figure. Dump cooling reduces the wall area that needs to be regeneratively cooled and simplifies the TC cooling

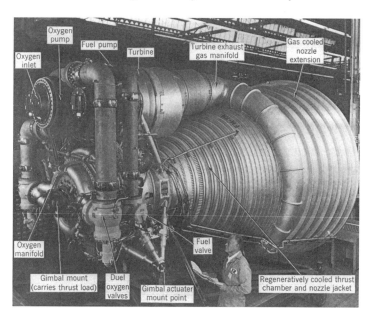

Fig. 7.8-10 Early version of the F-1 LPRE. It was the highest thrust engine (1.5 million pounds) in the world, when it first flew in 1967 (220 in. long and 144 in. wide). Courtesy of The Boeing Company, Rocketdyne Propulsion and Power; copied from fourth edition of Ref. 12, p. 8.

---

*Personal communications, V. Wheelock, The Boeing Company, Rocketdyne Propulsion and Power, 2000–2004.

jacket. Dump cooling can result in a very small increase in specific impulse; however, it does not cause a loss in performance, compared to an engine with a separate turbine exhaust duct. As mentioned earlier, dumping minimizes blowback of turbine exhaust gas at altitude. The large single-shaft turbopump is next to the chamber and is oriented vertically with the cold oxygen pump at the top, the two-stage hot turbine at the bottom, and the fuel pump in the middle. High-pressure fuel was used for the hydraulic power to open or close the main valves and for the actuators of the gimbal motion.

In June of 1962, an F-1 engine being tested at Edwards Air Force Base in California destroyed itself as a result of combustion instability. This event set off a feverish round of design changes and tests on this program.[20] Task forces were formed to investigate this phenomena, and they examined instabilities in other large LPREs. There was a tremendous pressure on Rocketdyne to solve this instability problem, and even the NASA administrator became involved. It took more than two years and many tests of potential solutions to eliminate this F-1 instability. A total of 14 different baffle configurations were built and tested; most of them were tested several times.[20] At least a dozen changes in the injection holes and their distribution were evaluated. At least two-thirds of all TC component tests and a major share of the complete engine tests were associated with the effort to achieve stable combustion. The selected final baffle configuration, which withstood all attempts to artificially induce instability in different modes, is shown in Fig. 4.10-3. It was an expensive crash effort, prolonged in part by an incomplete understanding at that time of the combustion behavior and instabilities.

The F-1 TP was large, used a single shaft, had inducer impellers ahead of the main pump impellers, and used a conventional design. The LOX pump of the F-1 TP experienced five major pump failures between 1963 and 1965.[15-18] Four of these occurred during the 107th and 111th s of planned runs for 165 s, and one occurred during startup. Investigations showed that it was initiated by metal contact and local metal heating between the rotating LOX pump assembly and the stationary pump housing causing a LOX/metal oxidation reaction. Design changes to increase clearances, eliminate fretting of the bearing surfaces, and strengthen the pump impeller vanes to reduce deflection under load solved the issue. The gas generator, which provided the warm gas to drive the TP, is shown in Fig. 4.5-8.

Large LPREs require a unique *start sequence*, and it seems to be different for every major type of LPRE. During the 1950s and 1960s, it was learned what to do, and through failures it was learned what not to do. The *start sequence of the F-1* is mentioned here as an example.[3,12,15,*] A preflight dry nitrogen gas purge is used to purge air from the oxidizer lines and the injectors of the main chamber and the gas generator, while the vehicle is still on the launch stand in order to remove atmospheric moisture, which might freeze on valve stems.

---

*Personal communications, V. Wheelock, The Boeing Company, Rocketdyne Propulsion and Power, 2001–2004.

Pressurization of the fuel tank is started by admitting pressurized helium from a storage tank. During engine operation, the helium is heated in a heat exchanger in the turbine exhaust duct, and this reduces the mass of required helium. The oxidizer tank is pressurized by evaporated oxygen, initially evaporated from the oxygen in the tank, and during operation taken from the oxygen pump discharge and evaporated in a heat exchanger in the hot turbine exhaust duct. These were the main prestart events.

The start sequence is shown in Fig. 7.8-11. After receiving the start command signal, the F-1 main oxidizer valve and the GG oxidizer valve are opened first, and oxidizer starts to flow slowly under gravity.[15] This flow is induced by the pressure of elevation of the liquid level above the pump augmented by the ullage gas pressure in the propellant tank. A pyrotechnic igniter is fired in the GG, the GG fuel valve is opened, fuel begins to flow slowly, and the GG starts to burn and create hot gas. This gas accelerates the turbine, and both pump discharge pressures begin to rise. In one version the GG oxidizer and GG fuel valve are opened simultaneously with the igniter. When the fuel pump discharge pressure reaches a predetermined value, it ruptures a disk to a hypergolic fuel cartridge, and this hypergolic fuel (a mixture of triethyl aluminum and triethyl boron) starts to burn with the low flow of oxygen in the chamber.

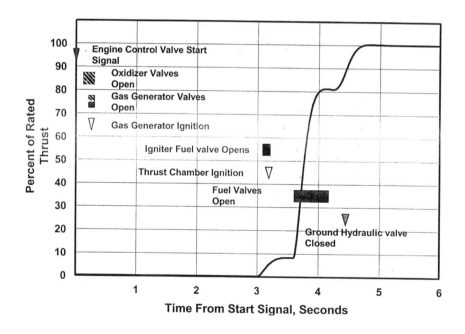

Fig. 7.8-11 Start sequence for the F-1 LPRE. Courtesy of The Boeing Company, Rocketdyne Propulsion and Power; copied with permission from Ref. 15.

Once combustion is sensed, the two main fuel valves are opened to allow main-stage combustion. This start procedure has been called a tank head start, which is simpler, but slower than a start with a solid-propellant cartridge or an initial GG propellant supply from separate small pressurized propellant tanks. With the tank head start it takes about 5 s (most of this time at very low flow) to reach full thrust. During this engine start, a hydraulic high-pressure fluid is supplied for valve and gimbal actuation from a ground source, which is disconnected at liftoff.

The F-1 was based in part on the experimental E-1 LPRE of 400,000 lbf thrust with LOX/RP-1 propellants.[3,*] Figure 7.8-12 shows its TC being ground tested. It was designed two years before the F-1 design began and tested before the F-1 design was frozen. It used a gas-generator engine cycle, a TP that resembled somewhat the F-1 TP, and a tubular thrust chamber. It was very similar in its fea-

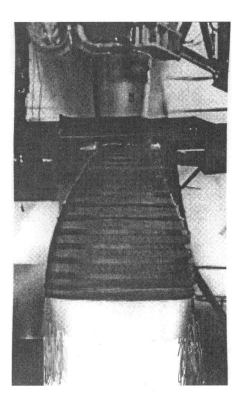

Fig. 7.8-12 Test firing of the experimental E-1 LPRE of 400,000 lbf thrust. The test stand platform obscures some of the engine. Courtesy of the Boeing Company, Rocketdyne Propulsion and Power (from personal communications, V. Wheelock, 2002).

---

*Personal communications, V. Wheelock, The Boeing Company, Rocketdyne Propulsion and Power, 2001–2004.

tures to the F-1; however, it did not aspirate the turbine exhaust into the nozzle diverging section, and it had a somewhat different start sequence.

There were 3248 single F-1 engine tests for a cumulative duration of 280,587 s, more than 4000 TC development tests, plus 34 tests of a five engine cluster and 13 flights.[15,19,20,*] There were a number of major failures during development prior to December 1965. In a formal reliability demonstration program with 336 declared tests, there were no engine failures. Altogether 98 production engines were delivered, and 65 of these have flown successfully at 100% reliability. The first engine tests were in 1961, the first test of a five-engine cluster in 1965, the first flight in 1967, the first manned flight in 1968, and the last flight (Skylab) in 1973.

The F-1A was an uprated version of the F-1 with a record sea-level thrust of 1,800,000 lbf, which is 20% higher than the F-1. As far as this author can determine, it was the highest-thrust LPRE ever to be ground tested. It was a potentially higher-thrust engine for the Saturn V booster stage. Design of the F-1A started in 1967 just before the first flight of the F-1. The engine was very similar to the F-1, operated on a gas-generator engine cycle, and used the same propellants. It had a higher chamber pressure (1161 psia vs 982 psia), a more powerful and heavier TP with slightly different impellers, required a strengthening of the TC, and had better thermal insulation. Test were conducted at the Air Force Rocket Propulsion Laboratory at Edwards, California, in early 1970. The specific impulse was 270 s (SL) and projected as 303 s (vac) at the same nozzle area ratio of 16 to one. NASA decided that this increased takeoff thrust was not needed in the Saturn V, and the program was not continued.

The *J-2 engine* was the first large engine to use liquid oxygen and liquid hydrogen as propellants and had 230,000 lbf thrust at altitude with a specific impulse of 425 s (vac).[3,7,16,19,†] Rocketdyne had extensive prior experience with liquid oxygen in several of their engines (described before) and with liquid hydrogen in connection with their nuclear rocket engine program and their research department efforts. This included work on a liquid-hydrogen pump, which had bearings and seals lubricated by LH_2 and cold-flow tests and chilldown tests with LH_2 of a Thor-type TC equivalent to a 135,000 lbf thrust level. This work was done in the 1950s.

The J-2 LPRE had three flight applications as already mentioned: as a single engine in the second stage (S IVB) of the Saturn I SLV, five J-2 engines were in the second stage (S II) of Saturn V, and a single engine in the third stage (also a S IVB) of the Saturn V SLV. Some of its data are listed in Table 7.8-1, and the engine is seen in Fig. 7.8-13. Design started in 1960. It was the first large Rocketdyne production engine with two separate direct-drive turbopumps, one for the oxidizer pump and one for fuel pump. This is shown in the

---

*Personal communications, V. Wheelock, The Boeing Company, Rocketdyne Propulsion and Power, 2001–2004.
†Personal communications, V. Wheelock, The Boeing Company, Rocketdyne Propulsion and Power, 2000–2004.

Fig. 7.8-13 J-2 LPRE installed in the S-IVB stage of Saturn SLV. It was the first high-thrust LOX/LH$_2$ engine (133 in. long, 80.5 in. wide). Courtesy of The Boeing Company, Rocketdyne Propulsion and Power, from Ref. 3.

schematic diagram of this engine in Fig. 7.8-14. The bearings of the TPs were lubricated and cooled by LOX and LH$_2$, respectively. The oxygen pump used a centrifugal impeller preceded by an inducer impeller. The fuel pump is different; it is a seven-stage axial-flow pump with an inducer impeller, and it is shown in Fig. 4.4-7. This type of axial-flow pump is more efficient than a centrifugal pump for this range of flow (and this helps to reduce the GG flow), but it is not as adaptable to off-design flow conditions. Thereafter all future high-flow main pumps were designed by Rocketdyne with mixed-flow centrifugal impellers and fewer stages. This fuel pump has the largest number of stages of any known flying rocket pump.

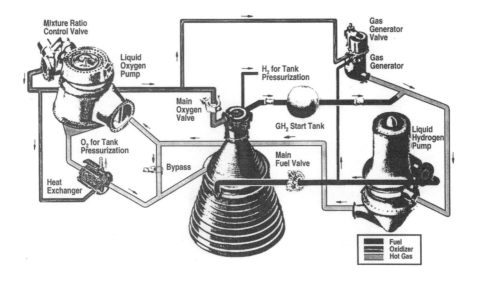

Fig. 7.8-14 Simplified flow diagram of the J-2 LOX/LH$_2$ LPRE shows two independently driven turbopumps and a gas generator to supply hot gas to two turbines in series. This engine can be restarted in a vacuum, and it has a propellant utilization system. Courtesy of The Boeing Company, Rocketdyne Propulsion and Power; copied from Ref. 16.

The TC had a tubular cooling jacket and multiple coaxial injection elements. The lower portion of the nozzle-exit section was dump cooled by turbine exhaust gas. The fuel-rich gas from the GG went through the two turbines in series and then was dumped or aspirated into the diverging section of the exhaust nozzle, where this gas cooled the single nozzle wall. It is common practice with all cryogenic propellants to initially bleed some of the cold cryogenic liquids through the piping, pump, and valve hardware in order to cool them and minimize making gas bubbles, which could reach the injector and chamber during the start transient period. The first long duration ground test (250 s) was conducted in September 1962. The first flight with the Saturn I happened in late 1967. Engine production proceeded for about eight years.

The starting of the J-2 engine was the same in the Saturn II upper stage (five J-2 engine) and the S IVB upper stage. An onboard high-pressure gas tank of hydrogen gas was used to initiate the rotation of the two turbines. There were redundant dual electronic igniters (the spark igniter had an electronic amplifier and oscillator) in the GG and the main TC. The igniter chamber, located in the center of the TC injector, had a small flow of gaseous propellants. They are shown in Fig. 4.8-7 and 4.8-8. The J-2 engine in the S IVB was restartable in space.*

During one of the Saturn V flights, one of the five J-2 engines in the second

---

*Personal communications, V. Wheelock, The Boeing Company, Rocketdyne Propulsion and Power, 2000–2004.

stage gave signals of an out-of-tolerance condition, which caused its premature (one minute before normal cutoff), but safe automatic shutdown.[16]* Although this was a useful demonstration of the engine-out capability of this five-engine cluster, it was a failure of the engine, and it caused a major investigation. During this same flight, the single J-2 engine in the third or S IVB stage would not restart for its second programmed burn. The engine worked properly during its first burn (there were no anomalies), and nothing unusual happened during the coast period before the second burn. Fortunately these two J-2 failures did not cause a vehicle failure or a loss of astronauts. However these malfunctions triggered a major eight-months long investigation of possible causes. Ground tests of this LPRE were made trying to duplicate the failures encountered in the flight. Eventually it was found that the cause of the two engine failures was the same, namely, a vibration failure of a small bellows supplying fuel to the augmented spark igniter chamber, which is located in the center of the main injector. The fuel line broke and allowed hot gas to leak out of the combustion chamber. The duplicate ground test of this event showed substantial burning and erosion of the igniter chamber. The remedy was to replace the bellows with a rigid, welded fuel line. On an earlier flight a J-2 engine in the five-engine cluster of the Saturn II was shut down prematurely, before it had fired for its full duration. However this thrust termination was intentional and intended to test the engine-out capability.

During the development period, some of the turbines of the J-2 fuel TP burned out as a result of excessive GG exit temperatures.[16] The following items were believed to have contributed to this condition: possible stall on an axial-flow impeller of the seven-stage fuel pump, initial flow with hydrogen temperatures above its boiling point (might have had bubbles), and possible fuel lead-time variations in the GG valves. More extensive thermal cooling of the hardware and a better timing sequence of the GG valves seemed to fix the problem. The start process and the overtemperature problem were analytically simulated, and this helped with the understanding of the events and the potential solutions.

The J-2 development program was continued with an experimental engine called the J-2X. It was an active experimental program for about three or four years beginning with design in 1964. It served well in allowing various suggested J-2 improvement to be tested, such as higher nozzle-exit area ratio and simplified controls.† The program continued with the J-2S, a simplified and enlarged version of the J-2, but with a different engine cycle. Its thrust (265,000 lbf) and chamber pressure (1245 psia) were higher than in the J-2. The engine used a topping engine cycle or tap-off cycle, and a simple diagram of this cycle is shown in Fig. 4.2-5. Hot chamber combustion gas was tapped off the chamber at the injector, diluted with hydrogen gas from the cooling jacket, which reduced the bleed gas temperature, and this gas provided the

---

*Personal communications, V. Wheelock, The Boeing Company, Rocketdyne Propulsion and Power, 2000–2004.
†Personal communications, V. Wheelock, and E. Schuster, The Boeing Company, Rocketdyne Propulsion and Power, 2002–2004.

power to drive the turbines. This tap-off cycle eliminated the gas generator, its controls, and several valves. The distribution of temperature and gas properties of the warm tapped-off gas were very uniform and reproducible. The exhaust gas from the turbines was introduced into the main nozzle flow at the diverging section of the nozzle. The performance of the J-2S was essentially the same as an equivalent J-2 engine at the same chamber pressure, area ratio, and mixture ratio. There was no continuation of this experimental program.

The Saturn-I and Saturn-V space launch vehicles had illustrious histories. It has already been mentioned that the Saturn-I was successfully launched by the H-1 engines 19 times beginning in October 1961. It served to investigate design features of the upcoming Saturn-V, to launch Skylab crews, and to fly to the first orbital international rendezvous and docking of the Apollo (United States)–Soyuz (Soviet Union) spacecraft during July 1975. The Saturn-V was launched with the F-1 and J-2 engines 13 times in connection with the world-renowned American moon circumnavigation, landing, and return program and the Skylab orbital station.

The *lunar ascent engine for the Apollo lunar excursion module* was acquired by Rocketdyne in a most unusual manner.[17,*] Even though it was called an engine, it was really a thrust-chamber assembly with valves, and its mission was to return the astronauts from the moon back to the command module, which was orbiting the moon. The pressurized-gas feed system for this engine was the responsibility of Grumman, the prime contractor for the lunar excursion module, which is shown in Fig. 7.8-15. The key objective for this propulsion system was 100% reliability because the lives of the astronauts depended on it. In 1963 the development of this TC assembly was awarded to Bell Aerosystems, the predecessor of ARC, which is covered in Chapter 7.11. The thrust was 3500 lbf, and the propellants were NTO and Aerozine 50 (50% hydrazine with 50% UDMH) at a feed pressure of 165 psia. An ablative TC design was chosen with multiple ablative layers in the nozzle-throat region and the chamber, but a single low-density ablative material in the nozzle-exit region. Bell Aerospace used triplet impinging injection elements similar to those in their successful Agena engine. A redundant set of propellant valves was developed. This ARC (formerly Bell Aerosystems) engine is shown in Fig. 7.11-13 in the version, which was modified by Rocketdyne.

According to NASA, the Bell Aerosystems people had three significant R&D problems and they were related to the injector. The combustion would at times become unstable, there were hard starts, and the erosion of the ablative material was not symmetrical, causing potential thrust misalignment. These problems continued for a couple of years, even though Bell tried several other injection elements or patterns and at the suggestion of NASA added baffles, which were supposed to suppress the instability.

In 1967, four years after ARC had started, NASA asked Rocketdyne to develop an alternate injector for the lunar ascent engine. The job was done in

---

*Personal communications, V. Wheelock, and E. Schuster, The Boeing Company, Rocketdyne Propulsion and Power, 2002–2004.

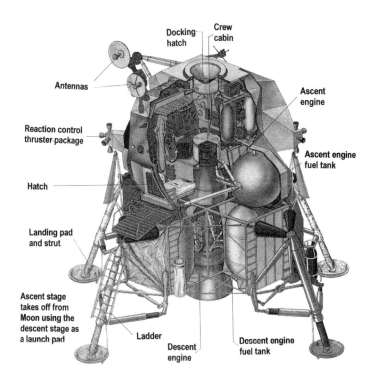

Fig. 7.8-15 View of partially sectioned Apollo lunar excursion module showing the location of its rocket engines. Courtesy of NASA, the Amber Press, and Northrop Grumman,; copied from Tim Furniss, *The History of Space Vehicles,* Thunder Bay Press, San Diego, CA, 2001, p. 1.

record time with the first firing test in one month after program go-ahead. The Rocketdyne injector, shown in part of Fig. 7.8-16, was made of aluminum, had three baffles, and acoustic cavities and proved to be stable when bomb tested. The uneven erosion of the ablative material was found to be connected with the flow distribution in the Bell injector manifold and was remedied by injecting film-cooling fuel at the periphery of the injector in an unsymmetrical distribution pattern. There were also hard starts (overpressure at ignition) and detonations, which were caused by a fuel lead. This was remedied by adding an extra fuel duct volume, which would delay the fuel entry to the chamber and ensure an oxidizer lead.* This additional fuel volume can be seen in the picture of the ascent LPRE also in Fig. 7.11-13. The Rocketdyne injector was qualified in 1968. Shortly thereafter NASA decided to give Rocketdyne the responsibility to assemble, test, deliver, and do the field service of the TC, with a valve package from Bell, an ablative TC from a subcontractor, and the Rocketdyne injector.

---

*Personal communications, E. Schuster, and V. Wheelock, The Boeing Company, Rocketdyne Propulsion and Power, 2003.

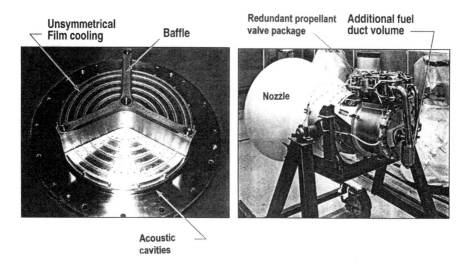

Fig. 7.8-16 Apollo lunar ascent engine and its injector with baffles and resonance cavities. Courtesy of ARC, Rocketdyne, and NASA; copied from Refs. 15–18.

Large Engines Using Liquid Oxygen and Liquid Hydrogen

The historic *space shuttle main engine (SSME)* was designed in 1972, and its design is a radical departure from prior Rocketdyne engines.[3,21–23]. It runs on liquid oxygen and liquid hydrogen and is still in limited production. Three of these engines are used in a Space Shuttle Orbiter, and they provide the main portion of the total impulse to the flight. Figure 2-11 is a sketch of the Space Shuttle Orbiter, and it shows the placement of the three SSMEs. Figure 7.8-17 shows two views of the engine, Table 7.8-1 indicates it has high performance, and Fig. 7.8-18 shows a simplified SSME flow diagram. It is the first flying U.S. LPRE with a *staged combustion engine cycle*. It gives a higher performance by about 5 to 7% compared to a GG cycle engine of the same chamber pressure and nozzle area ratio. It is the only *reusable man-rated U.S. large pump-fed LPRE* currently flying. It can be throttled between 67 and 109% of rated power level. The SSME flow sheet shows four turbopumps, namely, a three-stage high-pressure fuel pump, a single-stage oxidizer pump with a dual inlet impeller, and two booster pumps. It was the first flying U.S. engine with booster pumps. The oxygen booster pump was driven by a hydraulic turbine with liquid oxygen and not a gas turbine. The fuel booster pump is driven by gasified hydrogen, which has been evaporated and heated in the cooling jacket.

The chamber pressure of 3319 psia (initial version of block I) is the highest of any U.S. pump-fed LPRE. This allows a relatively small chamber and nozzle, a high nozzle-exit-area ratio (68.8) without excessive nozzle flow separation, and a high specific impulse (453 s at altitude). However it has very high heat-transfer rates and is therefore more difficult to cool. For this reason a copper-alloy inner wall, and a milled channel design is used in the cooling jacket at the chamber and nozzle-throat region. This milling operation is seen in Fig. 4.3-6. The copper alloy was developed by Rocketdyne to obtain higher strength at elevated temperatures, yet retain some of the good conductivity of copper. This high chamber pressure was reduced in the next version of the SSME, block II, as mentioned in the following.

The throttled or lower thrust condition (down to 67% of rated thrust) is required for two reasons. During ascent of the upper atmosphere, full thrust would cause excessive aerodynamic loads (dynamic pressure) on the flight vehicle structure (thereby requiring a higher vehicle structure weight). Furthermore during the last roughly 70 s of powered flight, when the onboard propellant mass is low, the acceleration of the vehicle and the astronauts needs to be limited to 3 g by throttling. The SSME design has its heritage in part on the J-2, the J-2X, the J-2S, and the BORD program described later in this chapter.

Illustrations of other SSME components can be found in other chapters. The initial version of the injector can be seen in Fig. 4.10-2, and it has coaxial bipropellant injection elements and baffles to prevent certain combustion oscillations. Later, after resonance cavities were built into the TC the baffles were no longer needed and were removed. The preburner for the high-pressure fuel TP is shown in Fig. 4.5-9, and its injector concept with coaxial injection elements is similar to the main injector. The unusual main oxygen valve, with a very low pressure drop, is seen in Fig. 4.7-6, the gimbal mount in Fig. 4.9-3, the electronic controller in Fig. 4.7-2, and a flexible high-pressure gas joint in Fig. 4.7-12. The two preburners and the main SSME TC each had two electronic controlled igniters.

The space shuttle's large liquid-hydrogen tank is pressurized (to a low tank pressure) by a small flow of warmed hydrogen gas (heated in the cooling jacket), which is tapped off the turbine discharge of the hydrogen booster pump. The oxygen tank is pressurized by a small flow of oxygen (tapped off from the discharge side of the main oxygen pump), which has been heated in a coiled pipe heat exchanger wrapped around the turbine of the high-pressure oxygen TP. The schematic flow diagram of the SSME does not include the pipelines leading the pressurizing gases to the propellant tanks, but it does show two flanged pipe stubs and the heat exchanger in the main oxygen TP.

The *health management system* of the SSME, which was an integration of advanced technologies jointly developed by Rocketdyne and the NASA Marshall Space Flight Center, monitors engine parameters, identifies anomalies, and its computer makes decisions on remedial action.[3,21-23],* It includes measurements of various temperatures, pressures, valve positions, shaft speeds, real-

---

*Personal communications, V. Wheelock, S. Claflin, and J. Halchak, The Boeing Company, Rocketdyne Propulsion and Power, 2002–2004.

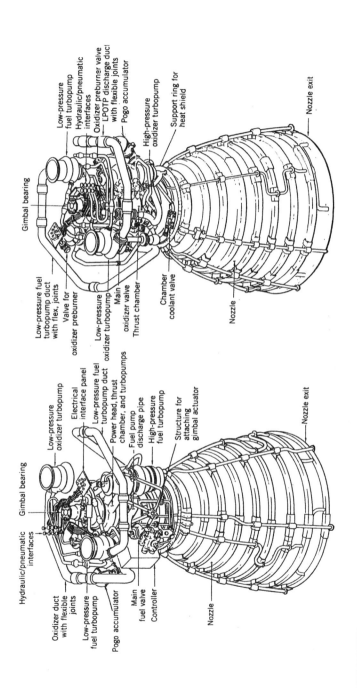

Fig. 7.8-17 Simplified sketches of two views of the space shuttle main engine. Courtesy of The Boeing Company, Rocketdyne Propulsion and Power; copied with permission from Ref. 12.

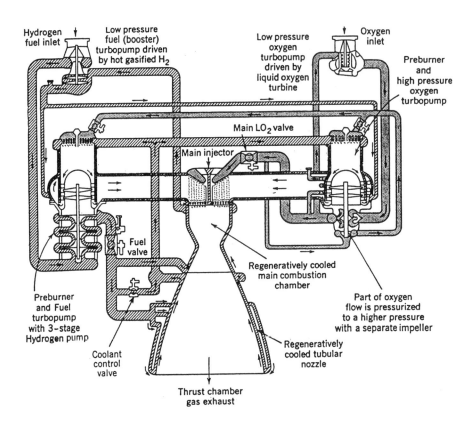

Fig. 7.8-18 Simplified flow diagram of the SSME showing four turbopumps, two preburners, and a staged combustion engine cycle. Courtesy of The Boeing Company, Rocketdyne Propulsion and Power; copied from Ref. 12.

time vibrations monitors on the TC and the TPs, and in some versions also a real-time optical plume exhaust spectroscopic analysis (e.g., detection of nickel or titanium metal species in exhaust). The computer compares the actual measurements with the calculated or expected values, determines the difference of the two, and evaluates the severity/possible effects of the anomaly. If the evaluation indicates a severe out-of-tolerance condition or an unacceptable anomaly, the computer will select and initiate a remedial action, such as a quick, safe shutdown or continued operation, but a lower thrust. The health management system includes various sensors, requires changes of the engine controller (discussed and shown in Chapter 4.7), and requires an additional health management computer. It was estimated that this health monitoring and management system will substantially improve the reliability of the three-engine cluster. The health monitoring system had not been called upon during a flight up to late 2002, but it has caused a number of shutdowns during ground tests and has certainly saved hardware.

The unique *power head* design shown in Fig. 7.8-19 consists of two forgings (made of nickel based superalloy 718), which are electron beam welded and machined together, so as to form a single piece.[21–23] It contains the main injector assembly with a novel oxygen/hot-gas heat exchanger. The power head is the backbone or key structure to which are attached the three-stage high-pressure fuel turbopump with its preburner and the single-stage oxygen pump with dual impeller inlets and its preburners. This design avoids a number of high-pressure joints and flexible hot-gas piping, which are potential sources of leaks. The figure also shows sections of the two TPs. The materials selected for the SSME are unusual and contain a large amount of superalloys.[24,25]

Improvements were made to the SSME in 1995 and again in 1996–1997 (Refs. 15–18,22). There were some problems with turbine blade cracking and shaft whirl in turbopump component tests and in a few static firing tests. The Rocketdyne-designed TP version seemed to have difficulties in overcoming these failures. One way for making the turbopumps and the engine less critical was to reduce the chamber pressure, which in turn would reduce the TP power and TP stresses. In the block II version of the SSME, the chamber pressure was reduced from 3317 to 2747 psia, and this reduced the feed pressures,

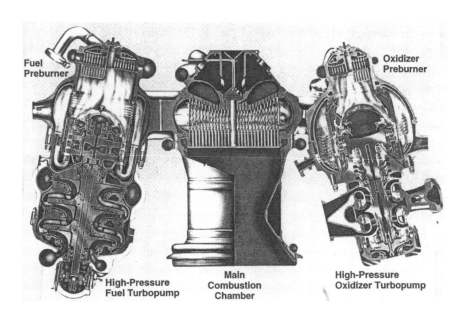

Fig. 7.8-19 Power head is the single-piece "backbone" nickel alloy structural element of the SSME, and it holds the engine's TC, turbpumps, and preburners. Courtesy of the Boeing Company, Rocketdyne Propulsion and Power (from personal communications, V. Wheelock, 2002.

many of the stresses, and the heat transfer.* A larger throat was built into the thrust chamber to accomplish the chamber pressure reduction. With this same block II change NASA required that two new robust and heavier new main turbopumps, developed by Pratt & Whitney (P&W), replace the original Rocketdyne turbopumps. NASA had gone to P&W and had asked them to review Rocketdyne's work on the SSME TPs and funded P&W to develop a set of backup TPs. The new TPs from P&W have the same interface as the Rocketdyne TPs, are interchangeable, and are quite similar in concept. These TPs are discussed briefly in Chapter 7.10. The two P&W turbopumps weighed 550 lb more than the Rocketdyne pumps they replaced, even though their pump discharge pressures were 20% lower. Another change in the block II SSME was the redesign of the power head to reduce pressure drops and increase its strength. The reliability of some sensors and controllers was also improved. Rocketdyne now assembles the P&W turbopumps into the SSME.

The *RS-68 booster engine* was designed for low cost in 1997 and 1998.[26-28] Propellants are LOX/LH_2. It flew first in October of 2002. A single RS-68 engine launches the first stage of the Delta IV family of SLVs as shown in Fig. 2-9. The initial version of the Delta IV has a single booster stage with a RS −68 LPRE, but for heavy payloads it is planned to use three such booster stages strapped together. This LPRE is shown in Fig. 7.8-20, and key data are in Table 7.8-1. At a thrust of 650,000 lbf, it is the highest thrust existing LOX/LH_2 engine and is throttleable to 58% of full power. The engine design was unusual because it was designed for low cost and not for maximum engine performance or minimum inert weight. The gas-generator engine cycle was selected because it is relatively simple, weighs less, and costs less than a comparable staged combustion cycle engine, but its performance is lower. The ablative nozzle-exit section is lower in cost that other nozzle-exit designs. The turbopump is a scaled-up version of a J-2 TP, has been simplified to have 40 unique parts per pump (as compared to many more parts in the SSME TP), and it can be assembled in one-third the time. There were a few development issues, which were resolved, with cracks in the turbine disk and the turbine blades and with extensive ablation rates in a portion of the ablative nozzle-exit skirt. A systematic approach to determine safe operating limits was taken, and the operating margins are shown in Fig. 7.8-21. These kinds of operating margins are typical of modern large LPREs.

The RS-68 is started by feeding helium from a ground system into the GG to start the spin of the turbines. A single pyrotechnic igniter is used in the TC and dual igniters in the GG. The engine could be throttled, and the thrust could be varied from 57 to 102% of rated value as shown in Table 7.8-1.

This RS-68 is the first LPRE that was fully designed on a computer using computer-aided design and analysis programs.[26-28] The database of this engine was then used for computer-aided manufacturing, field service,

---

*Personal communications, V. Wheelock, S. Claflin, and J. Halchak, The Boeing Company, Rocketdyne Propulsion and Power, 2002–2004.

Liquid Propellant Rocket Engines in the United States (Summary)

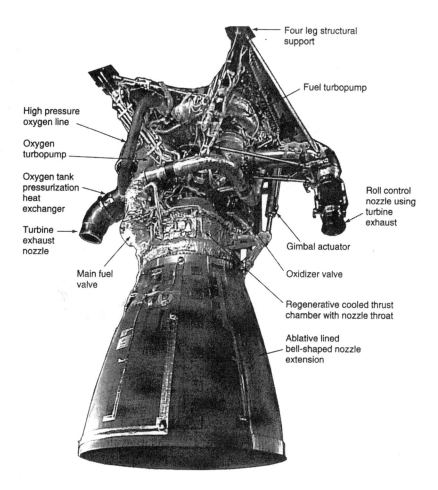

Parameter	Thrust chamber	Engine
Specific impulse at sea level (max.), sec	368	364
Specific impulse in vacuum (max.), sec	416	410
Thrust, at sea level, lbf	640,700	650,000
Thrust in vacuum lbf	732,400	745,000
Mixture ratio	6.74	6.0

Fig. 7.8-20 View of one side of the RS-68 LOX/LH$_2$ rocket engine for the Delta IV space launch vehicle. Courtesy of The Boeing Company, Rocketdyne Propulsion and Power copied with permission from Refs. 3, 26, and 28.

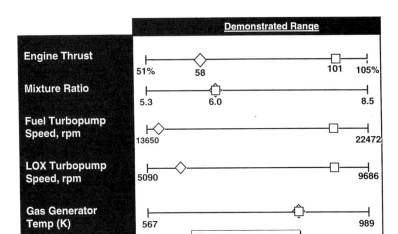

Fig. 7.8-21 These operating margins of the RS-68 LPRE are typical for a large pump-fed LPRE and the data are usually obtained from development and qualification tests. Courtesy of The Boeing Company, Rocketdyne Propulsion and Power; data from Refs. 26–28.

inventory control, information management, and other computer programs throughout the company. The common product database included a precise geometrical description of the engine and all of its parts, the geometry of the manufacturing tooling, bill of materials, NC (numerical control) data/programming, inspection records, and some test data. The computer system had a network of interactive work stations for the designer, the analysts (stress, thermal, or performance), manufacturing planners, NC programmers, etc. For a pump impeller or a throttle valve stem an accurate image of the part was developed describing the geometry and a finite element image for stress analysis. There also was a significant reduction in parts count, in the cumulative lengths of welds, and in touch labor in the manufacturing plant, when compared to earlier LPREs. The number of days of testing and the number of tests were also dramatically reduced.

Other Large Liquid Propellant Rocket Engines

In this book a large engine is defined as having more than 1500 lbf thrust. The other large engines listed here were all experimental engines, which were not intended to fly. In the preceding sections of this chapter, several such experimental engines had already been mentioned. They were the X-1 (advanced version of Thor and predecessor of the H-1 LPRE), E-1 (a 400,000 lbf predecessor of the F-1 at 1.5 million pounds thrust), the J-2X and J-2S (simplified versions of J-2), and the F-1A (larger version of F-1). The following listing is not complete.

In 1961 the X-8 experimental booster engine was designed using LOX/LH$_2$ propellants. It was rated at 90,000 lb thrust and a specific impulse of 408 s (vac) (Ref. 3). It had two separate turbopumps, a two-stage fuel pump and a single-stage oxygen pump, each with its own turbine. This feature of separate TPs was used in the J-2. It also had a gas-generator engine cycle, and the X-8 was tested before the J-2 design was finalized.

The NOMAD was an early program intended to become a future upper stage with the high-energy propellant combination of liquid fluorine and hydrazine.[3,*] It was rather unusual for Rocketdyne to not only develop the LPRE, but also to develop a complete upper stage that might fit on top of the Atlas, Titan, or Thor SLVs. Design started in 1958, and the stage is shown in Fig. 7.8-22. The engine had a nominal thrust of 12,000 lbf, a specific impulse of 357 s (vac), and was about 6 ft in diameter. A clever feature was the liquid-nitrogen jacket that covered the top half of the spherical oxidizer tank. The cold nitrogen served as a refrigerant, prevented boil-off of fluorine during storage or launch holds, and allowed rapid tanking of the fluorine without vapor escaping, which is what normally happens with a cryogenic propellant being loaded in a tank, which is at ambient temperature. The liquid nitrogen also surrounded the helium tanks, which reduced its size, and helium was used for tank pressurization. The TC was regeneratively cooled and gimbal supported to give yaw and pitch control of the upper stage. Small helium-gas roll control jets were also provided. Even though this program achieved its objectives, there was no follow-on work. As with other programs using high-energy propellants, there were concerns of possible spills of this highly toxic oxidizer at the launch sites, during a launch accident, or during transport.

Rocketdyne also developed two engines for which competitors were concurrently also developing a very similar engine to the same requirement.[3,*] In both cases the competitor was selected for completing the development and production. The Rocketdyne SE-10 engine was an Apollo lunar landing engine (10,500 to 1080 lbf thrust, deep throttling, with NTO and 50% hydrazine-50% UDMH, specific impulse 305 s), and it was started in 1963. Because the lunar landing engine was critical to the Apollo moon landing mission, NASA decided to develop two engines and then select one for qualification, installation, and flight. The version developed by TRW (today Northrop Grumman) was selected for this mission, it flew successfully, and it is described in Chapter 7.9.

The other LPRE, called RS-23, was for the orbit maneuver engine (OME) of the Space Shuttle Orbiter (same propellants, 6000 lbf thrust, specific impulse 313 s at vacuum), and Rocketdyne started this job in 1972, mostly on company money. Rocketdyne developed a good engine with a clever injector.[†] Aerojet's engine was developed to the same requirements, had a platelet injector, was selected by NASA and the prime shuttle contractor, and has flown successfully in all shuttle missions for many years. It is shown in Fig. 7.7-13.

---

*Personal communications, V. Wheelock, The Boeing Company, Rocketdyne Propulsion and Power, 2002–2004.
†Personal communications, V. Wheelock, and M. Yost, The Boeing Company, Rocketdyne Propulsion and Power, 2003, R. Thompson, retiree of Rocketdyne, 1990–2004.

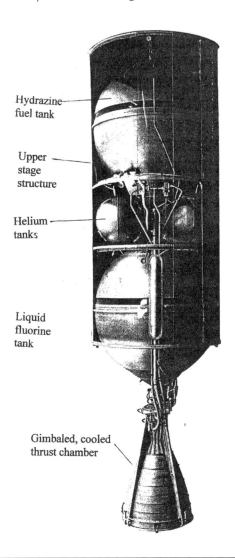

Fig. 7.8-22 NOMAD upper stage was developed to operate with liquid fluorine and hydrazine propellant, but it has not flown. Courtesy of the Boeing Company, Rocketdyne Propulsion and Power; copied with permission from Ref. 3.

The basis for the selection of Aerojet for the OME LPRE was most unusual. The prime contractor for the Space Shuttle Orbiter was the Space Systems Division (SSD) of North American Rockwell at that time. The shuttle contract with NASA required that major shuttle subsystems be obtained on a "make or buy decision." Such decisions affected other divisions of the company. For example, a make decision of the guidance subsystem by SSD would mean that the Autonetics Division of the company would make the guidance system in house. The propulsion decision by NASA and the Space Systems Division was made early and was a buy decision on the OME, which means that Rocketdyne, being part of the same North American Rockwell organization, would automatically be excluded from competing. However this decision on the OME was not communicated to Rocketdyne until much later, after Rocketdyne had developed and tested a good engine. Incidentally the space shuttle main engine was awarded by NASA directly to Rocketdyne, before the Space Shuttle Orbiter contract, and was not subject to the make or buy decision of the company's Space Systems Division. There were people in NASA and SSD who knew about the buy decision on the OME, but both NASA and SSD encouraged Rocketdyne to do the work on the OME, and Rocketdyne spent considerable effort and company money to develop such an engine. NASA even made available the NASA test facility at White Sands Proving Ground for tests of this engine. The delay in telling Rocketdyne was a dubious management practice, which might have resulted in creating some competition to Aerojet, the contractor to SSD for the OME.

In 1970 Rocketdyne developed and tested an advanced space engine identified as RS-30. It was a new upper-stage experimental engine supported by NASA, used LOX/LH$_2$ propellants at a nominal thrust of 20,000 lbf, and the measured specific impulse was projected to be a respectable 473 s in a vacuum.[3,*,†] The turbopump was located on the side of the TC. With a radiation-cooled nozzle extension the nozzle-exit-area ratio was 400, which—for a large engine—was perhaps one of the largest area ratios at that time. It had a staged combustion cycle, and it was the only other Rocketdyne or U.S. engine with this efficient cycle besides the SSME. It had a chamber pressure of 2000 psia, which was higher than any of the prior Rocketdyne engines, but not as high as the SSME. It was used in part as a test bed for evaluating scaled-down features and scaled-down components of the SSME. Mostly it was used to demonstrate the feasibility of a higher pressure upper-stage LOX/LH$_2$ engine. It is shown in Fig. 7.8-23 at the mouth of a diffuser (for conducting simulated altitude tests) with its large radiation-cooled nozzle-exit section.

The RS-44 was also an experimental upper-stage LPRE using LOX/LH$_2$, and it was called the advanced expander cycle engine. Designed in 1981, it used not only the expander engine cycle, but it also featured an extendible nozzle.[‡] The basic flow disagram for an expander engine cycle is shown in Fig. 4.2-4 and the concept of an extendible nozzle in Fig. 4.3-12. When extended to a nozzle-exit-area ratio of 320, the vacuum specific impulse was an estimated 481 s at a thrust of 15,000 lbf. During the initial tests, there was insufficient

---

*Personal communications, V. Wheelock, and M. Yost, The Boeing Company, Rocketdyne Propulsion and Power, 2003, R. Thompson, retiree of Rocketdyne, 1990–2004.
†Yost, M. C., "No Riding on or in the Rockets or my thirty years in the Rocket Factory," unpublished personal biographical account about activities and project assignments while employed at Rocketdyne and United Technologies Corp., 1992–1994.
‡Personal communications, V. Wheelock, The Boeing Company, Rocketdyne Propulsion and Power, 2002.

Fig. 7.8-23 View of advanced space engine (with a nozzle area ratio of 400) at the mouth of a diffuser for simulated altitude tests. Courtesy of The Boeing Company, Rocketdyne Propulsion and Power; copied with permission from M. C. Yost, *No Riding on or in the Rockets or my Thirty Years in the Rocket Factory,* unpublished personal biographical account, 1992–1994.

thermodynamic energy in the turbine inlet flow or really not enough power to drive the turbine, and this was remedied by lengthening the combustion chamber and thus increasing the heat-transfer area to the gaseous hydrogen in the cooling jacket. Test used a diffuser to simulate altitude. This effort on an expander engine cycle was not continued at Rocketdyne.

In 1967 Rocketdyne conducted tests of a new German LOX/LH$_2$ thrust chamber designed by the Bölkow Engineering Bureau for very high chamber pressure and very high heat transfer.* The BORD project (abbreviation for Bölkow-Rocketdyne) was jointly financed. The BORD TC was very unique for its time and is described as a derivative of another German TC in Chapter 9.5. The TC had the first actual milled channel cooling-jacket configuration seen by Rocketdyne. The Soviets had a milled channel about 15 years earlier. Because

---

*Personal communications, V. Wheelock, The Boeing Company, Rocketdyne Propulsion and Power, 2002.

the Germans did not have a facility for this propellant combination at the required high pressures, it was mutually convenient to use the new Rocketdyne Reno Test Facility, which was able to provide the high pressures needed for this TC testing. The maximum chamber pressure reached during about a dozen tests was 283 bar or about 4200 psi. This confirmed the German predictions. The work helped Rocketdyne to gain experience with high-pressure cryogenic propellants and their heat transfer; this was helpful to Rocketdyne in the SSME competition, which came shortly thereafter.

Expansion/Deflection Nozzles

The investigation of these expansion/deflection (E/D) nozzles between about 1960 and 1966 was thorough, and included design studies, cold-flow tests, developing an analytical model, and the construction and testing of several experimental engines. It covered a wide range of thrust sizes, nozzle area ratios up to 100 to one, different centerbody geometries, and gas from the turbine discharge into the chamber. The basic concept for this E/D nozzle can be seen in Fig. 4.3-9. It has a central cylinder-shaped injector, which can protrude into the chamber, and in some versions the injection and the initial combustion gas flow are essentially radially outward, but the supersonic flow is turned by the outer wall, which has a specially designed curved shape.[12,*] In other versions the injection is essentially axial. The E/D nozzle allows altitude compensation, that is, it operates at optimum expansion at all altitudes.

The three experimental engines are briefly described next. In 1961 Rocketdyne developed two booster engines, which had an E/D nozzle. These two engines operated at 50,000 and 10,000 lbf thrust, respectively, used a pressurized feed system, and ran on NTO with A-50 fuel. The 50,000-lbf engine was uncooled, and its short duration of a couple of seconds demonstrated its feasibility at a chamber pressure of 300 psia, good combustion stability, and a performance somewhat better than an equivalent bell-shaped nozzle. The 10,000-lbf engine had cooling to allow prolonged operation, a large area ratio of 300, a chamber pressure of 225 psia, and the TC length was about 40% of an equivalent TC with a conical nozzle. It was tested at the altitiude test facility at the Arnold Engineering Test Center in Tullahoma, Tennessee, at pressure ratios between 400 and 10,000. Data correlated well with theoretical prediction, and performance was slightly higher than an equivalent nozzle with a bell-shaped exit. The third set of TC hardware was a cooled design, had a vacuum thrust of 9900 lbf, and operated with a different propellant combination, namely, LOX/LH$_2$. Test were performed at the NASA Lewis Research Center in a simulated altitude facility. Test data covered a range of simulated altitudes from sea level to 100,000 ft. Nozzle performance was again better than an equivalent TC with a bell-shaped nozzle exit. At one altitude (6000 ft) it actually gave a 20% improvement in nozzle thrust coefficient

---

*Personal communications, V. Wheelock, The Boeing Company, Rocketdyne Propulsion and Power, 2002.

compared to an equivalent bell nozzle. At some other altitudes, there was also an increase, but it was much less. A centerbody bleed of turbine exhaust gas provided a 5.8% increase in nozzle thrust coefficient at 6000 ft altitude.

The theoretical predictions and the cold-flow tests agreed and correlated very well with the hot-firing test data. In general the TCs with E/D nozzles performed slightly better than an equivalent TC with a bell-shaped nozzle exit. E/D nozzles were much shorter. However the performance of an aerospike nozzle was usually better than that of an E/D nozzle, and some of the cutoff aerospike designs were shorter than a cutoff E/D nozzle. One study indicated that an engine with an E/D nozzle would be somewhat heavier than an aerospike engine or an engine with a conventional nozzle. Therefore Rocketdyne in 1966 began to concentrate its efforts aerospike designs and developments, and they are described near the end of this chapter. No further work on E/D nozzles was done after about 1966. The only other known LPRE with an E/D nozzle was developed and ground tested by the Chemical Automatics Design Bureau in Russia in the 1990s, and this is described in Chapter 8.5. There is no known engine with an E/D nozzle that has actually flown.

Prepackaged LPREs

It was already mentioned that in the decades leading up to the mid-1940s the solid-propellant rocket motors had difficulties in meeting the environmental temperature limits, which were then from −40 to + 140°F for ground-launched weapons and −65 to + 165°F for air-launched missiles. Prepackaged liquid propellant rocket propulsion systems could meet these limits, but they were more complex than a comparable solid rocket motor. Other U.S. companies, such as RMI (Chapter 7.2), also developed prepackaged LPREs, but in larger sizes.

Beginning in 1951, the NALAR (acronym for North American Liquid Aircraft Rocket) engine development started for an air-launched unguided missile.[3] It was 2.75 in. diam, 49 in. long, was launched from tubes carried by the aircraft, and had four folding fins. These fins were retracted in the launch tube and unfolded in free flight. Propellants were IRFNA and UDMH, and it had a solid-propellant cartridge for tank pressurization. The nominal thrust was 1900 lb for a duration over ½ s, producing a burnout flight velocity of 2300 ft/s. It was the smallest diameter prepackaged production LPRE, and a cross section is shown in Fig. 7.8-24. The central inner fuel tank was separated from the outer oxidizer tank by a long, thin, flexible aluminum metal bladder, which was cylindrical in its expanded condition and folded in a starshape during storage. Only the fuel tank was pressurized directly with hot high-pressure gas from the solid-propellant gas generator, and the oxidizer tank was pressurized by expansion of the flexible folded metal bladder. The ullage for both propellants was contained in a separate flexible bag inside the fuel tank only. The oxidizer tank had zero ullage. As changes in ambient storage temperature occurred, it caused changes in the density and volume of the propellants, and the flexible bladder would move to compensate with the expansion or contraction of the confined fuel ullage volume only. It had the advantage that both

Liquid Propellant Rocket Engines in the United States (Summary) 451

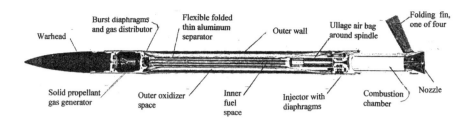

Fig. 7.8-24 Section through the NALAR aircraft launched unguided air-to-surface missile of 2.75 in. diam. with an aluminum bladder between the two storable propellants. Courtesy of The Boeing Company, Rocketdyne Propulsion and Power; copied and adapted with permission from Ref. 3.

propellant tanks were pressurized simultaneously, there was no propellant lead, and the start was repeatable from missile to missile and safe. The start of NALAR was fast, about 25 ms from electrical signal to 90% thrust.

A good air-launched unguided missile has a high hit probability, if its time to target and also its terminal velocity are essentially constant, even though the ambient propellant temperature and the air temperature can vary widely. A clever NALAR ullage compensator device inside of the slender fuel tank allowed a nearly constant time to target even though the ambient temperature was varied. With the increase in air temperature (and a reduction of air density and drag), the total delivered impulse needed to be reduced to achieve the same terminal burnout flight velocity. A typical solid-propellant rocket and a typical LPRE without a compensator would have increased thrust and a higher terminal velocity at high ambient temperatures, which is not what is wanted. In the NALAR the compensator trapped some of the fuel at higher ambient temperatures, reducing the amount of fuel burned and thus reducing the total impulse. The compensator's flapper valve trapped a quantity of fuel at the higher ambient temperatures, thus reducing the total impulse but keeping the burnout velocity and the time-to-target essentially the same. Over 3400 rockets were built, and most were flight tested from a stationary tube or launched from an aircraft. Some of the NALAR rockets were flight tested in a 10-ft-diam tunnel at the Santa Susana Field Laboratory, specially built and instrumented to record a zero-velocity launch and the initial powered flight portion. However the program was canceled in 1954 because the solid-propellant version had overcome its grain cracking problems at extremely low ambient temperatures.

The LAR (acronym for liquid aircraft rocket) was actually developed by the Naval Ordnance Station at China Lake, California, but Rocketdyne contributed to its design and built all of the hardware.[3] The LAR was a 5-in.-diam unguided air-to-air missile, also had folding fins and was tube launched. Propellants were IRFNA/hydrazine with some ammonium thiocyanate as a freezing point depressant. It had 17,000 lbf thrust for about 1/2 s. It had a pressurized feed system using a solid propellant, and it had a clever device (water bags) to limit reactions between the fuel-rich hot pressurizing gas and the liquid propellants, particularly the oxidizer. Rocketdyne built and delivered about 250 units.

Lance Missile Rocket Engine

The Lance LPRE development for a surface-to-surface missile started in 1964 as a subcontract from LTV (Ling, Temco Vought), the prime contractor, and it resulted in a most unique thrust chamber. Propellants were IRFNA/UDMH. The 20.5-ft-long missile, developed by the LTV Aerospace System's Division, had integral propellant tanks, a piston-type expulsion device in the oxidizer tank, and a solid-propellant gas generator. The unique dual-thrust-chamber assembly with its valves and thrust vector control was developed by Rocketdyne.[3,*] A simplified sketch of this unique concentric configuration in Fig. 7.8-25 shows an annular outer booster thrust chamber (nominally 50,000 lb thrust for roughly 6.1 s) and an inner throttleable sustainer thrust chamber (5000 lb maximum thrust at altitude operating for about 114 s). The ablative liners were used in both thrust chambers and were fastened to a forged steel piece; the outer layer was fastened to the aluminum casting. The cast aluminum injector contained the propellant distribution passages, five annular grooves for the booster injector with rings (not shown) that have the injection holes drilled into them, and the main fuel and oxidizer valves. The pitch and yaw TVC was accomplished by pulsed liquid fuel side injection at four places on the outer nozzle exit. It is the only known flying LPRE with liquid fuel side injection. This liquid injection TVC mechanism was chosen in part because it could be accommodated in less vehicle volume than any other TVC scheme. Cutoff of the booster TC is achieved by quickly closing the main fuel and main oxidizer valves (they are normally open) using a pyrotechnic squib. The variable thrust of the sustainer (from 5000 to 14 lbf) was achieved by a movable pintle (with an ablative face) controlled by a servocontrol valve and actuated by fuel pressure. This large throttling ratio of 357 to 1.0 is the highest known anywhere in LPRE history. Cutoff of the sustainer TC was achieved by moving the sliding pintle into the closed position.

The chamber pressure at full flow was about 950 psi, and the nozzle-exit-area ratios and specific impulses for the booster were 5.7 and 238 s and for the sustainer 4.0 and 227, respectively. The thrust-chamber assembly dry weight was 173 lb, and it was 19.4 in. long and 21.4 in. diam. The first full-duration rocket engine test was in early 1965, about one year after the contract started. Shortly thereafter, tests with the missile tanks and the pressurizing system began, and the first flight tests were in late 1965. The Lance missile had a minimum required shelf life of five years, when filled with propellants.

A total of 3229 engines were produced.[†] As with all LPREs that are produced in reasonable quantity, the costs went down as manufacturing person-

---

*Personal communications, V. Wheelock, The Boeing Company, Rocketdyne Propulsion and Power, 2002.
†Personal communications, V. Wheelock, The Boeing Company, Rocketdyne Propulsion and Power, 2002–2004.

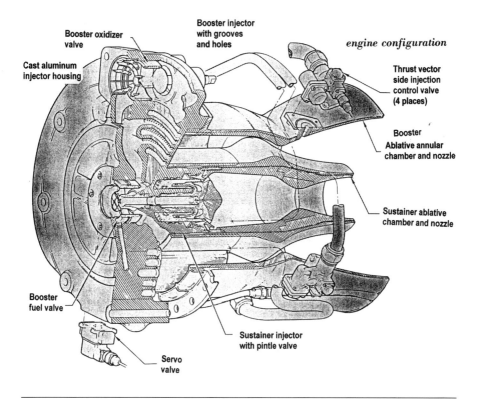

Fig. 7.8-25 TC assembly of the Lance surface-to-surface missile has an outer annular booster TC (with fuel side injection into the nozzle for thrust vector control) and a central smaller sustainer TC, which has the widest known throttling range of 327 to 1. (19.4 in. high and 173 lb mass). Courtesy of the Boeing Company, Rocketdyne Propulsion and Power copied with permission from Ref. 3 (also from personal communications, V. Wheelock, 2002).

nel became more efficient and as production automation was introduced. The initial 103 Lance development engines required approximately 1300 man-hours per engine, the initial manufacturing engines between 680 and 480 man-hours/engine and the final 1000 engines required 320 to 380 man-hours each. Deployment with the Army was in the early 1970s, and some Lance batteries were sent oversea. It replaced some short-range missiles that had used solid-propellant rocket motors, such as the Honest John. The Lance missile was decommissioned between 1990 and 1992. After the demilitarization some Lance missiles became available as target missiles.

A modified version of the Lance design was flown several times in the Homing Overlay Experiment, which was performed to gather data for future missile defense systems.[3,*] In these flight tests an experimental three-stage interceptor missile was aimed to hit a reentering ballistic target missile. The third or terminal stage of the interceptor vehicle had to maneuver to achieve impact with the incoming fast moving target. In 1979 Rocketdyne designed a scaled-down version of the Lance engine for the third stage of the experimental interceptor missile. It had an axial annulus-shaped TC with a thrust of 12,500 lb and a central sustainer TC with a thrust of 2083 lb; this configuration is similar to the Lance engine. The propellants were changed from IRFNA/UDMH to NTO/MMH (easier to store), the fuel side injection TVC system was removed and replaced by a gimbal mount, and the overall total impulse was greatly reduced. The sustainer thrust chamber could be throttled and restarted.

Rocketdyne Reaction Control Systems and Small Thrusters

The development of small thrusters for spaceflight vehicles is really a different business than those discussed before in this chapter. Information about small thrusters and their RCSs is given in Chapter 4.6. Rocketdyne has produced and delivered about a dozen different small thrusters, all of which have been flown, and most of them are mentioned in this chapter. In addition Rocketdyne developed about 30 different experimental thrusters, and only a few of these are discussed here.[3,†]

Rocketdyne developed low-thrust LPREs with three kinds of thrust chambers. The company pioneered the use of ablative cooling for LPREs, which up to the 1960s had been used only on solid-propellant rocket motors.[12] The basic concept of ablation cooling is shown in Fig. 4.6-11. The use of beryllium for conductive cooling (1966–1990s) was another new approach. They also developed small thrusters with radiation cooling using carbon fibers in a carbon matrix (beginning about 1980). Many of these small LPREs have requirements for multiple pulsing (up to several hundred thousand starts in some qualification tests), small pulse durations, and/or rapid start transients. All used pressure feed systems, storable propellants, and closely coupled propellant valves. All RCSs operate at low chamber pressures (90–150 psia) in order to minimize the inert weight of the propellant tanks and pressurizing-gas tanks.

The first *ablative lined thrust chamber* was designed in 1960 for the auxiliary propulsion system for the Agena upper stage, and it had 45 to 50 lbf thrust each.[3,†] The propellants were NTO and a mix of 75% hydrazine with 25% MMH. As explained in Chapters 4.3 and 4.6, ablative small thrusters are particularly well suited for applications, where there is a need for throttling and or restart as compared to regenerative cooling. Some of the lessons on how to make a good ablative were learned from solid-propellant rocket motors, but

---

*Personal communications, V. Wheelock, The Boeing Company, Rocketdyne Propulsion and Power, 2002–2004.
†Personal communications, E. Schuster, and V. Wheelock, The Boeing Company, Rocketdyne Propulsion and Power, 2002–2004

much had to be learned for making liners and nozzles for application to LPREs. The best materials and fabrication processes were identified, the best angles of the fibers relative to the axis of the TC and the avoiding of delamination (opening of cracks or gaps in the lay up of the ablative liner) during storage and firing were determined.

The *Gemini manned spacecraft* used 32 ablative lined thrust chambers, and they were used to achieve the correct velocity (in magnitude, position, and angular alignment) for proper and safe reentry into the atmosphere.[3] The piloted Gemini capsule itself had two redundant propulsion systems with eight 25-lb thrust chambers, with their own feed systems for attitude control and reentry alignment. The support module, which was attached to the Gemini reentry capsule, had eight more 25-lb thrust units plus two at 85 lb and six at 100 lb thrust with their own feed system for orbital maneuvers and attitude control. Design started in 1962, and the first flight was in 1964. Beginning with Gemini all subsequent Rocketdyne ablative units used NTO with 100% MMH. A cross section of a Gemini ablative thruster and an external view of another are shown in Fig. 7.8-26. The thick ablative liner is protected from excessive erosion by a ceramic sleeve and a ceramic nozzle insert and by extra layers of ablative material wrapped so that its thermal conductivity would be low.

Two sizes of ablative thrust chambers (45 and 25 lb thrust) were used on the *Titan Trans-stage* for the flight control of the payload after the two trans-stage engines had been shut off; these small thrusters were designed in 1964.[3] This stage is mentioned in Chapter 7.7. The *Apollo command module* had 12 ablative thrust chambers at 93 lb thrust each for attitude control and reentry alignment control of the Apollo Command Module.[3,18,*] There were two redundant LPREs with six thrusters each. These were also designed in 1964 and operated at 137 psia nominal chamber pressure with a steady-state specific impulse of 274 s at altitude. The ceramic liner used in prior designs was eliminated by precharring a sleeve in the thruster. Figure 4.6-12 shows one of these Apollo command module thrusters cut open for postflight inspection; the discolored layer caused by intense heat is difficult to see in a black and white photo. The thick ablative sleeve near the nozzle exit is needed because this thruster is close to the heat shield, which becomes very hot during reentry. In 1965 the 72-lb thrust units were designed for ullage control or *propellant settling of the Saturn IVB upper stage*.[18] This settling of the liquid propellants is needed in gravity-free space prior to the restart of the J-2 engine. The central gimbaled axial-velocity control engine of the Peacekeeper postboost control system also used ablative liners and was restarted several times in space. Altogether Rocketdyne developed 10 different small thrusters with ablative liners and three larger TC also with such liners. Within a few years the small ablative thrust chambers, used in reaction control systems, were replaced by lighter and cleaner burning beryllium and refractive metal chambers.

---

*Personal communications, E. Schuster, and V. Wheelock, The Boeing Company, Rocketdyne Propulsion and Power, 2003.

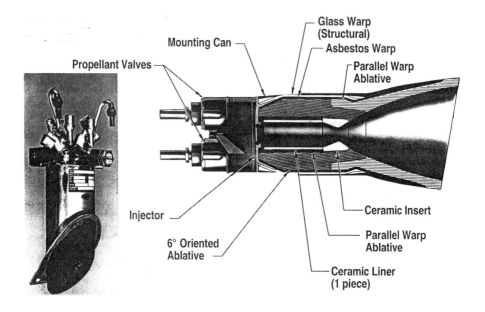

Fig. 7.8-26 Section and external view of two ablative thrusters used on the Gemini manned spacecraft. The 25-lbf thruster on the left was used for attitude control; its nozzle is cut off at an angle in order to fit the vehicle skin contour without protruding. The 100 lbf thruster was used on the Gemini orbital maneuvering module, which is detached from the manned capsule just prior to reentry. Courtesy of The Boeing Company, Rocketdyne Propulsion and Power from Ref. 3.

The second approach to small bipropellant thrust chambers was a unique Rocketdyne contribution to the state of the art of propulsion technology. It uses a thick *beryllium wall* in the chamber and also in the nozzle-throat section.[3,*] Figure 4.6-18 shows schematically how the heat is transferred by heat absorption and conduction within the beryllium from the hotter nozzle-throat region and conducted to the film-cooled chamber wall region. There the heat is then transferred from the hot chamber wall to the cooler fuel-rich film-cooling gas layer. Because heat is returned to the chamber gas, this method has been called *interregenerative cooling or interegen cooling*. Beryllium has some unique properties that make this possible, and Table 4.6-1 shows good heat conduction, high heat absorption per unit mass, low density or low weight, and a substantially higher melting point than aluminum. Rocketdyne developed and tested at least nine different thrust chambers using intergen cooling.

The first thruster with this cooling method was designed in 1966 for the postboost reaction control system of the Minuteman III ballistic missile. Postboost control systems (PBCS) are used in ballistic missiles to impart an accurate terminal velocity to each of several warheads. This was discussed in

---

*Personal communications, E. Schuster, and V. Wheelock, The Boeing Company, Rocketdyne Propulsion and Power, 2003.

Chapter 4.6, and the LPRE for a PBCS is illustrated in Fig. 4.6-20. The first interregen thruster for the attitude control of the PBCS had 23 lbf thrust (vacuum) at a specific impulse of 259 s and used NTO/MMH at a low nozzle area ratio. A derated version at 18 lbf thrust was also developed, but it had poor performance. These two thrusters were not selected, but instead a radiation-cooled version with thermal insulation developed by ARC was chosen. One can be seen in Fig. 4.6-14.

This interregenerative cooling concept is still being used on the axial thrust chamber (316 lb thrust) of the Minuteman III PBCS. This TC has been given the company designation RS-14.[3,*] Its design started about 1966, had a nominal thrust of 316 lbf (vac), and the thrust chamber was fabricated from a single billet of beryllium. This axial thrust chamber is shown on the left in Fig. 7.8-27 with an insulation blanket, together with two other interregen thrusters. The injector for this thruster is shown in Fig. 7.8-28 and is discussed further in a separate paragraph. The engine and its PBCS stage are discussed in Chapter 7.11. Designed in 1966, its specific impulse was 288 s (vac), it was gimbal mounted, had an external thermal insulation bracket, and it used NTO/MMH. This axial thruster is delivered to ARC, where it is integrated into the Minuteman postboost control system. A total of 996 of these systems have been delivered, and more than 150 have flown. Its injector has resonance cavities, which were discovered by accident as explained next.

This interregen type of TC is also still used on the Peacekeeper (stage four) postboost control system in eight small thrusters, each at 70 lbf thrust. Figure 4.6-20 shows a schematic diagram of this propulsion system and its fourth stage, both of which were built by Rocketdyne. Beryllium TCs were also developed for the Mars Mariner and Viking planetary space probes, each with a vacuum thrust of about 300 lbf. They can also be seen in Fig. 7.8-27. They had the beryllium wall only in the chamber and the nozzle-throat region and used a radiation-cooled niobium skirt for the lower part of the nozzle. The Mars Mariner started in 1968 and flew in 1971, and the Viking Orbiter started later and flew first in 1975. Both had relatively long durations (1940 and 2980 s, respectively), used the same propellants, and both had a gimbal ring outside of the nozzle throat. All three of the interregen thrusters in this figure were put into production, and all three have flown successfully.

In the development program of the Minuteman RS-14 thruster mentioned in the preceding paragraph, it was difficult to achieve a TC configuration with stable combustion. It led to the accidental discovery of acoustic cavities, which subsequently have become very effective in controlling combustion instability. The tests were made with a small injector and a heavy-duty beryllium chamber/nozzle as shown in Fig. 7.8-28. Test were inconsistent, mostly stable combustion, but some tests and some hardware had unstable behavior.[†] It

---

*Personal communications, E. Schuster, and V. Wheelock, The Boeing Company, Rocketdyne Propulsion and Power, 2003.
†Personal communications, V. Wheelock, and M. Yost, The Boeing Company, Rocketdyne Propulsion and Power, 2003.

Fig. 7.8-27 Three thrusters using interregen cooling with beryllium. Courtesy of the Boeing Company, Rocketdyne Propulsion and Power copied with permission from Ref. 3 (also from personal communications, V. Wheelock, and E. Schuster, The Boeing Company, Rocketdyne Propulsion and Power, 2002–2004).

required considerable effort to identify the cause, and various modifications were evaluated. The test hardware had a cylindrical gap on the inside of the joint between the injector and the chamber, and this gap can be seen in Fig. 7.8-28. The development engineers noticed that this gap was not always the same. They decided to widen the gap. They learned that the TC (with an injector known to cause instability) had stable operation when the gap was large (about 0.060 in. wide or wider) and sometimes unstable operation when the gap was small (0.030 in. or less). So the investigators repeated the tests and found the same results. This then led to further experiments with different gap widths, various shaped gaps, deeper gaps, and compartmented or interrupted gaps. It turned out that this gap acted as a quarter-wave resonator for the radial mode of combustion instability, and it also influences the tangential mode. Acoustic cavities are discussed further in Chapter 4.10. In this program the astute observations came before the theory was developed for cavities. Resonant cavities have become one of the more effective methods for avoiding

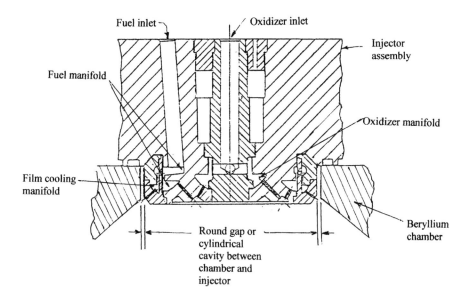

Fig. 7.8-28 Partial section drawing of the RS-14 thruster (axial engine of the Minuteman III PBCS) showing complex injector manifolds. The cylindrical cavity between the injector and the beryllium chamber led to the discovery of acoustic cavities for controlling combustion instability. Courtesy of The Boeing Company, Rocketdyne Propulsion and Power copied with permission from Yost, M. C., "No Riding on or in the Rockets or my thirty years in the Rocket Factory," unpublished personal biographical account, 1992–1994.

combustion instability and have been used on many other TCs. To this day the RS-14 Minuteman thruster has been an active program, and it still has the acoustic cavities.

The Germans also have developed a small thruster with interregen-type cooling in the early 1990s. The material for absorbing and conducting the heat is a platinum alloy containing some rhodium. It has good conductivity, but is very heavy compared to beryllium. It is shown in Chapter 9.5.

The small thrusters used for the *terminal maneuvers* of the final or top stage of an antiaircraft or antimissile defensive vehicle are somewhat different.[3,12] It is sometimes called kinetic energy weapon. Its defensive maneuvers are aimed to produce an impact with the fast-moving approaching intended target. With the high velocities of the fast-approaching threat (missile or aircraft) and the fast-moving defensive interceptor vehicle, the time for flight trajectory changes (to achieve a hit) is usually very short (often less than a minute). Also space for propellant tanks and thrusters is usually very limited in a terminal stage of a small interceptor vehicle. This application needs small thrust chambers for pitch, yaw, and roll control in order to keep the target seeker pointed at the target. It also needs a relatively high pulsing thrust (300–900 lbf) for sideways translation maneuver, also called divert maneuver. This divert thrust has to go through the center of gravity of the defensive vehicle, so that there would be no turning or rotation of the flying interceptor because otherwise the target seeker on the interceptor would lose its target. Pulsing is usually also required. Therefore these LPREs operate at higher thrust than ACS thrusters (to achieve high side accelerations), at a higher chamber pressure (200–800 psia) and lower nozzle area ratio (so that the TCs are small and can be placed into a tight space) than most attitude control LPREs for satellites. The interregen cooling scheme and the radiation-cooling scheme for small thrusters is not suitable because they are not able to operate at the higher chamber pressures and at the higher heat-transfer rates.

The third type of small thruster developed at Rocketdyne uses oriented carbon fibers in a matrix of carbon or silicon carbide as the wall material.[3,12,29] It can withstand the higher heat transfer rates and the higher wall temperatures, which result from higher chamber pressures used in defensive missions. One type is shown in Fig. 7.8-29. As seen in Fig. 4.6-13, carbon can operate at a much higher wall temperature than any of the metals (e.g., niobium) commonly used in small thrusters. To protect the carbon from the rapid oxidation by the oxidizing species in the hot propellant reaction products, it is necessary to provide an oxygen-free reaction gas or an internal protective coating. By plating or vapor depositing an external metal ring on the forward end of the carbon thrust chamber, it is then possible to weld or braze a metal injector to the carbon chamber. This type of thruster can better accommodate the higher heat transfer at the higher chamber pressure as needed in interceptor terminal stages. Rocketdyne has developed several sizes of these carbon thrust chambers (between 5–500 lb thrust) and has also investigated other materials such as a silicon carbide-carbon fiber combination. These kinds of thrusters are likely to be useful in defensive system terminal stage applications.* Rocketdyne has developed more than one complete engine system for defensive missions using higher chamber pressure carbon-type thrusters, and some have flown in a hover facility mentioned in the next paragraph. Information of specific defense systems or their engines/thrusters was not available.

These terminal maneuver systems can undergo flight tests in a unique indoor facility.* The Hover Test Facility at Edwards Air Force Base allows such limited free flighttests of terminal vehicles without restraints or umbilical cords to the vehicle. "Hovering" is achieved, when the divert thrust is equal to the test vehicle's weight. The hover test facility allows limited translation and rotation maneuvers within a large hangar building with safety steel nets and special instruments. Rocketdyne provided inputs to the design of this unique facility.* A Rocketdyne experimental terminal vehicle engine system with ACS thrusters and divert thrusters was the first to use this facility in 1987 and 1988.

---

*Personal communications, E. Schuster, The Boeing Company, Rocketdyne Propulsion and Power, 2002–2004.

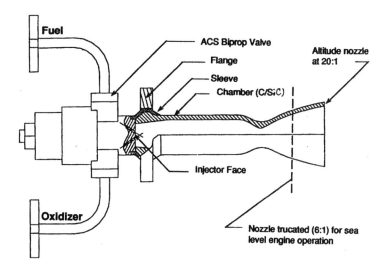

Fig. 7.8-29 Simplified section through a 30-N (6.7-lbf) thruster using a carbon fiber with silicon carbide material. Courtesy of The Boeing Company, Rocketdyne Propulsion and Power, copied with permission from Ref. 29.

A number of different maneuvering terminal vehicles (with multiple TCs developed by different companies) have since been tested there.

Rocketdyne Aircraft Superperformance Engines.

Rocketdyne developed, built, and did preliminary flight rating tests (PFRT) on five different LPRE for aircraft applications using 90% hydrogen peroxide and jet fuel (JP-4 or JP-5) as propellants.[3,*] Two of these have actually flown. The five LPRES are shown in Fig. 7.8-30, and work on these LPREs was done between 1955 and about 1969. They are the only U.S. aircraft superperformance LPREs using hydrogen peroxide and jet fuel. The fuel came directly from the aircraft's fuel tanks, and the oxidizer had separate tanks inside the aircraft. All used a silver-screen-type catalyst at the injector end of the chamber, where the 90% hydrogen peroxide was decomposed into steam and hot oxygen. The hot decomposition gas was injected into the combustion chamber, where it readily ignited with the jet fuel and burned in an efficient combustion. All of the TCs were of double-wall steel construction with expansion bulges in the outer wall. The oxidizer was the coolant, and it entered directly from the cooling jacket into the catalyst bed. The first three engines shown in the figure (AR-1, AR-2, and AR2-1) used pressurized-gas feed systems, and the

---

*Personal communications, V. Wheelock (2003) and J. M. Cumming (1980) The Boeing Company, Rocketdyne Propulsion and Power.

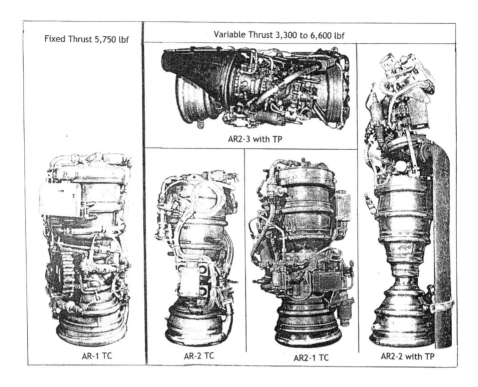

Fig. 7.8-30 Five aircraft superperformance rocket engines were developed, and two of these (AR-1 and AR2-3) were flown. Courtesy of the Boeing Company, Rocketdyne Propulsion and Power obtained with permission from Ref. 3.

AR2-2 and AR2-3 used a turbopump feed system with a monopropellant hydrogen-peroxide gas generator. This allowed lightweight propellant tanks and thus better aircraft performance. The last four of these LPREs could be throttled down to about 50% of rated thrust. Table 7.8-2 lists some data of the AR2-3 LPRE, which is the engine that has the most flight hours. It is also shown in a perspective line drawing in Fig. 7.8-31 (Ref. 3).* A simple flow diagram of the AR2-3 is shown in Fig. 7.8-32. The turbopump, shown in cross section in Fig. 7.8-33, is simple, has a single shaft, the two bearings are lubricated and cooled by jet fuel, and the TP is driven by its own gas generator with its own catalyst bed.

The AR-1 LPRE was a monopropellant engine using 90% hydrogen peroxide and had a pressurized feed system. It was the first Rocketdyne aircraft engine to be flight tested on a Navy FJ-4F fighter aircraft, and the installation of this engine with its pressurized feed system is shown in Fig. 7.8-34 (Ref. 30). The tanks were heavy, and the aircraft performance improvement was not spec-

---

*Personal communications, V. Wheelock (2003) and J. M. Cumming (1980) The Boeing Company, Rocketdyne Propulsion and Power.

Table 7.8-2 Performance and Weight of the AR-2-3 Aircraft Super Performance Rocket Engine

Nominal thrust at 100,000 ft, lbf	6600
Minimum thrust, as percent of nominal	50
Nominal specific impulse @ 100,000 ft, s	246
Chamber pressure at full thrust, psia	560
Nozzle area ratio	12:1
Mixture ratio	6.5
Dry weight, lb	225
Overhaul/service requirements after cumulative firing duration of	
For servicing, min	60
For overhaul, min	120

tacular. The next several flight tests were done with the pump-fed AR-2-3 rocket engine, and it allowed a much thinner and lighter oxidizer tank, and the fuel came directly from the aircraft's fuel tanks. The FJ-4F airplane was later modified and flight tested with the AR2-3 engine for 103 flights with a total rocket engine operation of 3.5 h. In level flight at altitude, this aircraft flew at supersonic speed at Mach number 1.31. An Air Force F-86 fighter aircraft was also flight tested with an AR-2-3 for 31 flights, and it flew horizontally at Mach number of 1.22. A total of 302 test flights were made with the AR2-3 engine in the Lockheed NF-104A fighter aircraft for a cumulative duration of 8.6 h, and the aircraft was able to reach an altitude of 121,000 ft. Three NF 104A aircraft suitable for rocket installation have been built. After the flight tests two of the aircraft have been flown periodically with the AR2-3 engine, and they have been used since the 1960s for the training of U.S. astronauts in a temporary zero-gravity environment.* None of these five LPREs were deployed with the Air Force.

The original aircraft rocket engines with hydrogen peroxide were developed in Germany in the period of the 1930s and the early 1940s. They used hypergolic fuels and turbopump feed systems, and they are described in Chapter 9.3. In Britain (Chapter 12) aircraft rocket engines were also developed with 90% hydrogen peroxide as an oxidizer using independent pressurized-gas feed systems. Many foreign-developed superperformance engines in Russia and France depended on a jack shaft from the turbojet engine to drive the rocket engine's turbine, which in turn drives the propellant pumps; however, usually the jet engine does not have enough power at high altitude to do so adequately. The Rocketdyne aircraft engines are self-contained, have their own feed system, and do not depend on power from the turbojet engine. The 90% hydrogen peroxide does not give off toxic fumes (in case of a spill or leak) as do RFNA or NTO, which are the propellants used by the Soviets and the French in their superperformance aircraft LPREs, which are discussed in Chapters 8.2 and 10.

---

*Personal communications, S. J. Smith, aircraft consultant formerly of Lockheed.

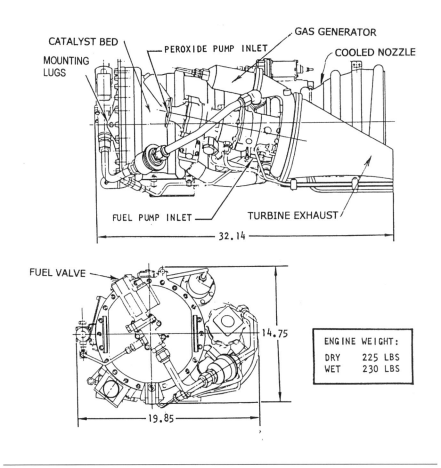

Fig. 7.8-31 Simplified drawing of the AR-2-3 superperformance aircraft rocket engine, Courtesy of The Boeing Company, Rocketdyne Propulsion and Power; copied with permission from Ref. 3.

Aerospike and Linear Aerospike

The aerospike nozzle, also called a plug nozzle is explained and compared with other nozzles in simplified sketches in Fig. 4.3-9, and the pressure distribution and flow patterns are shown in Fig. 4.3-11. The pressure distribution over the ramp surface of spike changes with altitude.[3,12,31–33] An aerospike nozzle has the advantage of operating at optimum nozzle expansion at all altitudes. If the spike is truncated (cut off), the performance diminishes only very slightly, but engines with a truncated aerospike nozzle will be much shorter than engines with more conventional nozzles.[12] The shorter length can reduce the inert mass of the flight vehicle. When discharging the turbine exhaust of an aerospike engine (with a gas-generator engine cycle) through the open surface of the cut-off spike, a small amount of additional performance can be gained. This compensates for the small loss caused by cutting off the spike. General Electric

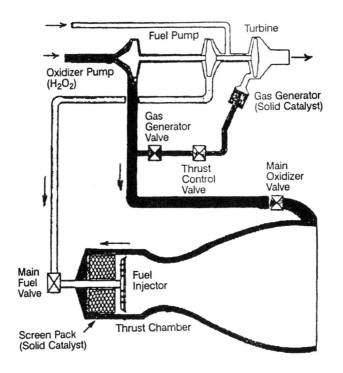

Fig. 7.8-32 Simplified flow diagram of the AR2-3. Courtesy of The Boeing Company, Rocketdyne Propulsion and Power, adapted from Ref. 30.

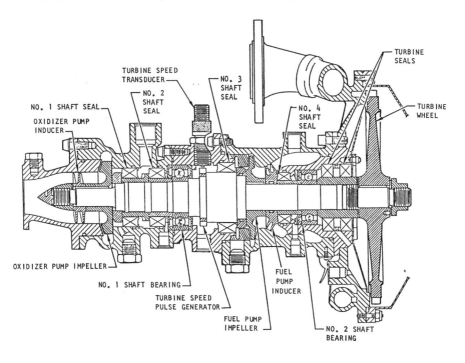

Fig. 7.8-33 Simple small TP for AR2-3 aircraft LPRE. Courtesy of The Boeing Company, Rocketdyne Propulsion and Power; copied with permission and adapted from Ref. 3.

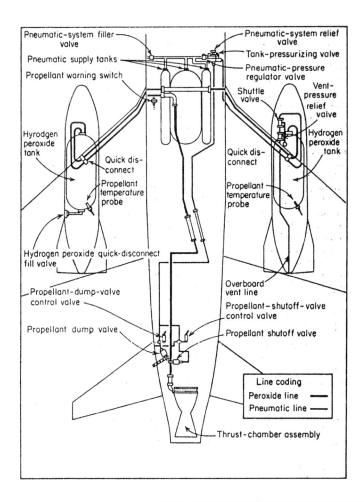

Fig. 7.8-34 Simplified diagram of AR-1 monopropellant hydrogen-peroxide engine with a gas-pressurized feed system installed in a Navy FJ-4 aircraft. Courtesy of North American Aviation and Rocketdyne, copied from Ref. 30.

Company was the first to work on an annular Aerospike LPRE at thrust levels of 16,000 lbf (H_2O_2) and then 50,000 lbf (LOX/LH_2) in the early 1960s. More information and pictures of these engines can be found in Chapter 7.3. However this company voluntarily went out of the LPRE business a few years later.

Rocketdyne has been working on aerospike LPREs since about 1965. Six different configurations of LPREs with axisymmetric aerospike nozzles were tested in the 1967 to 1972 period.[3,*] The first one used the decomposition products of hydrogen peroxide at a thrust level of roughly 4000 lbf. Pressurized feed systems were used initially at thrust levels of 18,000 lbf (vac)

---

*Personal communications, V. Wheelock, The Boeing Company, Rocketdyne Propulsion and Power, 2003.

and 50,000 lbf (vac) with LOX/LH$_2$. Two configurations of aerospike LPREs were then run at higher chamber pressures up to 250,000 lbf thrust, again using LOX/LH$_2$ propellants. There also were small units that were tested with LOX/RP-1, a type of kerosene, at thrust levels between 1500 to 2000 lbf, 3000 to 4000 lbf, and up to 18,000 lbf. The combustion chamber was annular and included some baffles. Almost all aerospike nozzles were cooled, were truncated, or had a cutoff spike and thus were very short.[31–33]

In 1967 Rocketdyne opened a second major test facility for large LPREs and particularly for cryogenic propellants. It was located about 25 miles north of Reno, Nevada, in a sparsely populated area and was known as the Reno Test Facility. It was established in part because the Santa Susana Test Facility near Los Angeles was predicted to be very busy and also because management was concerned about the gradual encroachment of new homes close to the Santa Susana test areas. This new Reno Facility is mentioned here because many of the tests on the aerospike TCs were done here. One of the test stands had heavy large high-pressure propellant tanks suitable for TC tests with cryogenic propellants at high pressures. A ground test of a cooled TC with a cutoff aerospike nozzle at about 40,000 lbf (SL) thrust is shown in Fig. 7.8-35 using LOX/LH$_2$.

The high-pressure testing of the BORD, a regeneratively cooled high-pressure German-designed TC at chamber pressures up to 4200 psi, was performed here in 1968. It was mentioned earlier in this chapter. Furthermore this site was used for operating the cooled experimental German TC at 3000 psi, and these tests supported Rocketdyne's proposal for the SSME. This Reno facility was closed in the early 1970s because of contract cancellations and a major decrease in business.

Beginning in 1970, a *linear version of the aerospike engine* was being developed at Rocketdyne.[12,32,33] This program was started because some vehicle prime contractors expressed interest in a long thin rocket engine rather than in a conventional axisymmetric round engine. It was a pump-fed LPRE and was tested at 100,000 lbf thrust with LOX/LH$_2$. There were two sets of 10 small thrust modules (also called thrust cells) on the top of each of the two opposed two-dimensional curved ramps. The nozzle exits of each of these small modules were rectangular. The two rows of modules permitted a continuous flow of hot supersonic gas over the two ramps, where the gas flow deflected and turned. The exhaust gases from the turbine are discharged through a porous plate or a perforated plate at the base of the cutoff spike, providing a small additional thrust (1–2%). The geometry of this linear aerospike can be seen in the next two figures, which show a more advanced version of this particular linear LPRE. The ramps and the small thrust chamber modules were regeneratively cooled. At the ends of the linear aerospike engine, there were hydrogen-cooled endplates, which confined the hot gases and kept them from spreading, which would have decreased performance. None of these aerospike or the linear aerospike LPREs found an application in the 1960s and 1970s, the interest in aerospikes diminished, and the work was temporarily discontinued.

In the 1990s NASA wanted to build and fly an experimental vehicle for exploring features for a future single-stage-to-orbit SLV. The development of the X-33 blended wing experimental flight vehicle was started. The linear aerospike engine could be tailored to fit into the trailing edge of this flight-test vehicle and would give a better performance than other types of LPREs. Rocketdyne was given the job of developing the XRL-2200 linear aerospike LPRE for this application.[12,32,33] The engine is shown in Fig. 7.8-36, and a ground test at the NASA facility at Stennis Test Center is seen in Fig. 7.8-37. There were no cooled side plates in these tests. This LPRE had a sea-level thrust of 204,400 lbf and a specific impulse of 339 s (SL) or 436 s (vac) at a nominal chamber pressure of 857 psia. Two engines (in line and bolted together) were intended to propel the vehicle. Another merit of this linear aerospike was differential throttling TVC, as shown in Fig. 4.9-6, and it was validated by full-scale engine tests.

Each of the 20 thrust cells is regeneratively cooled by hydrogen fuel, uses a milled-channel-type copper cooling jacket, and has a round nozzle throat, but a rectangular nozzle exit.[3,12,32,33] By throttling selected banks of small thrust modules by about ±15%, more than adequate torques could be applied to the vehicle in pitch, yaw, and roll. Such throttling was accomplished in one of the engine development tests of two engines coupled together, and it was also tested by General Electric in the 1960s.

The *combustion wave ignition system* used with linear aerospikes was clever and unique because warm gas from a single igniter combustion chamber was used to ignite two banks of 10 small thrust cells each. In principle it had a cen-

Fig. 7.8-35 Test firing of an axisymmetric aerospike (with annular slot nozzle) at about 40,000 lbf (SL) of thrust. Courtesy of The Boeing Company, Rocketdyne Propulsion and Power copied with permission from Ref. 3.

tral small gas generator, which generated fuel-rich gas at an elevated temperature, but well below the melting point of steel. This warm gas was sent through an array of insulated steel tubes to each of the 20 modules or cells. In each of the modules, the warm gas was injected through a coaxial type of ignition tube similar in concept to a coaxial injection element. The warm gas was in a tube in the igniter's center, gaseous oxygen was injected through an annular orifice around the center tube, and another outer injection annulus (around both of the other two) injected some gaseous hydrogen. The ignition flame from this combined igniter/injector element was hotter than the warm gas because of the chemical reaction between the fuel-rich warm gas and the hydrogen with the oxygen. In actual practice the start events in this simultaneous ignition in the 20 thrust modules of each of the two linear aerospike engines had to be precisely timed, measured, and controlled through an electronic set of controls. It turned out to be a rugged and reliable ignition system. An earlier version of this combustion wave igniter was used in the first linear aerospike, developed in the 1970s as just discussed.

A small-scale model of the RS-2200 was developed and flight tested several times in a SR-71 high-altitude aircraft. It used gaseous oxygen and gaseous hydrogen from pressurized tanks in a blowdown mode, four thrust cells on each of the two curved ramps, and the type of the ignition system just mentioned. Design thrust was 7000 lbf, and the nominal specific impulse was 430 s. This model engine was installed at the trailing edge of the vertical stabilizer of the aircraft. Flight tests demonstrated the aerodynamic validity of the linear aerospike at different altitudes and flight speeds. These were the only known flights of an aerospike engine.

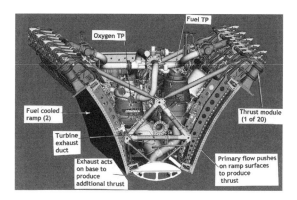

Fig. 7.8-36 Computer-generated drawing of one version of the XLR-2200 LPRE with a linear aerospike nozzle with two ramps and 10 thrust cells, but without the cooled side plates. Courtesy of The Boeing Company, Rocketdyne Propulsion and Power; copied with permission and adapted from Ref. 3 (also from personal communications, V. Wheelock, 2003).

Fig. 7.8-37 XLR-2200 linear aerospike LPREs (two sections coupled together) being tested at NASA's Stennis Space Center without cooled side plates. Courtesy of NASA and The Boeing Company, Rocketdyne Propulsion and Power; copied with permission from Ref. 3.

In 2001 NASA canceled the X-33 experimental vehicle program and with it also the linear aerospike LPRE effort. The reason for the termination had nothing to do with the Rocketdyne engine effort. The engine development was proceeding well at the time with several successful static firing tests. The actual ground-test performance was 1 or 2% below the design performance, and the changes necessary to attain full performance had not then been made.

Research

Rocketdyne had an active, first-class research department, which was engaged in a large variety of R&D tasks. It started its work in 1947.* They were one of the first in the industry to perform analytical simulation of combustion and nozzle flow, and in the 1950s they provided early data on the theoretical

---

*Personal communications, R. Thompson, The Boeing Company, Rocketdyne Propulsion and Power, 1970.

performance of many different propellant combinations. Some of these data is shown in Table 4.1-1. They performed laboratory tests of different propellants, determining their physical and/or chemical properties, including their ignition characteristics or compatibility with materials of construction. They undertook the successful synthesis of new energetic propellants, but they did not find their way into LPREs. They test fired a variety of propellants including those mentioned in Chapter 4.1 using small thrust chambers. The experimental work on mixtures of liquid fluorine and liquid oxygen (called FLOX) with various fuels demonstrated the feasibility of this approach to high-energy propellants. They deserve the credit for the idea in 1952 of adding soluble fluorine compounds to nitric acid to inhibit the corrosion of tank wall materials, which then became inhibited red-fuming nitric acid. As already mentioned in Chapter 4.1, this research department also synthesized a new hydrocarbon, diethylcyclohexane, which gave a little more specific impulse than RP-1 and would have been readily manufactured in a petrochemical plant. The timing of this discovery was inappropriate because the government had just made the decision to standardize RP-1 as the hydrocarbon fuel for U.S. LPREs and was not interested in funding other chemical plants for this new organic compound. The research department also undertook R&D programs on ignition, heat transfer, combustion instability, or improved instrumentation. When funding for research related to LPRE became essentially unavailable, this research department was effectively dissolved in the 1970s.

Concluding Comments

Rocketdyne has been the largest U.S. company in this LPRE field and around 1970 it probably was for a time the largest such company in the world. It has had an active illustrious history of developing, building, and flying booster LPREs for MRBMs, booster and sustainer engines for an ICBM, several different SLVs, especially the Saturn/Apollo SLV, several tactical missiles with storable propellants, a variety of small thrusters, aircraft superperformance, and with all the common types of liquid propellants. They also developed a number of nonflying experimental engines. Rocketdyne did some groundbreaking work, such as contours for nozzle exit sections, inhibited red fuming nitric acid, aerospike nozzles, interregen cooling, or design by computer. The limited production of the space shuttle main engine is scheduled to terminate, when the planned phase out of the space shuttle will happen in the near future. The limited production of the RS-68 for the Delta IV SLV, some continuing work in small thrusters, and various studies seem to keep the company going, however at a greatly reduced level of business.

References

[1]Winter, F. H., "Rocketdyne—a Giant Pioneer in Rocket Technology: the Earliest Years, 1945–1955," International Academy of Astronautics, Paper 97-IAA.2.2.08, 1997.

[2]Winter, F. H., "The East Parking Lot Rocket Experiments of North American Aviation, Inc., 1946–1949," International Academy of Astronautics, Paper 99-IAA.2.2.07, 1999.

[3]"Atlas," "NALAR," "H-1," "J-2," "F-1," "Navaho," "Apollo ACS," "Gemini ACS," "Redstone," "SSME," "RS-68," "LAR," "AR2-3," "NATIV," "Lance," "Aerospike," "Jupiter," "Health Management," "Liquid Propellant Rocket Propulsion Systems," and several other LPREs, Company Data Sheets, Pamphlets, Brochures, The Boeing Co., Rocketdyne Propulsion and Power, Canoga Park, CA, 2000–2004.

[4]Bullard, J. W., "History of the Redstone Missile," History Division, Army Missile Command, Huntsville, AL, Oct. 1965.

[5]Brennan, W. J., "Milestones in Cryogenic Liquid Propellant Rocket Engines," AIAA paper 67–978, 1967.

[6]Iacobellis, S., "Liquid Rocket Engines, Their Status and Their Future," AIAA Paper 66–828, 1966.

[7]Murphy, J. M., "Liquid Rocket Propulsion System Advancements, 1946–1970," 14th History Symposium of the International Academy of Astronautics, Sept. 1980.

[8]Dixon, T. F., "Development Problems of Rocket Engines for Ballistic Missiles," *Interavia*, LXXI, 1959, pp. 818–828.

[9]Heppenheimer, T. A., "The Navaho Program and the Main Line of American Liquid Rocketry," *Air Power History*, Vol. 44, No. 2, Summer 1997.

[10]Gibson, J. N., *The Navaho Missile Project*, Schiffer Publishing, Atglen, PA, 1996.

[11]Smith, R. A., and Minami, H. M., "History of the Atlas Engine," American Astronautical Society, Paper 89–555, 1989.

[12]Sutton, G. P., and Biblarz, O., *Rocket Propulsion Elements*, 4th and 7th eds., Wiley, Hoboken, NJ, 1976, 2001.

[13]Fuller, P. N., and Minami, H. M., "History of the Thor/Delta Booster Engines," American Astronautical Society, Paper 89–554, 1989.

[14]Braun, J. H., "Development of the Jupiter Propulsion System," *History of Rocketry and Astronautics*, Vol. 20, American Astronautical Society History Series Univelt, San Diego, CA, 1991, Chap. 5, pp. 133–144.

[15]Biggs, R., "Development of the F-1 Rocket Engine," presented at the *Joint Propulsion Conference of AIAA/SAE/ASME*, July 2004.

[16]Coffman, P., "Development of the J-2 Rocket Engines," presented at the *Joint Propulsion Conference of AIAA/SAE/ASME*, July 2004.

[17]Harmon, T., "Lunar Excursion Module Ascent Engine," presented at the *Joint Propulsion Conference of AIAA/SAE/ASME*, July 2004.

[18]Harmon, T., "Saturn S IVB Ullage, and Apollo Command Module Reaction Control Engines," presented at the *Joint Propulsion Conference of AIAA/SAE/ASME*, July 2004.

[19]Vilja, J., "Rocketdyne Advanced Propulsion System Overview," AIAA Paper 97–3309, 1997.

[20]Oefelein, J. C., and Yang, V., "Comprehensive Review of Liquid Propellant Combustion Instabilities in the F-1 Engines," *Journal of Propulsion and Power*, Vol. 9, No. 5, 1993.

[21]Biggs, R., "Space Shuttle Main Engine, the First Ten Years", "History of Liquid Rocket Engine Development in the United States, 1955–1980," American Astronautical Society History Series, Vol. 13, Univelt, San Diego, CA, 1992 Chap. 4, pp. 69–122.

[22]Harris, S. L., "Block II: The New Space Shuttle Main Engine," AIAA Paper 96–2853, 1996.

[23]Colbo, H. I., "Development of the Space Shuttle Main Engine," AIAA Paper 79–1141, July 1979.

[24]Jewett, R. P., and Halchak, J. A., "The Use of Alloy 718 in the Space Shuttle Main Engine," *Proceedings of the International Symposium on the Metallurgy and Application of Superalloys 718, 625 and Various Derivatives*, edited by E. A. Loria, TMS, Warrendale, PA, 19 April 1991, pp. 749–760.

[25]Lewis, J. R., "Materials and Processes for the Space Shuttle Engines," *Metal Progress*, March 1975, pp. 41–51.

[26]Conley, D., Lee, N. Y., Portanova, P. L., and Wood, B. K., "Evolved Expendable Launch Vehicle System: RS-68 Main Engine Development," International Astronautical Congress, IAC-02.S.1.01, Oct. 2002.

[27]Wood, B. K., "RS-68: What and How," AIAA Paper 98–3208, 1998.

[28]"RS-68 -First Flight Edition," *Threshold* No. 20, Winter 2002.

[29]Hodge, K. T., Allen, K. A., and Hemmings, B., "Development and Test of the ASAT Bipropellant Attitude Control System Engine," AIAA Paper 93-2587, 1987.

[30]"Integrated Propulsion Systems," *Handbook of Astronautical Engineering*, edited by H. Hermann Koelle, McGraw–Hill, New York, 1961, Sec. 20.06, pp. 20–136–20–150.

[31]Kinkaid, J. S., "Aerospike Evolution," *Threshold*, No. 18, Spring 2000, pp. 4–13.

[32]Fuller, P. N., "Linear Rocket Engine Design-Fabrication-Testing," Society of Automotive Engineers, Paper 730944, Oct. 1973.

[33]Cannon, I., Kotake, A., and Jones, D., "Assembly of the XRS-2200 Linear Aerospike Rocket Engine," AIAA Paper 99-2183, 1999.

7.9 Propulsion Products Center, Northrop Grumman Corporation

Today the LPRE work is done in this Propulsion Products Center (PPC), which is part of the Space Technology Sector of Northrop Grumman (NGC) in Redondo Beach, California. For many years the LPRE effort was part of the Space and Electronics Group of TRW, Inc. In the fall of 2002, the aerospace-related part of TRW was acquired by the Northrop Grumman Corporation. The work on LPRE has been focused on small monopropellant and bipropellant thrusters and also on large and small thrust chambers with ablative liners and/or pinle-type injectors.[1–3,*]

The LPRE work had its origins in the work done at the NASA Jet Propulsion Laboratory (JPL) in the mid-1950s. Beginning about 1956, JPL investigated the reaction rates and combustion phenomena of hypergolic storable liquid propellants. This included tests with an injector where flows of propellants entered the combustion chamber in annular sheets similar to the pintle-type injector shown later in this chapter. It also included experimental JPL studies of catalysts for decomposing hydrazine. Around 1960 several of the JPL engineers, who had worked on hydrazine monopropellant engines and the predecessors of the pintle injector, moved to the Space Technology Laboratory (the predecessor organization of TRW), and they continued their work in these areas.* NGC has been a builder of spacecrafts and satellites, a business that was larger than the LPRE business. This PPC-NGC is the only U.S. rocket propulsion organization that works not only on the LPREs, but also on their installation and the propulsion related issues of the spacecraft. It has a close relationship with the TRW spacecraft designers, and its propulsion systems can be found in most of the company's satellites and spacecraft.

The Rocket Products Center has worked on a variety of different LPREs, different propellants, and different applications. They have available a series of flight-proven small thrusters, an experience background of ablative liners in TCs, and have done more on pintle injectors than other LPRE development organizations. All of their LPREs use pressurized-gas-type feed systems. The only exception is one set of small thrusters supplied with propellant from gear pumps, which were developed by the prime contractor. The PPC-NGC has a small hot-fire test facility at their Redondo Beach headquarters and a larger one near San Juan Capistrano in California.

Gaseous Propellants

It has already been said that inert cold-gas attitude control systems (ACS) were historically the first TVC method used for steering space vehicles. Several of the initial spacecraft (1958–1959) built by the predecessor of NGC Space Technology used a reliable simple cold-gas system from their own LPRE group

---

*Personal communications, G. Dressler, and R. L. Sackheim, PPC, Northrop Grumman Corp., 2003–2004.

for attitude control, for example, in Pioneer 1 and 2. Their cold-gas TVC systems also flew on the Nimbus or the Earth resources satellites. Flights with PPC's cold-gas systems continued on other spacecraft until about 1974.* As discussed in Chapter 4.1, the performance of cold-gas thrusters was poor (typical specific impulses were about 70 s). On some of the Vela satellites, the PPC used the excess electrical power available during part of the flight to heat the cold gas and thus increase its specific impulse, which provided an extra margin.*

One novel approach used ammonia stored as a liquid in a small high-pressure tank.* A small amount was discharged through an orifice into a lower-pressure storage tank, where the heat of the metal hardware evaporated it into an ammonia gas. This system gave a specific impulse somewhat higher than most cold-gas systems, and it had less inert mass because there were no large heavy high-pressure gas storage tanks. This vaporized ammonia system was used on the LES-8 and –9 and the initial Vela satellites between 1969 and 1976.

Hydrazine Monopropellant Thrusters

The PPC-NGC was the first U.S. LPRE organization to fly hydrazine monopropellant LPREs with a gas pressure feed system. It was first flown in the Able 4A (also known as Pioneer P-1) in 1959. Because a good catalyst did not exist in 1959, PPC engineers used a slug start, that is, an initial injection of a small amount of the hypergolic oxidizer nitrogen tetroxide (NTO); the initial combustion reaction created the high temperature necessary for the subsequent thermal self-decomposition of the hydrazine. The next spacecraft Able 4B carried enough NTO for six slug starts. These early monopropellant thrusters were suitable for a few relatively long-duration firings, such as an orbit injection or orbit-transfer maneuver, but they were not able to perform multiple short-duration pulsing operations. In the 1960s the Shell 405 catalyst became available, and a series of hydrazine monopropellant LPRE systems (with multiple thrusters and a gas pressure feed system) were then developed. Each thruster had its own small bed of solid catalysts. These LPREs were then capable for multiple pulsing operations for attitude control, spread out over a long period of time. The chambers and nozzles were made of steel alloyed with cobalt and some were made of Haynes 25 alloy.[1,2] Thrusters in sizes from 0.1 to 150 lbf thrust were developed and qualified. Four of these monopropellant hydrazine thrusters are shown in Fig. 7.9-1 and described in Table 7.9-1 (Ref. 2).

A 1-lbf hydrazine monopropellant thruster was actually demonstrated for 550,000 cycles.* Although these thrusters have a modest performance (specific impulse between 220 and 250 s depending on the particular design parameters), the engine systems are less complex and simpler than a higher-performing bipropellant engine. They are often preferred for appli-

---

*Personal communications, G. Dressler, and R. L. Sackheim, PPC, Northrop Grumman Corp., 2003–2004.

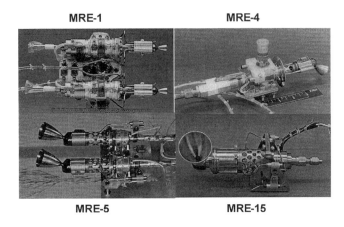

Fig. 7.9-1 Photos of four different hydrazine monopropellant rocket engines (MRE). Two are shown as modules with two thrusters. Courtesy of Northrop Grumman Space Technology from personal communications, G. Dressler, PPC-NGC, 2002–2004.

Table 7.9-1 Data on Selected Small Hydrazine Monopropellant Thrusters

Engine designation	Thrust, N or (lbm)[a]	Specific impulse, s (steady state)	Minimum impulse bit, lbf-s	Weight, single-thruster/dual-thruster module, lbm
MRE-1	3.4 (0.77)	218	0.017	1.2/2.1
MRE-4	9.8 (2.2)	217	0.054	1.1/NA
MRE-5	28 (6.3)	232	0.11	NA/3.2
MRE-15	66 (14.8)	228	0.27	2.5/NA

[a]At 275-psia feed pressure.

cations with a modest total impulse. The thrusters designed by this company are quite similar to the designs developed by other LPRE organizations. The first pulsing hydrazine monopropellant engines were used on Pioneer 6, 7, 8, and 9 spacecraft between 1965 and 1968. These hydrazine thrusters have been used sporadically ever since in different military and NASA spacecraft.

It is difficult to have pulses of very low total impulse when using a catalytic decomposition bed in the thruster. For extremely low thrust and short duration they used a system that was novel at the time. A small catalyst bed (equivalent to those used on a small thruster of approximately 0.1 lbf thrust) received a small slug of hydrazine propellant, and the hot reaction gas products (initially at about 2000 F) were discharged into a tank, where they were stored under pressure. When small torques on the vehicle (or a small thrust between 0.01 and 0.05 lbf with durations up to 20 ms) were needed, a small amount of this gas was discharged from the tank through hot-gas valves and through one or two of the uncooled alloy metal nozzles of the engine. When the pressure in the gas tank became too low, the next slug of hydrazine was catalyzed to recharge the tank with more hot gas.* Such a propulsion system was flown beginning in the 1970s on the defense support satellites with an average pulsing thrust of roughly 0.035 lb. Some years later the Russian Experimental Design Bureau Fakel came up with a monopropellant scheme that is somewhat similar and also achieves very low impulse bits. It is described in Chapter 8.12.

The Japanese company Ishikawa Harima Heavy Industries (Chapter 11) wanted to obtain a proven hydrazine monopropellant thruster. After surveying products from several development organizations, they selected TRW (today PPC-NGC) for obtaining a licence through a Japanese Trading Company as a means for acquiring this technology.

Bipropellant Thrusters, with Ablative Liners or Radiation Cooling

The PPC-NGC developed and built several types of small bipropellant thrusters. The earliest (1960) used ablative liners in the TCs, which at the time was the best available technology for pulsing operations. Such an ablative-type thruster was developed for the auxiliary propulsion system (APS) of the Saturn IVB stage (the third stage of the Apollo/Saturn V and also the second stage of the Saturn I SLVs).* Six thrusters were needed for each S-IVB stage. About 300 of these were delivered, and Fig. 7.9-2 shows one of them. The Saturn S-IVB was driven by the large restartable J-2 (LOX/LH$_2$) LPRE, which is described in Chapter 7.8. This APS provided not only roll moments during the powered flight of the gimbaled J-2 engine, but complete pitch, yaw, and roll control during the unpowered portions of the S-IVB flight. The APS thrusters had unique "capillary tubes" in the injector, which featured impinging unlike doublet injection patterns. Capillary implies long small injection holes, (0.020 in. diam and 2 in. long with an L/D of 100); this creates a substantial friction pressure loss and provides stable, clean, laminar flow injection streams and stable combustion. The chamber and nozzle used ablative material (silicon fibers at 45 deg in a phenolic cloth) and a coated molybdenum nozzle insert. Propellants were NTO/MMH. Design thrust was 150 lbf for each of the six ACS thrusters. Deliverable vacuum specific impulse was 281 s, and the total cumulative operating durations that had to be demonstrated were in excess of 500 s. During the long-duration ground-test firings, there was a development problem (which was later remedied) with heat soak-back causing excessive char formation in the ablative material in contact with the metal throat insert. This

---

*Personal communications, G. Dressler, and R. L. Sackheim, PPC–NGC, 1998–2004.

Fig. 7.9-2 External view of the thruster (about 15 in. high) of 150 lbf thrust with an internal ablative liner for attitude control of the Saturn S IVB stage. Courtesy of Northrop Grumman Space Technology.*

insert can then become loose, particularly with vibration, and allow the gas to bypass the nozzle insert.*

The next generation of bipropellant thrusters used walls made of niobium (often still identified by its former name columbium) with radiation cooling for the chamber and nozzle.[4,*] They were initially intended for apogee and orbit maneuvers. Typically with NTO/MMH a TR-306 produced 105 lbf thrust and a specific impulse of 316 to 320 s, depending on the specific design. Such a thruster is shown in Fig. 7.9-3. It has flown successfully in several satellites. One application, for the AXAF-I satellite, is described next.[5]

---

*Personal communications, G. Dressler, PPC–NGC, 2002–2004.

The predecessor company TRW furnished a modified version of the 100-lbf bipropellant thruster for the integrated secondary propulsion system for the upper Agena stage, which was developed by the Lockheed Missiles and Space Company (today part of Lockheed Martin) of Sunnyvale California. Its propellants, high-density nitric acid (with 44% dissolved nitric oxide), and UDMH containing 1% silicon oil were the same as the Agena main engine and came from the same propellant tanks. When burned, the fuel forms a fluffy deposit of silicon oxide on the inside of the TC walls, and this insulating layer reduces the heat transfer and the maximum wall temperatures of the niobium walls. Two small gear pumps, driven by an electric motor through magnetic couplings, were used for pressurizing the propellants. This pumping system was developed by the prime contractor Lockheed and is described in Fig. 4.4-10 and in Chapters 4.4 and 7.11. The TRW TC operated at a higher chamber pressure than the other bipropellant thrusters, and 28 thrusters were delivered to Lockheed.

The PPC also developed and flew TCs made out of rhenium coated on the inside and outside with iridium for oxidation protection.[4] Rhenium melts at 5733°F and niobium at 4380°F. With this rhenium metal a higher heat transfer and higher combustion temperatures can be accepted. It could provide up to 325 s with NTO/MMH and 330 s with NTO/N_2H_4, which is about 4 to 6 s more than with niobium. This apogee engine was used to place a Chandra spacecraft into its final orbit in 1999.

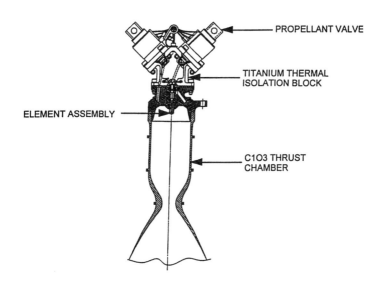

Fig. 7.9-3 Upper portion of a typical bipropellant thruster using a niobium wall. Courtesy of Northrop Grumman Space Technology (from personal communications, G. Dressler, PPC-NGC, 2002–2004).

This organization claims to have been first to deploy a pressure-fed dual-mode LPRE with both bipropellant and monopropellant thrusters. It is called SCAT (secondary combustion augmented thruster).[1,7,*] The bipropellant mode had higher specific impulse and more thrust, was partly cooled by propellant, and was used for maneuvers; it usually consumes the major share of the propellants. The monopropellant mode gave lower thrust and lower specific impulse. As seen in Fig. 7.9-4, the chamber and nozzle-throat region were cooled by NTO, which was vaporized in the nickel cooling jacket. The cooling jacket consisted of simple drilled holes with spiral inserts to increase the cooling velocity and absorb more heat. The hydrazine was decomposed in a separate catalyst bed to form warm gas, and this heated gas was injected into the combustion chamber, where it reacted with the NTO, which was also injected after it had been vaporized in the cooling jacket. It was a gas-to-gas reaction, and the combustion was very efficient. The monopropellant mode of operation had a lower mass flow and gave a lower thrust. It was usually used for attitude control maneuvers. SCAT flew for the first time in the GEOLITE satellite in 2002. Its bipropellant thrust was between 14 and 4 lbf and its monopropellant thrust between 4 and 0.75 lbf. The continuous thrust decrease was the result of a blowdown pressurized-gas feed system. The SCAT also had a pulsing capability.

An example of one of their complete systems is the AXAF-1 (Advanced X-Ray Astrophysics Facility), and its flow diagram is shown in Fig. 4.6-17. Whereas many small-thrust LPREs have either monopreopellant on bipropellant thrusters, this sytem had both. There were three different sets of TCs.[5] The bipropellant TCs (105 lbf) provide for major maneuvers, such as moving the spacecraft from a parking orbit into its final ellipical orbit. The hydrazine monopropellant thrusters constitute a reaction control system (RCS); the larger thrusters (20 lbf) provide pitch, yaw, and roll control during and after the orbit-transfer maneuvers. The momentum unloading propulsion system (MUPS) provides small monopropellant thrusters (0.1 lbf) for desaturating the momentum fly wheels from time to time during the long-duration orbit operation. Cavitating venturis are used to minimize the water-hammer-induced transients during system priming and to equalize engine startup transients.

Pintle Injectors

Northrop Grumman's PPC has refined the pintle injector technology over a period of 43 years and has designed and tested over 60 different pintle configurations covering a thrust range between 5 lbf and 650,000 lbf thrust.[1,3,6,*] Pintle injectors were defined and discussed in Chapter 4.3. Fig. 7.9-5 shows three sketches made in 1961 of different pintles. The one on the left has a fixed pintle or a constant area for injection. The other two sketches show a variable area injection scheme with a movable sleeve. Generally the oxidizer is injected into the chamber as a flat sheet or radially outward. The fuel is

---

*Personal communications, G. Dressler, PPC–NGC, 2002–2004.

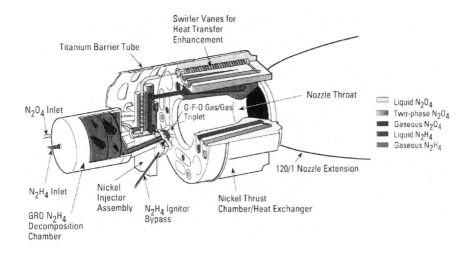

Fig. 7.9-4 This secondary combustion augmented thruster (SCAT) can operate as a bipropellant thruster (for major flight maneuvers) or at lower thrust as a monopropellant hydrazine thruster (for reaction control) in a spacecraft. Courtesy of Northrop Grumman, Space Technology.*

injected as an annular sheet flowing roughly in an axially downward direction. In one of the sketches, the oxidizer spray is not a continuous sheet, but interrupted by cogs in the movable sleeve.

As of October 2001, there has not been any combustion instability incidence with a pintle over a wide range of chamber pressures and with 25 propellant combinations. It gives good performance, but requires a relatively large combustion chamber volume with a characteristic length L^* between 30 and 100 in. It is claimed to be less expensive to produce than an injector with many small injection holes, particularly at higher thrust levels. In recent years TRW has tested low-cost LOX/LH$_2$ ablative thrust chambers with a single pintle-type injector at thrust levels of 13,000 and 650,000 lbf. It demonstrated that it had stable combustion, good performance, and relatively low cost. Eight different thrust chambers with pintle injectors have flown, most of them with fixed (nonmovable) sleeves or pintles.

Another feature of the pintle injector is its ability to shut off the propellant flow at the entry location in the chamber or at a flat injector surface. This reduces the "dribble volume" after shutoff essentially to zero. It is achieved with a two-position pintle sleeve as shown in Figs. 7.9-5 and 7.9-6 (Ref. 6). Most other small pulsing TCs have a small volume of propellant trapped between the injector surface and the valve seat. This trapped propellant dribbles out after shutoff and causes some afterburning and a small but unknown amount of extra thrust for a short but unknown timer period. The two-position

---

*Personal communications, G. Dressler, PPC–NGC, 2002–2004.

482 History of Liquid Propellant Rocket Engines

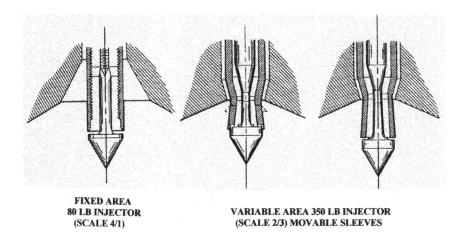

Fig. 7.9-5 Three early sketches (1966) of two types of pintle injectors, a fixed pintle type and a movable sleeve type with face shutoff. Courtesy of Northrop Grumman Space Technology; copied from Ref. 6.

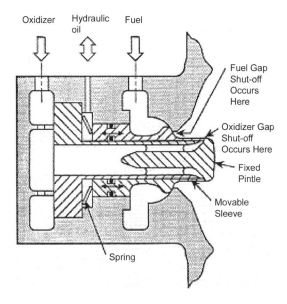

Fig. 7.9-6 Simplified section of injector with a central fixed pintle and a movable sleeve (actuated by hydraulic oil) for face shutoff, used in a pulsing experimental pilot escape propulsion system. Courtesy of Northrop Grumman Space Technology; copied from Ref. 6.

pintle sleeve acts as a dual valve right at the entry location of the propellants. This results in essentially no afterburning, gives more precise pulsing operations, but can also give sharp pressure rises at start.

The pintle face shutoff feature and a pulsing operation were used in the divert thruster of an experimental multistage defense missile named ERIS (exoatmospheric reentry interceptor system). ERIS was launched twice from Kwajalen Island in the Pacific Missile Range against an incoming ICBM in 1991, and it was successful in intercepting the incoming warhead.[6,7] The ERIS TC has a nominal thrust of 910 lbf, used NTO/MMH propellant, and operated at a relatively high chamber pressure of about 1600 psia. ERIS is discussed again in the next section of this chapter.

Pintle-type injectors have been developed and used earlier in other countries. The British Alpha LPRE used a three-pintle injector with a face shutoff feature beginning in 1947. It is shown in Fig. 12-4. The German HWK 109-509 LPRE had nine pintle-type injection elements (also with face cutoff) in its injector assembly, and its testing began in 1941 as described in Chapter 9.3. These foreign multiple-pintle injection elements are different because PPC always uses a single-pintle assembly in the center of the injector.

The movable pintle injector has been used successfully by the PPC to achieve deep throttling.[6] With conventional multihole injectors it has been possible to reduce or vary the thrust by a factor of two to one (by using throttle valves in the propellant lines); throttling down by three to one can become risky. Such modest throttling is used in a number of large LPREs. At the low flow the injection pressure drop is very low, resulting in poor atomization, inefficient combustion (loss of 5 to 20% of specific impulse), and often in combustion instability. With a variable-area pintle the pressure drop across the injection element and the injection velocity are essentially constant as the flow is throttled.

There are not very many flight applications that need deep throttling. Examples are planetary or lunar landing, terminal stages of defensive missiles, some tactical missiles, and aircraft superperformance propulsion. The historic U.S. lunar module descent and landing engine of the PPC had such a pintle injector and was capable of a 10-to-1 thrust reduction with only a 4% loss of specific impulse.[8,9,10] The TC had an ablative liner with a metal nozzle extension. A section of this engine and an enlarged section of its pintle injector are shown in Fig. 7.9-7. The trick in deep throttling is to maintain a high injector pressure drop and the proper mixture ratio. This is accomplished by varying the annular injection area (by hydraulically positioning the movable pintle sleeve) and at the same time varying the throat area of the two cavitating venturis with movable centerbodies. These venturis are inside valves and are labeled as flow control valves in the figure. The cavitating venturis control the propellant flow and the mixture ratio at any particular thrust level. An experimental PPC engine called Sentry, demonstrated a 19-to-1 throttle range using a variable-area pintle.* Tests showed acceptable performance and stable combustion.

---

*Personal Communications, G. Dressler, PPC–NGC, 2002–2004.

Large Thrust Chambers with Ablative Liners

Ablative liners were described in Chapter 4.3. The *lunar module descent and landing engine* had such an ablative liner because regenerative cooling would be unsatisfactory at the low thrust level. Fig. 7.8-15 shows a simplified section sketch of this lunar module and identifies the location of the descent/landing engine. The thick ablative TC liner is shown in Fig. 7.9-7, but parts of the injector face were regeneratively cooled. Supplementary film cooling, injected through holes at the periphery of the injector face, reduced the erosion and abrasion of the ablative material. The rated thrust of the engine was 9850 lbf at a chamber pressure of 103 psia, and the vacuum specific impulse was 303 s at a nozzle-exit-area ratio of 47.5 to 1.0. This modest chamber pressure

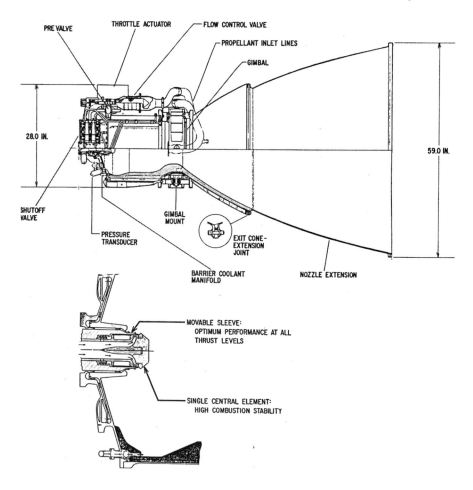

Fig. 7.9-7 Simplified section of the gimbal-mounted lunar descent/landing engine with an enlarged section of its throttling pintle-type injector. Courtesy of Northrop Grumman Space Technology; from Refs. 8 and 10.

implies a relatively low heat transfer and little erosion or abrasion. The propellants were NTO and A-50. When throttled down to 10% of its rated thrust, the specific impulse would be reduced to about 290 s, which is a relatively small loss. The gimbal frame is placed aroung the nozzle throat. This gimbal location requires less vehicle volume for gimbaling than a gimbal mount located on top of the injector. The propellant feed lines from the propellant tanks go through the gimbal assembly. A total of 84 lunar descent/landing engines were produced, and the engine flew several times satisfactorily in the 1960s and 1970s.

Ablative liners were also used in a series of thrust-chamber assemblies for upper-stage applications, such as the Delta launch vehicle. The ablative chamber of this TR-201 is seen in Fig. 7.9-8, and the ablative liner is in the chamber and initial segment of the nozzle. The nozzle-exit extension is metal

Fig. 7.9-8 View of a TR-201 thrust chamber for the upper stage of a Delta SLV. Courtesy of Northrop Grumman Space Technology (from personal communications, G. Dressler, PPC-NGC, 2002–2004).

and radiation cooled. It has also a pintle injector. Nominal thrust level was 9800 lbf using storable propellants. At the end of the blowdown of the pressurizing gas, the thrust was reduced to about 8000 lbf. About 77 of these were delivered, and they performed well and reliably. When high-performance upper-stage LOX/LH$_2$ engines became available and proved to be reliable, there was no longer a need for upper-stage engines with storable propellants and pressurized-gas feed systems.

An ablative liner was also used in the ERIS, which was mentioned in the preceding section.* It had a lateral or divert TC with the nozzle axis at right angle to the chamber axis in order to accommodate the tight space limitations of that experimental interceptor vehicle stage. It had a constant thrust of 910 lbf (vac), an ablative liner, and a chamber pressure of about 1600 psia. It flew successfully eight times in tests of intercepting incoming ballistic missiles.

A similar TC was developed for the Army's SENTRY missile program, also with an ablative liner. The nozzle-exit section was at 90 deg to the chamber axis, had a slot-exit section, and the exhaust gas formed a sheet squirting out of the side of the missile. The interaction between the TC's exhaust gas sheet and the airflow around the missile supposedly provided an augmented side force when compared to the exhaust jet only. This LPRE could be throttled over a 19 to 1.0 range, was restartable, and could be pulsed. It had approximately 8200 lbf thrust at sea level, ran on NTO/MMH, and had a low nozzle-exit-area ratio of about 4.0 to 1.0. The design chamber pressure was 2200 psia, which caused the TC to be very small for its thrust level (about 15 in. high) and the feed system to be very heavy. It also caused a much higher heat transfer than other ablative TC with pressurized feed systems, and the erosion and abrasion rates of the ablative material were much higher than in the lunar descent/landing engine or the TR 201 TCs. This SENTRY TC was ground tested, but the thrust augmentation was never tested, and the LPRE has never flown. The SENTRY and ERIS TCs are unique. Although there have been quite a few metal TCs with nozzles at right angle to the chamber axis, these two are the only ones known by the author to have ablative liners and high chamber pressure.

This rocket propulsion group undertook several experimental projects aimed at a large low-cost pump-fed LPRE. In the late 1960s and the early 1970s the company developed and tested at the Air Force Rocket Propulsion Laboratory at Edwards Air Force Base an NTO/UDMH TC of 250,000 lbf thrust with a pintle injector and an ablative liner.[6] The tests achieved initially 92% of theoretical performance, but after some improvements it became 95%. In the 1990s several experimental LOX/LH$_2$ TCs were developed and tested at three progressively higher thrust levels, namely, 16,300 lbf, then 40,000 lbf, and finally at 650,000 lbf (Ref. 11). All had a single pintle injector, a relatively large chamber volume, and ablative liners. Some of the liners used low-cost silica fibers and a phenolic resin. A turbopump for the 650,000-lbf LPRE was also designed and

---

*Personal communications, G. Dressler, PPC–NGC, 2002–2004.

built through an agreement with Allied Signal, who developed this TP. The TP was given some preliminary component testing. Hot firings of the large 650,000-lbf thrust chamber were conducted at the NASA Stennis Space Center in Mississippi. Unfortunately for the PPC-NGC none of these experimental demonstration tests and none of the accompanying studies resulted in a contract to develop a pump-fed low-cost LPRE for a large flight vehicle.

In 1967 and in 1968 the company tested a high-energy propellant combination, which was, at that time, a strong interest of the government. Most U.S. companies in the business did some work on one or more of these exotic propellants. The company tested FLOX as the oxidizer; it is a mixture of LOX and liquid fluorine.* The fuel was methane and alternatively a mixture of methane and ethane. Thrust levels were up to 3000 lbf, and the TC had a pintle-type injector. The resulting combustion efficiency was high at 99%. As discussed elsewhere in this book, the work with these highly toxic propellants proved feasible, but was not continued.

Gelled Propellants

Northrop Grumman's PPC and the U.S. Army have pioneered the use of gelled propellants and investigated different propellant formulations and operating characteristics. Gelled propellants have additives that make them thixotropic (jelly-like) materials. Most propellants (including the cryogenic propellants) have been gelled successfully in the laboratory, and a number have been tested in TCs, mostly on a small scale. The merits and disadvantages of gelled propellants have been briefly discussed in Chapter 4.1. This PPC was the first to build and test an experimental TC at 8000 lbf thrust with gelled fuel and gelled oxidizer in 1983.* In the following are two examples of advanced development programs using gelled propellants. Gelled propellants have now been investigated experimentally and analytically for several potential applications in the past 35 years. To date, none has as yet been selected for a production LPRE.

Figure 7.9-9 depicts a typical "breadboard installation" for testing a LPRE with gelled propellants and a pressurized feed system.* This kind of installation allows good visibility and access. This particular experimental assembly was tested in 1989 and was aimed at a pilot escape seat module, whereby a pilot can eject himself (together with the rocket-propelled seat) from a flying aircraft in an emergency. The final propulsion system, called advanced crew escape seat—experimental or ACES-X, was intended to be built into the bottom (underneath the seat) and the backrest of the ejectable pilot seat. The LPRE would expel the pilot while strapped into the seat. The LPRE gave a pulsed operation (down to 8 ms) for four thrust chambers (not shown in their final correct position in this breadboard) of 1500 lbf each to stabilize the flight path of the pilot, who is strapped to the seat module and exposed to a sudden air blast. The high thrust also prevented the pilot from an impact with the

---

*Personal communications, G. Dressler, and R. Sackheim, PPC–NGC, 1998–2004.

Fig. 7.9-9 Breadboard installation of a prototype LPRE for pilot escape with gelled propellants. Items are identified in text. Courtesy of Northrop Grumman Space Technology (from personal communications, G. Dressler, PPC-NGC, 2002–2004).

vertical tail of the aircraft. The system uses a solid-propellant GG shown on top, for expelling gelled IRNA and gelled MMH fuel with 60% carbon particles by weight, from two piston tanks, shown near the bottom. Pressure regulation was by means of a pressure relief valve. Operating time was less than 2 s with about 4500 psi in the solid-propellant gas generator and 2500 psi in the chambers. At this high chamber pressure the specific impulse was 281 s, but it was lower during rapid pulsing. Each thruster has a pintle-type injector similar to the one shown in Fig. 7.9-6 with face shutoff and servovalves to hydraulically actuate the pintles. The hydraulic oil tank (in the upper left) was also pressurized by the hot gas from the solid-propellant GG, and the oil was used to open and close the pintle valves. Because of the high heat transfer, there were a couple of burnouts toward the end of the burning period, but this problem was solved. The short pulses were commanded by a flight control system aimed to

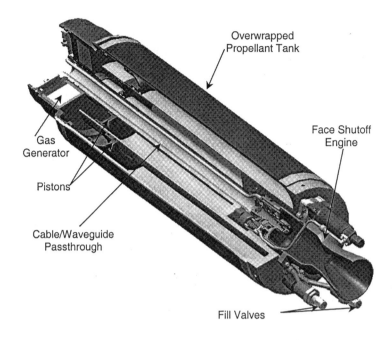

Fig. 7.9-10 Simplified diagram of a compact, preloaded LPRE for propelling a smart maneuvering ground-to-ground missile. The engine is 6 in. in diameter and 23.5 in. long. Courtesy of Northrop Grumman Space Technology; copied with permission from Ref. 12.

keep the pilot and his seat from tumbling. The final version was designed to give pitch, yaw, and roll moments while ejecting from an airplane flying at supersonic or transonic speeds and also while flying upside down close to the ground. The attitude control, position sensing, and firing command control were rather sophisticated. Although these breadboard system tests were successful, this crew escape design was not selected for production. Later the requirement for a LPRE for pilot escape was canceled.

Figure 7.9-10 shows a second potential application for jelled propellants. It is a unique prepackaged LPRE for a smart missile to be launched from an Army vehicle.[12,*] The objective was to have the shortest practical essentially constant time to target over a range of ambient temperatures. It uses gelled propellants and a face shutoff-type pintle injector capable of some throttling and restarting.[12] Work started about 1996. It uses a solid-propellant gas generator (with multiple grains that can be selectively ignited) for an "adjustable" tank pressurization, and it has pistons in the two concentric tanks for positive propellant expulsion. The gelled propellants are IRFNA and MMH; the gelled fuel has been loaded with small carbon particles making it denser and more

---

*Personal communications, K. Hodge, PPC–NGC, circa 2001.

energetic. The system allows control of the chamber pressure and powered flight duration, has provisions for the expansion and contraction of the propellant with changes in the ambient temperature, and minimizes variations in the time to target, which improves its kill probability. This test missile with its gelled propellants has flown successfully in two experimental launches from a military vehicle in March 1999 and in May 2000. There has not been a follow-up effort for a LPRE for a smart missile. Nevertheless the two flights of this experimental missile are believed to be the only flights with a rocket engine that has gelled propellants.

Concluding Comments

In 2004 the PPC NGC was still building various versions of monopropellant hydrazine thrusters and storable bipropellant thrusters and most have flown in company-developed satellites. In the past this center has developed and built small ablative-lined thrusters and several larger ablative-cooled TC including the Apollo lunar descent engine with deep throttling. The company pioneered in pintle type injectors and in gelled propellants. The center undertook innovative work aimed at a low cost large LPRE and at engines with gelled propellants, but these efforts did not result in a production program.

References

[1]Sackheim, R. L., "TRW's Family of High Performance Spacecraft Engines: Status and Impact," JANNAF, Paper, Dec. 1996.

[2]"Monopropellant Hydrazine Thrusters," TRW, Data Sheet, TRW, Redondo Beach, CA, 1996.

[3]Sackheim, R. L., "Rocket Propulsion," *Quest*, Summer 1983.

[4]Mayer, N. L., "Advanced X-Ray Astrophysics Facility—Imaging (AXAF-I) Propulsion System," AIAA Paper 96-2896, July 1986.

[5]Chazen, M., and Sicher, D., "High Performance Bipropellant Engine," AIAA Paper 98-3356, 1998.

[6]Dressler, G. A., and Bauer, J. M., "TRW Pintle Engine Heritage and Performance Characteristics," AIAA Paper 2000-3871, July 2000.

[7]Fritz, D., Dressler, G., Mayer, N., and Johnson, L., "Development and Flight Qualification of the Propulsion and Reaction Control System of ERIS," AIAA Paper 92-3663, July 1992.

[8]Elverum, G., Staudhammer, F., Miller, J., Hoffman, A., and Rockov, R., "The Decent Engine for the Lunal Module," AIAA Paper 67-521, July 1967.

[9]"Lunar Module Descent Engine," TRW, Data Sheets, TRW, Redondo Beach, CA, 1979.

[10]Gilroy, R., and Sackheim, R. L., "The Lunar Descent Engine—a Historical Perspective," AIAA Paper 89-2385, 1989.

[11]Gavitt, K., and Mueller, T., "TRW LCPE 650 klbf LOX/LH$_2$ Test Results," AIAA Paper 2000-3853, July 2000.

[12]Hodge, K. F., Crofoot, T. A., and Nelson, S. "Gelled Propellants for Tactical Missile Application," AIAA Paper 99-2976, June 1999.

7.10 Pratt & Whitney, a United Technologies Company

This company's main product line has been and is its turbojet engines. The name of the company is abbreviated as P&W in this book. The management decision to enter into the LPRE business was made in 1957 and serious work on rocket engines started in 1958. It was based in part on their extensive prior experience on rotating turbojet machinery and on having used, handled, pumped, and burned liquid hydrogen in a special turbojet engine (1956), which at the time was a secret project.[1]

RL10 Engines

P&W is the developer of the first flying liquid-oxygen/liquid-hydrogen LPRE designated RL10. With an active engine life of more than 42 years and with more than 350 engines, which have flown (and most of these were restarted during their spaceflights), the RL10 was the most successful, profitable, and historic rocket engine of P&W. The initial engine version had 15,000 lbf thrust. The initial application was the Centaur upper stage with two of the gimbaled RL10 engines.[1–6] This dual-engine configuration allowed vehicle control in pitch, yaw, and roll during engine operation without a separate attitude control rocket engine.

Design of the RL10 LPRE started in 1958. The first static engine firing occurred in 1959, preliminary flight rating tests in late 1961, and the first successful flight was launched in a Centaur upper stage in November 1963. One version of this engine is shown in Fig. 7.10-1 and described later in Table 7.10-1. It fulfilled a dream of Tsiolkowsky and Goddard, who more than two generations ago had identified LOX/LH_2 as the best practical high-enery propellant combination for spaceflight. It was the first flying LPRE with these propellants and the first in the world to use an *expander engine cycle*, where the liquid hydrogen is heated (from about −423°F to about −165°F) and evaporated as it flows through the cooling jacket of the thrust chamber.[1,3,7] This cycle is shown schematically in Fig. 4.2-7, and it does not have a gas generator.

P&W used formed, double tapered, and flattened 347 stainless-steel tubes for the cooling jacket of the thrust chamber and the nozzle.[1,6,7,*] It is basically similar to the tubular TC concepts developed years earlier by RMI, Aerojet, or Rocketdyne. P&W used the same subcontractor (LeFiel) who already had experience with double tapered tubes. Type 347 stainless steel was the tube material, which has good elongation properties, can be brazed, and is rust resistant. Silver was the brazing material. Brazing was done for 20 h in a special furnace with a reducing atmosphere (to prevent oxide contamination). The tubes and external rings were assembled and held in the furnace by a special rotary fixture. Because the TC axis was horizontal in the furnace, the TC was rotated

---

*Personal communications, C. Pignoli, and P. N. Mills, Liquid Space Propulsion, Pratt & Whitney, a United Technologies Company, 2000–2004.

Fig. 7.10-1 Early version of RL10 LOX/LH$_2$ LPRE for upper-stage applications. Courtesy of Pratt & Whitney, a United Technologies Company.

Table 7.10-1 Characteristics of two Versions of RL10 Rocket Engines[4,6,8,9,a]

Characteristics	Early Version	A Late Version
Design year	1958	1997
Designation	RL10A-3	RL10B-2
Thrust in vacuum, lbf	15,000	24,750[b]
Chamber pressure, psia	300	633
Nozzle expansion area ratio	40	290[b]
Specific impulse (vacuum), s	427	465.5[b]
Mixture ratio	5.0:1	5.88:1
Design life (no. firings/cumul. duration)	100/1.25 h	300/10 h

[a]Personal communications, C. Pignoli and P. N. Mills, Liquid Space Propulsion, Pratt & Whitney, a United Technologies Company, 2000–2004.
[b]With extendable diverging nozzle segment.

slowly to prevent distortion and achieve an even distribution of the brazing material.

The injector featured three curved sheets and multiple concentric coaxial injection elements with the oxygen coming through the small inside tubes and the gasified fuel coming through the annular space around it.[1,6] It is shown in Fig. 7.10-2. It was patterned after a similar injector developed by the NACA in Cleveland (today NASA Glenn Research). An innovation attributed to P&W was the porous stainless-steel material (called Rigimesh); it was used for the injector face; hydrogen oozed through the pores of this material providing a transpiration cooled surface. The porosity of the material could be controlled in its fabrication process. This basic injector design with the Rigimesh was used on subsequent larger U.S. LOX/LH$_2$ injectors.

In an expander cycle the gasified hydrogen, which had been heated in the cooling jacket by about 270°F above its boiling point, powers the turbine of the turbopump, and the hydrogen turbine exhaust gas is then injected into the combustion chamber, where it is burned with the liquid oxygen.[1-3,6,7] The turbopump, shown in Fig. 7.10-3 is positioned vertically. The shaft with the two-stage turbine and the two-stage hydrogen pump (with open impellers) is connected to the lower-speed oxygen pump (with a shrouded single impeller) by a gear case. The gears and bearings were initially lubricated and cooled by oil as was the custom in P&W jet engines. P&W developed the first bearing cooled and lubricated by liquid hydrogen in 1958, it was in the fuel pump for the LH$_2$ turbojet engine program mentioned earlier. P&W found that the bearings of the RL10 pump could be cooled and lubricated by the cryogenic hydrogen, which is a good coolant, but not a very good lubricant.

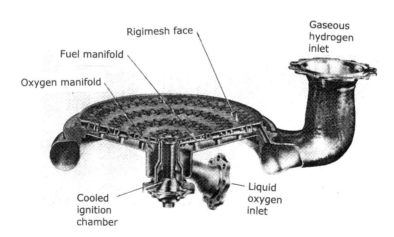

Fig. 7.10-2 Injector of the RL10 LPRE with a Rigimesh porous sheet at the combustion chamber face. Courtesy of Pratt & Whitney, a United Technologies Company; copied with permission of American Astronautical Society from Ref. 1.

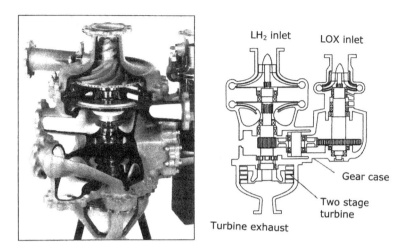

Fig. 7.10-3 Partial cutaway and simplified cross section of the turbopump of the RL 10 showing back-to-back hydrogen impellers, inducer, and gear case. Courtesy of Pratt & Whitney, a United Technologies Company; copied with permission from Ref. 6.

Cooling the bearings in oxygen pumps is more difficult because of the potential oxidation of the rubbing metals. The first LOX-cooled-and-lubricated bearing of P&W was operated in 1966 in a large TP, which is mentioned in the next section of this chapter. Robert Goddard was actually the first to submerge a bearing in liquid oxygen, but the details of this information might not have been available to P&W at that time. With the original gear material Waspalloy® gears would sometimes weld together. By changing to a gear alloy 6260, which was coated with molybdenum disulphite, the self-welding was eliminated.[6] The low hydrogen gas temperature allowed the turbine, its blades, and housing to be made of aluminum alloy.

To prevent the overspeeding of the turbopump rotors and to ensure a good start, it was and is necessary to cool the pumps (and avoid pumping gas) down to cryogenic temperatures prior to start. This is done systematically on all LPREs using cryogenic propellant by bleeding the cold liquid propellants through the pumps, piping, and valves prior to start and thus cool the hardware. This is critical in second-stage engines because the propellants consumed in the cooldown, just prior to start at altitude, reduce the amount of propellant available for combustion, reduce the mass ratio of the stage, and reduce the payloads. After considerable effort P&W achieved a 20-s cooldown by precooling the pumps with liquid helium on the ground prior to takeoff and using a special cooling bleed valve.[1,6] The turbine power and the shaft speed were controlled by an automatic bypass of a small amount of the gaseous hydrogen around the turbine as seen in Fig. 4.2-7. The first satisfactory RL10 restart in space occurred in October 1966.

A spectacular failure of the Centaur dual engines occurred on a new vertically firing test stand at P&W's Florida Research and Development Center in November 1960 on the first simultaneous firing of two engines.[1,6] It illustrated the importance of achieving safe ignition and a reliable start with more than one engine. One engine started properly, but the other did not. The hot flame from the working engine lit the unburned mixed propellant flowing out out of the nozzle of the other engine, and this resulted in an explosion of the premixed, unignited propellants accumulated in the chamber. The cold engine was demolished and busted into small pieces, and the firing operating engine was also damaged beyond repair. The test operator had failed to turn on the electrically driven booster pumps, which were mounted on the propellant test tanks. The inlet pressures of the main pumps were below the expected values. Pump cavitation probably caused a lower and irregular flow, and it possibly might also have caused the absence of an ignitable mixture at the spark plug of the unignited engine.

Injector failures occurred early in the RL10 development program, both on a horizontal and a vertical test stand, when starting under altitude conditions.[1,6] Apparently some of the oxygen, which precedes the fuel in the start sequence, ran backwards through the annular tubes of the multiple injection elements and got into the passages of the fuel manifold of the injector. When gaseous hydrogen fuel was then admitted to the fuel manifold, an explosive mixture was formed inside the fuel passages of the injector. When ignition occurred, this mixture exploded. The remedy was to pull an adequate vacuum prior to start (with a steam injector and diffuser) and to redesign the igniter (in the center of the injector) so that it would have a reliable small flow of gaseous oxygen at the central igniter with a spark plug. This small flow continued until after the fuel valve was opened near the igniter and the ignition sign-off signal was given. No explosion occurred when the steam ejectors of the test facility ran long enough to reduce the pressure below the value where gas would flash back.

The initial RL10 control system relied on a timer, which inititiated the opening or closing of the various valves and the operation of the spark igniter. Here each step of the start was initiated by a timing sequence without knowing that the previous step had been accomplished satisfactorily. Because of failures, like the one described in the preceding paragraph, the control was switched to a "ladder-type system."[6,7] It required that each commanded event had to be confirmed by a sensor reading or a measurement before the next step could be initiated. This type of ladder control had previously been used by Goddard and the other engine developers for at least a decade and also had been used by the Centaur vehicle designer (Convair).

There have been at least nine different, improved or upgraded versions of this RL10 engine.[1-6,*] The thrust was increased in steps until it reached the value shown in Table 7.10-1. The chamber pressure was increased, the nozzle-exit-area ratio was increased, and the performance was thus also increased. It has seen service as a dual engine in the Centaur upper stage with either the Atlas booster

---

*Personal communications, C. Pignoli, and P. N. Mills, Liquid Space Propulsion, Pratt and Whitney, a United Technologies Company, 2000–2004.

or the Titan booster as a first stage of the launching vehicle. The Centaur was used in unmanned lunar landings, planetary flyby/orbiters, Mars lander, astronomical observatories, and a number of other programs. A variable-thrust version (throttled down to 30%) with a cluster of four reusable engines was developed for the McDonnell Douglas experimental Delta Clipper (DC-X), a NASA-sponsored vertical takeoff and vertical landing test vehicle. Furthermore a cluster of six RL10 engines was successfully operated for an experimental version of the Saturn S IVB upper stage, which has been used in both of the Saturn I and Saturn V space launch vehicle. However this six-engine configuration was not selected by NASA for this application. Instead a larger single Rocketdyne J-2 LOX/LH$_2$ engine was chosen. There were RL10 versions that allowed a low-thrust idle mode, restarting, improved suction conditions (lower propellant tank pressures), higher chamber pressures, and larger nozzle area ratios.

P&W was the first in the world to fly an *extendible nozzle* with a LPRE.[4,6,7] The version for the RL10B-2 can be seen in Fig. 7.10-4, and data are in Table 7.10-1. This extendible nozzle concept was invented by a man in the United Aircraft Research Laboratory in 1948. It had been used by P&W in ground tests of their experimental RL-129 engine as early as 1967 and had been flown in the United Technologies Corporation (UTC) solid-propellant upper-stage rocket motors beginning around 1990. The nozzle extension is stowed around the engine during ascent and then lowered into position (by synchronized ball-screw actuators) after the drop-off of the lower vehicle stage, but before the second-stage engine is started at altitude. Thus the nozzle length and expansion area ratio could be greatly increased during the flight in space. The high nozzle area ratio allowed an improvement in performance at altitude, and the shorter initial vehicle length allowed a reduction in the inert mass of the vehicle. The specific impulse of 465.5 s is the highest for any flying U.S. LPRE. The radiation-cooled extendible nozzle segment was made of a three-dimensional weave of strong carbon fibers in a matrix of carbon[4] and was first flown in 2000. This particular nozzle extension uses three-dimensionally oriented carbon fibers in a carbon matrix and is relatively light in weight, when compared to metal nozzle-exit extensions. This woven carbon-fiber extension was developed by SEP in France (see Chapter 10), was nonporous, and did not leak gaseous hydrogen. Two earlier versions of an uncooled nozzle extension were made of the refractory metal niobium, but at much lower nozzle area ratios. They were the RL10A4 (tested in the early 1990s) and the RL10A4-1, first flown in 1995. All used three synchronized ball screws for moving the nozzle extension into working position, and they are driven by a single electric motor and connected by a toothed flexible belt.[4]

High-Chamber-Pressure Efforts

Operation at high chamber pressure (and attaining the benefits of this high pressure) was an aim of P&W soon after their LPRE effort was started.[6]

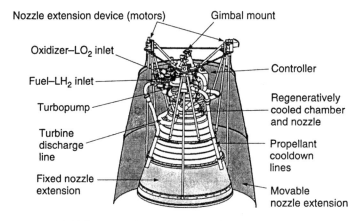

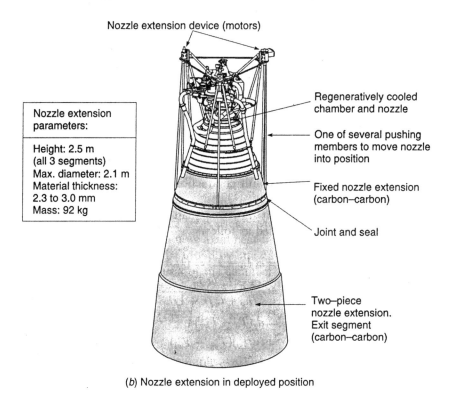

Fig. 7.10-4 Simplified sketch of the RL10B-2 LPRE with an extendable carbon nozzle skirt. Courtesy of Pratt & Whitney, a United Technologies Company; copied with permission from Ref. 7.

Beginning in the 1960s, P&W worked on several sets of high-chamber-pressure components, namely, TCs, TPs, and preburners at several different thrust levels, and ran a makeshift engine at 50,000 lbf. The advantages and disadvantages of high chamber pressure were discussed in Chapter 3.2. An Air Force interest in a classified manned glider project, proposed by McDonnell Douglas, was an added impetus to develop a new high-pressure LPRE.

In about 1959 or 1960 P&W studied methods for increasing the chamber pressure of their expander cycle engine and found that it would be necessary to greatly increase the turbine inlet temperature (from −165 to +1400°F) to accommodate the required higher pump power. This could be accomplished by burning the cold hydrogen with a small flow of LOX in what later was called a preburner, and so the concept of the first U.S. staged combustion cycle evolved from the expander cycle and from jet engines with afterburners. P&W actually subsequently tested a heavy-duty engine with a simulated staged combustion cycle. Because they did not have a suitable high-pressure turbopump available at that time, they used an orifice to simulate the pressure drop across the turbine, and so P&W conceived and ground tested a heavy-duty engine with a simulated staged combustion cycle well before Rocketdyne thought of it seriously. Much later Rocketdyne developed an experimental engine and the SSME with this engine cycle, and the SSME has been flown many times.

The heat transfer increases almost linearly with the chamber pressure, and a very good cooling method is needed to keep the chamber/nozzle walls from becoming too hot. In the 1960s P&W developed a unique TC design using *transpiration cooling*, a form of *film cooling*.[6,7] This concept was based on stacking up a large number of flat ring-shaped thin sheet-metal wafers fastened together. Each wafer had many small grooves, through which the cooling fluid (fuel) flowed slowly. These grooves were not straight, but curved or arc shaped. A piece of one of these wafers with multiple grooves is shown in Fig. 7.10-5. The coolant regeneratively cools the wafer, is injected (nearly tangentially) at low velocity from all these grooves into the thrust chamber, and forms a cool boundary layer along the chamber and nozzle walls. The film coolant entering from all of the many grooves reduced the heat flux to the walls. Around 1963 a heavy-duty TC with this cooling method was built and tested at a 10,000-lbf thrust level at chamber pressures up to 3300 psi. Subsequently TCs using this unique transpiration cooling scheme were developed and tested at 50,000 lbf (1964–1966) and 250,000 lbf (1966–1968). Figure 7.10-6 shows a simplified drawing of the transpiration-cooled TC at 250,000 lbf thrust level with the thin wafers, cut in the shape of annular rings and stacked up to form the inner wall of the TC. The TC was tested as a component in 1967. The amount of fuel needed for this wall transpiration cooling turned out to be small, sometimes as low as 1% of the fuel flow. Aerojet came up with this same concept, apparently independently of P&W, using their patented platelet technology with thin etched plates, which were stacked and brazed together. This is discussed in Chapter 7.7. This grooved wafer concept has technical merit, but it was never adopted or accepted for a production LPRE.

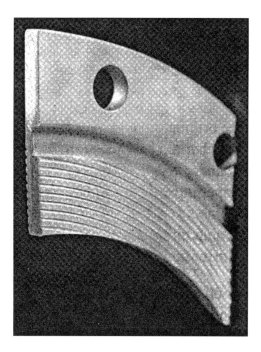

Fig. 7.10-5 Transpiration-cooling scheme for thrust chamber using stacked-up, ring-shaped wafers with small grooves in the shape of arcs. A piece of the wafer with grooves is shown. Courtesy of Pratt & Whitney, a United Technologies Company; copied with permission from Ref. 6.

In 1964 P&W embarked on a program to develop a preburner for their intended staged combustion cycle suitable for a 10,000-lb LPRE thrust engine and subsequently for a 50,000-lb engine.[6] Because the preburner is located very close to the turbine, which is driven by the hot gas, it is critical that the gas temperature distribution in the preburner be uniform. This is often difficult, particularly when the engine has to be throttled. This uniformity of the temperature over the cross section was achieved after extensive testing and some redesign of the preburner.

The high-pressure LOX and the LH_2 turbopumps for the staged combustion tests were separate assemblies, each with one pump per turbine. The initial experimental TPs (intended for a 50,000-lbf LPRE) were of the "battleship type" or heavy-duty design, and they had significant technical problems.[6] For a chamber pressure of 2500 to 3000 psi, a pump discharge pressure of between 4500 and 5500 psi would be needed. For example, the fuel TP had high stresses in the turbine blades and could not reach the desired high pump discharge pressures. It was a good learning experience. The experimental heavy-duty engine with the two heavy-duty TPs and the transpiration-cooled TC were tested for several firings, the longest being 17 s.

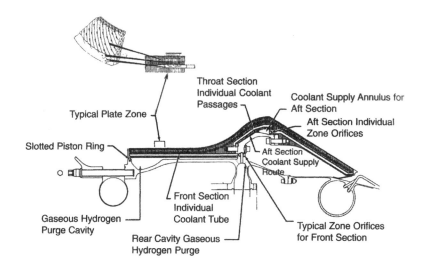

Fig. 7.10-6 Simplified section of a transpiration cooled thrust chamber (250,000 lbf thrust) with multiple wafers or stacked annular thin plates, each with grooves for film coolant flow. Courtesy of Pratt & Whitney, a United Technologies Company; copied with permission from Ref. 6.

Two large TPs, one for fuel and one for oxidizer, were developed in the 1966–1968 time period.[6] They were intended to ultimately supply propellants to a future LPRE of 350,000 lbf thrust. Figure 7.10-7 shows the rotor of the oxidizer turbopump, which incidentally was the first P&W LOX pump, which

Fig. 7.10-7 Rotor of the high-pressure oxygen turbopump sized for a potential 350,000-lb thrust engine, showing inducer, shrouded impeller, a balance piston, and a two-stage turbine. Courtesy of Pratt & Whitney, a United Technologies Company; copied with permission from Ref. 6.

used LOX as a bearing coolant and lubricant. During the development, there were several problems, and they provided a learning experience.[6] Some of the materials were subject to hydrogen embrittlement and had to be changed. To obtain a "stiff" support for the high-speed rotating assembly, P&W used a special type of roller bearing, on which they filed a patent. A very small clearance gave this bearing the necessary high stiffness and a slight curving of the rollers allowed the bearing to be self-centering. However these roller bearings could not take axial loads. The rotating assembly was sensitive to small changes in axial position, and the axial forces changed magnitude and direction during start or shutdown. To limit the axial shaft travel, the design included a balance piston, which in effect was the equivalent of two face-to-face axial hydrostatic bearings. This was probably the first crude hydrostatic bearing in a U.S. TP. In 1968 and 1969 a more advanced set of two turbopumps (at an equivalent thrust level of 250,000 lb) was developed. Figure 7.10-8 shows external views of the two fuel TP, one intended for a future 350,000-lb thrust LPRE and a later one for a possible 250,000-lbf thrust LPRE. They were based in part on the experience with the larger TPs already mentioned. The fuel pump (for the 250,000-lbf LPRE) delivered a pressure rise of 6700 psi, then the highest value in any U.S. TP. Neither of these two sets of TPs was operated with an engine.

Fig. 7.10-8 Two high-pressure liquid-hydrogen turbopumps. The left TP is for a future 350,000-lbf LPRE with high-strength aluminum pump housings, and the right TP is for a future 250,000-lbf LPRE using supermetal 718 for the housing. Courtesy of Pratt & Whitney, a United Technologies Company; copied with permission from Ref. 6.

Because the government canceled the contracts for the planned vehicles, the programs for both of these two sets of high-pressure TPs were stopped, and they were never operated with a full engine. However this extensive TP experience did pay off for P&W as explained in the next paragraph.

Other Projects

P&W developed alternate versions of the high-speed multistage high-pressure liquid hydrogen turbopump and the high-pressure liquid oxygen turbopump for the space shuttle main engine (SSME).[6,8] This engine had been developed and is still being built by Rocketdyne. When Rocketdyne had technical problems with their own design of these turbopumps (such as excessive bearing wear or turbine blade cracking), NASA decided to have P&W evaluate the Rocketdyne turbopumps, and a report was issued by P&W. Thereafter NASA gave P&W the job to develop alternate TPs for this engine. This TP job was made easier because Rocketdyne reduced the SSME chamber pressure by 17%, and this in turn reduced the pump discharge pressure by about the same percentage. These alternate TPs were more rugged, had a lower pump discharge pressure, and of the two TPs were about 550 lb heavier than the original Rocketdyne versions.[6,8,9,*] They were designed to be interchangeable with the existing Rocketdyne pump and to fit into the SSME. NASA decided to use the P&W large turbopumps in the SSME, and they are still being built by P&W and delivered to Rocketdyne for assembly into the SSME, beginning with block II of that engine. The exterior of these two TPs and their main features are almost identical to the Rocketdyne original versions. Figure 7.8-19 shows the powerhead frame of the SSME to which the TPs are attached.

In the early 1990s P&W believed that a new larger, higher thrust LOX/LH$_2$ LPRE was needed for future applications in upper stages at about 50,000 lbf thrust for spaceflight applications with heavy payloads. In about 1996 work started on components aimed at a 50,000-lbf thrust (later extended to 60,000 lbf) upper-stage oxygen-hydrogen LPRE.[8,*] This engine is designed for a modified expander cycle engine with two turbopumps, one for pumping oxygen and one for hydrogen. The first component development started that year for a novel and very unique compact two-stage liquid hydrogen turbopump shown in Fig. 7.10-9 (Ref. 10). It has a single-piece titanium rotor, which is very stiff, with integrally machined vanes for the two centrifugal pump impellers and a matching pump inducer impeller as well as a machined set of unusual turbine vanes or blades. The radial inflow turbine allows a compact design. Radial inflow turbines have not been used often in LPREs. The only other such turbine known to the author was developed by Curtiss Wright (Chapter 7.4) starting in 1948, for the X-2 experimental aircraft, which flew in 1955. The P&W experimental TP has split hydrostatic bearings for accepting both radial and axial loads, and these bearings are lubricated and cooled by a

---

*Personal communications, P. N. Mills, Liquid Space Propulsion, Pratt & Whitney, a United Technologies Company, 2000–2004.

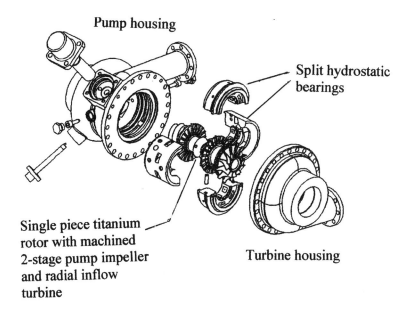

Fig. 7.10-9 Exploded view of an advanced experimental two-stage liquid-hydrogen fuel turbopump with a single-piece titanium rotor and with an inducer impeller, driven by a radial inflow turbine. Courtesy of Pratt & Whitney, a United Technologies Company; copied with AIAA permission from Ref. 9.

small flow of filtered liquid-hydrogen fuel. The novel stiff one-piece rotor design, the inherent stiffness of the bearings, and the good damping qualities of the fluid-bearing films minimize rotor excursion or wobble, and thus allow small clearances and high efficiencies. When operating at the 50,000-lbf thrust level and a liquid flow of 1600 gpm, the turbopump rotates at about 166,700 rpm, and the 3-in.-diam impellers give a head rise of 136,700 ft at a tip speed of 2180 ft/s and a pump efficiency of 64% (Ref. 10). The radial inflow turbine has a diameter of about 3.2 in., a power level of 5900 hp, an efficiency of 78% at a pressure ratio of 2.16. Cast inconel 718 was used for the pump housing and weldable Waspalloy for the turbine housing. The total number of parts of the novel TP was 29, which compares favorably with 135 parts for the turbopump of the earlier RL10-3-3A version.

By 1999 there was enough information available to seriously plan an experimental engine, which was identified a year later as RL60. It is shown in Fig. 7.10-10 without the nozzle extension. The LH_2 TP can be seen on the left side of the chamber, and the LOX TP is on the right side. In late 2000 it was decided to make it into an international project. The LOX TP was developed by the Chemical Automatics Design Bureau (CADB) of Russia,[10,11] the fuel pump was developed by the Japanese company Ishikawa Harima Heavy Industries (IHI),[12] certain valves by Techspace of Belgium (United Technologies

Fig. 7.10-10 RL60 experimental LPRE at 60,000 lbf nominal thrust has components developed by non-U.S. development organizations. Courtesy of Pratt & Whitney, a United Technologies Company (provided through personal communications, P. N. Mills, 2005).

owns a major share of this company), and a fuel-cooled nozzle extension with a clever-weld technique was given to Volvo Aero of Sweden.[13] Work on an experimental TC in this same thrust size was initiated at P&W, and two TC were built in 2003.[15] All of these components were tested and delivered to P&W. An upper-stage demonstrator engine with an LOX/LH$_2$ expander engine cycle at about 50,000 to 65,000 lb thrust was assembled and was about to be tested at the time this part of the manuscript was written.[9,14] This demonstrator engine did not have some of the features planned for the future final engine development, such as booster pumps or an extendable nozzle-exit skirt. As far as this author can determine, there were no immediate prospects for a vehicle application for this RL60 engine.

Like other rocket engine developers, P&W/UTC had its share of R&D projects and experimental engines (or components) that were not continued or were not selected for flight. For example, this included some work on firing a RL10 engine using liquid fluorine and liquid hydrogen as propellants.[6] The extreme reactivity and toxicity of fluorine, the potential serious consequences of a spill or accident, and the relatively modest performance improvement stopped any further work. Another project was a nozzle-exit skirt with two concentric nozzle-exit walls (one inside the other), and the walls had large perforated holes. The inner wall was fixed, and the adjacent outer wall could be rotated through a small angle. When the outer wall was rotated into a position where its holes lined up with and matched the holes of the inner nozzle-exit cone wall, atmospheric air could be aspirated through the holes into the flow of the overexpanded exit section of the nozzle. This concept was statically tested on one version of the RL-10. The idea was for the airflow to augment the nozzle mass flow and increase the thrust at low altitudes; furthermore, it would prevent flow separation (which can cause destructive side forces on the nozzle). Further work was stopped because some parts were subjected to the high combustion stagnation temperature, there were some mechanical design problems, and the improvement in vehicle performance was relatively minor.

Russian Rocket Engines

In 1996 P&W reached an agreement with the largest and most experienced Russian developer of large LPRE, NPO Energomash at Khimki (Moscow Region), to market some of the proven Russian LPREs for propelling U.S space launch vehicles.* This included the RD-170 (1,777,000 lbs thrust) liquid-oxygen-kerosene booster engine, the RD-180 (described next), the RD-120 (187,400 lbf vac thrust) liquid-oxygen-kerosene upper-stage engine, and a tripropellant LPRE.[15] All are described and shown in figures in Chapter 8.4. The RD-170 is the largest thrust existing LPRE uses a staged combustion cycle, four thrust chambers, two preburners, and a single main turbopump. The RD-120 was originally a second-stage LOX/kerosene LPRE, but a first-stage version has also been developed. It was test fired at the P&W LPRE test station in Florida in 1995 and was the first such test of a large Russian LPRE in a U.S. rocket engine test facility.[16] It is shown in Fig. 8.4-15.

Of these several available Russian rocket engines only the RD-180 found a U.S. application as a booster for advanced Atlas space launch vehicles. P&W and the Energomash have formed a joint venture organization called RD AMROSS; it is responsible for the delivery, servicing, and launch support of the RD-180 and will eventually be responsible to help produce it in the United States.[17] Detailed documentation of this RD-180 has been given by Energomash to the joint venture and thus to P&W personnel in 2003.

This LPRE is a scaled-down version of the flight proven RD-170 (with four thrust chambers), but the newer RD-180 has only two thrust chambers. Figure 7.10-11 shows this engine, and Table 7.10-2 gives some data.. It was developed by NPO Energomash (see Chapter 8.4) with inputs from P&W and has now

---

*Personal communications, P. N. Mills, Liquid Space Propulsion, Pratt & Whitney, a United Technologies Company, 2000–2004.

Fig. 7.10-11 RD-180 has two gimbal-mounted TCs (pointing up), a central common TP, two booster pumps, a high-pressure turbine exhaust pipe (bottom), which feeds gas to the injectors. A TVC actuator is on the right. Courtesy of NPO Energomash and Pratt & Whitney, a United Technologies Company (provided through personal communications, P. N. Mills, Liquid Space Propulsion, Pratt & Whitney, 2003).

Table 7.10-2 Selected Data on the RD-180 Liquid Propellant Rocket Engine

Thrust, lbf	860,200 at SL and 933,400 in vacuum
Chamber pressure, psia	3722
Specific impulse, s	311.4 at SL, 337.8 in vacuum
Nozzle-exit expansion ratio	36.87 to 1.0
Propellants	LOX/kerosene
Mixture ratio, O/F	2.72
Height/diameter, in.	140/118
Dry weight, lbm	11,898

been accepted as the new booster engine for the Atlas 3 and the uprated Atlas 5 space launch vehicles.

This LPRE uses a staged combustion cycle, an oxygen-rich preburner, an oxidizer lead during start with an initial flow of a hypergolic starting fuel, high-pressure helium for valve actuation and system purging, a single-shaft high-pressure main turbopump, and two booster turbopumps. The two thrust chambers can each be gimbaled through plus or minus 8 deg. The chamber pressure is the highest of any LPRE launched from the United States. The high heat transfer requires a combination of regenerative cooling and film cooling, and the TC design is essentially identical to the one used in the RD-170 and is described in Fig. 8.4-16. The flow sheet of the RD-170, shown in Fig. 8.4-17, is similar to the flow sheet of the RD-180. About 70% of the RD-180 components are reported to be the same as the flight-proven RD-170. About 30% had to be scaled down or modified. The RD-180 engine performance is superior to any U.S. developed LPRE using the same propellants, but most of these were developed much earlier.

The development and qualification testing of the RD-180 was done at Energomash in Russia.* During development, a total of 19 engines were built, and 95 engine tests were performed. Additional engines were built and operated during the qualification tests at Energomash. Four tests were conducted at the Marshall Space Flight Center in Alabama, with the engine mounted in the Atlas 3 booster stage. It was launched twice in May 2000 and in February 2002. Six additional engines were built, and 30 engine tests were conducted at Energomash to qualify the engine for the Atlas 5 configuration and operation. The first launch with the Atlas 5 was in August of 2002.

P&W also agreed to market the RD-0146 LPRE developed by the CADB of Voronesh, Russia.[8] It is shown in Fig. 8.5-23 and has never been installed in a flying vehicle. It has a nominal thrust of 22,000 lbf (vac), uses LOX/LH$_2$

---

*Personal communication, P. N. Mills, and AMROSS, Liquid Space Propulsion, Pratt & Whitney, 2003–2004.

propellants, and also uses an expander engine cycle. Except for integral booster pumps, a somewhat lower specific impulse, and an integral hydrogen roll control system, the engine characteristics are similar to the P&W RL10B-2 engine. P&W has not found a U.S. application for this engine. The RD-0146 engine is again discussed in Chapter 8.5.

Concluding Comments

The RL10 series of upper stage engines are still in limited production, 42 years after the first flight. There have been changes in design and increases in thrust level and performance. It was the first to fly with LOX/LH$_2$, the first with an expander engine cycle, the first LPRE with an extendible nozzle, a porous material to cool the injector face, an early TC with a unique film cooling scheme. Various attempts for P&W to sell a larger thrust level engine (RL60, high pressure TPs) did not result in an actual flight application. A joint venture with NPO Energomash (Russia) did find an application for the RD-180 LPRE in the United States as booster for the Atlas 3 and Atlas 5 SLV. In the future this engine may be produced in the United States.

References

[1]Tucker, J. E., "The History of the RL10 Upper Stage Rocket Engine, 1956-1980," *History of Liquid Rocket Engines in the United States, 1955-1980*, edited by S. E. Doyle, American Astronautical Society History Series, Vol. 13, Univelt, San Diego, CA, 1992, Chap. 5, pp.123–151.

[2]Brown, J. R.,"Expander Cycle Engines for Shuttle Cryogenic Upper Stages," AIAA Paper 83-1311, June 1983.

[3]Heald, D., "LH$_2$ Technology Was Pioneered on Centaur 30 Years Ago," *History of Rocketry and Astronautics*, edited by P. Jung, American Astronautical Society History Series, Vol. 21, Univelt, San Diego, CA, 1992, pp. 205-220.

[4]Santiago, J. R., "Evolution of the RL10 Liquid Rocket Engine for a New Upper Stage Application," AIAA Paper 96-3013, 1996.

[5]Richards, G. R., and Powell, J. W., "The Centaur Vehicle," *Journal of the British Interplanetary Society*, Vol. 42, March 1989.

[6]Mulready, D., *Advanced Engine Development at Pratt & Whitney, the Inside Story of Eight Special Projects, 1946-1971*, Society of Automotive Engineers, Warrendale, PA, 2001.

[7]Sutton, G. P., and Biblarz, O., *Rocket Propulsion Elements*, 7th ed., Wiley, Hoboken, NJ, 2001.

[8]"RL 10, RL 60, RL 10B-2, SSME turbopumps, RD-0146, and RD-180," Pratt & Whitney, Data Sheets, 2003.

[9]Minick, A., and Peery, S., "Design and Development of an Advanced Liquid Hydrogen Turbopump," AIAA Paper 98-3681, July 1998.

[10]Dimitrenko, A. I., Ivanov, A. V., Kravchenko, A. G., Mishin, A., Pershin, V., and Minnick, A.,"Development of an Advanced Liquid Oxygen Turbopump," AIAA Paper 2000-3852, July 2000.

[11]Dimitrenko, A. I., Ivanov, A. V., Kravchenko, A. G., and Minick, A. B.,"Advanced Liquid Oxygen Turbopump Design and Development," AIAA Paper 2000-3878, July 2000.

[12]Ohta, T., Kimoto, K., Kawai, T., Motomura, T., Russ, M., and Paulus, T., "Design, Fabrication and Test of the RL60 Fuel Pump," AIAA Paper 2003-5073, July 2003.

[13]Damgaard, T., Johansson, L., and Rydén, R., "Laser-Welded Sandwich Nozzle Extension for the RL60 Engine," AIAA Paper 2003-4478, July 2003.

[14]Bullock, J. R., Popp, M., and Santiago, J. R., "RL60 Demonstration Engine Design, Manufacture, and Test," AIAA Paper 2003-4489, July 2003.

[15]"RD-170, RD-180, RD-701, and RD-120," Pratt & Whitney, Data Sheets 1997.

[16]Croteau, M. C., Grabowski, R. C., and Minick A. C., "Live Fire Testing of the NPO Energomash RD-120 Rocket Engine at Pratt & Whitney," AIAA Paper 96-2608, 1996.

[17]Press release, Pratt & Whitney, West Palm Beach, FL, 24 Sept. 2003.

7.11 Atlantic Research Corporation (ARC), Liquid Rocket Division

This Liquid Rocket Division of Atlantic Research Corporation (ARC, since 1988 a unit of Sequa Corporation) is a relatively small organization that has used acquisitions to expand its product line. ARC obtained its LPRE technology from three sources.* The former *Bell Aircraft Company* started its work on LPRE in the 1950s; in 1960 Textron bought this aircraft company and established the *Bell Aerospace Division* as a separate organization; it contained the LPRE capability.[1] It developed a line of hydrogen-peroxide monopropellant engines, a line of low-thrust pulsing bipropellant engines, and a versatile upper-stage space launch LPRE called Agena. In 1983 *(Atlantic Research Corporation) (ARC)* acquired this liquid propellant operation (together with its personnel and facilities in Niagara Falls, New York) in order to supplement their own large solid-propellant rocket motor product line. In August 2004 Aerojet acquired ARC, mostly because of its solid-propellant motor capabilities. As a condition of this acquisition, Aerojet was required to divest itself of the liquid propellant operations of ARC in Niagara Falls, New York, and England because Aerojet already had two in-house groups, which had developed similar small thrusters. As of August 2004, when this chapter was written, this divestiture had not been finalized.†

The second source came to ARC in November 1997, when it acquired the *Royal Ordnance Company* of Westcott, England. With it they received some skilled people, facilities, a line of proven, storable bipropellant modern apogee LPREs operating with hydrazine as a fuel (most others used MMH), a number of smaller bipropellant LPREs for RCSs operating also with hydrazine as a fuel, and an entry into the European market. The third source was the *hydrazine monopropellant rocket engine product line* originally developed by *Hamilton Standard Division of United Technologies Corporation* in the late 1960s.[2] This line had been acquired by the Kaiser Marquardt Corporation a few years earlier, but when this Kaiser Marquardt merged in 2001 with Primex (which already had a similar monopropellant product line) the government required Kaiser Marquard to divest itself of this monopropellamt hydrazine product line. Therefore in May 2000 ARC was able to acquire the proven Hamilton Standard designs, manufacturing know-how, test data, customers, and inventory of this hydrazine monopropellant line. ARC and its predecessor organizations had developed a line of hydrazine monopropellant thrusters of their own; this set of products did not have a truly satisfactory technology and did not sell well. The Hamilton Standard line was well proven and rounded out ARC's small LPREs capability. These product areas helped ARC to now have qualified small LPREs of several major types, and each will be discussed briefly.

The family of *hydrogen-peroxide monopropellant thrust chambers* played a significant role in the early decades of this LPRE history.[3-8],* They were the

---

*Personal communications, Mayne Marvin, Atlantic Research Corporation, 2001–2004.
†In early October 2004, the Liquid Rocket Division of ARC was acquired by Ampac-ISP, which is part of the American Pacific Corporation, Las Vegas, Nevada.

first LPREs used for attitude control and spaceflight maneuvers of spacecraft and upper vehicle stages. Bell Aircraft, ARC's predecessor, produced them in approximately 20 different sizes between 1.0 and 500 lb thrust, and over 2600 were built during the 1950s and 1960s. Figure 7.11-1 shows one of them.[5,6] The nozzle with a conical exit section is at right angle to the centerline of the catalyst bed. This 14-lbf thruster had a steady-state specific impulse of 156 s with a nozzle area ratio of 15 to 1.0. A *silver screen catalyst* decomposed the 90% peroxide, and gas pressure expulsion was used to feed the propellant to the catalyst beds in the TCs. The thrust chambers were largely scalable, reliable, and in some applications also reusable. Many were pulsed, but with the valves then available and the placement of these valves the typical start delays were about 40 to 60 ms (and longer if the catalyst were cold or the thrust were large), and minimum pulse width was then about 200 ms. This delay compares to typically 10 ms for modern small thrusters with current propellants and modern valves. Hydrogen peroxide is no longer used since about 1965, primarily because of its low specific impulse of around 150 s for steady-state operation and as low as 90 s in short pulses. ARC and others learned early that performance (specific impulse) decreased with pulse width.

They also learned that the axis of the nozzle does not have to be in line with the axis of the catalyst chamber; it can be at an angle to it. The nozzle exit can also be scarfed or cut off at an angle (to fit the vehicle skin contour) with only a very small performance penalty. This geometry helps to place nozzles into crowded vehicle compartments, where there is no space for a longer straight nozzle.

The Bell/ARC hydrogen-peroxide LPREs with gas-pressurized feed systems controlled the flight path and attitude of the sound-barrier-breaking X-1B and the X-15 NASA research airplanes (needed at high altitude where there was little or no air to produce control forces on aerodynamic surfaces) and of the first manned U.S. flights with the Mercury space capsule in 1961. Figure 7.11-2 shows the placement of the components of two redundant LPRE systems,

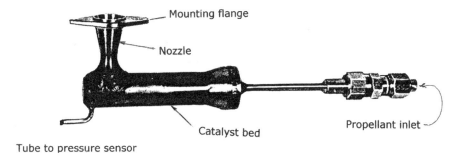

Fig. 7.11-1 Picture of a 14-lbf thruster using hydrogen peroxide as a monopropellant, circa 1951. Length was 9.2 in. Courtesy of ARC; copied from Refs. 7 and 8.

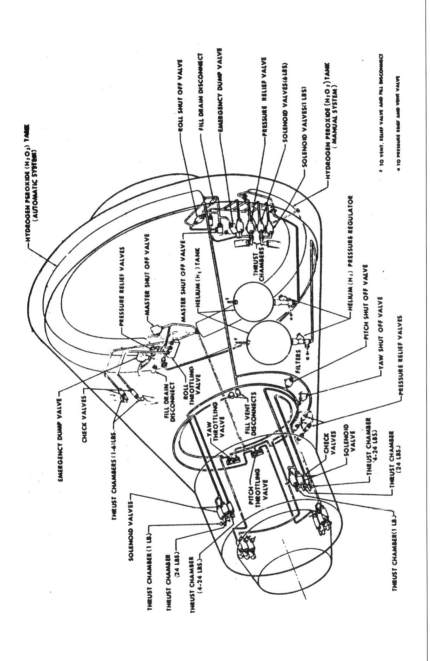

Fig. 7.11-2 Perspective view of hydrogen-peroxide monopropellant propulsion system of the Mercury manned space capsule showing an automatic and a manual redundant LPRE used for attitude control and small maneuvers. Courtesy of ARC; copied from Refs. 7 and 8.

each with its own tanks, valves, and thrusters (1, 6, and 24 lb thrust) within the envelope of the manned Mercury capsule. Bell/ARC hydrogen-peroxide thrusters were also used in the Centaur upper-stage vehicles for flight-path and attitude control (1.5, 3, and 50 lb thrust in 1962) and in the Dynasoar vehicle.[1,3-6] NASA developed five lunar landing training vehicles for astronaut flight training in preparation for actual landings of the larger Apollo lunar landing module. These vehicles each had several hydrogen-peroxide monopropellant thrusters, several at 90 lbf for attitude control and 500 lbf for controlling the rate of descent, and these were provided by ARC's predecessor.

The development of a good package of silver screens as a catalyst took some effort.[3,4] The German experience during World War II with liquid catalysts (Hellmut Walter Company) was not fully applicable because it was primarily for long durations and not for short multiple pulses of thrust. There were some problems with the catalyst bed. The thrust decreased with time. The causes were incomplete decomposition of the peroxide, uneven distribution of liquid flow across the cross section of the catalyst bed, and degradation of catalyst activity with use. Also there was some leakage along the edge of the bed because of improper design, inadequate sealing of the cylindrical container, and sometimes improper screen compaction or mesh sizes. Gradually these problems were all overcome.

The Bell/ARC "rocket belt" was a pressure-fed complete hydrogen-peroxide monopropellant rocket engine package, which was strapped to a man's back and hips. It enabled the man to fly over 50-ft obstacles and for short distances (200 ft) without wings, rotating aircraft machinery, propellers, or landing gear. Between 1961 and 1970 this rocket belt and its improved versions were flown at least 550 times, some at public events, such as at the 1964 New York World Fair.[1,3,4] It garnered a great deal of publicity at the time, it was considered by some people to merely be a toy, and it never found a market where it would be effective.

In the 1970s Bell Aerospace (ARC predecessor) started to develop a series of *good bipropellant low-thrust rocket thrusters* for reaction control, attitude control, stationkeeping, flywheel desaturation, or liquid propellant settling.[5-8] Almost all used nitrogen tetroxide and monomethyl hydrazine (NTO/MMH). Selected data are shown in Table 7.11-1, and they are discussed further in this chapter. The company tested different chamber/nozzle materials (such as tantalum, tungsten, molybdenum) and eventually settled on niobium, for good manufacturability, and relatively low weight. This metal's name has officially been changed to niobium, but the old designation columbium is still in common use in the industry. A silicate coating to minimize oxidation by the hot reaction gases and a process to apply this coating were selected from various alternatives. The injectors originally were also made of columbium and welded to the chamber. Today some of them use a titanium combined with columbium for the injector. The start delays and minimum pulse width were about the same as those already mentioned, namely, 40 to 60 ms. With new valves closely mounted to the injector, the time delay for valve opening or closing can be less than 4 ms, and the pulse duration can be as low as 10 ms,

Table 7.11-1 Selected Data on Several ARC Thrusters[a]

Engine	Nominal thrust, lbf	Nominal specific impulse, s	Propellants	Status	Typical application
Bipropellant thrusters					
ARC-5lb	5	293	NTO/MMH	Flown	Satellites
Leros 10	2.5	275	NTO/MMH	Flown	Satellites
ARC-20	23/18	230	NTO/MMH	Flown	Minuteman III
Leros 1B	145	318	NTO/hydrazine	Flown	Apogee maneuver
Leros 1C	103	325	NTO/hydrazine	Flown	Apogee maneuver
Leros 2B	90	320	NTO/MMH	Qualified	—
Monopropellant thrusters					
Monarc 1	0.2	226	Hydrazine	Flown	Attitude control
Monarc 5	1	230	Hydrazine	Flown	Attitude control
Monarc 22	5	235	Hydrazine	Flown	Attitude control
Monarc 445	100	235	Hydrazine	Flown	Apogee maneuver

[a]For a particular TC the thrust is essentially proportional to chamber pressure, and the specific impulse can vary with chamber pressure, mixture ratio and nozzle expansion ratio. (From Refs. 7 and 8.)

particularly for the low-thrust units. ARC now has a stable of small flight-proven thrust chambers assemblies in sizes between 0.2 and 350 lb and has been a leader in this field.[5–7] Some of the bipropellant thrusters were used in experimental defensive missile developments, such as a divert thruster. ARC has investigated other materials for these small bipropellant thrusters, such as platinum, rhenium, ablative liners, Haynes 25 alloy, carbon fibers, or nozzle throats made of tantalum (with 10% tungsten), but none has been fully developed, qualified, or selected for a production application.

In 1987 ARC acquired the Royal Ordnance organization, a British company, and with it a product line of *small thrusters, using NTO/hydrazine as bipropellants*.[7,8,*] Compared to MMH, which was used as a fuel in heritage ARC thrust chambers, pure hydrazine gives a few more seconds of specific impulse and a higher average propellant density, and these are needed in some high-performance spaceflight applications. Several are identified as the "Leros" thrusters in Table 7.11-1. Hydrazine, of course, has a high freezing point (34°F, almost the same as water), and all valves, catalysts, and pipe lines have to be heated in a cold environment such as in space, but electric heaters have been

---

*Personal communications, Mayne Marvin, Atlantic Research Corporation, 2001–2004.

accepted by vehicle designers and operators as a necessary adjunct in many space launch vehicles. Chapter 12 gives additional information.

The 1997 acquisition of a group of *hydrazine monopropellant thrusters*, developed originally by Hamilton Standard, helped ARC to fill a significant gap in its product line with reliable proven products.[2,10–14] These LPRE are relatively simple, and the exhaust products are noncontaminating and transparent. Hamilton Standard started in this field in about 1964 and had qualified thrust-chamber units with integral valves in sizes of 0.2, 1.0, 2.0, 5.0, 20, and 100 lb thrust. They are identified as the Monarc thrusters, and some of them are listed in Table 7.11-1. Figure 7.11-3 shows an early version of the 5-lbf thruster, which delivered a specific impulse of 231.5 s. The thermal standoff sleeve and the distance between the valve and the thruster minimize the heat transfer to the sensitive electric insulation of the valve assembly. Some of these thrusters demonstrated more than 300,000 restart cycles. In a set of referee tests run by the Air Force of hydrazine monopropellant thrusters from seven different development organizations, only the Hamilton Standard design and the Rocket Research design completed the rigorous endurance trials.[9] With the acquisition ARC also received information on complete engine designs, field support equipment, and some additions to their customer base. ARC had developed some of their own hydrazine monopropellant thrusters in the 1960s and 1970s, but this set of thrusters was not fully competitive and did not fly often. So the company welcomed obtaining the Monarc product line and customer base.[10–14]

Engine Systems

Since August 1965, ARC has designed, built, and delivered several different completely *integrated LPRE systems* with tanks, pressurizing system, bipropel-

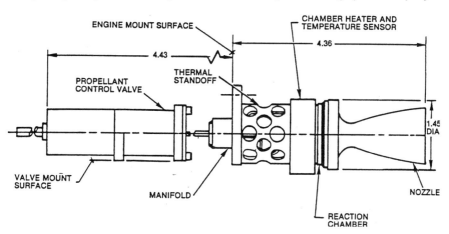

Fig. 7.11-3 This typical 5-lbf hydrazine monopropellant thruster originally designed by Hamilton Standard. Courtesy of ARC; copied from Ref. 2.

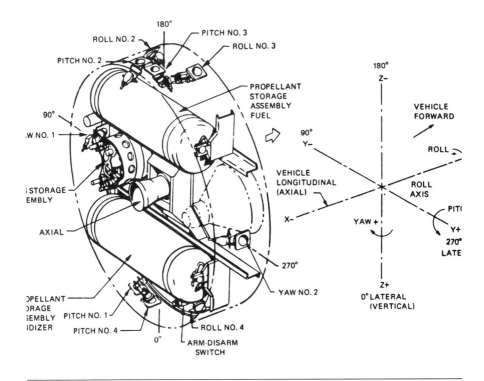

Fig. 7.11-4 Sketch of the Minuteman III postboost propulsion system. It has 11 small thrusters and two positive expulsion propellant tanks, which are preloaded at the factory. Courtesy of ARC; copied from Refs. 5 and 6.

lant thrusters, and support structures. For example the Minuteman III postboost propulsion system (PBPS) work started in 1965, and it has been built into a separate stage of the missile as seen in Fig. 7.11-4.[1,5,15,*] It actually constitutes the maneuverable fourth stage of this Minuteman ballistic missile. It was 52 in. diam and 11.75 in. high. A large number of these have been produced. The two positive expulsion propellant tanks have a double wall with a piston and bellows, as shown in a figure later; one contains nitrogen tetroxide and the other monomethyl hydrazine. It has a central 315-lb thrust gimbaled axial-velocity-control thrust chamber, which used beryllium interegen cooling technology (see Chapter 4.3) and was designed and is still produced by Rocketdyne (Chapter 7.8), and was delivered to ARC and assembled and serviced by ARC. The engine also has six ARC 23-lb pitch and yaw thrust chambers and four ARC 18-lb roll thrust chambers, with a niobium chamber/nozzle, radiation cooled and thermally insulated. These ARC thrusters are listed in Table 7.11-1, and one is shown in Fig. 7.11-5. Explosively activated isolation

---

*Personal communications, M. Marvin and F. Boorady, Atlantic Research Corporation, 2001–2004.

valves admit the propellants from the propellant tanks to the piping. Individual fast acting dual propellant valves (at each thrust chamber) control the start and stop of the pulsing operation. Originally designed for five years of storage, when loaded with propellants, it has been successfully tested 32 years after loading. ARC assembles the hardware and services the engines in the field. A total of 867 Minuteman III PBCSs have been delivered, and about 229 of these have flown, and some of these were stored for 15 years before flight. There has not been a single failure associated with this LPRE. The company received an Air Force award for the outstanding work and excellent reliability of this PBCS. The Minuteman III was still an active program at the time of writing this manuscript, and recent work included refurbishing old units.

Another auxiliary bipropellant propulsion system was developed for the target missile in the Gemini–Agena orbital rendezvous. It provides for orbit adjustments and propellant orientation.[7,8,*] Each had positive expulsion propellant tanks, a pressurized-gas feed system, a single small pulsed thruster at 16 lbf nominal thrust, a 100-lbf thruster for maneuvers, and used mixed oxides of nitrogen (NTO plus some dissolved nitrogen oxide) with UDMH as propellants.

ARC developed a complete engine system (called hydrazine actuation system or HAS) for application in the British submarine-launched Chevaline missile.[7,8,*] It had 12 hydrazine monopropellant thrusters (supplied by Rocket Research) for pitch yaw and roll control, used two ARC hydrazine storage tanks

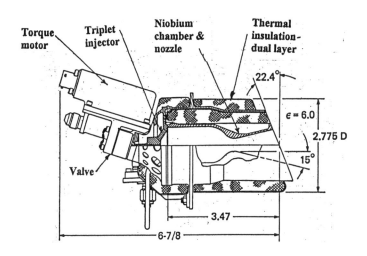

Fig. 7.11-5 Six of these 23-lbf TC assemblies are part of the Minuteman III postboost control system. The TC is insulated to maintain an outer-wall temperature below 400°F. Courtesy of ARC; copied from Refs. 7 and 8.

---

*Personal communications, M. Marvin, Atlantic Research Corporation, 2001–2004.

with rolling diaphragms for positive expulsion, and an ARC warm-gas monopropellant hydrazine pressurization systems. This warm gas comes from a gas generator, where hydrazine is decomposed into warm pressurizing gas, with a small propellant tank and its own inert gas-pressurization feed subsystem; this scheme was first developed at ARC in the 1970s. In addition the missile had a separate bipropellant system with higher-thrust bipropellant (NTO/hydrazine) thrusters, which were furnished by Royal Ordnance in England. Its feed system had two dual-piston pumps. Each pump assembly had an oxidizer piston pump, a fuel piston pump and a larger-diameter warm-gas actuating piston. The warm gas was produced in a separate ARC gas generator by again decomposing hydrazine, which came from the same gas-pressurized GG feed system as used by the other GG. It is the only flying dual-piston propellant pump known to the author. The first flight was in 14 months after go-ahead. Less than 200 LPRE assemblies have been delivered to the Royal Navy of Britain, and at least 35 of these (with the HAS) have been flown. Decommissioning of the missile started in 1997.

After World War II, Bell Aircraft hired General Walter Dornberger, the military commander of the German guided missile war efforts in Berlin and Peenemuende.* He became the chief scientist of Bell Aircraft and made significant contributions, such as in the Dynasoar boost-glide vehicle program. Bell also hired about 10 other German engineers, who had been involved in the V-2 effort. Dornberger and a couple of the other Germans had some experience with the German LPRE efforts. They were of considerable help in the design of the LPREs for the Rascal and Agena, which are described next.

Beginning in 1954, the predecessor Bell Aircraft developed and then produced a flying bomb called *Rascal* (GAM-63) to be carried and released by a B58 bomber aircraft. The rocket engine for this air-to-surface missile was also developed by this company. It is shown in Fig. 7.11-6, and some data are in Table 7.11-2. Its nonhypergolic propellants were IRFNA and JP-4 (aviation-grade gasoline). It had three fixed (not movable) rocket thrust chambers, a bipropellant gas generator, and a single-shaft turbopump. A solid-propellant start cartridge in each chamber was ignited by electric power, and the gas generator ahead of the TP had an electrical igniter. Although the engine passed qualification tests, in 1957 the government dropped the requirement to use rocket propulsion for the Rascal missile, and this engine did not fly. The Rascal missile, was deployed with the U.S. Air Force, but without a rocket engine.

Agena Rocket Engine

The *Agena rocket engine* was probably the best known, most historical, and the largest LPRE developed by ARC. Its principal applications were to drive an ascending upper stage (called Agena) and also to provide pitch and yaw torques for orbit maneuvers. This stage was developed by the Lockheed

---

*Personal communications, M. Marvin and F. Boorady, Atlantic Research Corporation, 2001–2004.

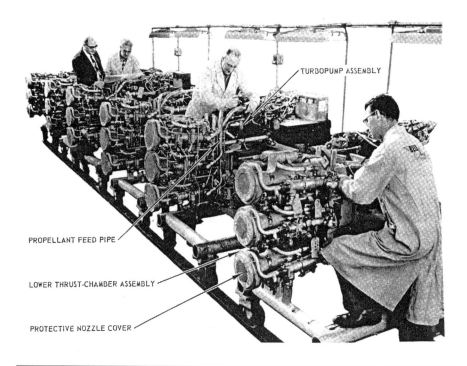

Fig. 7.11-6 View of the assembly line of the LPRE with three thrust chambers for the Rascal missile. Courtesy of ARC; copied from Ref. 16.

Table 7.11-2 Selected Data for four ARC LPREs

Characteristic	Rascal	Agena 8096	Agena 8096-39	Lunar ascent[a]
Thrust, lbf	4,000 (SL)[b]	16,000 (vac)	17,000 (vac)	3,500 (vac)
Propellants	IRFNA/JP-4	IRFNA/UDMH	IRFNA/UDMH	NTO/50% UDMH 50% hydrazine
Specific impulse, s	265	290.5	300	310
Number of TCs	3	1	1	1
Engine cycle	GG[c]	GG	GG	Pressurized gas
Mixture ratio, O/F	NA	2.8	2.94	1.51
Design year	1954	1965	1971	1963
Nominal chamber pressure, psia	NA	506	545	120
Flight proven	No	Yes	Yes	Yes

[a]Job was started by ARC, but completed by Rocketdyne.
[b]12,000 lbf total with 3 thrust chambers.
[c]GG means gas-generator engine cycle.

Corporation in Sunnyvale, California. The turbopump-fed engine used a gas-generator cycle and NTO/UDMH as hypergolic propellants.[17-21],*

Figure 7.11-7 shows the Agena engine, and a flow diagram is shown in Fig. 7.11-8. It has a unique aluminum (6061 T6) thrust chamber (oxidizer cooled) with a relatively thick wall, which contains long drilled holes (inclined to the axis) as cooling passages. Straight holes can be drilled into a double curved nozzle-throat section of the cooling jacket as can be seen in Fig. 7.11-9. Some versions had a five-compartment injector baffle to prevent the occurrence of unstable combustion. The radiation-cooled nozzle-exit section (between area ratio of 12 and 45) is made of titanium reinforced externally with molybdenum stringers and hoops, which can be seen as an egg-crate pattern in Fig. 7.11-7. To provide multiple restart capability in space, there are two positive expulsion piston tanks for the initial flow of propellants to the gas generator,

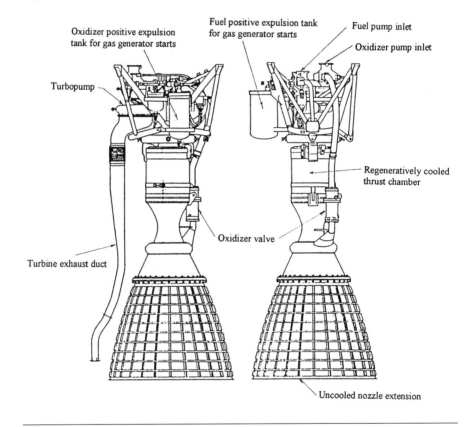

Fig. 7.11-7 Agena 8096 engine. Its earlier version was the first pump-fed LPRE used in upper stages of U.S. space launch vehicles in early 1960s. Courtesy of ARC; from Refs. 17 and 21.

---

*Personal communications, M. Marvin and F. Boorady, Atlantic Research Corporation, 2001–2004.

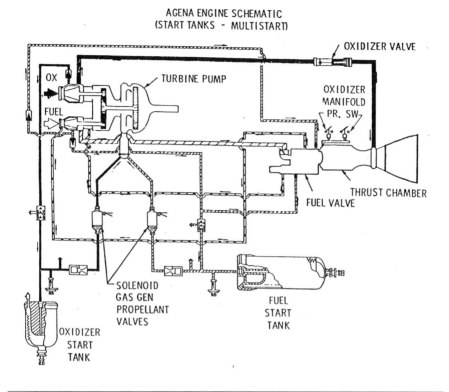

Fig. 7.11-8 Simplified flow diagram of one version of the Agena engine with restart capability. Courtesy of ARC; from Refs. 17 and 21.

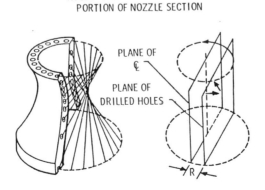

Fig. 7.11-9 Diagram to explain how straight cooling holes can be drilled at an offset and angle through the aluminum chamber/nozzle wall. Copied from Ref. 21.

and they can be identified in the schematic diagram. These tanks can be reloaded in flight with high-pressure propellant from the pump discharge lines and be ready for the next engine start. Cavitating venturis in the feed lines leading to the GG control the GG propellant flow (and thus the thrust) and the GG mixture ratio (and thus the GG's gas temperature). The single-shaft turbopump had an oil-lubricated gear case, which allowed the turbine to rotate faster than the propellant pumps, and this gave higher turbine and pump efficiencies. In turn this contributed to a lower GG propellant flow and raised the specific impulse by perhaps 1 or 2 s. A simplified section view of one of the versions of the Agena engine turbopump is shown in Fig. 7.11-10. It was based in part on the turbopump experience with the Rascal engine and used oil lubrication of the bearings and gears. Like the turbopumps of other LPRE organizations at that time, a geared turbopump was common; it helped to reduce the flow of GG propellant and raised the specific impulse slightly.

The first Agena engine was developed around 1956 and 1957, used IRFNA and JP-4 jet fuel, was not gimbal mounted, not restartable, and it did not fly. Five improved versions have been developed since that time, and three of those did fly. Agena engine Model 8096 had the largest production (300 units) and most flights. It had the performance characteristics given in Table 7.11-2

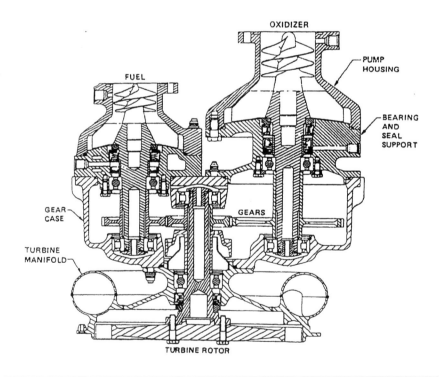

Fig. 7.11-10 Section view of one version of the Agena turbopump. Courtesy of ARC; copied from Ref. 21.

and also a dry weight of 296 lb, nozzle area ratio of 45, gimbal angle of ±5 deg, height of 83 in., and exit diameter of 32.5 in. It was the first major U.S. liquid propellant rocket engine to be test fired at a simulated altitude of about 90,000 ft at the Arnold Engineering Development Center in Tullahoma, Tennessee. The engine consisted of 26 subassemblies, which were individually tested. The total parts count was 1860.

The first two versions of this LPRE (1957–1958) had a lower nozzle area ratio (15), used the nonhypergolic propellants IRFNA and JP-4 jet fuel, and were started with solid-propellant cartridge ignition (restart was then not possible). The second version flew once. Because engineers were uncertain at the time that altitude ignition (at essentially zero ambient pressure) could be accomplished, the first flights used a nozzle plug, which was ejected once the chamber pressure begins to rise. Subsequent tests proved that such a plug was really not necessary. The subsequent engine versions used hypergolic UDMH as a fuel reacting with IRFNA (acid plus 14% N_2O_4, 0.7% HF and 2% water), and this simplified the engine system, and UDMH containing about 1% silicon oil. Some data are in Table 7.11-2. A total of 19 Agena 8096-39 have flown satisfactorily.

The silicon oil would burn and form silicon dioxide as small white solid particles, which would deposit on the inside chamber and nozzle walls as a fluffy white layer. This layer provided a high-temperature thermal insulation for the hot wall. In fact it reduced the overall heat transfer to the cooling jacket by about 33%. When the silicon-oxide layer became too thick (perhaps up to $\frac{1}{8}$ in.), pieces of the layer would break off or fall off or be dragged off the wall by the hot gas, but according to ARC, a new, but thinner protective layer would form almost instantaneously on the bare spot. This Agena engine is the only flying engine known to the author to successfully use this internal coating method of heat-transfer reduction.

Later ARC again changed propellants and went to a different oxidizer called high-density nitric acid, which consisted of 56% nitric acid and 44% dissolved NTO with an average density of 102.7 lb/ft^3. The propellant combination with UDMH fuel, containing 1% silicon oil, gave about 5% more specific impulse and about 4% more average propellant density.[22] It was extensively ground tested, but did not fly.

The concept of a self-renewing internal silica insulation coating was developed originally at General Electric Company in the early 1960s, but with LOX/kerosene or LOX/jet fuel as propellants. It was also tried by Reaction Motors, Inc., and Rocketdyne, but with mixed results. Although ARC liked the silicon coating and seemed to have good results, the other companies did not pursue it because it caused occasional local burn-throughs. There is a plausible mechanism for these failures. The silicon-oxide insulation layer at GE and Rocketdyne was occasionally relatively hard and crusty and not always as fluffy and low in density as the coatings at ARC. When a piece of the brittle insulation broke off (and was discharged through the nozzle), a sharp edge formed at the remaining hard insulation layer next to the chamber wall. At this edge the hot gas can reach temperatures close to the stagnation temperature in the combustion chamber. The local heat transfer would increase greatly, and the local wall temperature could at times quickly exceed the melting point of the wall material, resulting in a local wall failure. Also the coating caused the chamber and throat diameters to decrease, causing an unpredictable modest rise in chamber pressure and a change in engine performance. The author does not know what happened at RMI, but this idea of a self-forming thermal insulation silicate layer was abandoned at General Electric and Rocketdyne. It might have been the propellant combination of IRFNA/UDMH that gave a fluffy coating without sharp edges.

The Agena upper stage, developed by Lockheed, had an auxiliary or integrated secondary propulsion system for roll control during the firing of the main Agena engine and also for some attitude control during the subsequent coast period. Initially this secondary sytem had small thrusters, its own small high-pressure propellant tanks with positive expulsion features, and a pressurized inert gas tank of its own. Later the propellants were obtained directly from the main engine's propellant tanks, which were at a much lower tank pressure and were pumped by a small set of gear pumps driven by an electric motor through magnetic (nonleak) couplings. This eliminated the separate heavy pressurized tank system. A bipropellant radiation-cooled thrust chamber with a pintle injector (furnished by TRW) provided about 90 lbf thrust, and the thrust could be pulsed at 1-s intervals. The pump unit was developed by Lockheed Missiles and Space Company and is shown in Fig. 4.4-10. It is the only known U.S. small flying thruster using a gear pump feed system.

The Gemini–Agena rendezvous was an experiment between the manned Gemini spacecraft and an unmanned target vehicle spacecraft. Both had the Agena as their second stage. In the first space rendezvous the two spacecraft were supposed to dock with each other and then perform some space maneuvers together.[7] The auxiliary propulsion sytem for the target vehicle was briefly mentioned earlier in this chapter. In the first rendezvous attempt the target vehicle separated properly from its Atlas booster, but failed to reach its intended orbit because of an automatic premature shutdown during the restart of its Agena engine. Postflight analysis indicated that a hard start (explosion) was the cause of the engine failure. A careful evaluation and a duplication of the failure in ground tests were undertaken.[23] Simulated altitude start tests were made in the vacuum facility of the Arnold Engineering Test Facility. The cause of the hard start (really an explosion of accumulated unburned propellant) was identified to be a faulty starting sequence, with an unfavorable fuel lead and a relatively long ignition delay. It was believed that a fuel lead allows some explosive compounds to form in the chamber filled with fuel. Changes were made in the control system to ensure an oxidizer lead, which gave smooth starts.

The first flight of an Agena LPRE was made in 1959. A total of 418 Agena engines have been produced, and 363 have flown with only one known engine failure, and it was described earlier. This LPRE propelled a large variety of different illustrious payloads into space trajectories for almost three decades.[21] This included the first U.S. satellite in a circular low orbit, the first into a polar orbit, flyby space probes to Mars and the moon, and the first to perform a significant orbit velocity vector change. It also was the first large U.S. engine with a vacuum restart and significant on-orbit maneuvers. As already mentioned, it was the first to participate in a space rendezvous and docking operation with a manned spacecraft, namely, the Gemini.

Between 1985 and 1998 ARC improved the Agena engine extensively, investigated and demonstrated new versions with slightly more performance, a significant reduction in inert engine mass, a larger nozzle and a higher nozzle area ratio, and operation with monomethyl hydrazine as a fuel (instead of UDMH). In 1999–2000 ARC and Aerojet joined together to build and demonstrate a more improved version identified as Agena 2000, and it is briefly mentioned in Chapter 7.7 (Ref. 24). However none of the improved engines were selected for flights. By that time good new upper-stage LPREs with LOX/LH$_2$ had flown successfully, and these cryogenic propellants would permit a major increase in payload or mission velocity, if compared to the improved Agena engine.

Other ARC Projects

The highest possible specific impulse with chemical propellants was always known to be with a fluorine oxidizer. So in September 1958 the first static firing of a TC at about 1000 lb thrust using liquid fluorine as the oxidizer was successfully accomplished at Bell. In 1960 a version of an Agena pump-fed engine was also ground tested with liquid fluorine and ammonia as propellants.* These tests (and tests at other organizations) proved the feasibility of using fluorine as an oxidizer for an upper stage launch vehicle, but this toxic, corrosive oxidizer was never selected for a flight application as explained elsewhere in this book. Another technically very interesting accomplishment of ARC was the decomposition of MMH by a solid-pellet-type catalyst; however, this discovery did not find a production application.

ARC did some pioneering work with *positive-expulsion liquid propellant tanks*.[5–8],* Positive expulsion tanks are needed for starting or restarting -pumpfed engines in gravity-free space. It prevents bubbles from entering into the propellant feed lines. One type was ARC's piston expulsion tanks using a long metal bellows as a seal between gas and liquid; it is shown in a simplified presentation in Fig. 7.11-11. It has an elliptically shaped piston, which fits the contoured end of the tank. Such a tank expels 98 to 99% of the stored propellant, and during storage the piston moves to allow propellant volumetric (density) changes with ambient temperature. The company also built some positive expulsion tanks with internal pistons and curved tank ends. Two of these were used in the Mercury capsule shown in Fig. 7.11-2. Another type of positive expulsion tank design used an ARC "rolling diaphragm" inside a specially shaped tank. It uses a patented aluminum diaphragm depicted in Fig. 7.11-12, which has photos and a section indicating the successive positions of this

---

* Personal communications, M. Marvin, Atlantic Research Corporation, 2001–2004.

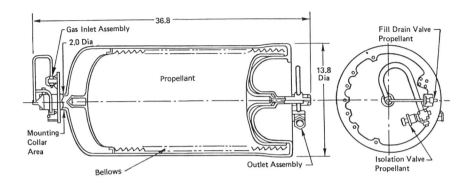

Fig. 7.11-11 Simplified section and end views of a typical positive-expulsion piston tank with a bellows seal many have flown. Courtesy of ARC; from Refs. 7 and 8.

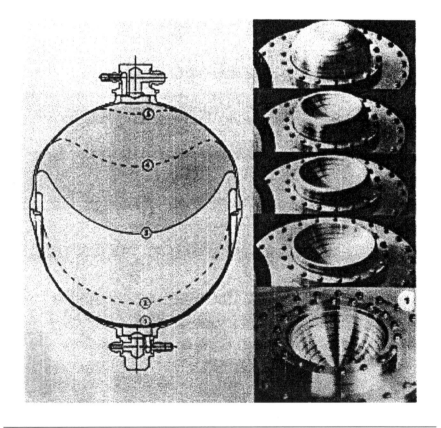

Fig. 7.11-12 Five successive positions of a rolling aluminum diaphragm, which provides positive expulsion of propellant; this type has flown successfully. Courtesy of ARC; copied from Refs. 7 and 8.

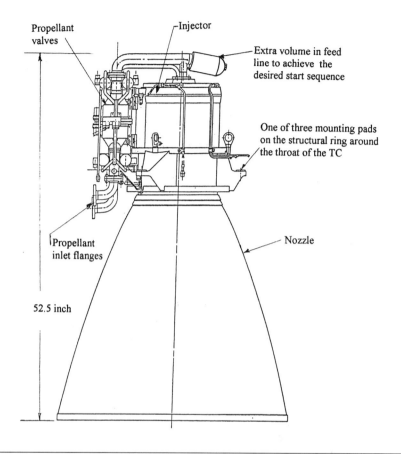

Fig. 7.11-13 Final version of the lunar ascent engine (3500 lbf thrust) was used to return astronauts from the moon's surface to the orbiting Apollo command vehicle. Courtesy of NASA, ARC, and Rocketdyne.

diaphragm. It has flown in several satellites and in a PBCS. The Apollo/SaturnV multistage space vehicle had a total of 31 ARC positive-expulsion tanks.

In 1963 ARC began the development and design of the Apollo mission *lunar ascent* thrust chamber and valve assembly. The pressurized feed system for it was developed by LTV Aerospace, the prime contractor for the lunar landing module. It has the mission of bringing astronauts from the surface of the moon back to the Apollo command module, which is orbiting around the Moon. The TC assembly can be seen in Fig. 7.11-13 and some data is in Table 7.11-2. The nozzle area ratio was 45.6 to one and duration was long, 465 seconds. Bell/ARC developed a TC with an ablative inner liner and a package of propellant valves. The TC failed to pass all the required demonstrations of inherent combustion stability. As discussed in Chapter 4.10 instability was triggered during demonstration tests by directed explosive charges. An unstable injector would continue to operate with large high frequency pressure waves, while a stable injector design would within a few milliseconds return to normal combustion. One requirement was to pass 12 consecutive bombardments, but the Bell/ARC TC could not pass this test. For about two years Bell/ARC tried various design changes and operating changes, including some suggested by NASA, to cure the problem and this combustion instability caused a delay in the overall program. Therefore NASA on an emergency basis had Rocketdyne design and build an alternate injector and it successfully passed all the required stability demonstrations.[24] This injector and the final TC configuration are shown in Fig. 7.8-17. So the final version, that was selected by NASA and successfully launched astronauts off the moon, had an ARC valve package, a chamber originally designed by ARC and then supplied by a subcontractor, a Rocketdyne injector, and the assembly and field service were provided by Rocketdyne. It was an example of a government initiated cooperation between two competitors.

Concluding Comments

Between 1956 and 1975 ARC developed and flew several versions of the first U.S. upper stage LPRE, the Agena engine, with thrust levels between 15,000 and 17,000 lbf and a total of 418 were built. The company was not able to find a suitable application for the more advanced Agena engine version, because by 1990 it was badly outperformed by new LOX/LH$_2$ engines. The main business has been and still is today in small thrusters (monopropellant as well as bipropellant types) and small positive expulsion propellant tanks. When Aerojet acquired all of ARC, the Liquid Propellant Operation was part of it. However the government required Aerojet to divest itself of the small ARC Liquid Propellant Operation, because Aerojet already had two groups in the same business. In 2004, when this chapter was written, Aerojet was in negotiation on this divestiture.

References

[1]"Chronology of Bell Products and Achievements," Informal Company Highlights, Bell Aerospace Systems, a Textron Co., Buffalo, NY, circa 1979.

[2]"*Hydrazine Propulsion Experience*," Hamilton Standard, United Technologies Corp., Windsor Locks, CT, Company brochure SSS-89-01, 1989.

[3]"*Hydrogen Peroxide Attitude Control Systems*," Informal Report 8500–953049, Bell Aerospace Systems, Niagara Falls, NY, Nov. 1971.

[4]Gribbon, E., Driscoll, R., and Marvin, M., "Demonstration Test of a 15 lbf Hydrogen Peroxide (90%) Thruster," AIAA Paper 2000–3249, 2000.

[5]*Reaction Control Experience, Bell Aerosystems Company*, Niagara Falls, NY, circa 1978.

[6]"*Propulsion Products and Experience*," Bell Aerospace Textron, Niagara Falls, NY, Internal Co. Rep. 0500–927020, circa 1980.

[7]"*ARC Space Propulsion*," Atlantic Research Corp., Niagara Falls, NY, Company Brochure, 2001.

[8]"*ARC Liquids Products and Capabilities*," Atlantic Research Corp., Niagara Falls, NY, 1999.

[9]Tolentino, A., and Grotter, R. J., "Advanced 5 lb Engine Demonstration," (AF-RPL report on life tests of different thrusters), AF RPL, Edwards AFB, Edwards, CA, March 1984.

[10]Polythress, C. A., and Conkey, G. H., "Development of a Long Life 125 lbf Hydrazine Thruster," AIAA Paper 80-1170, 1980.

[11]Marcus, M., "Historical Evolution of the Hydrazine Thruster Design," AIAA Paper 90-1838, July 1990.

[12]Sansevero, V. J., Jr., Garfinkel, H., and Archer, S. F., "On-Orbit Performance of the Hydrazine Reaction Control Sub-System for the Communications Technology Satellite," AIAA Paper 78-1061, July 1978.

[13]Francis, A. E., "Qualification Test for the Multiple Satellite Dispenser 0.2 lbf Rocket Engine Assembly," Technical Rep., Oct. 1973, May 1974.

[14]Saenz, J., Jr., "Life Performance Evaluation of 22 N (5 lbf) Hydrazine Attitude Control Engines," USAF/RPL (Rocket Propulsion Laboratory) Edwards AFB, Edwards, CA, Technical Report, Nov. 1980.

[15]"Minuteman, Post Boost Propulsion System," Bell Aerospace Textron, Niagara Falls, NY, data sheet, circa 1980.

[16]Herrick, J., and Burgess, E., (eds.), *Rocket Encyclopedia, Illustrated*, Aeropublishers, Los Angeles, CA, and Burgess, E., 1959.

[17]DeBrock, S. C., and Rudey, C. J., "Agena Primary and Integrated Secondary Propulsion Systems," *Journal of Spacecraft and Rockets*, Vol. 11, No. 11, 1974; also AIAA Paper 73-1212, Nov. 1973.

[18]Yaffee, M., "Bell Adapts Hustler Rocket Engine for Varied Missions," *Aviation Week*, 21 Nov., 1960, p. 52–61.

[19]Albanese, D. T., Ferger, T. M., Loftus, H. J., and Schmidt, C. M., "Development of Improved Agena Engine Injector," AIAA Paper 76–684, July 1976.

[20]DiFrancesco, A., and Schmidt, C. M., "Bell Model 8096 Agena Engine—Hot Restart Demonstration," AIAA Paper 76–683, 1976.

[21]"The Agena Story, Bell Agena Rocket Engine, 1959–1986," Bell Aerospace, Informal Co. Rep., circa 1986.

[22]Loftus, H. J., "Applications of High Density Nitric Acid Oxidizer and UDMH with Silicon Additive Fuel to the Agena Engine," Bell Aerospace, a Textron Co., Niagara Falls, NY, Internal Co. Rep., 1965.

[23]Boorady, F. A., and Douglass, D. A., "Agena-Gemini Rocket Engine—Hard Start Problem Resolved During Project Sure Fire," AIAA Paper 67–259, 1967.

[24]Gribben, C., Driscoll, R., Marvin, M., Wiley, S., Andersen, L., and Fischer, T., "Design and Test Results on Agena 2000 Main Axial Engine," AIAA Paper 98–3362, 1998.

Liquid Propellant Rocket Engines in Russia, Ukraine, and the former Soviet Union

The Soviet Union has done more work on LPREs than any other country. It is not possible to include all of the many LPRE or R&D achievements, and only selected efforts will be described here. A few historical events or interesting rocket engines will be described in somewhat more detail. All LPRE work in Russia and formerly in the Soviet Union have been supported by the government, and even the amateur societies had some government funding.

As mentioned elsewhere in this book, the LPRE design bureaus in the Soviet Union have designed and built more than 500 different liquid propellant rocket engines, more than any other nation by a wide margin. This number is based on known products from 11 different design bureaus and is the author's best estimate. This number does not count the early experimental TCs or the modified versions of an engine. One particular LPRE has nine different versions, but they are counted here as one.

The early LPRE efforts, aircraft rocket engines, and the organizations working on LPRES are discussed in the next three chapters. The next 10 chapters discuss specific rocket engines and their technologies sorted by individual design bureaus or R&D organizations. The final chapter on Soviet/Russian LPRE activities describes a summary of key LPRE events, accomplishments, and findings.

An abbreviated summary version of Chapters 8 to 8.13 on LPREs in Russia was published earlier. It contains only some of the material described here.[1]

Reference

[1]Sutton, G. P., "History of Liquid Propellant Rocket Engines in Russia, formerly the Soviet Union," *Journal of Propulsion and Power*, Vol. 19, No. 6, 2003, pp. 1008–1037.

8.1 Early History (1929–1944)

The visions of Tsiolkowsky (Chapter 5.1) and the early amateur effort in the Soviet Union (Chapter 5.5) were the preludes to steady and continuing government supported efforts in LPRE. In 1919 Nicolai I. Tikhomirov, a spaceflight enthusiast, wrote a letter to Lenin, the Communist leader of the country at that time, asking for a government grant to do serious research on solid- and liquid propellant rocket propulsion.[1] This request was reviewed by the Soviet military and other government bureaus and was approved two years later. Originally this government group was founded in Moscow in 1921 as Tikhomirov's Jet Propulsion Laboratory, and later in 1925 it was moved to Leningrad. (This city is now again known as St. Petersburg.) Initially it concentrated on propulsion theory and solid-propellant rockets.

In June of 1928, Tikhonmirov's Laboratory was reorganized and called the Gas Dynamics Laboratory (abbreviated as GDL). In 1929 it began to work on LPREs.[1-4] In 1931 Valentin P. Glushko, a young engineer and space enthusiast, became the leader of the LPRE section within the GDL. He had studied LPREs since he was 15 years old, and he had exchanged letters with Tsiolkowsy. He and his group (1931–1938) are credited with the development of several early Soviet thrust chambers and early engine concepts, which are described in this chapter.[5-9,*] Also he proposed and implemented the idea of hypergolic ignition around 1931; this was earlier than in Germany or the United States. In 1939 Glushko became the head of his own organization and between 1940 and 1945 worked on aircraft superperformance rocket engines.[2,3,6,*] Glushko is identified with many of the advances in LPREs and with the development of most of the larger sized LPRE in the Soviet Union.[7-9] More information about this outstanding man and his engines will be found in Chapter 8.4.

In August 1931 the Moscow Gruppa Isutcheniya Reactivnovo Dvisheniya (abbreviated as GIRD) was set up with government support.[1-4] A simple translation would be the Moscow Group for the Study of Reaction Motions. A number of the Moscow amateur society members joined this Moscow GIRD. Initially it had groups for LPREs, rocket-propelled test flight vehicles, ramjets, and jet engines. The work on LPREs at the Moscow GIRD proceeded independently of the LPRE work done at GDL in Leningrad, but both of these LPRE organizations started LPREs at about the same time. Fridrikh Tsander became the leader of the LPRE section within the Moscow GIRD. Other GIRD groups were also set up, such as those in Leningrad, Gorky, Kiev, Krahkov, Baku, Tbilisi, or Rostov. They were controlled from Moscow. Sergei P. Korolev, an engineer and pilot, became the leader of the Moscow GIRD in 1932.[2-4] He personally also worked on several vertically launched rocket test vehicles and rocket propelled aircraft. Only occasionally was he involved directly with the design or development of LPREs. Korolev later became the first Soviet leader to work on early Soviet ballistic long-range missiles and space-launch-vehicle

---

*Personal communications, V. K. Chvanov, first deputy of general director and general designer, NPO Energomash, 2002–2003.

development efforts (RNII explained later). He became a chief designer of his own design bureau, which developed ballistic missiles and space launch vehicles, but also included groups for the development of LPREs. A photo of Korolev is shown in Fig. 8.2-11. He became the leader of all of the Soviet ballistic missiles and space-vehicle development work. He was responsible for initiating work on ICBMs, Sputnik (the world's first satellite), and all of the other early USSR satellites.[10] He influenced the development of LPREs and selected the most suitable engines and the most suitable LPRE development organizations for his long-range missiles and space vehicles. Korolev was made a corresponding member of the Soviet Academy of Sciences in 1953 and a full member in 1958. His key job as the leader of the Soviet ballistic missile and space program and his membership in the Academy were kept secret. He was known as the "chief designer" and not by his name. Only after his death did his powerful role and his relationship with the top Soviet hierarchy become known.

Technical Advances in Thrust Chambers at the Gas Dynamics Laboratory

The technical LPRE issues addressed by the Moscow GIRD are quite similar to those undertaken by GDL, and both efforts are described next. Altogether these two development organizations designed more than 120 different TCs, all with pressurized feed systems. Each one incorporated some improvements, different or new features. Of these less than 80 were actually ground tested, and some were retested after the engine was modified. Only about eight of them were actually flown in experimental vehicles.[2,3] Several of these engines, but not all of them, will be described here. Some of the historic engines are described in more detail, and the technical changes, progress in specific technologies, and some failures will also be discussed. Many of these engines demonstrated or provided technical proof of useful, novel design features, and these found their way into subsequent production engines that were adopted for a specific application. As is typical in R&D programs, many of these experimental designs had features, which were not effective, were not adopted by later developments, and today are obsolete. Although GDL and GIRD aimed the development programs of most of their engines at specific flight applications, none of these engines found an actual military or space-related application, and none were mass produced or deployed with the military forces.

Early in the work Glushko and his department at GDL evaluated and selected the propellant combination for his early TCs. Between 1931 and 1933 he investigated a number of propellants, namely, nitric acid, nitrogen tetroxide, nitromethane, hydrogen peroxide, toluene, gasoline, alcohol, and kerosene. He also demonstrated for the first time in LPRE history the phenomenon of self-ignition using nitric acid and a fuel consisting of an organic phosphorous compound and toluene. A special apparatus was built to test ignition and combustion qualities. At first they tried homemade monopropellants made of a mix of nitrogen tetroxide with gasoline, or toluene, and kerosene.

They thought monopropellants would make a simpler engine feed system. Although these mixed monopropellants ran smoothly much of the time, there were several violent explosions of the premixed propellants, and this approach was abandoned. So the GDL investigators decided to abandon monopropellants and to try bipropellants. They preferred storable propellants because there was no evaporation loss and no ice formation. By 1933 GDL settled on nitric acid and kerosene because these propellants were readily available and resulted in a reasonably high performance. It was the favorite propellant combination at GDL for more than a decade. They also liked nitrogen tetroxide (NTO), but this oxidizer was not readily available in the Soviet Union in the 1930s. A small plant was built to produce it, and TCs were tested with this oxidizer. In the 1950s NTO began to replace nitric acid in the Soviet Union. Korolev liked LOX/kerosene because of its good performance. Some work proceeded with this combination, mostly at other LPRE organizations, but a couple of decades later Glushko used these propellants in large LPREs.

In 1931 Glushko's section at GDL built the first experimental Russian bipropellant TC called ORM-1 and shortly thereafter also an improved version ORM-2.[2,3,*] The abbreviation ORM stands for opytnyi raketnyi motor or experimental rocket engine. In the early years the word "motor" was often used in the Soviet Union to designate any kind of rocket propulsion, but they switched to the word "apparatus" or "engine" for LPREs. The ORM-1 was repeatedly static fired beginning in mid-1931 and also in 1932. The OMR-1 configuration is very different from modern TC versions, and it was cut off at the nozzle throat. Figure 8.1-1 shows an annular uncooled combustion chamber and a nozzle pointing upward. It did not have a supersonic diverging nozzle-exit section because its size and shape would influence the performance. The chamber and nozzle were uncooled, but coated with copper (supposedly to conduct away the heat) and a thin layer of gold (to prevent corrosion of the copper with nitric acid). The TC was supported on a perforated piece of pipe. The external open tank allows a pool of water to surround the TC. The investigators hoped that this external water bath would cool the hot parts, but this cooling by stagnant water proved to be insufficient. The fuel flowed through a check valve into three injection tubes, which were parallel to the nozzle axis. Only one injection tube is shown in the cross section of the figure. The oxidizer was supplied from the bottom, and it also went through a check valve into three injection tubes. (Only one is shown.) The other pipe was for measuring chamber pressure. Ignition was by wadded cotton rags, soaked in alcohol, and put into the chamber prior to firing. It was initiated by a fuse, whose wires came out of the nozzle. The OMR-1 in its first few tests developed a maximum thrust of 20 kg with nitric acid and toluene as the initial propellants. This TC was later fired with LOX/kerosene and other propellants. It was equipped with three interchangeable steel nozzles (with throat diameters of 10, 15, and 20 mm,

---

*Personal communications, V. K. Chvanov, first deputy of general director and general designer, NPO Energomash, 2002–2003.

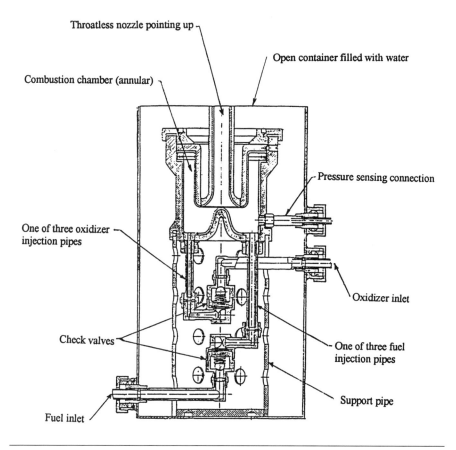

Fig. 8.1-1 Cross section of the ORM-1, the first Soviet thrust chamber tested in 1932. It had a cutoff nozzle on top. Drawing provided courtesy of NPO Energomash.

respectively) also copper plated and without a diverging section. The propellants were supplied from thick-walled heavy-duty test stand propellant tanks and were forced into the TC by high-pressure stored gas. The duration of firing of this first TC was limited by the heat capacity of the thrust chamber walls; it could run only for about 2 to 8 s without burnout. Specific impulse values were low by today's standard, about 140 to 170 s, but these numbers were very good for a first effort.

OMR-2 had swirl injection elements, which sprayed conical formations of propellants into the chamber, and simple open injection tubes were no longer used. It also had "dynamic cooling"; presumably the water in the open container surrounding the chamber was stirred or made to flow and was not static. OMR-3 was designed to use hypergolic propellants, eliminating the requirements for an ignition system and simplifying the start process.[3] Also one version was designed with a variable throat area by moving the nozzle against a fixed central inner tapered cone. The OMR-3 was one of several GDL

TCs that were never tested; the control of the variable throat area and the sealing of the moving hot parts turned out to be difficult, and at that time the need for throttling was not urgent.

Between 1932 and 1937 a large variety of TC designs, designated as ORM-1 to ORM-102, were investigated at GDL in Leningrad and also a good number at GIRD in Moscow.[3,8,9] Some were actually tested, some were built, but not tested, and some were merely design studies. By 1933 GDL had TCs that could run more than a minute because of advances such as those mentioned in the following. In that year the first vertical flight of an experimental small test vehicle with a LPRE was accomplished.

These TCs had different injection locations and different elements (various spray patterns, multiple holes, different direction of the injection elements, slotted jets, alternate swirl-type sprays, and different numbers or patterns of injection elements, etc.). In general, GDL preferred to use swirl-type injection spray elements, which emit a thin conical sheet of propellant, and this sheet then breaks up into droplets. Many of theses spray injection elements were initially located at the chamber wall, pointing toward the center of the chamber. Beginning in 1936 and 1937 the spray injection elements were located in the forward end or head of the TC, and this is where they still are today. Some of the TCs had spray injection elements with an integral check valve, which prevented backflow from the chamber into the injector manifold or supply tubes. The check valve also prevented propellant dribble at shutoff.

The first eight ORM TCs had annular chambers similar to ORM-1 shown in Fig. 8.1-1. GDL abandoned this annular design, because it had more surface to cool and a higher total heat transfer than a straight cylindrical or oval chamber.[3] They then designed and tested different chamber configurations or shapes, including cylindrical, spherical, oval or football shaped, conical with a decreasing cross section and reverse conical, that is, with an increasing chamber cross section.[3] Figure 8.1-2 shows one type of a GIRD 10 TC with such a reversed cone shaped chamber; it was developed in 1937. They also designed chambers with different chamber volumes in order to investigate the effects on performance and heat transfer. Another feature that was included in several OMR and GIRD TCs was a precombustion chamber, which was thought to help with mixing of propellants and with a smooth start procedure. One version is seen in Fig. 8.1-2. They built and tested several types and sizes of precombustion chambers. None of the early TCs in the Soviet Union or in other countries had design provisions for compensating for the thermal growth of the hot wall.

GDL soon abandoned the use of alcohol-soaked pieces of cloth for ignition as was used in ORM-1. Instead they went to electrical ignition (spark plugs) with their nitric-acid/kerosene propellants. When spark ignition proved to be unreliable, they went in 1933 to a chemical ignition using a charge of metal powder mixed with nitrate crystals and an electrical fuse. In 1934 they also used pyrotechnic (solid propellant) ignition. The concept of a hypergolic (self-igniting) fuel reportedly was conceived by Glushko around 1930 or 1931 and experimentally investigated in a laboratory. Hypergolic propellants were tested in a container with an electrical timer. The first self-igniting fuel that was suc-

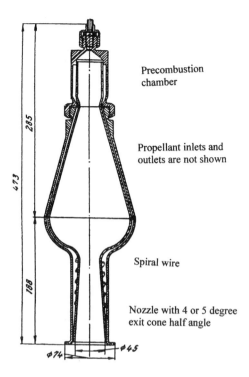

Fig. 8.1-2 Cross section of a cooled version of the GIRD 10 engine with a pear-shaped combustion chamber and a precombustion chamber. Dimensions are in millimeters. It was the first Soviet bipropellant engine to fly (around 1934). Copied from Ref. 2.

cessfully operated in a TC was a mixture of an organic liquid, which contained phosphorous, with turpentine and some gasoline. These early hypergolic liquids were not used as fuels because they did not give a high performance and some were very viscous at low temperatures. Therefore kerosene continued to be the principal fuel of choice. By about 1936 they had developed slug starts with a small initial quantity of hypergolic fuel. Originally it was injected from a separate gas-pressurized tank, but later the hypergolic start liquid was loaded into a piece of the fuel line and confined by two burst diaphragms, upstream and downstream of the hypergolic fuel slug. During start, the fuel pressure would break the diaphragms.

Cooling of the thrust chambers was a major issue in almost all of the test programs of GDL and GIRD, and a variety of approaches was tried.[2,3] The first few experimental TCs used uncooled walls, which would melt locally when the heat absorbing capacity of the wall would be exceeded and the wall would become too hot. Then GDL tried several different designs of external fins to provide *air cooling*.[3] One set of fins around a nozzle, made of cuprite (a high-conductivity metal), can be seen in Fig. 8.1-3 of the OMR-26 TC. The airflow, which cooled this finned radiator at the nozzle, was aspirated by the exhaust jet. Different fin

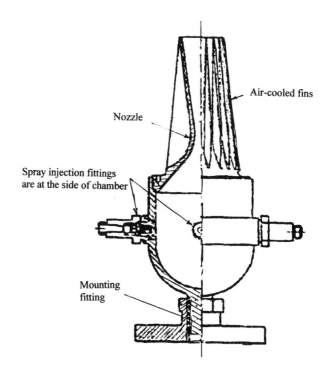

Fig. 8.1-3 Half-section of the ORM-26 thrust chamber with air-cooled fins around the nozzle and injection from the side. Courtesy of Energomash; from Ref. 3, p. 91.

geometries were tried. However air cooling was not aequate to handle the high heat release of the very hot gases, did not prevent burnouts, and was abandoned. *Film cooling* was used later in many TCs of both GDL and GIRD with different designs, and it usually was effective and avoided burnouts. But it was soon learned that film cooling decreased the specific impulse or performance, which was undesirable. By 1934 GDL designers used partial regenerative cooling, which they called 'external dynamic cooling'.[3] Beginning with ORM-34 *regenerative cooling* was designed into the nozzle only because that was the location where most of the burn-outs had occurred. Some TCs used the oxidizer and some the fuel as the coolant in the nozzle cooling jacket. They learned that the flow velocity of the coolant had to be increased to absorb all of the heat without excessive metal temperature or damage to the inner wall. The GDL had many TCs where the nozzle only had a cooling jacket, but the chamber was usually uncooled or cooled only by film cooling. One example is the ORM-48 (1933), and it is shown in Fig. 8.1-4. The spiral-shaped ribs in the cooling jacket cause an increase in the cooling velocity, which augments the heat transfer. This nozzle-only-cooling concept can also be seen in Fig. 4.3-2, which has a partial section of the ORM-50 TC. It shows a fuel cooled jacket around the nozzle only and an uncooled chamber with side injection (two oxidizer and two fuel spray elements). It was not until

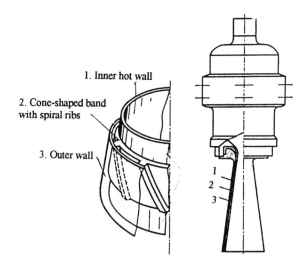

Fig. 8.1-4 ORM-48 engine (1933) was one of the early Soviet TCs with a cooling jacket around the nozzle. The combustion chamber was uncooled. The four injection spray elements are not shown here. Copied with American Astronautical Society permission from Fig. 6 of Ref. 5, p. 93.

1935 that they tested a TC with two separate cooling jackets, one around the chamber and one around the nozzle. By that time they had injection elements mounted at the forward end of the chamber, and they no longer used side injection from the chamber wall. Neither ORM-48 or ORM-50 were flown.

Another way to keep TC walls cool enough to carry the internal loads and stresses was to reduce the heat transfer by applying *a low-conductivity insulation layer or ceramic liner* on the inside of the hot walls. At first the LPRE department at GDL evaluated different thermal insulation prepared from zirconia, magnesia, alumina, and other ceramics in tests using hot gas from available smokeless solid propellant. The Soviet engineers experimented with ceramic nozzle inserts and ceramic sleeves in chambers and they also tried ceramic coatings and layers of dried ceramic paste, which was applied to the inner wall and baked. The ceramic-lined TC 12K is shown in Fig. 8.1-5 and was tested in 1935.[3,4] It ran on LOX/alcohol and used aluminum oxide for the pear-shaped chamber liner and a magnesium oxide formulation in the nozzle. This set of ceramic liners were often heavy, the ceramic components frequently cracked, and TC with ceramics were not considered to be reliable. An improved ceramic version with some regenerative cooling, called 02, was used for propelling some flight-test vehicles in 1936 and 1937; one was a winged vehicle. This 02 engine is mentioned again later. Ceramics are usually brittle materials, and they can form cracks caused by excessive thermal stresses. These cracks usually become larger with each successive rocket operation. Larger cracks in ceramic coating and spalling (falling off of small local pieces of the liner or ceramic coating) can cause excessively high local temperature

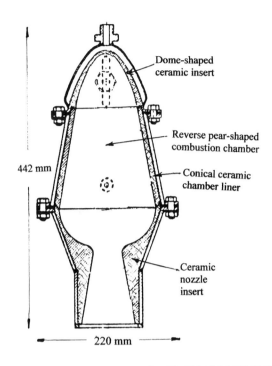

Fig. 8.1-5 12K experimental uncooled thrust chamber with a ceramic liner and a ceramic nozzle. Copied from Ref. 6.

regions, which in turn can cause local wall failures. Thereafter the Soviet engineers limited the ceramic chambers to a single use of relatively short duration, typically no more than 30 s. Ceramic liners/inserts have been used successfully in some applications, but today it is not a favored TC design for LPREs.

In 1933 and 1934 the ORM-52 was then the best LPRE with reasonable performance and life. It can be seen in Fig. 8.1-6 (Ref. 3). In 1933 tests it produced a thrust of 250 to 300 kg or 550 to 660 lbf, a specific impulse of 210 s, and one unit had been demonstrated in 29 consecutive runs and a cumulative operation of 535 s. Like all of the early engines, it used a pressurized gas feed system. The ORM-52 was intended for propelling a vertically launched experimental rocket vehicle and if successful also possibly a marine torpedo or an experimental airplane. The TC had a nozzle cooled by nitric acid, a film-cooled chamber, six spray injection elements with integral check valves, injection from the side of the chamber (three each for fuel and oxidizer), and an uncooled hemispherical forward dome. For starting the kerosene fuel was preceded by a small quantity of hypergolic liquid. In 1935 this ORM-52 LPRE flew in a small rocket test vehicle, which was launched vertically.

The idea of a *curved diverging nozzle exit contour* was mentioned by V. P. Glushko in a letter to K. E. Tsiolkowsky in 1930.[1] Several of the many ORM

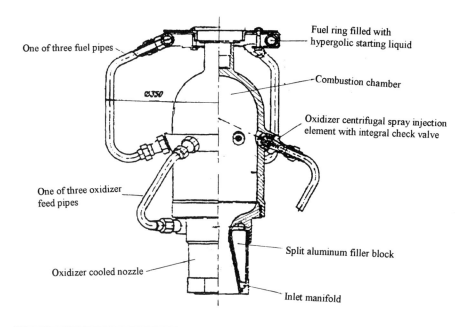

Fig. 8.1-6 Section of the ORM-52 using nitric acid and kerosene propellants. It was the GDL's LPRE to fly in 1937. Courtesy of NPO Energomash; copied from Ref. 3.

TCs developed by GDL in 1932 to 1935 had a curved nozzle skirt, but most used a conical nozzle-exit shape. It has been known in the Soviet Union that a properly shaped nozzle-exit contour diminishes divergence losses and can increase performance by 1 or 2% compared to a cone of the same length and a half-angle of 15 degs. Also it allows a slightly shorter, and thus lighter nozzle without performance loss. During the 1930s, some analysis was performed in the Soviet Union on the curved nozzle exit. GDL undertook some experimental work comparing different nozzle-exit lengths, exit cone half-angles, and exit contours using a pendulum as a balance with two opposed nozzles as shown in Fig. 8.1-7.* With this pendulum method small differences in thrust between two nozzles could be accurately sensed, and the cone half-angle (and some years later also the curved or bell-shaped contour) could be optimized. Different common nozzle-exit configurations are shown in Fig. 4.3-9. In the United States Rocketdyne also used a pendulum with two opposed nozzles to optimize the nozzle contour, but not until 1955 and 1956.[†] Little work was done in the Soviet Union on curved nozzle exits between 1935 and about 1949, when the concept was revived. Conical exits were used on the early

---

*Personal communications, V. K. Chvanov, first deputy of general director and general designer, NPO Energomash, 2002–2003.
[†]Personal communications, R. B. Dillaway, 2003–2004, formerly of Rocketdyne Propulsion and Power.

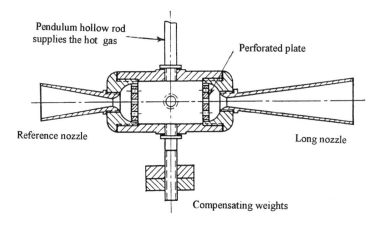

Fig. 8.1-7 Simplified section through a pendulum balance used to compare the thrust produced by two nozzles with different exit contours. Courtesy of NPO Energomash.

large Soviet TCs (1946–1950), and curved nozzle-exit contours appeared in large Soviet TCs in 1952 to 1953. In the United States large bell-shaped nozzle exits appeared later, namely, in 1957 to 1960.

GIRD Organization

As already mentioned, the GIRD organization became the second LPRE development organization in the Soviet Union. F. A. Tsander, a disciple of Tsiolkowsky, one of the early Russian spaceflight visionaries and a pioneer in LPREs, played a key role in getting GIRD started. He was originally employed at TsAGI (Central Aerohydrodynamic Institute, the key research organization for airplanes). His personal interest was in spaceflight, but his proposals at TsAGI to work on rockets or spaceflight were essentially ignored. On his own he had worked on an analyses for LPRE, such as performance, heat transfer, or comparison of different propellants. In 1924 Tsander published an imaginative popular booklet on flight to other planets. He had given public lectures on spaceflight, which greatly contributed to the popularizing of the subject and the belief that spaceflight could indeed be accomplished.[2,3]

In December 1930 Tsander had put an advertisement into a Moscow newspaper asking for responses from people interested in "planetary communications" or spaceflight. About 150 responded, and with them Tsander organized an informal amateur group to investigate rocketry and space exploration (1931). The group managed to obtain some government funds for their activities. This support came from Osoaviakhim (a government society for the promotion of defense, aviation, and chemical production); it sponsored amateur and paramilitary activities among Soviet youths and interest groups in fields such as gliders, automobile racing, or hot air balloons.[10] In mid-1932 this amateur group's name was changed to GIRD, and this new group obtained some more government funding and was no longer merely an amateur organization.[2,3,10] Sergei P. Korolev had joined the amateur group, and shortly after GIRD was founded Korolev became the leader of the Moscow GIRD.[11] There were four sections; Tsander became the section head for the development of LPREs and his section worked independently from the Gas Dynamics Laboratory in Leningrad, where Glushko and others were also working on LPREs. Korolev, in addition to being the GIRD leader, also headed the section that developed experimental launch vehicles for the initial flights of rocket engines. Korolev also became the individual who reviewed the activities of all of the other GIRD organizations (located in other cities). GIRD and some of these other cities were mentioned in the third paragraph of this chapter.

In Moscow, Tsander's section test fired its first engine in 1932 known as OR-1 (12 lb thrust) running on compressed air and gasoline.[2,3] It is shown in Fig. 8.1-8. Because OR-1 used air, people considered it to be more like a blowtorch or a forerunner of a jet engine than a true rocket engine. It used a spark plug for ignition. The air was admitted into a cooling jacket at the nozzle and entered the chamber through slots at the forward end. The slots cannot be

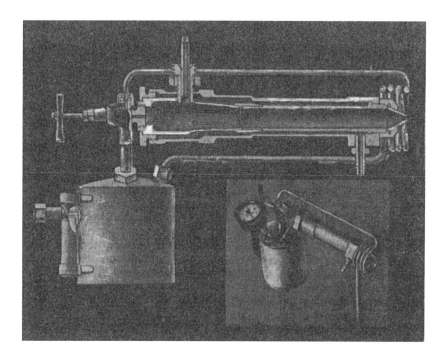

Fig. 8.1-8 Simplified section of the OR-1, the first engine designed and tested by Tsander. It used compressed air and gasoline. (around 1931). Adapted from Refs. 2 and 3.

seen in the figure. Many people do not count these early tests because they were done with air as the oxidizer and fuel was the only liquid propellant.

Tsander was interested in metal fuels because he had the idea of using the structural metal from the rocket engine and from the vehicle stage as extra propellants, once the regular fuel was exhausted.[3] He believed this metal hardware could be transformed into a form suitable for combustion, then the inert mass of the vehicle could be reduced, and the mass ratio would be improved. He considered several potential methods, such as changing the metal into powder form (to be mixed with the fuel) or to melt it and inject it into the combustion chamber in liquid form. The first step would be to test propellants with suspended metal particles. He made a couple of preliminary tests with the OR-1 engine using metal particles suspended in the fuel. He never was able to perform more work on such tests because he had burnouts and explosions on some of his other LPREs, and thus his time was spent fixing these more urgent problems. Others at GIRD later investigated metal additives to fuel, but the results were not encouraging, and no further work was done with metal fuels at GIRD.

The next LPRE model of Tsander, identified as OR-2, is shown schematically in Fig. 8.1-9. It was a very complex LPRE for its time and was originally intended to be installed in an experimental glider.[2,3] Design thrust was 50 kg or 110 lbf, and propellants were liquid oxygen and gasoline. It used liquid nitrogen, which was evaporated, for gas pressurization of the propellant tanks and water cooling of a cavity built around the nozzle. As seen in the figure, it had heat exchangers for evaporating the liquid oxygen and the liquid nitrogen with warm water from the nozzle cooling jacket. Water was used to cool the nozzle and oxygen gas for cooling the chamber. There was a closed water circuit, with the flow returning to the nozzle jacket by a separate pump, which was driven by an electric motor. The ignition was with a spark plug. The first few tests of the OR-2 engine in 1933 resulted in failures. There were several problems, including the bursting of the TC and combustion instabilities. The design had to be modified.

Tsander did not live to see the tests or the modifications of the OR-2 LPRE. He was ill and went to a sanitarium to regain his strength. He died unexpectedly in early 1933 when he contracted typhoid fever at the age of 46. After his untimely demise his group of engineers continued the work on OR-2. They changed the engine dramatically to make it work. The water circuit was eliminated, the TC was redesigned with a ceramic insert as shown in Fig. 8.1-5, and the fuel was changed from kerosene to 85% ethyl alcohol. Simple compressed gas was used (instead of evaporated liquid nitrogen) for the pressurization of the propellant tanks. It basically was a new engine with about twice the thrust (100 kg or 220 lbf), and it was given a new designation, namely, 02 LPRE.[3] It took three years to be modified four times, debugged, retested, and finally prepared for flight approval tests. It was flown successfully in two applications as described in the next section of this chapter. This 02 engine was flown successfully in 1936 as mentioned next.

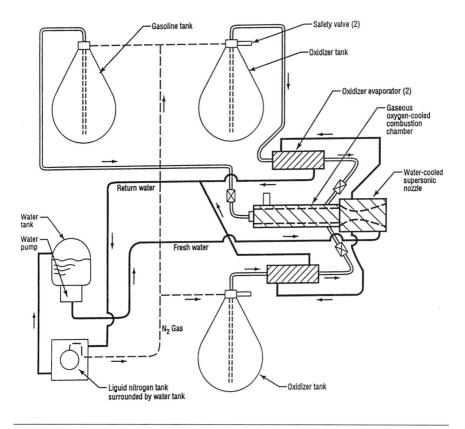

Fig. 8.1-9 Schematic diagram of an early version of the OR-2 LPRE. Adapted from Ref. 3.

Initial Flights of Soviet LPREs

A small number of the new LPREs were flown in experimental fight vehicles because they needed to be tested under actual flight conditions (accelerations, thrust variation with altitude, effect of ambient pressures or dynamic temperatures, etc.). Both vertically ascending vehicles (sounding rockets) and winged vehicles were specially developed for these tests. Some of the GIRD groups outside of Moscow also launched vehicles propelled by a LPRE. This section will mention only some of the more prominent flight tests. There were other flight test vehicles that will not be discussed.

The 09 engine was developed in 1932 and early 1933 for the GIRD 09 rocket test vehicle. It was actually a hybrid rocket engine using liquid oxygen as the oxidizer and a solidified (gelly or soft rubber like) gasoline as a fuel.[3,10] This type of hybrid was selected because at that time its designer Mikhail K. Tikhonravov believed it to be more reliable and faster to develop than a bipropellant engine. However there were problems and burnouts during development, and with three redesigns the development took longer than expected.

The fuel gel was fabricated by dissolving an organic resin in the gasoline. All of the gelled fuel was contained inside the chamber between the outer (uncooled) wall and a much smaller cylindrical screen. The oxygen flowed through the holes in the screen to the solid gelled gasoline, and the combustion products went through the screen into the central cavity and to the uncooled nozzle. The mixture ratio, thrust (25–33 kg or 55–73 lbf), and chamber pressure (5–6 atm) were not constant, but varied during the run. The oxygen tank pressurized itself by LOX evaporation caused by absorbing heat from its surroundings, and the maximum tank pressure was limited by a safety valve. The initial pyrotechnic igniter was replaced by a spark plug with its power supply. After trying several materials, copper was selected for the uncooled nozzle and brass for the chamber wall, which had an asbestos insulation layer on the inside. The engine team working on this 09 program was headed by M. K. Tikhonravov and the flight vehicle by S. P. Korolev, who was personally involved in the testing and launching. Tsander was not concerned with this project. The GIRD 09 was launched vertically on 17 August 1933, a key date in Soviet LPRE history. It was the first flying Soviet rocket test vehicle propelled by a rocket engine with a liquid propellant. This engine only had one liquid propellant, namely, the liquid oxidizer. The first two flights failed. On the third flight the vehicle reached an altitude of about 400 m. This flight helped to get the attention of top Soviet officials and to get more funding. Six GIRD 09 launches were made, and one reached 1500 m altitude.

The 02 engine mentioned earlier was installed in the 216 winged unmanned rocket test vehicle and in 1936 was flown four times with the engine working satisfactorily. The test vehicle was again developed by Korolev's department. This 02 LPRE will be discussed again later in this chapter.

The development of the "10" rocket engine was started in early 1933 at the Moscow GIRD while Tsander was still alive and able to participate. It was to use LOX/85% alcohol and provide a thrust of 60 to 70 kg for a duration of 20 to 30 s at a chamber pressure between 8 and 10 atm or roughly 120 to 150 psi (Refs. 3 and 10). The first test failed. The GIRD engineers changed the design, but each of the four major redesigns failed in the tests. Figure 8.1-2 shows the fourth revision of the 10 TC, which had mostly minor problems. It was the fifth redesign that finally seemed to meet the requirements. It had a precombustion chamber, an inverted conical (pear-shaped) combustion chamber, oxygen cooling in the cooling jacket around the nozzle and chamber, and a gas-pressurized feed system. The TC also had a ceramic coating, but it is not shown in the figure. Fuel was injected through axial holes (not clearly visible) at the forward face of the precombustion chamber, and oxygen was injected radially through holes in the side wall of the precombustion chamber (also not visible). The measured specific impulse was between 162 and 175 s, indicating a poor combustion efficiency. However the people working on this LPRE were pleased, when it lasted for more than 20 s and did not fail. The experimental rocket vehicle driven by the 10 LPRE was also identified as the "GIRD 10" or the GIRD X test vehicle. It was launched in early 1934, and this was the very first flight of a true bipropellant LPRE in the Soviet Union.

There were problems with the launching of rocket vehicles; they often flew in an unintended direction and on the wrong flight path. For unguided military bombardment rockets the accuracy of hitting its target depends critically on flying the desired flight path. Immediately after the vehicle lifted off its launchpad, the flight velocities were very low, and thus the initial aerodynamic forces on the flight control surfaces were insufficient to stabilize the flight. One solution would be to do away with aerodynamic fins and spin stabilize the vehicle by rotating it on its axis. Engineers at RNII investigated such a spinning rocket, which had several predictable development problems. The project did not proceed. Another solution, independent of the engine, was to build long launch ramps with long guide rails, so that the vehicle would attain a reasonably high velocity before the vehicle entered free flight. However these long launch structures were not desired by the military authorities. Yet another solution was to give the vehicle a very high initial acceleration or a high initial thrust for a short time followed by a longer-duration low thrust. This can be achieved by filling the combustion chamber of the LPRE with fast-burning solid propellant (e.g., double-base propellant), which would yield a high thrust at very high chamber pressure for a very short time (typically 0.5 s), thus needing only a short guide rail. The liquid propellant engine would be started after the solid propellant is consumed, and it would propel the vehicle to an even higher velocity (or higher altitude), and the flight would be aerodynamically stable. At least two such combination solid/liquid propellant rockets were developed and flight tested. The KRD-600 had an initial thrust of about 5000 kg (11,000 lb) for about 0.5 s, and the sustaining thrust of the LPRE was about 1100 kg (2420 lb) for a much longer duration. The engineers solved the problem of preventing the high pressure gas of the burning of solid propellant from entering through the injector into the liquid propellant feed system. Also they designed the combustion chamber to take the high initial chamber pressure of 220 atm or about 3230 psi. Soon it was learned that two separate solid-propellant rocket motors on the same vehicle could achieve the same result with a much simpler design. So this approach of using a combination solid–liquid propulsion system was abandoned.

Reorganization

At that time the country had two different groups doing work on LPREs. So the suggestion was made to combine the informal Moscow GIRD group, which was supported by the Ministry of Heavy Industry, and the Leningrad Gas Dynamics Laboratory (GDL), which was supported by the military (Marshall M. N. Tukhachevskiy) and form a single government propulsion research and development organization.[3,6,10] This amalgamation was completed in October 1933. The new organization was known as the Reaktirnyi Nauchno Issledovatel'kii Institut or RNII (Reaction Propulsion Research Institute), and it reported to the People's Commissariat for Heavy Industry. RNII worked not only on LPREs (with Glushko as its section leader), but also had departments on solid-propellant rocket motors, ramjets, and flight-test vehicles. This institute was subsequently renamed several times and became NII-1 (Scientific Research Institute) in 1944. It was historically important because it developed some unique and historical LPREs and because several of the key design bureaus working on propulsion and flight vehicles were spawned there and later spun off as separate organizations. RNII continued the work on improved ORM engines and the launches of experimental vehicles propelled by LPREs.

The first RNII Chief, Ivan T. Kleimentov (a former chief of GDL), was appointed by Marshall M. N. Tukhachevskiy. For a short period his deputy was Sergei P. Korolev, who was mentioned earlier.[11] In January 1934 Georgiy E. Langemak, an artillery officer and solid-rocket-motor specialist, replaced Korolev and became the deputy chief of all of RNII. Then Korolev became the chief of the section in RNII dealing with test vehicles for flight testing of LPREs. In 1936 or 1937 he was demoted from chief to senior engineer. It is not clear why Korolev was twice reduced in rank, but it might have involved a dispute about the allocation of resources. Kleimentov and Langemak wanted to apply more resources to the solid-propellant motor work, which was eminently successful and resulted in useful military weapons, such as the Katyusha, a simple solid-propellant rocket used by the infantry for multiple surface-to-surface bombardment. But Korolev badly wanted more resources for LPRE and his flight-test vehicles, which at the time did not have good prospects for a military application. The individuals listed in this paragraph will be found again at the end of this chapter, but in a totally different context.

RNII and its successor organizations investigated and tested many different ideas.[2,3] They continued to develop and test different and better TCs and their pressurized feed systems. But they also investigated gas generators, positive displacement pumps, the first centrifugal propellant pumps, the first turbopump, and the first Soviet engine a with gas-generator cycle. RNII built the first heavy duty TC firing test stands with remote control from a protected control station. They designed, built, and ground tested LPREs for experimental sounding rockets, an experimental aerial torpedo, prototype tactical military missiles, and experimental rocket airplanes. Many of these were then flight tested. Some of the rocket engines for aircraft applications were developed at RNII, and they are treated in the next chapter.

Gas Generators and Turbopumps

The first gas generators (GG) were designed and static fire tested at RNII in 1935 to 1936, and they were officially "certificated" as an engine component in static tests in 1937. Three propellants were used in this GG: nitric acid, kerosene, and water.[2,3] As seen in Fig. 8.1-10, the first GG had two chambers: one water-cooled chamber for the initial hot combustion and a second with a sharp-edged baffle for mixing and making the gas composition and temperature more uniform. The water was the regenerative coolant for the first chamber and also the diluting agent or gas cooling agent. The output of this GG was 40 to 70 liters per second of gas (1.4–2.5 ft^3/s) at 20 to 25 atm (about 300–368 psi) and 450 to 500°C or 842 to 932°F. Later some restartable versions had a

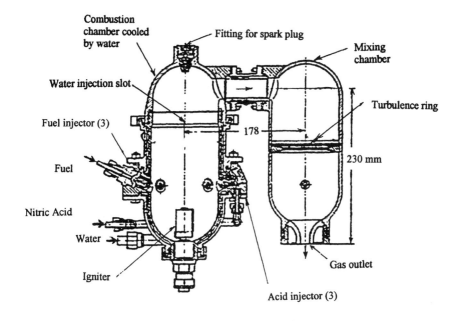

Fig. 8.1-10 Simplified cross section of one version of the first Soviet gas generator. It had two chambers and used water for cooling and dilution. This poor quality sketch was the only section drawing available. Adapted from Ref. 3.

small ignition chamber with an electrical igniter attached to the combustion chamber. At least one other tripropellant GG of a somewhat different configuration and higher gas output was also developed. They were intended to drive a turbopump for a rocket engine. However Soviet GGs were not operated with a turbopump (TP) until much later, namely, in 1945. This delay was in part because the Soviets had success in flying aircraft rocket engines with feed systems using gear pumps driven by a piston engine. For an illustration of typical gear pumps, see Fig. 8.2-7. They allowed a big savings in tank weight and did not need a gas generator. So there was apparently no urgency to develop a TP. But the engineers understood that a piston engine or a jet engine loses power at altitude (which implies a lower thrust of the LPRE), while a gas generator/turbopump does not lose power. These aircraft rocket engines with TPs and GGs will be described in the next chapter.

In the mid-1930s investigations of various positive displacement pumps were undertaken. Gear pumps had the advantage of accurately controlling the mixture ratio, even during throttling, and this was probably part of the reason why gear pumps were selected by Glushko for the aircraft LPREs discussed in the next chapter. By 1945 there was enough evidence for centrifugal pumps to be preferred. In the United States Goddard performed similar investigations of different positive displacement pumps (pistons, bellows, gears, etc.) and came to the same conclusion, but about 10 years earlier.

A few studies of TPs and apparently some component investigations of TPs were performed at RNII in the mid- and late 1930s. For example in 1939 and 1940 a GG and small turbopump were planned for a torpedo drive. There might have been some component tests, but the project was not completed. In 1940 design of a new engine of 300 kg thrust with a turbopump was studied and analytically investigated, but not built. Its turbopump had a single-stage high-speed turbine, a gear speed reducer, and pumps for oxygen, kerosene, and oil. In 1942 to 1945 the development of a nitric-acid/kerosene LPRE with four TCs with a total thrust of 1000 to 1200 kg and a turbopump feed system was undertaken. It seemed to have a low priority because the project was not completed and was displaced by work on gear pumps, which were then used in experimental aircraft rocket engines. In 1945 there was the first ground test of a Soviet turbopump with a tripropellant gas generator, but no detail was found to describe the engine or the TP. The development of a bipropellant GG came several years later.

More Experimental Engines and more Test Flights

In May 1936 a winged rocket-propelled unmanned RNII projectile, known as Model 216, was launched from inclined rails for the first time. It had a modified version of the 02 LPRE using alcohol and liquid oxygen delivering 984 N or 220 lb thrust at takeoff. The vehicle had a takeoff weight of 79.4 N or 175 lb, a takeoff speed of 118 ft/s, and a maximum speed of 180 m/s or 590 ft/s (Ref. 3). There were several versions of the 02 LPRE, including one with ceramic liners and one with a cooling jacket. It was originally developed for a test vehicle, which was launched vertically upward in the 1933–1934 period.

Glushko's LPRE department team continued to develop a series of LPREs designated ORM using various propellants, mostly storable propellants with nitric acid and kerosene. The goal was to develop a good enough engine for flight. Beginning with ORM-34, all TC nozzles were cooled by one of the propellants flowing through some type of cooling jacket around the nozzle section. Several of the new ORM TCs were aimed at different materials and different fabrication processes.

The ORM 63 TC had a thrust of 300 kg and had improved spiral spray elements and improvements in the fabrication processes.[3] It was produced in limited quantities to learn how to build it properly. It featured both roller welding at the expansion joints and high-temperature soldering of several other joints. It also had regenerative cooling in both the nozzle and the chamber. In 1936 the ORM-64 had a metal-nitrate mix in the igniter, a little better specific impulse of 216 s, and was operated in both the vertical and the horizontal attitude. This engine was specifically aimed at the RP 318 experimental aircraft and the 212 remotely controlled winged rockets, and both of these vehicles had these dual firing attitude requirements. However neither the ORM-63 nor the ORM-64 was qualified or flown.

The ORM-65 engine, first tested in 1936, was one of the more successful pressure-fed LPREs with storable propellants. It was tested in three experimental applications, as explained in the following.[1-3] Some of the data are in Table 8.1-1, and its TC configuration is shown in Fig. 8.1-11.

This engine is described here in some detail. There were three steel parts in the TC: the uncooled head with the igniter, the cooled chamber with its cooled nozzle (with integral spiral ribs), and the outer housing. They were threaded together with asbestos packings for sealing the joints. The space between the nozzle wall and the housing was filled with a split aluminum insert, pinned, and fastened with a locking screw. The thermal expansion of the inner wall was compensated by a lead ring gasket near the nozzle exit, which was tightened by a threaded annular nut. During operation, some of the lead extruded into a gap. The nut had to be tightened after every run. The injector had three spray injection heads of fuel injected at right angles to the TC axis and three spray heads for nitric acid inclined at 60 deg to the axis. The spray heads had swirl inserts (helical passages) and were sealed with aluminum packings. The nitric acid was heated as it flowed through the spiral channels of the cooling jacket prior to injection. The front end of the chamber did not have a cooling jacket. The electrical squib had a resistor that burned out when the circuit was closed, igniting the small gunpowder charge inside the squib; the hot gases ignited the solid-propellant grain, which in turn ignited the liquid propellants. The solid propellant was a mixture of nitrates with powdered metal fuel, a favorite in Soviet ignition systems, that had been proven on the ORM-64. The ORM-65 was the first engine put into serial production and after it was thoroughly tested in many static test firings. One particular unit of this LPRE was started 49 times and had an accumulated duration of 30.7 min.

Table 8.1-1 ORM-65 Rocket Thrust Chamber

Propellants	Nitric acid and kerosene
Thrust, SL	155 kg (341 lbf) rated, 50–175 kg possible
Chamber pressure, nominal	22 atm (323 psia), pressure fed
Specific impulse, SL	210 s (215 s in some tests)
Duration	38 s typical, longest test run 230 s
Propellant tank pressure	35 atm
Cooling of chamber and nozzle	With nitric acid as regenerative coolant; film cooled at head end only; jacket had spiral passages in chamber and nozzle
Outside temperature of head	300–400°C during operation
Nozzle expansion ratio	4.0
Weight of TC, total	14.26 kg
Length of TC	465 mm
Inside chamber diameter	102 mm
Thrust chamber, outer diameter	175 mm (max)

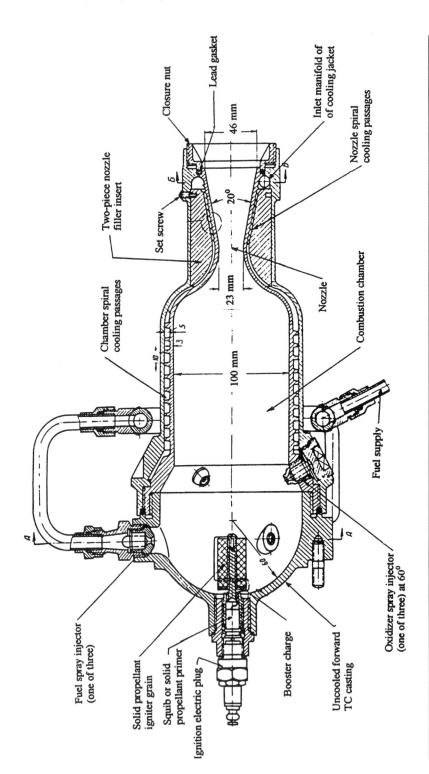

Fig. 8.1-11 Thrust chamber of the OMR-65 LPRE (1936) was regeneratively cooled, except for the upper dome with the injection elements. Courtesy of NPO Energomash. Copied from Ref. 3 and provided with callouts.

The ORM-65 was put into a limited production and used in three experimental flight vehicles. It was installed in the unmanned experimental winged guided rocket vehicle 212, an aerial torpedo, and the experimental rocket glider aircraft RP-318. The first two applications are shown in Fig. 8.1-12 and 8.1-13. These two vehicles were designed by Korolev's department. In 1937 and 1938 there were 13 static tests of the ORM-65 engine mounted in the 212 test vehicle, and in 1940 a modified ORM-65 LPRE propelled the 212 winged rocket in two flights. The 212 test vehicles failed to reach the predicted flight paths. In 1937 and 1938 there were more than 21 runway ground tests of the ORM-65 engine installed in the rocket aircraft RP-318. However this airplane did not fly with the ORM-65 engine, and the sources do not give an explanation why the flight tests were canceled. The cause might be that another aircraft rocket engine (RDA-1-150 discussed in the next chapter) was then man-rated and ready to fly; it did replace the ORM-65 in this RP-318 airplane, and it did fly in 1940. The ORM-65 engine also launched and propelled an aerial torpedo twice in 1939, but there were no details available about the torpedo or its operation.

The ORM-66 to ORM-70 were improvements to the ORM-65, and they were developed between 1936 and 1938. They included putting a cooling jacket around the head of the chamber, improving the igniter, using aluminum for some of the components, and reducing the complexity and the weight of the engine. None of these engines were flown.

Fig. 8.1-12 Guided winged test vehicle 212 flew with a modified ORM-65 LPRE, a solid-propellant rocket motor booster, and a launching track in the snow. Copied with permission of NPO Energomash from Ref. 9.

Fig. 8.1-13 This RP-318 rocket glider had high wings and a skid for landing. The nozzle of the ORM-65 is visible at the tail. Copied with permission of NPO Energomash from Ref. 9.

RNII also pioneered in the development of *ground test facilities*. Figure 8.1-14 shows a very small test stand built in 1932 for firing vertically downward. The ORM-50 TC can be seen in this figure, and the propellant tanks and the gas pressurization system was taken from one of the test vehicles, intended for vertical launching. This TC is also seen in Fig. 4.3-2. RNII subsequently built larger and heavier test stands with protective reinforced concrete walls and brick construction. It had a separate test bay, which could be closed off and heated in the winter, a control room protected by thick walls, two separate bays for the heavy-duty propellant tanks, and one bay for a heavy-duty gas-pressurized feed system with a high-pressure gas supply.

Summary Findings

The technical progress in the Soviet Union during the early period of this LPRE history was remarkable. It started with uncooled crude TCs, which could run just a few seconds without failure, and ended with fully cooled TCs, which could run as long as the propellant supply would allow. Chamber pressures

Fig. 8.1-14 Early test facility with an ORM-50 TC in position to fire vertically down. Copied with permission of NPO Energomash from Ref. 9.

started at 3 to 8 atm or approximately 50 to 115 psia and were advanced to about 25 atm or 365 psia. Thrust levels increased from about 89 kN or 20 lb to more than 4450 N or 1000 lbs. All used pressurized-gas feed systems during the period up to 1944. They tried and abandoned a number of propellants. The Moscow people used mostly liquid oxygen and kerosene or alcohol, and the Leningrad people used mostly nitric acid and kerosene. Reliable ignition and startup techniques were developed. Specific impulses started at about 140 sec and progressed to a high of 216 s. Initially there were few injector spray elements in each TC, and they were located in the middle of the chamber wall. During this period (up to 1944), injector spray elements were improved and placed in larger numbers into the forward end of the TCs. The design of cooling jackets advanced so that all exposed hot walls would be regeneratively cooled. Some of the engines were installed in experimental unmanned vehicles, and many of them were actually flight tested.

RNII has been the key governmment agency to spawn several LPRE organizations in the Soviet Union.[10] Groups or departments of RNII and its successor organizations were spun off and grew into independent design bureaus or independent R&D institutes. The first was Glushko's Design Bureau, which became independent around 1939 and later was called NPO Energomash (NPO means

Scientific Production Association). Others were the Isayev's Design Bureau, which started in 1942 (today called Chemical Machinery Design Bureau), Dushkin's Design Bureau (it existed only briefly), Lyulka Design Bureau (aircraft piston engines and jet engines), and the Keldish Research Institute. Several will be discussed in Chapters 8.4, 8.6, 8.8, and 8.13. Not only did RNII develop the first gas generator, the first flying TCs, and the first TP, but later (1958) it ground tested the first LPRE with a staged combustion cycle and hatched new engine control systems. There were a couple of other government funded R&D organizations, which worked on LPREs during the 1930s and 1940s, but they were shut down after several years of operations. Their engines are not mentioned here, and they did not fly.

Purges

In 1937 and 1938 an unexpected disaster happened. Stalin, the ruler of the Soviet Union, initiated an aggressive campaign against the intelligentsia of the country. He targeted key government officials, top military officers, and leading personalities in all kinds of endeavors and occupations. Thousands were arrested on baseless charges and were sent to prisons or labor camps, and a good many were actually executed.[4,10] People just disappeared. These purges or detentions and murders were conceived by Stalin to eliminate possible opposition to his regime and were undertaken by the secret police known as the NKDV or Peoples' Commissariat for Internal Affairs.

In the field of aeronautics and rocketry, a number of key personnel were victims of this purge. Field Marshall M. N. Tukhachevskiy, who was the deputy commissar for military and naval affairs, was arrested in 1937, and after a short trial he was convicted. He was severely beaten to make him confess, and he was then executed. His name was mentioned earlier in this chapter. He was the key military authority in the aeronautical and missile field, and he controlled the policies, activities, and the budgets of several government organizations including the GDL, and he had approved the merger of GDL into RNII. He was a strong supporter of rocket activities and had many discussions with the key people at those government organizations, for which he was responsible. Tukhachevshiy was accused on false charges of passing military intelligence information to a foreign power (Germany), sabotaging the Red Army, and promoting capitalism in the USSR. His mother, sister, and two brothers were accused as accomplices and were also executed.

The first man to become chief of RNII, I. T. Kleimentov, who was appointed by Tukhachevskiy, and his RNII Deputy Georgii E. Langemak were arrested on false charges in 1938 and executed, in part, perhaps because they had business dealings with Tukhachevskiy or might have been implicated by forced testimony of some of those that were beaten into confessing. These men were also mentioned before in this chapter. One of the chiefs of GDL Nicolei Il'lin also was arrested and executed. Valentin P. Glushko, the leader of the LPRE group at GDL and later a department at RNII, was arrested in 1938 and indicted on drummed-up accusations of being an enemy of the people, responsible for errors, omissions, and disruptions of work on test stands. This might have been based on false testimony obtained under torture, presumably from Kleimentov, or Langemak and others. Glushko was sentenced in absentia to eight years in prison.

Sergie P. Korolev, the head of GIRD and later the deputy chief of RNII and subsequently chief of its rocket vehicle department, was also arrested on phony charges and sent to prison. The fact that he was twice demoted in rank about a year and a half before his arrest might have saved him from a worse fate.[10,11] One of the accusations was that Korolev destroyed the RP-318 rocket airplane, but that airplane was at the time in a hangar at the airfield, where tests were being done. Several others at RNII, particularly in the solid-rocket-motor department, also suffered this fate. There is no list or count of these victims.

These purges sent a shudder through all people in responsible positions in the Soviet Union. At RNII most people had direct contact with some of those that were arrested or had disappeared. They were afraid of being identified as an accomplice or of possibly being accused of unknown subversive activities, like spending the people's money foolishly. It dramatically changed the way business was done at RNII and also at many other Soviet organizations. Many decisions (design, purchasing, test plans, etc.) were postponed. All decisions or conversations were stifled and had to be done in such a way that they could not be interpreted as being antigovernment. Documents that could show possible waste, incrimination, design decisions, certain photos, or dealings with the people who were arrested disappeared. Engine failures or unfavorable test data were hushed up. This was one of the reasons why many of the development and test operation documents and photos of the 1930s have not been preserved. The business tempo at RNII slowed down, and the working climate was very unsettling. No major new LPRE project was started in 1938 or 1939; only some of the existing projects such as the RP-318 rocket-propelled aircraft were continued. In 1939 Andrey G. Kostikov, who was suspected to have collaborated with Stalin's secret police and might have caused some of the arrests, became the chief of RNII. Several key LPRE developers, such as Tikhonravov or Dushkin, for reasons that are not clear, escaped arrest and were allowed to continue the work at RNII. Only in the early 1940s did the business climate change.

In the early 1940s some of the imprisoned experts were encouraged to continue their work, but in prison. This change of behavior toward the political prisoners might have been because of the advance of the German Army into the Soviet Union and the need for developing new Soviet weapons. For example Andrey N. Tupolev, the famous Russian airplane designer, had been arrested and incarcerated. However he was allowed to put together a team to design bomber airplanes. Tupolev was allowed to pick 25 people to assist him, and this list included S. P. Korolev. In 1940 Korolev was taken out of a brutal labor camp and transferred to work for Tupolev. The Tupolev TU2 bomber was designed in captivity and was put into production later.

Valentin P. Glushko was also allowed to do work on rocket engines for aircraft applications in prison.[7] His team was under the guard of the secret police in a makeshift facility in Kazan, about 450 miles east of Moscow. When Glushko tried to put together a team of experienced engineers, he had difficulty finding enough of them (they were in different prisons) and arranging to have them come to work for him. He was able to also find some engineers and academics experienced in related technologies, and he trained them in LPRE design. This team developed the RP-1 and other aircraft engines described in the next chapter. At Glushko's request Korolev was able to leave Tupolev prison design group and joined Glushko's prison team developing rocket engines for airplanes. Korolev became the deputy chief in charge of aircraft flight tests, which are described in the next chapter.[11] When Korolev was deputy chief of RNII, Glushko worked for him. Ironically in the captive LPRE design group Glushko was the chief designer, and Korolev worked for Glushko.

References

[1] Stoiko, M., *Soviet Rocketry, Past, Present, and Future*, Holt, Reinhart, and Winston, New York, 1970.

[2] Blagonravov, A. A., and others, *Soviet Rocketry, Some Contributions to Its History*, Moscow, 1964; translated by Israel Program for Scientific Translations, Jerusalem, 1966.

[3] Moshkin, Ye. K., *Development of Russian Rocket Technology*, Mashinostroyeniye Press, Moscow, 1973 (in Russian); also NASA Technical Translation TT F 15,408, March 1974 (in English).

[4] Winter, F. H., *Prelude to the Space Age; the Rocket Societies 1924–1940*, National Air and Space Museum, Smithsonian Inst. Press, Washington, D.C., 1983.

[5] Prishchepa, V. I., "History of the First Space Rocket Engines in the USSR," *History of Astronautics and Rocketry*, edited by E. I. Ordway, American Astronautical Society History Series, Vol. l0, Univelt, San Diego, CA, 1989, Chap. 8.

[6] Dushkin, L. S., "Experimental Research and Design Planning in the Field of Liquid Propellant Rocket Engines Conducted Between 1934 and 1944 by the Followers of F. A. Tsander," *History of Rocketry and Astronautics*, American Astronautical Society History Series, Vol. 7, Univelt, San Diego, CA, 1972.

[7] Pakhmanin, V. F., and Sterpin, L. Ye. (ed.) *Once and Forever—Documents and People on the Creation of Rocket Engines and Space Systems of Academician Valentin Petrovich Glushko*, Mashinostroyeniye Press, Moscow, 1998 (in Russian).

[8] Kulagin, I. I., "Developments in Rocket Engineering Achieved by the Gas Dynamics Laboratory in Leningrad," *First Steps Toward Space*, edited by American F. C. Durant III and G. S. James, Astronautical Society History Series, Vol. 6, Univelt, San Diego, CA, 1985, Chap. 10.

[9] Glushko, V. P., *Rocket Engines of the Gas Dynamics Laboratory—Experimental Design Bureau, (Rakenye Dvigateli GDL-OKB)*, USSR Academy of Sciences, Novosti Press, Moscow, 1975; also NASA Technical Translation TT F 16847, Feb. 1976.

[10] Siddiqi, A. A., *Challenge to Apollo, the Soviet Union and the Space Race (1945-1974)*, NASA SP-2000-4408, Washington, D.C., 2000.

[11] Harford, J., *Korolev, How One Man Masterminded the Soviet Drive to Beat America to the Moon*, Wiley, New York, 1997.

8.2   Rocket Engines for Piloted Aircraft

In the late 1930s and the 1940s the Soviet military forces perceived a need for increasing the rate of climb and the altitude speed of fighter aircraft in order to intercept enemy bombers and outmaneuver potential enemy interceptor aircraft. Because LPREs offered a potential solution, the Soviet government initiated several R&D programs at different LPRE organizations. As a result, the Soviets built and flight tested a large variety of superperformance LPREs for fighter airplanes. More than 16 such aircraft superperformance engines were developed, and this effort was performed at four different LPRE design organizations.[1-5] The early work at RNII (Reaction Propulsion Research Institute) mentioned in the preceding chapter included some effort on LPREs for unmanned winged vehicles and was the basis for work on piloted rocket-propelled aircraft. Although Russia no longer produces rocket engines for manned military aircraft today, the development of these engines helped to build a viable LPRE capability and constitutes a significant accomplishment in early Soviet LPRE history.

The RDA-1-150 aircraft rocket engine was historically significant, was the first to propel a manned Soviet aircraft, and was conceived at RNII. Nitric acid and kerosene were the propellants. It was developed in 1938 and 1939 (during the purge) by a team led by Leonid S. Dushkin, who later became a chief designer of his own design section.[1,6] Glushko, who would have been responsible for the development of all LPREs, was away in jail at the time. However he contributed to this design before he was arrested. The TC of this engine is shown in Fig. 8.2-1 and had a few of the features of earlier TCs, such as the injection elements. This TC used dual regenerative cooling, that is, the kerosene cooled the nozzle at relatively high cooling velocities and the acid was circulated through spiral channels around the inner chamber wall. Although it is not discernible in the figure, the nozzle cooling jacket had double spiral ribs integral with its outer wall, which increased the local cooling velocity and improved the heat absorption by the coolant. The injector assembly occupied the forward dome (a novelty in the Soviet Union at the time) and had a central ignition device and eight spray heads or spray injection elements (four each for nitric acid and for kerosene); the axes of these injection elements were pointed toward the center of the chamber. The propellant distribution manifolds in the injector provided some cooling for the injector assembly, again a new design in Soviet LPRE technology. This is different from prior designs, such as the ORM–65, which had the injection spray elements mounted on the side of the chamber and an uncooled forward chamber dome. Furthermore there were two injection elements for startup. It had a separate ignition stage at a reduced and controlled low initial propellant flow. The injector, chamber, and nozzle were separate subassemblies, which were screwed together. The TC and the nozzle assembly each had a sealed joint for thermal expansion in the outer wall. It had a simple gas pressurized feed system with heavy tanks. Some performance parameters of this LPRE are given in Table 8.2-1.

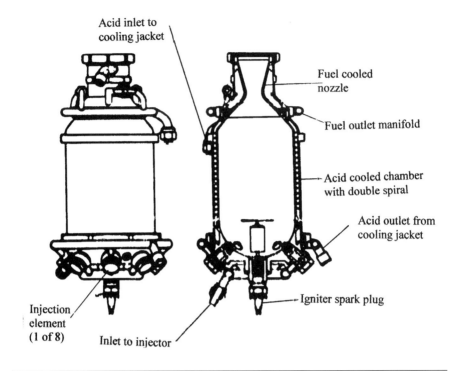

Fig. 8.2-1 View and section of the thrust chamber of the RDA-1-150 engine, the first Soviet LPRE to propel a piloted aircraft. It was in a glider airplane in 1940. Copied from Ref. 1.

The RDA-1-150 engine was first flight tested on 28 February 1940 as a powerplant for the RP-318 glider aircraft, which was towed to altitude by another aircraft. This was a memorable day in the history of Soviet aviation because it was the first time a Soviet pilot was propelled solely by a LPRE. The glider was developed by Korolev's section of RNII, and it was shown in Fig. 8.1-13 in the preceding chapter. As mentioned, the ORM-65 engine had been installed and

Table 8.2-1 Data of the RDA-1-150 LPRE, the First to Propel an Aircraft

Engine Parameters	As designed	Measured in ground tests
Thrust, kg	80–150	Up to 140
Specific impulse, s	150–198	145–186
Chamber pressure, atmospheres	8–18	Up to 18
Longest firing duration, s	NA	200
Weight of thrust chamber, kg	10.5	NA
Nozzle-throat/exit diameters, mm	25.5/50	Same

ground tested in runway tests with this glider before the RDA-1-150 LPRE was substituted. This new engine was considered to be superior and more reliable. However the RP-318 glider never flew with this earlier ORM-65 engine. The flight tests with the RDA-1-150 showed that the engine was reliable, but the flights did not give the desired results, mostly because the thrust was too low and it took a long time to accelerate the aircraft. During flight, the thrust would not usually exceed 100 kg (220 lbf), well below engine ground test values.

Nevertheless the flight test results were sufficiently encouraging to have a major influence on future superperformance rocket engines. It was decided to increase the thrust for future LPREs. The Soviets did not know that this RP-318 flight happened after the Germans had already flown rocket-propelled aircraft, the Heinkel He 112 (1937) and the He 176 (1939) with a sophisticated Walter monopropellant turbopump-fed LPRE. However the RP–318 flew several years before the United States accomplished a rocket-powered glider flight in 1943 with a small simple Aerojet LPRE.[6]

Beginning in 1941, the design of the D-1-A-1100 LPRE was undertaken by Leonid S. Dushkin and his team at RNII, and it was originally intended to become an aircraft super performance engine for a military fighter aircraft and also an assisted takeoff engine for military bomber aircraft.[1,7] It also used nitric acid and kerosene as propellants and a pressurized gas feed system with heavy tanks. The thrust was much higher than the RDA-1-150, could be varied from 1400 down to 350 kg (or 3080 to 770 lbf), a specific impulse of 203 to 156 s, and the LPRE could run for 3 min uninterrupted. By 1941 RNII had developed and tested TCs with the injector at the forward end and with internal cooling of the injector assembly. One version of the TC of this engine is shown in Fig. 8.2-2. The nozzle was cooled by the fuel, and the chamber had a nitric-acid cooling jacket, similar to earlier TCs. The injector had a central igniter and 63 individual injection spray nozzles arranged in a five-ring circular pattern. The two main propellant valves were actually downstream of the cooling jackets, thus allowing a short startup and minimizing the postcutoff total impulse caused by the dribbling out of trapped propellant downstream of the valves. During the ground tests with the aircraft, there was an engine explosion that sent the pilot and a mechanic to the hospital and severely damaged the airplane. After some test firings with a tied-down aircraft and some runway hops, the first rocket-powered takeoff and historic flight took place on 15 May 1942, a key date in Soviet LPRE history. The aircraft was the BI-1 fighter, which was designed by A. Ya. Beresnyak and A. M. Isayev under the guidance of aircraft experts.[8,9] Isayev later became a LPRE developer and in 1942 established his own LPRE design bureau as explained in Chapter 8.6. Only seven BI-1 aircraft were built. One is shown in Fig. 8.2-3. This aircraft was constructed mostly of wood. To prevent a reaction between the acid and the wood, the engine compartment was lined with stainless-steel sheet and insulation. In one test it took only 50 s to go from ground level to 5000 m or 16,400 ft altitude, a record for its day. Flight testing with the D-1-A-1100 engine continued until 1945, when the end of the war stopped further work. The airplane and also the engine were never put into production.

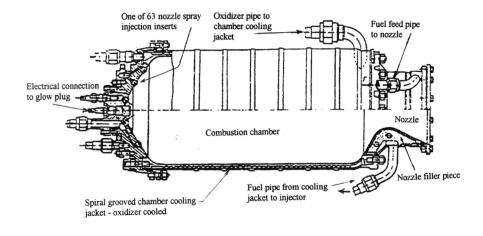

Fig. 8.2-2 Cross section of the thrust chamber of the D-1-A-1100 engine, which flew in the BI-1 rocket-powered aircraft beginning in May 1942. Copied from Ref. 1.

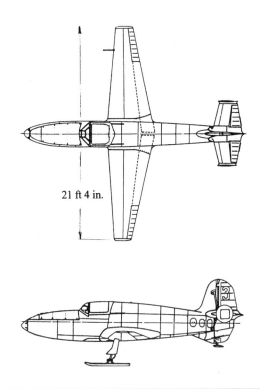

Fig. 8.2-3 Simplified sketches of the BI experimental glider. It is exhibited in the Moscow Air Museum. It was the first Soviet aircraft to take off and fly on rocket power only. Adapted from Ref. 8.

Engines in Russia, Ukraine, and the former Soviet Union 563

In one experimental model the hypergolic igniter fluid was supplied through a rotating arm, which was placed through the nozzle into the chamber during start. The arm was then swung out of the way as shown in Fig. 8.2-4. No other model is known to the author to have this method of introducing the hypergolic start liquid. This variable-thrust model had two thrust levels with

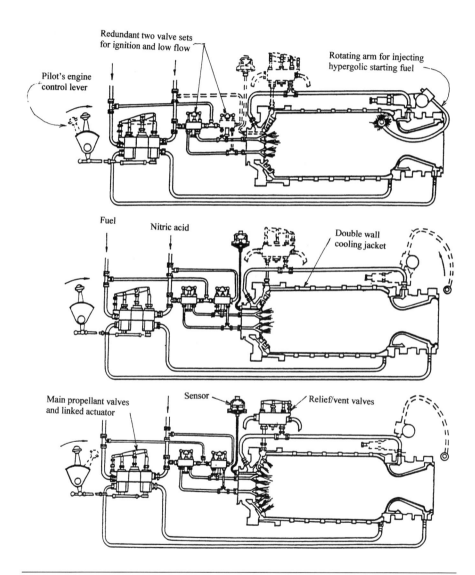

Fig. 8.2-4 Three sketches of an experimental thrust chamber assembly showing 1) ignition (with hypergolic liquid injection from the rotating arm), 2) cruise mode with only part of the injection elements operating (rotary arm has been retracted), and 3) full flow mode with all injector elements supplying propellant to the chamber. Courtesy NPO Energomash.

propellants flowing to two different parts of the injector and a configuration similar to the D-1-A-1100. The top two sketches show the low-thrust operation with injection of propellant only through injection elements located in the central portion of the injector. At full flow and full thrust all injector elements have propellant flow.

Another set of aircraft LPREs (RD-1–RD-4) were designed and developed during the early 1940s by a new group, which Glushko was allowed to form in prison.[1,3,4,*] This group used the services of other prisoners, who had suitable technical backgrounds. Design started with a few technical personnel in 1939. In 1940 this organization moved to Kazan, a city 300 miles east of Moscow. The group officially became a separate 'experimental design bureau,' and Glushko was named its chief designer. The initial ground tests of the new RD-1 aircraft rocket engines were performed in Kazan in 1942. The team and its testing operations were still under guard by the secret police, which was officially called the People's Commissariat for Internal Affairs. Glushko and most of his team members were released from prison in 1944. The team moved to Khimki near Moscow in 1946 and was renamed Energomash in 1967.

Glushko's aircraft rocket engines were the only ones in the Soviet Union, where the propellants were supplied and metered by gear pumps.[1,3,*] As stated before, a pumped feed system allows low propellant tank pressures and thinner tank walls; this permits a major reduction in the inert mass of the tanks and improves aircraft performance. The RD-1 engine was the first in this series, and it was modified several times. One version is shown in Fig. 8.2-5 and a flow diagram in Fig. 8.2-6. It used nitric acid and kerosene, had a nominal sea level thrust of 300 kg (660 lb), a chamber pressure of 22.5 atm (331 psia), and a specific impulse of 200 s. The first version had spark ignition, but later versions used a chemical start with a hypergolic start liquid. The only other known small gear pump for feeding small thrusters was built and flown in the United States, as described in Chapter 7.11.

The two propellants were supplied by a gear pump, which was driven by a shaft through a clutch from the aircraft's piston engine. Figure 8.2-7 gives a section through the gear pump assembly. The gear pump, being a positive displacement pump, ensured the intended mixture ratio (by the width and dispacement of the gears) over a wide range of thrust or flow. With centrifugal pumps there usually is a change in mixture ratio as the thrust is decreased. The gears of the acid pump had to have a generous clearance, so that there would not be a direct contact of mashing gears, which would cause local heat generation and potentially also a strong chemical reaction between the acid and the metal of the gear teeth. The gears on the fuel pump had tight clearances and were used to drive the oxidizer pump gears. By 1944 they had developed a special lubricant (which was called nitro-oil) that did not react with the acid, and it was used in valve and pump seals and bearings in several of these aircraft LPREs.

---

*Personal communications, V. K. Chvanov, first deputy of general director and general designer, NPO Energomash, 2002–2003, and C. Ehresman, Purdue University, IN, 2002–2005.

Engines in Russia, Ukraine, and the former Soviet Union **565**

Fig. 8.2-5 View of one version of the RD-1 aircraft superperformance aircraft engine. Copied with permission from photo supplied by NPO Energomash.

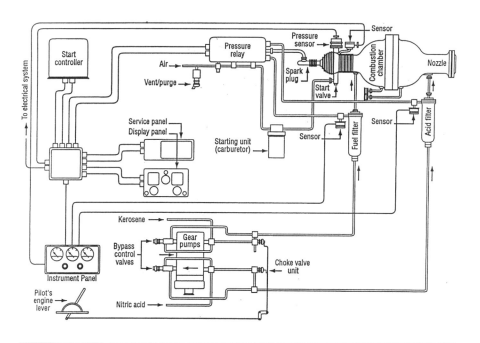

Fig. 8.2-6 Simplified schematic diagram of one version of the RD-1KhZ. Copied with permission from data supplied by NPO Energomash.

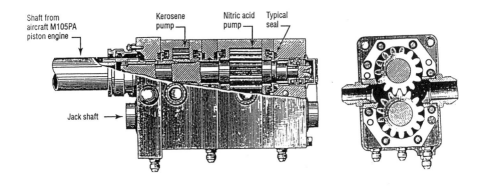

Fig. 8.2-7 Section through gear pumps of the RD-1, RD-2, and RD-3 LPREs. These were the first Soviet flying propellant pumps. Copied with permission from drawing of NPO Energomash.

Different versions of the TC of the RD-1 were tested.[1,3,*] All had regenerative cooling (by the nitric acid) in both the nozzle and the chamber. Early TCs had an ignition chamber in the center of the forward end of the TC as shown in Fig. 8.2-8. The top part of this ignition chamber was air cooled and had external cooling fins and a central spark plug. The bottom part of this ignition chamber was regeneratively cooled by fuel. An ether-air start mixture was supplied to the ignition chamber through a separate starting valve. Injection of the liquid propellants was through multiple injector spray element inserts, attached to the head or forward end of the chamber. In some versions the main propellant valves were actuated through a pilot valve and closed by springs. The pilot's control lever actuated the choke valve units shown in the schematic flow diagram and through electrical connections on the lever (not shown in the flow diagram) also the start controller, spark plug, and start valve. The relief valves or safety valves on the discharge side of the gear pumps opened automatically when the pressure was excessive, allowing some of the propellants to return to the suction side of the pumps.

Another version called RD-1KhZ was very similar, but it had chemical ignition with a hypergolic slug start with enough stored hypergolic fuel for a limited number of restarts.[1,3,4,*] The hypergolic starting fuel was 75% carbonal (a highly reactive material) with 25% gasoline, and it was injected from the top face of the injector. One version of this TC is shown in Fig. 8.2-9. The container holding the hypergolic starting fuel in the figure sometimes was called a carburetor. This RD-1KhZ version was qualified in 1943 and put into limited production in two versions in 1945 and was flown in experimental military aircraft beginning in 1945 as listed in the following.

---

*Personal communications, V. K. Chvanov, first deputy of general director and general designer, NPO Energomash, 2002–2003.

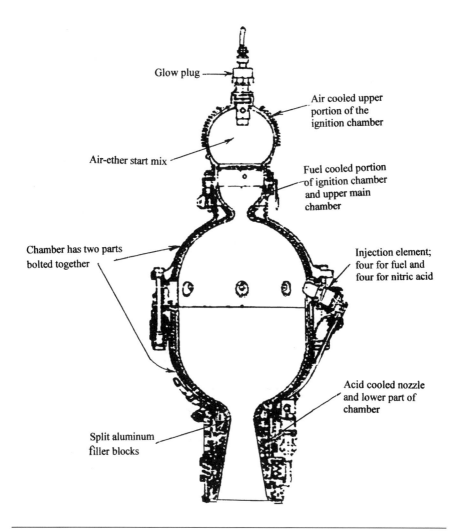

Fig. 8.2-8 Simplified section of RD-1 early thrust chamber with side injection and a partly cooled ignition chamber mounted on top. Courtesy NPO Energomash; adapted from Ref. 1.

One unit of the RD-1 was fired in the Kazan test facility for 70 min in 1942, qualified and first flown in 1943, and put into limited production in 1944. The first flight of the RD-1 engine was on a Pe-2 tactical bomber airplane (designed by V. M. Petlyakov) in 1943, and there were more flights in 1944. The version of this airplane Pe-2R, which was modified for the rocket engine, is shown in Fig. 8.2-10; it had a two man crew, namely, the pilot in the front and a gunner/radio man in the back.[3,8,*] During the test flights, the second position was

---

*Personal communications, NPO Energomash, 2002–2003.

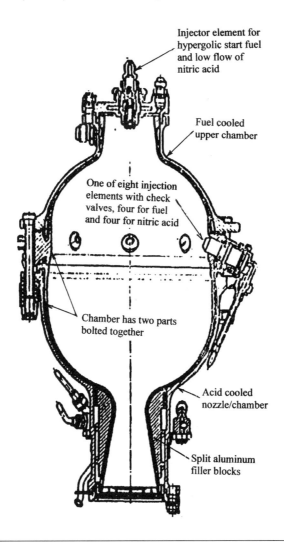

Fig. 8.2-9 Simplified section through one of the RD-1KhZ thrust chambers with injection of a hypergolic propellant during the start and with side injection of the propellants. Courtesy NPO Energomash; adapted from Ref. 1.

used for the flight engineer, who could observe the rocket engine plume or flame and some engine instruments. Both aircraft positions had the lever controls for the rocket engine. S. P. Korolev, who at that time was in charge of the flight tests for the RD-1 engine team, personally flew in the engineer's position on a number of the initial key test flights.[3] Fig. 8.2-11 is a picture of Korolev in his flight suit. At the time no one would have known that this man would become the most powerful leader of Soviet long–range missiles and spaceflight efforts. Some of his accomplishments are mentioned in the preceding chapter and in Chapter 8.11.

The initial ground tests of this Pe-2R aircraft with its RD-1 rocket engine were aimed at getting good engine reliability, avoiding failures, and having smooth ignitions (no hard starts) and reasonable performance. There were some engine problems and explosions that damaged the aircraft, resulting in

Fig. 8.2-10 Enlarged frame from a film record of the twin-engine Pe-2 tactical bomber airplane flying with rocket power. The exhaust plume is slightly smoky. Adapted from Refs. 3 and 7.

Fig. 8.2-11 S. P. Korolev in flight suit just before one of his test flights as engineer on the Pe-2R rocket propelled airplane. Copied with permission from Ref. 3.

aircraft repair, destroyed engines and engine design modifications. The first flight with the Pe-2R airplane propelled by a RD-1 engine was in 1943. In 1944 and 1945 this particular LPRE was used as a superperformance engine in ground tests and flight tests in several other experimental military aircraft. The RD-1 LPRE was also flown in the La-7A fighter aircraft (designed by S. A. Lavochkin in 1944), the Yak-3 fighter aircraft (designed by A. S. Yakolev and flown in 1945), and the Su-6 high altitude fighter aircraft (designed by P. O. Suhkoy).[1,3,8,9] The RD-1KhZ engine (with the hypergolic start fluid) had flown in the Pe-2R, Yak-3, La-7A, La-120R, and Su-7R in 1945.[1,3,8,9] Figure 8.2-12 shows the Su-7R firing during a ground test. Altogether there were more than 400 ground and flight tests with these six experimental airplanes. Altogether the RD-1 and the RD-1KhZ flew approximatey 169 times. There were a number of engine malfunctions, failures, and explosions. In 1945 shortly after the war had ended, an Su-7R with a RD-1KhZ engine was making a practice run in preparation for an air show. The LPRE blew up, and the airplane crashed, killing the pilot.[8,9] The reasons for these explosions could not be found. In 1946 a Lavochkin 120R aircraft was shown at a Toshino air show in a successful spectacular flyby with the rocket engine being fired.[3]

The extensive flight tests showed some flight performance improvement could be achieved. For example the Yak-3 fighter airplane had a 182-km/h (113-mph) speed increase at 7800 m (25,600 ft) altitude, and the climb to altitude was greatly shortened.[8,9] However, the extra weight of the engine and the propellants seriously reduced the airplane's range and payload, diminishing its military defense capability. Neither the RD-1 or the RD-1KhZ were deployed with the USSR Air Forces.

Fig. 8.2-12 View of Sukhoi Su-7 attack modified aircraft during ground test with the RD-1KhZ engine firing. Copied with permission from NPO Energomash from Ref. 3.

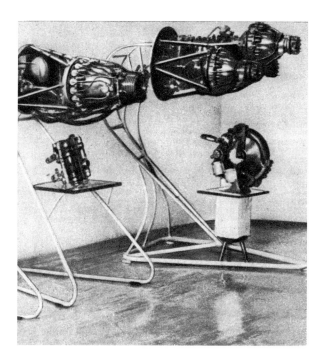

Fig. 8.2-13 Display of RD-2 and RD-3 engines. The turbopump below the RD-3 was the first in the Soviet Union ground tested with a LPRE in 1945. Copied with permission of NPO Energomash from Ref. 3.

The RD-2 LPRE used the same propellants, but had variable thrust initially between 200 kg (440 lbf) and 600 kg (1320 lbf) and used a chemical ignition (hypergolic initial fuel slug).[1,*] It also used gear pumps. The maximum firing duration was limited to a few hours by the wear on the gear pumps. One version of the RD-2 is shown on the left side in Fig. 8.2-13 with the gear pump assembly exhibited underneath the TC assembly with its controls. The bypass valves shown in the flow diagram allowed a change in propellant flow and thrust of the LPRE while the aircraft's main piston engine and the gear pump were operating at constant speed. The RD-2 passed its qualification tests in 1947, but apparently it did not fly.

The TCs of all of these engines were generatively cooled by nitric acid or by acid and kerosene. The thermal expansion of the outer chamber wall on all of TCs of these Glushko LPREs was accomplished by using belleville springs on the bolts holding the chamber and nozzle portions together with a ring seal in the outer wall of the chamber and also the nozzle.[1] Injection was by spiral

---

*Personal communications, V. K. Chvanov, first deputy of general director and general designer, NPO Energomash, 2002–2003.

spray element inserts made of stellite, and they included small spring-loaded check valves to minimize the dribble volume and prevent backflow.

The next engine RD-3 (1944–1945) had three TCs and a variable thrust between 100 kg and 900 kgf (1980 lbf), which then was a relatively large variation.[1] Individual TCs could be turned on or off and also could be throttled. The initial version had a small ignition chamber at the head of each TC. This small antichamber had a spark plug and spray nozzle inserts for acid, fuel, and hypergolic start fluid. It could be restarted several times. Initial versions used gear pumps driven by a piston engine from the aircraft. The TC was regeneratively cooled by nitric acid. This RD-3 engine did not fly.

One version of the RD-3 engine was modified to use turbopump feed systems with centrifugal pumps and a separate gas generator instead of the gear pumps driven by the airplane's engine.[1,3] It can be seen on the right in Fig. 8.2-13; the assembly of the three TCs with their controls is on the top shelf of this exhibit, and the turbopump assembly with its gas generator is lower on a pedestal. Maximum thrust again was 900 kg. The TP had a single-stage turbine, a reduction gear, an oil lubrication unit, an acid pump, a fuel pump, and a water pump. The gas generator was developed independently at an earlier date and was supplied with three propellants, namely, oxidizer, fuel, and a water alcohol mixture (or sometimes just plain water) from three separate small pressure-fed propellant start tanks. The water-alcohol mixture was used in cold weather because it had a lower freezing point than ordinary water. The water was added to the GG to reduce the temperature of the gas admitted to the turbine, which in turn drove the centrifugal oxidizer and fuel pumps. The GG had three chambers: a small ignition chamber, a combustion chamber, and a mixing chamber similar to the GG shown in Fig. 8.1-10. The engine came in two packages: a TC/valve package and a TP/GG package with the three propellant start tanks and their gas-pressurizing system; they could be installed in different parts of the airplane. With the addition of the TP and the GG, the engine dry mass went from 77 kg (RD-2) to 190 kg (RD-3), but the propellant tanks in the airplane were reduced by a larger mass. The gas temperature of the GG and thus the thrust were established by controlling the flow of the water-alcohol mix to the GG. The thrust was controlled by the pilot's engine lever with mechanical linkages to the main propellant valves and the GG water-alcohol valve. One experimental version of these aircraft rocket engines used a hydrogen-peroxide monopropellant GG; this feature was picked up from the Germans and simplified the engine system. No details were found about this turbopump. An experimental version of the RD-3 engine with the tripropellant GG (oxidizer, fuel, and water) and with the turbopump was first static tested in 1945. This was one of the Soviet Union's first self-contained (independent of external power) aircraft rocket engine tests with a TP, multiple restarts, and throttling. This engine with a TP feed system was not fully developed and did not fly.

The RD-4 LPRE had a nominal thrust of 1000 kg (2200 lbf), and its development was started in 1946.[1] Its development was never completed because by that time Glushko had personally ground tested the German V-2 engine, learned from the German technology, and devoted his design bureau's new work to much larger new LPREs.

In the early 1940s, while Glushko's team was working on his aircraft rocket engines in his design bureau at Kazan, a very similar LPRE aircraft rocket engine project (but with a turbopump feed) was given to NII-1, the successor of RNII. A team, which was headed by Leonid S. Dushkin, was to develop aircraft engines that used a pump feed system, and was self-contained (no external power source like a piston engine) with automatic controls.[1,7] Studies began in 1940, design of the first version was in 1942 and the first experimental work in 1943. Two engines were chosen, each used nitric acid and kerosene, featuring a variable-thrust, multiple start capability, a turbopump feed system, and automatic control of start, ignition, and stop operations. The TCs were similar to the one developed for the D-1-A-1100 with all of the injection spray elements at the head of the chamber. Figure 8.2-14 shows the thrust chamber had for the first time a flat face and an orderly pattern of spray injection elements. The TCs had sliding expansion provisions in their outer wall to allow for the thermal expansion of the hot inner walls. The TP might well have been the first in the Soviet Union, but no details of this TP were found. The engine performance was good. The gas generator was based on the work done at RNII in the mid-1930s and used three propellants: nitric acid, kerosene, and water or water with alcohol. The starting propellants for the GG came from

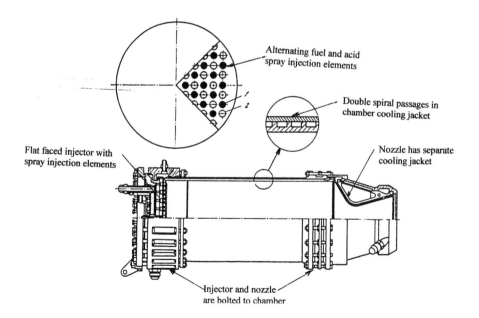

Fig. 8.2-14 Section through the TC of a RD-2M, a more advanced TC with a flat injector face. A partial view of the injector shows multiple injection spray elements: 1, identifies oxidizer injection elements or posts; and 2, identifies for fuel injection elements. Adapted from Ref. 1.

Table 8.2-2 Parameters of Selected Aircraft Superperformance Engines of RNII (1942–1947)

Designation	RD-2M	RD-2M3V
Thrust, kg force	1,450/400	1,500/100
Number of TCs	1	2
Chamber pressure, atm	20.5/6.7	20/6
Specific impulse, s	200/142	200/145
Assembled engine weight, kg	200	223
Turbine horsepower and rpm	75/12,000	80/12,000

small separate gas-pressurized tanks. The control system had some new valves and was rather complex. The identification and characteristics of these two LPREs are given in Table 8.2-2.

Ground testing of a complete engine with a multiple start capability was accomplished in 1945. These were flight tested in an I-270 experimental aircraft, which was designed by A. I. Mikoyan specifically for investigating rocket-engine-propelled interceptors.[8,9] The first satisfactory flight with the RD-2M 3V was in September 1947. It was the first flight with a Soviet-designed turbopump. This was approximately four years after Glushko's first flight on the RD-1 LPRE with gear pumps and a piston engine drive. A sketch of the aircraft is shown in Fig. 8.2-15. Only two I-270 aircraft suitable for rocket test flights were built. After both airplanes were destroyed during the tests, the program was stopped in 1947. There was no follow-up work on either of these two NII-1 engines.

The Kosberg Design Bureau (mentioned again later as the Chemical Automatics Design Bureau) also undertook the design and development of several other aircraft super-performance LPREs in the late 1940s and early 1950s. Several of their engines are noteworthy, and they are discussed in and a picture is shown in Chapter 8.5. The RD-0101 using ethyl alcohol and LOX and a turbopump feed system was flown in the E-50A experimental aircraft. The RD-0102 with a thrust of 39.2 kN (8813 lbf) using LOX/kerosene and chemical ignition was a successful LPRE. It was produced in limited quantities and thoroughly flight tested in the Yak-27V fighter airplane.

LPREs for aircraft superperformance were also developed by the Isayev Design Bureau (later renamed KB Khimmash). The aircraft rocket engine was identified as the RD-1, but it was different than the earlier Glushko engine, which had the same designation. Isayev's RD-1 was flight tested in a BI-1 aircraft, an aircraft previously used to flight test a different LPRE. This RD-1 had a nominal thrust of 1100 kg (2420 lbf) and was developed from 1941 to 1943. This Isayev engine and the Dushkin D-1-A 1100 engine were also flight tested in a Flovov 4302 experimental aircraft. The flight tests ended in 1947. More information about Isayev's aircraft rocket engines can be found in Chapter 8.6.

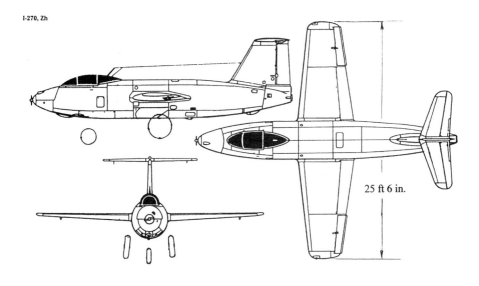

Fig. 8.2-15 Three simplified views of the I-270 experimental rocket-propelled aircraft. Adapted from Ref. 8.

Several of these aircraft rocket engines were also used as jet-assisted-take-off (JATO) engines, which is an easier task, because it does not require a restart. None seemed to be droppable or reusable JATO units, which were developed in other countries. The Yuzhnoye Design Bureau discussed in Chapter 8.10 has developed a LPRE specifically for a JATO application. Some of these aircraft engines were used as JATO units with modified bomber aircraft. However there was little interest to use rocket thrust assist for overloaded military aircraft, and none were requested or deployed with the Soviet Air Forces.

The intended reliability (99+%) and design life (typically 45 h of flight and 100 starts) of the newly developed aircraft superperformance rocket engines was actually never really reached. The Soviet Air Force was never really satisfied with these aircraft rocket engines or their reliability. The addition of an auxiliary LPRE helped the aircraft to climb to altitude quickly and allowed some altitude maneuvering improvements, but the range and the payload of the aircraft were greatly reduced, thus severely limiting its operational capability and usefulness. Furthermore the military people did not like the corrosion problems with nitric acid or the injuries it caused to personnel. None of the fighter airplanes equipped with rocket engines were ever used by the Soviet Air Force in combat. In retrospect this big effort (4 design/development organizations, approximately 16 LPREs for aircraft, installed in about 12 different military airplanes, and a lot of (flight testing) has been called a poorly conceived set of programs and a waste of resources. As said at the beginning of this chapter, these Soviet aircraft rocket engines were indeed historic achievements in the LPRE technology, served to train qualified personnel, to advance the state of the art, and helped to build up a good LPRE capability. This new capability was effectively used in developing LPREs for other applications, as described in the subsequent chapters.

References

[1]Moshkin, Ye. K., *Development of Russian Rocket Technology*, Mashinostroyeniye Press, Moscow, 1973 in Russian; also *NASA Technical Translation TT F-15,408*, March 1974 (in English).

[2]Blagonravov, A. A., and others, (eds.), *Soviet Rocketry, Some Contributions to Its History*, Moscow, 1964; translated by Israel Program for Scientific Translations, Jerusalem, 1966.

[3]Glushko, V. P., *Rocket Engines of the Gas Dynamics Laboratory—Experimental Design Bureau, (Raketnye Dvigateli GDL-OKB)*, Novosti Press, Moscow, 1975; also NASA Technical Translation TT F 16847, Feb. 1976; also NPO Energomash, Russia (abbreviated English version).

[4]Siddiqi, A. A., "Rocket Engines from the Glushko Design Bureau: 1946-2000," *Journal of the British Interplanetary Society*, Vol. 54, 2001, pp. 311–334.

[5]Winter, F. H., *Prelude to the Space Age; the Rocket Societies 1924–1940*, National Air and Space Museum, Smithsonian Inst. Press, Washington, D.C., 1983.

[6]Ehresman, C. M., "Liquid Rocket Propulsion Applied to Manned Aircraft", AIAA Paper 91-2554, June 1991.

[7]Dushkin, L. S., "Experimental Research and Design Planning in the Field of Liquid Propellant Rocket Engines Conducted Between 1934 and 1944 by the Followers of F. A. Tsander," *History of Rocketry and Astronautics*, American Astronautical Society History Series, Vol. 7, Univelt, San Diego, 1986.

[8]Gordon, Y., and Gunston, B., *Soviet X-Planes*, Midland Publishing, Hinckley, England, 2000.

[9]Gordon, Y., and Sweetman, B., *Soviet X Airplanes*, Motorbooks International Publishers, Osceola, WI, 1992, Chaps. 15 and 17, pp. 38–54.

8.3 Organizations Working on Liquid Propellant Rocket Engines

Most of the countries with a LPRE capability perform the development and production in the same organization. This is not the case in the Soviet Union or today in Russia. The design bureaus do the development, and the serial production is usually done in a separate set of production plants.

Design Bureaus

A design bureau (DB) in the Soviet Union is really a company that designs, develops, and usually tests new or improved hardware, machinery, or equipment. These DBs are unique technical organizations responsible for development, building, and testing of experimental hardware, but not usually the serial production of the hardware. The Russian words for DB are "konstruktorskoye buro," which is abbreviated as KB. There are several grades of KBs, such as SKB (special DB), OKB (experimental DB), or a full fledged KB. They were identified by an alphanumeric designation and often named after their chief designer, but name changes and reorganizations of these bureaus were quite common. A number of these design bureaus specialized in developing LPREs, either as a main product line or as a supplementary or secondary product line. Each LPRE DB normally also had the personnel and the facilities to conduct laboratory tests and flow tests and to manufacture and hot-fire ground test its own experimental LPREs. Some had special facilities for very large LPREs, for static firing of engines installed in a stage of a vehicle, or for engine firing tests at simulated altitude. Sometimes the testing of an engine was done at a different organization and location, where more suitable test facilities were available. Some certification and quality control testing of LPREs is conducted at different (off-site) facilities. Many a DB had its own housing, gymnasium, and medical facilities for the employees and their families.

Beginning around 1939, the Soviet government wanted to expand their large missile capability and thus also broaden the LPRE capability of the country. The RNII R&D organization, mentioned in Chapters 8.1 and 8.2, by itself was not considered suitable for such expansions. Therefore several new LPRE design bureaus were established in the 1940s, some of them as spin-offs from RNII. Some of the LPRE experts from RNII went into several of the design bureaus. The list of design bureaus developing LPREs is given in Table 8.3-1.[1-3,*] This list is not complete because there were some others during periods of this history. Of those listed, only one is outside of Russia today. Some of the LPRE DBs are or were totally committed to the LPRE field, but in other DBs the LPREs were a minor product. Some universities, government R&D

---

*Personal communications, J. Morehart and others, The Aerospace Corporation, 2001–2004, and M. Coleman, Chemical Propulsion Information Agency, 2001–2003.

Table 8.3-1 Design Bureaus and R&D Institutes

Organization name	Location	Area of specialty
1) NPO Energomash (named after Academician Valentin P. Glushko, Formerly OKB 456)	Khimki, Moscow Region	High-thrust LPREs for 1st and 2nd stages of ballistic missiles and space launch vehicles
2) Chemical Automatics Design Bureau (CADB), (S.A. Kosberg Design Bureau) KB Khimautomatiki	Voronezh, Russia	Mostly upper stages of SLVs; flew one SLBM engine, and one LOX/LH$_2$ LPRE
3) KB Khimmash (Isayev Design Bureau or Chemical Machinery Design Bureau)	Korolev (was called Kaliningrad) Moscow Region	Upper-stage LPREs for orbit maneuvers satellites or SLVs; for SLBMs, short-range ballistic missiles; reaction control thrusters/engines
4) NII Mashinostroyeniya (R&D Institute of Mechanical Engineering)	Nishnaya, Sverdlovsk Region, Russia	Reaction control thrusters for spacecraft and orbiting stations (bipropellants)
5) A. M. Lyulka Engine DB (today Saturn Science and Production Association)	Moscow, Russia	Gas turbine aircraft engines; power supplies; In the past also developed large LOX/LH$_2$ engines
6) NK Engines (N. D. Kuznetsov Design Bureau), (now Samara Science and Technical Complex)	Samara, (formerly Kuibychev), Russia	Jet and aircraft engines; in past some large LPREs with LOX/kerosene
7) Southern Machine Building Production Association, incl. Yuzhnoye Design Bureau	Dnepropetrovsk, Ukraine	Space launch vehicles, missiles; vernier and upper-stage LPREs for SLVs, spacecraft, and missiles
8) RKK Energiya (was Korolev's Design Bureau), RKK means Rocket and Space Corporation, formerly OKB-1	Korolev, Moscow region, formerly Kaliningrad	Launch vehicles and space craft; but also some upper-stage LPREs and vernier LPREs and postboost vehicle control
9) OKB Fakel	Kaliningrad, Russia	Mostly electric propulsion and some hydrazine monopropellant small thrusters

(continued)

Table 8.3-1 Design Bureaus and R&D Institutes (continued)

Organization name	Location	Area of specialty
10) Keldysh Institute for Applied Mechanics (originally part of RNII)	Moscow	R&D and analysis on all types of propulsion and flight vehicles, mission analyses/optimization
11) State Institute of Applied Chemistry	Leningrad	Develop, characterize, synthesize new propellants and their production processes
12) Tumansky DB, also Machine Building Scientific Production Association Soyuz	Tushino, Russia	Primarily jet engines, small nuclear reactors, small thrusters for satellites
13) Isotov DB now closed	Leningrad	Upper-stage LPREs, lunar lander, auxiliary engines,
14) Stepanov DB (NPO Soyuz) Turayevo Machine Building	Turayevo, Russia	Small thruster for ACS, other unrelated products

Based on personal communications, J. Morehart and others, The Aerospace Corporation, 2001-2004, and M. Coleman, Chemical Propulsion Information Agency, 2001-2002. NPO Energomash, 2002; the Chemical Automatics Design Bureau, 2002; and NII Mash, 2002, 2003. Also on Refs. 1-3.

institutes/laboratories also conducted R&D related to LPREs, and they worked with the DBs.[4] Two such institutes are listed in this table. Almost all of the organizations in Table 8.3-1 have had more than one name or designation, were reorganized, relocated, merged, or otherwise changed. Only some of the names of each DB are listed in the table.

The history of each of the first 11 organizations in the table will be discussed in separate chapters later in this book. The last three organizations are known to have worked on some LPREs, might no longer be active, and valid information of their LPRE work could not be obtained. Therefore these three will not be described in this book.

Each DB usually had expertise in specific types of LPREs with different applications, propellants, or engine sizes. Each LPRE DB has its own test facilities and a shop for building experimental engines and test equipment. The DBs are not of the same size in terms of product variety, facilities, engine size, or personnel. Competition between DBs was at times fierce, particularly when two of these DBs were developing engines for the same application requirement. There also was cooperation when a LPRE was designed and developed by one DB, and the design information and know-how was transferred to another DB, which then built newer or uprated versions of the same basic engine and dealt or coordinated with the production plants.

The first four design bureaus are still in the LPRE business today and were established between 1940 and 1945. In 1959 Nikita Kruschev (the Soviet leader at that time) decided to reassign some of design bureaus in the aircraft and aircraft engine business to work on rocket vehicles and rocket propulsion tasks.[1] His decision might have been influenced by the Cold War and the desire for the Soviet Union to be first in space. The Kuznetsov DB, Lyulka DB, and Tumansky DB are examples, were engaged primarily in aircraft or jet engines, but in the 1960s they were required to also develop some rocket engines. They have since been phased out of the LPRE business.

How did a design bureau get work? Rocket engine work usually originated with a "requirement" for a new engine or for a preliminary study of a new engine, and these were issued by one of the vehicle (missiles or spacecraft) design bureaus, who themselves were working on a primary new mission. So the development of a LPRE or of design studies for a LPRE were always aimed at specific requirements. The decisions on awarding vehicle and LPRE development jobs to specific design bureaus was often the result of interactions between lead personalities.[1,*] Once a vehicle or system's chief designer was comfortable with a chief designer of a LPRE design bureau and satisfied with the prior engine products he had received, it was not unusual to see continued reliance on that LPRE bureau for that class of engines. For example, all but one of the LPREs for submarine-launched missiles were developed by the same design bureau for the same vehicle development prime contractor.

The Soviet LPRE activity increased greatly in the 1960s and reached a peak in the 1970s. This was the period of the Cold War between the Soviet Union and the United States and the competition for space-flight achievements. At that time between 14 and 17 Soviet design bureaus and R&D institutes were involved in LPREs. In the last 25 years the engine development effort has diminished not just in the Soviet Union, but also in other countries. The number of Soviet or Russian organizations seriously working on LPRE development has shrunk to about half, and several of the currently existing design bureaus are working with less than half their former LPRE staff levels.[*,†] In some design bureaus only some of the current personnel are engaged in LPREs because they have started "conversion products" (such as commercial pumps or oil field equipment). They have recently been encouraged by their government to export LPRE technology and hardware.[†]

Although the early investigations of LPREs in the Soviet Union employed a variety of different storable liquid propellants, the government and the industry seems to have standardized on a single storable propellant combination,

---

*Personal communications, J. Morehart and others, The Aerospace Corporation, 2001–2004, and M. Coleman, Chemical Propulsion Information Agency, 2001–2004.
†Personal communications, NPO Energomash and the Chemical Automatic Design Bureau, 2002, and NII Mash, 2002, 2003.

namely, NTO/UDMH. Since about 1965, these propellants have been used for ballistic missiles, SLVS, orbit maneuvers, attitude control, and submarine-launched missiles. The only exceptions are LOX/kerosene engines and LOX/LH$_2$ engines for spaceflight. There is an obvious national benefit in the logistics of the supply of a single set of storable propellant to the R&D establishments, the military missile forces, launch ranges, or production test facilities. It saves costs in development, compatible materials, or personnel training. Furthermore it avoids costs for producing other propellants.

Production Plants and Test Facilities

Production of LPREs usually took place in one of the independent serial production plants, dedicated to this purpose.[1-4,*] DBs were usually not allowed to undertake serial production of the products they developed. If the quantity of a batch of production LPREs was small enough, then occasionally such batches were allowed to be manufactured in the experimental shop of a DB. A partial list of LPRE production plants is in Table 8.3-2. Many of these plants were known by other names, have been reorganized, and have been diverted to other work. The cognizant design bureau, which developed the engine, maintained a branch or a group of engineers and fabrication experts at the production plant. In the USSR a serial production plant has the right to redesign parts of a LPRE in order to make the hardware more producible, but the design bureau has to agree that the proposed change does not detract from the engine's performance, safety, or function. Some of the production plants have a long history of manufacturing LPREs and have built several different rocket engines, but other production plants have only built a few or only one type of engine or only for a limited period of time. Only one of the listed LPRE production plants is at this time in the Ukraine, outside of Russia.

As an example of a production plant, the Krasnoyarsk Machine Building Plant will be briefly discussed here.[5] It is located in the city of Krasnoyarsk on the trans-Siberian railroad about halfway between the Ural mountains and the Sea of Okhotsk. The company was started in 1932, but work on large missiles, spacecraft, and their components only began in the 1950s. They worked on rocket engines as well as vehicles. This company, abbreviated as Krasmash, produced an early ICBM (R-9), at least two versions of submarine-launched missiles, and participated in the production of the Proton and Kosmos launch vehicles. In the field of LPREs, they produced engines developed by KB Khimmash, NPO Energomash, and RKK Energiya. This production plant is equipped with various laboratories (materials, mechanical properties, welding, etc.), modern inspection equipment (X-ray, die penetrant, surface inspection), test stands for testing of hardware under loads or pressure, vacuum furnaces, investment casting equipment, vapor deposition equipment, specialized

---

*Personal communications, J. Morehart and others, The Aerospace Corporation, 2001–2004, and M. Coleman, Chemical Propulsion Information Agency, 2001–2004.

Table 8.3-2 Examples of Production Plants for LPREs[2,3,a]

Name	Location	Example of assignment
1) Metallist Samara OAO[b]	Samara, Russia	TCs for RD-253
2) Ust-Katav Wagon Building Plant	Ust Katav, Ural Region, Russia	KB Khimmash engines, some work for CADB.
3) Perm Motor Production Plant	Perm, Russia	RD-214, RD-216
4) NPO Motorostroitel	Samara, Russia	RD-107/108 family
5) Voronezh Mechanical Plant	Voronezh, Russia	Production of engines designed by CADB
6) Krasnoyarsk Machine Building Plant (Krasmash)	Krasnoyarsk, Russia	RD-119, RD-243, SLBMs
7) Yuyzhmash Plant or Southern Machine Building Production Plant	Dnepropetrovsk, Ukraine	RD-101/103, RD-120, RD-261/262, vernier engines
8) PO Polyot	Omsk, Russia	RD-170/171
9) Krasnaya Oktobr	Leningrad, Russia	SAM's LPREs, RD-0235/RD-0236

[a]Personal communications, J. Morehart and others, The Aerospace Corporation, and M. Coleman, Chemical Propulsion Information Agency, 2001–2004.
[b]OAO is a Russian abbreviation for an open joint stock company, as opposed to a joint stock company or one exempted from privatization.

welding equipment (including electron beam welding), and facilities for production ground firing tests of LPREs. They cooperate with a number of R&D establishments, such as the Krunichev Research and Production Space Center and state R&D organizations specializing in automation, welding, or instrumentation. In recent years they also produce commercial hardware, such as industrial refrigerators, because the production of large missiles, space launch vehicles, or LPREs has been dramatically reduced.

Most of the testing of LPREs for R&D or qualification has been done at the test facilities of each of the DBs listed in Table 8.3-1. Even some of the R&D institutes such as the Keldysh Institute had some rocket test facilities. However the Soviets also had independent separate organizations for testing very large engines or for testing a cluster of LPREs installed in a part of a vehicle. There are extensive facilities for these tests at NII Khimmash (this means Science and Technolgy Complex—Chemical Machinery) at Sergiyev Posad north of Moscow and at NII Mach in Nishnaya Saldo in the Ural mountains. The test facility NII Khimmash is not to be confused with KB Khimmash (Chemical Machinery Design Bureau), a developer of LPREs, and it is not related, is located elsewhere, and is discussed in Chapter 8.6. The large LOX/LH$_2$ RD-0120 LPRE was tested at NII Mash in the Ural mountains. NII Mash is an abbreviation for NII Mashinostroyeniya (R&D Institute of Mechanical Engineering), and it is directly involved in the development and production of small thrusters, as explained in Chapter 8.7.

German Influence

At the end of World War II, the Soviet rocket missile and propulsion effort was given a big boost, when they gained access to the German wartime work, which in 1945 was perhaps 10 years ahead of the work in the Soviet Union. Shortly after the end of the war, the Soviet government sent expert teams to Germany for gathering technical know-how, test German secret weapons (and also LPREs), and collect military and production hardware. For example, the German V-2 production plant, which was located in the Soviet zone of Germany, was refurbished, rebuilt, and reactivated. Facilities to build V-2 missiles and V-2 rocket engines were then built in the Soviet Union. A large number of these V-2 missiles (with a slightly modified LPRE) were actually built in Russia, and many were launched in the Soviet Union, beginning in 1947. A slightly modified version of the German V-2 was actually deployed with the Soviet military forces as their first short-range ballistic missile. The Soviets also copied, built, and tested several other German missiles and propulsion units, which had not been fully developed before the war ended. This included building a small number of the German antiaircraft missiles Wasserfall, Schmetterling, and Taifun. The Soviets managed to extract a lot of information from the Germans.

The Soviet Union also obtained the assistance of experienced German engineering and design personnel, which were transported to the Soviet Union and paid a good salary. They were sent to different plants or locations. For example a group of about 24 German engine designers, fabrication managers, and engineers was assigned to the Glushko DB in Khimki to assist with rocket engines.[4] They provided advice, conducted studies, and prepared preliminary designs of advanced engines for the Soviets. The largest group of Germans consisted of production and launch personnel, and they helped with the Russian versions of the V-2 production, launches, and flight testing. The Germans never learned what missiles and engines the Soviets themselves were developing on their own. After four to six years in Russia (at times under uncomfortable or harsh conditions), the groups of Germans were allowed to go home.

The Germans did not receive any feedback about the benefit or the use of their studies to the Soviets. Many of the returning Germans downplayed the value of their contributions to the Soviet missile and space effort, but today it is believed that it was considerable.[6] The Russians appreciated the help with the V-2, but many said that the studies and preliminary designs of the Germans were of no value to the design of USSR hardware. However today there is evidence that the contributions and benefits to the Soviet programs were considerable.[6]

Injuries and Fatalities

The Soviets also have the dubious distinction of having injured and killed more people in accidents or malfunctions of LPREs than in any other country. There were a number of unfortunate explosions and launch accidents, which caused the deaths of many of the workers. One of the most dramatic disasters was the 1960 accident of the R-16 ballistic missile on the launchpad at the Tyuratam launch facility.[1,7] It caused the deaths of 124 engineers and military personnel including Marshal M. I. Nedelin, a former deputy minister of defense. The second-stage rocket engine of the R-16 missile suddenly began to fire, which broke up the missile that had been fully loaded with propellant. It resulted in an explosion of the mixed hypergolic propellants. The failure was eventually blamed on the missile control unit not having a protective circuit to block spurious electric signals, which induced a start signal. Although the immediate cause of the accident was not caused by the engines themselves, the spill, ignition, burning, and exploding of the hypergolic propellants (nitric acid with additives and UDMH) were responsible for the tragedy. This and other fatal LPRE accidents were hushed up and remained secret for about 30 years. A false story was officially released that Marshall Nedelin had died in an airplane accident. There were fatal accidents in some other launches, in several of the LPREs test operations, and in a few of the flights of rocket-propelled aircraft.

References

[1] Siddiqi, A. A., *Challenge to Apollo, the Soviet Union and the Space Race (1945-1974)*, NASA SP-2000-4408, Washington, DC, 2000.

[2] Lardier, C., "Liquid Propellant Rocket Engines in the Soviet Union," Conference paper International Academy of Astronautics, Paper 99-2-3-09, Oct. 1999.

[3] Lardier, C., "Les Moteurs-Fusées Soviétiques de 1946 a 1991 (Soviet Liquid Propellant Rocket Engines from 1946 to 1991)," International Academy of Astronautics, Papers 97-2-3-03, 1997, 98-2-3-09, 1998, and 99-2-3-04, 1999.

[4] Koroteev, A. S., and Demianko, Y. G., "RNII—the Keldysh Research Center as a part of the History of Home Rocket Manufacturing," 10th International Symposium on the History of Astronautics, June 1995.

[5] www.fas.org/spp/civil/Russia/Krasmash.htm (cited May 2005).

[6] Przybilski, O., "Die Deutschen und die Raketenantriebswerkentwicklung in der USSR (The Germans and the Development of Rocket Propulsion in the USSR)," *Luft und Raumfahrt*, 1999, No. 2, pp. 30–32 and No. 3, pp. 28–32, and No. 4, pp. 33–40 (in German); also *Journal of the British Interplanetary Society*, Vol. 55, No. 11/12, 2002, pp. 404–427 (in English).

[7] Bolokin, A., *The Development of Soviet Rocket Engines (for Strategic Missiles)*, Delphic Associates, Inc., Falls Church, VA, 1991, (translated from Russian).

8.4 NPO Energomash

Organization's History

This organization is best known for developing more successful large LPREs than any other LPRE establishment in the world. Its history can be traced back to the liquid propellant rocket-engine section (1930) of the Gas Dynamics Laboratory (GDL) and the liquid propellant rocket-engine section of the Reaction Propulsion Research Institute (RNII), the prewar rocket-engine organizations mentioned in Chapter 8.1. This LPRE organization, which was headed by Valentin P. Glushko, went through several reorganizations and relocations as will be described here.[1-5] From its beginning in 1931, Glushko was the leader of the LPRE section and since 1944 he became a chief designer, a very prestigious title with major responsibility in Soviet technical organizations. His portrait is in Fig. 8.4-1. He was the head of Energomash and later became the leader of most of the Soviet missile and spaceflight developments. He was selected to be an associate member of the Soviet Academy in 1953 and a full member in 1958. He is considered by many as the key individual in Soviet large LPRE developments.

As reported in Chapter 8.1, the Stalin purges claimed a good number of senior military officers and technical intelligentsia. These people were arrested, uprooted, deported, or killed. In 1938 Glushko and several of his coworkers were arrested on trumped-up charges of espionage and sent to labor camps. About a year later Glushko and several of his associates were allowed to continue their development work with rocket engines, but as prisoners in a factory in Tushino, near Moscow. In 1940 this group was transferred to another prison factory at Kazan, which is about 450 miles east of Moscow. There they developed aircraft rocket engines with thrusts of 300 to 1100 kg while under guard. They built crude test facilities and started with the development of the RD-1 aircraft rocket engine, described in Chapter 8.2. In 1944 the group was elevated to an OKB, which means an Experimental Design Bureau, and Glushko was made its chief designer as its leader, but they had to operate in a prison. Glushko and about 33 of his men were released from captivity only in 1944, and the rocket-engine work continued without the prison confinement.

Shortly after the war with Germany had ended in 1945, V. P. Glushko, S. P. Korolev (later leader of Soviet ballistic missile and spaceflight efforts), A. M. Isayev (later became chief designer of his own LPRE design bureau), and others were put into military officer uniforms and sent to Germany to gather information and hardware of German rocket missiles and their propulsion systems.[1,2,6] During this period, Glushko was put in charge of testing German V-2 rocket engines at Lehesten, a production test firing facility located in the Russian zone of Germany. He assembled a staff of Soviet engineers at Lehesten (mostly from his own design bureau) to examine the design, materials, fabrication processes, and operation of the V-2 LPRE and to study possible improvements.

In July of 1946, Glushko's development team (then identified as Experimental Design Bureau OKB 456) was put into a former aviation plant at Khimki in the

Fig. 8.4-1 Valentin P. Glushko, key individual in large liquid propellant rocket engines. Photo courtesy of NPO Energomash.

Moscow region.[1,2,6] This factory had been vacated during the war, as the German Army came close to the capital city. The assignment for OKB 456 was to develop LPREs, and they started with a German LPRE, which was about 15 times larger than anything they had done before in the Soviet Union. To accomplish this job, the old factory was renovated, new buildings were erected, and the proper facilities and equipment for developing, building and testing experimental rocket engines had to be created or acquired.[1] Although most of the improvements and additions of the facilities were done between 1946 and 1950, expansion and upgrading has continued to the 1990s. Because of the cold winter weather, the facilities are indoors and heated. The design bureau has four unique engine test stands for large LPREs.[1] The engine's exhaust gas is captured, sent through a large (vertical axis) vortex chamber with intensive water sprays. The cooled and cleaned gas is conducted away from the test stands (by means of ejectors) in large ducts and is exhausted through a tall chimney. Figure 8.4-2 shows two exhaust chimneys and a vortex chamber. The design bureau obtained modern equipment for the usual kinds of LPRE tests, such as pressure testing, TC firing, pump testing with water and an electric drive of 37,000 kW or roughly 50,000 hp, turbine and turbopunp testing, cold-flow testing for various components and complete engines, strength testing of structures under static and dynamic loads, and equipment for testing control apparatus and monitoring

Fig. 8.4-2 Two exhaust chimneys and a vortex gas cooling chamber of the large LPRE tast stands at the Khimki test facility. Copied with permission from NPO Energomash from Ref. 1.

functions. Altogether in about 1998 Energomash had by its own count 83 separate test facilities for LPREs their components.[1] The existing manufacturing plant at Khimki also obtained the appropriate machinery and inspection equipment to build LPREs. The first production, starting in 1946, was a series of copies of the German V-2 rocket engine.

In 1954 the production plant at Khimki and the Experimental Design Bureau (OKB-456) were put together into one organization under Glushko. In 1967 the OKB-456 was elevated to a full design bureau (in Russian it is abbreviated as KB for konstruktorskoye buro), and the name Energomash was adopted.[1,2,5] It was then known as KB Energomash. A major reorganization of the Soviet aerospace industry took place in 1974. A new vehicle/propulsion conglomerate was created under Glushko's leadership.[2-5] The design bureau of S. P. Korolev called KB Energiya (ballistic missiles, space launch vehicles, and some LPREs) was combined with KB Energomash (large LPREs) into a single entity called Energiya Scientific Production Organization or NPO Energiya.[2,5] Korolev, who was the chief designer of KB Energiya and who had also presided over the council of chief designers, had passed away, and Glushko was selected to replace him. Glushko became the leader and administrator of the new combined organization, and he picked Victor P. Radovsky to be the chief designer of the Energomash suborganization. This new organization stayed in place until Glushko died in 1989.[2,7] In 1990 Energomash was separated from the NPO Energiya. In 1991 Boris I. Katorgin became the new general designer and head of KB Energomash, and shortly thereafter it was officially renamed NPO (Scientific Production Association) Energomash after V. P. Glushko. In 1998 it had about 6500 employees with a work space of 282,000 m^2 or about 3,000,000 ft^2 (Refs. 1 and 6). At that time it might well be the largest LPRE development organization in the world.

Table 8.4-1 (Ref. 8) gives some data on 30 of the Energomash principal rocket engines.[1,3-5] The data of some engines do not always agree with similar data in some of the references because it is not always clear which version of the engine is quoted. The code for the numbering of LPREs in this design bureau is this: the RD-100 series is for LOX (mostly with kerosene); the RD-200 series is for storable propellants; the RD 300 and 500 series for exotic propellants, and the 700 series for tripropellant LPREs. There were other Energomash engines (not listed in table) that have also flown, some engines whose intended vehicle never flew, some that were experimental engines (which were not intended to fly), and then there were concept LPREs, where only analyses, design studies and sometimes some static component testing was done. By Energomash's own count (up to about 1998) there were altogether 95 LPREs; of these some 35 passed qualification tests, and 30 have flown.[1] If other experimental engines and early work are also included, the total number could be considerably larger. Siddiqi[5] counted 120 engines, and this does not include early LPREs.

After the end of World War II, the Soviets had acquired German hardware, facilities, deployment details, and manufacturing know-how, some German rocket experts, and data of several German missiles (V-2 missile, Wasserfall

Table 8.4-1 Data on selected LPREs of Energomash

Designation	Thrust (SL) vacuum (V), kN	No. of TCs per engine	Specific impulse, s	Propellants	Chamber-pressure, kg f/cm²	Engine-mass, kg	Height/diameter m	Development period, years	Application and comments
ORM-65	1.72 (SL)	1	215 (SL)	HNO₃/kerosene	25.5	14.3	0.46/0.38	1936	RP-318 glider, winged rocket
RD-1	2.94 (SL)	1	200 (SL)	HNO₃/kerosene	21			1941–1945	Flew on Pe-2, La-7, Yak-3 and Su-6 aircraft
RD-1 KhZ	2.94 (SL)	1	200 (SL)	HNO₃/kerosene	20.4	14.3	0.85/0.4	1941–1946	Flew on Pe-2R, Yak-3, La-7R, La-120R, Su-6, and Su-7 aircraft
RD-100	257 (SL)/304 (V)	1	199 (SL)/233 (V)	LOX/75% alcohol	16.2	1,209	3.7/1.65	1946–1950	R-1 SRBM, copy of V-2
RD-101	363 (SL)/404 (V)	1	210 (SL)/237 (V)	LOX/92% alcohol	21.6	1,178	3.54/1.65	1946–1951	R-2 (SS-2) SRBM
RD-103M	432 (SL)/500 (V)	1	232 (SL)/300 (V)	LOX/92% alcohol	21.1	1,867	3.22/1.65	1946–1953	R-5M(SS-3 Mod 2) MRBM
RD-107	814 (SL)/1000 (V)	4	256 (SL)/313 (V)	LOX/kerosene	60	1,190	2.86/1.85	1954–1957	First-stage R-7 ICBM, and R-7A for SLVs
RD-108	745 (SL)/941(V)	1	250 (SL)/316 (V)	LOX/kerosene	52	1,625	286/1.85	1954–1958	Second stage for same vehicles
RD-111	1412 (SL)/1628 (V)	4	275 (SL)/317 (V)	LOX/kerosene	80	1832	2.1/2.74	1959–1962	First stage R-9A (SS-8) ICBM
RD-119	105 (V)	1	352 (V)	LOX/UDMH	80	107.2	2.17/0.96	1958–1962	Second stage of Kosmos 2 SLV
RD-120	833 (V)	1	350 (V)	LOX/kerosene	166	1,125	3.87/1.95	1976–1985	Second stage of Zenit and Zenit 3 SLVs
RD-120 K	740 (SL)/870 (V)	1	295 (SL)/356 (V)	LOX/kerosene	187	1,037	2.53/1.54		Improved RD-120 (smaller nozzle) has not flown
RD-170	7257 (SL)/7904 (V)	4	309 (SL)/337 (V)	LOX/kerosene	250	10,750	4.0/4.0	1976–1987	First stage of Energia SLV
RD-171	7257 (SL)/7904 (V)	4	309 (SL)/337 (V)	LOX/kerosene	250	NA	4.0/4.0	1976–1987	RD-170 adapted for Zenit 2, Sea launch

(continued)

Table 8.4-1 Data on selected LPREs of Energomash (continued)

Designation	Thrust (SL)[a] vacuum (V)[b], kN	No. of TCs per engine	Specific impulse, s	Propellants	Chamber-pressure, kg f/cm²	Engine-mass, kg	Height/diameter m	Development period, years	Application and comments
RD-180	3824 (SL)/4148 (V)	2	311 (SL)/338 (V)	LOX/kerosene	261.7	5,330	3.58/3.2	1992–1998	First stage of Atlas 3 and 5 SLVs two-chamber derivative of RD-170/171
RD-191	1922 (SL)/2085 (V)	1	310 (SL)/337 (V)	LOX/kerosene	263.4	2,200	4.0/1.45	1998	Single-chamber derivative of the RD-170, not yet flown, proposed for Angara SLV
RD-214	636 (SL)/730 (V)	4	230 (SL)/264 (V)	HNO₃/kerosene	44.5	645	2.38/1.5	1952–1957	R-12 (SS-4) MRBM; first stage of Kosmos (SL-7)
RD-216	1481 (SL)/1677 (V)	4	246 (SL)/289 (V)	HNO₃/UDMH	75	1,350	2.19/2.26	1958–1960	R-14 (SS-5) IRBM; first stage of (SL-8) Kosmos SLV
RD-218	2221 (SL)/2608 (V)	6	246 (SL)/289 (V)	HNO₃/UDMH	75	1,960	2.2/2.8	1958–1961	First stage of R-16 (SS-7) ICBM
RD-219	883 (V)	2	293 (V)	HNO₃/UDMH	75	760	204/2.2	1958–1961	Second stage of R-16 ICBM
RD-251	2363 (SL)/2648 (V)	6	270 (SL)/301 (V)	NTO/UDMH	85	1,729	1.7/2.52	1961–1965	First-stage R-36 (SS-9) ICBM
RD-252	902 (V)	2	317.6 (V)	NTO/UDMH	91	NA	NA	1961–1966	Second-stage R-36
RD-253	1471 (SL)/1638 (V)	1	285 (SL)/316 (V)	NTO/UDMH	150	1,080	3.0/1.5	1961/1965	First-stage Proton SLV (6 engines)
RD-264	4168 (SL)/4521 (V)	4	293 (SL)/318 (V)	NTO/UDMH	210	3,600	2.15/3.0	1969–1973	First-stage R-36M (SS-18 Mod 1–4) ICBM

Engine		Thrust (SL/V), kN	Propellants	Isp (SL/V), s	Mass, kg	Mixture ratio	Years	Application
RD-268	1	1147 (SL)/1236 (V)	NTO/UDMH	295.6 (SL)/318.5 (V)	230	770	2.15/1.08	1969–1976	First stage of MR-UR 100 ICBM
RD-270	1	6276 (SL)/6717 (V)	NTO/UDMH	301 (SL)/322 (V)	266	4,770	4.85/3.3	1962–1971	Development not completed
RD-275	1	1590 (SL)/1745 (V)	NTO/UDMH	289 (SL)/316 (V)	163	1,070	3.0/1.5	1996–2001	Improved RD-253 for Proton-KM.
RD-301	1	96.1 (V)	LF$_2$/NH$_3$	400 (V)	120	183	1.89/0.98	1969–1977	Upper-stage experimental engine; program abandoned
RD-701	2					3,800	5.0/5.0	1988	Tripropellant experimental engine, not yet qualified
Mode-1		3920 (SL)	LOX/LH$_2$/kerosene	415 (SL)	300				
Mode-2		1590 (V)	LOX/LH$_2$	460 (V)	122				
RD-704	1					2,000	5.0/2.3	1988	Tripropellant experimental engine, not yet qualified
Mode-1		2040 (SL)	LOX/LH$_2$/kerosene	415 (SL)	300				
Mode-2		810 (V)	LOX/LH$_2$	461 (V)	122				

Explanation of Abbreviations

SL = sea level condition, V = vacuum conditions, SRBM = short-range ballistic missile, MRBM = medium-range ballistic missile, ICBM = intercontinental ballistic missile, NTO = nitrogen tetroxide, UDMH = unsymmetrical dimethyl hydrazine.

Engine Configuration or Application

RD-107 and RD-108 are also used in the Vostok, Voshkod, Molniya and Soyuz SLVs;
RD-216 consists of two RD-215 engines, each with two TCs and single TP, using a GG engine cycle;
RD-218 consists of three RD-217 engines, each with two TCs and one TP. RD-217 is an improved version of RD-215.
RD-251 consists of three RD-250 engines, each with two TCs and one TP. RD-251 is also used in Tsyklon SLV.
RD-264 consists of four RD-263 engines, each with one TC and one TP.
RD-268 was derived from the RD-263.
Data from Refs. 1–5 and J. Morehart, The Aerospace Corporation.

antiaircraft missile, Schmetterling, Taifun, etc.). Between 1946 and 1950 about 24 German V-2 engineers were in residence at Khimki and were paid a salary to prepare new LPRE designs, give advice, and assist with the V-2 engine effort in the USSR.[5,7] The V-2 LPRE at 25.4 tons or 56,000 lbf was an order of magnitude larger and more advanced than anything the Soviets had developed on their own before 1945. The Soviets produced copies of the German V-2, and Glushko was in charge of the V-2 LPRE production and testing. Other Russian organizations made copies of the German Wasserfall antiaircraft missile and flew them.

Energomash originally worked on a slightly modified German V-2 engine, which they called RD-100.[1,3-5,9] They started with the same propellants used by the V-2, namely, LOX/75% ethyl alcohol. This engine was put into production for the Soviet R-1 short-range ballistic missile and was produced at the Khimki plant. The Soviet version was not an exact copy of the German hardware.

The next two engines, the RD-101 and the RD-103, were uprated and improved versions of the V-2 engine.[1,3-5,9] They had higher thrusts, higher chamber pressures, and more performance than the copy of the V-2 engine (RD-100) as shown in the table. The fuel was changed from 75% ethyl alcohol to 92.5% ethyl alcohol, and this improved the performance. The liquid catalyst (permanganate solution) for the decomposition of the hydrogen peroxide gas-generator propellant was replaced by a solid catalyst, which was mostly made of silver, and this simplified the engine feed system. A film-cooling ring slot, which can be seen in Fig. 8.4-3, was placed into every one of the 18 injection heads of the TC to help with the increased heat transfer caused by the higher chamber pressures.[7] The control scheme was simplified and modernized. These two engines were placed into production and used in the short-range ballistic missiles R-2 and R-5, which were deployed by the military forces.

Thereafter they used kerosene because it was available and gave higher performance.[5-7] Before making this decision, Glushko operated a slightly modified version of the RD-103, called RD-103K, where kerosene was substituted for the alcohol fuel. As discussed later, these tests gave mixed results. Thereafter they developed several engines specifically designed for LOX/kerosene; they were at higher thrust and represented a significant improvement over the V-2. Instead of the 18 separate injection domes of the V-2, the next Glushko engine (RD-110) had 18 flat injector plates instead of the domes, a curved injector face, and a smaller combustion chamber. This engine is discussed later in this chapter. Thereafter they developed a single dished multiple plate injector and a more slender chamber.

Experimental Thrust-Chamber Investigations

Between 1946 and 1951 this DB designed and tested at least two sizes of experimental TCs to explore a variety of LPRE parameters. The small TCs, identified as the KS-50 series (with 100 to 50 kg thrust or 220 to 110 lbf), were used to determine performance and heat transfer of different propellant combinations and to investigate a variety of different injector elements. One version is shown in

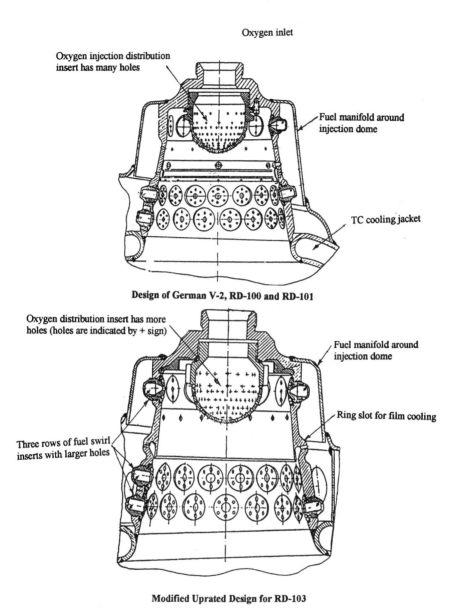

Fig. 8.4-3 Comparison of the injection heads of the RD-101, which is very similar to the German V-2, and the uprated RD-103, which has more flow, more injection holes, a higher chamber pressure, and a new film-cooling injection slot. Courtesy NPO Energomash; copied with permission of the British Interplanetary Society from Ref. 7.

Fig. 8.4-4. There must have been more than 10 different versions of this small TC. The cooling jacket of the nozzle was separate from the cooling jacket of the combustion chamber, which was typical of earlier ORM TCs described in Chapter 8.1. The cooling fluid was usually water, and the two separate cooling jackets allowed changing the cooling velocity of one without changing the other. The inner wall was copper in some versions and bronze alloys in others. Some versions had wires in the cooling passages, and some used milled channels. Some had straight channels in the cooling jacket, and some had spiral passages, which can often absorb more heat. The most common fuel was kerosene, but they also ran with other propellants, such as fluorine-type oxidizers with various fuels and beryllium hydride suspensions in fuels. In 1949 and 1950 a series of 194 tests were made with 45 different single injection elements. Many design variations were explored. It has a single central coaxial bipropellant injector element. It has a spiral insert in the central hole and tangential injection in the

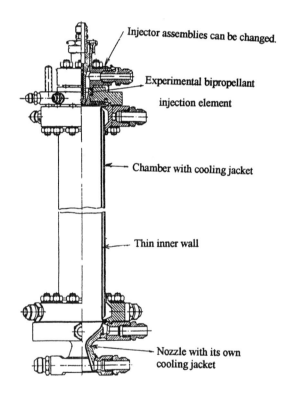

Fig. 8.4-4 One version of the KS-50 experimental TC used for evaluating 45 different injection elements and several propellants. Courtesy NPO Energomash; copied with permission of the British Interplanetary Society from Ref. 7.

outer annulus, thus giving a swirl to both propellants, which emerged as conical sheets of propellants that readily formed small droplets. These types of injector elements gave good performance.

The larger experimental set of TCs was identified as ED-140, and they had a 240 mm diam and a nominal thrust of 7 tons or 15,400 lbf (Refs. 1, 5, and 7). One version of this TC is shown in Fig. 8.4-5. It had a flat injector face with an array of multiple injection elements of a single-propellant spray type. It also had a curved nozzle-exit contour, which was one of the earliest Soviet applications of a bell-shaped nozzle with a reasonably high thrust level in the Soviet Union. It was relatively easy to substitute different chamber segments, alternate cooling

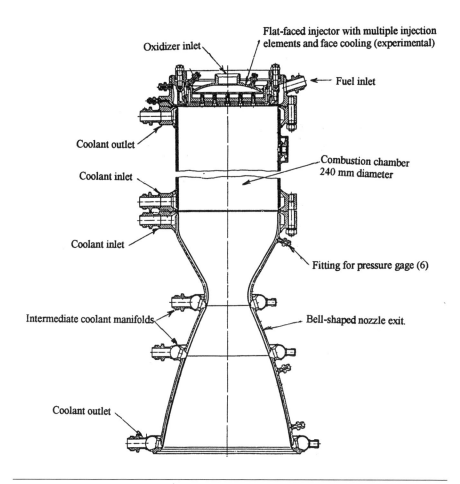

Fig. 8.4-5 One type of the ED-140 experimental TC used to investigate effects of changing chamber pressure and/or mixture ratio on heat transfer and to evaluate cooling-jacket designs. Courtesy NPO Energomash; copied with permission of the British Interplanetary Society from Ref. 7.

jackets, and different nozzle segments. The ED-140 was used to investigate the influence of chamber pressure and mixture ratio on performance or heat transfer and explore different hypergolic start liquids, different injection elements and patterns, different cooling jacket designs, and some different film-cooling configurations.[7,10] It also confirmed Glushko's conviction that a cylindrical chamber and a flat-face injector were appropriate. The development of the cooling-jacket design and construction, which was later adopted for the new larger Energomash engines, was materially aided by these tests. This jacket design is innovative because it used a thin inner high-conductivity wall that substantially reduced thermal stresses in that wall. The Germans had suggested using Cuprodur for the inner-wall material; it is a bronze copper alloy with 2% chromium. It had reasonably good strength at high temperature and a relatively high thermal conductivity and had been tried successfully by the Germans. It took a while for the Soviets to duplicate this material because minor impurities cause significant changes in physical properties. Finally a good bronze material was produced in sheet metal form, manufactured into a test article and tested in the ED-140 TC.

Brazing and Welding of Cooling Jackets

Concurrently with these experimental TCs, several techniques were explored for brazing together the inner wall, the intermediate channel walls or ribs, and the outer TC or cooling-jacket wall.[1,2,10] Several cooling jacket designs schemes can be seen in Fig. 4.3-4. The surfaces to be joined are coated with the brazing material. The key to obtaining a good brazed joint between these three flexible sheet components or between the top of the ribs (on the milled channels) and the outer wall lay in the method of supporting these parts in the brazing furnace. A good joint will form if all of the surfaces to be brazed are either in contact with each other or are only a few thousandths of an inch apart. The surface tension in the liquified brazing material allows filling small gaps (between components, which are to be joined) with braze material. One process used by the Soviets was pressure brazing. The flexible sheet metal components are assembled on a brazing fixture, put in a vacuum furnace, and are pressure brazed, that is, the metal sheets are pressed together by air-inflated bags, which are pushing on the metal assembly. A thin metal inflatable bag can push against the inner wall, or alternatively one or more contoured sleeve-type bags can push against the outer wall or both the inside and outside bags push together against the cooling-jacket assembly in the furnace. Different brazing materials were also investigated. In addition to the brazing, there has to be also some welding, such as welding the inner and outer wall of the nozzle to those of the chamber or the welding together of the injector plates or the welding of the injector to the chamber. Different welding and brazing sequences and different brazing materials have been investigated. Some of the TCs for the next several Soviet LPREs have been considered to be a scale-up of the ED-140 TC. This type of lightweight cooling jacket was used in the majority of large Soviet TCs thereafter.

Engines with Liquid Oxygen

The design bureau decided to replace the alcohol with kerosene as the fuel because it offered 10 to 20 s more specific impulse. The initial experience with the first few LPREs using LOX/kerosene was not encouraging. The RD-103 LOX/alcohol LPRE was modified and operated with kerosene, but the results were not decisive. Nevertheless, Glushko persisted. One of the first large engines designed to use kerosene as a fuel instead of alcohol was the RD-110. Its thrust (140,000 kg or 308,000 lbf vac) was the highest of its day, and its development started in 1947.[5,7] It was originally intended to propel the booster stage of the R-3 missile. The chamber was relatively very small and of a spherical shape as recommended by the resident Germans. The top of the chamber had 19 flat plate injector pieces as shown in Fig. 8.4-6. The individual plate-shaped injectors had been developed on the ED-140 experimental engine mentioned earlier. The RD 110 engine was assembled, but not tested at full operating conditions because its cooling with a double-walled cooling jacket was marginal, and it did not fly. The cooling jacket used thick inner and outer walls, similar to the German V-2 designs. At the higher chamber pressure and higher combustion temperature of LOX/kerosene, this cooling jacket was marginal and did occasionally fail. The engine was not expected to be fully reliable. This was one of the reasons why it was not tested at full power. Korolev himself proposed canceling the development program of the R-3 and with it the RD-110 engine program.[2,7,11]

The RD-105 booster engine (61 tons or about 134,000 lbf thrust vac) and the RD-106 sustainer or second-stage engine (slightly less thrust, but higher area ratio) also used LOX/kerosene and were intended for the R-6 ballistic missile[5,11] For thrust vector control a set of four graphite jet vanes was developed, as used satisfactorily on prior large Soviet engines. The development of these engines was started in 1952 and had a number of problems. The most difficult issue was the destructive high frequency gas vibration in the combustion chamber, which caused the engine to fail immediately after reaching full thrust level. At this early time the understanding of combustion instability was not adequate to fully cure this problem. It was thought that the combustion was incomplete, and it was decided to lengthen the chamber, so that there would be more volume and time for the combustion reaction to be more complete. This helped, but occasionally combustion vibrations and failure did still occur. Furthermore the long duration of 250 s caused serious erosion of the jet vanes, so that they would not be effective during the last part of the engine operation. In 1954 the payload of the missile was increased, and the engines did then not have sufficient thrust for takeoff. The programs for the missile and these two engines were canceled. By that time Glushko and his engineers had not yet solved all of the key engine problems. These engines (RD-105, RD-106, and the RD-110) had their programs canceled and are not included in Table 8.4-1.

The first major, truly new, and successful rocket engines, which were put into mass production, were the RD-107 and the related RD-108 engines. The RD-107 is shown in Fig. 8.4-7 and the flow diagram for either engine in

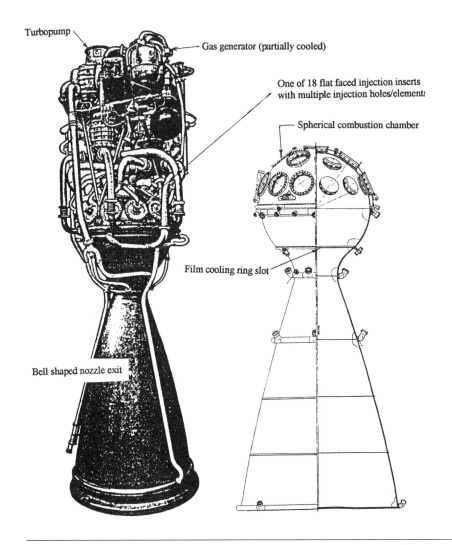

Fig. 8.4-6 View of RD-110 engine and a simplified cross section of its TC, not drawn to same scale. Courtesy of NPO Energomash; copied with permission of the British Interplanetary Society from Ref. 7.

Fig. 8.4-8 (Refs. 2,5,9,12,13).* The external appearance and the flow diagram for the RD-108 are essentially the same. These engines had some new technology, are truly historic, had the highest thrust of any flying LPRE at that time, and deserve a more detailed discussion. The experience with the ED-140 and the unfinished engine developments just listed provided a background to the TC designs. The RD-107 has four fixed (not movable) main TCs and two

---

*Personal communications, J. Morehart, The Aerospace Corporation, M. Coleman, Chemical Propulsion Information Agency, and A. Siddiqi, author.

Fig. 8.4-7 RD-107 and the RD-108 (very similar but not shown here) LPREs have propelled more large missiles and SLVs than any other LPRE in history. Courtesy NPO Energomash; copied from Ref. 17.

smaller hinged vernier TCs supplied by the same feed system, one turbopump (positioned horizontally) and one monopropellant hydrogen-peroxide gas generator with a solid catalyst. The RD-108 was very similar, was developed in parallel, had four vernier thrusters, and had a somewhat lower chamber pressure and slightly lower thrust in order to allow a longer burning duration. Propellants were liquid oxygen and kerosene.

Four RD-107 engines, one in each of four strap-on boosters, and one RD-108 in the center core of the flight vehicle have been originally used on the R-7 ICBM

RD-107 SCHEMATIC

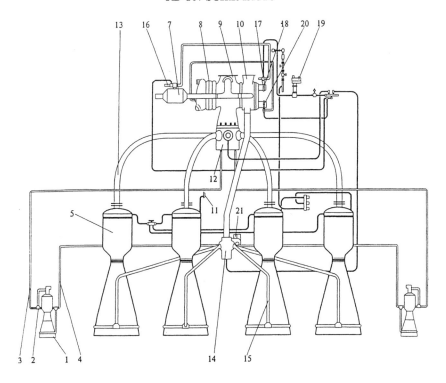

Fig. 8.4-8 Simplified flow sheet of the RD-107 or the RD-108 LPRE: 1, one of two vernier TCs; 2, rotary joint for hinge of vernier TC; 3, 4, LOX and kerosene supply lines to vernier TCs; 5, one of four main TCs; 7, gas generator (H_2O_2) with solid bed catalyst; 8, turbine; 9, LOX pump; 10, fuel pump; 11, pressure sensor of system for regulating engine thrust; 12, main LOX valve; 13, one of the main LOX distribution pipes; 14, main fuel valve; 15, one of four high-pressure fuel pipes; 16, explosive cutoff valve for hydrogen peroxide; 17, pressure reducer valve; 18, hydrogen-peroxide pump; 19, electrically driven gas pressure reducing valve; 20, liquid-nitrogen pump; 21, electrically driven fuel flow throttle subsystem. Courtesy of NPO Energomash; copied from Refs. 12, 16, and 17.

and later on the Sputnik, Luna, Vostok, Molniya, Voshkod, and Soyuz SLV families.[1,2,5,9,12] The R-7 ballistic missile was deployed in the military forces of the Soviet Union, and some of the derivative SLVs have had a long life and are still used today. Engine development started in 1953/1954. The first test with a single TC happened in 1955 and the test with four TCs in 1956. Combustion instability was encountered during development, but it was remedied, mostly by injector modifications. The first successful spaceflight was in October 1957 with the launch of Sputnik, the first satellite to reach orbit. It is probably the world's longest living rocket-engine project (50 years) and the world's most utilized satellite booster rocket engine. At the end of 2001, over 1630 vehicles had been launched with the production models at a success rate of 97.5%. There were at least eight versions of the RD-107 and nine versions of the RD-108. In 2002 the RD-107 and the RD-108 were still being modified to fit the latest version of the Soyuz SLV.

The four-TC configuration was selected for this engine (and also for other Soviet engines) because it resulted in a shorter engine (and thus saved some vehicle length and vehicle inert mass) and it was easier to build (smaller, simpler TC manufacturing equipment). However it was more complex (more piping and start fluid distribution) and had a larger vehicle diameter, which translated into approximately 40% more vehicle drag than an equivalent LPRE with a single TC. Also the designers knew from experience that combustion vibration in a larger diameter single TC would be more difficult to remedy than in a smaller one.

The compact turbopump (TP) of Fig. 8.4-9 has two in-line shafts with a coupling, a concept that came from the German V-2 TP.[12,*] The merits and heritage of the two in-line shaft design are mentioned in Chapter 4.4. The TP has a steel turbine with two rows of blades (with an exhaust heat exchanger at the turbine discharge end), a shrouded double-inlet oxygen impeller, a single-sided fuel impeller, a gear case, to which small pumps for the hydrogen peroxide and liquid nitrogen are attached. Each of the two in-line shafts rides on two ball bearings. The coupling is a hollow floating cylindrical sleeve with internal splines; this type of coupling is smaller and lighter that the coupling used on the German V-2 TP. The hydrogen peroxide is decomposed by a solid catalyst in a gas generator, and the resulting hot steam, containing oxygen gas, drives the turbine. The cold liquid nitrogen is gasified (by flowing through a heat exchanger located in the turbine exhaust gas assembly) and then used for pressurizing the main propellant tanks.

The four TCs are regeneratively cooled by kerosene. As already mentioned, they allow a shorter engine than an equivalent single TC. Each TC has two different cooling-jacket designs.[10,12,*] At the nozzle throat region, where the heat transfer is the highest, it has an inner wall with integral vertical passage channels. In the United States it is called milled slots. It is brazed to the outer stainless-steel wall. The geometry of these milled channels is such that the cooling velocity is highest in the throat region and upstream of the throat. This milled slot design has been used also in Britain, the United States, France, or Japan. In less critical regions (chamber or nozzle exit) the cooling channels are made of corrugated sheet metal between the inner and the outer wall. Figure 4.3-4 gives examples of some of several cooling-jacket cross sections, including the milled channel type and the corrugated intermediate sheet design. In the high heat-transfer regions the inner wall is made of a bronze-chromium alloy of relatively good conductivity and high-temperature strength. The outer walls are made of high-strength, nonmagnetic stainless-steel sheet. All of these precision-formed sheets are coated with brazing material at those surfaces that

---

*Personal communications, J. Morehart, The Aerospace Corporation.

602 History of Liquid Propellant Rocket Engines

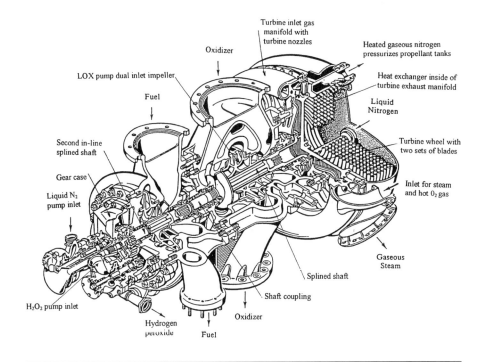

Fig. 8.4-9 Turbopump of the RD-107 LPRE with two in-line shafts. Courtesy of NPO Energomash, (from personal communications, J. Morehart, Aerospace Corporation, M. Coleman, Chemical Propulsion Information Agency, and A. Siddiqi).

need to be bonded and are then assembled in a fixture and brazed together inside a special vacuum furnace. The braze material is manganese based. This combination welding and brazing technique for building good TCs is described in Chapter 4.3, and the early experimental work was tested in experimental TCs, such as the experimental ED-140 mentioned earlier in this chapter. A similar lightweight cooling-jacket design and injector design are used in most other large Soviet TCs. This corrugated middle sheet design is a contribution to the state of the art of LPREs and is used only in the Soviet Union.

The vernier TCs or steering TCs have about 38 kN thrust each in the RD-107, and they are hinge mounted to allow full attitude control of the vehicle during engine operation.[2,6,*] The specific impulse from a vernier TC has been given as 248 s at sea level. The propellants are supplied from the same single TPs of this LPRE. The development of these vernier TCs for the RD-107/108 was supposed to have been done at OKB-456 in Khimki, but Glushko declined to accept this job because he did not want to spend efforts on low thrust engines and because his DB was then busy with problems on the RD-105 and

---

*Personal communications, J. Morehart, The Aerospace Corporation.

RD-110. Korolev, the designer of the vehicle for these engines, then assigned the development work of the vernier TC to his LPRE section in his own design bureau. Here the S1.35800 thrust chamber was developed to fit this vernier need. The TC work was transferred from Korolev's bureau to the CADB. When the requirements changed (switch from an ICBM to a SLV), the vernier engine was no longer suitable. Glushko did not like the design, and decided to develop the higher-performance vernier TC. Glushko then had the development of the complete engine assembly in his organization. The technology and the S1.35800 hardware were transferred to CADB (Chapter 8.5), where it was uprated and eventually became the RD-0105 LPRE.

The injectors of the RD-107 and the RD-108, seen in Fig. 8.4-10, feature an outer welded dome-shaped metal cover (on top of the oxidizer distribution cavity), a middle bulkhead plate (which separates the fuel and oxidizer passages), and a flat inner wall made out of a more conductive material. Many short steel tube assemblies (or bipropellant injection elements) at right angles to the injector face are brazed to the middle and inner walls. Spiral inserts into the center tube and tangentially drilled holes allow for tangential flows of the oxygen and the fuel inside each injection element. The propellants then flow in spirals and are "whirled." At the exit of each injection element the propellant flows form two conical propellant sprays. There are 337 such tubes or injection spray points, arranged in 10 rings. The outer rings spray only fuel on the inner

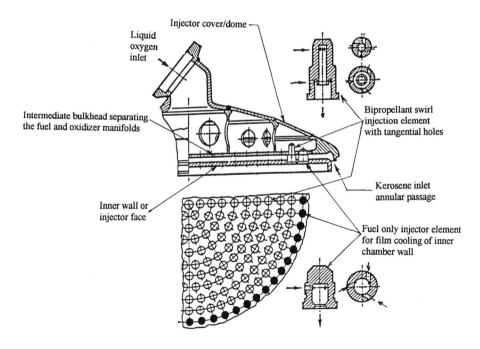

Fig. 8.4-10 Sections of the injector of the RD-107 and the two types of tangential swirling injection elements or spray inserts. Courtesy of NPO Energomash.

chamber wall (film cooling) and keep the wall temperature below 380°C or 716°F. Combustion instability was encountered and was eventually remedied by design changes in the injector during the development of this LPRE.

Figure 8.4-11 shows a view of the aft end of the Soyuz SLV. There are four strap-on boosters, each equipped with an RD-107 engine and a center core RD-108 engine. The strap-on boosters are dropped off in flight, while the center core of the vehicle continues to be accelerated by its rocket engine. Altogether there are 20 large TCs and 12 vernier TCs for a total of 32. All are started at the same time using hypergolic ignition start fluid. As already mentioned, there are as many as 17 different versions of these two engines, and

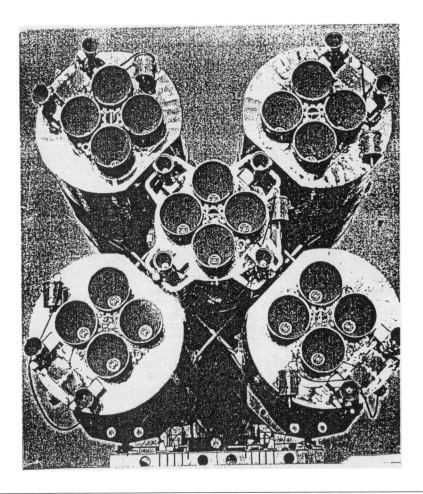

Fig. 8.4-11 View of the aft end of the Soyuz SLV shows four strap-on boosters each powered by a RD-107 LPRE with two hinge-mounted vernier TCs and a vehicle core stage powered by a RD-108 LPRE surrounded by four hinge-mounted vernier TCs. Courtesy of NPO Energomash.

some of them have slightly different performance numbers than what is stated in Table 8.4-1.

The RD-111 LPRE was a new, improved LPRE with four TCs and a single large TP running on LOX/kerosene. It was used for the booster stage of the intercontinental ballistic missile R-9A.[5,6,12,13] Engine development started around 1958/1959. It resembled the RD-107, but had about 75% more thrust. It was the first Glushko engine with four hinge-mounted TCs, and therefore there no longer was a need for separate hinged vernier TCs. Figure 8.4-12 shows the TP of the RD-111. There are two in-line TP shafts, each with two ball bearings, coupled together with a floating spline shaft. The centrifugal pumps have dual inlet impellers to balance axial hydraulic pressure loads in the rotating assembly and also to reduce the impeller inlet velocity (which gives an extra margin for cavitation). Sophisticated inducer impellers were also used. The RD-111 used a bipropellant fuel-rich gas generator, and this eliminated the third propellant (hydrogen peroxide) used in earlier engines. The fuel-rich mixture is unusual because the majority of Soviet gas generators and preburners have an oxidizer-rich mixture ratio. This engine experienced combustion

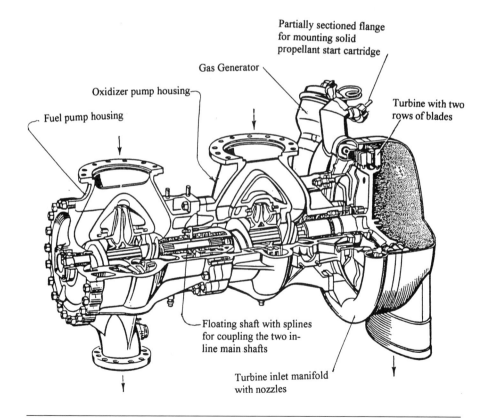

Fig. 8.4-12 TP for the RD-111 LPRE has two in-line shafts, four bearings, and a coupling. Courtesy of NPO Energomash from Ref. 14.

instabilities during development, which caused a delay in the program. The first flight was a failure; the engine disintegrated because of high-frequency gas vibrations, and the propellants mixed and exploded. The RD-111 entered military service in 1965. It was the last LPRE that Glushko's LPRE organization developed for a missile or SLV developed by Korolev in his KB Energiya. For a brief period of time, Korolev was so concerned about the engine problems that he tried unsuccessfully to eliminate Glushko's RD-111 engine from the program.[2] This caused a rift between these two key personalities in the Soviet aerospace business.

The RD-119 LPRE from Glushko's design bureau was originally intended for the upper stage of the R-7 ballistic missile, but was replaced by the RD-0105 engine (developed by CADB), which was available earlier. This decision, made by Korolev, contributed to a worsening of the relationship with Glushko. The RD-119 was then designated for second stage for the Kosmos SLV and developed between 1958 and 1962. Propellants were LOX/UDMH because UDMH gives a somewhat higher specific impulse (vac 352 s) than kerosene. The thrust with a large nozzle-exit-area ratio of 116 was 105 kN (23,600 lbf) in a vacuum. The single fixed (nongimbaled) TC was built mostly of titanium. As seen in Figs. 8.4-13 and 8.4-14, the attitude control of the second stage was obtained by routing and pulsing the turbine exhaust gases to four large pitch and yaw nozzles and four small vernier roll nozzles. The flow of hot gas to each pair of these nozzles is controlled by one valve with three positions: flow to the right nozzle, flow to the left nozzle, or equal flow to both nozzles. The turbopump, which is shown in Fig. 4.4-5, is very long and has two separate in-line shafts connected with a spline coupling. The unique GG is the only one known to use thermal decomposition of UDMM. Engine starting and stopping was unique and accomplished with fast acting explosive valves. The RD-119 was retired in 1977.

Glushko had experienced destructive combustion instability problems during the development of most of his LOX/kerosene engines, and he had a strong prejudice against developing more such LPREs with this propellant combination. When Glushko was asked by Korolev to develop four new LOX/kerosene engines for his gigantic lunar flight N-1 vehicle, Glushko refused to accept the job, in part because of difficult combustion stability problems with this propellant.[2,5] Korolev wanted LOX/kerosene partly because it gave slightly better performance, and he went instead to a less experienced LPRE design bureau (Kuznetsov) for developing these four engines. Energomash did not start the development of another LOX/kerosene engine for about 15 years. By that time combustion instabilities were better understood, and more remedies for preventing this destructive high-frequency vibration were then known. So Glushko changed his mind and developed two outstanding new engines, and they are described next.

The RD-120 second-stage engine (shown in Fig. 8.4-15) has high performance and uses liquid oxygen and kerosene, a staged combustion cycle, and ignition by the injection of an initial small quantity of hypergolic fuel. In the second stage of the Zenit launch vehicle, this particular engine is mounted on

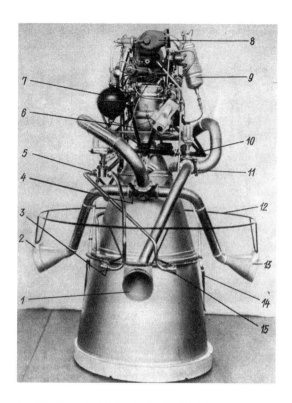

Fig. 8.4-13 View of the RD-119 LPRE for the second stage of the Kosmos SLV: 1, vernier nozzle for pitch in front of main nozzle; 2, 13, nozzles for yaw control; 3,15, small roll control nozzles; 4,5,11, electrically driven gas distribution valves; 6, combustion chamber; 7, high-pressure gas tank; 8, turbopump; 9, gas generator; 10, mounting structure; and 12, assembly tool to support nozzles. Courtesy of NPO Energomash from Ref. 1.

a fixed thrust structure in the middle of a toriodal tank. Its fixed single TC has a high nozzle area ratio of 106:1, and the engine uses an oxygen-rich preburner and a helium system for valve actuation and purge. This engine had two small booster pumps allowing low tank pressure. The Zenit SLV second stage needed four vernier TCs (called RD-8) for flight-path control, and they were developed by the Yuzhnoye Design Bureau, discussed in Chapter 8.10. The first flight of the RD-120 was in 1985. NPO Energomash also developed a version called the RD-120K (with a shorter nozzle) for the booster stage of an undefined small SLV. It had a smaller nozzle area ratio, higher chamber pressure, and a little more thrust. It comes with either a gimbal suspension or with vernier TCs. It has not yet flown. The RD-120 was one of the engines that has been licensed to Pratt & Whitney, and this company tested and demonstrated the engine at its Florida facility as mentioned in Chapter 7.10. However no suitable application has been found in the United States.

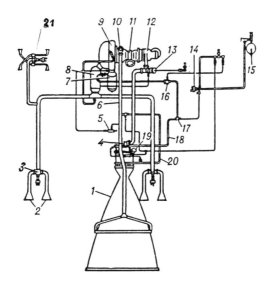

Fig. 8.4-14 Simplified flow diagram of the RD-119 LPRE: 1, combustion chamber and nozzle; 2, turbine exhaust nozzles for pitch and yaw; 3, electrically driven gas distribution valve; 4, cutoff valve—explosive actuated, 5, GG mixing valve (thrust control); 6, high-pressure fuel pipe; 7, gas generator; 8, heat exchanger for vaporizing oxygen; 9, two-stage turbine; 10, 16, cutoff valves—explosive actuated; 11, fuel pump; 12, LOX pump; 13, LOX valve—explosive actuated; 14, electrically driven nitrogen control valve; 15, compressed nitrogen tank; 17, GG oxidizer pressure control valve; 18, oxidizer pipe for GG; 19, TC fuel cutoff valve; 20, film-cooling pipe; and 21, roll control nozzles with three-way control valve. Courtesy of NPO Energomash.

The RD-170 is a historically significant engine with the highest thrust in the world (7904 kN or 1.77 million lbf in vacuum and 7256 kN or 1.63 million lbf at sea level) at a very high chamber pressure of 250 atm or 3674 psia. Much has been written about this LPRE.[1-6,9,12,13,15-17] It used LOX/kerosene propellants, which Energomash had not promoted for about 15 years. It has four gimbaled TCs, a single large 14,500 rpm TP of about 250,000 hp, two booster turbopumps, and uses a staged combustion cycle with two preburners, one on each side of the turbine. The preburners operate oxidizer rich and burn all of the oxygen and a portion of the fuel. The RD-170 is shown in Fig. 8.4-16, and a simplified flow sheet is in Fig. 8.4-17. The booster TPs allow a lower propellant tank pressure (saves tank weight), a higher shaft speed of the main TP (smaller, lighter TP assembly), and they provide a margin against cavitation in the main pump impellers. The turbine of the fuel booster pump is driven by liquid fuel tapped off the discharge of the main fuel pump, and the turbine of the oxidizer booster pump is driven by oxidizer-rich preburner gas, which is tapped off the turbine exhaust manifold and cooled with helium in a heat exchanger. This oxidizer-rich gas flow to the oxidizer booster pump is con-

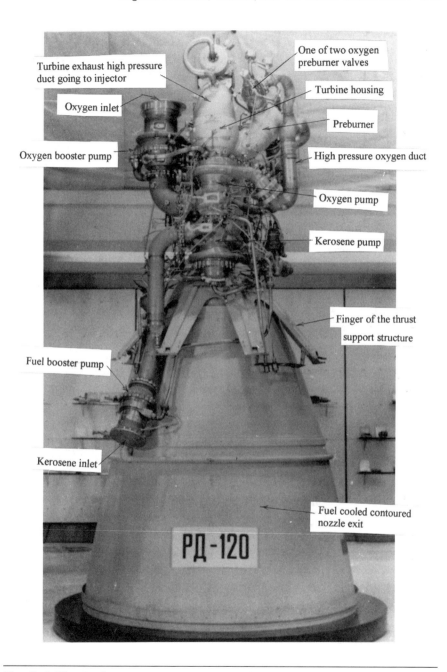

Fig. 8.4-15 LOX/kerosene RD-120 engine drives the upper-stage engine for the Zenit SLV. Courtesy of NPO Energomash from Ref. 1.

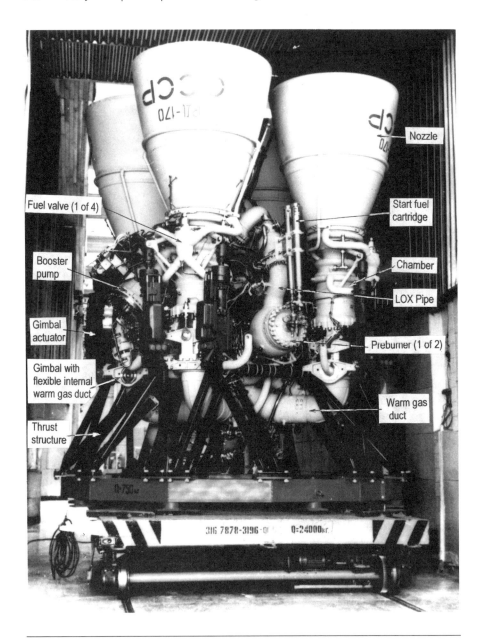

Fig. 8.4-16 RD-170 LPRE has an altitude thrust of 1.77 million pounds, the highest in the world. One version is still in production. The engine is upside down on a factory cart. Courtesy of NPO Energomash (from personal communications, NPO Energomash, 2001–2002).

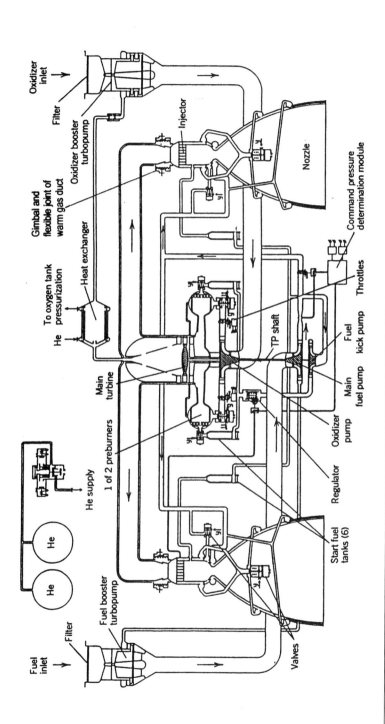

Fig. 8.4-17 Simplified flow sheet of the RD-170 LPRE. To simplify the diagram, only two of the four main TCs are shown, and the vernier TCs are omitted. Courtesy of NPO Energomash (adapted from personal communications, NPO Energomash, 2001–2002; also available in Sutton, George P., *Rocket Propulsion Elements*, 7th ed., John Wiley & Sons, Inc., Hoboken, NJ, 2001, p. 395).

densed or dissolved in the liquid oxygen, which then flows to the main oxygen pump. The RD-170 and the RD-120 are the first large Soviet LPREs with LOX/kerosene to have booster pumps. The helium subsystem, which is only partly shown in the schematic flow diagram, supplies inert gas to actuators and control valves and purges. It is indicated by the symbol y in four places of the figure. Ignition is by injecting hypergolic fuel slugs from separate containers into the two preburners, vernier thrusters, and the four TCs.

Figure 8.4-18 shows a thrust chamber of a type used in the RD-170 and its derivative engines. It is one of the more advanced LOX/kerosene TCs that has ever been developed. It is regeneratively cooled with fuel, uses supplementary film cooling through three slots in the combustion chamber, and its assembly is done both by brazing and welding of the parts.[1,5,18] As seen in Fig. 8.4-18, the kerosene fuel at ambient temperature is fed into the cooling jacket just upstream of the nozzle, where the heat transfer is the highest. A portion of the fuel cools the nozzle-exit segment. The heated fuel is then transferred by external pipes to the bottom of the chamber cooling jacket, where it absorbs more heat; from there it flows directly into the injector. The intensive heat is absorbed without the formation of detrimental carbon deposits, which would impede heat transfer and cause excessive wall temperatures. The cooling channels, as seen in cross section in Fig. 8.4-19, are essentially rectangular and straight, except in the converging nozzle section, where the rectangular channels are arranged in a helical flow pattern. Figure 8.4-19, shows the 271 injection elements, which are of two basic types as described in Fig. 8.4-20. All of the injection elements are brazed into injector face plate and the intermediate plate. One type of injection element protrudes beyond the injector face. A look at Fig. 8.4-19 reveals that 54 of these protruding elements (shown by simple small circles) form antivibration baffles, namely, a circular central baffle and six nearly radial baffles. The baffles are not in a continuous wall, but have gaps between protruding injection elements. These gaps cause extra damping to an oscillating cross flow. The other injection elements shown in Fig. 8.4-20 each end at a countersunk recess of the injector inner face plate. There are four slightly different detail designs with four different mixture ratios. The amount of black color in the circles of Fig. 8.4-19 (one-quarter, one-half, or three-quarter black) designates different mixture ratios; the more black area, the more oxidizer rich is the mixture ratio of that element. The oxidizer-rich gas flows through the injection elements in an axial direction, but the fuel is injected through tangential holes, forming a vortex in the annular space of the injection element and emerging from bottom of the element in a conical spray. The distribution of the different types of injection elements on the injector's face is not random, but generally follows spiral-type patterns, as can be seen on Fig. 8.4-19. The variations of the four groups of injection elements in their flow rates, their mixture ratios, and their placement on the injector face are needed to avoid high-frequency combustion instability over the operating range.[18] The inner-wall material is a copper alloy with relatively high thermal conductivity, and the outer walls, manifolds, and pipes are stainless steel. The tops of the ribs of the inner chamber and nozzle walls are brazed to the outer

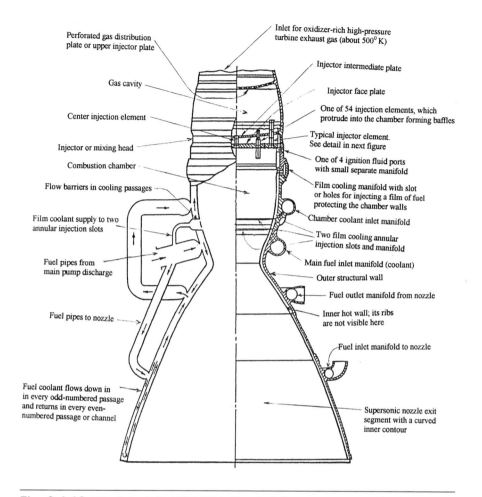

Fig. 8.4-18 TC of the RD-170 LPRE. The fuel flow circuit is on the left and a cross section on the right. Adapted from NPO Energomash patent drawing from Ref. 18.

walls in a special vacuum furnace using special tooling. The RD-170 has a relatively high nozzle expansion area ratio for a booster engine, namely, 36.8. It is just small enough so that it will not cause an undesirable nozzle flow separation when launched at sea level. As with other Soviet engines, the nozzle-exit section is bell shaped.

The RD-170 has a smooth start transition to full thrust in over 3 s, using tank pressure feed only. Slugs of hypergolic fuel (aluminum triethyl enclosed in long cylindrical containers) are injected into the main and vernier thrust chambers and also into the two preburners. The pressurized fuel expels the hypergolic fuel into the combustion chambers, where they react with oxygen. After a fully automatic check of all of the TCs, the main turbopump and the two booster turbopumps are brought to full speed. At shutoff the thrust

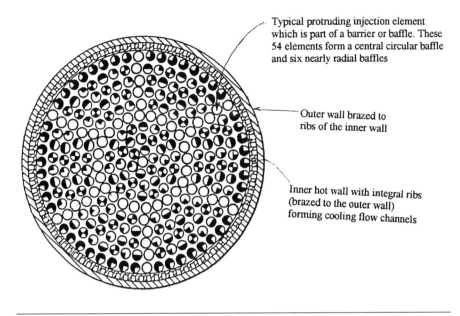

Fig. 8.4-19 Front view of the injector face and section of the cooling jacket of the TC shown in Fig. 8.4-18. Adapted from NPO Energomash patent drawing from Ref. 18.

drops first to 70% for a few seconds, then to 50% for the next 2 s, where it can be held for up to 10 s and is then shutoff in 0.5 s. This gradual shutdown avoids engine damage, which can occur if the coolant flow is stopped too quickly.

Although engine development started in 1976, the first satisfactory full thrust engine ground tests took place in 1982. Initial tests showed combustion instability and other problems.[2-5,9,16,*] A special group of experts was convened to analyze the problems, the program was delayed, but successful ground tests were eventually made in 1984. The first flight was in 1985. This RD-170 LPRE was used as a booster on the Energiya SLV. A slightly modified version is being used for the booster stage of the Zenit SLV, as described in the next paragraph, and also the U.S. Sea-Launch program. In the serial production every one of these RD-170 engines was statically fired prior to delivery. However the individual TCs and some other components were not tested individually.

The RD-171 is very similar to the RD-170, and it was tailored to fit the Zenit 2 SLV and the U.S. Sea-Launch applications (ship-based launch platform), which was called Zenit 3 in Russia.[6,14,16] The gimbals in the RD-170 were replaced by hinges in the RD-171. The first ground test of this engine in a Zenit 2 booster stage in June of 1982 resulted in a spectacular unfortunate explosion, which was diagnosed as a probable turbine failure. The test was run at a facility of NII Khimmash (not connected to the KB Khimmash Design

---

*Personal communications, J. Morehart, The Aerospace Corporation.

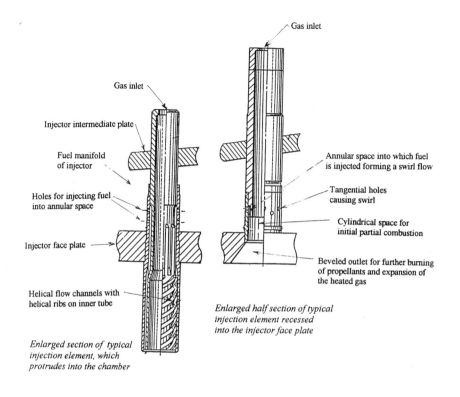

Fig. 8.4-20 Two typical injection elements of Fig. 8.4-19. Adapted from Energomash patent drawing from Ref. 18.

Bureau), and it destroyed the engine and the vehicle stage structure and caused severe damage to the test stand. The remedy was to strengthen the turbopump so as to reduce turbine vibrations and to install a filter to catch small particles, such as machine shavings or pieces of scale from tank or pipe walls. The impact of such small particles had been observed to cause dents and stress concentrations on the surface of turbine blades and had caused some turbine blade failures.

There are two derivative versions of the RD-170/171. There is a single TC version (RD-191), which was fully developed, but has not yet flown.[1,2,*] The two TC version (RD-180) was developed with inputs and support from Pratt & Whitney of the United States, and it has a little better performance and somewhat more thrust per TC.[1,2,19,*] It was selected for the booster of the Lockheed Martin U.S. advanced versions of the Atlas SLV. It has been manufactured and production tested at Energomash in Russia and serviced and sold through a

---

*Personal communications, J. Morehart, The Aerospace Corporation.

joint venture company formed by NPO Energomash and Pratt & Whitney. The RD-180 LPRE is shown in Fig. 7.10-11, and it was first flown with the Atlas III in May 2000 and with the Atlas V in August 2003.

In the mid-1990s Energomash also designed or developed at least four LPREs using LOX with *methane* as a fuel.[3-5] The fuel was actually liquified natural gas, which is about 98% methane. It is not clear to this author just how far this development had progressed. This propellant combination gives an engine performance and an average propellant density, which is intermediate between engines using kerosene and hydrogen as fuels. Methane is not as cold as hydrogen, and the exhaust products are also environmentally benign. These engines were intended for first and second stages, the orbit insertion maneuver, and the ACS of the Riksha space vehicle, but the engines and this vehicle have not yet flown.

Engines with Storable Propellants

Development of large LPREs with storable propellants began at Energomash in the late 1950s and proceeded concurrently with the engines using liquid oxygen.[20,21] These storable propellants were used on Soviet ballistic missiles because they allowed long time storage in the vehicle and missiles were ready to launch on very short notice. The Soviet military people did not like the operational problems created by cryogenic propellants (long launch preparations, no long term storage, ice formation, etc.). Early versions of these LPREs used nitric acid and kerosene, but later versions used nitrogen tetroxide (NTO) and UDMH as propellants.[1,3-5,12]

The RD-214 and 216 were the first large LPREs production engines with storable propellants using gas-generator engine cycles. The RD-214 engine had a monopropellant hydrogen peroxide GG with a solid catalyst, used nitric acid and kerosene propellants, and jet vanes for steering (like the V-2). It is shown in Fig. 8.4-21. The RD-214 was used as the first stage of the R-12 intermediate-range ballistic missile or IRBM and later for the first stage of the small version of the Kosmos satellite launcher. The RD-214 development started shortly after the RD-107 development was started, and it bears a resemblance to some of the features of the RD-107.

The RD-216 was larger and more advanced. It consisted of two identical RD-215 engines, each of which had two TCs and a single TP. They used a bipropellant GG and ran on RFNA (concentrated nitric acid containing 27% NTO) and UDMH as propellants. The RD-216 originally was installed in the booster stage of the R-14 ballistic missile. A modified version, identified as RD-216M, was later adapted for the Kosmos 3 SLV. The unique injector of the RD-215/216 is shown in Fig. 8.4-22. The first orbital Kosmos flight of the RD-214 was in March 1962 and for the RD-216 in August 1964. The RD-218, which was larger than the RD-216, and the RD-219 upper stage engine were developed for the first and second stages of the R-16 ICBM.[1,9,16,20,21] Figure 8.4-23 shows the RD-219 upper-stage engine, which was a very short engine, because its TP with a horizontal shaft was installed between two TCs There were additional, larger LPRES with storable propellants for ballistic missiles.

Engines in Russia, Ukraine, and the former Soviet Union

Fig. 8.4-21 RD-214 was the first Glushko engine with storable propellants, nitric acid (with 27% NTO) and kerosene: 1, thrust chamber with bell-shaped nozzle; 2, gas generator; 3, turbine; 4, oxidizer pump inlet; 5, fuel pump inlet; 6, air pressure regulator; 7, hydrogen-peroxide flow control valve; 8, one of oxidizer pipes; 9, pressure relay; 10, support structure; 11, oxidizer shutoff valve; and 12, one of fuel pipes. Copied with permission of NPO Energomash from Ref. 17.

The relationships between these engines are shown in simplified form in Table 8.4-2, which was copied with permission and adapted from Ref. 5. The table shows the design principle of combining two or three engine blocks into larger engines, and this simplifies production, assembly, servicing, and spare parts. By increasing the chamber pressure, the TCs could be smalller, and the performance would be slightly better. The first stages with six thrust chambers in the RD-251 (Fig. 8.4-24) and RD-261 had hinge mounting of these TCs. The Tsiklon and Kosmos were space launch vehicles, and the other vehicles identified in this table were all ballistic missiles.

The RD-253 is a historic LPRE, and it featured a staged combustion engine cycle and was developed between 1961 and 1965.[1,5,9,13] It was not the first Soviet application of this efficient engine cycle, but the first for Energomash, and it was with a large engine using storable propellants. One version is shown

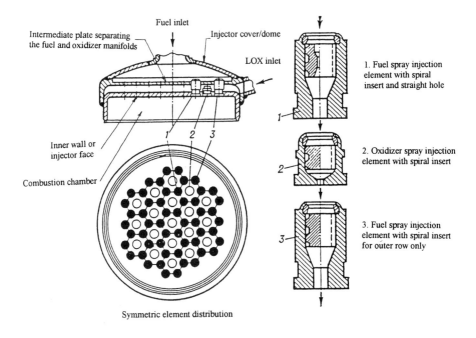

Fig. 8.4-22 RD-216/RD-215 injector, using red-fuming nitric-acid spray elements (white circles) and UDMH spray elements (black circles), has a symmetrical honeycomb pattern of injection elements. Courtesy of NPO Energomash (from personal communications, NPO Energomash, 2001–2002).

in Fig. 8.4-25, and a flow diagram is in Fig. 4.3-8. As explained elsewhere in this book, the first staged combustion cycle engine was achieved at OKB-1 in 1958, and the first one that flew (engine S-1.5400) was attributed to the Korolev Design Bureau in 1960 and is described in Chapter 8.11. As explained in Chapter 4.2, engines with this staged combustion cycle give somewhat better performance (2–9% depending on engine parameters) than those with an equivalent gas-generator cycle. The few percent of extra performance are particularly effective in ambitious or high-velocity missions.

Six RD-253 engines are used in the first stage of the heavy-lift Proton SLV. Each LPRE features an oxidizer-rich preburner and a hinge-mounted TC. The early version had the TP protrude above the TC, but later engine versions were shorter, had a side-mounted TP, and somewhat improved performance. An unusual design feature is the annular ejector in the low-pressure oxidizer feed line. Although ejectors have been known for many years, this is the first application disclosed in a USSR LPRE. By injecting a small directed flow of high-pressure liquid (from the pump) into the suction pipe, it is possible to slightly raise the pressure of the main flow. It raises the suction pressure at the main NTO pump inlet and provides more margins against cavitation problems. Several Russian and non-Russian engines have used ejectors instead of booster

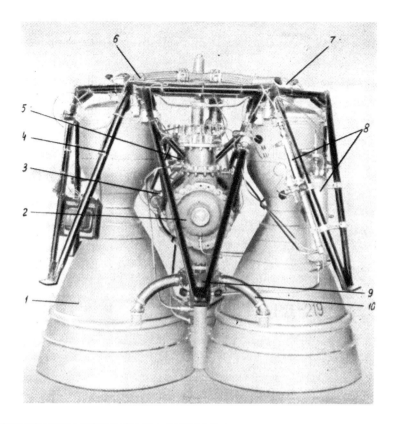

Fig. 8.4-23 View of RD-219 short LPRE with two TCs and one TP between the TCs. It was installed in the second stage of the R-16 ballistic missile: 1, thrust chamber; 2, fuel pump; 3, gas generator; 4, mounting structure; 5, oxidizer pump; 6, one of oxidizer pipes; 7, oxidizer shutoff valve; 8, tanks for starting fuel; 9, fuel valve; and 10, one of fuel pipes. Courtesy of NPO Energomash from Ref. 1.

pumps. Ejectors are certainly simpler than a booster pump, but they are notoriously inefficient. With a staged combustion cycle the penalty is more pump power and a somewhat heavier engine.

The RD-253 engine control regulates the engine thrust by causing changes in the fuel flow to the preburner and regulates mixture ratio by changing the fuel flow through the fuel cooling jacket. This was also one of the first large production engines known to have chemical gas pressurization of the propellant tanks; one small gas generator produced a fuel-rich gas mixture, which is supplied to the fuel tank ullage space, and one produces an oxidizer-rich gas, which flows into the oxidizer tank ullage space. This chemical tank pressurization scheme was used about 10 years earlier in the U.S. Bomarc missile. In the 1990s the development of an updated and uprated version of the

Table 8.4-2 Families of Glushko LPREs with Storable Propellants for Several Ballistic Missiles and SLVs[5]

Step	Engine Units	Engine Block Assembly	Variants
Development steps in engine family based on RD-215 (2 TCs and 1 TP)			
1	RD-215	—	—
2	2 × RD-215	= RD-216 (R-14, stage 1)	Modified to R-216M (Kosmos)
3	RD-215 improved	= RD-217	—
4	3 × RD-217	= RD-218 (R-16, stage 1 and Kosmos 3)	—
5	High-altitude RD-217	= RD-219 (R-16, stage 2)	—
Development steps in engine family based on RD-250 (2 TCs and 1 TP) higher performance than RD-215			
1	RD-250	—	—
2	3 × RD-250	= RD-251 (R-36, stage 1)	Modified to RD-261 (Tsiklon, stage 1)
3	High-altitude RD-250	= RD-252 (R-36, stage 2)	Modified to RD-262 (Tsiklon, stage 2)
Development steps of engine family based on RD-263 (1 TC and 1 TP) (higher performance than RD-250)			
1	RD-263	—	—
2	4 × RD-263	= RD-264 (R-36M, stage 1)	—
3	RD-263 modified	= RD-268 (MR-UR-100, stage 1)	—
4	RD-263 modified	= RD-273	—
5	4 × 273	= RD-274 (R-36M2, stage 1)	—

RD-253 called RD-275 was initiated for use with the Proton K space launch vehicle. It has, according to a release by the Krunichev Space Center, about 7% more thrust and a higher chamber pressure than the RD-253. The first flight with the new RD-275 with the Proton SLV was in April 2001.

The RD-270 LPRE was the highest thrust (640 tons or 1.4 million lbf at SL and 685 tons or 1.5 million lbf in vacuum) storable propellant engine, and it had a novel engine cycle, which is a variation of a staged combustion cycle.[3-5,22] It was based on injecting only gaseous propellants into the TC, using oxidizer-rich gases from an oxidizer-rich preburner and also fuel-rich

Fig. 8.4-24 View of the clustered RD-251 LPRE, which consists of three RD-250 engines, each of which has two TCs and one common TP. Courtesy of NPO Energomash (photo from personal communications, J. Morehart, The Aerospace Corporation).

gases from a fuel-rich preburner. It was hoped that a gas/gas injection scheme, which eliminates the droplet formation and the droplet evaporation from the combustion process in the chamber, would give better mixing and diffusion, superior performance, and stable combustion. As seen in Table 8.4-1, it had the highest thrust of all listed storable propellant rocket engines, a very high chamber pressure of 266 atm or about 3900 psi, was being developed between 1965 and 1971, and used NTO/UDMH. The RD-270 is shown in Fig. 8.4-26. It was originally intended to use nine RD-270 engines for the booster stage and three more in the second stage of the large UR-700 lunar launch vehicle, which was proposed as competition to the N-1 Moon SLV. This installation was not favored by the vehicle designer because the LPRE was too large and too long in size, and the high thrust caused structural vehicle load distribution problems. It was also rejected by Korolev for his N-1 lunar launch vehicle because he wanted LOX/kerosene engines. This dispute between Glushko and Korolev about using LOX/kerosene or NTO/UDMH caused a further deterioration in their relationship and might have affected the progress of Soviet rocketry. The RD-270 went through 29 firing tests (many of which had technical problems) and 21 experimental engines before work was stopped. The UR-700 vehicle program, for which the RD-270 was originally intended, was canceled. The expected engine reliability and performance improvement were apparently not validated in this initial brief development effort. It seems Energomash did not try to develop another LPRE with gas-to-gas injection.

Engines with High-Energy Propellants

Glushko had undertaken some exploratory work with liquid fluorine since the 1950s. The experimental KS-50 TC, mentioned earlier in this chapter, was tested with this oxidizer. Of course the attraction was the higher specific impulse and the relatively high density of the oxidizer (smaller vehicle), and studies showed that it would allow substantially higher payloads. After tests at higher thrust levels (500 kg and 1500 kg or 3300 lbf), after analyzing alternate fuels, and after investigating the implications to the flight vehicle, the propellant combination of liquid fluorine and liquid ammonia was selected as the most promising. The intended application for such an engine was as an upper stage of a space launch vehicle because this would allow much heavier payloads in orbit. A development program for an engine with 10 tons of thrust (22,000 lbf) was started in 1960, and the design was identified as the RD-303.[5] The tests were conducted at the Primorsk Branch of the Design Bureau, which was located east of Leningrad. Between 1963 to 1967, about 100 test firings were conducted. A large launch vehicle was planned and investigated by the Yangel vehicle design bureau, and it included a high-energy upper stage. The design requirements of this application caused some change in the engine requirements, and the modified LPRE version was then identified as the RD-302 LPRE.[5] About 309 ground firings and over 40,000 s of operation were accumulated. After the Yangel project collapsed, an upper stage on the Proton Space Launch Vehicle became the likely application for this high-energy LPRE. More specifically it was a heavy communications satellite payload, and the development work of a new upper stage, to be propelled by the new fluorine/ammonia engine, was begun. The modified engine was identified as the RD-301, and it was tested extensively. It had a TP feed system with a gas-generator cycle and a large nozzle area ratio (108.7:1). The thrust was the

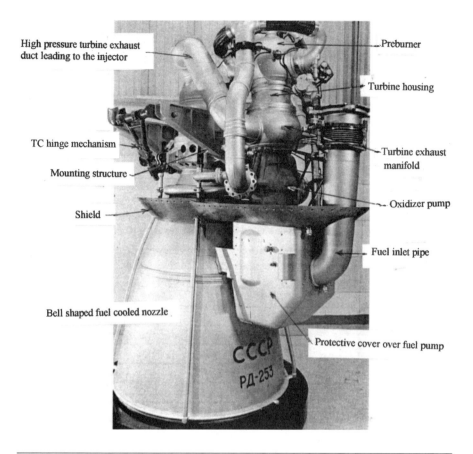

Fig. 8.4-25 View of one version of the RD-253 LPRE with storable propellants and a staged combustion cycle used on the Proton SLV. Here the TP (vertical shaft) is located next to the TC; in another version the turbopump is partly above the TC. Courtesy of NPO Energomash from Ref. 17 (also personal communications, NPO Energomash, 2001–2002).

same as the prior versions, and the specific impulse at altitude was expected to reach 420 s. The RD-301 is shown in Fig. 8.4-27. The material problems for bearings, seals, valves, gaskets, and chamber walls were severe, but they were mostly solved. Satisfactory test procedures (including sensing, purging, or decontamination) were worked out, but firing preparations and launch preparations did take a longer time than with other propellants. Altogether 274 engine had been built, and the cumulative firing time was over 200,000 s (56 h) spread over more than the 15 years of testing.

In 1977 the program was terminated, just before tests of the engine installed in the vehicle upper stage were to begin.[5,22] The official reason was a redirection of the communications satellite programs. The unofficial reason probably was the extreme toxicity and corrosiveness of fluorine and the drastic consequences of a potential spill, launch accident, flight abortion, or exposure of people, equipment, or the environment. The safety issues for the vehicle, surrounding communities, and the launch range had not been satisfactorily addressed by the time the program was canceled. The Primorsk branch of NPO Energomash was deactivated and closed in 1989.

This is a historically remarkable and unique achievement, and no one else has developed a complete new pump-fed engine with this extremely toxic, very corrosive, and highly reactive oxidizer. The high-energy propellant program was a major effort, and at one time it was given a high government priority. Other countries (Germany, France, United States, including this author) have worked with fluorine as an oxidizer, but at much lower thrust and chamber pressure levels, using mostly simple gas pressure feed systems or a modified existing LPRE. Testing with fluorine and other high-energy propellants is reported in Chapters 7.7, 7.8, 7.10, 7.11, 9.8, and 10. In the United States breathing fluorine is considered a serious health hazard in concentrations greater than 1 part

Fig. 8.4-26 View of RD-270 large LPRE (1.51 million lb vac thrust) used a novel gas-gas injection scheme. Engine development was not completed. Courtesy NPO Energomash.[22]

Fig. 8.4-27 Experimental high-energy propellant RD-301 LPRE using liquid fluorine and liquid ammonia. Courtesy of NPO Energomash (from personal communications, NPO Energomash, 2001–2002).

per million for a relatively short-term exposures. For the same reasons the work on engines with highly toxic propellants was also stopped in the United States and other countries.

During 1960 and 1966, Energomash also worked on the pump-fed RD-501 engine using concentrated hydrogen peroxide and pentaborane as propellants. It is the only known complete pump-fed engine with this very toxic fuel. It is not known how far the Soviet development effort had progressed. Work was stopped well before the engine was qualified. In the United States General Electric developed and tested TCs with the same propellants and also tested diborane as a fuel. The U.S. work was discontinued because of very poor combustion efficiencies of the boron compounds and the high toxicity of the fuels.

Tripropellant Rocket Engines

In 1989 Energomash started the development of two tripropellant large booster engines.[3-5,22] The RD-701 had two gimbaled TCs, and the RD-704 had a single TC. Both used a staged combustion engine cycle. The aim was to drive a single-stage-to-orbit launch vehicle using a combined first- and second-stage engine. Data given in Table 8.4-1 show a very high chamber pressure of 4260 psi for its boost phase. For the initial period of the flight (boost phase or ascent through atmosphere), the engine burns both kerosene and LH_2 with LOX at a high thrust level and a high chamber pressure. For the remainder of the flight (which is a sustainer phase), it burns only LH_2 fuel with LOX at a much lower thrust and chamber pressure. The advantage of this dual fuel concept is a somewhat higher average fuel density, which allows a smaller total propellant tank volume, a slightly lower vehicle structure mass, a single TC for two propellant combinations, a lower drag, and a slightly improved vehicle performance. Another design bureau, KB Khimautomatiki, worked on a third tripropellant engine based on their LOX/LH_2 RD-0120 LPRE. The tripropellant concept was originally investigated and might have been originated by Rudi Beichel (originally of Peenemünde, Germany and later of Aerojet in the United States). The United States and other countries supported tripropellant engine studies and a few component tests, but never built the hardware for a complete tripropellant engine. The engine would be more complex, have probably six TPs (including booster turbopumps) and a more complex TC, be very expensive, and the vehicle performance improvement has been estimated to be small (maximum of 3%). The Russians have done some engine testing and were seeking international partners to share the cost of further engine development, estimated at more than $500 million. This is another example where the Soviets undertook three parallel developments for an application, which did not have a firmly designated future vehicle and which did not have a compelling mission advantage. None of these Soviet tripropellant engines has been fully developed.

Energomash had its share of unsuccessful efforts and LPRE developments that were cancelled before they could be completed. Nevertheless the historical record of successful LPREs of Energomash is very impressive.

Concluding Comments

NPO Energomash developed more large LPREs than any other LPRE organization in the world. It includes engines for the first and sometimes the second stage of several ballistic missiles (short range, several at medium range and several at ICBM range), for space launch vehicles (SLV), which were converted from military retired long range missiles (restartable for second stage engine), or for SLVs which were specifically intended for this purpose. Most of the work was with LOX/kerosene or NTO/UDMH. This Design Bureau was the first to use four large TCs with a single TP, the first with LPREs in strap-on boosters, turbine exhaust for attitude control, and hinge-mounted vernier engines for attitude control. They get credit for the first large LPRE with a staged combustion cycle using storable propellants. They employed experimental TCs for developing new cooling jacket designs and new injectors. Energomash worked on engines with TPs using high energy very toxic propellants (several F_2/NH_3 engines and also one with H_2O_2/B_5H_9) and one of these was qualified, but did not fly. They solved complex combustion instability problems in several of these engines, some at very high chamber pressures. They also learned how to use a good single engine in families and install them as clusters of 2,3,4, or 4 engines in a booster stage. Their indoor engine or TC firing test stands (for firing vertically down) were the first large ones to be built in the Soviet Union for high thrust. They are located in a densely populated part of the Moscow region and are heated for cold weather.

References

[1] *"NPO Energomash,"* illustrated brochure NPO Energomash, Khimky, Moscow Region, circa 2000 (in English).

[2] Siddiqi, A. A., *Challenge to Apollo, the Soviet Union and the Space Race (1945-1974)*, NASA SP-2000-4408, Washington, DC 2000.

[3] Lardier, C., "Liquid Propellant Rocket Engines in the Soviet Union," International Academy of Astronautics, Paper 99-2-3-09, Oct. 1999.

[4] Lardier, C., "Les Moteurs-Fusées Soviétiques de 1946 a 1991 (Soviet Liquid Propellant Rocket Engines from 1946 to 1991)," International Academy of Astronautics, Papers 97-2-3-03, 98-2-3-09, and 99-2-3-04 (in French).

[5] Siddiqi, A., "Rocket Engines from the Glushko Design Bureau," *Journal of the British Interplanetary Society*, Vol. 54, 2001, pp. 311–334.

[6] Pakhmanin, V. F., and Sterpin, L. Ye. (eds.), *Once and Forever—Documents and People on the Creation of Rocket Engines and Space Systems of Academician Valentin Petrovich Glushko*, Mashinostroyeniye Press, Moscow, 1998. (in Russian).

[7] Przybilski, O., "Die Deutschen und die Raketenantriebswerkentwicklung in der USSR (The Germans and the Development of Rocket Propulsion in the USSR)," *Luft und Raumfahrt*, 1999, No. 2, pp. 30–32, No. 3, pp. 28–32, and No. 4, pp. 33–40 (in German), *Journal of the British Interplanetary Society*, Vol. 55, No. 11/12, 2002 (in English).

[8] Sutton, George P., "History of Liquid-Propellent Rocket Engines in Russia, Formerly the Soviet Union," *Journal of Propulsion and Power*, Vol. 19, No. 6, 2003, p. 1018.

[9] *Russia's Arms Catalogue*, Vol. 4, edited by the Ministry of Defence of the Russian Federation, Military Parade, Ltd., Moscow, 1997, Parts 3–6, 9, 12–14.

[10] Salkutdinov, G. M., "The Development of Methods of Cooling Liquid Propellant Rocket Engines (ZhRDs), 1903 to 1970," *History of Rocketry and Astronautics*, American Astronautical Society History Series, Vol. 10, Univelt, San Diego, CA, 1990.

[11] Biriukov, J. V., "The R-3 Rocket Project Developed in the USSR in 1947 – 1959 as a Basis for the First Soviet Space Launchers," *History of Rocketry and Astronautics*, edited by J. D. Hunley, American Astronautical Society History Series, Vol. 19, Univelt, San Diego, CA, 1997, pp. 193–199.

[12] Bolokin, A., *The Development of Soviet Rocket Engines (for Strategic Missiles)*, Delphic Associates, Inc., Falls Church, VA, 1991. (translated from Russian).

[13] Wade, M., www.astronautix.com, (cited May 2005).

[14] Prishchepa, V. I., "History of the Development of the First Space Rocket Engines in the USSR," *History of Rocketry and Astronautics*, edited by F. I. Ordway, American Astronautical Society History Series, Vol. 9, Univelt, San Diego, CA, 1989.

[15]Katorgin B., and Sternin, L, "Pushing Back the Missile Technology Frontiers," *Aerospace Journal*, No. 5, Sept.–Oct. 1997.

[16]Gubanov, B. I., "USSR Main Engines for Heavy Lift Launch Vehicles, Status and Direction," AIAA Paper 91-2510, 1991.

[17]Glushko, V. P., *Rocket Engines of the Gas Dynamics Laboratory – Experimental Design Bureau, (Raketnye Dvigateli GDL-OKB)*, Novosti Press, Moscow, 1975; NASA TT F 16847, Feb. 1976.

[18]*"Liquid Propellant Rocket Engine Chamber and Its Casing,"* U.S. Patent 6,244,641, 12 June 2001, assigned to NPO Energomash.

[19]Performance Data Sheets on the RD-120, RD-180, and RD-701, Pratt & Whitney Div., United Technologies Corp, and Energomash, circa 1993.

[20]*Jane's Space Directory,* revised biannually, Jane's Information Group, Coulsdon, Surrey, England, U.K.

[21]Hindley, K. B., *Handbook of Russian Rocket Engines*, 1st ed., Technology Detail, Clifton, York, U.K., 1999.

[22]Haeseler, D., "Soviet Rocket Motors on View," *Spaceflight*, Vol. 35, Feb. 1993, pp. 40–41.

[23]Gakuna, G. G., *Construction and Design of Liquid Rocket Engines*, Mashinostroyeniye Press, Moscow, 1989 (in Russian).

8.5 KB Khimautomatiki or Chemical Automatics Design Bureau

Originally this organization worked on auxiliaries for aircraft engines (fuel-injection systems, starters) and was founded in Moscow during World War II under the leadership of Semion A. Kosberg, who headed the work between 1941 and 1965. He became the chief designer of the experimental design bureau named after him. Its roots can be traced to a Moscow carburetor factory of 1940. In 1941 the bureau and factory were evacuated to Berdsk in the Novosibirsk Region of Siberia. In 1945/1946 it was relocated to Voronezh (about 300 miles south of Moscow), and in 1954 the bureau was asked to begin work on LPREs and components for aviation engines.[1-8,*] In 1966 it was renamed as KB KhimAutomatiki or the Chemical Automatics Design Bureau (abbreviated as CADB). It is still working on LPREs, but today at a lower level than 40 years ago. To compensate for the reduction in the LPRE business, CADB has started other work, such as the development of oil and gas equipment, commercial pumps, agricultural equipment, or medical devices. They also are developing a new supersonic jet engine using liquid hydrogen as a fuel, and it has been flight tested.

CADB and its predecessor organization developed these principal types of LPREs:

1) In the 1950s it was LPREs for superperformance of fighter aircraft.

2) In the late 1950s it developed and put into mass production LPREs for the upper stages of antiaircraft missiles.

3) CADB is best known for developing a variety of upper-stage engines and then developed several booster engines for ballistic missiles and SLVs, and this work is still being pursued today.

4) A historic engine running on LOX/LH$_2$ was developed and launched (1987), and several other engines using LOX/LH$_2$ were also investigated, but did not fly.

5) CADB developed a unique booster engine for a submarine-launched ballistic missile between 1977 and 1985, and it was deployed in the fleet.

6) Other engine types were investigated, such as a tripropellant engine, an engine with an unusual expansion/deflection nozzle, an engine with an expander cycle and an orbital maneuver engine.

This design bureau has developed more than 60 different LPREs; of these about 35 were qualified, and about 30 were put into production. Table 8.5-1 shows data for several of these.[1,2,*] Some of the experimental LPREs mentioned in this chapter are not included in this table and some data on the listed engines was not available.

---

*Personal communications, V. S. Rachuk, general designer and general director of CADB, and J. Morehart, The Aerospace Corporation.

Table 8.5-1 Data on Selected LPREs of CADB

Designation	Engine Cycle	Thrust (SL)/vacuum (V), kN	Specific impulse, s	No. of TCs per engine	Height/diameter, m	Chamber pressure, MPa	Propellants	Engine mass, kg	Development period	Application
RD-0101	GG	19–39(SL)	240(SL)	1	1/0.4	4.4	LOX/alcohol	—	1954–1957	E-50A aircraft
RD-0102	GG	15–39 (SL)	250(SL)	1	—	4.25	LOX/kerosene	—	1954–1957	Yak-27V aircraft
RD-0105	GG	49.9 (V)	316 (V)	1	—	4.6	LOX/kerosene	—	1957–1958	Luna SLV, third stage
RD-0106	GG	294(V)	—	—	—	—	LOX/kerosene	—	1958–1960	Second stage, R-9A ICBM
RD-0107	GG	298 (V)	324(V)	4	—	—	LOX/kerosene	—	1958–1962	Third stage, Molniya
RD-0109	GG	54.5(V)	323(V)	1	1.57/1.10	5.0	LOX/kerosene	121	1958–1960	Third stage, Vostok SLV
RD-0110	GG	298(V)	326(V)	1 + 4V	1.57/2.40	6.8	LOX/kerosene	408.5	1963–1967	Third stage, Soyuz and Molniya SLVs
RD-0120	SC[a]	1962(V)	455(V)	1	4.55/2.42	21.8	LOX/LH$_2$	3450	1967–1983	Energiya SLV with four sustainer engines
RD-0124	SC	294(V)	359(V)	4	1.575/2.4	15.5	LOX/kerosene	450	1996–1999	Third stage of improved Soyuz SLVs
RD-0200	GG	6–59(SL)	230(SL)	2	—	6.7	NTC/UDMH	—	1957–1961	V-1100 SAM missile
RD-0202[b]	SC	4 × 559(SL)	316 (V)	1	—	14.5	NTC/UDMH	—	1961–1963	First stage, UR 200 ballistic missile
RD-0210	SC	589(V)	326(V)	1	2.33/1.47	—	NTO/UDMH	565	1963–1967	Second stage, Proton SLV
RD-0212 consists of	SC	612.6(V)	324(V)	1 + 4V	3.00/3.78	14.7	NTO/UDMH	638	1963–1970	Third stage, Proton SLV
Main RD-0213	SC	581.7(V)	326(V)	1	3.0	14.7	NTO/UDMH	—		
Vernier RD-0214	GG	30.9(V)	293(V)	4	—	5.49	NTO/UDMH	—		

Engine	Cycle	I_{sp} (s)	Chambers	Thrust (kN)	Propellants	Mass (kg)	Years	Application
RD-0215[c]	SC	—	—	3 + 1	NTO/UDMH	—	1963–1966	First stage of RS-10 (SS-10) ICBM
RD-0225	PG[d]	3.92(V)	291(V)	1	NTO/UDMH	23	1966–1971	Almaz orbital station with two engines
RD-0228[e]	SC	1 + 4V	—	1.0/0.45	NTO/UDMH	—	1967–1974	Second stage of RS-20A and 20B (SS-18) ICBM
RD-0232[f]	SC	—	—	—	NTO/UDMH	—	1969–1974	First stage of RS-18 (SS-19) ICBM
RD-0235	SC	Main	—	1 + 4V	NTO/UDMH	—	1969–1974	Second stage of RS-19 for (SS-19) ICBM
RD-0236	—	Vernier	—	4	NTO/UDMH	—	1969–1974	
RD-0237	PG[d]	Postboost control	—	2	NTO/UDMH	—	1969–1974	Third stage or payload maneuver engine
RD-0243[g]	SC/SC	814 (estimated)	—	1 + 4V	NTO/UDMH	27.8	1977–1985	First stage of RSM-54
RD-0255[h]	SC/SC	—	—	1 + 4V	NTO/UDMH	—	1983–1987	Second stage of RS-20V (SS-18 mods. 5/6) ICBM
RD-0750				1			Circa 1989	Tripropellant version of RD-0120 (2 modes of operation)
Mode 1	SC	418(V)	1720		LOX/kerosene/LH$_2$			
Mode 2	SC	451(V)	780		LOX/LH$_2$			

[a]Staged combustion. [b]Has three RD-0203 and one RD-0204, which also provides task pressurization. [c]Consists of three RD-0216 and one RD-0217, which has gas generators for propellant tank pressurization. [d]Pressurized gas feed cycle. [e]Consists of one main RD-0229 and one vernier engine with four TCs (GG), [f]Consists of three RD-0233 and one RD-0234, which provides for pressurization of propellant tanks. [g]Submarine missile engine, consists of the RD-0244 main engine, which is submerged in the propellant tank and the RD-0245 vernier engine with four hinged TCs, [h]Consists of one RD-0256 main engine and one RD-0257 vernier engine with four TCs; both engines submerged in propellant tanks. [Data Refs.1,2,5,6,8, and from personal communications with V. S. Rachuk (CADB), J. Morehart (The Aerospace Corporation), and M. Coleman (CPIA)].

Early Efforts

In 1954 CADB started work on an aircraft rocket engine using a monopropellant, namely, ethyl nitrate, because it was thought to become a simple engine. The DB also worked on a second design of such an aircraft superperformance engine with this monopropellant. Although they managed to make it work, they had a few explosions and problems and abandoned this monopropellant approach. The next few engines (1956 and 1957) used bipropellants and were also intended to become superperformance rocket engines for military interceptor aircraft.[1,2,5-7,] * The RD-0100 and the RD-0101 used LOX/alcohol propellants and they are not listed in Table 8.4-1. The RD-0102 and RD-0103 used LOX/kerosene. Some featured variable thrust. They used a gas-generator cycle and a turbopump feed system. Figure 8.5-1 shows the RD-0102, which was restartable and reusable. A communication from CADB indicated that the development of these engines was not completed. However the *Russian Arms Catalogue* indicated that the were flown in experimental versions of military aircraft (Mikoyan E50-A, Yakovlev 27V fighter airplane, and a Sukhoy fighter/interceptor). Altogether with aircraft rocket engines from CADB, the Glushko DB, RNII, and KB Khimmash, there were more than 14 such engines, and they are described in Chapter 8.3. None were used in combat.

Beginning in 1946, the Soviet authorities wanted to develop a surface-to-air defensive missile, and several design bureaus became involved in developing these missiles and their propulsion systems. Between 1957 and 1960 Kosberg's Bureau developed the RD-0200 and RD-0201 LPREs with storable propellants for the upper stages of antiaircraft missiles.[1,2,4,7,8,]* Both used nitric acid with kerosene and had a maximum thrust of about 13,000 lbf.

Fig. 8.5-1 RD-0102 aircraft superperformance LPRE was restartable and throttleable. It used LOX/kerosene. Copied with CADB permission from Ref. 1.

---

*Information for this chapter was provided by V. S. Rachuk, general designer and general director of CADB, and J. Morehart, Aerospace Corporation.

The RD-0200 is shown in Fig. 8.5-2, had two TCs, one TP, one GG, and its thrust could be throttled by a ratio of 10 to 1.0. The cooling jacket of the thrust chamber used the corrugated intermediate sheet described in Chapter 4.3. During development, combustion vibrations were encountered, both in the TCs and the GG. The designs were changed until the problems were remedied. The RD-0200 was designed for the second stage of the antiaircraft missile 5V11, which was developed by the Lavochkin DB. The RD-0201 was similar to the RD-0200, but the thrust chambers were hinge mounted and actuated to achieve full thrust vector control during powered flight. It was designed for the third stage of the antiaircraft missile V-1100. Other LPREs for antiaircraft missiles were developed by KB Khimmash and are described in the next chapter. These missiles and their LPREs were mass produced and

Fig. 8.5-2 Turbopump-fed RD-0200 LPRE was used in the second stage of an antiaircraft missile, and more than 2000 missiles were deployed. Copied with CADB permission from Ref. 1.

deployed around several major Soviet cities. The production quantities for these rocket engines (such as those designed by CADB and KB Khimmash), which were used in surface-to-air missiles, were the largest in history of any class of pump-fed LPREs.* The total quantity was estimated to be between 12,000 and 20,000 missiles.

Engines for Missiles and Space Flight

CADB successfully developed various upper-stage LPREs and a couple of booster LPREs for guided ballistic missiles and SLVs, and many are listed in the table and mentioned next. The early engines used gas generator engine cycles, and in the 1960s the DB shifted to staged combustion cycles. The complimentary vernier TCs were initially supplied with propellant from the main turbopump. In the mid-1950s Kosberg's Experimental Design Bureau got the first job for a spaceflight application. The vernier TCs for the Glushko RD-107/RD-108 LPRE were transferred from Korolev's DB Energiya, where they had been largely developed, to OKB Kosberg for completing the development and for guiding the impending production.* This engine was mentioned in the previous chapter. This thrust chamber had the lightweight cooling-jacket design with a corrugated intermediate wall as described in Chapter 4.3. Subsequent engine designs of this DB used this or a related construction method. When the vehicle design requirement changed, the current vernier TCs were no longer satisfactory, and the development of new vernier TCs was then undertaken, but this time by Glushko at his DB. Thus Kosberg no longer had to build and deliver these vernier TCs, which might have had some technical problems. However the work was helpful to Kosberg and formed the basis for the design of the engine described in the next paragraph.

The RD-0105 was the first LPRE developed for spaceflight in this bureau. It had a single TC, supplied by a single TP, operated on LOX/kerosene, and operated with a gas-generator cycle. It is shown in Fig. 8.5-3. This engine was the first at this experimental DB to use the lightweight thin inner-wall construction with brazing and welding for the cooling jacket of the TC. This construction technique is discussed in Chapter 4.3 and was originally poineered by Isayev and tested in Glushko's EB-140 experimental TC. It was an improved version of the vernier engine just mentioned. The turbopump was based in part on the TP experience of their aircraft LPREs. The RD-105 flew successfully in the third stage of the Luna (moon) vehicle, boosted by an adaptation of the two-stage R-7 ICBM. It made unmanned flights to and around the moon.[1,2,4-6,*]

The RD-0106 was the first production engine of CADB with a single TP feeding four TCs. The TC design was based on the RD-0105. The length of the engine was reduced by going to four TCs. The RD-0106 was put into production for the second stage of the R-9A ICBM. Little information seems to be available about this engine. An improved version, identified as RD-0107, was developed, again using LOX/kerosene as the propellants. It also had four

---

*Personal communications, J. Morehart, The Aerospace Corporation, 2001–2004.

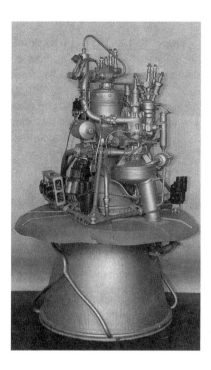

Fig. 8.5-3 RD-0105 was CADB's first engine in a space vehicle and was used in several unmanned "Luna" (moon flight exploration) missions. Copied with CADB permission from Ref. 2.

smaller vernier TCs, which were supplied with propellants from the main TP. The RD-0107 has been used in the third stage of the Molniya SLV, a version of the Soyuz SLV family. A few years later another modified version, identified as the RD-0108, was developed and used on the third stage of the Voshkod SLV series also part of the Soyuz vehicle family.

The RD-0109 engine (thrust is 54.8 kN), shown in Fig. 8.5-4, is an improved version of the lower-thrust RD-0105.[1,2,5,6] It had a single TC and about 2% more engine performance, in part because of improved bipropellant injection elements. The RD-0109 was used for propelling the orbital stage during the six manned launches of the Vostok program, and flights began in 1960. This historical engine helped to launch the first man, Yuri Gargarin, into a space orbit. The cooling-jacket portion of the nozzle-exit section had an open corrugation (no outside wall) in order to reduce weight. A cross section of such a cooling jacket was shown as item d in Fig. 4.3-4. This design idea of omitting the outer wall is attributed to Semion Kosberg, the first chief designer.* He probably did not know that a similar design had been tested by the M. W. Kellogg Company in the United States in 1947, about 12 years earlier; it is depicted in Fig. 4.3-7.

---

*Personal communications, V. S. Rachuk, general designer and director of CADB.

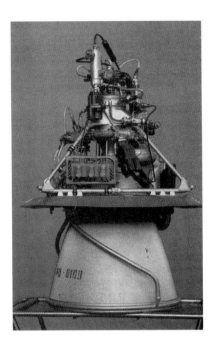

Fig. 8.5-4 RD-0109 propelled the third stage of the Vostok Space launch vehicle. Copied with CADB permission from Ref. 2.

CADB overcame combustion vibrations in the RD-0109 during ignition by locking a throttle valve in the chamber's oxidizer manifold into a temporary intermediate valve position during the start, which was a novel approach.[2]

The RD-0110 LOX/kerosene engine was a reliable upper-stage LPRE. It is shown in Fig. 8.5-5 and one of its four TCs in Fig. 8.5-6. It replaced the RD-0107 engine, which had about the same thrust and specific impulse, on the second stage of the Molniya and Soyuz SLVs, and it first flew in 1964. The RD-0110 uses a gas generator engine cycle, has four fixed TCs, and a single TP. Its four hinged vernier TCs provide flight stability control and are supplied with hot gas from the GG.[1–8,*] The Russians do not call them vernier TCs, but "nozzles" because they do not have a combustion chamber and merely accelerate the warm pressurized gas to supersonic velocities. Gas from a solid-propellant cartridge starts the spinning of the turbine, and igniters start combustion in the four TCs and the GG. The engine has a limited mixture ratio adjustment (+11 to −14%) by throttling the fuel flow and a limited thrust control (+7 to −9.5%) by throttling the oxidizer flow to the GG. The original welded construction of the turbine had some cracks and had failed and was replaced by a stronger cast turbine.[2]

---

*Personal communications, V. S. Rachuk, general designer and general director of CADB.

Fig. 8.5-5 RD-0110 LPRE and its predecessor (RD-0107) have propelled more than 1200 upper stages of Soviet launch vehicles. Copied with CADB permission from Ref. 1.

The RD-0110 TC uses a cooling jacket construction similar to other Soviet TCs as described in Chapter 4.3. A corrugated intermediate sheet metal is used in the chamber, the converging nozzle section, and the upper part of the nozzle-exit section (see Fig. 8.5-6). All cooling jacket parts are made of stainless steel (for corrosion resistance) and are brazed and welded together. Only in the throat region, where the heat transfer and the inner wall temperature are usually the highest, is the inner wall made of a high-conductivity copper (with a small amount of strengthening alloy) with milled channels brazed to an outer steel curved wall as shown in Fig. 8.5-7. The throat piece is pressure brazed together in a vacuum furnace and then joined together to the chamber and nozzle-exit pieces. The lower nozzle-exit segment does not have an outer wall, which contributes to a lower engine weight. Here the corrugated wall confining the coolant passages are exposed. Film coolant is supplied through a separate manifold, as seen in Fig. 8.5-6, and is injected through a circular slot upstream of the nozzle throat.

Occasionally unstable combustion occurred during the TC development testing, and a program to enhance the combustion stability was initiated. This

638 History of Liquid Propellant Rocket Engines

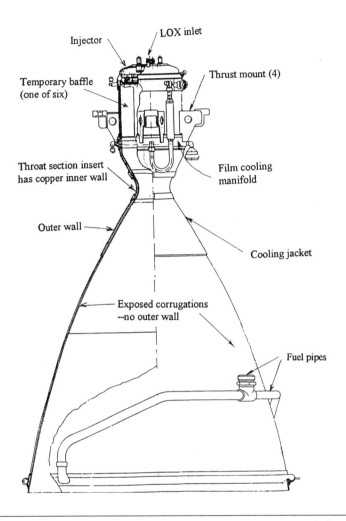

Fig. 8.5-6 Thrust chamber of the RD-0110 has different types of flow passages in its cooling jacket. The lower two-thirds of the cooling jacket of the diverging nozzle segment does not have an outer wall, and the corrugations are exposed. Courtesy of CADB; copied with AIAA permission from Ref. 9.

program is described in detail in Ref. 9. Tests were done with experimental injectors using multiple injection elements of various designs. The injector and a cross section of one of its injection elements are shown in *Fig.* 8.5-8. Several types of bipropellant swirling concentric-tube-type injection elements (propellant forms conical spray sheets) and an injector element using holes and impinging liquid streams were tested for stability. The final injector design was stable during TC operation and had 91 bipropellant coaxial spray injection elements of two types arranged in a pattern of five concentric circles. The length L in the section of the injector element (Fig. 8.5-8) was critical in determining the spray impingement location and the combustion behavior. In

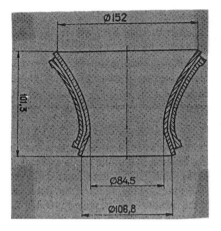 

Fig. 8.5-7 Throat piece of the RD-0110 cooling jacket has a thin inner wall with integral ribs made of a copper alloy. The top of the ribs are brazed to the outer steel wall. The inner and outer walls are subsequently joined to the other pieces of the cooling jacket. Courtesy of CADB; copied with AIAA permission from Ref. 9.

the center region the injector elements are designed for the combustion to begin in the exit cavity of each injection element. In the injection elements near the periphery of the injector, the combustion occurs mostly in the combustion chamber just outside of each element. However occasional instability still happened during the start transient. The solution was to glue six consumable internal fins or baffles to the combustion chamber wall as seen in Fig. 8.5-9. These baffles are made of felt-like porous material that is consumed or burned off in the first few seconds of operation. These temporary barriers are clever unique antivibration devices, which are effective only during the start, and are not found in other countries. The thrust chamber in this figure is a low-altitude test TC with a shortened nozzle of lower nozzle-exit area ratio.

This LPRE first flew in 1964, as was already mentioned. By 2004 (40 years later) about 2100 of the RD-0110 and its predecessor RD-0107 LPRE had gone through static firing tests, and more than 1400 of them had propelled the upper stage of a SLV. The demonstrated reliability had a 0.9984 success rate, which is indeed a high rate.

A replacement engine for the workhorse RD-0110 engine has been under development at CADB in recent years.[1,2] This RD-0124 has the same vehicle interface and fits into the same vehicle space. However it has a much higher chamber pressure (15.5 MPa vs 6.8 MPa), a larger nozzle area ratio of 198 to 1.0, and about 9% more specific impulse. Qualification tests were in progress in 2003. This new engine allows up to 950 kg or more than 2000 lb more payload in a Soyuz-2 SLV.

CADB provided the engines for ballistic missiles using the storable propellants NTO/UDMH beginning in the early 1960s.[1,2,4,6–8] The RD-0202 was a

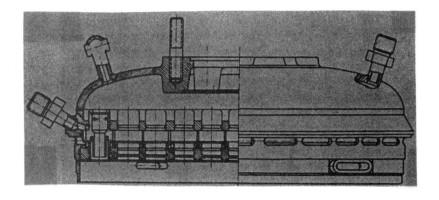

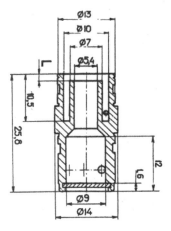

Fig. 8.5-8 Injector of the RD-0110 has 91 brazed-in injection element inserts, but only one is shown in the half-section. An enlarged section of one of these coaxial bipropellant spray elements is shown. Courtesy of CADB; copied with AIAA permission from Ref. 9.

---

four-engine cluster for the first stage of the UR-200 ballistic missile, and it consisted of three RD-0203 single TC engines and one RD-0204 engine, which was nearly identical, except it provided the warm gases for pressurizing the two propellant tanks of the first stage to a moderate pressure. A single RD-0203 engine is shown in Fig. 8.5-10. The RD-0204 provided oxidizer-rich gases for the pressurization of the NTO tank by tapping off some preburner gas downstream of the turbine. It also provided fuel-rich gas for pressurizing the UDMH tank from a gas-generator/mixer.[1,2,5,6,*] The RD-0205 LPRE was

---

*Information from V. S. Rachuk, general designer and general director of CADB, and J. Morehart, The Aerospace Corporation.

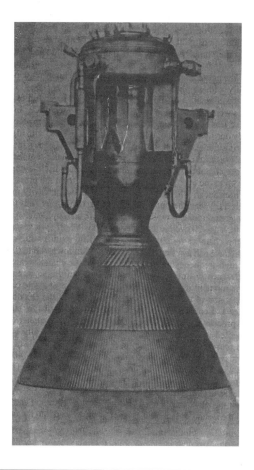

Fig. 8.5-9 Consumable baffles are visible in this cutaway view of the TC of the RD-0110 LPRE. Courtesy of CADB; copied with AIAA permission from Ref. 9.

for the second stage of this UR-200 missile and consisted of the RD-0206 main engine with a single TC and with its own TP and the RD-0207 vernier engine with four smaller hinge-mounted vernier TCs and its own turbopump. These engines provided for the pressurization of the propellant tanks of the second stage of the missile. Each of these two engines was a self-contained independent unit. The RD-0203, -0204, and -0206 LPREs all used staged combustion cycles with oxidizer rich preburners. The RD-0203 engine was claimed to be the first at this DB to use a staged combustion engine cycle with storable propellants; it might have occurred slightly earlier than the RD-253 developed by Glushko. This cycle allows a higher chamber pressure and usually a higher nozzle area ratio and thus a higher performance than an equivalent engine with a gas generator cycle. The UR-200 missile was never deployed with the Soviet military forces.

Fig. 8.5-10 RD-0203 LPRE is one of four engines developed for the UR-200 ICBM, which was not deployed. These engines did not fly. Copied with CADB permission from Ref. 2.

CADB developed several LPREs for the second- and third-stage engines of the Proton SLV. All of these engines used NTO/UDMH propellants, the main engines and oxidizer-rich preburners. The second stage of Proton used a cluster of four hinge-mounted LPREs, each of which could be deflected by +/− 3 degs. Three were the RD-0210 engine, and one was a RD-0211 engine, which was very similar, except it also provided the warm gases for propellant tank pressurization.[1-8,*] Figure 8.5-11 shows a single RD-0210 LPRE. Cooled oxidizer-rich gas from the turbopump discharge of the RD-0211 engine was injected into the NTO tank and UDMH-rich gas from a gas generator/mixer of the RD-0211 was supplied to the UDMH tank of this second stage. If oxidizer-rich gas would have been injected into the fuel tank, there would have been a vigorous combustion in the tank, and the ullage temperature could have been excessive. If the pressurizing gases are too hot, they can cause a heating of the fuel in the tank, possible weakening of the tank structure, and if UDMH gets too hot, it can lead to a decomposition of this fuel; therefore, the tapped-off fuel-rich gas is usually cooled before it is admitted to the fuel tank or

---

*Personal communications, V. S. Rachuk, general designer and general director of CADB, and J. Morehart, The Aerospace Corporation.

Fig. 8.5-11 U-shaped large pipe on this RD-0210 LPRE feeds high-pressure turbine exhaust gas to the injector. Three are used in Proton second stage. Copied with CADB permission from Ref. 1.

alternatively the gas might be generated at a very oxidizer-lean mixture ratio. Thrust is controlled by a throttle/regulator installed in the preburner's fuel manifold. A throttle valve in the main fuel manifold of the combustion chamber provides the mixture-ratio control, which is aimed at simultaneous emptying of the two propellants in the vehicle's propellant tanks. The RD-0210 and RD-0211 replaced an earlier version identified as RD-0208 and RD-0209 in the same application. Serious problems were discovered and remedied during the RD-0210 engine development.[2] There were fires in the turbine manifold with oxidizer-rich hot gas reacting with the metals and high-frequency gas vibrations occurred in the main TCs. The manifold burning was remedied by changing materials of several parts and design changes in the TP to avoid metal contact. The combustion instability was fixed by closing some of the openings in the fuel injectors.

This concept of having one of the engines in a four-engine cluster provide for the chemical tank pressurization is unique; it was originally used for the RD-0203 and −0204; it has since been used with several other four-engine-cluster applications. This dedsign concept seemed to be found only at CADB.

Fig. 8.5-12 RD-0212 is used to propel the third stage of one version of the Proton SLV. The vernier system is a separate LPRE with four TCs. Copied with CADB permission from Ref. 1.

The RD-0212 was used to propel the third stage on a version of the Proton SLV, and it can be seen in Fig. 8.5-12. This LPRE consisted of two separate autonomous LPREs, the RD-0213 and the RD-0214.[1-8,*] The sustainer or main engine RD-0213 has a single TC, which is a modification of the RD-0210 TC, and uses a staged combustion engine cycle with an oxidizer-rich preburner. In the figure a large U-shaped exhaust duct leads from the turbine outlet of the preburner to the injector of the main engine (RD-0213), and several of the vernier TCs are visible. The vernier engine RD-0214 represents a novel approach. It uses a GG engine cycle, has four hinged vernier TCs (total vernier thrust is 30.9 kN), but the vernier propellants are not supplied from the TP of the main engine, but from a separate small single vernier TP, which is supplied with warm gas from its own fuel-rich gas generator. Gas tapped off downstream

---

*Personal communications, V. S. Rachuk, general designer and general director of CADB, and J. Morehart, The Aerospace Corporation.

of this small turbine is used to pressurize the oxidizer tank of the second vehicle stage. Fuel-rich gas from the separate fuel–rich gas generator in the RD-0214 engine supplies the gas for pressurizing the second-stage fuel tank. The hinged vernier TCs can be rotated through a large angle, namely, +/− 45 deg. The TP of the vernier engine is not visible in Fig. 8.5-12. The author's concept of a likely simplified flow diagram of a typical main engine with an auxiliary, separate vernier engine and the two tank pressurization schemes is shown in Fig. 8.5-13, but the valves, restriction orifices, or burst diaphragms are omitted. This unique combination of a main engine and a separate vernier engine with two GGs has some advantages over competitive engine designs (simpler R&D and manufacture, possible weight saving, and option to run verniers before or after the main LPRE operation). It was used by CADB and has only been observed in Russia. Several of the engines discussed next also have this feature.

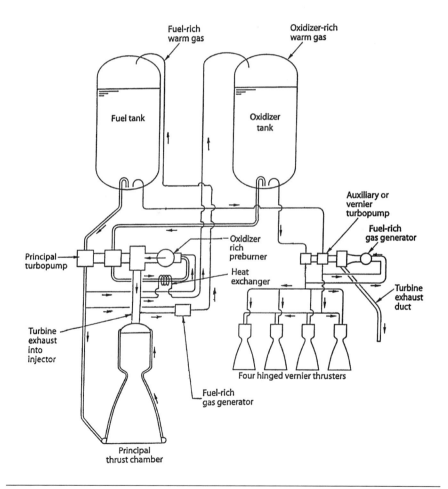

Fig. 8.5-13 Simplified possible flow diagram of RD-0212 (prepared by author).

Fig. 8.5-14 RD-0233 is a large booster engine. Three RD-0233 and a similar RD-0234, but with tank pressurization features, one of the four engines propel the first stage of the RS-18 ICBM. Copied with CADB permission from Ref. 1.

The RS-18 long-range ballistic missile was propelled by a similar set of four engines as used for the Proton SLV. The RD-0232 engine consists of three RD-0233 engines and one RD-0234, which is similar, but also provides tank pressurization.[1-7] This engine cluster propels the first or booster stage of the missile. Figure 8.5-14 shows the RD-0233 LPRE. The RD-0235 main engine and the RD-0236 (vernier engine with four hinged TCs) propel the second stage of this ICBM. The main TP and the smaller, horizontal vernier TP are visible in Fig. 8.5-15, which depicts the RD-0235/RD-0236 engine combination for the second missile stage.

The RD-0225 LPRE is a low-thrust engine and it is different from other CADB LPREs. Two of these were used for space maneuvers and orbital corrections of the "Almaz" manned orbital complex.[1-6,*] It is one of the few CADB

---

*Personal communications, V. S. Rachuk, general designer and general director of CADB.

Fig. 8.5-15 Combination of the RD-0235 main engine and the RD-0236 vernier engine propels the second stage of the RS-18 ICBM. The small TP for the four vernier TCs is clearly visible. Copied with CADB permission from Ref. 1.

engines with a gas-pressurized feed system. It has a high nozzle area ratio. It was designed for 100 starts in space, one year in orbit, and a cumulative firing duration of 20 min. The combustion chamber and nozzle-throat region are regeneratively cooled, and the large radiation-cooled nozzle-exit skirt (single wall) is coated on the outside with a chrome-nickel alloy of high emissivity, which helps to reduce the wall temperatures. To provide long-life electromechanical propellant valves and to avoid friction between the moving and stationary parts during restarting, the valve stem and its magnetic anchor are mounted on flexible membranes. Engine development began in 1967, and its first flight with the orbital station was in 1974.

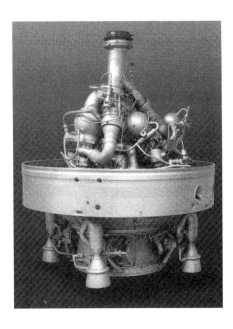

Fig. 8.5-16 RD-0243 engine is partly submerged in the tank of the first stage of a submarine-launched three stage missile. The bottom of the lower propellant tank is integrated with the engine. Copied with CADB permission from Ref. 1.

Submerged Engines

This design bureau built and delivered only one LPRE (RD-0243) for a submarine-launched ballistic missile, and this engine was installed inside a missile propellant tank and submerged in hypergolic propellant. However another design bureau, KB Khimmash, built more than 10 submerged engines; they (together with figures and a discussion of launching from a submarine) are presented in the next chapter. The RD-0243 engine in Fig. 8.5-16 is the LPRE for the first stage of a submarine-launched three-stage ballistic missile RSM-54, which has the range of an ICBM.[1-7,*] The RD-0243 consist of a main engine RD-0244 with a fixed TC and a vernier engine RD-0245 with four hinged TCs. The engine is mostly submerged inside the UDMH tank of the missile's first stage, and the nozzle skirt is integrated with the bottom part of the tank structure. The four hinge-mounted vernier TCs are outside of this fuel tank. The main engine used a staged combustion engine cycle with an oxidizer rich preburner for driving the main turbine. The vernier engine also uses a staged

---

*Personal communications, photo from Ref. 1 supplied by V. S. Rachuk, general director and general designer of CADB, and some data courtesy of J. Morehart, The Aerospace Corporation, 2002.

combustion cycle with an oxidizer-rich preburner. The chamber pressure in the main combustion chamber is high and has been estimated at a nominal value of 275 kg/cm^2 or 3911 psia.[5,6] One source reports it to be as high as 4600 psi, which was probably achieved during ground–based limit testing.[5,6] At either of these chamber pressures, the heat transfer is very high, requiring a special cooling-jacket design.

The gas for pressurizing the main oxidizer tank is obtained by bleeding oxidizer-rich preburner gas at the turbine exhaust of the main TP and cooling of the tapped-off gas with oxidizer propellant in a heat exchanger. The warm gas for pressurizing the UDMH tank comes a tap-off of fuel-rich gas at the turbine discharge of the RD-0245. It is only a small portion of the gas from a fuel-rich gas-generator/mixer of the RD-0244 vernier engine. The upper part of the RD-0244 engine was submerged in the bottom propellant (fuel) tank and fastened (and sealed) to the tank bottom with a skirt at the middle of the nozzle-exit section. The thrust force was carried by the tank wall. The TP, preburner, valves, and all of the hot and cold piping were submerged in the hypergolic propellant of the fuel tank. The joints of the engine components were welded and extensively pressure proof tested. All of the engine components inside the propellant tank had to be absolutely leakproof for long periods, while the submarine cruised the ocean with liquid hypergolic propellants loaded into its missiles. The RD-0245 was the complimentary smaller engine with its own TP and four smaller movable (hinged) vernier TCs, which were mounted outside of the missile's first-stage propellant tank. It provides the flight-path control for this missile stage. This vernier engine was outside of the propellant tank and obtained its propellants from the same missile tanks as the main engine. It was integrated with the RD-0244 submarine-launched main booster engine, and both engines started simultaneously. The vernier engine (RD-0245) did operate for a slightly longer period than the main engine (RD-0244) allowing some flight corrections after the main engine cutoff. The relatively large hot turbine exhaust pipes (characteristic of a staged combustion cycle) can be seen in Fig. 8.5-16 going to each of the four vernier injectors and the main injector. A simplified flow sheet of the RD-0243 would be similar in concept to the one shown in Fig. 8.5-13 and a diagram for a submarine missile LPRE is shown in Fig. 8.6-6 in the next chapter. The engine development period is given as 1977 to 1985, and the first flight was in 1986. CADB claims that the RD-0243 engine is the most advanced in terms of specific mass (probably engine mass per unit of thrust) among all existing engines of a similar class.

In 1983 CADB provided the RD-0255 engine for the second stage of the RS-20 ICBM also called the SS-18, Mods. 5/6 heavy-lift long-range ballistic missile.[1–7] It is shown in Fig. 8.5-17 and consists of a main engine RD-0256 and a vernier engine RD-0257 with four TCs as described in Table 8.5-1. The integral tank bottom can be seen as a part of the engine. Even though this is not a submarine missile application, the main engine was also partially immersed in the propellant tank. This allowed more propellant to be put into a given missile stage length and this extra propellant in turn improved the missile

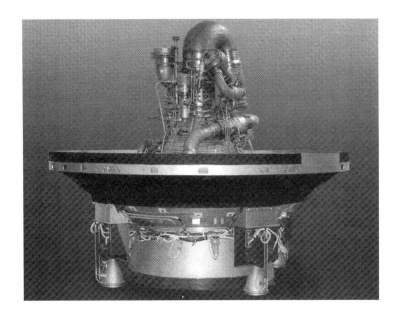

Fig. 8.5-17 RD-0255 LPRE consists of a large main engine (RD-0256) and an autonomous vernier engine (RD-0257). These are submerged in the bottom tank of the second stage of the R-20 ICBM. Copied with CADB permission from Ref. 1.

performance. The RD-0228 engine, which was used on earlier models of this ICBM RS-20A and RS-20B, (SS-18 Mods. 1–4) preceded the RD-0255.

LOX/LH$_2$ Rocket Engine

An exception to the stable of flight-proven CADB engines is the RD-0120 engine, which uses LOX and LH$_2$. It can be seen in Fig. 8.5-18, and a simplified flow diagram is in Fig. 8.5-19. Four of these gimbaled engines were used as the main propulsion for the second stage of the Energiya SLV.[1–6,8,10–12,*] The RD-0120 is the first and probably the only Soviet LOX/LH$_2$ LPRE that has flown. It had a staged combustion cycle and is one of the few Soviet LPREs with a fuel-rich preburner. Only about 22% of the hydrogen is used for cooling the TC; the largest portion goes directly into the preburner. The thrust is the highest of all of the LOX/LH$_2$ engines developed in Russia. The vacuum specific impulse of 455 s is the highest of any flying Soviet LPRE in history.

The RD-0120 has some interesting features and development issues. For example, electric plasma igniters were developed for the thrust chamber and the preburner, allowing restarts in space. The hydrogen pump impeller is

---

*Personal communications, photo and data from publications sent by V. S. Rachuk, general director and general designer of CADB.

Fig. 8.5-18 View of the RD-0120 is the only one of about seven or eight Soviet LOX/LH$_2$ LPREs that has flown. Four RD-0120 were used in each Energiya SLV. Copied with CADB permission from Ref. 1.

strong enough to rotate at the relatively high tangential velocity of 930 m/s or 3060 ft/s. Some of the engine materials were investigated for adequate strength and elongation at cryogenic temperatures and at other operating conditions. The turbine was fabricated by powder metallurgical processes to allow light weight at hot gas temperatures. A special antifriction coating, called Aftol with a low friction coefficient between 0.01 and 0.018, was developed for some moving parts of this engine. The design indicates an advanced state of the art, in part because it benefited from the earlier experiences of such LOX/LH$_2$ LPREs in other countries, from the earlier D-57 LOX/LH$_2$ engine developed by the Saturn/Lyulka Design Bureau (see Chapter 8.8), the KVD-1 LPRE of KB Khimmash (see Chapter 8.6), and from earlier experimental work with this propellant combination at CADB.

The single shaft of the main TP has three hydrogen impellers, two oxygen impellers, and a two-stage turbine. Two booster pumps helped to allow a higher shaft speed of the main TP, and this reduced the main TP's size,

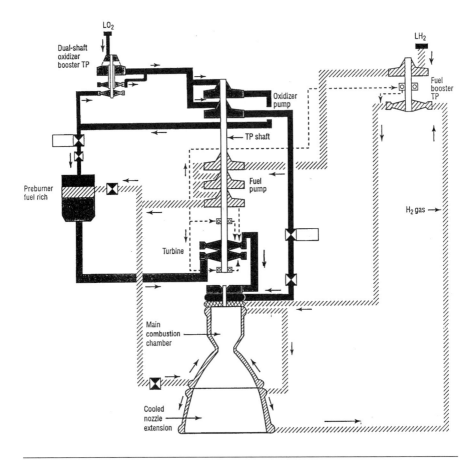

Fig. 8.5-19 Simplified flow diagram of the RD-0120 LPRE showing a staged combustion cycle with booster turbopumps. Courtesy of CADB; copied with AIAA permission from Ref. 10.

reduced propellant tank pressure (reduced tank inert mass), and provided more cavitation margin. An example of advanced design is the lightweight twin-spool oxygen booster pump, which is shown in Fig. 8.5-20 (Ref. 13). There is no other known rocket engine TP with two concentric shafts and two pump impellers running at different rotational speeds. The first impeller, which runs at the lower shaft speed, is driven by a single liquid flow turbine, supplied with high-pressure LOX from the main LOX pump. The second impeller consumes more power, is driven by two turbines, and also has an axial-flow pump impeller. The liquid-oxygen exhaust discharge from the booster turbines is mixed with the pumped oxygen flow. The pressure rise in the lower-speed booster pump stage allows the second-stage booster pump to be smaller and lighter, and it also allows the main oxidizer pump to run at a

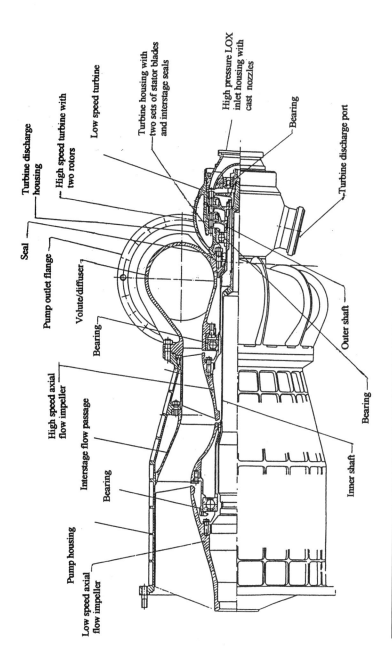

Fig. 8.5-20 Cross section of the oxygen booster pump of the RD-0120 with two concentric shafts (each with one pump the impeller, and its own turbines) rotate at different shaft speeds. Courtesy of CADB; adapted from Ref. 13.

higher shaft speed, thus reducing the size and weight of the main TP. This does a better job of suppressing excessive cavitation in the first main oxidizer pump impellers, than a single shaft booster pump. This dual-shaft booster turbopump is believed to have a superior performance compared to a single-shaft booster TP, but it is more complex and more expensive. The fuel booster pump is driven by warmed gaseous hydrogen, which was heated in the nozzle cooling jacket. Warm hydrogen gas is used for pressurizing the fuel tank and helium heated by the engine for pressurizing the oxidizer tank.

The LPRE electronic control system is advanced and serves to ensure adequate propellant supply under all transient and steady-state operating conditions.[1,2,10,11] This engine has a limited thrust control and a mixture-ratio control. It automatically provides for mixture-ratio changes during ignition and start. There is also an automatic safety system that monitors a series of diagnostic sensors.[11] This includes sensors giving readings of turbopump shaft speeds, gas temperatures at the main turbine inlet and the two booster turbine inlets, temperatures in the igniter region of TCs or preburners, axial displacement of oxidizer pump rotors, pressure changes in the cavities between the LOX and the H_2 pumps, or helium-gas control pressure in critical valves. It then uses special algorithms to evaluate the anomalies of these measured parameter, and a computer then can decide what went wrong and what remedy, if any, should be implemented (such as safe shutdown). Of course such a system cannot foresee all of the potential failure modes, and during development has falsely triggered several engine shutdowns. During the flights of Energiya, there were no problems with the RD-0120 engine safety system, but during some static tests the safety system shut the engine down and saved hardware.

The Soviet's first flight with this LOX/LH$_2$ LPRE was in 1987 in the Energiya SLV. There was a second flight (each flight with four RD-0120 engines) in 1988 before the SLV project and the RD-0120 engine program were stopped. If compared to other nations, the Soviets were late in flying a LOX/LH$_2$ engine. The United States' first LH$_2$ engine started flying in 1963, the European Space Agency (France) in 1979, China in 1984, and Japan in 1986. The RD-0120 engine was designed and built by the Chemical Automatics Design Bureau, but static tests were run at NII Khimmash at Sergiyev Posad, which has large static firing test facilities for large cryogenic LPREs in Russia. Some development tests were also run at a large test facility of NII Mash at Nizhnaya Salda in the Ural mountains. The engine was exhibited in Moscow in 1990.

CADB built and evaluated several different experimental LOX/LH$_2$ rocket engines, which were aimed at a possible LOX/LH$_2$ booster stage and an interorbital transfer stage.[2] Beginning in 1989, after the successful flights of the RD-0120, this DB worked on engines with an expander engine cycle (no preburner or gas generator), where the hydrogen temperature was raised solely by heat absorbed from the TC cooling jacket. There were experimental engines with a single turbopump, one with two separate turbopumps, one for pumping the liquid oxygen and the other the liquid hydrogen. The chamber length had to be increased to obtain sufficient heat-transfer surface for raising the hydrogen gas to a high enough temperature to provide the needed power to the turbines. In one version CADB put some ribs on the inner wall of the combustion chamber to increase the heat-transfer surface and raise the hydrogen temperature. They also tried an annular chamber, which was shorter, but had a large inner-wall surface. There also were two experimental engines with staged combustion cyles, but different component configurations. There also was a version with two booster pumps (not directly connected to the LPRE), one each in the bottom of the two propellant tanks. Another engine version had four TCs and was about 35% shorter. None of these LOX/LH$_2$ engines seem to have been fully developed or qualified, and none was selected for an actual flight vehicle.

Tripropellant Rocket Engines

CADB has investigated several possible versions of tripropellant rocket engines, has tested key components, and operated a bench model of a partial engine.[1,2] Like the Energomash tripropellant engines discussed in the preceding chapter, this work was aimed at a potential single-stage vehicle to be placed into a low Earth orbit. The engine concept is based on a modification of the LOX/LH$_2$ RD-0120 LPRE, to which they added kerosene during the tripropellant boost phase. CADB had built an experimental engine and had started work on engine tests. A key novelty is a new preburner, which had been operated as a component at full flow in a bipropellant or tripropellant mode. Figure 8.5-21 shows the setup for the bench test of a partial engine. All of the oxygen and much of the hydrogen were supplied by the turbopumps of the RD-0120, which were proven and available. All of the kerosene and some of the hydrogen were supplied from pressurized tanks of the test facility in lieu of appropriate pumps. Data for the planned engine are given in Table 8.5-1.

As was mentioned in the preceding chapter, Energomash worked on two other tripropellant engines with one or two gimbal mounted TCs. The Russians were looking for a foreign partner to finance the completion of the engine development. Further work on all three Russian engines was stopped. As already stated, this tripropellant concept was also investigated in the United States, but without building any full-sized hardware, and further work was also stopped because the predicted payload increase (1 to 3%) was modest and the engine would be quite complex. Russian studies showed a 10% predicted improvement, which might have been optimistic.

Other CADB Projects

CADB has undertaken some interesting projects in the last few years. In 1998 they have designed, built, and tested an experimental LPRE (RD-0126) with an expansion-deflection nozzle.[1,2] The TC is shown in Fig. 8.5-22 and the expansion-deflection nozzle flow concept is sketched in Fig. 4.3-9. It has an outer cooled curved nozzle exit and a cooled centerbody. At the lower altitudes the exhaust gas clings to the outer wall, and the middle section has essentially atmospheric pressure. At high altitudes the nozzle-exit section flows full, and

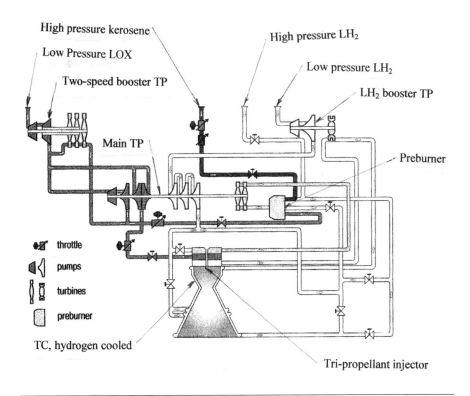

Fig. 8.5-21 Simplified schematic diagram of a bench test setup of an experimental tripropellant LPRE based on the RD-0120 LOX/LH$_2$ engine. Copied with CADB permission from Ref. 2.

the exhaust gas is distributed over the whole nozzle-exit area. An expansion-deflection nozzle can be substantially shorter than most other nozzles and operates at optimum gas expansion at all altitudes. No CADB data on the performance, propellants, or test results were available, and, as far as the author can tell, it has not yet been fully developed. The first known LPREs with expansion-deflection nozzles were designed and tested at Rocketdyne in the early 1960s. Engines at thrust levels up to 50,000 lb thrust were developed and tested as described in Chapter 7.8. At that time the performance benefits of an engine with such a nozzle were not considered to be sufficiently better than other nozzle types to warrant further effort.

CADB also investigated and demonstrated that engines, which were originally designed for NTO/UDMH, can be converted to an environmentally benign propellant combination.[2] They tested the RD-0256 second-stage engine and the RD-0244 submarine engine with LOX/kerosene without any significant redesign of the engines and with a small performance enhancement. The propellant combination of LOX/liquified natural gas (really 98% methane) was investigated by design studies. It gave more performance than LOX/kerosene, but less than LOX/LH$_2$. Methane is readily available in Russia. A version of the RD-0110, and

Fig. 8.5-22 Partial section of the RD-0126 experimental TC with an expansion-deflection nozzle. Copied with CADB permission from Ref. 1.

possibly also a version of the RD-0120, were operated with LOX/methane in ground tests. Other countries have also explored the use of methane as a rocket fuel and it has some avid supporters. As far as the author can tell, no one has used or flown methane as a fuel in a production LPRE.

CADB has done work in improving the fabrication of turbopumps.[13] This includes processes for making turbines, pumps, and other engine parts by using powder metallurgy to fabricated materials, which have good strength and adequate ductility at high as well as low temperatures. The French have developed titanium parts by a powder metallurgy process for their Vinci LPRE pump impellers.[14] Presumably CADB parts are fabricated in the same way. CADB also made precision aluminum castings for pumps housings, valve housings, and other parts with good surface finish (important for low friction losses in pump impellers or pump housings) and with good strength. In some CADB engines precision castings can account for at least 30% of the engine weight. The author has not found any published information on these processes by the Russians.

International Cooperation

An agreement was reached in the 1990s for Aerojet (USA) to market CADB rocket engines, including the LOX/LH$_2$ RD-0120 engine and their potential derivatives to future U.S. applications, but to date no requirement and no customer have materialized. Under a Europe–Russia cooperative agreement (initiated by SNECMA in France and its Vulcain engine European partners), rocket-engine demonstration tests of this RD-0120 engine were conducted at NII Khimmash, Russia's largest test site, in August 1995. The purpose was to validate a mathematical model for future such engines.

There were other agreements not related to the RD-0120. For example, CADB agreed to develop and to provide Pratt & Whitney a newly developed LOX turbopump for an experimental upper-stage P&W LPRE with a nominal thrust of 60,000 lbf. This pump was developed at CADB, delivered to P&W, and integrated with an engine as mentioned in Chapter 7.10. There were also cooperative agreements with Astrium in Germany and with Volvo in Sweden (TPs).

CADB also developed and delivered to Pratt & Whitney one RD-0146 LPRE with an expander cycle.[1,15] In 1997 CADB started a project (in cooperation with the Krunichev Space Center) on the RD-0146 for future use with an advanced version of the Proton SLV or an advanced version of the Angara SLV, which was then under consideration. This engine design was based in part on prior experimental expander cycle engine of CADB. It has features for restart in a vacuum and two separate TPs, one for fuel and one for oxidizer, and could be fitted with an extendable uncooled nozzle skirt. Testing was completed in Russia. This engine is shown in Fig. 8.5-23 and is discussed again in Chapter 7.10. In accordance with an agreement with Pratt & Whitney, one RD-0146 LPRE was delivered to them in 2003, and P&W attempted to market it, but did not find a customer. The thrust level (about 22,000 lbf), specific impulse (455 s vac), and engine cycle are similar to a LPRE of P&W's own flight-proven LOX/LH$_2$ RL10 engine (24,500 lbf). However there are some differences. The RD-0146 has integral booster turbopumps driven by gaseous oxygen and gasified fuel respectively, and they allow lower tank pressure and substantial savings in tank weight. Furthermore this engine provides warm fuel gas for roll control and does not require a separate set of small thrusters. The specific impulse (vac) of the RD-0146 is predicted to be about 451 to 455 s, but the P&W RL10B-2 has an actual value of 465.5 s.

Concluding Comments

CADB is perhaps best known for having developed more upper stage LPREs than anyone else. More than 1400 launches have been made with one type of their LPREs, each using four thrust chambers. This organization also built engines for third-stage applications, a couple for the first stages of flight vehicles, and for an orbital space station. Their RD-0120 LOX/LH$_2$ engine is the only one with these propellants that has actually flown in the Soviet Union; it had one of the first health monitoring systems and a novel efficient oxidizer booster pump, which had two pump impellers rotating at different speeds on two concentric shafts. This DB created several unique schemes for pressurizing the propellant tanks with warm gases not found elsewhere. Another amazing innovation was the use of consumable temporary baffles in the chamber to counteract combustion instability during start-up. This DB developed only

Fig. 8.5-23 RD-0146 operates with an expander engine cycle and has separate booster pumps. The TC is longer than needed for good combustion performance, but long enough to transfer enough heat to the coolant fuel for driving the TP. Courtesy of CADB; copied from Ref. 15.

one engine for submarine-launched missiles; it consisted of a separate vernier engine with four hinged TCs and a main engine, which was submersed in the fuel tank and had the highest known chamber pressure (about 4200 psi). In the late 1950s and the early 1960s CADB developed aircraft superperformance rocket engines, which were flight tested, and also upper stage LPREs for anti-aircraft missiles, which went into mass production. CADB tested several experimental LPREs. One was a breadboard modification of their LOX/LH$_2$ engine configured as a tripropellant engine with kerosene as the third propellant; it was aimed at a possible single stage to orbit application. They also developed an experimental TC with an expansion/deflection nozzle and an engine with an expander cycle and two booster pumps. In order to compensate for the lack of LPRE business, this DB has started R&D work on commercial equipment and products for other businesses.

References

[1]"*KB Khimmmautomatiki* (Chemical Automatics Design Bureau)", Russian Space Agency, Brochure in Russian and English, CADB, Voronesh, Russia, 2001.

[2]Chemical Automatics Design Bureau, *Russian Space Bulletin*, Vol. 4, No. 3, 1997 (English and Russian).

[3]Ratchuk, V., "Best Rocket Engines from Voronesh," *Aerospace Journal*, No. 6, Nov.-Dec. 1986, pp. 30–33.

[4]Wade, M., www.astronautix.com, (cited May 2005).

[5]Lardier, C., "Liquid Propellant Engines in Soviet Union," International Academy of Astronautics, A Paper 99-2-3-04, Oct. 1999.

[6]Lardier, C., "Les Moteurs-Fusées Soviétiques de 1946 a 1991 (Soviet Liquid Propellant Rocket Engines from 1946 to 1991)," International Academy of Astronautics, Papers 97-2-3-03, 98-2-3-09, and 99-2-3-04.

[7]The Ministry of Defence of the Russian Federation (ed.), *Russia's Arms Catalogue*, Vol. IV, Military Parade, Ltd., Moscow, 1997, Parts 3–4, 9, 12–14.

[8]"CIS/Russia: Launch Vehicle Propulsion," *Jane's Space Directory*, Jane's Information Group, Alexandria, VA, 1994–1995, 1999–2000.

[9]Rubinsky, V. R., "Combustion Instability in the RD-0110 Engine," *Liquid Rocket Engine Combustion Instability*, edited by V. Yang, and W. Anderson, Progress in Astronautics and Aeronautics, Vol. 169, AIAA, Washington, D.C., 1995, Chap. 4, pp. 89–112.

[10]Rachuk, V., Gontcharov, N., Matrinyenko, Y., and Fanciullo, T. J., "Evolution of the RD-0120 for Future Launch Systems," AIAA Paper 96-3004, July 1996.

[11]Rachuk, V. S., Goncharov, N. S., Matrinyenko, Y., Barinshtein, B. M., and Sciorelli, F. A., "Design, Development and History of the Oxygen /Hydrogen Engine RD-0120," AIAA Paper 95-2540, July 1995.

[12]Anufriev, V. S., Goykhingberg, M. M., Kalmykov, G. P., and Sirachev, M. K., "From the History of Research and Design of Russian LOX/ LH$_2$ Rocket Engines," *Acta Astronautica*, Vol. 43, No. 1–2, 1998, pp. 19–21.

[13]Demyenenko, Yu. V., Dimitrenko, A. I., and Kalitin, I. I., "Experience of Developing Propulsion Rocket Engine Assembly Feed Systems Using Boost Turbopump Units," AIAA Paper 2003-5072, 2003.

[14]Guichard, D., and Du Tetre, A., "Powder Metallury Applied to Impellers for Vinci Turbopump," *6th International Symposium for Space Transportation of the 21st Century* [CD-ROM], Association Aéronautique et Astronautique de France, 2002.

[15]RD-0146 LPRE, Data Sheet Pratt & Whitney, a United Technologies Co., West Palm Beach, FL, 2003.

8.6 KB Khimmash or Chemical Machinery Design Bureau

Isayev

Alexei Mikhailovich Isayev was one of the Soviet Union's most famous development engineers of LPREs, and he became the founder and initial chief designer of his own rocket-engine design bureau, which later was named after him. In 1967 his DB became identified with the name just listed. His portrait is shown in Fig. 8.6-1. However he did not start his professional career in rocket propulsion.[1–5] As a young, bright, and hard-working engineer, his first job was designing airplanes in a small experimental design bureau (OKB-129) headed by Viktor F. Bolkhovitinov, an experienced aircraft developer. In the late 1930s two of his young and eager engineers, Alexander Bareshnyak and Isayev, designed and built the first Soviet rocket propelled airplane. It was identified as BI-1, and the "BI" stands for the first letter of the names of the two designers. The airplane was shown in Fig. 8.2-3, and it was the first Soviet aircraft capable of flying fully under rocket power. After some flights as a glider and after some powered runway tests, the BI-1 first flew with its rocket engine in February 1940. As described in Chapter 8.2, it was propelled by the D-1-A-1100 rocket engine developed at RNII, and it operated on

Fig. 8.6-1 Alexei M. Isayev, chief designer. Photo provided by Christian Lardier of Air & Cosmos, Paris.

nitric acid and kerosene propellants. During the design and testing of this rocket engine, Isayev became intimately familiar with it.

Beginning in about 1942, he worked on installing propulsion systems in aircraft and on improving the engine designs. In February 1943 he became the leader of a small group of propulsion engineers within Bolkovitinov's Experimental Design Bureau.[1-4] This group was set up in part because the leaders of this design bureau did not like the engines then available. Isayev's tasks were to help engine developers make better engines and to develop and install some good new rocket engines. This small propulsion group was the seed from which a new capable LPRE design bureau would grow.

In 1944 the aircraft development team of Bolkhovitinov was merged with a successor of RNII (discussed in Chapter 8.1) into a new organization NII-1 or Scientific Research Institute #1. Isayev's group developed a new aircraft rocket engine, which had been started in 1943, a LPRE JATO, and other engines as described in the following.

When World War II was over in early 1945, Isayev was conscripted to go to Germany and help collect information on the V-2 and other German weapons.[1,2] He was put into an Army officer's uniform and given the rank of major. The Soviet military at that time had a scavenging team headquartered at Bleichenrode, Germany, trying to make sense out of captured German hardware, reports, and interview results. Isayev became the specialist on rocket propulsion issues because he was then the only man on the team with hands-on experience in this field. He visited Peenemünde and other key places of German rocket-engine work, saw and examined German wartime LPRE hardware, and worked on reestablishing a German production line of V-2 engines. When about 50 V-2 engines were found in storage, Isayev set up a plan and organization to test some of these engines at Lehesten, at a local production test facility. Glushko, who had 10 years more experience with LPREs, was called in to supervise the testing at Lehesten, and Isayev was sent home to use his newly acquired information about German LPREs in the development of Soviet engines.

In 1947 Isayev was promoted and given the prestigious job of chief designer of his LPRE section within NII-1. In 1948 Isayev's group of about 21 people was transferred to NII-88 in Kaliningrad, a suburb of Moscow.[1] This was a missile design (and later a space-launch-vehicle design) organization, and it was headed by S. P. Korolev between 1946 and 1966. The two men formed a close working relationship, and, as will be seen later, it resulted in a good number of future LPREs. Isayev's group became a special design bureau (SKB) within Korolev's organization. In 1952 it officially became an OKB or experimental design bureau, identified as OKB-2 of the NII-88 plant. In 1957 OKB-2 separated from Korolev's organization and became an independent experimental design bureau.[1] In 1967 the name Khim Mash (Chemical Machinery) was adopted, and it became a full-fledged design bureau known as KB Khimmash. To this day this design bureau has continued to work with LPREs in Kaliningrad. The city's name was changed to Korolev in recognition of Korolev's leadership of the Soviet missile and space efforts.

Isayev was a clever inventive engineer, and he has been given credit for several significant technical innovations.[1,4,5] He was the original inventor (in 1946) of the cooling-jacket design for large thrust chambers. It is an integrated welding-brazing design, and it has been improved by others. It features an inner thin hot sheet metal shell (stressed by thermal gradients and the pressure of the cooling fluid) and an outer cooler shell (takes the pressure loads in tension), and they are connected by the intervening corrugated sheet, which forms the cooling channels. It is described as item a in Fig. 4.3-4. He is also credited for conceiving the "submerged" or "wet" engine concept used on submarine-launched multiple stage missiles as discussed separately in this chapter. He was the first to use four TCs in lieu of an equivalent single TC, which allowed a shorter engine and a lighter vehicle. In 1950 Isayev tried a cruciform-cooled injector baffle (about 50 to 100 mm deep) protruding into the TC that had combustion instability incidences. This baffle seemed to cure what later was identified to be the tangential mode instabilities. Isayev used this cross-shaped baffle in some of his other engines, such as the S2.253 for the R-11FM missile discussed later in this chapter. Similar cruciform injector baffles were used for solving stability problems in thrust chambers of other Soviet LPREs. For example, the RS-17 SRBM engine used such a cruciform baffle. A decade or so later injector baffles were discovered in the United States, but with a somewhat different design and for much larger TCs. A picture of a U.S. baffle is shown in Fig. 4.10-2.

Isayev's Design Bureau developed and built 1) aircraft LPREs in the 1940s; 2) small pulsing thrusters, mostly for attitude control, both monopropellant and bipropellant types; 3) small spacecraft engines for space maneuvers, attaining Earth orbits, lunar missions, or deep-space missions mostly between 1950 and 1972; 4) LPREs for submarine-launched ballistic missiles beginning in 1956; 5) LPREs for tactical missiles (surface to air or SAM, surface to surface, and air to surface) beginning in 1948; and 6) one LOX/LH$_2$ engine starting in 1962 for an advanced version of the Soviet N-1 lunar program vehicle.[1-7] This bureau has worked on an estimated 50 to 80 different LPREs. There are not enough reliable data available to the author to prepare a table of key LPREs of KB Khimmash. This DB seems to have been reticent about revealing technical information of its activities or its products and also did not answer letters from the author.

Isayev's Design Bureau developed a number of JATO engines and small aircraft rocket superperformance engines in the 1940s. The RD-1 aircraft LPRE (same engine designation as another engine of Glushko) had a thrust of 1100 kg (2420 lbf) and a pressurized feed system. It was flight tested in the BI experimental aircraft around 1944 or 1945. This aircraft was mentioned earlier in this chapter. There reportedly were also JATO engines for the bomber aircraft Tupolev Tu-4 and Ilyushin Il-28. None were adopted by the Soviet Air Force.

Small Thrusters for Spaceflight Applications

There were several simple small attitude control systems with uncooled nozzles for spacecraft in the 1950s and 1960s. Initially they used pressurized inert gas and later the catalytic decomposition of concentrated monopropellant

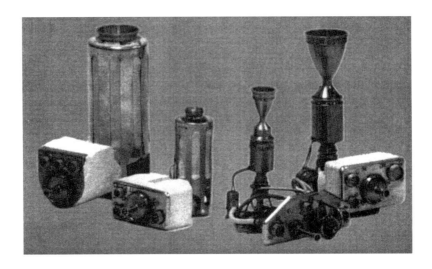

Fig. 8.6-2 Several hydrazine monopropellant thrusters with heated catalyst beds and electrical heaters. Some have thermal insulation. Courtesy of KB Khimmash (photo from personal communications, C. Lardier, Air & Cosmos, Paris, 2004).

hydrogen peroxide. The space capsule of the first manned orbital flight piloted by Yuri Gargarin had two redundant systems, each with eight small uncooled thrusters using inert high pressure gas or decomposed monopropellant for flight maneuvers and for reentry alignment.* An early hydrogen peroxide system was installed in the Soyuz-spacecraft in 1966 for the alignment of the spacecraft for reentry. There were two thruster sizes: one was rated at 1.5 kg thrust or 3.3 lbf and the other at 10 kg or 22 lbf. These hydrogen peroxide monopropellant thrusters were replaced by better performing propellants.

Beginning in the late 1960s the development of hydrazine monopropellant and NTO/UDMH bipropellant low thrust, pulsing thrusters was undertaken at OKB-2. It resulted in a series of available qualified spacecraft engines for flight control with higher performance than hydrogen peroxide. *Hydrazine monopropellant thrusters* were in sizes of 5, 10, 25, and 50 N thrust (roughly 1 to 11 lbf) with vacuum specific impulse values between 228 to 235 s. All used gas-pressure feed systems, a Soviet-developed solid catalyst, and fast-acting valves.[5] Figure 8.6-2 shows several hydrazine monopropellant thrusters developed by KB Khimmash. They have been used in several Soviet spacecraft, satellites, space stations, missiles, and SLV for providing attitude control, stationkeeping, minor trajectory maneuvers, small flight-path corrections, etc. An example is the DOK-10 hydrazine monopropellant engine with a thrust of 10 N or 2.2 lbf, a vacuum

---

*Personal communications, J. Morehart, The Aerospace Corporation, 2001–2004.

Fig. 8.6-3 Several bipropellant thrusters with valves mounted on the injector. Courtesy of KB Khimmash.*

specific impulse of 229 s, a nozzle area ratio of 46, using an iridium-based solid catalyst, which was heated to enhance fast starting, repeatability, and stability. The DOK-50 (50 N or 11 lbf thrust) was similar. The DOT-5 and DOT-25 (5 and 25 N thrust) used a wire catalyst preheated to 620°K or about 347°C.

Figure 8.6-3 shows several different bipropellant thrusters with closely coupled valves.* An example of a bipropellant TC is the DST-100A, with NTO/UDMH, with a vacuum thrust of 100 N or 22 lbf, and a specific impulse of 304 s (vac) at an area ratio of 100 to 1. One unit reportedly was tested satisfactorily for 450,000 ignitions. For restarts in space, this OKB-2 developed radiation-cooled and ablative small bipropellant thrusters. Initial radiating thrusters had lower performance and used stainless-steel walls, before refractory metal chamber/nozzle walls were available. A glass fiber plastic laminate with a steel outer shell was used as the ablative.

Regeneratively cooled thrusters are not suitable because in the small size the heat capacity of the coolant is usually not sufficient to absorb all of the heat transfer from the combustion gases to the thrust chamber. Furthermore restart becomes unpredictable because the soak-back of heat to the cooling

---

*Personal communications, photo from C. Lardier, Air & Cosmos, Paris, 2004.

jacket after shutoff can cause vapor formation of the stagnant coolant and a breakdown and decomposition into gas.

Space Rocket Engines

Beginning in the 1950s, Isayev's Design Bureau also developed at least 30 different rocket engines for lunar and planetary flight missions, and for apogee and other orbital maneuvers, retrorockets, space-launch-vehicle upper stages, and planetary or moon descent.[2-4,6,7,*] The first assigned task was to create a flight-control LPRE for the Vostok manned SLV, and engine development started in the late 1950s. At that time the rocket experts doubted that reliable ignition could be obtained in a gravity-free vacuum. Isayev's first solution was to avoid the vacuum and ignite under pressure, by putting a temporary closure in the nozzles of the thrusters for the Vostok SLV. After some experiments and after a flight, it was learned that it was possible to ignite propellants in a vacuum, and therefore the nozzle closure was no longer necessary. In tests it was learned that the ingestion of gas bubbles by the propellant in the first 2 to 8 s of operation could lead to drastic engine failures. So as to avoid gas ingestion and to facilitate multiple starts in space, Isayev concluded that some means of positive feed system was needed. So Isayev was the first in the Soviet Union to develop positive expulsion devices for operation in space and invented what was called an elastic compensator, which means enclosing the propellant inside the tank in a separate flexible bladder, thus separating the propellant liquid from the pressurizing gas. He also developed a design where the ullage or gas volume of the tank was enclosed in a flexible, expandable bladder surrounded by propellant. Together with some chemical companies, he developed new plastic sheet materials compatible with the stored propellants for long periods of time.

Figure 8.6-4 shows an example of a Khimmash space propulsion system; it is the KDU-414 LPRE with nitric acid and UDMH, with a thrust of 1.96 kN (445 lb), and a nitrogen-pressurized feed system.[2-4,7] This LPRE was developed between 1958 and 1960. It was used for orbit maneuvers in planetary missions, such as Venera, Kosmos, and in two versions of the Molniya SLV. It was also used with Soviet communications satellites. The propellant tanks were inside the conical structure shown in the figure, the nozzle is seen on top, the high-pressure gas tanks are near the base of the cone, and the controls and hydraulic equipment are mounted to the outside of the cone. It had positive expulsion bladders in the propellant tanks and could be restarted. Another example is the lunar soft landing engine KTDU-417 used in unmanned moon missions. It has two engine systems, both with variable thrust, a pressurized feed system, and it uses nitric acid and UDMH propellants. The larger single TC (18.25 to 7.35 kN or 4100 to 1650 lbf) is restartable, runs for 10 or 11 min, and was used for slowing down the flight velocity during the main lunar

---

*Personal communications, J. Morehart, The Aerospace, Corporation, 2001–2004.

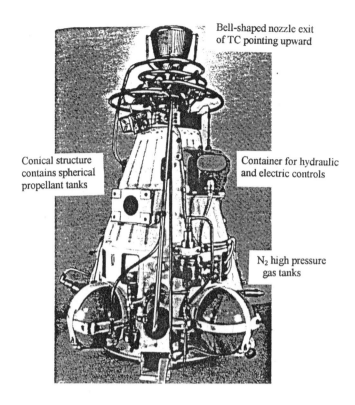

Fig. 8.6-4 KTU-414 LPRE is used for orbit maneuvers and has a conical structure around internal spherical propellant tanks (not visible). Courtesy of KB Khimmash; copied from Ref. 6.

decent trajectory. The second smaller dual-TC unit runs for up to 30 s just before moon touchdown; its thrust varied from 3.4 to 2.0 kN or 770 to 456 lb.

OKB-2 developed most of the rocket engines for space maneuvers used in the third stages or satellites for the Suyuz, Vostok, Voshkod, or Salyut SLV and for a space station. The S5.53 LPRE is one of these.[2–4,7] It used nitric acid (with 27% NTO) and UDMH propellants and had a thrust of about 425 kg or 923 lbf. It was used on satellites lifted by the Soyuz SLV for apogee and orbital maneuvers. The engine system also had eight small monopropellant thrusters (1.5 kg or 3.3 lbf thrust each) for attitude control and reentry alignment. The 880 lbm of propellants were contained in four spherical tanks with a gas-pressurized feed system.

For larger thrusts, long durations, and certain space maneuvers it was possible go to a turbopump feed system and save considerable inert mass in the propellant tanks and the pressurizing gas tanks.[5,7] In turn this allowed more payload in the spacecraft. Isayev might have been the first to develop a spacecraft engine with a TP and a gas-generator cycle capable of restart in space.

He used a combination of positive expulsion bags in the propellant tanks and a purging or blowing out (prior to restart) of any propellant residues in the propellant lines or injectors of the TC and the GG. He also designed the pipes and components, where propellants are flowing, not to have pockets or cavities, where propellant could remain trapped. The first such spacecraft engine was used on the Vostok's return from its satellite orbit to a selected landing site.[5] It had a pump feed system, storable hypergolic propellants, a gas-generator engine cycle, and the turbine exhaust gases had swiveled exit nozzles for stabilizing the flight of the spacecraft during engine operation. The tanks used plastic bags to confine the ullage gas and expel the propellants. The propellant tanks had a propellant management system, which was based on surface-tension devices at the outlet of these tanks, to ensure a propellant flow without gas bubbles. This complex type of a space-capable LPRE with a turbopump feed system did not exist at this time in other countries.

The KTDU-35 was another turbopump-fed LPRE used for orbital maneuvers, including the deorbit maneuver, of manned Soyuz spacecraft.[2–4,7] A diagram is shown in Fig. 8.6-5. Originally the propellants were nitric acid and UDMH, the

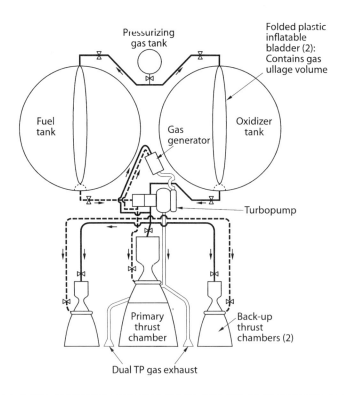

Fig. 8.6-5 Simplified diagram by the author of the KTDU-35 used for orbital maneuvers and deorbiting of Soyuz spacecraft. Some valves and diaphragms are not included. This LPRE was developed by KB Khimmash.

thrust was 4.09 kN or less than 1000 lbf, and the long cumulative burning time was over 15 min. The KTDU-35 had a primary TC with a high nozzle area ratio. The backup engine had two TCs at a lower nozzle area ratio (and thus at a lower performance), but with about the same total thrust. During the Soyuz 33 flight, the primary TC malfunctioned, and the flight was completed using the backup TCs. The plastic folded bladders containing the ullage volume are indicated in the figure. There also were four small monopropellant attitude control thrusters, which are not shown in the figure; they used solid catalysts and a separate pressurized-gas feed system. This engine was first flown in 1966. Modified versions of this engine using NTO/UDMH and different amounts of propellants were used in other Soviet spacecraft.

The "rendezvous-vernier" LPRE for one version of the payload stage of the Soyuz SLV also had a turbopump feed system.[2-4,7] There was a main engine and a standby or redundant engine supplied from the same propellant tanks. Plastic bags in the propellant tanks ensured positive expulsion without gas bubbles and a reliable start in the gravity-free vacuum of space. This LPRE could be restarted and was used for space maneuvers, while approaching the target spacecraft or a space station for a rendezvous; it could also be used for a breaking impulse to transfer the Molniya payload from an orbit into an Earth reentry path. Such a TP system with a low thrust engine is very unusual because at low propellant flow the TPs are notoriously inefficient and the control system for restart can become quite complex. In the United States and in other countries, similar maneuvers have usually been performed with engines that have pressurized feed systems, which are heavier in inert mass, but simpler and perhaps more reliable. None of the other countries are known to develop a turbopump-fed LPRE for rendezvous maneuvers.

Figure 8.6-6 shows the S5.92 engine, which was used on the Frigate Space vehicle, the Phobos mission, and the Mars 8 mission. It had a thrust of 20 kN (about 5000 lbf), a vacuum specific impulse of 327 s, using NTO/UDMH, and a long burn time, reportedly 20,000 s (cumulative).[2-4,7] The KRD-61 LPRE of almost 4200 lbf was used in an upper stage of the Luna SLV for unmanned lunar exploration. It is shown in Fig. 8.6-7, and it had a high nozzle area ratio and a vacuum specific impulse of about 313 s.

LPREs for Submarine-Launched Missiles

The Soviet Union had a number of different submarines capable of carrying and launching ballistic missiles. Some of these submarine-launched ballistic missiles (SLBMs) were propelled by solid-propellant rocket motors, but there were problems with them. Historically the majority of Soviet SLBMs used LPREs with hypergolic storable propellants.[2-4,6,7,*] The Soviets developed single-stage SLBMs and several two stage ballistic missiles with ICBM range, all using LPREs. They also developed at least one three-stage missile, which had

---

*Personal communications, J. Morehart, The Aerospace, Corporation, 2001–2004.

Fig. 8.6-6 S5.92 LPRE was used for Venus and Mars flight missions. Courtesy of KB Khimmash (photo obtained through personal communication, J. Morehart, Aerospace Corporation, 2001–2004).

ICBM range, but with a respectable payload. No other nation ever developed a submarine-launched long range missile with LPREs.

The first ship-launched missile, the R-11FM, was a modification for ship application of the short-range, single stage, land-based ballistic missile R-11 with conventional vehicle structure and tank designs.[3,5] Isayev's organization modified the engine to become suitable for marine use. The S2.253 LPRE for the R-11FM had about 8 tons of thrust and was developed during 1952 to 1957. It ran on nitric acid (with 20% NTO) and kerosene. During development, there was some excessive combustion vibration, and therefore this engine was equipped with a cruciform baffle at the injector face. There also was a version with hypergolic propellants, which did not need a separate start system. The RD-11FM could be launched from the deck of a sub-

Fig. 8.6-7 KDR-61 was used on unmanned flights to the moon with the Luna SLV. Courtesy of KB Khimmash, from Refs. 2–4 (from personal communications, C. Lardier, Air & Cosmos, Paris, 2004).

marine or another navy surface vessel. The submarine had to surface to launch this missile, which made the submarine more vulnerable to potential enemy action.

Beginning with the R-21, the subsequent submarine-based missiles were designed so that they could be launched while the submarine was under water. The missile had to rise through the water to the ocean surface and then fly through the air to its target. All of the Soviet SLBMs were designed and developed in a vehicle design bureau headed by Victor Makeyev, and all of the LPREs (except one), which were used to propel these missiles, were developed by Isayev's design bureau. There were six classes of submarines capable of launching such missiles. Each of the stages of the different ballistic missiles was propelled by its own LPRE, and most also needed auxiliary or vernier engines.

The length of a missile carried by a submarine is limited by the height of the submarine. So to reduce this length, the designers soon used common bulkheads between the missile's oxidizer and fuel tanks, and then they submerged the engines inside the propellant tanks. Figure 8.6-8 shows a diagram comparing these two concepts for a typical two-stage submarine-launched missile. With these concepts the vehicle's length can be reduced by about one-third, more propellant can be carried in a missile of the same diameter, and the missile performance can sometimes be improved. Figure 8.6-9 shows

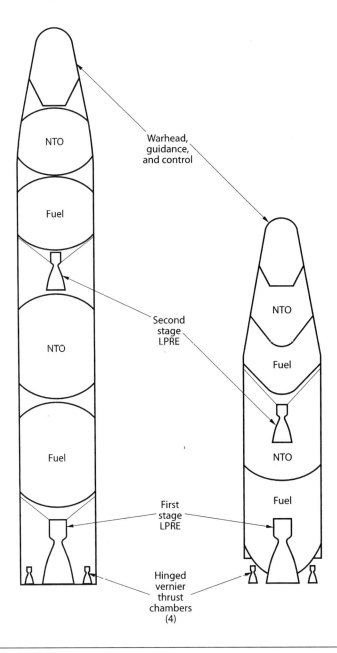

Fig. 8.6-8 Comparison of the missile lengths of two design approaches for tanks with and without common bulkheads and for open as well as submerged LPREs.

 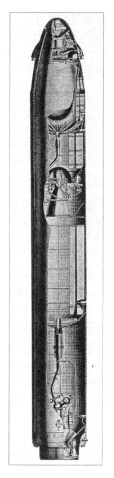 

Fig. 8.6-9 Partial simplified sections through a single stage and two two-stage submarine launched missiles with submerged LPREs. Not drawn to the same scale. Adapted from Ref. 6, part 6.

more detail for three specific SLBMs, but they are not exactly drawn to the same scale.[6] The early ship-launched missiles with LPREs were single stage, launched from a surface ship, and had conventional lightweight propellant tanks with elliptical ends and an intertank structure between tanks. The first underwater launch experiments were undertaken with an engine designed specifically for underwater launch. This engine was developed by Isayev's bureau in 1959–1962. It had four hinged TCs supplied by the same turbopump with limited thrust control and limited mixture-ratio controls aimed at simultaneously emptying both propellant tanks. Propellants were stored and sealed in the vehicle's propellant tanks, materials were compatible with long-term storage and ocean environment requirements, and the engine was

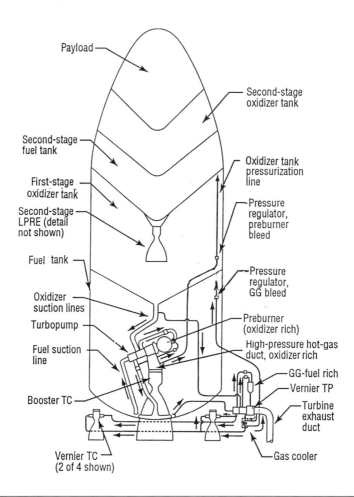

Fig. 8.6-10 Simplified flow diagram of LPREs in a two-stage submarine-launched missile with automatic tank pressurization. The valves and burst diaphragms are not shown. Courtesy of KB Khimmash; reconstructed by author.

made to be leakproof and able to withstand the high hydrostatic underwater pressures during a submersed launch.*

Most of the stages of these ballistic missile are really powered by two LPREs: 1) a main engine with a single large TC, which is fixed (or gimbaled in some second stages) and has either a GG cycle or staged combustion cycle with an oxidizer-rich preburner, and 2) a vernier engine with two gimbaled TCs or four hinged thrust chambers.* The author's idea of a likely flow diagram for many of these submerged engines is shown in Fig. 8.6-10. The turbopump, gas generator, and heat exchanger of the vernier engine are shown for clarity in enlarged

---

*Personal communications, J. Morehart, The Aerospace, Corporation, 2001–2004.

representations outside of the missile, but in reality they would be within the missile's envelope. The vernier TCs were supplied by a separate small turbopump feed system with either a fuel-rich gas generator (shown in the figure) or later with a fuel-rich preburner. The oxidizer tank is pressurized by tapping off some oxidizer-rich turbine exhaust gas from the main engine, and the fuel tank is pressurized by fuel-rich cooled turbine exhaust gas, which has been tapped off from the vernier engine turbine exhaust manifold. The pressurizing oxygen-rich gases are cooled to a manageable temperature in a heat exchanger or cooler (not shown in Fig. 8.6-10) with high-pressure propellant taken from the pump discharge. The pressure of pressurizing warm gases has to be controlled, and this is done by warm gas pressure regulators, which are shown.

The complete second-stage LPREs with a gimbal-mounted main engine and with their movable vernier TCs were placed inside the oxidizer propellant tank of the first stage. Figure 8.6-11 shows the engine originally designed by

Fig. 8.6-11 This engine has been used on the third stage of the RSM-54 submarine-launched missile. Its TP is at an angle and located near the nozzle throat of the TC. Courtesy of KB Khimmash (photo obtained through personal communications, C. Lardier, Air & Cosmos, Paris, 2004).

KB Khimmash for the third stage of the RSM-54 submarine-launched missile. Some second-stage submersed engines had a single gimbal-mounted main TC and roll nozzles using a bleed of turbine exhaust gas, and others used separate hinged verniers. The first-stage main LPRE was submerged in the UDMH fuel tank; only the nozzle-exit section protruded out from the tank, and the thrust force was transmitted through the nozzle skirt to the fuel tank walls. The hinge-mounted vernier TCs with their own TP feed system were outside of the first-stage fuel tank. At stage separation an explosive cord, placed on the propellant tank wall, separated the second stage from the first stage, and the electrical and hydraulic control lines between stages were also severed. A small portion of the cut off empty tank remained with the upper stage. This design concept resulted in a self-contained and very compact package and a short missile length, but was achieved at a penalty in inert vehicle mass and probably at a higher risk of failure.

In the third stage of the RMS-54 missile the propulsion was provided by a combination system of a third stage engine (with TP and a non-gimbaled TC) and a post boost control system with multiple thrusters. During the acceleration period of the third stage, thrust was provided by the fixed engine and the pitch, yaw, and roll controls were provided by the multiple thrusters; all the propellant was supplied for both systems from the turbopump. After the intended flight velocity had been attained, the large TC was shut down, severed from the vehicle and discarded. The TP then operated at a lower speed and flow and the same thrusters could then operate at a lower thrust during the post boost phase of the flight.

The first few sea-launched missiles were not designed to withstand the external water pressure and could not be launched under water from a submerged submarine. As already mentioned, the submarine had to rise to the surface to launch these missiles. When the dynamics of underwater launch were understood and when the propellant tanks could withstand the external water pressure and the LPREs were structurally strengthened, then a submerged launch from a submarine could be undertaken.

The development of the LPREs for submerged submarine launches had to meet several tough requirements. There are many places on a submersed engine where leaks can occur, such as at piping joints, valve stems, injectors, turbopumps seals, gas generators, rotary hinge joints of vernier thrusters, vent lines, drain lines, or sensors. Even a small leak of one hypergolic propellant into a space filled with the other propellant would cause a combustion and the generation of hot, high-pressure gas, which in turn could have disastrous effects on the missile and the submarine. The OKB-2 people have developed design features and fabrication techniques to positively prevent a leak of vaporized or liquid propellant into a space, where the other propellant (or its vapor) is present during storage or would be present during operation. The leak prevention also applies to hydraulic fluids (such as for the hinged TC actuators), and they should not be allowed to mix with the oxidizer or dilute the UDMH. Leaks of pneumatic inert gas (such as for valve operation) must also be eliminated to avoid bubbles in the propellant feed system. Furthermore there should be no leaks to the environment, such as into the submarine and to the ocean, particularly because UDMH and NTO are highly toxic. The Russians use the word "ampulization" for this isolation of the different fluids and the associated structural and seal provisions.

An all-welded construction with extensive pressure and leak testing during fabrication was developed for all of the engine components, piping joints, and connections. Tanks or containers were hermetically sealed during storage. Because hot surfaces of a submerged engine (walls of GG, turbine, TCs, hot gas pipes) could cause boiling or decomposition of adjacent liquid propellant, heat screens or insulation layers were used. The "warm" pressurizing gases cause heating of the propellant at the liquid surface inside the tank and cause thermal stresses in the tank walls. Also the propellant would rise in temperature (particularly the last portion of the propellant removed from the tank), and the resulting in lower density and higher vapor pressure could cause pump cavitation. To prevent this overheating and possible cavitation, the hot turbine exhaust gas is cooled with high pressure liquid propellant in a heat exchanger, one of which is seen in the flow diagram. Furthermore the turbine can be slowed down toward the end of operation, which reduces the thrust, but enhances the cavitation resistance. To avoid gas bubbles from entering the combustion chamber (and cause start problems or combustion vibrations), they devised a gas/liquid vortex-type separator (built into the pump), and the separated gas is discharged overboard. Both of these features are unique and historical and have not been seen elsewhere.

The propellant tanks of submarine-launched missiles are often odd in geometrical shape and therefore heavy, when compared to a spherical tank or to a cylindrical tank with elliptical ends. Visual inspections of the engine or certain prelaunch engine functional checkouts are essentially impossible. The start process must be very reliable and must occur after a long period of propellant storage. This is a unique class of LPRE that does not exist in other countries.

The RD-0243, developed by CADB and seen in Fig. 8.5-16, is the only submarine engine not developed by KB Khimmash. Little data could be found on any of these submerged engines.[6,7] Altogether KB Khimmash developed perhaps 10 different principal engines, about five vernier engines, and perhaps two or three postboost velocity control engine systems.* All of these were for the application of missiles launched from submerged submarines.

Tactical Missiles

The literature, which was obtained through library searches, provided little detail on the LPREs for tactical missile applications The Soviet regime and its military establishment wanted to take advantage of the Wasserfall antiaircraft missile, whose development had not been completed in Germany by the end of World

---

*Personal communications, J. Morehart, The Aerospace Corporation, 2001–2004.

War II. This Wasserfall missile and its engine flow diagram are shown in Figs. 9-10 and 9-11. The job of making copies and completing the development was assigned to another design bureau, and they managed to fly the USSR-made copies. It soon became evident that the payload and range of this single-stage surface-to-air missile were too small.

Isayev became involved and attempted to reduce the propulsion system weight and upgrade the performance of the LPRE of the Wasserfall.[1,2] In the late 1940s he tried to use chemically generated gas from a solid propellant or a liquid propellant gas generator) instead of the compressed nitrogen for tank pressurization. This would have resulted in a small weight savings, but the engine system would be more complex. This effort to use chemical pressurization was not successful and was not continued.

Isayev also tried a new TC, but it would not significantly upgrade the Wasserfall missile. However the application of a turbopump feed system would achieve a significant reduction in inert tank weight, which would allow a better Wasserfall missile performance. This occurred after RNII had already tested a TP in 1945. The development of an experimental TP was undertaken, and it is not known if this TP development was ever completed or if the TP was installed in a Russian Wasserfall. However this TP, which was intended for a LPRE of about 8 tons of thrust, became the basis for upper-stage LPREs of larger two-stage tactical missiles and later of new Soviet short-range missiles.

A few of the early Soviet designs for surface-to-air missiles were single-stage vehicles (like the Wasserfall), but most were two stage, with a solid-propellant motor in the booster stage and a LPRE in the upper stage. Isayev tried a new TC design with 8 tons of thrust, but it had some combustion instability in the TC. For the next LPRE Isayev went to four smaller TCs of 2 tons each.[2-4,7] This modification seemed to cure the instability; in retrospect it was a tangential vibration mode. He subsequently learned how to stabilize the combustion in the TC. The new S.09-29 storable propellant engine of 8-ton thrust (four chambers) was developed around 1950 and installed in the surface-to-air missile 205/R-113 or SA-1. It went into production, and many were built and deployed.

The S5.1A engine for a surface-to-air missile was developed during the 1954 to 1958 period. It had a variable thrust ranging from 17 to 5 tons or 37,400 to 11,000 lbf. It had a turbopump, used a nitric-acid oxidizer (RFNA) and a hypergolic amine-based fuel, and the engine weighed about 122 kg or 268 lb. A modified engine, called S3.42A with a variable thrust (17,000 to 5000 kgf or 37,400 to 11,000 lbf), was used on the upper stages of the 217 and 218 surface-to-air missiles (also known as SA-2 or V-750) and in an uprated version (18 ton) in the 5Ya24 surface-to-air missile. Figure 8.6-12 shows two LPREs of KB Khimmash used in upper stages of air-defense missiles.[6] All used a simple gas-generator engine cycle, a single-shaft TP, acid cooling, and some were restartable. The TC used the cooling-jacket design with corrugated intermediate metal sheets as discussed in Chapter 4.3.

In addition to KB Khimmash, the CADB (Chap. 8.5) also developed some LPREs for surface-to-air missile applications, as was mentioned in the preceding chapter. At least six different surface-to-air missiles were put into mass

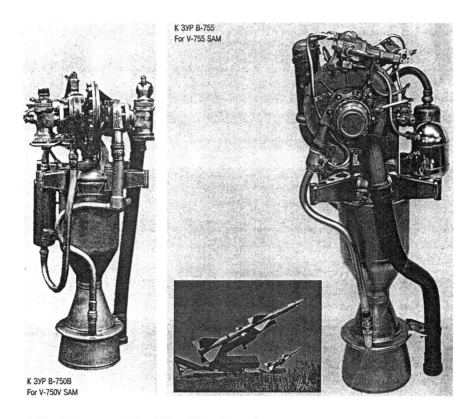

Fig. 8.6-12 These two LPREs, used in the sustainer stages of the V-750V and V-755 surface-to-air tactical defense missiles, were put into mass production. Courtesy of KB Khimmash; copied from Ref. 6.

production. Thousands were installed around Moscow, Leningrad, and other cities. They were in such a hurry that some of these missiles were produced and even deployed before they were fully qualified. As many as 2000 to 5000 LPREs were installed of just one missile, and of course more than this quantity of engines were built.[6,*] A version of the SA-2 missile was exported to perhaps 10 other countries for their air defense. Altogether between 10,000 and 20,000 antiaircraft missiles were built. These represent the largest production quantities of pump-fed engines in all LPRE history.

Korolev's organization (see Chapter 8.10) with some help from Iseyev developed a relatively small short-range or tactical surface-to-surface missile identified as the R-11, later as SCUD B, surface-to-surface missile. This missile did not fit into Korolev's plan to build ICBMs, and he did not give much personal attention to it.[1] Isayev built its S2.253 engine with a pressurized feed system. It used nitric acid and kerosene and had a sea-level thrust of 8.3 tons or about

---

*Personal communications, J. Morehart, The Aerospace Corporation, 2001–2004.

Fig. 8.6-13 KVD-1 LOX/LH$_2$ engine was originally intended to fly in an upper stage of the N-1 Moon SLV. It was sold to India and has flown in an Indian SLV first in 2001. Courtesy of KB Khimmash from personal communication, J. Morehart, The Aerospace Corporation, 2001–2004.

29,400 lbf. The missile was flown in three series of tests between 1953 and 1955. It was accepted by the military in mid-1955 and became a useful mobile military weapon with a long life. It soon replaced the R-1 missile, which was a Soviet copy of the German V-2. The R-11 quickly became operational in the Soviet Union. The SCUD B was larger, had an engine with a TP-feed, and 60% more range. SCUD B and some SCUD A missiles were exported to various countries including Egypt, Iran, Iraq, Lybia, North Korea, Slovakia, or Syria. It and its engine were modified several times. It is the only Soviet developed surface-to-surface missile that has been operated in combat, first by Egypt in the 1973 war against Israel. Eventually it became the renowned mobile Scud B missile (this name was given by NATO) used by Iraq against Israel and Saudi Arabia in the Persian Gulf war of 1991 and more recently by Russia in the

Chechen conflict. An uprated longer-range version (more propellant, but less payload), as modified by Iraq, was also used in this conflict. Copies of this missile and engine were built in several countries.

Isayev's DB also was reported to have developed several LPREs for air-to-surface tactical missiles.[2-4] For example, the S5.33 LPRE (83 kN full thrust, 5.9 kN cruise thrust) propelled the Kh-22 air launched missile.

Engines with Liquid Hydrogen

The benefits of LOX/LH$_2$ and its superior performance were well understood by the people responsible for the design of the N-1 moon exploration multistage vehicle. Initial plans visualized at least one or two of the upper stages to use this energetic propellant combination.[1] Korolev and his team pushed for LOX/LH$_2$ to gain some payload margin in the N-1 design. Contracts were given to OKB-2 (Isayev) for an engine of about 7.0 to 7.5 tons of thrust and to OKB-165 (Lyulka see Chapter 8.8) for an engine with about 40 tons.* The major initial problem was to get an adequate supply of liquid hydrogen, which did not then exist in the Soviet Union.[1] Existing suppliers had inadequate capacity and inefficient plants. Progress was slow. Also it was decided to build special facilities for testing with liquid-hydrogen fuel, and that took longer than planned.

Isayev developed a single TC engine with a staged combustion cycle identified as KVD-1 or 11D56, and it is shown in Fig. 8.6-13. It was to have a thrust of 73.6 kN or about 16,500 lb, at a chamber pressure of 58.1 atm, capable of five restarts in space over a period of 11 days.[1,7,8]* It was expected to provide a vacuum specific impulse of 463 s, which would be the highest performance in the Soviet Union. The two verniers TCs of 2 kN each using LOX and gasified hydrogen were part of this system. The program was started in 1962, and the first static test firing was five years later in 1967, partly because of inadequate funding and partly because of a lack of enough liquid hydrogen. The program continued through the early 1970s. The delays in this and other vehicle subsystem programs caused Korolev to change his plans for the N-1 SLV, and he took the LOX/LH$_2$ engines out of the first version (even though this greatly reduced the payload) and replaced them with LOX/kerosene engines. The LOX/LH$_2$ engines were rescheduled to fly with a more advanced, but subsequent N-1 SLV version. The N-1 vehicle program was canceled in 1974 after four failed flights, and with it the further development of this engine was stopped. The KDV-1 had not flown in the Soviet Union. The engine was revived, and it has been marketed for use on the fourth stage for a modernized version of the Proton SLV program.* It has also been sold to India to serve as an upper stage in their SLV, as discussed in Chapter 14. The contract called for seven KVD-1 engines to be delivered to India.[7-9] The first flight of the new KDV-1 propelling the upper stage of the India's GSLV took place in April of 2001.

---

*Personal communications, J. Morehart, The Aerospace Corporation, 2001–2004.

Other Items

In the 1950s Isayev developed a booster LPRE for the twin launchers of a supersonic ramjet. The project was called Burya, which means storm in Russian, and its rocket-propelled vehicle boosted the ramjet to its operating altitude and speed. The program was aimed at an eventual cruise missile with an intercontinental range. The S2.1100 engine has a cluster of four thrust chambers and was integrated with its propellant tanks forming a cylindrical assembly. Two such booster assemblies, one on each side of the ramjet, were needed to launch the test vehicle. Total rocket engine thrust at takeoff was 137 metric tons or about 300,000 lb. The program was not continued.

The design bureau KB Khimmash should not be confused with another entirely different organization in a related business, but with the same abbreviated name, namely, NII Khimmash. NII stands for Scientific Research Institute. It was located near Zagorsk about 90 miles north-northeast of Moscow. NII Khimmash is a rocket-engine test organization and has a series of large facilities for static testing of large LPREs and for tie-down rocket firing tests of large complete vehicle stages. Other design bureaus have been using these facilities for testing their larger engines. For example, the RD-0120 of CADB and the second and third stage of the N-1 SLV were tested here.

Concluding Comments

This Experimental Design Bureau was started by Alexei Isayev, one of the Soviet Union's most creative engineers. Amongst other inventions he reportedly was the originator of the cruciform baffles for controlling combustion vibrations and the cooling jacket with intermediate corrugated sheet metal. Both of these ideas have become widely used. KB Khimmash was active and still works on many types of small thrusters using cold inert gas, decomposition products of hydrogen peroxide (now obsolete), catalyzed hydrazine, or storable bipropellants. The bureau is perhaps best known for its several rocket engines for various space missions, such as for satellites, orbit maneuvers, retroaction, space rendezvous, lunar operations, or for third stages of planetary or lunar launch vehicles. A separate, most unique application, not found in any other country, is several LPRE immersed inside the propellant tanks of submarine launched ballistic missiles. Several second stage engines of anti-aircraft missiles were developed in the 1950s and they were deployed in large numbers. KB Khimmash developed and qualified only one LOX/LH$_2$ engine, but it has not flown in the Soviet Union. However it was sold to India and did fly for the first time as the third stage of an Indian SLV.

References

[1]Siddiqi, A. A., *Challenge to Apollo, the Soviet Union and the Space Race (1945-1974)*, NASA SP-2000-4408, Washington, DC, 2000.

[2]Lardier, C., "The Soviet Rocket Engines Beginning in 1945," International Academy of Astronautics, Paper 97-2-2-03, 1997 (in French).

[3]Lardier, C., "The Soviet LPREs of 1946 to 1991," International Academy of Astronautics, Paper 98-2-3-09, 1998 (in French).

[4]Lardier, C., "Liquid Propellant Engines in the Societ Union," International Academy of Astronautics, Paper 99-2-3-04, Oct. 1999.

[5]Tavzrashvily, A. D., "Engines and Propulsion Units for Space Vehicles Constructed by Alexey M. Isayev," *History of Rocketry and Astronautics*, American Astronautical Society History Series, Vols. 18–19, Univelt, San Diego, CA, 1984–1985.

[6]The Ministry of Defence of the Russian Federation, *Russia's Arms Catalogue*, Vol. IV, Military Parade, Ltd., Moscow, 1997, Parts 3–6, 9, 12–14.

[7]"CIS/Russia: Launch Vehicle Propulsion," *Jane's Space Directory*, Jane's Information Group, Coulsdon, Surrey, England, 1999–2000.

[8]Anufriev, V. S., Goykhingberg, M. M., Kalmykov, G. P., and Sirachev, M. K., "From the History of Research and Design of Russian LOX/ LH_2 Rocket Engines," *Acta Astronautica*, Vol. 43, No. 1–2, 1998, pp. 19–21.

[9]"India's GSLV Reaches Orbit, But Can It Be a Contender?", URL: http://www.jones.com/010420_1_n.shtml, (cited 20 April 2001).

8.7 NII Mashinostroeniya or the R&D Institute of Mechanical Engineering

In 1958 an R&D organization (abbreviated as NII Mash) for small thrusters was set up as a branch of the R&D Institute of Thermal Sciences, which has been a leading R&D institute on all types of propulsion technology in the Soviet Union. This institute can be traced back to RNII, where it originally started as a small analytical group. It has undergone reorganizations and name changes. In 1995 this institute was renamed the Keldish Research Center, and it is described in Chapter 8.12. A new branch was established specifically for developing and providing a family of small bipropellant thrust chambers or thrusters for the flight control of rocket-propelled vehicles.[1-3,*,†] It is located at Nizhnyaya Salda about 100 miles north of Ekaterineburg (formerly Sverdlovsk) in the Ural mountain region. In 1981 it separated from the parent organization, but kept its original name as listed in the preceding heading. It is one of several organizations in Russia that developed and delivered small thrusters. The applications for small thrusters were described in Chapter 4.6, which also includes an injector design of NII Mash. Most of the information and all of the detailed data in this chapter were generously furnished by NII Mash in 2002 at the author's request.*

NII Mash's main business is to develop and make small bipropellant thruster assemblies.[1-3,*,†] Other Soviet organizations also make these kinds of small thrusters, but not as a main product line in those companies. At NII Mash the small thrusters are identified as low-thrust rocket engines or LTREs. They have qualified about 30 different small thrusters or LTREs. They design and provide the propellant valves and sometimes include pressure transducers with their thrust chamber assemblies. Occasionally NII Mash develops a complete engine system with propellant tanks and a gas-pressurization system. Their LTREs have been flown in attitude control sytems, used for flight stabilization or minor maneuvers of spacecraft, and have been installed in vehicle stages, space stations, and military and nonmilitary satellites. Up through 2002 their thruster assemblies have flown in more than 800 Soviet rocket vehicles and with more than 16 families of spaceflight vehicles. For example, their thrusters were used successfully in several auxiliary modules of the Space Station *MIR*, the Space Station *Salyut*, several versions of the Soyuz manned spacecraft, and the Buran Space Shuttle. The missions cited by NII Mash seem to relate primarily to space exploration applications. Little data were available about the use of their thrusters in military applications. NII Mash has not produced or developed hydrazine monopropellant thrusters.

The first generation of small thrusters used radiation-cooled stainless-steel single-walled chambers/nozzles, a single central spray type injector element,

---

*Personal communications, Y. G. Larin, chief designer, NII Mashinostroeniya, Nizhnyaya Salda, Russia, 2002–2003.
†Personal communications, J. Morehart, The Aerospace Corporation, 2001–2004.

running on NTO and UDMH at a mixture ratio of 1.85. Available thrust sizes were 50, 100, 135, 200, and 400 N.[2,3] They have been serially produced, and they have flown. As an example, the 400 N (approximately 90 lbf) thruster is shown in Fig. 8.7-1, and it has flown in at least five different satellites. It weighs 2.5 kg (5.5 lbm) and operates at 95 psia chamber pressure. The chamber diameter in the figure is 51 mm or 2.0 in. The vacuum steady-state specific impulse for all five of these LPREs varies between 235 and 254 s, for nozzle area ratios of about 52 to 56. The valves operate nominally at 27 +/– 7 V. The four connections or fittings with protective covers in the figures are for oxidizer, fuel, electrical wiring, and a pressure sensor. The start delay to reach 90% of full thrust was about 30 ms for the smaller sizes and about 50 and 100 ms for the two

Fig. 8.7-1 LTRE-400 is typical of the first generation of small thrusters, and it was used on Kvant, Kristall, Spektr, Piroda, and Almaz spacecraft. Copied with NII Mash permission from Ref. 2.

larger sizes. Most of their first-generation thruster assemblies were endurance tested for 100,000 start cycles and some of them for more than 500,000 start cycles. To avoid damage of the stainless wall at the hot nozzle-throat region, the film-cooling design (discussed later) provides for the local wall temperature to remain well below approximately 1200° K or about 1700°F. This was achieved by using 30 to 40% of the NTO as a film coolant. This NTO absorbs heat energy, when it is vaporized and heated, and also chemically, when NTO breaks down into nitrogen oxides and hot oxygen. With the oxidizer-rich annular combustion gas zone near the periphery of the chamber/nozzle, the gas in the center is fuel rich, which gives the best performance. Comparable small thrusters in other countries typically used fuel for film cooling, which was then considered to be safer because it does not have much free hot oxygen, which under some circumstances can cause oxidation (burning) of the hot wall.

The 12-N thruster is shown in Fig. 8.7-2, and it is unusual. Its chamber and nozzle-throat region are partly regeneratively cooled with a fuel-cooled coiled-tube segment and also an oxidizer-cooled coiled-tube segment.[2] This cooling arrangement and the placement of the propellant valves close to the injector are shown by the simplified diagram in Fig. 8.7-3. The nozzle-exit section is radiation cooled. The nominal thrust is 12 N or 2.7 lbf, dry mass was given as 0.55 kg, the specific impulse was 279.6 s in steady-state vacuum operation, and the start delay up to 90% thrust was 0.015 s. It had flown satisfactorily in the Kwant, Kristall, Spektr, Priroda, Almaz, and other spacecraft.

With the propellants in the cooling coils in intimate contact with the hot inner wall, there is the potential danger of heat soak-back from the hot inner wall to the stagnant coolants after shutdown. The heated coolant will change temperature, density, and other properties, can contain vapor bubbles, and, if very hot, the coolant will locally decompose and form a gas. When restart occurs relatively soon after shutdown, then the initial propellants flowing into the injector are warm and can contain bubbles, which often cause problems. These effects depend on the time between pulses and the duration of each pulse. In other words this 12-N thruster will work well in many applications, but there are regions of pulsing operation where there might be problems.

The second generation of NII Mash small thrusters became available in the 1970s and 1980s, was developed in different sizes, and used a niobium (often still called columbium) chamber/nozzle wall with a protective disilicide internal coating.[2,3,*,†] Data for seven different sizes are given in Table 8.7-1. Because this refractory material allows a higher wall temperature, it requires less film cooling, it has higher nozzle-exit-area ratios, and the performance is much better than the first-generation LTREs. Higher nozzle-exit-area ratios will improve the specific impulse even further. All of the thrusters in the table run on NTO/UDMH at a nominal mixture ratio of 1.85 and at 27 V power. Start delays were about the same or slightly shorter than the first generation. Two or more

---

*Personal communications, Y. G. Larin, chief designer, NII Mashinostroeniya, Nizhnyaya Salda, Russia, 2002–2003.
†Personal communications, J. Morehart, The Aerospace Corporation, 2001–2004.

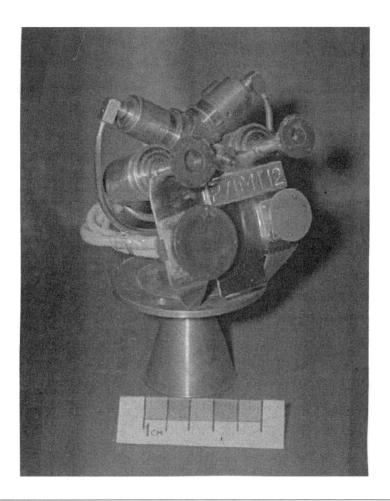

Fig. 8.7-2 LTRE-12 was different. It is partially cooled by propellant flowing through coils around the chamber. Thrust is 12 N or 2.70 lbf. Copied with NII Mash permission from Ref. 2.

individual thrusters are often packaged into a subassembly, which makes it simpler and easier to check and install. Figure 8.7-4 shows such a package.

In 2002 NII Mash had been operating some experimental thrusters at a higher performance or a lower inert mass than the data given in the table. They also experimentally investigated MMH (for better performance) and alcohol or kerosene running with gaseous oxygen (for enhanced environmental compatibility).[2,*,†]

---

*Personal communications, Y. G. Larin, chief designer, NII Mashinostroeniya, Nizhnyaya Salda, Russia, 2002–2003.
†Personal communications, J. Morehart, The Aerospace Corporation, 2001–2004.

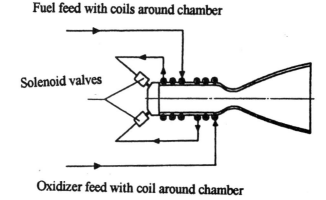

Fig. 8.7-3 Simplified diagram of the LTRE-12 showing the regenerative-cooling coils around the chamber. Copied with NII Mash permission from Ref. 2.

Figure 4.6-5 shows the basic injection scheme of using a coaxial cylindrical injection element with double conical sprays.[2] There is only a single injection element in each of their thrusters. Prerotation of the propellant flow is achieved by swirling vane inserts in each propellant passage. A conical fuel spray emerges from the lip of the center hole, and it is shown by a dashed line. The conical spray of oxidizer starts at the lip of the outer cylinder (shown by a dash-dot-dash line), and the spray forms small droplets and reaches the outer wall for film cooling. In addition there is a series of axially directed individual oxidizer injection holes, whose jets (shown by dash-dot-dash lines) break up the conical sprays (or the conical flow of burning atomized propellant droplets). This allows hot gas and vapors to flow between zone I and zone

Table 8.7-1 NII Mash Second-Generation Bipropellant Thrusters[2]

LTRE model number	50M	100M	135M	135MA	200A	500A/1	400M
Average rated thrust, N	54	98	130.5	130.5	196	490	392.4
[a]Average specific impulse, s	290.3	308.3	308.3	310.3	312.3	315.3	300.2
Inlet pressure, MPa	1.32	2.06	2.16	2.06	2.06	2.06	1.96
[b]Startup delay, s	0.030	0.030	0.030	0.030	0.037	0.050	0.050
Mass, kg	1.3	1.5	1.5	1.8	2.0	3.5	3.0
Nozzle expansion area ratio	52.35	100	100	120	100	150	67

[a]for steadystate operation
[b]from start signal to 90% of full thrust

Fig. 8.7-4 Cluster of five thrusters is part of the Russian-built cargo unit of the International Space Station. These are LTRE 135M and LTRE 12 thrusters. Copied with NII Mash permission from Ref. 2.

II of the figure and equalizes the pressures between these two zones. According to their chief designer, this feature allows stable combustion operation.[2,*] NII Mash and other Soviet companies prefer NTO as the film coolant, in part because it absorbs heat during decomposition into nitrogen oxides and it stores well. The oxidizer droplets from the upper cone spray shown in the figure reach the TC wall and provide the film cooling of the chamber and the nozzle-throat region. The wall temperature of the critical nozzle-throat region will not usually exceed 1200 K or about 1700°F. This type of injection element is different from most small thrusters developed in several other countries, where impinging streams or discrete oxidizer and fuel jets are used and where fuel is used for film cooling.

Two of the NII Mash designed electromagnetic valves are shown in Fig. 4.7-11. Both have an integral filter to remove small solid particles, such as rust flakes or machining chips.[2] The valve on the left uses a hemispherical soft Teflon® type plunger sealing against a conical seat. The valve is normally closed by a spring and opened or held open by an electromagnet. The enlarged detail indicates that the seal is made by deforming a small ring-shaped portion of the Teflon® (on the tip of the plunger) of depth b caused by the spring load. NII Mash believes that there is a possibility for the formation of oxides in the annular clearance space between the armature and the housing of the electromagnet, which would bind and prevent the movement of the

---

*Personal communications, Y. G. Larin, chief designer, NII Mashinostroeniya, Nizhnyaya Saldo, Russia, 2002–2003.

Fig. 8.7-5 View of one version of the Molniya communications satellite with a propulsion system using pressurized-gas and NTO/UDMH propellants. Copied with NII Mash permission from Ref. 2.

armature or plunger. NII Mash designed the valve so that the magnetic force would break most any binding spot. The other valve on the right side has an armature or plunger in the form of a disk held in place by a dual belleville-type spring. It opens and closes more rapidly than the other valve (0.003 to 0.005 s) and usually will not bind by the potential formation of oxide. NII Mash has an experimental small propellant solenoid valve that weighs only 8 g. It is used on a small 8-N thruster using gaseous propellant, and it opens in 1.0 ms at 10-MPa pressure and 2 g/s of flow. The industry's Coordination Council has recommended the use of such solenoids valves in all Russian small thrusters. One of the NII Mash solenoid valves is now a standard for all of the small thruster applications and also is used in thrusters made by other companies.[2]

To achieve the required high reliability, NII Mash has applied a common sense rigorous and stepped approach.[2,*] It has steps such as these: 1) minimum number of moving parts; 2) trustworthy, yet forgiving seals, which are not affected by changes in operating conditions (e.g., force, temperature, pressure); 3) duplication or redundancy (e.g., dual valves) and sometimes triplication of less reliable joints or contacts (e.g., they often used triple electrical contacts); 4) avoidance of excessive flow velocities in tubes or valves (choking); 5) firing tests with propellants of every thruster that is to be delivered, under test operating conditions, which are agreed upon with the customer; 6) qualification testing under overload or overstress conditions, such as higher chamber pressure, longer duration, off-mixture-ratio operation, or warm/cold propellants; 7) appropriate tooling and fixtures during fabrication and assembly (also proper inspections of key dimensions, surfaces, etc.) and checks during manufacture, such as pressure or flow tests and electrical continuity tests; 8) periodic certification of equipment, machinery, test instruments, and personnel; and 9) proper installation and functional checkouts in the flight vehicle. This includes proper loading of propellants and venting of propellant lines to eliminate bubbles. NII Mash seems to have carefully planned these and other steps, and they have delivered good products. They are aware of two known flight failures of small thrusters, but these thrusters were from another supplier and not from NII Mash.

One of the NII Mash thrusters (400 N or 90 lbf) was tested by Aerojet in Sacramento, California, in 1994. It had a stainless steel nozzle extension and a chamber that was film cooled by NTO.[4] The steady-state performance of this thruster with NTO/UDMH measured 253 s of specific impulse, just below the Russians nominal value of 255 s. It took 30 ms during start to reach 90% thrust, which is typical of NII Mash practice. But when Aerojet used MMH instead of UDMH as a fuel, there came a surprise: the specific impulse was higher by 40 s, namely, 293 sec (Ref. 4). In 1994 the cost of the TC was reported to be about 10% of a similar U.S. version at that time. In the United States MMH is commonly used as the fuel, thrusters have often higher nozzle-exit-area ratios (more than 200), start delays are shorter, and vacuum specific impulses can be 10 s higher in many cases. In Russia UDMH seems to be the standard fuel for all LPREs with storable propellants, is available at launch sites, and is used with the other engines in the vehicle.

NII Mash recently also has built some cold-gas thrusters.[2] A number of thrusters at 0.8 N or 0.18 lbf thrust each were used with nitrogen (specific impulse is approximately 73 s) and helium (specific impulse is about 189 s) in some versions of the Kosmos satellite, first flown in 1995. The 5-N thruster (1990) operating on compressed air is used on the man maneuvering unit fastened to a cosmonaut's back during space walks.

---

*Personal communications, Y. G. Larin, chief designer, NII Mashinostroeniya, Nizhnyaya Saldo, Russia, 2002–2003.

For the Buran space shuttle vehicle launcher they developed a 200-N thruster using gaseous oxygen and liquid kerosene operating at an oxidizer-rich, but variable mixture ratio of 3.5 to 6.0 (Ref. 2). Although liquid oxygen would have been available in this vehicle, cryogenic liquids have not usually been successful in small thrusters because it is essentially impossible to predict the density, amount of evaporation, or mass flow. Gaseous oxygen is more predictable and can be made to work. This LTRE flew first in 1988. Finally they have built a few complete LPREs with multiple thrusters and lightweight propellant tanks, storable propellants, using gas-pressurized feed systems. One such LPRE was developed for the Molniya satellite. It has a 600-N (135-lbf) sustainer thruster with a specific impulse of 305 s (steady state), four attitude control thrusters at 10 N each, several propellant and pressurizing gas tanks, and suitable controls and piping.

Figure 8.7-5 shows a version of the complete propulsion system that was developed for the Molniya satellite. The nozzle of the axial thruster is on top, the smaller attitude control thrusters are mostly hidden in this view, but the spherical propellant tank and the small spherical high-pressure gas tanks are visible. They have also experimented with a self-contained pressurized feed system using lightweight fiber reinforced plastic tanks and multiple thrusters.

In addition to their R&D facilities and manufacturing facilities for small thrusters, NII Mash has a large facility for use by other LPRE developers.[2,3] It was used in development testing of the large LOX/LH$_2$ CADB RD-0120 engine for the Energiya SLV (about 400,000 lbf thrust).

NII Mash is a relatively small organization dedicated primarily to bipropellant small thrusters using mostly NTO/UDMH. The first generation of the thruster family used stainless steel chambers and nozzles. The second used niobium and gave better performance. Typical TCs have a single coaxial injection element, oxidizer film cooling, and their own fast acting valves, capable of pulsing. At least 800 successful flights (without failure) have been made using their small reliable thrusters. NII Mash also developed thrusters with partial regenerative cooling, another with gaseous oxgen and kerosene, and a couple of complete propulsion systems with propellant tanks, structures and their own gas pressurizing system.

References

[1]The Ministry Defence of the Russian Federation (ed.), *Russia's Arms Catalogue*, Vol. IV, Military Parade, Ltd., Moscow, 1997, Parts 4, 13, 14.

[2]"Low Thrust Rocket Engines and Propulsion Systems of R&D Institute of Mechanical Engineering," NII Mash, Informal Company Document, Nizhnyaya Salda, Russia, 2002.

[3]http//users.ev1.net/ larin/spce_russia/niimash (cited Oct. 2005).

[4]Dornheim, M. A., "Aerojet Tests Russian Oxidizer Cooled Thruster," *Aviation Week and Space Technology*, 27 June, 1994, pp. 75, 76.

8.8 NPO Saturn, formerly OKB Lyulka

This experimental design bureau (OKB-165) was identified with Arkhip M. Lyulka, its first chief designer, and has traditionally developed gas turbine aircraft engines. OKB-165 was started in 1946 and resides in Moscow. The name NPO Saturn (Scietific Production Organization Saturn) was adopted in 1967 when it was amalgamated with other organizations.[1] In the 1960s and early 1970s this bureau was essentially required to work on LPREs, primarily on a high-energy LPRE for the N-1 lunar launch project. This launch-vehicle program was canceled in 1974, and this engine project and further work on LPREs were stopped in 1975.

The Lyulka's Design Bureau's only contribution to the technology of LPREs was a LOX/LH$_2$ upper-stage engine, which had several unique features.[2,3] It was the highest-thrust LOX/LH$_2$ engine tested in the Soviet Union at that period of time, and it is shown in Fig. 8.8-1. As mentioned before, the original design concept of the N-1 manned moon flight vehicle, which was designed

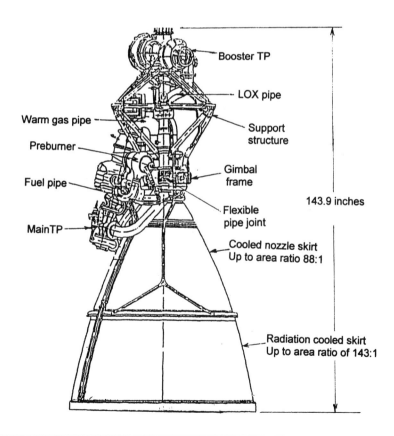

Fig. 8.8-1 Simplified line drawing of the 11D57 (or (D-57) LOX/LH$_2$ LPRE. Copied with AIAA permission from Ref. 4.

Table 8.8-1 Selected Parameters of the D-57 LPRE

Parameter	Value
Thrust in vacuum	19.35 kN (88,300 lbf)
Chamber pressure in combustion chamber	10.93 MPa (1,585 psia)
Engine mixture ratio, oxidizer/fuel	5.8
Estimated specific impulse in vacuum, s	456.5
Nozzle area ratio at nozzle exit	143:1
Nozzle area ratio for regenerative cooling	88:1
Engine weight	821 kg (1812 lbm)
Maximum diameter	1.58 m (72.5 in.)

by Korolev's DB, required the use of this cryogenic high-performance propellant combination in one or two of its upper stages. There were initial vehicle design configurations that used this engine in the second stage (cluster of six or eight hinge-mounted engines) and in the third stage (one or two gimbaled engines). When this and other the engine program were delayed and when technical problems surfaced, the plans for the N-1 program were changed.[1] It was decided to use conventional LOX/kerosene engines in all stages of the initial version of the N-1 vehicle. The oxygen/hydrogen upper-stage engines (Lyulka D-57 engine and Isayev's KVD-1 engine) were then designated to propel upper stages of a more advanced but later version of the N-1 vehicle. In making this decision, the designers were well aware that the payload was greatly diminished; this meant that there would be only two astronauts instead of three flying toward the moon and only one astronaut would land on the moon. It also required engines to use turbopumps, where possible, and the margins for unplanned extra inert mass were drastically diminished.

OKB Lyulka started by developing and testing several small LOX/LH_2 TCs. The original engine model, identified as 11D54, is a LOX/LH_2 engine with a modified staged combustion cycle.[2,3] A variation of this Lyulka engine, identified as the 11D57, was modified with a gimbal mounting for the fourth stage of the same later version of the N-1 vehicle.[1] In more recent years the designation D-57 has been given to this LPRE. A simplified outline drawing of the engine is shown in Fig. 8.8-1, and selected engine data are in Table 8.8-1. This version of the engine had a gimbal mount around the converging section of the nozzle, but it is only partly visible in this figure. The single preburner delivered a fuel-rich gas, as did all of the other Soviet LOX/LH_2 LPREs. The D-57 engine could be throttled to less than half of its rated thrust.

Figure 8.8-2 is a simplified schematic flow diagram. It shows two TPs in an unusual arrangement. One has at lower shaft speed and lower power and drives two booster pumps (one booster pump for fuel and one for LOX). The

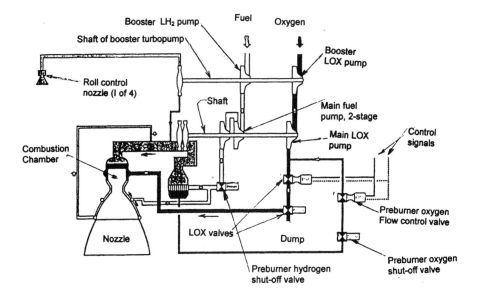

Fig. 8.8-2 Simplified schematic flow diagram of the 11D57 (or D-57) LOX/LH$_2$ LPRE. Adapted from Ref. 2; copied and adapted with AIAA permission from Ref. 4.

main TP operates at higher speed and much higher power and drives the two main pumps, a single-stage oxidizer pump and a two-stage fuel pump. This is a unique TP arrangement that is not known to exist elsewhere. It reduces the number of turbines, allows some simplification, and can be lighter than other feed systems. It is technically difficult to throttle the flow with two TPs in series without significant changes in mixture ratio, and therefore throttle valves and controls had to be installed in the lines feeding the TC and feeding the preburner. All of the hot preburner gas flows first through the main turbine, and most of the gas then flows into the main TC, but some of the exhaust gas is tapped off from the main turbine discharge pipe and used for driving the turbine of the booster pumps and then for supplying gas to roll control nozzles (eliminating vernier thrusters). Several different nozzle-exit versions were built and tested. The main nozzle's exit area ratio was originally planned to be 220:1, but that was too large for the vehicle. Some engines were built at a ratio of 170:1, but the flow separated from the nozzle wall in ground tests. The extendible radiation cooled nozzle had an area ratio of 143:1 with a base of a regeneratively cooled nozzle at 88:1. Most of the ground testing was done with a nozzle-exit-area ratio of 88:1 or less.

The main nozzle and a part of the chamber is cooled by a partial flow (30%) of liquid hydrogen; most of the hydrogen (70%) flows into the preburner. The upper portion of the chamber is cooled by liquid oxygen. The pressures in the cooling jackets are high. The control of the thrust is obtained by controlling the flow and the mixture ratio of the propellants to the preburner by means of two throttle valves.

This engine had several noteworthy features, and they are described next. Reference 4 gives more detail of these features, the engine's development, testing, materials, and development problems.

1) The engine is not using a conventional staged combustion cycle, where all of the preburner gas is injected into the combustion chamber. A small portion of the turbine discharge drives the turbine of the booster TP, supplies gas to the roll control nozzles, and is dumped overboard. This causes a small performance penalty, but eliminates separate roll control thrusters.

2) The two booster pumps are driven by the same turbine on a single shaft, which is different from other LPREs, where each booster pump has its own turbine and there are two booster TPs. This might well be easier to start, and it eliminates a turbine.

3) An extendible nozzle was used in one version of this LPRE. Its main merits were a shorter engine and a savings in vehicle mass. It was moved into its extended position while the engine was firing (but at 50% of full thrust), and this is unique. In most other countries the extendable nozzles are moved into their operating position before the engine is started. It is the first extendable nozzle known to the author that is placed into position while the engine is operating. There is a small performance advantage to a two-position extendable nozzle, if a significant portion of the flight is at relatively low altitude.

4) The D-57 has two different cooling jackets. One jacket in the chamber portion (next to the injector) is cooled by liquid oxygen. The lower part of the chamber and the nozzle throat region (down to an area ratio of 88:1) are cooled by some of the fuel in a separate cooling jacket; this warmed-up fuel flow then goes into the main injector. Although there had been some early Soviet TCs with two cooling jackets in the 1930s (nozzle and 100% of chamber), these were apportioned differently. This engine is the only large LOX/LH$_2$ LPRE known to the author with two cooling jackets in the chamber, one with oxidizer as the coolant and the other with fuel as the coolant.

5) The inner walls in the chamber and the nozzle-throat region (down to an area ratio of 4:1) are coated with zirconium oxide, and this insulation layer reduces the heat transfer and wall temperatures. The coating was applied to an inner wall of a milled channel cooling jacket made of copper, which was alloyed with a small amount of chromium. Several other LPRE organizations have tried to use such a zirconia coating, but the coating spalled off or peeled off locally and was not satisfactory. The Russians must have had a good coating technique.

6) The alignment of the main TP housing was achieved by axial dowels and line boring of the case prior to assembly.

7) Radial pins are provided for the thermal growth of the turbine housing.

8) For starting and restarting of the engine, a flow of high pressure helium is supplied to the turbine of the booster pump. This creates enough propellant pressure to start the initial combustion in the preburner and the initial rotation of the main pumps. For a fast start other engines have used high-pressure hydrogen (J-2 engine), and the use of inert helium gas is safer and different.

This Lyulka engine has been developed in several slightly different versions and ground tested extensively in component tests and engine tests. The development of this engine took much longer than anticipated because of three reasons. Initially (early 1960s), there was not enough liquid hydrogen available to do meaningful testing.[1] Furthermore the building of new plants to produce liquid hydrogen and the buiding of suitable test facilities took longer than anticipated. Secondly, the funding was stretched out and was at times inadequate. The third reason was technical problems, which required more time to solve than expected.[1,4] In 1975, when the engine work was stopped, a total of 53,000 seconds of ground-test operation had been recorded, a total of 105 such engines had been built, and most of them (about 85%) had been used up in the development effort.[2] This high number of R&D engines suggests that there were serious technical difficulties. This LPRE has never flown, but the work served as a lesson and a prototype for the subsequent development of the larger RD-0120 by CADB (see Chapter 8.5), and it did fly.

No known LPRE work seems to have been done at NPO Saturn/Lyulka since 1975. In the 1990s the engine was revived and redesignated as D-57, and it included some plans for improvements. In 1993 Aerojet teamed up with Lyulka to market and promote this D-57 engine in the United States for a potential single-stage-to-orbit SLV application.[4] This effort resulted in some serious U.S. vehicle studies, but no hardware sales.

References

[1]Siddiqi, A. A., *Challenge to Apollo, the Soviet Union and the Space Race (1945-1974)*, NASA SP-2000-4408, Washington, DC, 2000.

[2]"CIS/Russia: Launch Vehicle Propulsion," *Jane's Space Directory*, published every two years, Jane's Information Group, Coulsdon, Surrey, England, 1999–2000.

[3]Anufriev, V. S., Goykhingberg, M. M., Kalmykov, G. P., and Sirachev, M. K., "From the History of Research and Design of Russian LOX/LH$_2$ Rocket Engines," *Acta Astronautica*, Vol. 43, No. 1–2, 1998, pp. 19–21.

[4]Anderyev, A. V., Chepkin, V., and Fanciullo, T. J., "The Development of the D-57 Advanced Staged Combustion Engine for Upper Stages," AIAA Paper 94-3378, June 1994.

8.9 OKB Kuznetsov, Reorganized as NPO Samara

This is the development organization of Nikolay D. Kuznetsov (an influential chief designer), and it was established in 1946 as an OKB or experimental design bureau, identified as OKB-276 for the purpose of developing aircraft piston engines and since about 1950 gas-turbine engines. It is located at the city of Kuybyshev, which is now known as Samara, some 500 miles in an easterly direction from Moscow. In 1967 it became a full design bureau and was renamed KB Trud.[1-5] Thereafter it was merged with a production plant and was given the designation NPO Samara. (Samara Science and Technical Complex).

As stated in Chapter 8.3, the Soviet government wanted to broaden the capability for developing ICBMs and SLVs, and the government caused several design bureaus to undertake jobs in this new field. Kuznetsov's DB was more or less told to work on LPREs.[1] In 1959 they were assigned their first serious job: the development of two LPREs intended for the first and second stage of the R-9M (SS-8) intercontinental ballistic missile.[2-5] They used LOX/kerosene and a gas generator engine cycle. A list of 10 important LPREs developed by this organization is given in Table 8.9-1, and it is divided into three groups.[2,3,6] Each group is for a particular vehicle.

The initial 11 months of the two engine development programs for the R-9M ICBM were troubled. Out of 57 ground test firings, 26 had failures, 21 experienced periods of combustion instabilty, and only eight of the firings were successful during this time period. However all of the technical difficulties were resolved, and Kuznetsov then had two very satisfactory LPREs, the NK-9 (about 80,000 lbf thrust) for the first stage and NK-9V (larger nozzle and 88,000 lbf in vac) for the second stage of the R-9 ballistic missile.[1-5]

The letters NK stand for the initials of the chief designer. The gas generator for these two engines produced oxidizer-rich gas. However these two NK engines were not selected for installation into the actual ballistic missile. Instead the RD-111 LPRE of the Glushko DB (for the first stage) and the RD-0106 LPRE of the Kosberg DB (later called CADB) (for the second stage) were chosen for a slightly different version of the R-9 ICBM. These two engines are described in Chapters 8.4 and 8.5 and were developed in parallel to the two Kuznetsov engines. The engines developed by OKB Kuznetsov then became surplus.

Next Kuznetsov's OKB-276 was given the job for all of the four LPREs for the first four stages of the ill-fated large N-1 lunar and planetary SLV.[2-6] This was a very responsible and important job for a LPRE design organization that was relatively inexperienced compared to other LPRE DBs. Politics played a role in the award to Kuznetsov's DB.[3]

S. P. Korolev, the man in charge of the Soviet spaceflight program wanted to develop a six-stage vehicle, called N-1, for manned lunar expeditions. It was an enormous Soviet effort, and it was aimed at getting to the moon before the Americans with their U.S. Apollo/Saturn moon launch vehicle.[1,7] For good reasons Korolev had made up his mind that LOX/kerosene was the best propellant combination for this project, and he selected it for the first stage of the

Table 8.9-1 LPREs of the OKB Kuznetsov/NPO Samara

Designation	Thrust, tons	No.TCs in stage	I_s, s sec	Engine cycle	Development period	Application
NK-9	36	NA[a]	NA	GG[b]	1959–62	First stage, R-9A ICBM
NK-9V	40	NA	NA	GG	1959–62	Second stage, R-9A ICBM
NK-15	154 SL[c]	30	297 SL	SC[d]	1962–69	First stage, N-1 moon SLV
NK-15V	179 vac[e]	8	346 vac	SC	1962–69	Second stage, N-1 moon SLV
NK-19	41 vac	4	352 vac	SC	1962–69	Third stage, N-1 moon SLV
NK-21	41 vac	1	352 vac	SC	1962–69	Fourth stage, N-1 moon SLV
NK-33[f]	154 SL	30	297 SL	SC	1969–74	First stage, N-1 moon SLV
NK-43	175 vac	8	342 vac	SC	1969–74	Second stage, N-1 moon SLV
NK-39	41 vac	4	352 vac	SC	1969–74	Third stage, N-1 moon SLV
NK-31	41 vac	1	352 vac	SC	1969–74	Fourth stage, N-1 moon SLV

[a]NA = data not available. [b]GG = gas generator engine cycle. [c]SL = sea level. [d]SC = staged combustion engine cycle. [e]VAC = vacuum. [f]Improved version of NK-15.

N-1 vehicle.[1] As mentioned in the prior chapter, he initially considered LOX/LH$_2$ for two or three of the upper stages. He subsequently eliminated LOX/LH$_2$ in the first version of N-1 and wanted LOX/kerosene engines for the first four stages. Korolev originally wanted Glushko's organization, the most experienced development organization of large LPREs, to design and build these four engines. But Glushko felt strongly, also for good reasons, that storable propellants were the best for this application, and he refused to accept the job to develop LOX/kerosene engines for the N-1. So Korolev decided to ask Kuznetsov. He personally visited Kuznetsov in Kuybyshev and persuaded him to develop the four LOX/LH$_2$ engines for the N-1 moon program.

Beginning in 1962, the Kuznetsov DB modified the NK-9V for use in the third and fourth stage of the N-1 vehicle (identified as NK-19 and NK-21 engines in the table). The OKB-276 also started to develop the two larger engines needed for the N-1 SLV, the NK-15 at nominally 154,000 kgf thrust (SL) or 339,000 lbf for the first stage and the NK-15V with a higher nozzle area ratio and 175,000 kgf thrust (vac) or 385,000 lbf (vac) for the second stage. By 1964 the Bureau had a nonfunctional mock-up of the NK-15. All of these engines used LOX/kerosene propellants, and all had efficient staged combustion engine cycles with oxidizer rich preburners. Figure 8.9-1 shows one of these engines. The designs of the major engine components were quite similar in all four of these LPREs. There was good cooperation between Kutznetsov's engineers and engineers at some of the other design bureaus.[1] The literature cites visits by engineers to the Glushko DB for technical consultation.

In the first or booster stage of the large N-1 SLV, there were 30 NK-15 engines, the largest number of large LPREs in any rocket flight takeoff.[1-4] Of these 24 were in a large circle near the outside diameter of the first stage, and six were in a small circle in the center. The second stage had eight NK-15V engines, which were similar, but with a larger nozzle area ratio; the third stage had four NK-19 engines, and the fourth stage had a single NK-21 engine, which was gimbal mounted. The third- and fourth-stage engines each gave about 402 kN or 90,000 lbf (vac) thrust. The TCs were regeneratively cooled by fuel, which went from the cooling jacket directly into the injector. The hot oxidizer-rich gas from the preburner burned with the heated fuel in the combustion chamber.

The NK-15, NK-15V, and the NK-19, which were used in the first, second, and third stage, could each be throttled down to about 60% of full power. This throttling of the selected engine in a cluster was used to achieve pitch and yaw control of the flight vehicle.[4,5] This method of thrust vector control of an engine cluster is unusual and not usually found in engines outside of the Soviet Union. Figure 4.9-6 shows how throttling would work with a four-engine cluster. In the first stage of the N-1, there were 24 NK-15 engines in a circle, and a bank of six adjacent engines would be throttled at the same time to achieve pitch and yaw control. The fourth stage engine NK-21 was gimbal mounted and thus could provide the pitch and yaw control of the flight of the fourth stage. Smaller thrusters were used for roll control of this fourth stage. In addition to the cluster of large primary engines, the first and second stage had four roll control TCs,

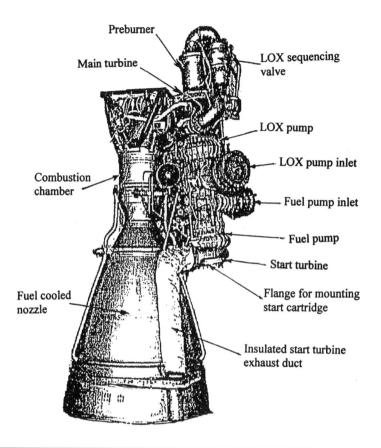

Fig. 8.9-1 View of NK-33 engine, which is very similar to the NK-15 engine. Copied with permission from NPO Samara.

and their engines were developed by Melnikov's department at Korolev's design organization. They are metioned again in Chapter 8.11. These roll control engines were identified as 11D121, and they used the same propellants. Each 11D121 had 7 tons or about 15,000 lbf of thrust (first stage) or 6 tons (second stage) of thrust and a vacuum specific impulse of 313 s.

All of these engines for the N-1 SLV were tested extensively and were then put into production. Testing included all sorts of overstress conditions, such as extended durations, extreme thrust variations (49 to 114% of power), excursions of chamber pressure or mixture ratio, propellant with different ambient temperatures, or variations in start and shutdown procedures. Test were also made in clusters of engines installed in the second, third, and fourth stage.

The large number of engines in the first and second stage allowed an engine-out capability. If one of the engines had certain predetermined kinds of abnormality in its sensed or measured parameters, which would indicate a potential impending failure, the vehicle controller would automatically shut off the engine with the "out-of-tolerance" parameter before the engine would fail and cause potential damage to the vehicle or a failure of the mission. Furthermore the controller would simultaneously shut off the engine on the opposite side of the vehicle, so that there would not be any large turning moment imposed on the vehicle. The remaining engines would continue to operate, but for a longer period of time. There also was a limited engine-out capability in the second and the third stages.

One of the handicaps of the engine development programs was the lack of a facility to ground test the complete first stage with all of its 30 large LPREs for their full duration.[1] It was too large and too heavy for any existing test facility. Because the Soviets at the time were trying to beat the Americans in a manned flight to the moon, the delay for creating such a facility and the delay for running the full-stage tests was not an option and was unacceptable. Instead Korolev depended on extensive analysis, some cold (nonfiring) test of the stage structure, and extensive ground tests of the engines. There were adequate facilities for hot testing each of the complete upper stages. Therefore the dynamics and interactions between the first stage structure, the full load of propellants, and the engines were never determined until the vehicle flew. In retrospect some of the flight failures described next would probably have been detected in ground tests of the first stage using dummy masses on top of the stage to simulate the upper stages and payload.

Another major handicap was that this set of engines could not be restarted. For example, there were some explosively actuated valves, which could not be restored to their original condition. Apparently some component of the engine could be damaged in the first firing, and it would be risky to attempt a second run. Quality was ensured by overstressing and testing two out of every seven engines selected at random from the production lots. If the two engines tested satisfactorily, they were then discarded, and the remaining five of the lot of seven were accepted and then installed in the flight vehicle. Of course there is a small probability that one of the five accepted engines would have an undetected flaw.

The first N-1 flight in February 1969 failed, and the three subsequent flights also failed to complete the powered flight portion of the trajectory.[1-7] On at least two of these flights and perhaps also on a third flight, the flight failure might have been triggered by a problem in one of the Kuznestov engines. However this has not been conclusively proven. It is not always possible to identify the exact cause of a flight failure or some of its consequences, but it is usually possible to identify several plausible or likely causes. One possible failure in the first flight was probably caused by a failure of a feed pipe leading to a preburner and the simultaneous failure of a small tube for an instrument. Then propellants spilled into the engine compartment and caused a fire, which in turn burned the electrical wiring insulation and affected adjacent engines. Another possible cause is a structural failure of the vehicle, when too many engines are started or stopped simultaneously. Good engineering practice requires a short time interval, say, 0.1 s, between starting banks of engines in a cluster. In the third flight a failure of one of the feed pipes occurred during engine shutoff, when very high dynamic pressures would build up in the long feed pipes.

After the first N-1 flight failure (and even more so after the second failure), an intensive modification and improvement program was undertaken on an urgent basis, with the objective to increase robustness, reliability, and durability. This included for example a freon fire extinguishing system in the engine compartments and additional thermal insulation of pipes and electric conduits. An engine improvement program of these four engines was also started at Kuznetsov's organization.[1,4–6,]* Restart and reuse capability was designed into the new versions and this allowed engine acceptance firings and/or vehicle stage check-out firing tests before the engine's flight. For example, valves with pyrotechnic actuation were replaced by solenoid valves or other valves, which could be operated again, and a gas purge was added to flush the propellant piping and avoid propellant accumulations in the chamber after shutoff. The turbopump castings were strengthened. An automatic control of the ignition process was installed, and it worked well. It included sensing of the ignition of the solid-propellant start cartridge before the hypergolic start liquid was released to the preburner and the sequencing of increasing oxidizer and fuel flows. Where possible, loose piping was eliminated and replaced by internal passages inside castings or tubing firmly fastened to components. Compared to other engines seen by the author, these engines looked relatively very clean with very little external small piping (such as those used for overboard dumping of seal leakage, valve actuators, purges, instruments, drains, vents, etc.), and almost no external loose wiring (for controls, valve actuation, or instrumentation).

The new improved engines had essentially the same performance, the same appearance, and the same interfaces as the four earlier engines, but they had been improved in several of the details and were given new designations. They were the NK-33 (booster), NK-43 (second stage), NK-39 (third stage) and NK-31 (fourth stage).[2–7,]* They are listed in the third group of LPREs in Table 8.9-1. Figure 8.9-2 shows a simplified schematic diagram. This diagram shows some details of the TP piping, the thrust is controlled by throttling the fuel flow to the preburner (which reduces the preburner gas temperature), and the mixture ratio is controlled by throttling two valves, one in each main propellant line. The minor valves are not shown in this diagram, and the purge lines are also omitted. Thrust termination is initiated by cutting off the fuel flow to the preburner. Table 8.9-2 shows some data for one of the four improved engines.

The R&D testing, qualification testing, and the production testing and inspection requirements were made tougher. These four improved engines passed qualification and overstress testing, and two sets of these engines were

---

*Personal communications, J. Morehart, The Aerospace Corporation, 2001–2004.

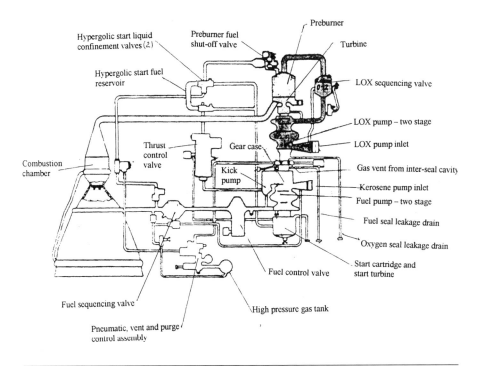

Fig. 8.9-2 Simplified schematic diagram of an NK variable thrust (throttling) engine. Courtesy of NPO Samara; adapted with AIAA permission from Ref. 7.

installed in two of the N-1 vehicles, but the N-1 program was canceled before the vehicles with these improved engines could fly.[6,7]

The turbopumps of all of the Kuznetsov LPREs for the N-1 SLV were different from other turbopumps, and a sectioned view of one TP in seen in Fig. 8.9-3. Each TP has two different turbines, one at each end of the TP assembly. One turbine is driven by oxidizer-rich preburner gas at about

Table 8.9-2 Selected data for NK-33 LPRE

Thrust	1501/1681 kN [339 (SL)/378 (vac) lbm]
Specific impulse	279/331 s
Propellants	LOX/kerosene
Nozzle-exit area ratio	27
Chamber pressure	14.5 MPa (2100 psia)
Engine dry weight	1242 kg (2738 lbm)
Engine height	3.7 m (146 in.)

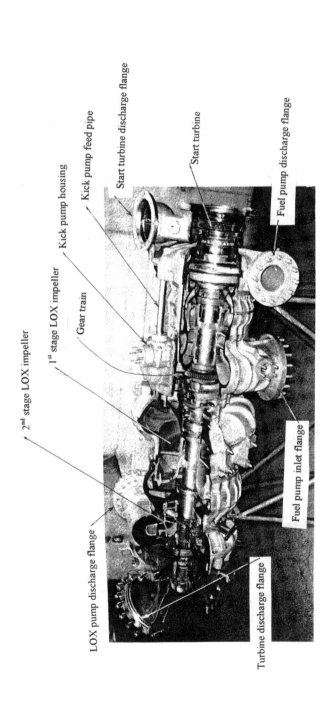

Fig. 8.9-3 Turbopump of the NK-33 LPRE with a turbine at each end, a two-stage oxygen pump, a two-stage kerosene pump, and a gear train driving a partial flow third-stage fuel pump. The main turbine cannot be seen clearly in this view. Courtesy of NPO Samara; copied with AIAA permission from Ref. 7.

671°F, and it operates for the full operating duration. The second turbine is used only during the startup and is driven by gases from a solid-propellant cartridge, and this enables a relatively fast start. For example the NK-33 reached 90% of its full thrust in 1.7 s. The extra startup turbine adds weight and lengthens the TP assembly and requires a solid-propellant start cartridge. A discussion of the reasons for using a separate start turbine is given in the next chapter on the Youshnoye Design Bureau, which also developed some TPs with two turbines. The main turbopump was rigidly fastened to the TC. The preburner assembly was attached to the main turbine, which was at the top of the TP assembly. The solid propellant start cartridge was attached to the start turbine, which was installed at the bottom of the TP assembly. These made the TP assembly relatively long, and it protruded above the TC injector. The exhaust from the start turbine was dumped overboard.

The nominal steady-state pump discharge pressures are high, namely, 5660 psia for the oxygen pump and 31.7 MPa for the kerosene main pump. A small portion of the fuel pump discharge is diverted to a small separate fuel "kick pump," and this pump raises the fuel pressure to 7865 psia, which is high enough to overcome valve and line pressure losses and the injector pressure drop of the preburner. The kick pump sticks out from the main fuel pump assembly and is driven (through a set of gears near the center of the TP) at a higher shaft peed (48,000 rpm) than the main pumps (17,440 rpm). The preburner consumes all of the oxygen and only a small flow of kerosene at a mixture ratio of 57.6. Most of the fuel flow goes through the cooling jacket. The chamber pressure in the combustion chamber could be varied between 1350 and 2200 psia.

Figure 8.9-4 presents a typical combustion chamber pressure vs mixture-ratio diagram for the NK-33 LPRE used commonly during many engine developments. It shows the engine was stable over a wide range of likely operating conditions well over the minimum range of the N-1 requirement.[6,7] It is part of the overstress testing to which these engines were subjected.

An appreciation of the size and complexity of the large N-1 space vehicle can be obtained from Fig. 8.9-5. The height was 105.3 m or 345.5 ft, and the base diameter was 16.8 m or 55.1 ft. The takeoff thrust was 4620 metric tons or 10.1 million lbf, the equivalent pull of perhaps 200 locomotives. The top picture is a bottom view, and it shows the 30 NK-15 engines. Of these eight were located near the center, and 24 were in a circle at the periphery. Selected banks of six adjacent engines were throttled to allow pitch and yaw control of the vehicle. Roll control was provided by four separate LPREs, but they cannot be seen in this figure. From the inclined criss-cross columns of the interstage support structures, it appears that a "hot separation" was used, when dropping off the first stage or dropping off the second stage. This hot-separation technique is discussed in more detail in Chapter 7.7 for the hot stage separation in the U.S. Titan missile and in Chapter 13 for the hot stage separation in the Chinese ICBM and heavy SLV.

The N-1 program was canceled in 1974, in part because there were four successive flight failures, which have already been mentioned.[1] The inert

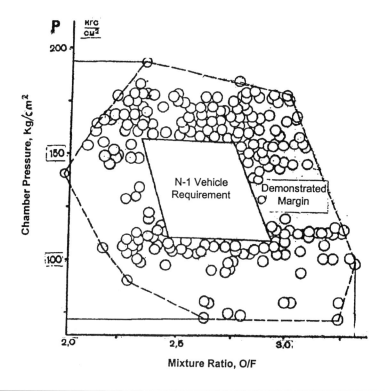

Fig. 8.9-4 Typical way to demonstrate stable operation over a range of mixture ratios and chamber pressures was to run many tests outside the normal range of operating conditions. Courtesy of NPO Samara; copied with AIAA permission from Ref. 6.

takeoff mass of the N-1 was the highest ever launched, considerably higher than that of the Apollo/Saturn V space-vehicle stack. However the payload, or the mass that could be placed on the moon, was substantially less with the N-1. This was because the Saturn V had LOX/LH$_2$ engines in the second and the third stage. Apollo could send three astronauts into space and land two on the Moon, but the N-1 could only carry two astronauts into space and land only a single one on the moon. The program cancellation of the N-1 might have avoided a public embarrassment of admitting a lesser payload capability, and this might have contributed to the decision to cancel the program. Furthermore the Americans had landed on the Moon five years earlier.

The engine programs were canceled shortly after the N-1 program was canceled in 1974. After 1975 this design bureau did essentially no further work on LPREs. At that time of cancellation, the government issued orders to destroy all N-1-related hardware.[1] Kutznetsov took a chance and put 70 new engines (not fired) and some 50 used experimental engines of his four N-1

708 History of Liquid Propellant Rocket Engines

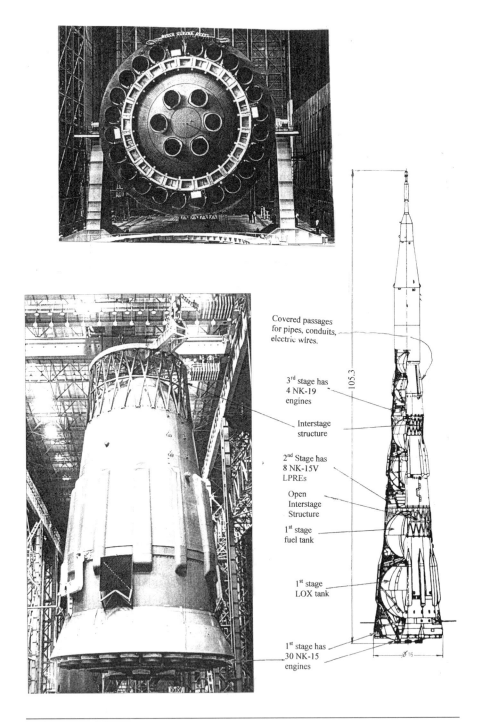

Fig. 8.9-5 Simplified drawing of the Soviet N-1 space launch vehicle for a moon mission; view of side and bottom of first stage. It is the largest and highest mass SLV that has ever taken off. Courtesy of NPO Energiya.

models into storage. This gamble paid off, when Aerojet became interested in the NK-33 believing that this old engine had improved technology (staged combustion cycle) when compared to existing U.S. LPREs using the same propellants.

Aerojet made an agreement in 1993 to market and eventually produce these engines in the United States. They sold the NK-33 engine (first stage N-1, has never flown) to Kistler Aerospace Corporation for use in the first stage (three engines) and second stage (one engine) of the unique Kistler reusable K-1 launch vehicle, which was being developed at the time this book was written.[6,7,*] This subject is also discussed in Chapter 7.7. Approximately 30 Russian NK-33 engines were taken out of storage in Russia and acquired for this U.S. program.

Kuznestov's development organization also did some work on a LOX/LH$_2$ LPRE, identified as the NK-35.[2,3] No information was available about its parameters or engine details. This work was not completed because priority had to be given to the LOX/kerosene engines for the troubled N-1 moon vehicle.

References

[1]Siddiqi, A. A., *Challenge to Apollo, the Soviet Union and the Space Race (1945-1974)*, NASA SP-2000-4408, Washington, D.C, 2000.

[2]Lardier, C., "Liquid Propellant Engines in Soviet Union," International Academy of Astronautics, *Paper 99 AIA-2-3-04*, Oct. 1999.

[3]Lardier, C., "The Soviet LPREs of 1946 to 1991," International Academy of Astronautics, *Paper 98-2-3-09*, 1998 (in French).

[4]*Jane's Space Directory,* Jane's Information Group, Coulsdon, Surrey, England, U.K.

[5]Hindley, K. B., *Handbook of Russian Rocket Engines*, 1st ed., Technology Detail, Clifton, York, U.K., 1999.

[6]Anisimov, N. D., Lacefield, T. C., and Andrews, J., "Evolution of the NK-33 and NK-43 Reusable LOX/Kerosene Engines," AIAA Paper 97-2680, July 1997.

[7]Lacefield, T. C., and Sprow, W. J., "High Performance Russian NK-33 LOX/Kerosene Liquid Rocket Engine," AIAA Paper 94-3397, June 1994.

---

*Personal communications, Carl Fischer, Aerojet Propulsion Company.

8.10   NPO Youzhnoye

This organization is located at Dnepropetrovsk, in the sovereign state of Ukraine, which used to be a part of the former Soviet Union. It was established in 1951 as a serial production plant (Plant 586) to produce and assemble ballistic missiles designed by Korolev's design bureau.[1] Originally it had a small cadre of engineers to deal with manufacturing issues. Today this plant still produces aerospace components and assembles missiles or space vehicle for the Russians.

In 1954 the top administrative people in the Soviet government were concerned about the monopoly of Korolev DB in the field of ballistic missiles and wanted to create additional organizations. Therefore in 1954 additional engineers from Korolev's DB were transferred, and the organization officially became an experimental design bureau OKB 586.[1] The design and development of their own missiles (and later also SLVs) started in 1954. Mikhail K. Yangel, originally a disciple of Korolev, was appointed the chief designer (1954–1971) of missiles and SLVs, and some were in competition with similar projects undertaken by Korolev. The uprating to a full design bureau and the name KB Yuzhnoye happened in 1966. In 1986 the organization's name was changed to NPO Youzhnoye. The abbreviation NPO is for Nauchno Proizvidstvennoye Obedienie or Science and Production Association. This organization has had a separate section devoted to LPREs since 1954. Most of the LPRE work mentioned next was actually performed when the Urkaine was still a part of the Soviet Union.

Most of the LPREs were in support of the vehicles developed at this DB, and LPREs were not the primary product of this plant. The LPREs were developed by a suborganization. Ivan I. Ivanov was the leader for the LPRE work, had been the deputy chief designer for LPREs under Yangel, and was responsible for most of the engines.[2–4,*] They developed LPREs in four areas: 1) vernier engines for flight path control of vehicles and/or stages of vehicles, 2) single-operation engines for upper stages of ballistic missiles and SLVs, 3) post-boost control rocket engines in payload stages of a ballistic missiles allowing accurate velocity control of each of several multiple warheads, and 4) one lunar landing engine. Table 8.10-1 has a list of 18 selected rocket engines developed by this DB.* It does not include small thrusters or experimental engines.

Only a few of the engines listed in this table will be discussed here. The RD-857 was a historic LPRE because it was an early Soviet engine with a staged combustion cycle, because it operated at two different thrust levels, and because it had an extraordinary gas-injection thrust-vector-control (GITVC) scheme. The engine is shown in Figure 8.10-1. The program started in 1963, but was terminated in 1967. It was intended to become a second-stage engine for a ballistic missile. This RD-857 had a single TC, was started only once, and

---

*Personal communications, J. Morehart, The Aerospace Corporation, 2001–2004.

Table 8.10-1 LPREs Designed by OKB-586/NPO Youzhnoye

Designation	Application/type	Development/production	Cycle	Propellants
RD-851	R-16 [SS-7] stage 1, vernier	1958–1963; serial produced	GG	IRFNA/kerosene
RD-852	R-16 [SS-7] stage 2, vernier	1958–1963; serial produced	GG	IRFNA/kerosene
RD-853	R-26, stage 2, main engine	Began 1960; terminated 1963	GG	IRFNA/kerosene
RD-854	R-36 [SS-9 Mod 3] orbital stage	1962–1967; serial produced	GG	NTO/UDMH
RD-855	R-36 [SS-9] stage 1, vernier SL-11/14, stage 1, vernier	1962–1965; serial produced	GG	NTO/UDMH
RD-856	R-36 [SS-9] stage 2, vernier SL-11/14, stage 2, vernier	1962–1965; serial produced	GG	NTO/UDMH
RD-857	RT-20P [SS-X-15] stage 2, main	Began 1963; terminated 1967	SC	NTO/UDMH
RD-858	N-1 lunar lander, main engine	Began 1965; terminated 1972	GG	NTO/UDMH
RD-859	N-1 lunal lander, backup engine	Began 1965; terminated 1972	GG	NTO/UDMH
RD-860	NA, low thrust restartable engine	NA, experimental	—	NTO/UDMH
RD-861	SL-14 Tsyklon, stage 3, main	1968–1972, serial produced	GG	NTO/UDMH
RD-862	RS-16 [SS-17], stage 2, main	1969–1972, serial produced	SC	NTO/UDMH
RD-863	RS-16 [SS-17], stage 1, vernier	1970–1973; serial produced	GG	NTO/UDMH

(continued)

Table 8.10-1 LPREs Designed by OKB-586/NPO Youzhnoye (continued)

Designation	Application/type	Development/production	Cycle	Propellants
RD-864	RS-20B[SS-18 Mod 4], PBCS[g]	1976–1978; serial produced	GG	NTO/UDMH
RD-8	SL-16 Zenit, stage 2, main	1976–1983, serial produced	SC	LOX/kerosene
RD-866	RS-22 [SS-24], PBCS	1980–1983; serial produced	PF[h]	NTO/UDMH
RD-868	Restartable apogee stage	Began 1983	GG	NTO/UDMH
RD-869	RS-20V [SS-18 Mod 5/6] PBCS	1983–1985; serial produced	GG	NTO/UDMH

This table was provided and assembled by James Morehart, The Aerospace Corporation, 2C03.
Symbols: GG = bootstrap gas generator, SC = staged combustion engine cycle, NA = not available or not applicable, PF = pressurized feed system, PBCS = post boost control system.
Engine relationships/comments: RD-855 and RD-856 were improved versions of RD-851 and RD-852 respectively; RD-860 was an experimental LPRE with a pneumatic feed system, RD-862 was an improved, simplified version of RD-857 and both had warm gas injection on sides of nozzle diverging section for thrust vector control, vernier thrust chambers were usually hinge-mounted, the RD-869 PBCS was based on the RD-864 PBCS. The nitric acid for the first three listings was inhibited by iodine.

had a fuel-rich preburner. At full power its thrust was 14 metric tons or about 30,800 lbf, and in the low-thrust mode it was 1.3 tons or 2860 lbf. The corresponding specific impulses were 329 and 250 s using the storable propellants NTO/UDMH.

Operation at the low-thrust mode was difficult. Almost all LPREs with a staged combustion cycle and storable propellants have a preburner that is oxidizer rich, but this engine is an exception. Fuel rich gas, which was tapped off from the preburner, was injected through hot-gas valves at one of four places at the diverging section of the exhaust nozzle, and this created pitch and/or yaw forces on command. Two of the four injection manifolds can be seen at the diverging section of the nozzle. The maximum side force was about 300 kg force (660 lbf) at full power and 8 kgf (about 18 lbf) in the low-thrust mode. An engine with side injection is more compact and has a smaller diameter than an engine with vernier TCs, has fewer problems with the blowback of hot exhaust gases at altitude, and does not need rotary joints, actuators, or power for rotating vernier TCs. Whereas the United States built a LPRE with liquid propellant for side injection, this design bureau is the only one in the world known to have developed engines with warm preburner gas side injection. Warm-gas injection is theoretically more effective than using liquid injection, because the warm gas scheme needs less side injection mass for the same side force. Roll control of the RD-857 was obtained through four nozzles using gas that was bled from the exhaust duct of the turbine, and the gas flow was controlled by hot-gas valves, which were normally closed.

The RD-857 turbopump had a starter turbine, which was a separate turbine in addition to the normal operating turbine. The start turbine was driven by hot gas from a solid propellant cartridge for a few seconds during the start period only. The exhaust gas from the start turbine is dumped overboard. The reasons for using a start turbine are discussed in a subsequent paragraph. The fuel tank is pressurized by bleeding warm gas from the turbine exhaust at a reduced pressure. The oxidizer tank is pressurized with oxidizer-rich gas in a pressurization gas generator. The low-thrust mode is achieved by using only the center portion of the injector elements for injecting oxidizer thus operating at a fuel-rich mixture ratio of 1.1 as compared to a ratio of 2.6 during full thrust. Thrust and mixture ratio had automatic controls. Development had not been completed by 1967, when the program was canceled.

On the basis of the experience of this program, Youzhnoye developed (1968–1972) the RD-862 engine, which was produced in quantity and has flown satisfactorily in the second stage of the R-16 (also identified as SS-17) ballistic missile. In external appearance the RD-862 was the same as the RD-857 shown in Fig. 8.10-1. This engine also uses a staged combustion engine cycle, has a single TC, and is suitable for a single operation. It had a thrust of 14.5 metric tons (31,900 lbf) and a specific impulse of 331 s (vac) at a chamber pressure of 135 kg/cm^2 (1984 psia). The gas side-injection TVC system and the staged combustion cycle features of the RD-862 were based on the RD-857 design; however, it did not have a low-thrust mode, and this helped to simplify the engine design. Roll control is achieved by bleeding fuel-rich gas from

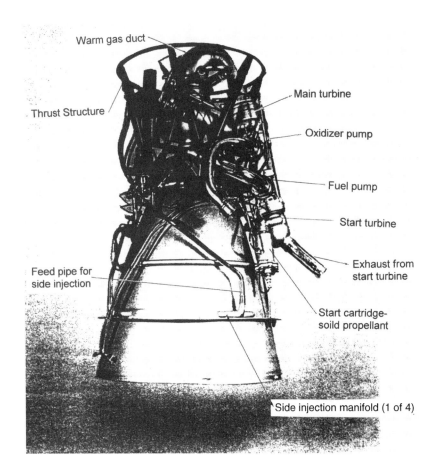

Fig. 8.10-1 View of the RD-857 rocket engine with warm-gas side injection into the nozzle for pitch and yaw control. Two of the four nozzle injection manifolds are visible. Courtesy of NPO Youzhnoye (from personal communications, J. Morehart, The Aerospace Corporation, 2001–2004).

the turbine exhaust manifold and sending this gas to two out of four nozzles, which are equipped with hot gas valves, which are normally closed and have no leakage. The scheme for pressurizing the propellant tank is similar to the scheme used on the RD-857. Several of the valves are limited to a single operation because they are actuated by pyrotechnic devices. The premature firing of this engine on the launch stand was the cause for a major accident, which is described in the summary (Chapter 8.13) and at the end of Chapter 8.3.

Start turbines seem to have been used in the Soviet Union only. A separate start turbine was used in addition to the regular turbine of the turbopump in the RD-857 and the RD-862. The start turbine makes the TP assembly more complex, longer, and heavier. Although it is not confirmed, the author postulates the following reasons for having a start turbine. The running turbine for the operation of the engine ran on clean gas for the duration of the LPRE's operation, had a relatively large gas flow and a relatively small pressure drop, and ran long enough for thermal equilibrium to be reached. Under these conditions this turbine had relatively thin blades with a relatively good turbine efficiency. The thin blades are often fragile and will not always withstand an impact of small solid particles. The impact of solid particles on highly stressed turbine blades has in the past caused turbine blade failures in the Soviet Union. Even if the particles did not damage the turbine blades, they have plugged up injector holes and caused problems of the combustion in the main TC.

The start turbine was driven by exhaust gas from a solid-propellant cartridge, had a large pressure drop across the turbine blades, and the duration would be relatively short, typically a few seconds. The best blade configuration or shape is different from the blades of the running turbine. The cartridge's gas is usually fuel rich, is smoky, and often contains some solid particles. Early solid-propellant cartridges usually had particles in their exhaust gas. These particles in the gas flow (from a solid-propellant grain) could come from small pieces of ignition wire, small hunks of thermal insulation, slivers of grain support brackets, or igniter squib pieces. Furthermore, deposits of solid carbon particles have been found on parts of the turbine blades and turbine nozzles. The start turbines have fairly thick, relatively inefficient blades, but they are robust and can withstand impacts of solid particles or buildup of carbon deposits without blade failures. The use of start turbines therefore has minimized possible damage to the running turbine and allowed better efficiency of the running turbine.

The earliest start turbine, known to the author, was with the turbopump of the S1.5400 LPRE developed by Korolev's Design Bureau as described in the next chapter. It flew satisfactorily in 1961. The Kuznetsov Design Bureau had about eight LOX/kerosene engines with start turbines as described in the preceding chapter, and one start turbine in a turbopump assembly can be seen in Fig. 8.9-3. As solid-propellant start cartridges were improved and had cleaner exhaust gas with essentially no solid particles, the need for having a separate start turbine diminished, and they were abandoned. The warm gas from solid-propellant start cartridges has become cleaner and has routinely been used for starting TPs of many LPREs in many countries, including the Soviet Union.

Between 1976 and 1978 the RD-864 was developed for the postboost vehicle (PBV) of the RS-20B ballistic missile, and after development it was placed into serial production.* Its mission was to impart an accurate terminal velocity (in magnitude and direction) to each of several warheads in the payload of the ICBM. A simplified schematic diagram with four TCs is shown in Fig. 8.10-2. Four TCs were mounted on swivel arms, but were retracted inside the vehicle during the powered ascent of the ICBM. Once the payload stage (or postboost vehicle) had been separated from the missile, the thrust chambers were rotated into the firing position as seen in the figure. Each thruster was hinged and could be rotated by +/− 55 deg. Rotary actuators cause the hinge rotation of the TCs out of the module's envelope. Rated thrust was 2060 kg

---

*Personal communications, J. Morehart, The Aerospace Corporation, 2001–2004.

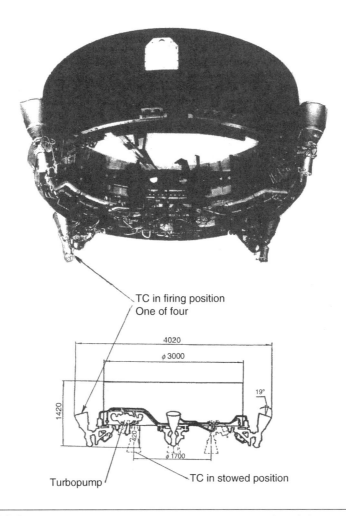

Fig. 8.10-2 Simple sketch and photo (not the same scale) of the RD-864 propulsions system of the post boost vehicle of RS-20B ICBM with four hinged TCs in their operating position, not the same scale. The dashed outlines of the TCs shows the retracted or stowed position during the ascending flight of the vehicle. Courtesy of NPO Youzhnoye (from personal communications, J. Morehart, The Aerospace Corporation, 2001–2004).

force (4532 lbf) with all four thrusters, and the specific impulse was 309 s when running in the full thrust mode and 862 kgf (1896 lbf) and 298 s respectively in the low or throttled mode. This RD-864 engine had the ability to cycle between high- and low-thrust modes up to 25 times. This LPRE had a turbopump, which can be seen in the figure, and ran on a gas-generator engine cycle continuously for up to 10 min. The turbopump speed is reduced when operating in the low-thrust mode. A solid-propellant start cartridge is used to initiate the rotation of the the turbine in the TP and thus start the engine. This scheme of control of a postboost vehicle is very different than those used in the United States; one U.S. system is described in Chapter 7.8.

The RD-858 and RD-859 LPREs were developed between 1964 and 1972 for landing on the moon. The engines were intended for the lunar landing module of one version of the N-1 SLV. Chief Designer M. Yangel turned down the opportunities to develop several engines for the lunar module of the N-1 vehicle, but finally agreed to develop these two LPREs, which incidentally were some of the few rocket engines that were not installed in a Yangel-designed vehicle. The RD-858 was the primary or main lunar landing engine, with a rated thrust of about 2 metric tons (4440 lbf), and it could be throttled down (during lunar descent before landing) to 0.858 tons. The chamber pressures were 80 to 34 kg/cm^2, respectively. The RD-859 was integrated with the RD-858, had two smaller TCs, one on each side of the main TC, and it was the backup or reserve engine, in case the primary engine failed to operate properly. It could not be throttled. A view of the combined two LPREs is shown in Fig. 8.10-3. The assembly also had four small nozzles operating with turbine exhaust. These two engines were part of the lunar module on the N-1 moon flight vehicle. In the four launches the N-1 space launch vehicle did not get

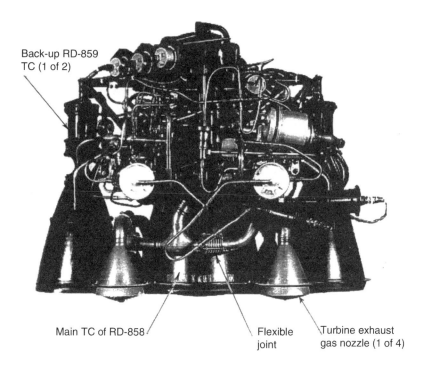

Fig. 8.10-3 View of the lunar landing engine (combined RD-858 and RD-859) for the planned astronaut landing on the moon. Courtesy of NPO Youzhnoye; photo credit: Wade, Haeseler.

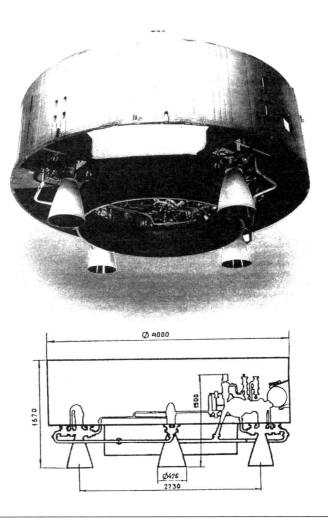

Fig. 8.10-4 Simplified sketch and photo of the RD-8 LPRE with four hinge-mounted TCs, a preburner, and one TP; it operates with a staged combustion cycle. Courtesy of NPO Youzhnoye; copied from a company publication through personal communications, J. Morehart, The Aerospace Corporation, 2001–2004.

close to the moon, and therefore the lunar module and these dual landing engines have never been operated in a spaceflight.

The RD-861 for the third stage of the Tsyklon was a single-operation, fixed LPRE with a single main TC. Thrust was 78.9 kN or 17,700 lbf at a chamber pressure of about 1330 psia. The specific impulse of 317 s corresponds to a nozzle area ratio of 111. Like most of the LPREs of this design bureau, it uses NTO and UDMH as propellants. The engine has several small thrusters for thrust vector control. The first flight was in 1977.

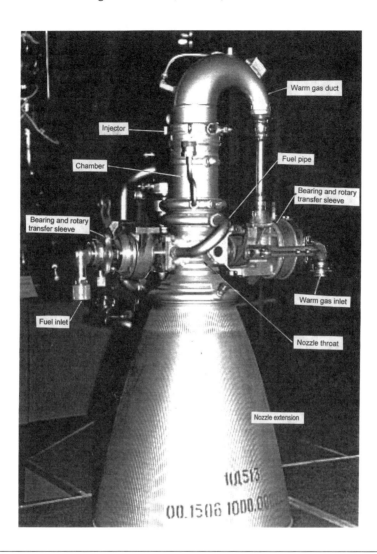

Fig. 8.10-5 One of four hinged thrust chambers of the RD-8 LPRE with its two rotary transfer joints for oxidizer-rich warm turbine exhaust gas and fuel.* Courtesy of NPO Youzhnoye.

A vernier engine of Yuzhnoye is the R-8 LPRE.[2-5,*] It is shown in Fig. 8.10-5. There are four hinged thrust chambers (swing is +/− 33 deg) and a total vacuum thrust of about 78.4 kN or 17,600 lbf. The specific impulse is 342 s for a nozzle area ratio of 104, and the relatively high value is caused in part by the more energetic propellant combination, the high area ratio, and the staged combustion

---

*Personal communications, J. Morehart, Aerospace Corporation, 2001–2004.

engine cycle. This performance is higher than most other LPREs with this propellant combination. As seen in Table 8.10-1, the propellants are different from other Youzhnoye engines, and they are the same (LOX/kerosene) as the propellants for the main engine of this stage (Energomash RD-120). The warm high-pressure turbine exhaust gas is supplied through long insulated pipes to the injector of each of four hinged TCs. Fig. 8.10-5 shows one of the TCs with its rotary joints, through which hot gas and liquid propellant were supplied to the TC. The nozzle-throat region and the chamber are regeneratively cooled, and the height of the TC is a little more than 1 m or 3.3 ft. The RD-8 vernier LPRE first flew in 1985 in the second stage of the Zenit SLV.

Yuzhnoye also had developed thrusters with gaseous propellant (air) at thrust levels of 0.196 and 0.98 N (0.043 and 0.215 lbf), and they have flown on spacecraft as late as 1994. A series of small NTO/UDMH bipropellant thrusters has also been developed. They provide a steady-state specific impulse between 220 and 240 s, and for short pulses it was between 150 and 160 s. These have been used on several spacecraft.

Today this organization is the only major aerospace flight-vehicle design bureau and production plant outside of Russia, but Yuzhnoye continues to produce or assemble some vehicles and vehicle stages for the Russians. As with other aerospace operations, the volume of work in vehicles and in LPREs has gone down appreciably causing concern about the future. However they seem to continue some work on rocket engines. For example Youshnoye has teamed with Fiat Avio of Italy and is offering an improved version of their RD-861 as an upper stage engine for the Vega SLV.[2] It uses a solid-propellant booster stage. The Vega is a small new multistage space launch vehicle of the European Space Consortium with a modest payload. The Italians have the lead role with this Vega SLV.

References

[1]Siddiqi, A. A., *Challenge to Apollo, the Soviet Union and the Space Race (1945–1974)*, NASA SP-2000-4408, Washington, DC, 2000.

[2]Lardier, C., "Liquid Propellant Engines in Soviet Union," International Academy of Astronautics, *Paper 99 AIA-2-3-04*, Oct. 1999.

[3]Lardier, C., "The Soviet LPREs of 1946 to 1991," International Academy of Astronautics, *Paper 98-2-3-09*, 1998 (in French).

[4]*Jane's Space Directory*, Jane's Information Group, Coulsdon, Surrey, England, U.K, new version issued every two years.

[5]Hindley, K. B., *Handbook of Russian Rocket Engines*, 1st edn., Technology Detail, Clifton, York, U.K., 1999.

8.11   Korolev's Design Bureau, Later NPO Energiya

This design organization was the first in the world to develop long-range ballistic missiles and heavy space launch vehicles. In 1946 it was started by Sergei P. Korolev, who is a famous man in Soviet astronautics and dominated the development of ballistic missiles and space launch vehicles in the Soviet Union for two decades.[1,2] Korolev was the chairman of the Council of Chief Designers, which included Glushko, Kosberg, Kuznetsov, and the heads of several other design bureaus involved with missiles and SLVs. He had good personal relationships with Nikita Krushchev and other top government officials in the Soviet Union. His proposals for new ballistic missiles or new space programs were readily accepted, and his recommendations for the allocation of needed state resources were generally followed. He was a very powerful man, humble and very persuasive, when he wanted something done. Even though he was the recognized aerospace leader and the top man in ballistic missiles or SLVs, his role was kept secret and was not acknowledged until after his death in January 1966. Korolev did not seek an official high-ranking job title, and he was never addressed by name. Instead people in contact with him, such as the Soviet astronauts, design bureau officials, secretaries or military officers, referred to him as the "chief designer." Some comments about him are in Chapters 8.1, 8.2 (including a photo), and 8.3.

Korolev's vehicle development organization became known as OKB-1 or Experimental Design Bureau number 1, and it was located in Kaliningrad, a suburb of Moscow. The main effort was the development of missiles and space launch vehicles, and it started with copying the German V-2 ballistic missile, which was the first such weapon deployed by the Soviet military.

The development of LPREs was a very minor, but important effort in this key vehicle development organization. This captive rocket-engine department developed several selected engines for application to its own vehicles. Some of these, for various reasons, could not be contracted to LPRE design bureaus and were undertaken in-house. The vast majority of all of the major rocket engines needed for Korolev's own missiles and own SLVs were contracted by him to one of the LPRE design bureaus, such as those headed by Glushko, Kosberg, Kuznetsov, or Isayev, and they were desribed in prior chapters.

Korolev's vehicle development organization was originally established in 1946 as a department #3 in an existing manufacturing plant (NII–88) in Kaliningrad. One year before, in 1945, the Soviet Union became aware of the details of the V-2 ballistic missile in postwar Germany, and Korolev's bureau was a logical national response to build their own Soviet missiles.[1-4] In 1950 this department became an experimental design bureau, identified as OKB-1, but within the framework of the manufacturing plant. In 1956 the OKB-1 organization was separated from the NII-88 plant and became an independent organization. After Korolev's death in 1966, OKB-1 was renamed Central Design Bureau of Experimental Machine Building and V.P. Mishin became its leader. He was fired in 1974. Work on LPREs was continued within this organization. Valentin Glushko, the general designer and general director of

Energomash, was selected to take over. Glushko's new organization, called NPO Energiya, included not only Korolev's former DB, but also Energomash (OKB-456). As mentioned elsewhere in this book, the name of the town of Kaliningrad, where the OKB-1 design bureau was located, was changed to Korolev. It was a posthumous honor in the great man's memory.

Korolev always wanted to have a group of engineers knowledgeable in propulsion to be a part of his organization. Initially he arranged for a group under Isayev to do some LPRE work for him In 1948 he arranged for the transfer of Isayev's rocket propulsion design organization to plant NII-88, and the two men collaborated together on many projects between 1945 and 1966. As outlined in Chapter 8.6, Isayev broke away to form his own independent development organization, which grew into a design bureau of its own. Korolev then established another in-house LPRE department in his own design bureau under Mikhail V. Melnikov, who became deputy chief designer for LPREs.[3-6,*] At least 10 different LPREs were developed by this department, and the majority of these have flown. This in-house capability also helped Korolev to evaluate the proposals, performance, and progress of the LPRE design bureaus, such as those of Glushko, Kosberg, or Isayev, who were then working to build engines for him.

When Glushko refused to develop the vernier thrust chambers for his historical RD-107 and RD-108 engines in 1954, this development task was assigned to Melnikov's in-house propulsion group. This was mentioned in Chapters 8.4 and 8.5. A good hinge-mounted TC using the latest construction techniques was developed in OKB-1 (engine S1.35800). At a later time, when the vernier TC requirements changed (different thrust level), the task of developing the new version of the vernier was given to and then accepted by Glushko. The work done by Melnikov's group was no longer needed, and the information and some hardware were transferred to the Kosberg DB, as explained in Chapter 8.5, where it became the basis for the development of other TCs.

The first Soviet LPRE with a staged combustion engine cycle to fly was the S1.5400 LPRE developed at Korolev's DB between 1958 and 1960. It has also been identified as the 11D33 engine. The principle of that staged combustion engine cycle was originally demonstrated in ground tests beginning in 1958. The benefits of a staged combustion engine cycle are explained in Chapter 4.2. Some data for two versions of this LPRE are listed in Table 8.11-1 (Ref. 7).* It flew on top of Molniya SLVs and Venera space vehicles, and they in turn were lifted by a variation of the two-stage R-7 ICBM. A picture of this engine is in Fig. 8.11-1. It shows that the horizontal turbopump had two turbines. On the left of the TP is a start turbine, which uses gases from a solid-propellant short-duration grain (in a cartridge located just underneath the TP), to bring the TP up to speed. The turbine on the right end of the TP is driven by gas from a preburner (not visible), and the high-pressure turbine discharge gas is then supplied to the injector of the thrust chamber through the large pipe on the right. This engine appears to be the earliest to have a separate startup turbine

---

*Personal communications, J. Morehart, The Aerospace Corporation, 2001–2004.

Table 8.11-1 S1.5400 and S1.5400A1 Liquid Propellant Rocket Engines

Parameter	S1.5400	S1.5400A1
Thrust, kg/lbf	6,500/14,300	6,800/14,960
Specific impulse, s	338.5	340
Propellants	LOX/kerosene	LOX/kerosene
Dry engine mass, kg/lbm	153/337	153/337

in addition to the usual turbine of the turbopump. Start turbines are discussed in Chapter 8.10. The first flight attempt failed because of a malfunction in a lower vehicle stage. It first flew successfully in a Venera vehicle in 1961. Between 1961 and 1964 the engine was improved, the new version was identified as S1.5400A1, and there were some changes in the performance as seen in the table.

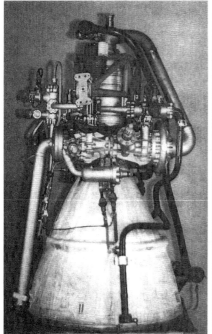 

Fig. 8.11-1 Front and side view of the S1.5400 LPRE, which was the first with a staged combustion engine cycle. Its first satisfactory flight was in an upper stage in 1961. Courtesy of NPO Energiya; photo obtained through J. Morehart, The Aerospace Corporation, 2004.

In the late 1960s Melinkov also developed the 11D58 LPRE for the top (fourth) stage of one version of the Proton SLV.[2,5,6] It had a thrust of 85 kN or about 19,100 lbf and operated with LOX/kerosene and a chamber pressure of 1122 psia. It was used for injection of payloads into an Earth orbit or into a planetary flight path. This engine was also installed in the fifth stage of the ill-fated large N-1 lunar SLV. One version of this rocket engine that used a special hydrocarbon fuel, identified as Syntin in lieu of ordinary kerosene. This synthetic fuel was discussed in Chapter 4.1. Compared to kerosene, this special fuel allowed an increase of the specific impulse of about 10 s or about 3%, and this would allow a significant increase in the payload. The engine version with Syntin was identified as 11D58M and had a slightly different thrust and chamber pressure.

Korolev's OKB-1 also started development of a short-range ballistic missile, which was later identified as the R-11 missile. Korolev was busy with larger missiles and SLV and really did not devote much time to this project. Isayev, who at the time worked for Korolev, worked on its LPRE. V.P. Makeyev, who then also was in OKB-1, worked on getting this missile through development and ready for production. When Makeyev and his group broke away from Korolev's organization and formed their own design bureau, the further development and full production of this R-11 project went with him. The specific engine is discussed in Chapter 8.6. It was a very successful project and quickly became operational with the Soviet military forces, and several versions were developed.

The 11D121 engine, developed at OKB-1, was used in pairs on the N-1 large first stage for roll control. It could swing ± 45 deg and use gaseous oxygen and hydrocarbon fuel. It was developed between 1962 and 1972. It had 7.0 tons (15,700 lbf) of thrust and a specific impulse of 313 s. One source[4,5] also reports that engines 17D11 (ACS with multiple thrusters, LOX/hydrocarbon) and 17D12 were developed (1985/1988) for the Buran SLV. LOX could be stored in space for at least a year by using a special gas cooler.

In conclusion this design bureau did groundbreaking work on missiles and space vehicles. The development of LPREs was really a sideline in this bureau, but they still managed to develop some new and historic LPREs, such as the first large roll control engines or the first flight of a staged combustion cycle engine.

References

[1] Siddiqi, A. A., *Challenge to Apollo, the Soviet Union and the Space Race (1945-1974)*, NASA SP-2000-4408, Washington, D.C., 2000.

[2] Harford, J., *Korolev—How one Man Masterminded the Soviet Drive to Beat America to the Moon*, J. Wiley, NewYork, 1997.

[3] *Jane's Space Directory*, Jane's Information Group, Coulsdon, Surrey, England, U.K.

[4] Hindley, K. B., *Handbook of Russian Rocket Engines*, 1st ed., Technology Detail, Clifton, York, U.K. 1999.

[5] Lardier, C., "Liquid propellant Engines in Soviet Union," International Academy of Astronautics, Paper 99 AIA-2-3-04, *Oct. 1999*.

[6] Lardier, C., "The Soviet LPREs of 1946 to 1991," International Academy of Astronautics, *Paper 98-2-3-09*, 1998 (in French).

[7] Varfolomeyev, T., "Soviet Rocketry That Conquered Space, Parts 4, 5, and 6," *Spaceflight*, Vol. 40, Jan./Mar./May, 1998.

8.12 OKB Fakel

This design bureau has been known primarily for electrical rocket propulsion developments. However it also has developed unique small hydrazine monopropellant thrusters and electrocatalytic (electrically heated or augmented) hydrazine monopropellant thrusters. It is located in Kaliningrad and was originally established in 1959 as a laboratory under the USSR Academy of Sciences. It was reorganized in 1971 as OKB Fakel and has been in this monopropellant business since the 1970s. They have done the development, testing, as well as much of the production of these small thrusters and their propulsion systems.

One of Fakel's specialties is hydrazine microthrusters with thrust values between 0.1 and 0.001 N (0.022 to 0.00022 lbf). They do not use a conventional decomposition catalyst, but instead use thermal decomposition of the hydrazine or the evaporation of the hydrazine by electric heat. These small thrusters and their feed systems are used in satellites or space stations for onboard projects that need a microgravity environment. More than 2000 such miniature thrusters have flown since 1982, mostly in the Kosmos satellites.

Fakel also has developed several small thermocatalytic hydrazine thrusters and their propulsion systems for attitude control of small spacecraft. Although it is not really a pure liquid propellant unit and is outside the scope of this book, it is mentioned here for characterizing the Bureau's product line. One has thrust levels of 5.9 to 1.0 N (1.3 to 0.23 lbf) with a specific impulse of 227 to 208 s. They require electric power (10 to 14 W) for about an hour prior to operation for slowly heating the hydrazine. The system operates under pressure of 0.9 to 0.2 MPa, weighs about 0.6 kg (1.3 lbm), and also has a radiator.

A propulsion sytem with eight small monopropellant thrusters (four main and four redundant spares) of 0.55 to 0.1 N each requires 14.6 W for the heaters, has a radiator area of 180 cm^2, and weighs 7.2 kg or 15.8 lbf. The life was given as 10.5 years. It has flown in the SECAT satellite since 2000.

Reference

[1]Murashko, V. M., Koriakin, A. A., Vinogradov, O. I., Rybalchenko, L. V., and Niatin, A. G., "Main Results of 20 Year Operation in Space of Monopropellant Liquid Propellant Rocket Engines of Experimental Design Bureau Fakel," *6th International Symposium for Space Transportation of the 21st Century* [CD-ROM], 2002.

8.13 R&D Institutes

There are a number of government R&D organizations that participated and have had an influence on LPRE in the Soviet Union and subsequently in Russia. Two of these are briefly described in this chapter. There were others, who were occasionally involved with LPRE, such as those concerned with welding technologies, aerodynamic R&D, launch complexes, or R&D on military applications of spacecraft. They are not discussed here.

Keldysh Research Center

An offspring of RNII grew into the Keldysh Institute of Applied Mathematics, and during its history it was also known by several other names including the Keldysh Research Center and formerly the Institute of Thermal Processes.[1,2] It started out as a small analytical section within RNII in the 1930s and was under the leadership (1946–1955) of the applied mathemetician Mstislav V. Keldysh. They did analysis and performed research on propulsion performance, heat transfer, combustion, combustion instability, stresses, and other areas related to propulsion with LPREs, solid-propellant motors, several types of electric propulsion, and later nuclear propulsion. They worked on a method for analyzing the theoretical performance of different propellants, different propulsion systems, and methods for optimizing the performance and staging of rocket-propelled flight vehicles. They performed some laboratory work and tested some small TCs. This R&D center also developed some small thrusters with NTO/UDMH (10 to 1000 N thrust), and this technology was later transferred as a spin-off to an organization identified as NII Mash (see Chapter 8.7). Also they developed a small hydrazine monopropellant microthruster of 0.5 N. People from this Institute claim to have done crucial work and proposed technical decisions on the first rocket-propelled airplane, the first ICBM, the engineering of LPREs with LH_2, the flight control of rocket-driven vehicles, or the reentry of spacecraft into the atmosphere. Experimental products included ion and Hall effects electrical thrusters, and also a solar powered propulsion system.

The Institute was reorganized often and became independent from NRII and its successor organizations. Besides propulsion-related projects, this Institute worked on a broad range of related technologies, on mission analysis, system integration, and ballistic flight-path computations. It continued its propulsion analyses and has worked closely with most of the design bureaus working on propulsion. Keldysh (1911–1978) became a member of the Soviet Academy, was selected to be its secretary (1953–1954), then a member of the presidium, and later its president (1961–1975). His opinions were very influential in the development of ballistic missiles, spaceflight, and thus also LPREs.

Russian Scientific Center for Applied Chemistry

This Center was started as the State Institute for Applied Chemistry in Leningrad (today St. Petersburg) in 1946. Today it is known by the name in the preceding heading, which is Rossiyskiy Nauchniyi Tsentr Prikladnaya Khimiya.[3] It supported the work of major LPREs DBs by developing and synthezising new propellants, by investigating the physical and chemical characteristics of propellants, and by developing the processes needed for the efficient production of certain propellants. In the late 1940s this Center reproduced and helped to put into production some of the German propellants, such as Tonka 250. In 1949 they synthesized dimethyl hydrazine and shortly thereafter unsymmetrical dimethyl hydrazine (UDMH), which at the time were not well known in the Soviet Union. Some years later they developed the process technology for the industrial production of UDMH, which has been used extensively in many Soviet LPREs. They investigated liquid hydrogen, its properties, handling, and compatible materials. They also worked on different hydrocarbons for the fuel of the RD-107 LPRE. They investigated special types of kerosenes, such as those that gave more specific impulse (such as Syntin, see Chapter 4.1) or better heat transfer characteristics. In the late 1950s and 1960s they explored liquid fluorine and related compounds in anticipation of the RD-301 LPRE (fluorine/ammonia) development, which is described in Chapter 8.4.

References

[1]Former Web site of the Federation of American Scientists on the Keldysh Research Center, circa 1999. No longer available.

[2]"RNII -the Keldysh Research Center as a Part of the History of Home (Domestic) Manufacturing," *Acta Astronautica*, Vol. 43, No. 1–2, 1998, pp. 13, 14.

[3]Web site of the Federation of American Scientists on the Russian Scientific Center for Applied Chemistry, circa 1999. No longer available.

8.14 Summary of Soviet or Russian Efforts in Liquid Propellant Rocket Engines

This summary describes the key events, accomplishments, and general findings of the work on LPREs undertaken in the Soviet Union and more recently in Russia. The items listed here are not in any particular order. It contains some comments and subjects that were not mentioned explicitly in the preceding 13 chapters.

Russia and the former Soviet Union was and still is the world's foremost nation in developing and building LPREs. The Russians have some very capable organizations in this field. Their efforts and their LPRE work were more extensive, and they were more active than those of any other nation. During 1950 to 1998, their organizations developed, built, and put into operation a larger number and a larger variety of LPRE designs than any other nation. The author estimates that at least 500 different LPREs have been developed. For comparison the United States has developed slightly more than 300. The Soviets also had more rocket-propelled flight vehicles than any of the other nations. They had more liquid propellant ballistic missiles and more space launch vehicles (SLVs) derived or converted from these decommissioned ballistic missiles than any other nation. They also had newly designed or dedicated SLVs such as the Energiya or the N-1 moon flight vehicle. As of the end of 1998, the Russians (or earlier the Soviet Union) had successfully launched 2573 satellites with LPREs or almost 65% of the world total of 3973 such launches up to that time. A major share of these Russian satellites was presumably for military purposes. All of these vehicle flights were made possible by the timely development of suitable high-performance reliable LPREs.

Several historical first flights with LPREs occurred in the Soviet Union. This includes the flight of the world's first ICBM (R-7) in 1956. It preceded the U.S. Atlas first ICBM flight by about five years. The Soviets launched the world's first satellite (Sputnik—4 October 1957) and the world's first manned orbiting spacecraft (Y. A. Gargarin—12 April 1961). It used a modified R-7 ICBM as a launch vehicle. These spectacular achievements could not have been accomplished without the rocket engines tailored to these applications.

The Soviet Union developed more long-range ballistic missiles (propelled by LPREs) than any other country. The Soviet military forces never used any of them in actual combat. The ballistic missiles in other countries have also never been fired against an unfriendly target. However, some versions of the R-11 Soviet short range surface-to-surface missile, which was exported by the Soviet Union, was actually used in combat. Egypt used them in its war with Israel in 1973, and Iraq used it and a modified version against Israel and Saudi Arabia in the Desert Storm conflict in 1991.

The Soviet Union developed in the late 1970s and is still building a version of the RD-170 LPRE, which has the highest thrust (1.777 million pounds vac) in the world. The highest thrust flying U.S. engine, the F-1 at 1.746 million pounds (vac), was developed about 15 years earlier, but it is no longer an active engine program.

The Soviets have also developed engines with the highest known chamber pressures. Several of their historic engines had chamber pressures between 3800 and 4250 psi, and one engine reportedly was tested at 4600 psia. The higher chamber pressure allows a higher nozzle area ratio, a somewhat higher performance, and the design of a smaller size nozzle and thrust chamber, which can be easier to package into a flight vehicle. Disadvantages are major increases of the heat transfer and usually a higher inert mass of the engine.

The Soviets were the first to develop and use strap-on droppable booster stages with liquid propellant engines (1950s) in several of early ICBMs and their SLVs. These were designed and built as a first stage and not as an add-on or afterthought to increase the payload capacity, as with a number of U.S. solid-propellant strap-on boosters. The liquid propellant version could deliver a higher specific impulse and a relatively clean environmentally acceptable exhaust gas, but the solid motors produced heavy smoke, had a higher density, and therefore somewhat less drag.

They were the first to develop a number of engines with four large thrust chambers (TCs) being fed by a single turbopump (TP), beginning in the 1950s. This engine is 20 to 35% shorter than a single chamber engine of the same total thrust and chamber pressure. This reduces flight vehicle length and usually some vehicle structural mass, but a multiple TC cluster is more complex, has a larger diameter, or a larger vehicle cross section than a single engine, which leads to a higher aerodynamic drag. In at least two engines the Soviet designers went to a four-chamber version because they found it to be more difficult to remedy combustion instabilities in the larger single-chamber version. The four-TC scheme has since been used by other countries, such as China or Britain.

Since 1960, the Soviet Union built and has flown at least 30 LPREs with a high-performance staged combustion engine cycle. In contrast the United States has flown only one about 20 years later, and Japan has flown three different versions of their large booster engine. With storable propellants almost all Soviet engines used an oxidizer-rich propellant mixture in the gas generators and preburners, a feature not found elsewhere. Only one of these engines (using NTO and UDMH) and also a Soviet LOX/LH_2 engine (with this engine cycle) are known to have flown with a fuel-rich turbine gas mixture in the preburners. The Soviets have not flown an engine with an expander engine cycle.

The first Soviet piloted LPRE-powered piloted glider flight was in 1940. The glider was towed to altitude by an aircraft, and the LPRE was fired after the glider was released. The first Soviets piloted experimental aircraft powered solely by a rocket engine took off under its own rocket power in 1942. Both engines had gas-pressure feed systems with heavy walls in the propellant and gas storage tanks. These propulsion systems were so heavy that they caused a major decrease in the useful flight range or payload of the aircraft. By 1945 Soviet engineers had ground tested an experimental turbopump-fed, aircraft engine. The LPRE with a TP was flight tested in 1947. It allowed some, but not enough, improvement in the aircraft performance. The Soviet Air Force lost interest in aircraft driven totally by rocket power, and the work was not continued. The Germans had pursued this concept vigorously, and their rocket-propelled fighter airplane, the Me-163, first engaged in combat in 1944, several years before the Soviets abandoned this approach.

Between 1938 and 1956 about 16 different aircraft-type LPREs were developed in the Soviet Union. They were intended to provide aircraft superperformance (where a LPRE augments a piston engine or a jet engine) and allow rapid climb and fast maneuvers at altitude. Some of these were also used for jet-assisted takeoff (JATO). Most of these LPREs were installed and flight tested in one of about 12 different military Soviet airplanes, which had been modified for flight testing of a rocket engine. Some of these rocket engines were tested in two or three of these test aircraft. The Germans flew their first superperformance engine in 1936 and had some in development when the war ended in 1945. None of these Soviet rocket-assisted aircraft were fully successful in greatly improving the military capability of the aircraft, none were placed into serial production, and none were used in combat. The Soviet aircraft rocket engines were not fully reliable, caused several flight failures, the deaths of several pilots, and were not really accepted by the Soviet Air Force. In retrospect the huge effort put into LPREs for Soviet aircraft and into their flight tests did not result in a single useful serial production or a really useful application.

The Russians have known about and have tried many different liquid propellants. A good number of different liquid propellants have been investigated in laboratories, a number of these were tested in small TCs, and a few in complete LPREs. They synthesized and tested a new synthetic kerosene (called Syntin) with a higher performance than ordinary kerosene and flew it in a few selected upper stages of SLVs, such as with engine for the third stage of the Proton SLV. With subcooled LOX and this synthetic Syntin fuel, the specific impulse was as much as 20 s higher than ordinary LOX/kerosene; this substantially increased the payload. They developed and ground tested engines with methane or pentaborane as a fuel and other engines with liquid fluorine as the oxidizer, but they did not fly. In the last 30 years they seem to have settled on a few specific propellant combinations, each for a specific category of applications. They are LOX/kerosene for the LPREs of some space launch vehicles and NTO/UDMH for military applications, SLVs, upper stages of SLVs, spacecraft, and multithruster flight control engines. They also developed good LOX/LH$_2$ LPREs for upper stages of SLVs and flew one such engine in 1987 and 1988. At the time of this writing, there did not seem to be an active flight program in Russia with this cryogenic propellant combination.

The Soviet TC technology matured early. The first hypergolic ignition was discovered in early 1931, investigated at a small TC scale in 1931, and was demonstrated in larger TCs in about 1935. Hypergolic ignition was discovered in Germany in 1935, and in the United States it was 1940. The Soviets had partially cooled TCs (cooled by fuel in the nozzle region only) as early as 1933 and 1934. The first fully regeneratively cooled TC dates back to about 1937. The Germans did it around 1935 or 1936, and the American Rocket Society tested one in 1938. As described in Chapter 4.3, they developed and refined a clever cooling jacket design with a corrugated intermediate sheet metal, which is unique to the Soviet LPREs. The Soviets had excellent tooling for fabricating complex cooling jackets out of sheet metal, using selective welding and brazing. This TC cooling-jacket construction is usually lighter than other types. The Soviets preferred multiple spray injection elements (both a simple and a coaxial type) brazed into their injectors. Good combustion efficiencies and good efficiencies for converting the energy of the hot high-pressure gases into the kinetic energy of the nozzle exhaust flow were demonstrated early.

High-frequency combustion vibrations have occurred during the development of a good number of Soviet large LPREs starting in the 1940s, and these instabilities have delayed program schedules and changed several of their engine development projects. A major multiyear high-priority analytical and experimental effort, which involved several organizations, has helped to gain a better understanding of this complex phenomena and has led to several methods that were successful in eliminating this destructive gas pressure oscillation in specific LPREs. These remedies were discussed in Chapter 4.10. A unique solution was the temporary consumable baffle that can control certain combustion vibrations during the start transient.

The Soviet Union was the only country known to put LPREs into submarine-launched ballistic missiles (SLBM). About 15 different LPREs (with hypergolic storable propellants) were developed for application in single-stage, two-stage, or three-stage SLBMs. These engines were immersed in hypergolic storable propellant inside a propellant tank, while the submarine cruised the oceans for long periods. These LPREs were leakproof, had essentially no prelaunch checkout, and could be started quickly. Their novel submerged engines, unconventional propellant tank designs, and novel tank pressurization schemes allowed a substantial reduction (up to about 35%) in missile length, which made it possible to put powerful multistage missiles into submarines. This submerged engine technology is unique in the Soviet Union and has also been applied to at least one of the upper stages of a ground-launched Russian intercontinental ballistic missile, allowing it to carry more propellant and/or more payload with a limited missile length.

In the area of TP technology, the Soviets made some significant technical contributions. For high-chamber-pressure LPRE operations it was necessary to develop TPs with pump discharge pressures of 450 kg/cm^2 or more (6600 psi or more), which is the highest of any known flying TPs. Although high-pressure TPs were developed in the United States, they were not integrated into an engine, and they did not fly. The first simple Soviet turbopump (intended for aircraft rocket engines) was statically tested in 1945. They learned about the two in-line shafts concept of the German V-2 TP, and in the 1950s the Soviets designed a number of TPs for some of their early large LPREs with this in-line two-shaft concept. Compared to an equivalent single-shaft TP concept, this allowed smaller lighter shafts, more durable smaller bearings, and sometimes a lighter TP inert mass, but it resulted in longer and more complex TP assemblies. Most of the TPs developed outside the Soviet Union and many in the Soviet Union had a single-shaft design concept. The Soviet Union gets the credit for developing the most powerful turbopumps; the power was approximately 250,000 horsepower for the RD-170 LPRE in the 1950s and more than that for the RD-270 in the late 1980s, but this last engine did not fly. They also developed a novel lightweight booster TP with two concentric hollow shafts (one inside the other) driving two different pump impellers, each rotating at a different shaft speed. This improved the cavitation resistance and might have slightly reduced the booster and main TP masses, but it is more complex and more expensive. None of the booster TP designs of other countries are known to use these dual concentric shaft features. They were early users of axial-flow inducer impellers (a concept obtained from the Germans), which reduced the required propellant tank pressure. The Soviets seem to be the only known users of small turbopumps, which can be very inefficient, for supplying propellant to small auxiliary TCs of certain spacecraft or of certain low thrust third stage engines. Other nations have used pressurized gas feed systems, which often have a much higher inert mass.

The world's largest production of turbopump-fed LPREs was for the second stages of several defensive Soviet antiaircraft missiles during the 1950s. They were deployed in large numbers to defend Soviet cities. Several different LPREs were produced in quantities between 1000 and 5000 each and put into surface-to-air missiles.

Their engineers had early experience with several methods of positive expulsion for emptying propellant tanks in a zero-gravity environment. For a number of restartable LPREs used in spacecraft applications, they built flexible bladders, which separated the liquids from the pressurizing gas in the propellant tanks. This was one way for preventing gas from being sucked into the propellant flow to the injector. They even developed special plastic bladder materials, which could withstand prolonged contact with nitric acid. In some of the tanks filled with storable propellants, the Soviet engineers would confine the ullage gas in the tank in a separate flexible inert plastic or sheet metal bag. This was a different method for obtaining positive expulsion, and it is different from tanks with flexible diaphragms. They also developed tanks with the propellants confined inside a plastic bladder. These bags would expand (or collapse) as the ambient storage temperature varied or as the pressurizing gas pushed out the propellants from their tanks. Both methods prevent the absorption of the gas in the propellant and ensured positive expulsion for restart in space.

Most of the known common designs for thrust vector control (TVC) with LPREs have been developed, produced, and flown by the Soviets (and also by the United States). They put into production TCs with jet vanes (on a few early LPREs only), hinges (rotation about a single axis), and gimbals (rotation about two axes). They also used hinged or gimballed nozzles for the turbine exhaust gas discharge to achieve flight-path control. The most common TVC system for large Soviets LPREs used multiple smaller hinged auxiliary or vernier TCs, which were placed next to one or more larger fixed (not moving) main TCs. The vernier TCs typically provided about 10 to 20% of the total thrust. In the earlier LPREs the vernier TCs were supplied with propellants from the main engine's TPs. In some of their more recent LPREs, the vernier TC propellants were supplied from a separate small turbopump feed system. In effect it was a separate independent small LPRE typically with four small vernier TCs and its own gas generator (or preburner); the propellants came from the same propellant tanks that also supplied the main engines. In the non-Soviet world the verniers or the reaction control systems usually have a lower thrust (1 to 5% of total thrust), and usually the verniers were supplied from separate heavy gas pressurized feed systems. The Soviet Union was the only nation to implement and fly a unique TVC scheme on a Yuzhnoye LPRE with warm-gas side injection (tapped off the preburner) squirted on command into the sides of the diverging nozzle section of a thrust chamber. In the United States a unique liquid propellant side injection scheme was produced in quantity, but none of these liquid injection schemes were found in the Soviet LPRE inventory. The Soviets also used throttling of selected engines in a cluster of four or more engines for pitch and yaw control. In the large Soviet N-1 SLV six adjacent engines were throttled during flight. In the United States such selective throttling was tested satisfactorily in two aerospike LPREs, but they were not flown.

The early investigators in Russia knew about the performance benefits (about 1 to 3%, depending on the engine parameters) of a curved or bell-shaped nozzle-exit section in the 1930s and conducted tests with curved nozzles in small TCs. They began to use curved nozzles in large LPREs in the early 1950s, about six or eight years earlier than in any other country. The Soviets have, according to the author's available information, not as yet flown an extendible exhaust nozzle, although they ground tested the concept in the 1970s (D-57 of Lyulka and later on another engine). In the United States extendable nozzle configurations were flown with solid-propellant rocket motors in the 1980s and with a LPRE beginning in the 1990s.

The Soviets put a large number of thrust chambers into the base of several of their launch vehicles and managed to start them all essentially at the same time. The R-7 ballistic missile and its space launch versions, the Vostok, Molniya, or Soyuz SLVs, each had 20 larger TCs with an additional 12 vernier TCs (total of 32 nozzles in the cluster). There was an engine with four TCs in the vehicle's core and also in each of the four strap-on liquid propellant boosters. The N-1, the largest SLV, had 30 large LPREs in its first stage, and it also had four additional vernier thrust chambers for roll control (total of 34). Outside of the Soviet Union the largest number of large LPREs was in the U.S. Saturn I booster with eight large H-1 LPREs and the eight nozzles on the British Black Knight.

Several of the design bureaus developed a unique engine system consisting of a main larger engine and a separate pump-fed vernier engine, typically with four smaller TCs. As mentioned before, the thrust of all of the four vernier TCs is typically 10 to 20% of the total thrust. There are some advantages in development and fabrication, and it allows the vernier LPRE to operate before, during, and/or after the main engine operation. Such a dual engine configuration was later used by the Chinese.

The Soviets have developed and flown a number of different schemes for propellant tank pressurization and many are similar in concept to those used in other countries. However three of these pressurization systems are unique to the Soviet Union and have not been seen elsewhere. One system uses two LPREs, which usually operate simultaneously. The larger engine with an oxidizer rich turbine gas, has a tap-off of some of this gas, which is cooled, reduced in pressure, and used to pressurize the oxidizer tank. The smaller engine, often called vernier engine, with multiple thrust chambers has fuel rich gas in its turbine and this gas is cooled and reduced in pressure for pressurizing the fuel tank. In another system promoted by CADB there are four engines driving a given stage, but only one of these four has provisions for generating both fuel rich gas and oxidizer rich gas for pressurizing the respective main propellant tanks. The third scheme uses direct injection of one of the hypergolic propellants into the tank with the other propellant. This system of generating the warm gas directly in the propellant tanks was investigated in other countries, but the only known flying direct pressurization system has been in a Soviet ICBM and has been operated for the last 30 years.

The Soviet LPRE organizations had relatively steady support and funding because of high military or spaceflight priorities, while many other countries had more frequent changes in their annual budgets for LPREs. This relatively steady funding has allowed a steady development of new engine concepts and new LPRE manufacture methods. However it also allowed some duplication and the initiation of some questionable programs.

The cancellation and duplication of major programs caused a poor utilization of LPRE resources, particularly during the busy period of the 1960s and 1970s. The Soviet Union developed engines and put some of them into production, when they later were really not needed. For example, on urgent military requirements the Soviets would at times support two competing parallel engine development programs. This was done on two sets of engines for the R-9A ballistic missile. They would then select one set of the engines for the intended application. The engine programs that were not selected were then canceled. So they ended up with quite a number of perfectly good engines that had to be scrapped or put into storage for a possible future application. This duplication helped them sometimes to pick a somewhat better engine, but it was a luxury that other nations could not afford very often. For the heavy N-1 space launch vehicle for lunar and planetary missions, they actually fully developed and placed into production at least two engines for every one of the six stages of the vehicle. Furthermore they also developed several higher performance (LOX/LH$_2$) upper-stage alternate engines for a subsequent improved N-1 vehicle capability aimed at more ambitious missions. When the big N-1 program experienced four successive flight failures and when the U.S. astronauts got to the moon before the Soviets did, the program was canceled. The Soviet Union then had more than a dozen newly qualified engines that did not then have an application and had to be scrapped or put into storage for an unknown potential future application.

Many of the flight applications, for which LPRE have done well, are today obsolete. For example Russia no longer uses LPREs for JATO, aircraft super-performance, or sounding rocket vehicles. The emphasis in Russia has shifted. The emphasis today is on applications for space launch vehicles, spacecraft, or reaction control for the steering of vehicles. It is not known if there are some new LPREs for as yet undisclosed military applications.

This author looked, but could not find much information accessible to people outside of Russia on a substantial number of LPREs. Very little information was available on some old LPREs, on engines developed by certain design bureaus, LPREs for submarine-launched missiles (with the exception of two engines where photos were found), little on postboost control systems for multiple warheads, or data for certain LPREs for military applications. Details of the remedies or analyses for overcoming and remedying combustion vibrations are fragmentary, and almost nothing could be found about critical high-temperature materials or manufacturing processes. The history of Russian engines in this book is therefore based on partial information only and would be altered when and if the missing data become available.

The LPRE business in the Soviet Union has seen its peak in the late 1950s to the early 1980s. This was the period when the staff of the LPRE design bureaus, the rate of innovation of novel LPRE features, and the output or deliveries were at their highest. Although the volume of business has greatly diminished, there is today still a lot of activity. This decrease has caused some of the design bureaus to leave the field of LPREs. There are fewer Russian design bureaus engaged with LPRE today. Energomash or CADB had about half as many people in this area in 2002 than they used to have about 30 years earlier.

More people were injured or killed in failures of Soviet LPREs or in liquid propellant spills than in all of the other countries. Some were failures in ground-test facilities, and some at launch facilities were caused by failures in vehicle stages or in the engines.

Furthermore, there were failures of experimental LPREs in aircraft, and major accidents, explosions, or spills of hazardous propellants. About 30 years ago the USSR government suppressed the news about a major launch accident at Baikonur Cosmodrome Launch Complex (Tyuratam), where 124 people were killed including some high-ranking officials. About 30 min before the launch, the second-stage engines of a new experimental fully loaded two-stage long-range missile (R-16) suddenly began to fire. The missile broke apart, and the propellants of the first and the second stage spilled, burned, mixed, and exploded. Apparently the control system did not have a protective circuit to prevent a spurious electric currents from generating an unexpected start signal. Although this failure was in the vehicle control system and not directly the fault of the LPRE control systems, the resulting disaster involved the sudden spill of tons of toxic, hypergolic HNO_3/UDMH propellants. One of the victims was Marshall Mitrofan I. Nedelin, a former deputy minister of defense and the commander of the Strategic Missile Forces. A news release falsely attributed his death to an airplane accident.

In Russia the LPRE field is today essentially mature. The basic engine system and key components have been fairly well defined several decades ago, and reliable operations have been achieved by their LPREs many years ago. Today engines are used for spaceflight applications, and they will continue to be needed for spaceflight launches and satellites, but probably at a lower rate because new satellites seem to last longer. Existing models of LPREs seem to fulfill most of the Soviet need, and some improvement in reliability or innovation in materials or engine feature will continue. However, the opportunities for developing a new LPRE are today not as plentiful in Russia as they used to be.

It appears that the Russians have successfully preserved the LPRE capability and for the time being (2003) have maintained this capability in several of the current design bureaus. They have initiated programs to slowly replace existing engines with newer more up-to-date LPREs. There might be a question if some of these are really needed. Furthermore some of these design bureaus have started "conversion" efforts in other (non-LPRE) technical areas, such as commercial pumps, oil field equipment, or medical devices. In the last dozen years the Russians have instituted a policy of exporting LPRE technology and some of their LPREs. This has also helped to maintain the capability and to obtain some additional funding. More than one design bureau can at the time of this writing develop any kind of LPRE that might be needed. With the retirement of skilled personnel and the aging of equipment, it is unknown how long Russia will be able to maintain this capability.

The Soviet Union exported or provided more LPRE-driven missiles to other countries than any other nation. They exported many defensive missiles (such as the SA-2) and short-range ballistic missiles (such as Scud) to their allies and to countries they considered to be friendly to them. They sold an upper-stage LOX/LH_2 engine to India and sold two large LOX/kerosene engines to the United States.

CHAPTER 9

Liquid Propellant Rocket Engines in Germany

The Germans have made early outstanding contributions to the field of LPREs. They were early in starting significant LPRE work, and their government recognized the potential of rocket propulsion in the 1930s and supported R&D aimed at major military applications. There are several distinct periods in the history of LPREs in Germany: 1) the early pioneers, 2) before and during World War II (until 1945), and 3) the postwar period, which began around 1955. The first period starts with the pioneers, the rocket amateurs, and the selection of suitable propellants. The LPRE work before and during the war was performed by three different LPRE organizations and culminates with LPREs for the historic wartime deployment of the rocket-propelled V-2 missiles and the Me163 fighter rocket aircraft. These three teams have found a place in this history: the unique German Army team in Peenemünde (1932–1945), the successful Hellmuth Walter Corporation at Kiel (1934–1945), and the BMW company's rocket engine operation mostly near Munich (1938–1945), each developing a series of remarkable LPREs. Their work is briefly described in Chapters 9.2, 9.3, and 9.4. All of them did their work secretly before and during World War II, and their achievements became known to the rests of the world only after the war had ended. Because of Adolf Hitler's desire to win the war with new weapons, the LPRE work generally had good government support. When the war ended, all of these projects came to an abrupt halt.

After the war (1945) there was no LPRE activity in Germany for 10 years, while the country was rebuilt. A new beginning of LPRE work started in 1955. After several mergers and reorganizations a single company, MBB (Messerschmitt, Bölkow, Blohm) emerged in the late 1980s in the German LPRE field. After further name changes and reorganizations this company became part of the EADS (European Aerospace, Defence, and Space Company), a large pan-European organization. The postwar period is described in Chapter 9.5. All of the references for the five chapters on LPREs in Germany can be found at the end of Chapter 9.5.

9.1 Early Efforts and Early Propellant Evaluations

The most famous German early pioneer was *Hermann Oberth*, and his work has been discussed in Chapter 5.3. His writings were well known in the 1930s, and his participation in an epochal movie production (which included the launching of a realistic looking multistage space vehicle) made him the most well-known space guru of his day.[1,2] He performed some LPRE experimental work in the 1930s, but it was not a groundbreaking achievement. He and his publications had an influence on subsequent German LPRE developments.

The Verein für Raumschiffahrt (German Amateur Society for Space Travel) undertook experimental work in LPREs and launched simple rocket vehicles with these engines. It was described in Chapter 5.2. Some of their innovations were later adopted by the German LPRE industry, and some of their members became employed in the emerging German LPRE industry, as will be explained later.

The Germans did a very thorough job of investigating a *large number of different propellants* beginning in the 1930s.[3,4] Some of this work has already been described in Chapter 4.1. Otto Lutz, one of the leaders of this tremendous task, gives an account of his laboratory investigations of self-ignition characteristics of about 1100 different propellant combinations between 1935 and 1945.[3] This work was done initially at the Technical University of Stuttgart and then at the Government's Aeronautical Research Institute at Braunschweig (Brunswick). It has been discussed in Chapter 4.1.

This propellant research group originally thought monopropellants would be simple to implement, and so they started with several of those. They soon discovered that a monopropellant is really a liquid explosive, and it is very hard to keep it from exploding occasionally. Therefore they quickly switched their investigations to bipropellants. First they evaluated alternate fuels and mixtures of fuel for use with 80% hydrogen peroxide and thereafter alternate fuels with different types of nitric acid. The term "hypergolic" or "self-igniting" was coined in 1935 by Wolfgang Neuggerath, a key researcher of this Institute, and this word was generally used in the German LPRE industry beginning around 1936.

Concentrated hydrogen peroxide had been used as a monopropellant by the Hellmuth Walter Company (see Chapter 9.3) in torpedoes, small submarines, or aircraft powerplants, and this was then well known in German military circles. By finding a suitable hypergolic fuel, which will ignite readily with hydrogen peroxide (particularly at low ambient temperatures), and the starting and restarting procedures and the hardware could be vastly simplified. The final selection by the Brunswick researchers and the engineers at the Hellmuth Walter Corporation was what the Germans called "C Stoff" (C-substance), which was a blend of 30% hydrazine hydrate, 57% methyl alcohol, and 13% water. The oxidizer, called "T Stoff" was 80% H_2O_2 or hydrogen peroxide. The water was added to the fuel to reduce the combustion temperature to below 1750°C (3182°F), which was the limit established by the company based on prior tests with an oxidizer-cooled cooling jacket.

Next came investigations of the self-igniting qualities of several types of *concentrated nitric acid* with many different fuels.[3,4] The German government wanted to switch to nitric acids because the hydrogen-peroxide supply in the country was limited and the future production of peroxide was committed for use in torpedoes, JATOs, or the rocket for the Me163 airplane, as discussed next. The objective of the early investigations with nitric acid was to find a suitable hypergolic fuel that would have a very short ignition delay at the required low ambient operating temperature. One of their typical ignition delay diagrams, determined in Lutz's laboratory in 1936, is shown in Fig. 4.1-1, and it is for two types of aniline. Because nitric acid was so corrosive and oxidized the tank or pipe materials, various additives to the acid were also investigated, but this approach was not successful. Many of the chemicals were ruled out because they were not readily available in Germany during the war, were too expensive, had undesirable properties (too high a freezing point or too low a density), were too toxic, or unstable during storage.

The BMW Company worked during the war with LPREs using nitric acid as the oxidizer, and they participated actively in these propellant investigations. BMW's own laboratory tried about 2000 propellant combinations, and the chemical firm I. G. Farben in Ludwigshafen also participated in this search. Several of the chemicals were tested by all three of these laboratories. Some of the results of these extensive propellant investigations will be mentioned in subsequent chapters, and a summary of these propellant investigations can be found in Chapter 4.1. The references are listed at the end of Chapter 9.5.

9.2 The Army Research Station at Peenemünde

The Heeresversuchsanstalt (Army Research Station) in Peenemünde was a most remarkable organization. The development of the V-2, the first ballistic missile, was the major historic achievement of this organization. It was called A-4 by its developers, and the term V-2 was coined by the propaganda minister of the German government. It stood for Vergeltungswaffe No. 2, which can be translated as retribution or retaliation weapon number 2. The LPRE propelling the V-2 missile was the first large LPRE, and its 56,400 lb thrust was higher (by a factor of between 5 and 10) than that of any prior LPREs.[1,2,5,6] It was a unique and most important milestone in LPRE history. Contrary to the policy of having German industry develop new weapons, the government decided in 1932 to secretly undertake the work in this field in government arsenals or government R&D institutes. The Army recruited technical personnel, some of them came from the Verein für Raumschifffahrt (VfR;) (German Society for Space Ship Travel) as described in Chapter 5.5. One of their recruits was Wernher von Braun, who soon became the technical leader of this highly complex missile effort. He completed his work on a doctorate degree, while still working for the German Army. Incidentally the title of his Ph.D. dissertation was "Constructive, Theoretical and Experimental Contributions to the Problems of Liquid Propellant Rocket Engines."

At their test station in Kummersdorf, a small suburb south of Berlin, the Army organization began to develop and test thrust chambers with thrust levels up to 1.4 metric tons or about 3000 lb. They tested a variety of different designs, several injector schemes, and several propellants. One of the early TCs was made of aluminum alloy, had a cooling jacket, operated with LOX/75% ethyl alcohol, and was rated at 3000 N or 680 lb thrust. There were many failures, such as quick burn-throughs, combustion instability, or low performance. Most major LPRE developers had to go through a series of different injection schemes. Years later when the Germans and the Americans compared early R&D efforts on injectors, there was a remarkable similarity in some of the schemes, which were tested. However the Germans did this 10 years earlier.

The propellants of choice were liquid oxygen and 75% ethyl alcohol. The fuel with 25% water was selected because its flame temperature (4760°F measured) was lower than 100% alcohol, caused less thermal stress, usually burned smoothly, and was a better coolant than several other fuels, which had been investigated.* Also the amateurs of the VfR had a favorable experience with this propellant combination, when compared to gasoline, kerosene, and other propellants tested by this society. A poisonous denaturing agent and a foul-smelling mercaptan were added to the ethyl alcohol, so as to prevent its use as an alcoholic beverage. However during the war, when schnapps was difficult to get, some German soldiers did obtain and drink the doped fuel.

---

*Personal communications, W. Riedel, W. von Braun, A. Thiel, D. Huzel, K. Rothe, and H. H. Koelle, formerly of German Army Experimental Station, Peenemünde,1946–1955.

During the war, the V-2 launch regiment had more casualties of soldiers, who died after drinking the denatured alcohol, than from enemy action.*

The A-1 (Aggregat #1 translated as Apparatus #1) test vehicle was developed early. It used LOX/alcohol propellants with a nominal thrust of 660 lbf and a gas-pressurized feed system.[5,]* Its LOX tank was made of a fiberglass-type insulating material, and it was located within the fuel tank. The A-1 was prone to fires and excessive oxygen tank pressure. The first flight of the A-1 in the summer of 1933 exploded because of an ignition delay. It was followed by the development of the A-2 test vehicle, which had separate individual propellant tanks. Its TC was new, had 15 kN or 3400 lbf maximum thrust, and spiral-wound wires to guide the cooling flow inside the cooling jacket. The injector had a central mushroom-shaped multihole oxygen disperser and many fuel side injection holes or side injection spray elements, which were pointed radially inward. It was the model after which the V-2 injector domes were later patterned, and it is described next. Two A-2 test vehicles were launched successfully in 1934 from an island in the North Sea, and both reached altitudes of 6500 ft. This propulsion system had a small novel liquid-nitrogen tank, where the nitrogen was evaporated by a small electric heater and the resulting nitrogen gas was then used for pressurizing the propellant tanks. This controlled liquid evaporation device was lighter in weight than an equivalent high-pressure nitrogen gas tank.

This LPRE also included features to briefly limit the initial propellant flow during startup. This low-thrust phase during the start allowed a relatively smooth thrust buildup without a "hard start" (high-pressure surge) and limited the accumulation of unburned propellant in the chamber that can occur at initial high propellant flow. This low-thrust preliminary stage was a significant contribution to the state of the art of LPREs and was also later incorporated and copied in the design of a good many other rocket engines.

Progress toward a better LPRE and a better flight vehicle remained speculative, in part because the government rejected an early Army request for a ballistic missile.* In 1937 most of the operation was relocated to Peenemünde on the Baltic Sea; it became a large military R&D facility with laboratories, static firing test stands, launching platforms, an airport, an experimental shop, and later some limited LPRE production facilities. The V-2 and other missiles were launched from there in a northeast direction over the Baltic Sea. The Army kept the LPRE station at Kummersdorf, and much of the R&D on small TCs and small LPREs continued to be performed there. The A-3 (21-ft-long and 1640-lb takeoff mass) was developed and tested by the end of 1937. The A-3 test vehicle was a reduced scale model of the A-4, and it validated the aerodynamic configuration of the supersonic missile. The A-3 had the first automatic flight control. A pressure feed system supplied the propellants to the TC, and the engine had a thrust of 1.4 metric tons or about 3000 lb. Flight tests

---

*Personal communications, W. Riedel, W. von Braun, A. Thiel, D. Huzel, K. Rothe, and H. H. Koelle, formerly German Army Experimental Station, Peenemünde, 1946–1955.

were conducted at Peenemünde, with the missiles being launched over the Baltic Sea. Although the pressure-fed LPRE worked, the guidance system and other components experienced major development problems.

In 1936 this Army team under the technical leadership of Wernher von Braun also adapted a version of the A-3 LPRE (with LOX/75% alcohol) for an aircraft superperformance application.[7,8] It gave 1000 kg or 2200 lbf thrust for up to 30 s. This engine was installed in a Heinkel He112 aircraft in 1937. The engine malfunctioned and caused the plane to crash. After repairs the test flights were resumed, and eventually the engine performed as intended and flew satisfactorily. The LOX was really not suitable for aircraft application because the oxidizer hardware (injector, valves, and pipes) had to be cooled by bleeding oxygen through the hardware prior to operation, the formation of ice (from condensed moisture in the air) increased the aircraft weight, and the evaporation of the oxidizer limited the storage period. There were no further efforts to develop an aircraft rocket engine by von Braun's team at Peenemünde.

By 1938 the German Army got approval and funds to proceed with the development of a mobile-based ballistic missile (250- to 300-km range and and 1-ton warhead). When the preliminary designs of the A-4 were made, the engine requirements were established as a thrust of 25 tons or 55,000 lb and a minimum specific impulse of about 200 s in order to reach the specified target missile range of 250 km (155 miles).[2,*] The subsequent engine R&D focused on these requirements. Testing of full-scale engine components started in 1938. Tests of a subscale interim missile, called A-5, also began in 1938 and continued in 1939. Its engine used a pressurized feed system. There were about 25 A-5 flights, and many of them focused on the automatic steering system.

The historic A-4 flight vehicle, later renamed as V-2, is shown in Fig. 9-1, its engine is shown in Fig. 9-2 and in Fig. 7.8-1 engine data are in Table 9-1, and a schematic LPRE flow diagram in Fig. 9-3.[1,2,6,*] This important engine is described here in more detail than the other German LPREs. The development of the thrust chamber, rated at 25.4 metric ton or 56,000 lbf at sea level, was simplified by mounting 18 nonimpinging fuel-cooled injector heads on the upper side of the combustion chamber. This injector head had its heritage in the injector of the smaller 1.4-ton TC, and this nonimpinging stream injection pattern and the propellant combination gave stable combustion operation. The TC and a more detailed section through one of the injector heads are shown in Fig. 9-4. In each of the 18 injection heads, the oxygen came through a central brass fitting with rows of multiple holes (flowing axially down as well as radially outward), and the fuel came through three rows of brass spray injection inserts and some rows of holes (flowing toward the center of the injection head). The fuel flow in the cooling jacket was admitted to the fuel injector manifold through the main fuel valve, which was mounted on top of the thrust

---

*Personal communications, W. Riedel, A. Thiel, D. Huzel and H. H. Koelle, formerly German Army Experimental Station, Peenemünde, 1946–1960.

Fig. 9-1 V-2 missile carrying scientific instruments is shown here during launch preparations at White Sands Proving Ground in New Mexico, 1950. Courtesy of U.S. Army and General Electric Company; photo from General Electric Company.

chamber. The combustion was not very efficient, and the chamber volume was relatively large, and therefore the TC was heavy. A measure of the relative chamber volume is the so-called characteristic length L^*, which is the chamber volume divided by the nozzle-throat area. The A-4 TC has an L^* value of 113 in., and newer TCs with good combustion efficiency would have a typical L^* value of 30 to 40 in. If the design would not have been frozen for mass production, the propulsion people at Peenemünde had gathered enough data to make a smaller, lighter, and better performing TCs.

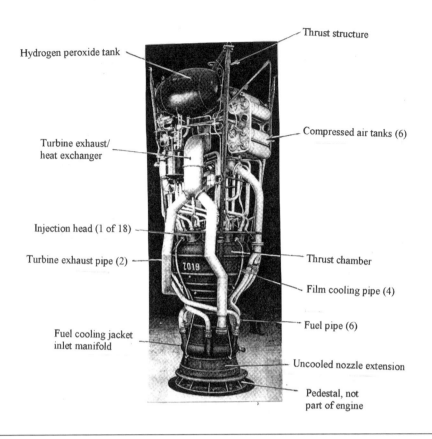

Fig. 9-2 V-2 LPRE had about 10 times more thrust than any prior rocket engine. From German Peenemünde Archives.

The regeneratively fuel cooled chamber was made of formed and welded low-alloy steel plates 4 mm thick. There were some burn-throughs of the cooled chamber or nozzle walls during development. These failures were remedied by adding three rows of film-cooling injection holes (each supplied through a separate manifold and external piping) upstream of the nozzle throat, thus augmenting the regenerative cooling, but at a slight loss in performance. One of the four pipes supplying the film-cooling fuel to the four manifolds, which supply film coolant to the four rows of film-cooling holes, can be seen in Fig. 9-4. The mass flows of these four film-cooling rows of holes are calibrated by flow restrictors. The chamber and nozzle grew about 4 mm in length when the inner wall became heated during operation. Four expansion folds were provided in the outer wall between each film-cooling ring to accommodate this dimensional growth of the hot inner wall.

Flight-path control of the V-2 is achieved by four movable aerodynamic fins at the lower ends of each of the four tail surfaces (they can be seen in

Liquid Propellant Rocket Engines in Germany **745**

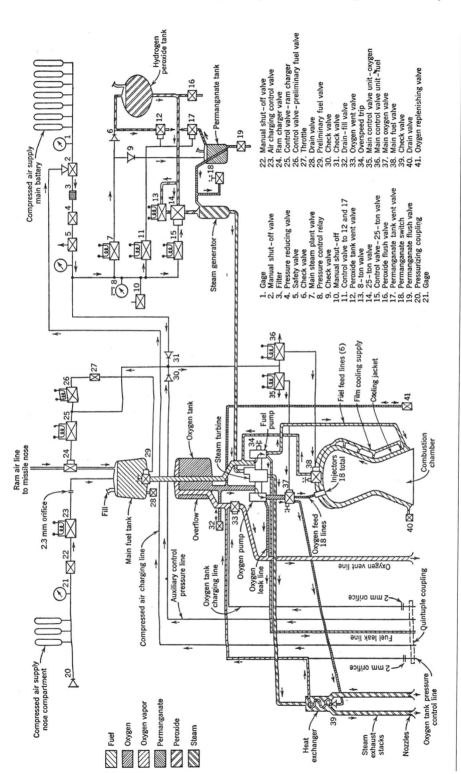

Fig. 9-3 Schematic flow diagram of the V-2 LPRE. Redrawn from German Peenemünde Archive data; copied with permission from Ref. 13, p.224.

Table 9-1 V-2 Rocket-Engine Data

Propellants, engine	LOX 75% ethyl alcohol, 25% water
Propellant, gas generator	80% hydrogen peroxide
Gas-generator catalyst	Aqueous solution of potassium permanganate ($KMnO_4$)
Thrust at sea level	25.4 metric tons or 56,000 lbf
Thrust at high altitude	64,000 lbf
Exhaust velocity at sea level	6540 ft/s
Specific impulse	203 s at sea level and 242 s in vacuum
Chamber pressure	220 psia
Propellant flow	276 lb/s
Mixture ratio, oxidizer to fuel mass flow	1.24
Average injection pressure differential	34 psi
Cooling-jacket pressure differential	63 psi
Number of injection heads	18
Nozzle-throat internal diameter	15.75 in.
Nozzle-exit area ratio	3.4
Film-coolant flow	13% of fuel flow
Characteristic chamber length L^*	113 in.
Engine weight, dry (incl. air tanks)	2050 lb
Propellant tank weights, dry	433 lb

Fig. 9-1) during the traverse of the atmosphere and by jet vanes during flight in the vacuum above the atmosphere. The aerodynamic fins and the jet vanes were coupled together and moved together. The lift forces on the aerodynamic fins and/or the jet vanes immersed in the exhaust jet allowed thrust vector control (TVC) of the V-2 missile in pitch, yaw, and roll. The jet vanes were mounted on rotary shafts just below the nozzle exit, as shown in Fig. 9.5, were made of sintered carbon and glowed white hot during firing operation. The oxidation of the carbon material (by the small amount of oxidizing gas species in the hot exhaust flame) caused the leading edges of the vanes to recede by about 1.0 or 1.5 in. during a typical powered flight. There was a short uncooled single-wall segment without a cooling jacket at the exit of the nozzle, but it was film cooled through a fourth set of film-cooling holes and a separate film-cooling manifold.

The 465-hp turbopump (TP) has been shown in Fig. 4.4-4 and was discussed in Chapter 4.4. It was the first TP for a large LPRE and the second flying TP in Germany. It was the highest power and the largest size TP of its day, exceeding prior TPs by a large margin. It also was the first flying TP in the

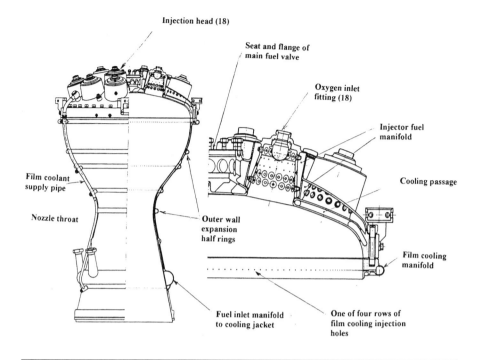

Fig. 9-4 Partial section through the thrust chamber and injector of the V-2 engine. It shows oxidizer injection holes in the oxygen inlet fitting, fuel film cooling and screw-in injector elements using a nonimpinging stream pattern. Adapted from German Peenemünde documents.

world with two in-line shafts.* The centrifugal fuel pump impeller and the two-stage aluminum turbine wheel were mounted on one shaft supported by two precision ball bearings. The oxygen pump impeller was on the other shaft supported by two oxygen lubricated journal bearings, and it had an axial bearing for counteracting hydraulic forces. The two shafts were connected by a flexible flange coupling. Shaft seals separated the regions of fuel, LOX, and turbine gases. The Hellmuth Walter Company in Germany had built TPs a few years before, but always with a single shaft and at a much lower power level. If the V-2 turbopump would have been designed with a single shaft, this shaft probably would have twice the diameter (for stiffness and rigidity), and the rotating parts would be larger in diameter and have considerably more mass; also the pump housings and the turbine casing would be larger in diameter and heavier. This TP had an overspeed safety feature that caused the engine to shut down automatically (by closing the GG valve) if the shaft speed became

---

*Personal communications, Kurt Rothe, formerly German Army Experimental Station, Peenemünde, circa 1965.

Fig. 9-5 Carbon jet vanes immersed in the rocket-engine's exhaust gas were the first means for thrust vector control of large LPREs. These V-2 graphite jet vanes provided flight control in pitch, yaw, and roll. From Ref. 6, p. 248.

too high. As stated elsewhere in this book, the Soviets later copied this two-shaft concept in several of their TPs in large LPREs.

The shaft was made of a heat-treatable carbon steel. The turbine inlet temperature of about 725 to 750°F allowed the turbine housing, rotor and stator turbine vanes to be made of cast aluminum alloys. The nozzle blocks were made of cast iron. The forged aluminum turbine disc was riveted to a two-piece steel center part, which had a six-pointed serration for fitting to the splined shaft.

For the pumps to be primed for starting and avoid cavitation, it is necessary to slightly pressurize the two propellant tanks with gas. The oxygen tank is pressurized by a small flow of oxygen, which is tapped off the oxygen pump discharge pipe and gasified in a heat exchanger, which is located in the hot turbine exhaust duct. The fuel tank is pressurized by compressed air during most of the flight and by ram air during the vehicle's traverse of the atmosphere. This LPRE was the first pump-fed engine with these tank pressurization features. The propellant tanks of the V-2 were separate and located within the missile's external structure. Since the late 1940s, the propellant tanks have usually been part of the structure.

The decomposition products of 80% hydrogen peroxide (a mixture of steam and oxygen at approximately 775°F) were created in a gas generator by catalysis with a solution of permanganate in water.[9,10] This method of creating hot gas for driving an uncooled turbine had previously been developed and proven by the H. Walter Company. This temperature was low enough to permit the use of aluminum alloys for the turbine, turbine blades, turbine housing, and steam piping. Compressed air was sent through a pressure regulator to pressurize the egg-shaped hydrogen-peroxide tank and the aqueous catalyst tank. Upon start the gas/steam mix quickly brought the turbopump up to rated speed, and this helped to shorten the start sequence.[6]

To reach a given target with a ballistic missile, the flight velocity at engine cutoff has to be exactly at a predetermined value. A short time before engine cutoff, the thrust of the engine was reduced to about 31% of full value (8 tons) because the uncertainty in total impulse (after a cutoff signal) and the uncertainty of the final missile velocity are smaller at the lower thrust level. Furthermore this diminishes a potential water hammer problem that would occur if the cutoff at full flow would occur too fast. This lower thrust is achieved by reducing the GG flow (by routing GG propellant through a smaller GG supply valve) and by bypassing some fuel flow from a port in the main fuel valve back to the fuel pump inlet. This bypass increased the fuel flow and helped to maintain a reasonable high cooling velocity in the cooling jacket. This flow reduction and the fuel bypass can be seen in the flow diagram in Fig. 9-3.

A unique, safe, and controlled start sequence was developed by this German team. After the pyrotechnic igniter had started to burn, the main propellant valves are opened, and propellants flow at a low rate (under gravity and the tank ullage gas pressure) into the chamber, where they are ignited by solid-propellant igniters and then burn; this was called the preliminary burning stage. The thrust during this low flow preliminary stage is about 17,000 lbf, not enough to lift the missile off its launch stand. When the flickering flame, which could be seen outside of the nozzle, appeared to be satisfactory, the launch operator gave the electric signal for main stage. The water-permanganate solution and immediately afterwards the hydrogen peroxide are then fed into the gas generator where hot gas is created by catalytic action. This gas accelerates the turbopump, which then increases the propellant flows and the feed pressures up to their full values. When the thrust equals the vehicle weight, the missile lifts off its launch platform. Variations of this start sequence have been used on other large LPREs since that time.

The flow diagram of Fig. 9-3 shows 34 different valves and pilot valves. Each valve has a specific purpose or function, which can be discerned by closely examining the flow diagram. This number of valves is typical of early large LPREs, which have a low-thrust preliminary operating stage, a reduced thrust period at cutoff, a separate propellant for the gas generator, and integrated propellant tank pressurization features. The number of valves can be reduced, if a new version of this engine were to be designed today using newer technology and if some changes in the functions or requirements would be allowed. The minimum number of valves would be for a simple bipropellant single-use only LPRE, which has a solid-propellant GG and five burst diaphragms. A burst diaphragm has been considered by some people to be a simple single-use valve.

The German government wanted an early V-2 missile production, and therefore it was necessary to freeze the design. If von Braun's engineers would have had an extra year, they would have been able to design a considerably better flight vehicle and a better, less complex engine, but this was then not possible. Design changes were strictly controlled, and only those that were critical or related to safer operations were allowed. Toward the end of the war, more changes had to be approved because many key materials were no longer available and a substitute material had to be selected, tested, and installed, and some of these were not effective. Also Allied bombing caused interference with production of some components, and alternate sources of supply had to be developed and qualified. The control of the manufacture during the war was very difficult. A total of 65,000 changes were required to get the V-2 missile through its final production.*

The key man for the development of the LPREs on this Army team was Walter Thiel, who reported directly to von Braun. He deserves the credit for several of the engine design features. By today's standards his engines were not the lightest or best performing pieces of propulsion machinery, but they were sensational achievements at the time, as the next few paragraphs will explain. Thiel was killed in 1943 during an Allied bombing raid on Peenemünde.

The historic V-2 LPRE was the first to have these *unique characteristics:*

1) It was the *highest-thrust LPRE of its time* by a wide margin.

2) It was the first really large LPRE with a *turbopump/gas-generator* feed system and a gas-generator engine cycle. It had the first large turbopump, and it used two shafts in line with each other.

3) The V-2 was the first rocket vehicle and the first LPRE to be *mass produced*. It has been the only large LPRE (over 30,000 lbf thrust) produced in relatively large quantity. More than 5000 have been built, and perhaps 3000 have flown. Of these about 600 were used in development and training flights. Toward the end of the war, the production rate reached 700 units per month. About 25%

---

*Personal communications, W. Riedel, formerly German Army Experimental Station, Peenemünde, 1946–1950.

of the V-2 missiles fired against England did not reach or explode on their target area. Some failed in flight control, some had warheads that did not explode, and some missiles broke apart during flight or reentry. Propulsion system failures did not occur very often.

4) It had a *unique engine start system*, which included gradual opening of main propellant valves and a preliminary operating stage at low propellant flow. It allowed a decision to proceed or to stop the launch based on visual observation of the exhaust flame during the preliminary stage. It also prevented the accumulation of excessive mixed unburned propellants on the chamber (avoids hard starts). It had a unique igniter, namely, four solid-propellant igniter cartridges rotating on a swastika-like wheel inside the chamber.

5) This engine was the first to be able to *run at reduced thrust* (about 31% of full thrust) for a short time during shutdown, while using the turbopump at low flow. This reduced the amount of afterburning during the shutoff procedure, helped the missile guidance system to achieve a more accurate terminal velocity for the warhead, and reduced the potential target error.

6) It was the first guided rocket vehicle with an *effective flight controller*, which commanded the movements of four movable fins and the four *jet vanes* to apply pitch, yaw, and roll movements to the missile during powered flight. The jet vanes caused extra drag and effectively reduced the engine performance by perhaps 1 or 2%. The vehicle also had four air rudders or fins to help control the portion of the flight in the atmosphere.

7) The engine and the missile were developed, produced, and fielded in *record time*. Detail design was in 1940, the first flight in June 1942 had a failure of the feed system, the second flight did not reach the target, but the third launch in October 1942 was a success. The first military use was in late 1944. Top government priority for obtaining materials and resources, adequate funding, and the freezing of the design made this schedule possible.

8) The V-2 was the first missile with a LPRE to use a mobile launcher. It could be launched from almost any place accessible to a large truck. The check-out of the engine and the loading of the propellants from tanker trucks had to be done in the field. In case of a launch problem, the missile's propellants could also be safely detanked or removed. This mobile launch technique has since been emulated by other mobile missiles in the Soviet Union, United States, or Iraq.

9) It served as a significant *lesson* for the design and development of other large LPREs. The United States, Soviet Union, Great Britain, and France obtained some V-2 hardware, German experts, and know-how. All of the other countries, which were interested in LPREs, ballistic missiles, or space launch vehicles, studied the V-2 design and literature.

10) *Ram air is used for the pressurization* of the fuel tank during a portion of the powered flight while the missile traverses the atmosphere; this saves some pressurization system mass.

11) The *empty fuel tank remains pressurized* after engine shutoff, so that the tank will not collapse during the missile's reentry into the denser atmosphere.

The United States and the Soviet Union considered the V-2 LPRE so significant that both countries copied this engine and manufactured it. Figure 7.8-1 shows a U.S. made copy. In addition, the United States, the Soviet Union, and Britain launched captured V-2 missiles. The Soviet Union was the only country to equip a factory for building complete, but slightly modified V-2 ballistic missiles and to deploy them in their military forces.

The German designers visualized and undertook the design of advanced versions of the V-2 flight vehicle. A longer-range development version of a V-2 vehicle equipped with sweptback wings was actually flown.[1,2] But others (larger vehicle, two- and three-stage versions, and even a piloted version) were mostly design studies. They had contemplated switching propellants from LOX/alcohol to storable hypergolic propellants, such as nitric acid (with a small amount of sulphuric acid) as an oxidizer and a mixture of vinylisobutyl ether and aniline as a fuel. Some component tests with these propellants were actually performed. This propellant switch would have increased the range and eliminated the launch delays caused by bleeding LOX through the engine for cooldown. It would also have eliminated the formation of ice on cold surfaces, which adds extra inert mass or the adding of extra LOX propellant just prior to launch; this topping-off was really needed for the longer-range missions. The end of the war stopped further advanced V-2 developments.

The Germans were slow to realize the need for antiaircraft missiles to defend against U.S. and British bomber raids during World War II. The team at Peenemünde and the BMW company worked together on a very simple unique prepackaged LPRE for the *Wasserfall (Waterfall) surface-to-air missile*. It is described in Chapter 9.4. Actually von Braun's German team did an antiaircraft missile study in 1942 and recommended the development of such a missile. However this program was delayed by the German military command in order to concentrate available resources on the V-2 development and production.* Because of this delay, the Wasserfall missiles were too late to be deployed during the war. There has been speculation as to what would have happened to the Allied bombing raids over Germany if the Wasserfall would have been deployed in quantity.

In January 1945, when it was obvious that Germany was losing the war, the personnel at Peenemünde were ordered to evacuate the facility, and Wernher von Braun and his team left the facility, but took many documents and draw-

---

*Personal communications, W. Riedel, formerly German Army Experimental Station, Peenemünde, 1946–1955.

ings with them. The majority of the key technical personnel surrendered to the American military, were transported to the United States, and eventually joined NASA, where they worked on launching U.S. satellites and on the Saturn/Apollo project. During the 1946 and 1971 period, the author had repeated meetings with von Braun and about 15 of the Germans who came with him from Peenemünde. These interchanges were in connection with the Rocketdyne development of the large LPREs for the NASA Saturn/Apollo SLVs, which were then designed by this team. A lot of information about German LPRE efforts came from this elite German team, and the data were very useful in the U.S. LPRE efforts. Several members of the former Peenemünde team joined various U.S. aerospace companies. For example, Walter Riedel, Dieter Huzel, and Kurt Rothe joined Rocketdyne; Rudi Beichel went to Aerojet; Kraft Ehricke went to Convair; and Walter Dornberger and a few others went to Bell Aircraft. Other German rocket experts went to the Soviet Union, France, and Britain.

It is worth repeating that the V-2 LPRE was an outstanding historic achievement. Its thrust was ten times larger than any other engine at or before that time. The engine incorporated novel features that were copied, adapted, or modified by many subsequent large rocket engines. All the other nations in the LPRE business studied the V-2 engine design and two (Soviet Union and the United States) build copies to learn the fabrication process.

9.3 Hellmuth Walter Corporation

Beginning in 1935, the Hellmuth Walter Kommanditgesellschaft (abbreviated as HWK and translated as Hellmuth Walter Corporation), located in Kiel on the Baltic Sea in Germany, based its business on what the Germans called "T-Stoff." It is 80% hydrogen peroxide with 20% water and a very small amount of phosphate stabilizer (to minimize self-decomposition). HWK used this monopropellant originally for torpedoes, small submarine drives, launching antisubmarine depth charges from a Navy ship, or emergency powerplants. Later it was applied to aircraft/missile catapults (such as those used to launch the V-1 missile) and to LPREs for aircraft and missiles.[9–11] This author met Mr. Walter after the war and found him to be a modest, serious, and very knowledgeable man. His company developed and flew more different LPREs than the other two German organizations. More than 18 different LPREs were developed by HWK from 1934 until 1945. What is so amazing is, that almost all of these HWK LPREs were produced in limited quantity, were qualified to fly, have flown in experimental vehicles, and more than half went into serial or mass production. None of the other LPRE organizations have such an outstanding record of success in having the majority of their LPREs adopted by the military forces for important applications.

The early HWK aircraft LPREs (1936 and 1937) were monopropellant hydrogen-peroxide units for assisted aircraft takeoff applications.[6,10] The first units had 100-kg (220-lb) thrust, but later developments went up to about 1500 kg (3300 lbf). The hydrogen-peroxide propellant tanks were pressurized by compressed air. The exothermic decomposition into steam and oxygen was first catalyzed with a paste-type catalyst, which was not fully reliable. A year later the simultaneous injection of a small amount of an aqueous catalyst solution of potassium permanganate (or alternatively sodium permanganate) together with the hydrogen peroxide was successfully developed, and this liquid catalyst was used in other German rocket engines and gas generators, such as the GG of the V-2 engine.

The first jet-assisted takeoff (JATO) flight with a HWK LPRE of 100-kg (220-lbf) thrust and hydrogen peroxide as a monopropellant was in February 1937. By comparison the first U.S. JATO was five years later, and the first Soviet JATO a year or two after the United States. The first German JATO flight test was with an overloaded Heinkel He-111 bomber-type aircraft. This is the first rocket-assisted takeoff with a LPRE in recorded history. Subsequent HWK JATO engines had 300 kg (660 lbf) and then 500 kg (1100 lbf) thrusts, and they flew in modified aircraft in late 1937. Figure 9-6 shows such a Walter JATO engine.[6,10,11] The flight tests of the small-scale A-1 models of the V-2 (mentioned in the previous chapter) were propelled by a HWK engine with a gas-pressurized feed system. These A-1 rocket vehicles reached a height of 18 km (11 miles) and were the first German rockets to exceed the speed of sound.

The 500-kg thrust monopropellant JATO (identified as HWK 109-501) was in production at HWK when World War II began.[6,9,10] It was first used experimentally for takeoff by the Heinkel He 111 and Junkers Ju 88 bombers. Several

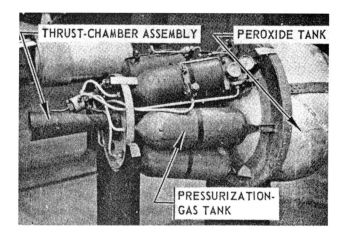

Fig. 9-6 Monopropellant hydrogen-peroxide assisted takeoff rocket engine of the Hellmuth Walter Company with a pressurized-gas feed system. The first aircraft takeoff with a rocket-assisted engine used a similar, but smaller LPRE in 1936. Copied from Ref. 6.

thousands of this JATO were delivered, and the German bomber airplanes were slightly modified to use this unit. Some JATO units were equipped with parachutes and were recovered and reused. Although the HWK JATO was designed for 60 flights, it actually seldom reached this number. During the war, there were thousands of rocket-assisted takeoffs of German Air Force planes on many different German air bases in Europe. This deployment is in stark contrast to other countries, which later also developed and flight tested liquid propellant JATO rocket engines (such as United States, Britain, or France), but did not use them in military operations. The use of HWK JATOs by the German Air Force was extensive during the early years of the war, but declined during the latter part of the war. This was because hydrogen peroxide was then more difficult to obtain and because many German pilots did not like the extra effort and the time needed for the fueling and installation of a JATO. They preferred to take off without the JATO, but with a lower payload.

One Walter JATO experimental model of 1000 to 1500 kg thrust used a bipropellant with gasoline as fuel to achieve higher performance. The hydrogen peroxide would be decomposed by a liquid catalyst, and then the gasoline fuel would be added to the hot oxygen-rich gas for further combustion. This greatly increased the specific impulse from about 148 to about 210 and up to 220 s. This JATO unit could be turned off at any time, and this was not possible with prior designs. All of the HWK JATO units mentioned so far have used pressurized-gas feed systems. Four of these new 1000-kg thrust JATOs were used in the first liftoff of a new bomber, the Junkers Ju287, which was the first German bomber equipped with jet engines. Although this bipropellant unit was produced in limited quantity, it was never deployed with the military aircraft operations. The Air Force preferred the simpler lower-thrust 109-501 monopropellant unit. A modification of this larger bipropellant unit propelled the experimental Enzian antiaircraft weapon just before the end of the war.

The first flight in the world of an airplane with a *superperformance or auxiliary LPRE*, in addition to its own aircraft engine, was made in a Heinkel He72 airplane with a simple modified JATO monopropellant Walter engine in 1936.[4-8] It augmented the airplane's propeller engine, used a gas pressure feed system, and could run for 45 s at 100-kg or 220-lbf thrust. It flew satisfactorily, but the flight-test data showed that the thrust was too small. There were earlier airplane flights with auxiliary solid-propellant rocket motors in Germany. However it was not possible to stop the thrust on command or to restart, and this severely limited the flight maneuvers. Solid propellants were not again considered for aircraft superperformance.

A new Walter monopropellant engine with a higher and variable thrust became available in 1938, and it was flown successfully in 1938 as power augmentation in a Heinkel He112 airplane.[7,8] Its maximum thrust was 950 kg or 2000 lbf. The flight-test data were encouraging and indicated that a new airplane could exceed 1000 km/h (621 mph), which would exceed established world records.

Heinkel designed and built such a new airplane, identified as the He176, to be powered solely by a LPRE.[6-8] It flew in 1939 with a Walter throttlable, restartable monopropellant hydrogen-peroxide engine of 600 kg or 1320 lbf. This was the very first historical takeoff and flight of an all-rocket-powered airplane. Figure 2-4 shows a sketch of this small historic airplane. It reached a speed of 700 km/h or 621 mph, which was much less than expected. One of the demonstration flights was observed by Hitler, Göring, and several of Germany's top generals. They were not impressed and disliked the loud rocket-engine noise. When war broke out, the He176 program was stopped. The airplane and its engine were put on display at the Berlin Air Museum, where they were destroyed by an Allied bombing in 1943.

The turbopump-fed HWK 109-509 engine series was probably the most famous Walter aircraft rocket engine. It was the primary and only powerplant for the Messerschmitt Me163 interceptor aircraft, the first rocket-propelled fighter airplane put into production and the only rocket fighter aircraft engaged in aerial combat during World War II.[6,7,9,10] One version of this historic rocket-propelled aircraft is presented in Fig. 2-7. A fleet of 364 airplanes and at least 470 LPREs were produced.

The Walter engine for the Me 163 was based on an earlier simpler LPRE, which was also qualified a couple of years earlier. It was a hydrogen-peroxide monopropellant engine version with a turbopump feed system, and this had a relatively low performance (specific impulse of about 147 s at SL). The TP was the very first TP in a flying LPRE. In 1940 the first test flights with this turbopump-fed monopropellant engine were made on a DFS 1944 experimental glider. It was towed to altitude, released from the towing airplane, and the engine, rated at 800 kg or 1760 lbf thrust, was started. In the first flight it was a surprise to find that the engine operated at about half its rated thrust value, and the desired flight performance was not attained. After a careful investigation it was then determined that cavitation in the oxidizer pump impeller reduced the flow of propellants. Nevertheless it is the world's first altitude flight with a pumped-fed rocket engine. Because of this incidence of cavitation, the Walter engineers paid special attention to having adequate net positive suction pressure in the pump feed line, which will prevent vaporization and thus avoid cavitation at the impeller inlet or a reduction in thrust. For this reason they developed inducer impellers ahead of the main pump impellers because this improves the cavitation resistance of the main pump impellers. The improved monopropellant version of the HWK 109-509 monopropellant engine (with inducer and modified TP) was put into production and used on the first series of successful test flights of several of the Messerschmitt Me 163 aircraft in 1941. This engine is shown in Fig. 9-7. The flat casting on the left is a gear case. On the other side of the gear case are the peroxide pump, the liquid catalyst pump, and the turbine.

The next version of the engine (developed in 1942) used bipropellants, namely, T-Stoff oxidizer and hypergolic C-Stoff fuel.[12] The nominal thrust was 15 kN or 3872 lbf. It was basically a new design identified as HWK 109-509. A-1. It first flew in early 1943. This C-Stoff fuel consisted of 57% methyl alcohol with 30% hydrazine hydrate and 13% water. It was hypergolic with the T-Stoff or 80% hydrogen-peroxide oxidizer. This fuel was based on the results

Fig. 9-7 HWK 109-509 LPRE was the initial monopropellant powerplant for the flight tests of the Messerschmitt Me163 military fighter aircraft. With improved bipropellant versions of this engine, the Me163 was engaged in combat operations in 1944 and 1945. Copied from Refs. 6 and 10.

of R&D work on propellants done at HWK and the Research Institute at Braunschweig, which was mentioned in Chapter 9.1. This LPRE did not need a liquid catalyst, its feed system was simpler, and restarting was easy. The early version (109-509.A-1) used an electric starter for initiating the rotation of the turbopump, but the next version (109-509.A-2) was similar, but used a monopropellant gas generator with a separate gas-pressurized small T-stoff tank for starting. This reduced the power drain on the aircraft electrical system. The next versions (109-509.A-2 and 109-509B) of this aircraft rocket engine had a second smaller TC (underneath the main chamber) for cruising, and this change augmented the range and effectiveness of the Me163 aircraft. It first flew in August of 1943. Its TP is shown in Fig. 4.4-3.

As seen in Fig. 9-7, all of the versions of this 109-509 LPRE had a long 6 in. diam tube as the principal structure.[6,9] The thrust chamber and in some models also a valve were at one end (at the tail of the aircraft), and a package consisting of the valves, turbopump, gas generator, and with some models also a gear case were at the other end of the 6-in. tube (inside the fuselage). The oval-shaped main thrust chamber and the nozzle are fuel cooled and can be seen in Fig. 9-8. The injector had three groups of three bipropellant pintle-type spray heads each at the flat injector face, and each group could be activated separately, thus achieving three levels of thrust.[6] Inside the structural tube was the propellant piping for each set of injector heads, for the cooling jacket, and some hydraulic lines.

The turbopump shown in Fig. 4.4-3 was historic because it was the first bipropellant TP (with fuel and oxidizer pumps) to fly, the first to power a combat fighter aircraft, and the first to use two axial-flow screw-type inducer impellers ahead of each of the two main pump impellers. The TP was designed by Bruckner and Kanis, a company that had developed turbines and hydrogen-

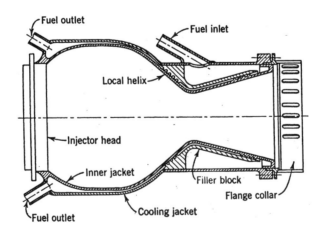

Fig. 9-8 This sectioned TC of the HWK 109-509 shows an oval chamber shape and the cooling jacket. The spray injection inserts in the injector face are not shown. Copied with permission from Ref. 13, p. 143.

peroxide pumps for torpedoes and submarines. The TP housing and impellers were made of aluminum and the single shaft from an alloy steel. These inducers improved the suction characteristics and limited cavitation, which had occurred with the earlier monopropellant TP model and plagued some early test flights. The inducers reduced the tank masses and probably slightly reduced the inert TP mass. This reduction in inert hardware mass improved the airplane performance. Although American, British, French, and Soviet personnel examined these inducers after the war in 1945, it was not until 7 to 10 years later that the inducer impeller concept was implemented in a flying LPRE in these countries.

The turbine was driven by a separate gas generator where 80% monopropellant hydrogen peroxide was decomposed into hot gas (775°F) by a catalyst. A disassembled view of the gas generator is shown in Fig. 9-9. The liquid catalyst used earlier had been replaced by a solid-type catalyst, which simplified the engine and eliminated a catalyst tank with its gas pressure feed piping and valves. Catalysts were either silver screen packs or cobalt impregnated pebbles confined by a wire basket. The development of a solid catalyst had problems in achieving good decomposition of the peroxide and long life. This was solved by a good flow distribution and/or proper injection element design and distribution and by proper packaging of the catalyst bed.

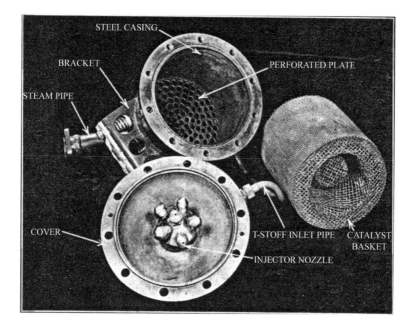

Fig. 9-9 Gas generator for a later version of the HWK 109-509 LPRE had its catalyst impregnated pebbles inside a basket. The output is superheated oxygen-rich steam. Copied with permission from Ref. 13, p. 217.

One interesting aspect of the oxidizer behavior came in developing a pump shaft seal for one of the early experimental oxidizer pumps used for a marine engine, as told by Hellmuth Walter. The stuffing box, which was initially sealed in a conventional manner with greased graphite impregnated fiber packing, blew out and stripped the bolts several times. Small quantities of hydrogen peroxide apparently leaked and reacted explosively with the hydrocarbons in the grease of the packing. A paraffin-based grease with talcum was then developed, and it was satisfactory. HWK later developed, dynamic shaft seals and Buna rubber seals.

The pilot's five-position engine lever (off, idle, low, medium, and high power) in the Me163 cockpit was connected by linkages to the centralized propellant feed control assembly, which in turn opened and closed valves.[6] The starting sequence did initiate a pumping action, but did not immediately admit propellant to the thrust chamber. Moving the pilot's throttle lever into the idling position opened the tank valves and initiated the start motor, which caused the pumps to rotate at low speed, thus filling the propellant lines.[6] A bypass line sent a small quantity of oxidizer to the gas generator, thus driving the turbine at low speed. An indicator in the cockpit showed that idling propellant pressures were attained and the pipe lines were filled with propellants. The pilot then could move the lever into the first power position. The gas generator now received a higher flow, and the pumps rotated faster. Three of the injector heads (designed with spring-loaded face-shutoff valves to open only when the pressure in the lines was sufficient) were supplied with propellant, which were then injected into the combustion chamber, where they ignited and burned. Hypergolic ignition was essential to allow a start and restart. When the pilot moved the throttle into the second power level position, three additional injector heads were supplied with propellants, and in the final lever position all injector heads were operating. When the engine was shut down, the oxidizer system was scavenged, sending residual catalyzed oxidizer gas through the turbine, and the fuel was scavenged through a scavenge valve at the cooling jacket of the thrust chamber.

The 80% hydrogen peroxide was considered to be safe, if proper cleanliness of containers, valves, and pipes were observed and handing rules were followed. This strong oxidizer will react with almost any organic material, and therefore any residues or traces of oil, grease, or fabric fibers had to be removed. All of the exposed surfaces had to be thoroughly cleaned, usually with acid and clean water. There were several accidents and some fatalities when procedures were not followed or when the oxidizer was accidentally mixed with a fuel or other organic material. After the war the United States and Britain used 90% hydrogen peroxide, and this has since become the standard concentration. In some TCs even 95% peroxide has been used experimentally. Although the higher concentration gives a better performance, it seems to be more sensitive to a slow self-decomposition and more difficult to store for long periods.

There were more than 1000 flights of the Me163, many in combat missions.[10] This reusable throttlable restartable LPRE therefore has flown more than any other aircraft rocket engine. The combat results of the Me163 were disappointing to the Germans because only a small number of Allied bombers were actually shot down. Even though hundreds of Me163 intercepter missions were flown, there were only 13 enemy bomber casualties. A version of this HWK 109-509 was also used as a JATO unit, but in experimental flights only. This 109-509 engine was skillfully copied (from incomplete drawings and fabrication documents) by the Japanese in 1944 and 1945 as explained in Chapter 11.

A controllable LPRE of about 750 kg (1650 lbf) thrust was developed by HWK and tested in the Messerschmitt Me 263 aircraft, but the engine development was stopped. A more advanced version with 2000 kg (4400 lb) thrust was then developed for this Me263, successfully flight tested, and several hundred were built. Thereafter the modified 263 aircraft were used as rocket-engine trainers for fighter pilots, and they were flown about 1000 times.

A modified version of the HWK 1000-kg JATO (gas pressure feed system) was used in the sustainer stage of the *Enzian* (German name for an alpine flower) swept-wing antiship missile under development by Messerschmitt around 1944 and 1945.[2] This program was just started, when the end of the war stopped it. Toward the end of the war, a similar version of this same engine was used in the development of the experimental Ba349 vertical takeoff aircraft called the *Natter* (German word for viper snake) air-defense vehicle. The work on this program was also stopped by the end of the war.[2]

After the war the experience with this oxidizer and these engines was used in Britain to build LPREs using hydrogen peroxide as an oxidizer, and some key personnel from HWK went to Britain. Hellmuth Walter himself spent several years in Britain. The relatively simple monopropellant hydrogen-peroxide gas-generator concept was used in the United States and in the Soviet Union for driving the TPs of some of the early large LPREs.

Walter LPREs were also used as a booster and in some models also as a sustainer power plant for several rocket-propelled *air-launched glide bombs*. The Germans developed at least four models of air-launched glide bombs with ranges of 4 to 10 miles, and one of these was designed to dive under water.[2,6,10] The pilot of the launching bomber aircraft guided this bomb visually by observing a flare or tail light on the flying weapon and sending radio signals from the aircraft to the flying bomb. The signals from the pilot's control unit activated the ailerons and elevators of the flying bomb. Walter delivered several different, but simple LPREs, for propelling the glide bombs, and this increased the effective range. Some LPRE had two TCs, and all used hydrogen-peroxide monopropellant, a gas pressure feed system, and a liquid catalyst. Engine design for the first model began probably around 1938, and the first successful flight tests were at Peenemünde in 1940. The glide bomb identified as the HS293 was the most successful, and it was deployed with the German Air Force; its hydrogen-peroxide monopropellant engine (109-507) was placed into production in 1941, and at least 1000 engines were built. It had a thrust of 300 kg (660 lbf) and a liquid-permanganate-type catalyst. This weapon, the HS293 radio-guided glide bomb, sank several British ships in the Bay of Biacayne in 1943.

Hellmut Walter is relatively unknown outside of German rocket engine circles. Yet the historic accomplishments of his company and of him are indeed outstanding. The first rocket powered flight of an experimental airplane was with a HWK engine. The first fighter aircraft, the Messerschmitt Me 163, that was mass produced, was propelled by a relatively sophisticated Walter LPRE and it was the only rocket fighter ever used in actual combat. HWK JATO units were the first JATO engines to fly and a couple of these models were the only ones put into a military service, namely with the German Air Force. The first rocket powered guided bomb used in combat was driven by a simple HWK LPRE. This company made great contributions to the LPRE technology. It included the first single shaft flying turbopump, the first inducer pump impellers for TPs, the first operational engine with hypergolic ignition, the first gas generator using concentrated hydrogen peroxide decomposed by a catalyst, the first variable thrust engine, and the first restartable rocket engine. Almost all the LPREs developed by HWK were produced in limited quantity, were qualified and have flown in experimental vehicles; more than half of their LPREs went into serial production and military service, which is a historic and most enviable record.

References are listed at the end of Chapter 9.5.

9.4 Bayrische Motoren Werke (Bavarian Motor Works)

The third significant LPRE developer in Germany before and during World War II was a special department of the automotive manufacturer BMW.[1] These three letters are an abbreviation of the names given in the chapter heading. This LPRE development group had the dubious distinction of having developed a number of good LPREs (and some of them were qualified or flight tested in experimental vehicles), but none were deployed with the German military services. Intense work on LPREs started in 1939. In 1940 Helmut Count (Graf) von Zborowski became the leader of all of the LPRE effort at BMW. He was a friend and coworker of Eugen Sänger, a well known pioneer mentioned in Chapter 5.4. After the war Zborowski went to France, became an employee of a French company, and among other contributions he helped to design new French test facilities for testing LPREs. The LPRE development and test work at BMW was done mostly in three locations: Zühlsdorf/Berlin, München (Munich)/Allach, and Bruckmühl (South of Munich).[1,2]

BMW's first LPRE contract was to build a JATO unit. They started by using the same propellants as the successful HWK engines, namely, T-Stoff or 80% hydrogen peroxide and C-Stoff, which is 57% methyl alcohol, 30% hydrazine hydrate, and 13% water. Because the T-Stoff was already committed for use in submarines, torpedoes, JATOs, and aircraft engines of HWK, there was not enough available for additional military applications, and BMW was asked to switch to nitric acid as the oxidizer. BMW did a lot of work on different alternate acid-type propellants, much of it in coordination with the Braunschweig Research Institute mentioned in Chapter 9.1.[3,4] Hypergolic ignitions were achieved with a type of turpentine and with aniline and were test fired at BMW in 1940. BMW investigated at least 1000 different propellants (one source says 2000) for reliable, fast hypergolic ignition, low freezing point, good energy release, and stable storage. They were probably the first ones to formulate mixed acid (contains 8 to 10% sulfuric acid and 2% water). It gave better ignition qualities, was easier to pump (lower vapor pressure) than RFNA (red-fuming nitric acid), and had a low freezing point, but poorer performance. This new acid was less corrosive than other acids, but only for a couple of months during the initial storage time period. They tried the addition of 6% hydrated ferric chloride to the acid as a catalyst; this helped with accelerating the ignition of fuels that normally would react more slowly, but it reduced the performance. It was not adopted in flying LPREs. By 1941 BMW had a JATO of 3000 kg thrust with nitric acid and a hypergolic fuel. It was flown as an experimental engine, but was not put into production.

Together with some of the experts from Peenemünde, BMW developed the packaged LPRE for the *Wasserfall* guided antiaircraft missile between 1942 and 1944. It had the ability to quickly climb to 50,000 ft altitude, and it had a range of 45 km or 28 miles (Refs. 1, 2, and 6). The Wasserfall missile is shown in Fig. 9-10. The four movable rudder surfaces provided the aerodynamic forces for controlling the flight path. The engine was the world's first prepackaged simple gas pressurized LPRE factory loaded with storable propellants. Its hypergolic storable propellants of nitric acid with 10% sulphuric acid and a "visol" fuel (vinylisobutyl ether plus aniline) were developed by BMW and I. G. Farben, a chemical company. Propellants were loaded into the missile at the factory, and these propellants stayed in the missile until it was fired or operated. The maximum thrust at sea level was about 78.5 kN or 17,600 lb for about 45 s. The injector of the Wasserfall thrust chamber had impinging stream injection elements, which is different from the nonimpinging patterns used in the V-2 TC. Its specific impulse was about 190 s. Its flow diagram in Fig. 9-11 shows a simple gas-pressurized feed system, with five burst disks or burst diaphragms. A burst diaphragm really is a single-use valve, which can only be opened once and then cannot be closed. Two diaphragms were in the gas lines (designed to open at 10 atm), two in the propellant lines (at 20-atm burst pressure), and one in the high-pressure gas line; this last diaphragm was opened by a pyrotechnically actuated piston. The safety valve had a pin, which had to be pulled manually just before the propulsion system could be operated. Similar safety valves with pins had been used earlier by the Walter company on their JATO units. The Wasserfall was the first German rocket vehicle, where the propellant tank's cylindrical walls were an integral part of the vehicle's structure. A clever unusual swiveled movable tank outlet with a flexible hose was inside both of the propellant tanks. It had a rotary joint and a weight at the end of the hose. This hose moved and rotated inside the propellant tanks during maneuvers or side accelerations of the missile. This feature allowed continuous propellant feeding during high acceleration side maneuvers without allowing bubbles of pressurizing gas to enter the propellant pipes leading to the TC. The first successful flight was in 1944 from Peenemünde. The Wasserfall engine was put into a limited production, and about 42 experimental flight tests were made; most were concerned with the guidance and control system. The war was over before the Wasserfall was ready for full production. Both the United States and the Soviet Union considered the Wasserfall to be an advanced and important military defense weapon at that time in history (1945–1946), and both countries built copies of the missile and learned by testing and flying it. In the United States the task of copying the Wasserfall missile was given to General Electric, and five flights were made as described in Chapter 7.3

BMW formulated the propellants and developed the engines that went into the German *Taifun (Typhoon)* antiaircraft barrage missile and the Schmetterling (Butterfly) antiaircraft missile. The *Taifun* was an unguided small missile (6.3 ft long weighing 66 lb) and intended as a barrage weapon to be fired in large quantities.[2] Originally the Taifun was to have solid-propellant rocket motors. Although there were problems of grain cracking at low ambient temperature, the development of the solid-propellant missile version was continued. The liquid propellant engine version used nitric acid and butyl ether as propellants. The LPRE developed by BMW had a rated thrust of 1850 lbf for 2.5 s and several features that were novel at the time. The oxidizer cylindrical tank was in the center and was surrounded by the fuel tank. Pressurization of the propellant tanks was by hot gas generated by a solid-propellant gas generator using

Fig. 9-10 The Wasserfall was the first medium-range antiaircraft missile. It was the first vehicle, where the propellant tanks were a part of the vehicle structure. The length was 25.8 ft or 7.85 m. Copied from Ref. 1.

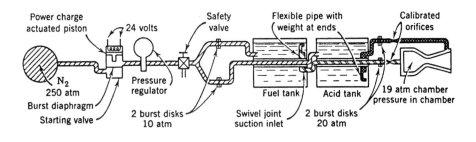

Fig. 9-11 Simplified flow diagram of the Wasser fall LPRE with prepackaged storable propellants, burst diaphragms instead of valves, and a swivel-mounted tank outlet. Copied with permission from Ref. 14, p. 263.

a double-base-type propellant. The gas entered the tanks after breaking burst diaphragms. The propellants entered into the combustion chamber after opening two other burst diaphragms. There was a special oxidizer throttle valve, which allowed a low initial flow and was opened by chamber pressure. The initial mixture ratio was fuel rich. As the chamber pressure built up during the starting transient period, the oxidizer valve opened further, permitting an increased flow of nitric acid. This valve was fully opened, when the chamber pressure exceeded about two-thirds of its rated value. Both the liquid- and the solid-propellant versions of the Taifun had started limited production, but the missile was not ready for deployment as the war ended.

The *Schmetterling* was a two-stage antiaircraft winged missile with two solid-propellant motors as boosters, which were dropped off, and a BMW engine in the sustainer stage.[2] The nitric acid was burned with a fuel, called Tonka 250, and it consisted of 57% xylidine and 43% triethylamine. The Taifun and the Schmetterling were not far enough along in their development or flight tests at the end of the war to be put into serial production.

BMW also developed an unusual LPRE for the Ruhrstahl company's *X-4 air-to-air wire-guided missile*. This air-launched missile was conceived to help German fighter airplanes shoot down enemy bomber aircraft. A partially sectioned view of this X-4 missile is shown in Fig. 9-12. The pilot guided the missile and remotely actuated the missile's aerodynamic control surfaces through two wires, which were unwound from bobbins in the missile and electrically connected the missile and its launching aircraft. The BMW LPRE, identified as 109-548, used an air-pressurized feed system with the hypergolic propellant combination of RFNA and Tonka, which was a mixture of xylidine and triethylamine. Because of the maneuvering of this missile, it was necessary to provide a positive expulsion of propellants from their tanks, without gas bubbles entering into the thrust chamber. BMW selected two sets of spiral-wound tubes, filled with oxidizer and fuel respectively, and the propellants were expelled by flexible movable pistons inside these spiral tubes. The pistons are not visible in Fig. 9-12. The maximum diameter of the fuselage was only 8.75 in. Although the concept of using a piston inside a tube was not new (Goddard

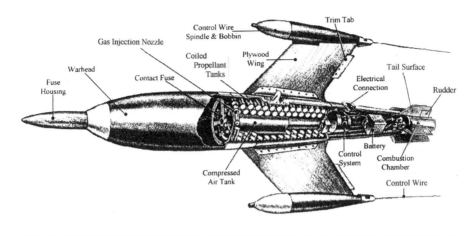

Fig. 9-12 Ruhrstahl X-4 air-to-air wire-guided missile with a BMW 109-548 LPRE using coiled tubes for propellant tanks. Adapted from Ref. 15.

tried it in the late 1920s), building the tubes in coils seemed to be a novel idea. Once activated, the high-pressure gas would pressurize the propellant tanks and move the pistons, the propellants would then break the burst diaphragms, enter the combustion chamber, and self-ignite. The acid-cooled TC had a nominal thrust of 1600 kg or 3520 lbf and ran for about 33 s at an oxidizer-rich mixture ratio of 3.7. By August 1944 about 140 missiles had been flight tested with a Fokke-Wulf 190 and a Junkers Ju88 as the launching aircraft. It was intended to deploy the missile with the Luftwaffe (German Air Force) in the Spring of 1945.

About 1000 to 1300 production airframes had been built at the Ruhrstahl factory in Brackwede, Germany, and were ready for the BMW engines to be installed. However the BMW assembly plant at Stargard was demolished by an Allied bomber raid, and all of the finished 109-548 rocket engines were destroyed. Of course, this seriously delayed the program and prevented the early deployment of the X-4 air-to-air weapon.

BMW also designed and tested a superperformance rocket engine for the Me 262 turbojet-driven aircraft with nitric acid as the oxidizer. It had an aluminum thrust chamber with a machined set of spiral cooling passages shown in Fig. 9-13. Because aluminum alloys melt at about 1100°F, there has to be a relatively cool boundary layer and considerable film cooling. This LPRE development and flight tests were also not completed when the war ended.

An alternate or backup aircraft rocket powerplant was also developed by BMW for the Messerschmitt Me163 rocket fighter aircraft, but with a different propellant.[1] It used concentrated nitric acid with 2% water and "gasöl" (a kind of diesel fuel) as propellants. It could be throttled, and it provided a higher thrust than the HWK 109-509 engine (with hydrogen peroxide and C-Stoff), which was in production for this application. Thrust of the BMW engine was

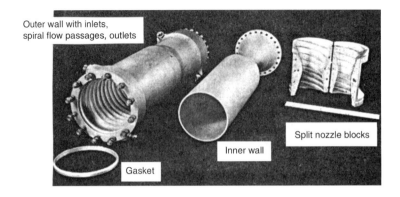

Fig. 9-13 Disassembled aluminum thrust chamber of BMW superperformance rocket engine for the He 262 aircraft. Copied with publisher's permission from Ref. 16.

78.5 kN (17,600 lbf), and the thrust chamber was acid cooled. The development of this engine was never completed.

The Liquid Propellant Rocket Engine Operation at BMW had a short life of only six years. It was disbanded after the war. During their brief existence they started about 10 different LPRE projects. BMW used mostly hypergolic propellants consisting of nitric acid as the oxidizer with various mixtures of self-igniting fuels. Some programs, such as the Wasserfall anti-aircraft missile went into limited production for flight testing, but not for actual defensive use. The time was too short and the war ended before any of these engines or their missiles could be deployed with the military services or used in combat. Therefore BMW is the only known LPRE developer, who saw none of its engines come to fruition.

References are listed at the end of Chapter 9.5.

9.5 German Liquid Propellant Rocket Engines Since 1945

Immediately after WW II all of the development or manufacture of LPREs (and in fact also all weapon systems) was stopped. The tremendous lead that Germany had in LPREs at the end of the war (estimated by one source to be 10 years ahead of all of the other countries) simply evaporated. As stated before, no development work on LPREs was done between the end of the World War II and about 1955.

Liquid Propellant Rocket-Engine Organizations

The history of the LPREs in postwar Germany is intertwined with the history of German aerospace organizations. It is a history of frequent company consolidations, mergers, and acquisitions. The next few paragraphs attempt to explain these organization changes.[1] In 1955, 10 years after the war, the engineering development bureau Bölkow, a group of just a few engineers, started a small LPRE effort at Stuttgart; initially it was mostly studies and a little experimentation aimed at a LOX/kerosene engine. Ludwig Bölkow, the leader, managed to get some government support. In 1958 the Bölkow organization was reorganized, moved to Ottobrunn (near Munich), and obtained some support for building rocket-engine test facilities. Bölkow secured a contract in 1959 for an experimental high-pressure LPRE, identified as the P-111.[1,2] It is a historic LPRE, and it is described next. In 1962 the German government officially decided to pursue developments toward the peaceful exploration and utilization of space and also the rebuilding of its aerospace industry.

Independently in July of 1961 three German aircraft firms [Focke-Wulf, Weser Flugzeugbau (translated as Airplane Building), and Hamburger Flugzeugbau] anticipated work for a joint European space project. They formed a cooperative working group called Entwicklungsring Nord (Development Consortium North), which was abbreviated ERNO. They obtained support from the German government to build experimental LPREs and also some test facilities. In 1967 this consortium of three companies became an independent single company called ERNO Raumfahrt Technologie (ERNO Space Travel Technology). It was the second German company, which had a group in the LPRE business at that time. In 1963 Bölkow and ERNO reached an agreement to cooperate in the development of the engine development for the planned Astris third stage of a European space launch vehicle, the first major postwar LPRE job in Germany.

Bölkow got more business, activated and operated a test facility at Leopoldshausen beginning in the early 1960s, and the organization continued to expand. There were further extensive consolidations of the aerospace industry in Germany.[1,17] In 1969 Messerschmitt (originally a military aircraft developer, who had already merged with Junkers, another aircraft maker) combined with Bölkow and Hamburger Flugzeugbau (Hamburg Aircraft Construction) and formed Messerschmitt-Bölkow-Blohm, Inc. and the company headquarters was still in Ottobrunn. The work on LPRE was continued

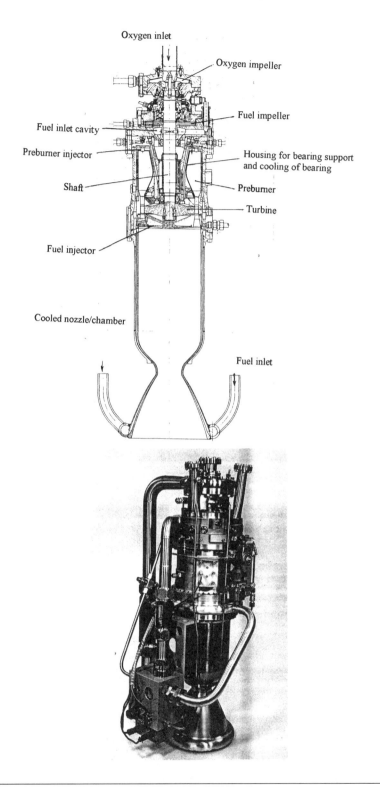

Fig. 9-14 P111 experimental LPRE (1959 to 1968) was ahead of its time. It combined the pumps, turbine, preburner, and thrust chamber into a single assembly. Courtesy EADS; adapted from Refs. 1 and 17.

in a department of this new company. This organization absorbed other aircraft companies and also absorbed ERNO, together with its LPRE group, in 1981. The consolidated company was called MBB, which was an abbreviation of Messerschmitt-Bölkow-Blohm. This then was and today still is the only German company with a department in the LPRE development business. MBB was also active in aircraft, missiles, and several other aerospace fields, and LPREs was not one of their larger activities. With additional consolidations [Dornier (aircraft), MAN (turbojets), and AEG (aviation electronics)], MBB was transformed into Deutsche Aerospace Aktiengesellschaft (German Aerospace Corporation, abbreviated as DASA) in 1989/1990. The LPRE operation within this conglomerate company continued its work on thrust chambers, small thrusters, propellant valves, etc. seemingly unperturbed by the organizational changes. Their key LPRE developments are described next.

In 1995 Mercedes-Benz, the automobile company, was persuaded to acquire Deutsche Aerospace, and the aerospace-related organization within Mercedes then was renamed Daimler-Benz Aerospace A.G. (also still abbreviated as DASA). In 1998 Mercedes-Benz acquired and merged with the Chrysler Corporation of United States, and the aerospace portion of the company was renamed Daimler-Chrysler Aerospace (still abbreviated as DASA). LPREs constituted only one of the activities pursued by this operation. In 1999 DASA spun off some of its aerospace business (including the LPRE work) as an independent company called Astrium. In 2002 Astrium merged with French aerospace firms and others into the new organization called EADS (European Aeronautics Defence Space). It is a huge international conglomerate and has organizations engaged in LPREs in France and Germany, solid-propellant rocket motors in France, the Ariane space launch vehicle, aircraft, guidance and control, satellites, aerospace structures, or military weapons.

LPRE Developments

Historically the most significant German LPRE effort in the 1960s was the development of the P111 experimental engine by Bölkow Entwicklungen KG (Bölkow Development, Inc.).[1,17,18] First preliminary design sketches were made in 1956, and the project ended around 1967. Propellants were LOX/kerosene. One version (1966) is shown in Fig. 9-14 and the cross-section drawing is of a different version. It had several unique and at the time novel features. The Germans called it a Hauptstromtriebwerk (mainstream propulsion equipment), but today it would be identified as a staged combustion cycle engine with an annular preburner as the first combustion stage and the main combustion chamber as the second combustion stage, just downstream of the turbine and main injector. The engine cycles were presented in Chapter 4.2. The P111 combined the preburner, TP and the TC into a single assembly. The nozzle and chamber were regeneratively cooled by the liquid oxygen, and it was the first known German application of the milled slot chamber wall described in Chapter 4.3. The preburner operated at an oxygen-rich mixture ratio to reduce the gas temperature. Goddard's gas generator (1930s) and this preburner were then the only such oxidizer-rich gas producers known outside

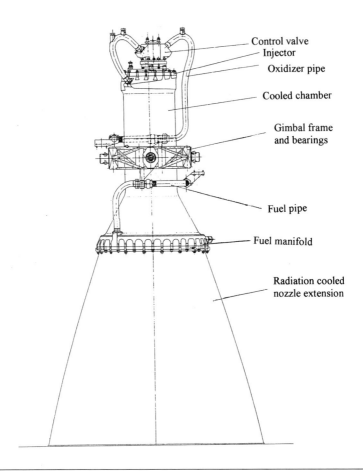

Fig. 9-15 TC of the Astris engine used in the third stage of the Europa SLV. It had a poor flight history. Courtesy of MBB/EADS; copied from Ref. 1.

of the Soviet Union. The P111 LPRE had a thrust of 49 kN (11,200 lbf), a relatively high chamber pressure (85 bar or 1250 psi) for this time, at a nominal mixture ratio of 2.7, and a sea-level specific impulse of 306, which was relatively high for a nozzle-exit-area ratio of 10.6 at this time in German LPRE history. One version of this engine could be throttled. The preburner had a nominal chamber pressure of 116 bars or about 1680 psi at a combustion temperature of 920° K or almost 1200°F.

The oxygen-free copper inner wall had milled slots, a novel feature that was then patented by Bölkow in Germany and also in the United States. The Soviets had developed this milled slot feature by about 1952 and used it for the first flights in the RD-107/RD-108 engines (developed in 1954–1957). However, this Soviet development was not then known to the Germans. The turbopump used a single shaft and had unshrouded vane-type impellers and dynamic shaft seals. The TP was not very efficient and was limited in the maximum turbine diameter. The P111 was a great learning experience; it never flew, and the Germans did not develop another staged combustion cycle engine or a turbine integrated with the injector. The engine was exhibited at a few professional meetings.

In 1966 to 1968 a joint U.S.–German project was undertaken namely to apply the milled slot copper inner-chamber-wall technology to LOX/LH$_2$ propellants at high chamber pressure.[18] Bölkow/MBB designed and built the TC for 130 kN (29,500 lbf) thrust, and Rocketdyne (USA) provided the high-pressure test facility at its new Nevada test site and ran 23 tests at different operating conditions. The project's name was BORD, which stood for the first two letters of Bölkow, and the RD was an abbreviation for Rocketdyne. The TC is almost the same as the TC shown in Fig. 9-14. The results showed that adequate LH$_2$ cooling could safely be obtained at a nominal pressure of 210 bars or 3045 psia and even at pressures as high as 286 bars or 4150 psia. This might have been the highest chamber pressure ever tested with this LOX/LH$_2$ propellant combination, and it was achieved without film cooling, but using a high-conductivity copper TC wall with milled cooling channels. A small temperature gradient across a relatively thin inner wall minimized the thermal stresses. The feasibility of a face-cooled injector face with multiple coaxial injector elements was also demonstrated at high chamber pressure. Rocketdyne used this technology of milled slots in a copper wall in its space shuttle main engine, and NASA paid the Germans for a license in 1975. The Germans did not know at that time that this milled channel cooling method and the coaxial injection elements were originally conceived in the Soviet Union about 25 years before. Milled channels have been used thereafter in other engines in the United States, Japan, and other countries. The coaxial injection elements have been used with other engines in the United States, Soviet Union, Japan, China, France, Germany, and India.

Participation in European Space Launch Vehicles

In 1962 Germany joined the European Launch Development Organization (ELDO) and began to support space-related projects. Germany participated in the R&D leading to a European SLV, called Europa, originally a British–French proposal. The United Kingdom was to provide the first stage (driven by the twin Blue Streak LPREs described in Chapter 12), the French the second stage (based on the propulsion of the Coralie described in Chapter 10), and the Germans were to develop the third stage (called Astris) for this SLV.[1]

The LPRE for this *Astris* upper stage was given to a cooperative working arrangement between Bölkow and ERNO.[1,19,20] Work on the LPREs started in 1963. The principal engine used NTO and a fuel consisting of 50% hydrazine and 50% UDMH, had a thrust of 22.56 kN (about 5,100 lbf), a vacuum specific impulse of 297 s, and a pressurized feed system. The Astris TC is shown in Fig. 9-15, was gimbaled, used regenerative cooling with fuel for the chamber and the throat region, and it had a radiation-cooled nozzle skirt. The gimbal frame was located around the nozzle-throat region, and this reduced the required vehicle space for the angular movement of the TC. There were also small TCs for roll control during the operation of the large LPRE of the third stage. Combustion instability problems in the main TC caused schedule delays and were remedied by changes in the injection pattern and the start sequence. The injector of the Astris main LPRE is shown on the left side in Fig. 9-16, and it had 72 sets of triplet impinging injection elements (1963–1967).[1] This Figure also shows the injector for the German Aestus LPRE, which was designed approximately 25 years later (1988–1995) and had 132 coaxial injection elements. The Aestus LPRE is discussed later in this chapter. By that time the preferred injector design for storable propellants had multiple coaxial bipropellant swirling spray injection elements. This improved injector raised the combustion efficiency and increased the specific impulse by about 10 s.

The other LPREs of this third stage were a pair of 400-kN (90-lbf) thrust gimbaled TCs for the attitude control of the stage.[21,22] This TC used the same propellants, with regenerative cooling of the nozzle throat and chamber (but augmented by film cooling), a radiation-cooled nozzle skirt, and a gimbal mounting. The TC is shown together with other small thrusters later in Fig. 9-21. Ground testing of both Aestus main and ACS LPREs was done at the company's facilities in Trauen and Leopoldshausen.

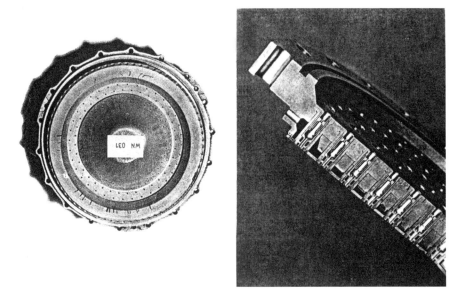

Fig. 9-16 Injectors of the Astris LPRE and the Aestus LPRE show the change in the German design approach of injector elements for storable liquid propellants over a period of about 25 years. (Photos are not to scale.) The Astris injector (left) with impinging stream injection elements was developed in 1963–1967 and the Aestus injector with coaxial injection elements in 1988–1995. Courtesy of MBB/EADS; copied from Ref. 1.

Of the 14 flights of the Europa SLV, only the last four were planned to operate and test the Astris third stage. The performance during these four flights was very poor. Three of these flights had problems with the propulsion ignition in a vacuum and the fourth had a computer problem. So neither the Astris stage or the three-stage Europa vehicle tests were successful. This set of failures was in part responsible for the abandonment of the Europa SLV. Germany and Bölkow/MBB were blamed in part for the demise of the Europa SLV.

When the French proposed the Ariane SLV in 1973 as a replacement for the Europa SLV, it found ready acceptance by other European nations (except for Britain). Several European nations agreed to cooperate and form a consortium.[1] Germany consented to participate and to provide technical and financial support. France assumed the responsibility for program management and overall system integration. French LPREs (Viking, see Chapter 10) were proposed for the first and second stages and a new LOX/LH$_2$ engine for the third stage, later identified as the HM-7 engine.[18] Although SEP (Sociéte Européenne de Propulsion) of France had done a lot of work with LOX/LH$_2$, they unselfishly agreed to let the German company MBB do the TC development for the new HM-7 LPRE. The overall responsibility for the engine was retained by SEP. Data for the HM-7 are given in Chapter 10, the engine is shown in Fig. 10-10, and its flow diagram in Fig. 10-11. MBB thus became responsible for the development of the TC (which was based on a French design and French experience with these propellants), did the design modifications and improvements, and conducted sea-level tests at its Leopoldshausen facility. Altitude simulation tests were done at a new facility at Vernon in France. The first flight of the new third stage with the HM-7 (without restart) and with the German-made TC was in December 1979 from the French launch facility at Kourou, French Guyana, in South America, located near the equator.

A clever feature of this TC is the spiral winding of the cooling tubes in the diverging nozzle section. The tubes have a variable spiral pitch, which allows a constant diameter of the initial tube prior to forming; the shaped tubes are e-beam welded together and are then provided with an external winding of strong fibers to contain the internal nozzle pressure. Advanced versions of the HM-7 were also investigated (more thrust, higher nozzle area ratio, etc).

When the dozen European countries supporting this SLV agreed to develop a more powerful new version of the SLV (Ariane 5), MBB participated in the new LOX/LH$_2$ engine for a bigger improved second stage. This engine, called *Vulcain*, was again under the overall responsibility of SEP in France, and the engine is discussed in Chapter 10. MBB worked on pieces of this engine, namely, the TC development and its sea-level tests, integrating the radiation-cooled nozzle-exit section, which was developed in Norway, with the TC,[1,18,20] the testing of the LOX TP (which was developed and built by Fiat in Italy), the testing of the gas generator (designed and developed by SEP), and the development of certain valves. The engine design, assembly, engine tests, and installation were done by SEP. This Vulcain engine is shown in Fig. 10-13, and the TC in Fig. 10-14. The Vulcain injector was largely developed by MBB/DASA and is shown in Fig. 9-17.

Fig. 9-17 Injector face of the Vulcain LOX/LH$_2$ rocket engine shows many coaxial swirling injection elements. Courtesy of Astrium/EADS; copied from Refs. 19 and 20, provided by G. Hagemann.

Its coaxial injection elements were based on experimental work done by MBB and SEP with prior injectors using LOX/LH$_2$. It gave excellent combustion efficiency and stable operation. A sea-level test of the Vulcain TC conducted in Germany is shown in Fig. 9-18. Its transparent flame shows the pear-shaped supersonics shock-wave pattern, and the white areas show the high-temperature regions in the exhaust gas plume outside of the nozzle. During launch, this engine is actually fired vertically downward, but the tests in the company's test facility were performed in a horizontal position because this can be done in a less complex, less costly, and available test facility.

The third stage of the new Ariane 5 SLV, called Aestus, has a storable propellant rocket engine for final orbit insertion and space maneuvers.[19,20,24–26] This was the first time that Germany (really MBB) had the responsibility for a complete stage. The LPRE has a thrust of 27.8 kN (6320 lbf) and is regeneratively cooled in the throat region (up to a nozzle area ratio of 10) and in the chamber region; its radiation-cooled nozzle-exit section is made of Haynes metal alloy. The propellants were NTO and MMH. The initial engine version used a pressurized feed system and a low chamber pressure of about 10.8 bars and had an altitude specific impulse of 306 s at an area ratio of 84. The fir-

Fig. 9-18 Diamond- or pear-shaped high-temperature regions and the diamond-shaped shock-wave pattern are visible in the largely transparent exhaust gas plume in a test firing of a large LOX/LH$_2$ Vulcain TC at the Leopoldshausen test facility. Courtesy of Astrium/EADS; copied from Ref. 23, p. 556.

ing duration was long, namely, 18.5 min. The Aestus LPRE or TC is shown in Fig. 9-19. A later version of the Aestus engine used a gas-generator engine cycle, a small TP, had a higher thrust (35 kN), a higher chamber pressure, and a smaller TC. This allowed a much lighter set of propellant and gas pressure tanks and over 700 kg more payload. Rocketdyne in the United States contributed to the turbopump design and development. The injector uses coaxial injection elements with storable propellants as shown in cross section in Fig. 9-16; it gives a comparison with an earlier German injector element design using impinging propellant streams. The Aestus injector design was the result of extensive testing and prior work. During the development, there were several incidences of high-frequency tangential combustion instability during the full thrust operation and also during the start transient. After detailed experimental investigations several changes were made that seemed to avoid these instabilities. In the fuel passage of the coaxial element, some additional slots provided improved atomization and distribution. Also there were changes in the valve opening rates and timing. Two small thrusters served to provide roll control for this third stage.

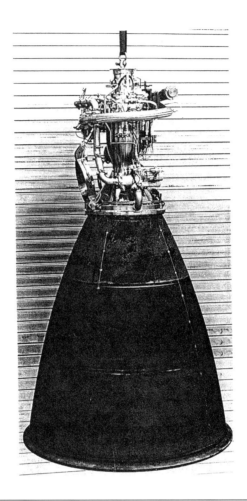

Fig. 9-19 Aestus TC was developed to propel a version of the third stage of the Ariane 5 SLV. Its injector is shown in Fig. 9-16. Courtesy of Astrium/EADS; copied from Refs. 19 and 20, provided by G. Hagemann.

Small Thrusters

In the early 1960s Bölkow developed the first German small bipropellant thrusters using storable propellants. One model at 400 N thrust is shown in Fig. 4.6-21. The chamber was regeneratively cooled, and the nozzle extension was radiation cooled.[1] It has been flown in some satellites.

In the late 1960s ERNO developed several monopropellant hydrazine thrusters in thrust sizes of 0.5, 2, 10, 20, and later at 400 N (Refs. 1, 27–29). The 0.5 N (0.1 lbf) thruster is shown in Fig. 4.6-9, and it is one of the smaller monopropellant thrusters. This ERNO thruster figure shows some of the several heat barriers and thermal insulation parts between the hot decomposition chamber and the heat-sensitive electromagnetic valve or the structural mounting pads. Furthermore it shows the electrical heater on the flow control valve (FCV) to prevent freezing of the hydrazine fuel and also on the catalyst bed to facilitate cold thrust starts. The minimum total impulse bit was between 0.005 and 0.015 Ns. The small monopropellant hydrazine thrusters were developed in Germany about a decade later than in the United States. Therefore the Germans had the advantage to copy or modify some of the features from earlier foreign thrusters. The first flight with a hydrazine monopropellant was with a 20 N (5 lbf) thruster in 1970 on the communications satellite ITELSAT III. It used Shell 405 catalyst obtained from the United States. The Germans did not want to be dependent on a foreign source of catalyst. So over a period of about two years, they developed their own catalyst, which was also based on iridium, but at 34% (Ref. 1). Another relatively new 400-N hydrazine monopropellant thruster actually has two different catalyst beds in series. The first one used iridium as the active ingredient, and the second used a iridium-ruthenium mix.

The company also developed (for a cryogenic upper stage of Ariane 5) a complete auxiliary engine system with titanium tanks including their internal bladders, valves, a gas-pressurization feed system and six small 400-N hydrazine monopropellant thrusters.[1,19,20] Development began in 1990. A package of three of these 400 N thrusters is shown in Fig. 9-20, and two of the nozzles were inclined at 65 deg to the thruster's chamber. Altogether the German monopropellant hydrazine TCs have flown on more than 25 satellites up to 2002.

The company also developed a series of bipropellant small thusters using mostly NTO and MMH between 1983 and 1997.[1,30-32] Some of these thrusters have used a single coaxial injection element and not an impinging stream injection element. Originally they had a small 400 N thruster with partial regenerative cooling jacket around the chamber and the nozzle-throat region. Figure 9-21 shows this thruster on the left side and illustrates the performance progress of German bipropellant thrusters over the years, as chamber pressure and nozzle area ratios were increased. The platinum alloy chamber/nozzles on the right side of the figure is a novel approach, but it is basically similar in concept to the intergen beryllium TC scheme developed earlier by Rocketdyne (beginning in 1966) as described in Chapters 4.6 and 7.8. The high heat energy transferred from the gas flowing at near sonic velocity into the throat wall is then conducted inside the platinum alloy wall to the chamber wall, where it gives up the heat to a film cooling layer. Initially this concept was tested in a 4 N thruster using a platinum-rhodium alloy for the TC walls (1983–1987), but it did not fly. Later (1989–1993) a platinum alloy was applied to a 10 N thruster and also to a 400 N thruster, and they found space application in satellites. As of 2002, more than 45 satellites successfully used the company's bipropellant thrusters.

Fig. 9-20 Assembled package of three hydrazine monopropellant thrusters at 400 N or 95 lbf thrust each used for flight control of upper stage in the Arianne 5 SLV. Courtesy of Astrium/EADS; copied with permission from Refs. 19 and 20.

Other Contributions

The Astrium company and its predecessors have also contributed to the state of the art of LPRE through studies and experimental work. This includes the analysis and testing of various kinds of advanced nozzles,[33,34] plumes, high-temperature components made from reinforced ceramics,[35] various injection schemes,[36,37] and feed system transients.[38]

Between 1967 and 1970 MBB's predecessor companies worked on small experimental TCs with liquid fluorine/liquid hydrogen at two different thrust levels. Initially they used a 300-N size and later a 4.9-kN (1100-lbf) TC size.[39] The highly toxic exhaust gas was scrubbed and cleaned. Further work was

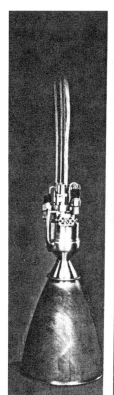  

Parameter	First generation	Second generation	Third generation
Chamber material	Ni-Mo-alloy	Ni-Mo-alloy	Platinum alloy
Vacuum thrust	392N	410N	415N
Chamber pressure	101 psi	101 psi	145 psi
Nozzle exit area ratio	77	150	220
Specific impulse, vac	303s	308s	318s
Cooling of chamber and nozzle throat	Regenerative	Radiation with some film cooling	Conduction, radiation, and film cooling
Cooling of nozzle exit extension	Radiation	Radiation	Radiation
Time period	Late 1960s	Late 1970s	Early 1990s
Flown in satellite	Symphony	TV-SAT	Eutelsat W-24

Fig. 9-21 Comparison of technical progress of three 400-N bipropellant TCs used for space maneuvers and attitude control. Courtesy of Astrium/EADS; adapted from Ref. 1.

stopped because of the high toxicity, increased costs, and potential damage to the environment. In the 1990s DASA undertook a joint German/Russian experimental thrust-chamber program to investigate transpiration film cooling with gaseous hydrogen.[40] Tests were performed at the Chemical Automatics Design Bureau in Voronesh, Russia. The company also experimented, on a small scale, with the pressurization of propellant tanks, which contain a storable propellant.[41] A small quantity of the other propellant was injected, so that hot-gas reaction products would be formed in the tank and these gases would then pressurize the tank. None of the results of these projects were incorporated into a new LPRE, and the work appears to have been discontinued.

Concluding Comments

In Germany, work on LPREs was restarted about 10 years after World War II (1955). It grew slowly at first. The ground tests of the P111 rocket engine beginning in 1963 with a staged combustion cycle was the most remarkable achievement of that time. When Germany decided to join the European space effort and signed the agreement to do so in 1963, the role of the German LPRE operation changed radically from an independent company/German government effort to the role of supporting a multinational space launch vehicle effort. This implied that the Germans' LPRE people would work on bits and pieces of an LPRE and that the work would be coordinated with, shared with, and reviewed by LPRE companies in other nations. Thus the Germans did projects like building and ground testing (but not designing) the TC of the HM-7, refining the design of the Vulcain injector and testing of the Vulcain TC, or the testing (but not the designing, building, or engine installation) of its TP and its GG. Finally in 1989 the Germans' LPRE organization DASA obtained the job for the design, building, and ground testing of the complete Aestus engine for the third stage of the Arione 5 SLV.

References

[1]Hopmann, H., *Schubkraft für die Raumfahrt, Entwicklung der Raketenantriebe in Deutschland (Propulsive Force for Space Travel, Development of Rocket Propulsion in Germany)*, Stedinger Verlag, Lemwerder, Germany, 1999 (in German).

[2]von Braun, W., and Ordway, F. I., III, *Space Travel, An Update of History of Rocketry and Space Travel*, Harper and Row, New York, 1969.

[3]Lutz, O., "A Historical Review of Developments in Propellants and Materials for Rocket Engines," edited by F. C. Durant, III and G. S. James, American Astronautical Society History Series, Vol. 6, Univelt, San Diego, CA, 1985, Chap. 11, pp. 103–112.

[4]Clark, J. D., *Ignition*, Rutgers Univ. Press, Rutgers, NJ, 1972.

[5]*The Story of Peenemünde*, Informal interview reports, Library of the Naval Postgraduate School, Monterey, CA, 1947.

[6]Herrick, J., and Borgess, E., (ed.), *Rocket Encyclopedia, Illustrated*, Aeropublishers, Inc., Los Angeles, CA, 1959.

[7]Myhra, D., "Project He 176," *Secret Aircraft Designs of the Third Reich*, Schiffer Publishing, Atglen, PA, 1998.

[8]Smith, J. R., and Key, A. L., *German Aircraft*, The Nautical and Aviation Publishing Company of America, Baltimore, MD, 1972.

[9]Walter, H., "Experience with the Application of Hydrogen Peroxide for Production of Power," *Jet Propulsion*, May–June 1954, pp. 166–171.

[10]Walter, H., "Report on Rocket Power Plants Based on T-Substance," *NACA Technical Memo 1170*, 1948 (English translation).

[11]McKee, L., "Hydrogen Peroxide for Propulsive Power; Production and Use by the Germans During World War II," *Mechanical Engineering*, Vol. 68, Dec. 1946, pp. 1045–1048.

[12]Burgess, E., "The HWK 109-509 Bi-fuel Rocket Unit," *Pacific Rockets*, Vol. 1, No. 4, March 1947.

[13]Sutton, G. P., *Rocket Propulsion Elements*, 1st ed., Wiley, New York, 1949.

[14]Sutton, G. P., *Rocket Propulsion Elements*, 3rd ed., Wiley, New York, 1963.

[15]http://www.luft46.com/missile/x=4.html.

[16]Sutton, G. P., *Rocket Propulsion Elements*, 2nd ed., Wiley, New York, 1956.

[17]Stöckel, K.,"Zur Entwicklungsgeschichte des Hochdruck Hauptstrom Raketenantriebswerk in Deutschland (History of the Development of the High Pressure Main-Flow Rocket Engine in Germany)," International Academy of Astronautics, Paper 8412, Oct. 1983 (in German).

[18]Rothmund, C., Hopmann, H., and Kirner, E., "The Early Days of LOX/LH$_2$ Engines at *SEP and MBB, History of Rocketry and Astronautics*, edited by P. Jung, Chap. 12, Vol. 21, Univelt, San Diego, CA, 1997, American Astronautical Society History Series, pp. 269–292.

[19]Daimler-Chrysler Aerospace, Company Brochures, Ottobrunn, Germany 1998.

[20]Astrium *Space Propulsion*, Brochure, Ottobrunn, Germany, 2001.

[21]Schwende, M. A., "Small Engines for European Satellites," AIAA Paper 97-3087, 1997.

[22]Teo, G. A., "Great Ideas to Power Spacecraft—40 Years of Propulsion Systems from Ottobrunn," International Academy of Astronautics, Paper 9910, Aug. 1999.

[23]Sutton, G. P., *Rocket Propulsion Elements*, 6th ed., Wiley, New York, 1992.

[24]Schmidt, G., Langel, G., and Zewen, H.,"Development Status of the Ariane 5 Upper Stage Aestus Engine," *AIAA Paper 93–2131*, June 1993.

[25]Butler, K., and Langel, G., "Storable Upper Stage Engine for Global Application—Aestus II, AIAA Paper 97-3347, 1997.

[26]Obermaier, G., Taubenberger, G., and Peyhi, D., "Coaxial Injector Development for Storable Propellant Upper Stage Turbopump Engines," AIAA Paper 97-3096, July 1997.

[27]Kollen, O., and Viertel, Y., "Development and Qualification of a Long Life 1N Monopropellant Hydrazine Thruster," AIAA Paper 96-2867, 1998.

[28]Hopman, H., "Über Raketenantriebswerke und Gaserzeuger mit Hydrazin als Monergol (Rocket Engines and Gas Generator with Hydrazine as Monopropellant)," *Luftfahrttechnik und Raumfahrttechnik*, Vol. 13, 1967 (in German).

[29]Rath, M., Schmitz, H. D., and Steenborg, M., "Development of a 400-N Hydrazine Thruster for ESA's Atmospherric Reentry Demonstrator," AIAA Paper 96-2866, July 1996.

[30]Schwende, M. A., Munging, G., and Schulte, G., "Bipropellant Thruster Family for Spacecraft Propulsion," AIAA Paper 92-3860, 1992.

[31]Barber, T. J., Krug, F. A., and Renner, K. P., "Final Galileo Propulsion System in-Flight Characterization," AIAA Paper 97-2946, July 1997.

[32]Pitt, R., and Rogall, H., "Design, Development and Evolution of the Ariane 5 Attitude Control System (SCA)," AIAA Paper 95-2808, July 1995.

[33]Hagemann, G., Immich, H., Nguyen, T., and Dumnov, G. E., "Advanced Rocket Nozzles," *Journal of Propulsion and Power*, Vol. 14, No. 5, Sept.–Oct. 1998.

[34]Öchslein, W., "Extendible Nozzle for the HM-60 Rocket Engine," European Space Agency, SP-265, Aug. 1986.

[35]Bayer, S., Strobel, F., and Knabe, H.,"Development and Testing of C/SiC Components for Liquid Rocket Propulsion Applications," AIAA Paper 99-2896, June 1999.

[36]Sternfeld, H. J., "Experimental Performance of Coaxial Injectors in LOX/GH$_2$ Rocket Engines," *Deutsche Forschungs-uns Versuchsanstalt für Luft-und Raumfahrt*, Rept. DLR-FB-71-56, June 1971.

[37]Schmidt, G., Langel, G., and Zeven, H., "Coaxial Injection Technology for Hypergolic Propellants," 2nd European Aerospace Conference on Progress in Space Transportation, European Space Agency, SP-293, May 1989, pp. 461–468.

[38]Obermaier, G., and Popp, M., "Dynamic Flow Analysis of Liquid Propellant Feed Systems," AIAA Paper 91-2281, June 1991.

[39]Seidel, A., Pulkert, G., and Wolf, D., "Development History of Several H_2/O_2 and H_2/F_2 Engines for Advanced ACSs, RVD and OMS Systems," AIAA Paper 91-3389, June 1991.

[40]Häseler, D., Mäding, C., Rubinskiy, V., Gorokhov, V., and Krisanov, S., "Experimental Investigation of Transpiration Cooled Hydrogen-Oxygen Subscale Chambers," AIAA Paper 98-3364, July 1998.

[41]König, M., "Full-Scale Experimental Investigation of Main Tank Injection for UDMH/NTO, UDMH/WFNA, and UDMH/RFNA Systems," *Deutsche Forschungs-und Versuchsanstalt für Luft-und Raumfahrt*, Rept. DLR-FB-73–65, Jan. 1973.

France's Liquid Propellant Rocket-Engine History

France has had an illustrious history in LPREs.[1] This country has developed a broad range of reliable engines for applications such as space launch vehicles, sounding rockets, antiaircraft missiles, small engines for flight-path control, or engines for aircraft.

Of its early pioneers the most notable was Robert Esnault-Pelterie (1881–1957), who was a dedicated visionary, inventor, and international proponent for astronautics.[2,3] His early efforts were devoted to airplanes and aircraft piston engines. His initial deliberations on spaceflight were first published in 1912, but his major work "L'Astronautique" came out in 1930 and was widely read. He first experimented in 1931 with a TC using the monopropellant tetranitromethane. An accidental explosion severed three fingers and damaged a fourth finger on his left hand, and this event convinced him to abandon monopropellants. He then selected the bipropellant combination of liquid oxygen and what was then called petroleum ether (a volatile hydrocarbon). Between 1934 and 1937 he ran thrust chambers with up to 100 kg (220 lbf) thrust for one minute at a specific impulse of 230, all very respectable values for that time. He was elected to the French Academy of Sciences, gave lectures and speeches on astronautics and rocketry, and was a very popular figure during his life.

There were several key LPRE organizations in France. In 1944 the Société d'Études de la Propulsion par Réaction (SEPR) was established, originally as a private company, but it soon became a government operation. Its principal operations have been and are in Vernon, France, near the river Seine about 70 km or 45 miles west of Paris. SEPR has been involved not only in LPREs, but also in solid-propellant rocket motors, but at a different plant location. In 1946 the Laboratoire de Recherches Ballistique et Aéronautique (LRBA) was established as a French defense department organization. Until 1971 these two organizations designed and developed almost all French LPREs; however, there were some others. In 1969 SEPR was merged with the Division de Engins et de L'Escape of the Société National d'Étude et de Construction de Moteurs d'Avion (SNECMA) into a single organization called Société Européenne de Propulsion (SEP).[4] This was done, in part, to accommodate the joint European efforts on a large European launch vehicle. In 1971 the propulsion activities of LRBA were transferred to SEP. The liquid propellant engine work of SEP in the 1970s and beyond has been largely concerned with LPREs for several versions of the Ariane, which became the European space launch vehicle. Right from Ariane's beginning this effort was undertaken with participation of other European countries. In 2003 SEP became a part of a new large European conglomerate called EADS (European Aeronautics, Defence and Space Company).* The German LPRE company called Astrium, the Ariane launcher vehicle organization, a defensive missile group, and others all became part of EADS. The French Solid Propellant Rocket Motor Organization, which used to be part of SEP, became a separate entity under EADS. The original manufacturing and LPRE test facilities were at Vernon (40 miles west of the center of Paris), and there are additional facilities at Villaroche and Villejuif.

There were a few clandestine developments and ground tests of simple gas-pressurized LPREs performed in France during the German occupation of 1940 to 1945. This was an amazing and risky business right under the noses of the Germans. They tested a TC using LOX and a liquid petroleum ether with thrust levels up to 1000 kg or 2200 lbf between. 1941 and 1943. One of the key investigators was Jean Jacques Barré, a pioneer not well known outside of France.[5]

Like other countries that at the time worked with LPREs, France learned a great deal about LPREs from the German WWII experience with their V-2, aircraft LPREs, and tactical missiles.[6] They obtained, studied, and tested German LPRE hardware and had the help of German rocket experts. For example, in 1946 LRBA had a staff of 60 German engineers and technicians, and 30 of them had LPRE experience. Other Germans went to other companies. The pioneer Eugen Sänger (see Chapter 5.4) and Helmuth von Zborowski (former leader of the BMW LPRE organization described in Chapter 9.3) were two of the more renowned Germans who assisted the French for a number of years. SEPR ground tested a captured German Walter 109-509 LPRE (with hydrogen peroxide). LRBA reconstituted a German V-2 from captured parts and conducted studies of larger V-2 type missiles. French efforts also had the benefit that many of their LPRE features and concepts (such as gimbals or LOX/LH$_2$ engines) had been tested and proven before in other LPREs in the United States or Soviet Union. The French had some access to these before they tried to develop their own versions of these concepts.

Flight Vehicles with Pressurized-Gas Feed Systems

All of the early French LPREs used simple gas pressure feed systems. The first one that was publicly released in 1945 was the SEPR-1, and it was designed for a jet-assisted takeoff application. The author does not know of it having flown. The second, the SEPR-2 LPRE, is shown in Fig. 10-1, and it was flown

---

*Personal communications, ESA and Astrium/EADS, data sheets from SEP/SNECMA, 1999–2004.

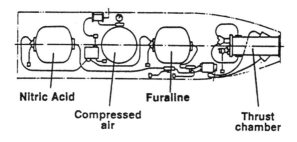

Fig. 10-1 Simplified sketch of the SEPR-2, representative of the early pressure-fed experimental LPREs of the late 1940s. Courtesy of SEPR/EADS; copied with AIAA permission from Ref. 4.

in an experimental antiaircraft missile. It had 1240 kg of thrust and used nitric acid and a hypergolic propellant called furaline, consisting of 41% furfural alcohol, 41% xylidine, and 18% ethyl alcohol. This engine was unveiled in 1947 (Refs. 1 and 4).

Between 1946 and 1953 a number of *tactical missiles* with LPREs were developed and flight tested; most were for surface-to-air applications.[1,4] Some of these LPREs were based on and inspired by German technology from World War II. Most were preloaded with storable propellants, such as nitric acid oxidizer and fuels like furaline, aniline with 20% furfural alcohol, or TX fuel (a mixture of triethylamine and xylidine). They had gas-pressurized feed systems, simple single-walled steel TCs, but some had graphite nozzle inserts. The early spherical propellant tanks did not use the available vehicle space efficiently. An example is the SE 4300 Guided Rocket Program, started in 1947, aimed at a potential antiaircraft weapon.[7] It became a two-stage experimental missile, had a solid-propellant booster motor and a liquid propellant sustainer engine in the winged upper stage. The LPRE of the upper stage had a pressurized-gas feed system, used nitric acid and furaline, had two spherical propellant tanks, the compressed air was in a spiral-wound tube, and the uncooled steel TC used a flat injector with doublet impinging injection holes. Thrust was 550 kg (1210 lbf) and could be throttled to 50%, and an experimental version went down to 15%. There were 80 different SE 4300 launches of 15 different vehicle versions between 1951 and 1957, and most were aimed at trying different guidance and seeker systems. None of the several experimental tactical missiles were truly satisfactory to the military authorities.

Between 1950 and 1964 France developed a series of *Veronique sounding rocket vehicles*, they were intended for atmospheric scientific exploration in support of the International Geophysical Year of 1958 and initially could carry 60 kg (132 lbm) of payload, such as typically meteorological instruments.[1,4,8,9] A total of about 100 vehicles were built and launched, using about five different versions of Veroniques; launches continued sporadically through 1975. Early versions had nitric acid and kerosene as propellants with 4000 N (899 lbf) thrust at sea level.[9] The maximum altitude was increased for different versions from 70 to 220 km, with different engine durations and different total propellant loading. The thrust chamber had a double-wall steel construction and used regenerative oxidizer cooling, augmented with some film cooling. All versions used a pressure feed system. Most flights had a gas generator to pressurize the propellant tanks with chemically generated warm gas instead of the cold inert pressurized gas, which was previously used in their LPREs. One type of this gas generator is shown in Fig. 10-2. It had three propellants in small pressurized tanks: nitric acid, kerosene fuel mixed with furfural alcohol, and water (or a dilute water solution of ammonium nitrate, which lowered the freezing point). This third water propellant was used to cool the GG walls, prior to being injected into the hot reaction gas through a spigot pointing toward the injector. This water diluted and cooled the gas to about 400°C. This chemical GG scheme was clever, seemed to be satisfactory for pressurizing both of the main propellant tanks, and had less inert mass than a high-pressure inert gas system.

The United States was successful with direct chemical pressurization with a single solid-propellant GG in the Bullpup missile, but for a very short duration, and the Soviet Union also investigated it; however, it was not pursued in

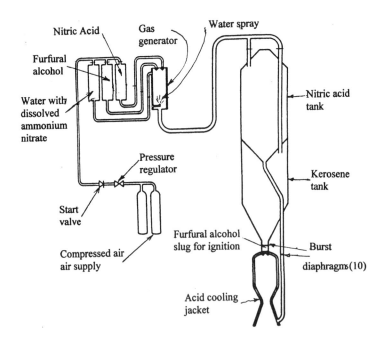

Fig. 10-2 Simplified flow diagram of the LPRE for the Veronique sounding rocket with a pressurized gas feed system using a chemically generated gas, diluted with water, for tank pressurization. Courtesy SEPR/EADS; modified from Ref. 8.

major LPREs. The United States and the Soviet Union went to a scheme using two liquid propellant GGs, one that is fuel rich and one that is oxidizer rich, all without water as a third propellant. This Veronique pressurization scheme with a single GG diluted with water has flourished in France, where the same GG scheme was used on a good number of different propulsion systems. The author suspects that there were some chemical reactions in the large propellant tanks between the warm gas and at least one or possibly both of the propellants. The propellant at the surface of the liquid in the tank was heated by the warm gas and also by the condensation of steam. The last portion of the propellants entering the TC has been heated, has a lower density (and therefore somewhat lower thrust), and has a higher vapor pressure, which can change the injection or the combustion characteristics. However none of these effects had been reported in the French literature accessible to the author.

The initial Veronique launch history in the 1950s was dismal.[8,9] The first two flight attempts in the Sahara desert in May 1952 were failures, and the subsequent six flights failed because of combustion instability. This was the first time in France that a key launch engine experienced high-frequency combustion vibrations, and it triggered a major effort not only at the LPRE company, but also at ONERA (research institute) and other organizations. Several changes were made. Instead of a flat face injector made of aluminum, the new injector was made of steel and had a cylindrical configuration similar in concept of sketch d in Fig. 4.3-16. In the old injector the direction of the injected propellant streams was basically directed downward toward the nozzle. In the revised design the injection of the propellant streams was essentially radial toward the centerline of the chamber. The injector face was an upper extension of the chamber wall. Fuel entered the injector from an external manifold. The oxidizer cooled the head end of the TC and entered the annular injector through holes, flowing vertically down. The actual injection into the chamber was through several circular rows of doublet self-impinging pairs of propellant holes. This type of injection design was carried forward to other French TCs using storable propellants and is unique to large French LPRE, as will be seen later. The next few flights were only partially successful, and therefore further changes were made. The double-wall regenerative cooled TCs were replaced by a single wall with a throat insert (made of ceramic or fiber-reinforced ceramic), and both were cooled by film cooling only. The fuel was changed from kerosene to turpentine, which at that time was believed to be less sensitive to instabilities. The last two models of Veronique used turpentine as a fuel. These actions seemed to have stopped further high-frequency combustion problems in this LPRE.

In its last version the Veronique was upgraded to 50% more thrust (60 kN) to allow a higher payload. The last flight of Veronique was in 1975 from Kourou, in French Guyana in South America. In the 1960s a larger sounding rocket vehicle was developed, called *Vesta*, capable of lifting 1000-kg (2200-lbf) payload to 300 km altitude. Its engine was similar to the Veronique engine, but it had more thrust.[8] Vesta only saw limited service, but it received lots of publicity, when it carried two small monkeys in 1967 (Ref. 4).

Space Launch Vehicles with Pressurized-Gas Feed Systems

The Coralie, shown in Fig. 10-3, was originally intended to become the second stage of the Europa I space launch vehicle, which was conceived in 1950 and initially supported by six European nations.[1,8] The British developed the Blue Streak engine for the first stage (see Chapter 12), the French developed the Coralie and its engine for the second stage, and the Germans developed the Astris third stage with its own engine as told in Chapter 9.5. The engine for Coralie was developed by LRBA and used NTO and UDMH as propellants. Some of the engine characteristics are listed in Table 10-1, which also lists engines that are discussed later in this chapter. It was based in part on the Veronique technology because it had a similar injector assembly and the same chemical warm-gas tank pressurization scheme. The injection of water, the third GG propellant, limited the gas temperature to 350°C or 662°F. As seen in Fig. 10-3, there were four hinge-mounted TCs because a single larger TC was too long for the vehicle. Also for the first time in French LPRE history, there was a low thrust phase (about 8.4 tons or 18,480 lbf) for a few seconds during

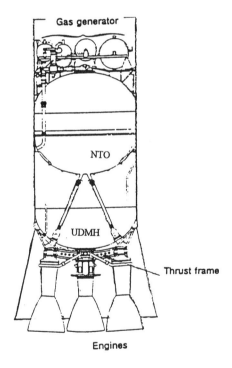

Fig. 10-3 Simplified section of the Coralie vehicle stage built and flown in the 1960s. It shows three of the four thrust chambers and a common bulkhead between the propellant tanks. Tank pressurization is by warm gas from a liquid propellant gas generator diluted with water. Copied with AIAA permission from Ref. 4.

Table 10-1 Selected Characteristics of the LPRES for the French and European Launch Vehicles

Vehicle or Stage	Engine name	First flight	Propellants	Thrust, kN/klbf	Chamber pressure, bar	Specific impulse, s
Europa SLV second stage	Coralie 4 TCs	1966	NTO/UDMH	70/15.7	13.7 vac	281 vac
Diamant A first stage	Vexin	1965	Nitric acid/turpentine	283/63.6 at SL	17.6	203 sea level
Diamant B first stage	Valois	1970	NTO/UDMH	344/77.3 at SL	19.6	219 sea level
Ariane I first stage	Viking V	1979	NTO/UDMH	617/138.7	53.5	247.6 sea level
Ariane II, III, IV, first stage	Viking V	1980	NTO/UDMH + hydrazine hydrate	678/152.4	58.5	278.4 vac
Ariane I second stage	Viking IV	1979	NTO/UDMH + hydrazine hydrate	721/162.1	52.6	295 vac
Ariane II,III,IV, second stage	Viking IV	1980	NTO/UDMH + hydrazine hydrate	805/181	58.4	295.5 vac
Ariane II,III,IV, boosters	Viking V modified	1988	NTO/UDMH + hydrazine hydrate	678/152.4	58	278.4 vac
Ariane III,IV third stage	HM-7A	1988	LOX/LH$_2$	63/14.1 vac	35	444 vac
Ariane VI third stage	HM-7B	1992	LOX/LH$_2$	64.2/14.4 vac	37	446 vac
Ariane V second stage	Vulcain 1	1997	LOX/LH$_2$	1145/276.0 vac	110	431.5 vac
Ariane V second stage	Vulcain 2	2000	LOX/LH$_2$	1350/303.5 vac	115	433 vac
Ariane V third stage	Vinci planned	2006	LOX/LH$_2$	180/40.5 vac	60	465 estimate vac

start and prior to going to full thrust of 28 tons or 61,600 lbf. This reduced thrust occurred also during shutdown for less than 1 s. The low-thrust phase helped to avoid excessive initial propellant flow, which has caused hard starts in the past) and reduced water hammer effects. These two short low-thrust periods are reminiscent of the German V-2 LPRE. The main propellant valves had two open positions, partly open for the reduced thrust phase and fully open for rated thrust. The locations of main propellant valves were unusual. There was a single fuel valve for the four TCs, but four individual NTO valves, each next to one of the TCs. The common fuel valve enabled essentially simultaneous ignition and simultaneous shutdown in the four TCs, and the close positioning of the oxidizer valves limited the NTO dribble volume. The single-wall TC had a zirconium-oxide coating and a graphite nozzle-throat insert. The chamber and nozzle were film cooled, and about 22.5% of the UDMH fuel was used for the film cooling.

On the first Europa SLV flight in November of 1966, the Coralie engine in the second stage failed to ignite. On the next flight the Astris cut off prematurely. The British were unhappy because their first-stage engine worked well in all of the 11 launches, whereas there were problems with the French second stage and the German third stage and the British had paid for the major portion of the Europa SLV cost. The French modified the Coralie stage into a first-stage experimental vehicle and called it *Cora*. It gave further tests to the Coralie LPRE. It was launched satisfactorily six times before the Europa SLV program, and with it the Veronique program was stopped. Some of the technology of Coralie was used later in other French LPREs. Because of the high tank pressure of 270 psi, the French second stage was heavy, and this fact detracted from carrying more payload in the SLV.

In 1959 the French government decided to establish a national deterrent using new ballistic missiles because the Soviet long-range missiles were perceived to represent a threat to France. Before developing a new ballistic missile, the French wanted to have an experimental large launch vehicle for testing reentry shields and novel guidance systems.[1] It was to become a two-stage experimental missile. The lower stage was to be propelled by the *Vexin* engine, which was developed by LRBA using a concept similar to Veronique. It had 28 tons (61,600 lbf) thrust and burned nitric acid and turpentine. The upper stage was called *Emeraude*, used nitric acid and turpentine, and was very similar in size and design to Veronique, but it was more sophisticated. The engine was gimbal mounted to allow pitch and yaw control, and the GG used a slow-burning solid propellant, whose gases were mixed with water to decrease the gas temperature, increase the flow by adding water vapor, and make it less reactive with the propellants in the tanks.[4]

In 1960 the government decided to use the Emeraude as a first-stage for a new three-stage launch vehicle, which was named *Diamant A*.[1] The second-stage engine was furnished by SEPR. The first-stage engine called Vexin had 28 tons (61,600 lbf) of thrust, was developed by LRBA, and used the same design concept of a warm-gas pressurization and the same propellants as Veronique.[4] Some of its characteristics are also in Table 10-1. The first launch

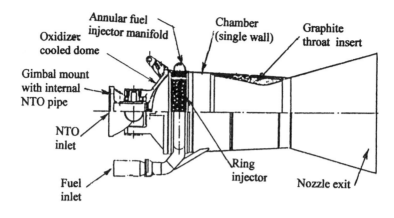

Fig. 10-4 Half-section view of thrust chamber of the Valois LPRE used in the Diamant B missile. Courtesy of SEPR/EADS; copied with AIAA permission from Ref. 4.

of Diamant A was in 1965. This Diamant A vehicle launched the first French satellite, named Astérix, in November 1965, and France became the third nation to send a satellite into Earth orbit. As this program progressed, a more powerful new *Diamant B* was initiated. The first stage used a different propellant than Diamant A, namely, NTO/UDMH. Diamant B was again chemically pressure fed by a single GG with water dilution, and the thrust was increased to 35 tons or 84, 700 lbf. The TC of this engine, now called *Valois*, is shown in Fig. 10-4, and the engine design concepts (film-cooled single-wall chamber, side injection, chemical tank pressurization) were very similar to the Vexin, Veronique, and to other French TCs, as will be seen later in this chapter. Diamant B was designed to be the first stage of a long-range ballistic missile.

The first stage of *Diamant B* was probably the largest and, for its size, the heaviest flying rocket-propelled stage with a pressurized feed system in all LPRE history.[1] The tank walls were thick and heavy, decreasing the mass ratio of the vehicle. The SEPR and LRBA people knew at the time that a turbopump-fed engine would substantially increase the payload or greatly reduce the size and inert weight of the vehicle. It is difficult to understand why the French LPRE hierarchy in 1960 persisted with conservative large heavy pressurized feed systems, when German engines with turbopumps had flown in the 1930s, when German TPs had been seen and tested by the French LPRE companies since 1945, when the Soviet Union and the United States had flown turbopumps 10 years earlier, and when the French had pump experience with their aircraft rocket engines, which are discussed in the next section of this chapter.

Aircraft Superperformance Rocket Engines

LPREs for aircraft were pursued by SEPR beginning in 1947. For military fighter airplanes of that period, these rocket engines greatly reduced the time to climb to high altitude, for intercepting intruding bombers, and improved the maneuverability and the speed at altitude. The first such LPRE was patterned after a German BMW rocket engine.[1,11] The SEPR-25 shown in Fig. 10-5 had an aluminum-alloy TC with acid regenerative cooling, a dry mass of 245 lbm, and a sea-level specific impulse of 196 s at a mixture ratio of 2.7. Table 10-2 gives a few characteristics of this superperformance engine. Its propellant pumps were driven directly through a jack shaft through a coupling from the engine of the airplane. The SEPR-25 engine was 60 in. long and only 11 in. high. The pump, gear case, and jack shaft connection are at the forward end. This engine flew 76 times on a modified Espadon military aircraft beginning in June 1952.

Between 1953 and 1963 eight different versions of similar, but improved aircraft superperformance LPRES were developed by SEPR and flight tested in different military aircraft. The seven engines listed in Table 10-2 have all flown.[1,4,11] The aircraft were modified to accept the rocket engine. Two of these LPREs became operational and were used in selected squadrons of the Mirage military fighter aircraft. Figure 2-3 shows a flying Trident fighter aircraft with the rocket engine in operation. All of these engines used a jack shaft from the aircraft's turbojet to drive the pumps, but the power available for the pumps became marginal at high cruising altitude. Both of the two deployed engines had a single thrust chamber of about 15,000 N or about 3400 lbf

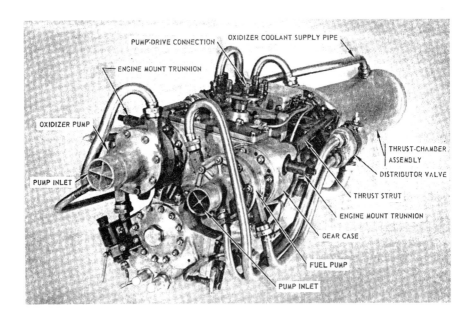

Fig. 10-5 Simplified sketch of the SEPR-25, the first French superperformance engine for military aircraft. Courtesy of SEPR/EADS (from personal communications, ESA and Astrium/EADS, data sheets from SEP/SNECMA, 1999–2004).

Table 10-2 LPREs used for Aircraft-Assisted Takeoff and/or Augmented Flight Performance

Engine	Propellants	Thrust, kN	No. TCs	Airplane for flight tests	First Flight
SEPR-25	Nitric acid/TX	15	1	Espandon	1952
SEPR 481	Nitric acid/furaline	15 and 30	3	Trident I	1954
SEPR 631	Nitric acid/furaline	15 and 30	2	Trident II	1957
SEPR 662	Nitric acid/furaline	15	1	Mystére IV	1956
SEPR 661	Nitric acid/furaline	7.5 and 15	2	Mirage I	1956
SEPR 841[a]	Nitric acid/TX	7.5 and 15	1	Mirage IIIC	1957
SEPR 844[a]	Nitrc acid/jet fuel	15	1	Miorage IIIE	1963

[a]These were produced and deployed in the air forces of France and several other countries.

thrust. The SEPR 841 used nitric acid and hypergolic TX2 fuel (triethyl amine and xylidine), and it could be throttled to 50% of full thrust. It became operational in 1961. The other, identified as SEPR 844 and shown in Fig. 10-6, was similar, but used kerosene as a fuel instead of TX2.[11] It could get its fuel directly from the aircraft's fuel tanks, but it also had a separate small TX2 tank to enable up to 12 hypergolic slug ignitions in flight. In an emergency the engine and the acid tank could be severed and jettisoned. To properly prime the pumps and avoid cavitation, ejectors were used in both propellant pump suction lines. The high-pressure flow to activate the ejectors came from a bleed at the pump discharge line. The SEPR 844 became operational in 1967 (Ref. 12). Servicing the rocket engine usually took about 15 min and was performed after every 50 ignitions. The Mirage aircraft (Model III, versions C and E) with the superperformance rocket engines were not only deployed with the French Air Force, but were also sold to Pakistan, Libya, Switzerland, Spain, and South Africa. The operation by the military service of these rocket engines was unique because in other countries their new auxiliary LPREs were only flight tested. In the Soviet Union, United States, or Britain the aircraft rocket engines were developed by several different companies. In France it was only one company that developed all of their aircraft rocket engines, namely, SEPR. As

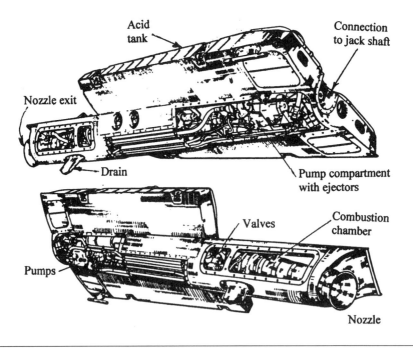

Fig. 10-6 Front and rear views of the SEPR 844 aircraft rocket engine, which was used in a squadron of Mirage III fighter airplanes. Courtesy of SEPR/EADS; copied with AIAA permission from Ref. 12.

far as the author can determine, none of these French aircraft rocket engines were ever used in combat. A total of 275 engines of the SEPR-841 and SEPR-844 were built. In the 1970s and 1980s turbojet engines became more powerful, particularly at altitude, and the rocket engines were no longer needed. In 1984 the use of these rocket engines was phased out of the French military service, but in Switzerland it was not until 1994. More than 10,000 rocket augmented flights have been completed with a 99% success rate, an accomplishment of which the French were proud.

Ariane Space Launch Vehicles

When the multistage *Ariane space launch vehicle* was proposed by the French in 1973 as a replacement for the Europa SLV, it was readily accepted by several other European countries. Britain, which originally provided the Blue Streak first stage for the Europa SLV, dropped out, in part because they did not want to incur or continue the associated high costs. For the initial version of Ariane, the French proposed providing four Viking engines for the first stage and one for the second stages.[1,4,12] This historical long-lived French Viking engine is discussed in more detail next. This French Viking engine is totally different from its namesake the U.S. Viking engine, which is mentioned in Chapter 7.3 and was earlier in the LPRE history.

SEPR had already worked on an experimental turbopump-fed large LPRE identified as M40, which became a predecessor of the French Viking engine. The designation of M40 relates to its thrust of 40 metric tons (88,000 lb). It allowed the testing of several features used later in the Viking engine, such as the design of the turbopump, the regulation of the gas-generator flow, which controlled the variable thrust, or the start sequence. The thrust chamber was patterned after the Valois TC shown in Fig. 10-4. Some of the gas from the fuel-rich gas generator was used to pressurize the fuel propellant tank, but with a TP it was at a reduced tank pressure level. It was first tested as an engine in 1969. The chamber pressure of this pump-fed experimental engine was two-and-a-half times higher than any of the prior gas-pressurized LPREs. An uprated version of the M40 was the M55, rated at 55 metric tons of thrust. At that time (late 1960s) the name "Viking" was first used for this engine.[1,4,13] All of the French engines from SEP/SNECMA/EADS have names beginning with V, which stands for Vernon, the location of the LPRE organization. Originally it was developed with the intent to use four of these engines on the first stage of a proposed improved Europa III SLV, an interim version, whose components were investigated, but this vehicle was never developed. Instead the participating European nations agreed to support the development of the Ariane, a French proposal for a versatile space launch vehicle. Its first and second stages were planned with Viking LPREs.

There are several different versions of the Viking engine.[1,4,13] The first was Viking II at 60 tons of thrust, an uprated version of the 55-ton engine. This Viking II was also later used as a ground-test engine. The initial Viking II LPREs had conical nozzle-exit sections, but the later versions all had bell-shaped nozzle exit contours, which gave up to 8 s more specific impulse and a corresponding flight performance increase. An assembly line of the Viking V is shown in Fig. 10-7 and a simplified schematic flow diagram in Fig. 10-8. The first Ariane SLVs had four gimbaled Viking V engines in the first stage, but the next version of this SLV had five. The Viking IV engine in the second stage had a nozzle area ratio of 30.8 to 1.0, which was about three times larger than that of the Viking V to give higher performance in a vacuum. The initial engine versions used NTO and UDMH for its propellants at a mixture ratio of 1.7 . As discussed next, incidents of combustion vibration later caused a change to a blend of 25% hydrazine hydrate with 75% UDMH as a fuel. There were three different versions of the Viking engine in the Ariane multistage vehicle For a five-engine first stage, there were four hinge-mounted Viking V engines, and the one in the center was fixed and not movable. In the second stage there was a single gimbal-mounted Viking IV engine with the larger nozzle exit.

The Viking *thrust chamber*, which can be seen in cross section in Fig. 10-8, was made of a cobalt alloy with an internal coating (silicate type for minimizing oxidation), and it had a graphite throat. Because the graphite nozzle insert

Fig. 10-7 Assembly line of Viking LPREs at SEP. Courtesy of SEP/EADS (from personal communications, ESA and Astrium/EADS, data sheets from SEP/SNECMA, 1999–2004).

occasionally suffered damage or excessive oxidation, it was replaced by a silica-fiber-reinforced phenolic resin material. The technology of heat-resistant fiber-reinforced plastics has been developed by SEP and is discussed later in this chapter. The light alloy injector (shaped like an upside-down cup) is cooled by the oxidizer flow, which then enters into the annular injector. A simplified sketch of this type of injector is shown in item d of Fig. 4.3-16. The injection of the propellant into the chamber is radially inward through 216 like-on-like doublet injection sets arranged in six rows, one on top of the next. Extra fuel is injected for film cooling. This side injection of the propellants with a film-cooled single-wall chamber and nozzle insert is unique to the French storable propellant TCs, and it is based on work with earlier French TCs. Since about 1970, Volvo Aero, a segment of the Swedish automobile manufacturer, has been the single source supplier of the Viking combustion chamber and nozzle, and they have delivered more than 1000 units. One version of the Viking LPRE with its chemical tank pressurization subsystem has been licensed to India, where it has been slightly modified and used to propel the second stage of the Indian SLVs. The engine was renamed as the Vikas engine by the Indians, and it is discussed in Chapter 14.

The turbopump assembly has a single shaft with a turbine and three pumps (oxidizer, fuel, and water). A simplified cross section of the TP can be seen as a part of the flow diagram in Fig. 10-8 and an external view of the TP in the upper part of each engine assemblies in Fig. 10-7. Water is added to the gas-generator flow to reduce the gas temperature to about 620°C. Water is also used to actuate the three valves that control the flow of three liquids to the gas generator. In a later version the pumps were equipped with inducer impellers, which allowed operation at a higher shaft speed and lower tank pressure and made the pumps on the second stage less critical. (The pump inlet head was marginal.) Both propellant tanks are pressurized by bleeding some of the 620°C (1148°F) gas from the three-propellant gas generator, reducing its pressure and diluting it with water to reduce the gas temperature.

In May 1980, during the second Ariane flight, a combustion instability occurred, and it disintegrated one of the Viking engines and disabled the launch vehicle. A big effort was made to correct the occasional combustion vibration problem. Flight records and old test records were reviewed, and a prior instability incidence was found and analyzed. Theoretical gas vibration studies were initiated, and possible alternatives were investigated. The injector was modified several times, and each change was tested, the acceptance criteria were changed, and 25% hydrazine hydrate was added to the UDMH fuel. These changes remedied the combustion problem. The next flight in June 1981 was successful, as were subsequent flights.

When the payload was increased, the design of the Ariane vehicle was changed to allow for more propellant, the thrust and chamber pressure of the LPREs were increased by 10%, and in one vehicle model three additional engines were added to the first stage. This was a version with eight Viking engines in the first stage. They also considered the addition of liquid propellant strap-on boosters. A single Viking engine powers each of the strap-on boosters,

800 History of Liquid Propellant Rocket Engines

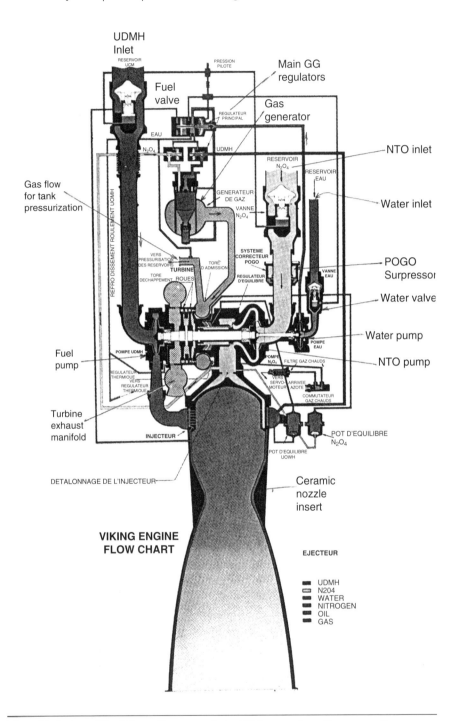

Fig. 10-8 Simplified flow diagram of the Viking engine. Courtesy of SNECMA/EADS (adapted from personal communications, ESA and Astrium/EADS, data sheets from SEP/SNECMA, 1999–2004).

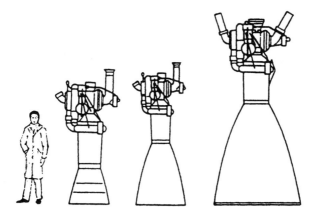

Fig. 10-9 Comparison of Viking II (low-area-ratio conical nozzle; used as sea-level test engine), Viking V (four or five used in first stage), and Viking IV (second stage, large nozzle exit). Courtesy of SEP/EADS; copied with American Astronautical Society permission from Ref. 13.

which are then dropped off. This engine is very similar to the first-stage engine, but had to be changed to allow for longer duration. Therefore there can be four different models of Viking engines in the same Ariane space launch vehicle: typically there would be five Viking V engines in the first stage (but the center engine was fixed and did not swing) and one Viking IV for the second stage. For heavy payloads and/ambitious missions some flights had one each of a modified Viking V for each of the two strap-on boosters. Figure 10-9 shows a size comparison of several different Viking engine versions.

LOX/LH$_2$ Engines

In preparation for a future upper stage with high-energy propellants, the government decided to support an R&D program, which included the firing of small hydrogen-cooled TCs (1962/1963), studies of various alternate engine configurations, and a new test facility suitable for cryogenic propellants. The first small TC tests in 1961 were actually performed with gaseous oxygen and gaseous hydrogen. When liquid hydrogen became available in adequate quantities, the French switched to small LOX/LH$_2$ TCs. Beginning in 1966, SEPR developed an experimental LOX/LH$_2$ LPRE identified as H-4 or HM4 (Refs. 1 and 14). It was rated at 40 kN or 9100 lbf thrust, had four small TCs, a single geared TP, and a LOX/LH$_2$ GG. The turbopump was the first one designed in France for a medium-sized engine; the two-stage turbine and the single-stage hydrogen pump were on one shaft, and the single-stage LOX pump was geared to a lower speed shaft. The gear case and bearings were cooled and lubricated by an unusual mixture of tributyl phoshate from a tank (lubricant) and gaseous hydrogen for cooling. The hydrogen was tapped off the cooling jacket outlet at about −150°C and heated in a heat exchanger to avoid freezing moisture. This is different from lubrication schemes in other known TPs. In 1967 a series of component tests were undertaken. The first test of a complete engine in March 1967 was a dramatic failure, caused by an improper cooling/ignition sequence. In July 1967 and in 1968, a small number of successful engine tests were performed, just before the program was canceled. At that time the government could not see a vehicle requirement that needed this engine, and the H-4 program was abandoned. However several years later this experience was applied in the design of a larger new LOX/LH$_2$ LPRE, as explained next.

The third stage of the Ariane SLV required a new engine using liquid hydrogen and liquid oxygen in order to attain the desired flight performance and payload. This engine, identified as HM7A, had features that were very similar to the H-4 engine and to earlier LOX/LH$_2$ engines in other countries. Figure 10-10 shows this HM7 LPRE.[4,15,*] The first static engine firing was in 1975, and the first installation in the third stage of the Ariane vehicle in 1978. It was the first flying LOX/LH2 LPRE developed in Europe. It had originally a vacuum thrust of over 60,000 N or 13,400 lbf, solid-propellant cartridges to ignite the gas generator and the thrust chamber, a fuel-rich mixture ratio of 4.5 to 1, and an altitude specific impulse of 421 s. The flow diagram of Fig. 10-11 shows a GG engine cycle and pyrotechnic (solid-propellant) igniters for the TC and the GG. The gimbaled engine provided pitch and yaw control, and a fuel tap-off from the regenerative cooling jacket provided roll control with heated hydrogen gas. (This was a novel feature in France.) The HM7 TP in Fig. 10-12 appears to be an improved larger version of the H-4 TP. Its turbine and the fuel pump were on the same shaft, but the oxygen pump speed was reduced through a gear case. A hard start or high chamber pressure peak occurred at altitude start, causing an unacceptable high momentary acceleration to sensitive vehicle components. This was remedied by increasing the opening rate of the main oxidizer valve and by reducing the helium purge on the oxidizer side. There was an uprated version HM7B with about 8% more thrust. Major components of this engine were developed and built in other European countries, but the engine assembly, and engine testing, and installation were by SEP. More than 85 HM7A and HM7B LPREs have been built. In five of the flights, the engine failed to perform properly.

The *Vulcain cryogenic-propellant engine* (originally identified as HM60) was developed to replace the second-stage Viking engine in order to increase the launch-vehicle's capability.[16–18,*] The thrust was increased by a factor of more than four. One version is shown in Fig. 10-13, and some data for two versions of this Vulcain engine are given in Table 10-1. It employs a gas generator cycle (because it was believed to be lower in cost and a smaller technology step forward) and was the first French LPRE with two separate high-pressure turbopumps, one for the oxygen and the other for the hydrogen. The first version

---

*Personal communications, data sheets from SEP/SNECMA, circa 2001.

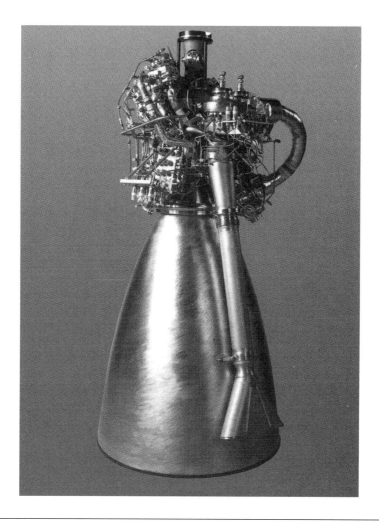

Fig. 10-10 HM7 LPRE uses LOX/LH$_2$ propellants and propelled the third stage of one version off the Ariane SLV. It is 2.0 m tall. Courtesy of SEP/EADS (from data sheets from SEP/SNECMA, 2002).

flew satisfactorily, and the next version shown in the table was uprated using an increased chamber pressure and increased nozzle area ratio (going from 45 to 61.7).[17–19] The altitude specific impulse of the Vulcain (433 s) is lower than HM7B LOX/LH$_2$ engine (446 s), in part because of nozzle area ratio of the Vulcain is lower (60 vs 83 to 1.0). Both TCs have bell-shaped nozzle-exit sections. The respectable chamber pressure of the Vulcain of 115 bars allows a relatively small thrust chamber and thus a somewhat shorter vehicle. The regeneratively cooled thrust chamber is shown partially in Fig. 10-14, has 360 longitudinal (variable cross section) channels in a copper alloy, which contains

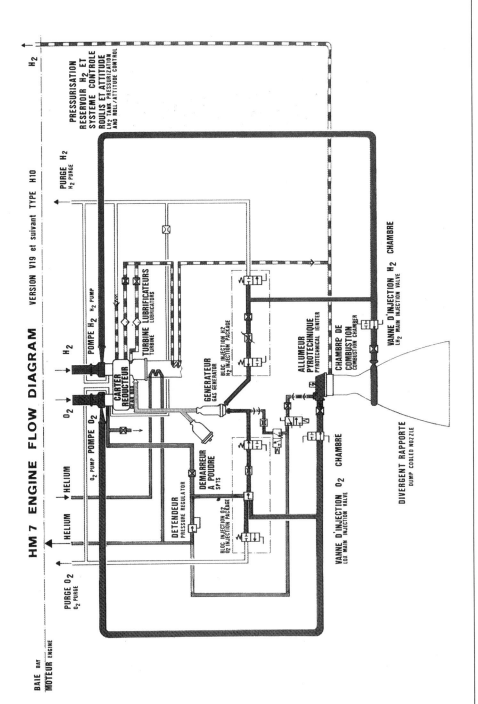

Fig. 10-11 Flow diagram of the HM7. Courtesy of SEP/EADS; copied with AIAA permission from Ref. 15.

Fig. 10-12 TP of the HM7 has inducer impellers, a shrouded impeller in the high-speed LH_2 pump, an unshrouded impeller in the lower-speed LOX pump and a heavy-duty gear case. Courtesy of SEP/EADS; copied with AIAA permission from Ref. 15 (also from personal communications, ESA and Astrium/EADS, data sheets from SEP/SNECMA, circa 2000).

some silver and zirconium, and an inconel injector with 516 coaxial injection nozzles. In the nozzle, beyond the copper section, a part of the hydrogen fuel is circulated through inconel spiral-wound formed tubes and dumped into the exhaust. This helical winding of the tubes, which are welded together, is a novel feature and allows ordinary (nontapered) tubes to be used, which are formed into rectangles of different height and width. The injector design and several of the key materials must have been inspired by earlier designs in other countries. In the Vulcain Mk II the exit portion of the diverging nozzle section used dump cooling with turbine exhaust gas to protect the wall. The Vulcain engine had two separate turbopumps. Its injector is shown in Fig. 9-17 and its exhaust plume in Fig. 9-18.

The fuel TP with a two-stage fuel pump in Fig. 10-15 and the oxidizer TP with a single-stage oxygen pump in Fig. 10-16 show an advanced state of the TP development at that time. It included inducer impellers, axial balance provisions, a complex shaft, good seals, relatively large ball bearings, and relatively stiff ball bearings supports.

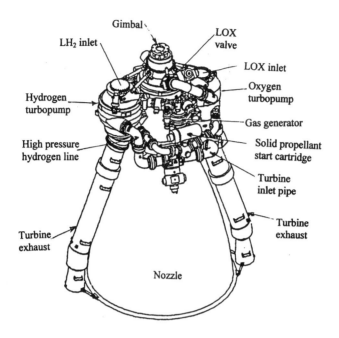

Fig. 10-13 Simplified sketch of one version of the Vulcain LOX/LH$_2$ LPRE for the second stage of the Ariane European SLV. Engine dry mass is 1850 kg or 4070 lbf, and height is 3.6 m or 11.8 ft. Courtesy of SEP/EADS (from personal communications, ESA and Astrium/EADS, data sheets from SEP/SNECMA, 1999–2004).

A new LOX/LH$_2$ third stage with a new engine for an improved version of Ariane 5 is called the *Vinci engine*, and it was promoted beginning in 1996 (Ref. 2.). It was a new LOX/LH$_2$ cryogenic engine, and its development was actually started in 1998 at SNECMA Moteurs. The intended (not its actual) performance is listed in Table 10-1. A model of the Vinci engine is shown in Fig. 10-17 and listed in Table 10-1. An expander engine cycle, the first for France, was selected because it gives 3 to 5% more specific impulse than a gas-generator engine cycle and allows a higher payload in GEO. The higher expected performance (465 s) is partly because of a high nozzle area ratio of 240 with an extendable uncooled nozzle-exit segment, which can be seen in the figure. It also has two separate TPs. The LOX TP operates 19,500 rpm and has 350 kW of power. The fuel TP runs at 90,000 rpm and has much more power, namely, about 2500 kW. In the development of this fuel–TP consideration was given to new ceramic ball bearings and pump impellers made titanium by a powder metallurgical process. By mid-2003 the fuel TP had been built and partially tested.[21] At the time of the writing of this book manuscript, the development of the Vinci engine had not been completed.

The joint development, manufacture, testing, and operation of the LPREs of the Ariane European Launchers are done by several countries (France,

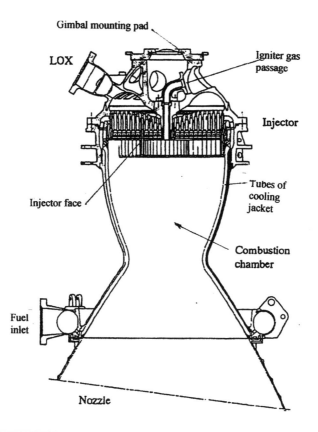

Fig. 10-14 Partial section of the Vulcain thrust chamber and its injector with part of nozzle cut off. Courtesy of MBB/DASA/EADS; copied with AIAA permission from Refs. 17–19.

Germany, Norway, Belgium, Italy, Netherlands, Sweden, Denmark, Spain, Austria, Switzerland, etc.) and have required unusual management methods and a will to cooperate. This process also requires a somewhat artificial breakdown of the tasks, so that each country has its own, well-defined specific contributions that happen to fall within the country's allocated resources and capability. For example, the development/testing of the Vulcain TC and its the gas generator were done in Germany, the turbine for the oxygen pump and the nozzle extension were developed in Norway, and some of the pumps in Italy, the hydrogen TP by France, the engine assembly, testing, and installation also by France, and components and some instruments in Spain, Belgium, or Austria. In Sweden the Volvo Aero organization has built torbopump rotor component for the Vulcain engine, stator components for the LOX TP, the TCs for the Viking engine, and the metal nozzle extension for Vulcain. It is an example of the heterogeneous pieces of the engines built by one of the countries in the consortium.

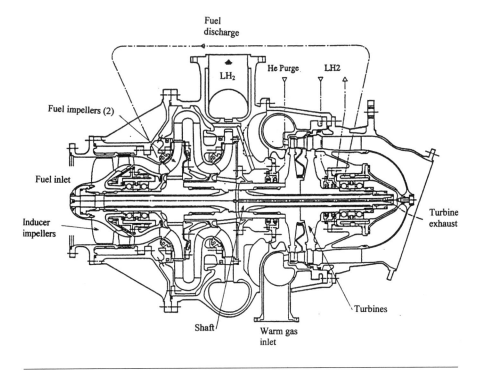

Fig. 10-15 Vulcain hydrogen turbopump. Courtesy of SEP/SNECMA/EADS; copied with AIAA permission from Ref. 19.

Composite Materials

Composite materials were originally developed for the French solid-propellant rocket motors. Several new materials have since been prepared for LPRE applications and also for reentry heat shields and various nonaerospace applications by the SEP operation at St-Medard-en-Jalles in France. Fiber-reinforced as well as homogeneous materials have been formulated. Materials with phenolic resins have been used for ablative materials and high heat-transfer and high-temperature applications, such as nozzle throats. Ceramic matrix composites use strong carbon fibers (or alternatively silica or ceramic fibers) in a multidirectional fiber pattern, and the filler material (between the fibers) can be aluminum oxide or silicon carbide. They have been used in hot-gas valves and small liquid propellant thrusters.[22,*] They have been successful at temperatures of about 3000°F or 1600 to 1700°C, and they can be designed to be resistant to thermal shock (sudden heating) and/or erosion. This means less film cooling and a somewhat higher specific impulse. The materials with carbon fibers are strong and can withstand high temperatures and high thermal

---

*Personal communications, data from SEP/SNECMA, circa 2001.

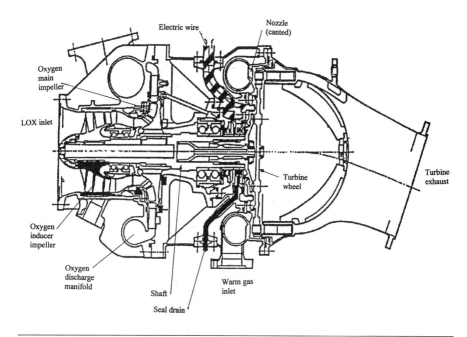

Fig. 10-16 Vulcain LOX turbopump. Courtesy of SEP/SNECMA/EADS; copied with AIAA permission from Ref. 19.

stresses, but the carbon can be readily oxidized (into CO or CO_2) if there are even small amounts of oxidizing ingredients in the combustion gases. Experiments have shown that the SiC fibers are more resistant to oxidation, but they are not as strong and are more likely to crack.

A variety of carbon materials has also been fabricated for use in nozzle-exit sections. A nozzle extension made of carbon matrix materials was used in the nozzle skirt of the HM-7. A special achievement was the development of a very large three-piece nozzle diverging section made out of a matrix of carbon fibers (oriented in three or four different directions), and the spaces between fibers are filled with additional amorphous carbon.[22],* The large major extendable nozzle-exit sections were built by SNECMA in three pieces for Pratt & Whitney's RL10B-2 engine, and it is shown in Fig. 7.10-5. Another two-piece extendable nozzle is shown in Fig. 10-17.

Small Thrusters

SEP also developed small hydrazine monopropellant thrusters (3 and 15 N) and some small bipropellant thrusters with NTO/UDMH (20 to 200 N); they are

---

*Personal communications, data from SEP/SNECMA, circa 2001.

Fig. 10-17 Three-dimensional model of theVinci LPRE with its extendable carbon nozzle in operating position. Height is 4.2 m or 13.8 ft. Courtesy of SNECMA/EADS (data sheets from SEP/SNECMA, circa 2003).

used for trajectory corrections and ACS on spacecraft, satellites, space payloads, and on military applications. The monopropellant thrusters required heaters for preventing the freezing of hydrazine and for preheating the catalyst bed. Specific impulses were between 225 and 235 s, and one unit had been pulsed more than 300,000 times.

A 200-N or 45-lbf bipropellant thruster with storable liquid propellants was selected for the attitude control and braking of the automated transfer vehicle flying to a space station or satellite.[23] SEP also developed a restartable orbital space maneuvering engine of 6 kN (1350 lbf) thrust with NTO and MMH. One version of this 6-kN TC has been tested with a matrix composite material for the chamber and nozzle.[24] These materials have been developed by SEP, and for the small thrusters they are basically carbon or silicon-carbide fibers in a matrix of silicon carbide. This LPRE is shown in one version in Fig. 10-18. Its basic injection elements are unusual. Each element has five impinging propellant streams. The four oxidizer jets were inclined to the chamber axis and impinged on the single axially directed fuel jet. It has an nozzle-exit area ratio of 100, a vacuum specific impulse of 315 s, and the height is 1.15 m or 3.77 ft. The injector is made of aluminum and it contains acoustic cavities for eliminating combustion instabilities. More recent versions worked stably without the original acoustic cavities. The engine uses a pressurized gas feed system. The use of ceramic materials allows a reduction in inert mass compared to a metal TC. A similar ceramic matrix material was demonstrated for a 20-N (4.5-lbf) bipropellant thruster. Some composite ceramic materials deteriorate with exposure to the hot combustion gases, and this can reduce the thruster's life and the safe number of restarts.

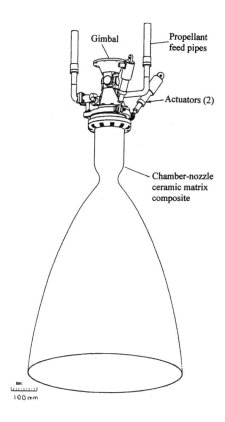

Fig. 10-18 Orbital maneuver thruster with 6 kN thrust made of ceramic matrix composites for the wall material. Courtesy of SEP/SNECMA/EADS; copied from Ref. 24 with AIAA permission.

Other Programs and Support

SEP/SNECMA/EADS has conducted a variety of analytical and experimental investigations related to LPREs. They experimentally investigated methane as a fuel, ran liquid-fluorine oxidizer in a small TC, and performed a number of evaluations of turpentine, kerosene, or TX fuel. They developed complex computer programs for heat transfer and stress analysis. They studied POGO instability in feed lines and vehicles, predictions of engine system parameters, combustion instability, engine system transient behavior, and engine health monitoring systems. At one time they studied and tested electromagnetic noncontact journal bearings for possible application to turbopumps, but did not pursue this concept.

A major technical support to the French LPRE organizations was the Office National d'Études et de Reserches Aérospaciale, which is abbreviated as ONERA. This R&D institute has conducted research programs in several aerospace fields, including subjects directly related to LPREs. They studied gaseous and liquid propellants and their interactions, analytical and experimental investigations of a number of different injection elements, or rocket combustion fundamentals (such as atomization, droplet formation, droplet burning, droplet evaporation). This included coaxial injection elements and different patterns of injection elements in the same injector. They developed analytical approaches to POGO oscillation in propellant feed lines and analytical treatments of high-frequency combustion instabilities. ONERA researchers have published learned papers on these subjects, and the list of such publications is quite long. They have gained international respect and have provided inputs and assistance to SNECMA/EADS and its predecessors; they also acted as advisors to French government agencies.

Conclusions

France has had a colorful history in its LPREs. They started with very simple engines, but more recently they have undertaken fairly complex LPRE designs. The French LPRE organizations have successfully developed and flown several reliable engines that have served them well. They have contributed to the advance of LPRE technology in some areas, such as composite materials, POGO vibration analysis, or operational superperformance rocket engines for military fighter aircraft. France is the only country that has developed and deployed aircraft rocket engines in squadrons of military aircraft.

They seem to have stayed steadfastly with proven technology even though there was then available more efficient and better performing technology. For example, they stayed with proven pressurized heavy feed systems for the first and second stages of larger flight vehicles for about 20 years; compared to other countries, they were probably 15 years behind in adopting turbopump feed systems for larger LPREs. Once they did accept turbopump feed systems, the French designs stayed with the gas-generator engine cycle because it was reliable, sometimes lighter in inert mass, and reportedly lower in engine cost. At that time in the LPRE history, the Soviet Union,

United States, and Japan had developed and flown LPREs with more advanced engine cycles, which gave better vehicle performance, were just as reliable, and their overall flight costs were not really so different. However at the time of the writing of the manuscript for this book, the European Consortium was developing a new upper-stage engine (Vinci) with an efficient expander engine cycle.

With storable propellants the French have used a unique side injector scheme with film cooling in single-walled TC for several applications, and it is still used today in the Viking. France seems to have done a creditable job of guiding and assisting the efforts in other countries in their European Space Organization and sharing the tasks and costs of development, fabrication, and operation.

References

[1]Villain, J., "The Evolution of Liquid Rocket Propulsion in France in the Last 50 Years," *History of Rocketry and Astronautics*, American Astronautical Society History Series, Vol. 17, Univelt., San Diego, CA, 1988–1989, Chap. 6, p. 87–98.

[2]Blosset, L., "Robert Esnault-Pelterie, Space Pioneer," *First Steps Toward Space*, edited by F. C. Durant and G. S. James, American Astronautical Society History Series, Vol. 6, 1985, Univelt, San Diego, CA, Chap. 2, pp. 5–31.

[3]Contensou, P., "The Contributions of Robert Esnault-Pelterie to Astronautics," *History of Rocketry and Astronautics*, edited by R. D. Launius, American Astronautical Society History Series, Vol. 11, Univelt, San Diego, CA, 1994, Chap. 16, pp. 195–199.

[4]Rothmund, C., "The History of SEP," AIAA Paper 91-2555, July 1991.

[5]Villain, J., "Jean Jacques Barré, French Pioneer of Rockets and Astronautics," *History of Rocketry and Astronautics*, edited by R. D. Launius, American Astronautical Society History Series, Vol. 19, Univelt, San Diego, CA, 1990, Chap. 13, pp. 277–282.

[6]Villain, J., "France and the Peenemünde Legacy," *History of Rocketry and Astronautics*, edited by P. Jung, American Astronautical Society History Series, Vol. 21, Univelt, San Diego, CA, 1997, Chap. 5.

[7]Jung, P., "The SE 4300 Guided Rocket Bomb," *History of Rocketry and Astronautics*, edited by R. D. Launius, American Astronautical Society History Series, Vol. 19, Univelt, San Diego, CA, 1997, Chap. 7, pp. 153–163.

[8]Rothmund, C., Moulin, H., Serra, J. J., and Lefon, J. L., "History of French Sounding Rockets Veronique and Vesta," International Astronautical Federation, Paper 99–2306, 1999.

[9]Corbeau, J., "A History of the French Sounding Rocket Veronique," *History of Rocketry and Astronautics*, edited by K. R. Lattu, American Astonautical Society History Series, Vol. 8, Univelt, San Diego, CA, 1974, Chap. 11, pp. 147–157.

[10]Rothmund, C., "Coralie: the Forgotten Rocket," *History of Rocketry and Astronautics*, edited by J. D. Hunley, American Astronautical Society History Series, Vol. 20, Univelt, San Diego, CA, 1997, Chap. 9, pp. 198–213.

[11]Rothmund, C., and Harlow, J., "A History of European Rocket Engines for Aircraft," AIAA Paper 99-2901, June 1999.

[12]Grosdemange, H., and. Schaeffer, G., "The SEPR 844 Reusable Liquid Rocket Engine for the Mirage Combat Aircraft," AIAA Paper 90-1835, July 1990.

[13]Rothmund, C., "The History of the Viking Engine," *History of Rocketry and Astronautics*, edited by P. Jung, American Astronautical Society History Series, Vol. 22, Univelt, San Diego, CA, 1998, Chap. 14, pp. 297–319.

[14]Rothmund, C., Hopman, H., and Kirner, E., "The Early Days of LOX/LH$_2$ Engines at SEP and MBB," *History of Rockets and Astronautics*, edited by P. Jung, American Astronautical Society History Series, Vol. 21, Univelt, San Diego, CA, 1997, Chap. 12, pp. 269–292.

[15]Pouliquen, M. F., and Gill, G. S., "Performance Characteristics of the HM7 Rocket Engine for the Ariane Launcher," *Journal Spacecraft and Rockets*, Vol. 16, No. 6, Nov.–Dec. 1979, pp. 367–372.

[16]Pauliquen, M. F., "HM60 Cryogenic Rocket Engine for Future European Launchers," *Journal of Spacecraft and Rockets*, Vol. 21, No. 4, July-Aug. 1984, pp. 346–353.

[17]DeBoisheraud, H., "Le Moteur Vulcain—Traveau Technologique sur la Chambre Propulsive et le Générateur de Gaz (The Vulcain Engine—Technical Efforts of the Combustion Chamber and the Gas Generator)," *ESA Bulletin*, No. 61, Feb. 1990, pp. 46–49 (in French).

[18]Brossel, P., Eury, S., Signol, P., Laporte-Weywada, H., and Micewic, J. B., "Development Status of the Vulcain Engine," AIAA Paper 95-2539, July 1995.

[19]Barton, J., Goupleau, C., and Jordant, P., "Vulcain Mk 2 Engine for Ariane 5 Evolution," AIAA Paper 95-2535, 1995.

[20]Frederic, J., and Ruault, J-M., "Vinci—a New Step for European Cooperation in Liquid Rocket Propulsion," International Academy of Astronautics, Paper 01-3205, Oct. 2001.

[21]Goirand, B., Alliot, P., Barthoulot, J-L., and Bonhomme, C., "Testing the First Fuel Turbopump of the Vinci Engine," AIAA Paper 2003-5069, July 2003.

[22]Mathieu, A., Monteuuis, B., and Gounot, V., "Ceramic Composite Materials for a Low Thrust Bipropellant Rocket Engine," AIAA Paper 90-2054, July 1990.

[23]Coutrot, A., "Storable Liquid Propellant Thrusters for Space Applications," International Astronautical Federation, Paper 96-S108, Oct. 1997.

[24]Melchior, A., "New Bipropellant Rocket Engine for Orbital Maneuvering," AIAA Paper 90-2052, July 1990.

Japan's Liquid Propellant Rocket-Engine History

Japan has concentrated its LPRE efforts on nonmilitary applications, namely, sounding rockets, space launch vehicles, and satellites for peaceful purposes. This emphasis was caused in part by treaty restrictions put on Japan after the war. Although there was some LPRE work done during the war period, nothing seems to have been done in Japan on LPREs between 1945 and about 1963. Since then, the progress has been steady and remarkable leading to successively more advanced LPREs and more sophisticated SLVs.

The government organization, called JAXA (Japan Aerospace Exploration Agency), was formed in 2003 by merging three existing government organizations, which are briefly discussed here.* JAXA is attached to the Ministry of Education and Science and has been considered to be the approximate equivalent of NASA in the United States. The National Space Development Agency (NASDA) was formed in 1969 and has been the primary government agency to the Japanese space program and awarded contracts until 2003. NASDA made the key decisions on LPREs, often with the advice of its contractors and the help of the National Aerospace Laboratory (NAL) mentioned next. There are two principal LPRE contractors. Mitsubishi Heavy Industries (MHI) had long been a designer and builder of aircraft and aircraft engines. This company started its work on LPREs during World War II and today concentrates on the large LPREs for SLVs. MHI develops, integrates, tests, and services these large engines. Ishikawajima Harima Heavy Industries (IHI) started later (1964) in the LPRE business and began with small thrusters and their LPRE systems. They built copies of the large TC of the Rocketdyne MB-3J. IHI has developed and built mostly small LPREs for ACS and orbit maneuvers, propellant tanks, and the turbopumps for all of the large LOX/LH$_2$ MHI engines. It has been the government's policy to split up the development and production work on large LOX/LH$_2$ LPREs between MHI and IHI. The National Aerospace Laboratory

---

*Personal communications, K. Kishimoto, retired chief engineer, Mitsubishi Heavy Industries, J. Nakamichi, Japan Aerospace Exploration Agency, and G. Suzuki, retired manager, Rocketdyne, 2001–2005.

(NAL), the second prior government organization, started in 1956, but did not get involved in R&D for LPREs until about 1962. Much of NAL's LPRE R&D work is at their laboratory at Kakuda in northern Honshu. Launches and vehicle static firing tests are conducted at the NASDA Tanegashima Space Center on Tanega island south of Kyushu. The LPRE facilities are discussed further at the end of this chapter. The Institute of Space and Aeronautical Sciences (ISAS), the third Japanese government organization, was a spin-off from Tokyo University, and it developed sounding rocket vehicles with meteorological or research payloads and small satellites with scientific payloads. Initially they used solid-propellant motors as the principal means of propulsion for sounding rockets and early SLVs. These early vehicles also had small LPREs with multiple thrusters for attitude control; these were provided by IHI or MHI.

These several organizations work closely with each other.[1] For example, the JAXA predecessor NAL undertook R&D and analysis projects to design, test, and define the development steps of the key large engines, which were then developed by MHI or IHI under NASDA sponsorship. NASDA used to define and develop the programs and justify them to the higher government levels, but it also did work on detailed technical subjects, such as shaft seals for turbopumps or the dynamic response of the LE-7 LOX pump. Many of the technical papers on the development of engines or components were written by JAXA engineers, even though most of the work was done by the two contractors. NAL has worked on materials problems of the hot thrust-chamber wall alloys or turbine-blade alloys, experimental heat transfer, nozzle configurations, alternate engine cycles, turbine-blade cracking, ball bearings, experimental studies of LOX/LH$_2$ staged combustion cycles, high-pressure gas seals in LOX TPs, vacuum ignition of LOX/LH$_2$, or the suction performance of pumps with two-phased flow. NASDA, NAL, and ISAS no longer exist as separate agencies, but are integrated under JAXA.

Prewar and World War II Activities

A little work on rocket thrust chambers was actually done in Japan in the 1930s before World War II. The Japanese Army undertook some rocket propulsion investigations.* Solid-propellant work started in 1931, and the very first liquid propellant efforts began in 1935. A simple rocket test facility was built on the grounds of the Army's Science Institute. Small thrust chambers were built and tested using LOX and alcohol with pressurization by nitrogen gas. No details were available. The work was stopped in 1939 because of the transfer of the key investigator to other work.

During the war, Japan was active in developing and building several LPREs for at least three expected military applications, but none were put into a military service.[2,3,*] In 1944 Mitsubishi Heavy Industries undertook its first effort

---

*Personal communications, J. Nakamichi, Japan Aerospace Exploration Agency, Rocketdyne, 2002.

in LPREs. They developed a simple LPRE for the I-go-1A and the I-go-1B radio-guided bombs, which really were air-launched missiles. The engine was called Tokuro-1 and used hydrogen peroxide as a monopropellant with sodium-permanganate-water solution as a catalyst. It used compressed air for tank pressurization. The engine development was followed by some initial flight tests. The work on the engine and the missile was done at MHI. However the Japanese Army changed its priorities and stopped further development.

Three copies of the German Messerschmitt Me163 rocket-propelled fighter aircraft and perhaps six or more copies of its LPRE were made by MHI.[2-4],* The Japanese called this aircraft "Shusui." The Germans, who were then war allies, shipped (as a part of their wartime assistance) documents and parts for a rocket-propelled airplane and its complete rocket engine by two Japanese submarines. Only one submarine managed to carry the information and some parts to Japan in July 1944. Illustrations of the German version of this aircraft and its LPRE can be seen in Fig. 2-7 for the Me163 and Fig. 9-7 for an earlier version of the engine that shipped. This German engine was identified as the 109-509-1A and was originally developed by Hellmuth Walter Company in Kiel, Germany, and is described in Chapter 9.3. Because the hypergolic bipropellants, namely, 80% hydrogen peroxide and a mixture of methyl alcohol, water, and hydrazine hydrate, had not previously been manufactured in Japan, the Mitsubishi Chemical Synthesis Company and the Edogawa Chemicals Company developed suitable equipment to produce these propellants in a very short time. The LPRE had a single thrust chamber (with multiple sets of coaxial injection elements) mounted on a tubular structure. It also featured a monopropellant gas generator and a sophisticated turbopump with inducer impellers. The maximum thrust was about 1500 kg or 3300 lbf; by shutting off different sets of injection elements, the thrust could be reduced.

This engine was copied by MHI, was named Tokuro 2, and initially built at the Nagoya Engine Laboratory and Motor Manufacture Plant of MHI.[2-4] Because of the threat of B-29 bomber attacks, the workforce on the engine was evacuated from Nagoya. Shortly thereafter a B-29 air attack hit the MHI Nagoya Works. An impromptu rocket-engine fabrication and test facility was established at Matsumoto in Nagano prefecture and another one at Yamakita-machi in the Hakone region.* Successful static test firings were accomplished at both places. Ground tests and runway taxi tests with the engine installed in the aircraft were mostly successful. The first attempt to fly in July 1945 was a failure. After a good takeoff and an initial climb, the engine suddenly stopped operating, Shusui crashed, was completely wrecked, and the pilot was killed.[4] Work leading to a second flight of a second aircraft had been started, but the end of the war stopped the program.

The LPRE technology learned from the German rocket engine was transferred to the sustainer stage of the Funryu 4 surface-to-air missile.[2,3] Earlier

---

*Personal communications, J. Nakamichi, Japan Aerospace Exploration Agency, circa 2002–2003.

versions of the Funryu missile used solid-propellant rocket motors. Funryu 4 supposedly was 13.1 ft long, 2 ft in diameter, and had a loaded weight of 4190 lb. It was designed to go to 20 mile altitude and a maximum speed of 650 mph. Called Tokuro-2, its LPRE was built by MHI, used the same propellants and essentially the same design concept as the engine for the Shusui aircraft. The test flight was scheduled for 16 August 1945, but the war ended one day earlier, and this LPRE has never flown.

Space Launch Vehicles

After years of inactivity, the Japanese government decided in 1962 to become active in the exploration and utilization of space. At that time the development of a series of sounding rockets with solid-propellant rocket motors and later a series of small space launch vehicles (SLV) was started. The first SLVs had solid propellants in all three stages and flew between 1975 and 1982. Compared to other nations, Japan had a late start for working on LPREs for space applications. However it had the advantage that the LPRE organizations in Japan could obtain literature and observe work done in other countries and obtain licenses of existing technologies. In this respect Japan was unique because its initial LPRE technology was based on several license agreements with several U.S. LPRE manufacturers.

In about 1970 NASDA laid out an ambitious spaceflight program involving a series of spaceflight launch vehicles of increased payload capacities.[1,5,6,*] They are described in Table 11-1, which identifies the various major LPREs. They used mostly LOX/LH2 engines because of their good performance, but they also used other propellants and some solid-propellant motors in the early SLVs. To get a good start on LPREs, the government of Japan made an agreement with the U.S. government to obtain state-of-the-art necessary technology. This then resulted in obtaining licences to build copies of U.S. LPREs and SLV stages, and some are mentioned later.

Mitsubishi Heavy Industries

As already mentioned, Mitsubishi Heavy Industries had started to develop LPREs during the war, but LPRE work was stopped in Japan for many years. Beginning in the 1960s, MHI resumed the development of LPREs and completed (and statically tested) several prototypes or experimental engines.[1] The company experimented with monopropellants, bipropellants, and hybrid-propellant rocket engines. MHI also furnished the hydrogen-peroxide monopropellant thrusters for the ACS of several of the sounding rockets and early SLVs developed under ISAS sponsorship. In the 1960s MHI also built a two-stage sounding rocket with a LPRE in the second stage. The sustainer stage of the two-stage LS-C used storable

---

*Personal communications, K. Kishimoto, retired chief engineer, Mitsubishi Heavy Industries, and J. Nakamichi, Japan Aerospace Exploration Agency, circa 2002–2003.

Table 11-1 Propulsion for Five Space Launch Vehicles

Vehicle designation	Third-stage propulsion	Second-stage propulsion	First-stage propulsion	Strap-on booster, (No.)	Payload, for GEO, kg	Launch period	No. of launches
N-1	Solid motor	LE-3	MB-3J	Solid (3)	130	1980–1986	7
N-2	Solid motor	AJ10-118FJ	MB-3J	Solid (9)	350	1981–1991	8
H-I	None	LE-5	MB-3J	Solid (9)	550	1986–1992	9
H-II	None	LE-5A	LE-7	Solid (2 or 4)	2000	1993–1999	7
H-IIA	None	LE-5B	LE-7A	Solid (2 or 4)	2200	Started 2001	6 up to 8/2004

propellants and a pressurized-gas feed system. MHI also investigated kerosene as a fuel with nitric acid and also with LOX.

The second series of SLVs, called N-1 and N-2, had a large LPRE in its first stage using LOX/kerosene propellants.[6,*] This engine was obtained through a license from Rocketdyne in the United States. It is a version of the MB-3 LPRE developed originally for the U.S. Thor missile (MRBM). A more advanced U.S. versions of this engine is shown in Fig. 7.8-6. The Japanese version was essentially identical. It had a thrust of 170,000 lb at sea level and a specific impulse of 262 s at sea level and approximately 300 s at altitude. The first engine and sets of components were supplied from Rocketdyne.[6] The servicing and assembly of the engine and later the building of Japanese copies of components were done by MHI and IHI. The thrust chamber was copied by IHI and the TP by MHI. This MB-3J engine was assembled and tested by MHI, and the first flight with the N-1 SLV took place in 1975 at a new test and launch facility at Tanegashima, an island south of Kyushu. The MB-3J engine was used for the first stages of the N-1, N-2, and H-I SLVs, and it proved to fly reliably.

The second stage of the N-1 SLV used the LE-3 LPRE developed and built by MHI.[6,7,*] Some of its engine parameters are given in Table 11-2. Values in this table are for a particular version of each of the listed engines. The LE-3 used NTO and A-50 fuel, a mixture of 50% hydrazine and 50% UDMH. Its nominal duration was 246 s. Its pressurized feed system used stored high-pressure helium gas. The LE-3 gimbaled TC is shown in Fig. 11-1. It was developed by MHI between 1970 and 1974 with the cooperation and some design assistance from Rocketdyne.[6] It is historically unique because it was the first indigenous LPRE and used three cooling methods: a tubular regeneratively cooled section for the chamber and the nozzle throat, an ablative nozzle segment, and a radiation-cooled nozzle-exit section. MHI also worked on another version of this upper-stage LPRE called LE-4; it had an all-ablative TC, but it was not fully developed.

The N-2 SLV used the same booster engine with larger propellant tanks or somewhat longer duration, but a different and higher thrust storable propellant second-stage engine, the Aerojet AJ10-118FJ. This upper stage became the responsibility of IHI and is discussed in the next section of this chapter.

The H-I SLV used the same LOX/kerosene first-stage engine as the N-2 SLV, but a Japanese-developed upper-stage LOX/LH$_2$ LPRE, called LE-5.[7–10,*] This engine is shown in Fig. 11-2, some engine parameters are in Table 11-2, and the engine was developed by MHI between 1977–1983. It had a gas generator and a gas-generator engine cycle. The engine flow diagram is in Fig. 11–3. An auxiliary turbine (located in the hydrogen high-pressure line upstream of the injector) provides power for the hydraulic pump to swivel the gimbaled engine. The tubes in the chamber and nozzle-throat section are made of a

---

*Personal communications, K. Kishimoto, retired chief engineer, Mitsubishi Heavy Industries, 2000–2005.

Table 11-2 Selected Data on Japanese-Developed Large Liquid Propellant Engines

Designation	SLV application	Stage	Thrust, kg	Specific impulse, s	Propellants	Nozzle area ratio	Chamber pressure, Mpa	Engine cycle	Length, m	Development
LE-3	N-1	Second	5,400 vac	290 vac	NTO/A-50	26	1.15	Pressurized helium gas	1.76	1971–1974
LE-5	H-I	Second	10,500 vac	450 vac	LOX/LH$_2$	140	3.65	GG cycle	2.65	1977–1984
LE-5A	H-II	Second	12,400 vac	452 vac	LOX/LH$_2$	130	3.98	Nozzle expander bleed cycle	2.67	1986–1991
LE-5B	H-IIA	Second	14,000 vac	450 vac 347.5 SL	LOX/LH$_2$	100	3.6	TC expander bleed cycle	2.54	1994–2002
LE-7	H-II	First	110,000 vac	446.5 vac 339 SL	LOX/LH$_2$	52	12.7	Staged combuston cycle	3.42	1984–1993
LE-7A	H-IIA	First	10,000	438 vac	LOX/LH$_2$	54	11.9	Staged combustion cycle	3.66	1994–2002

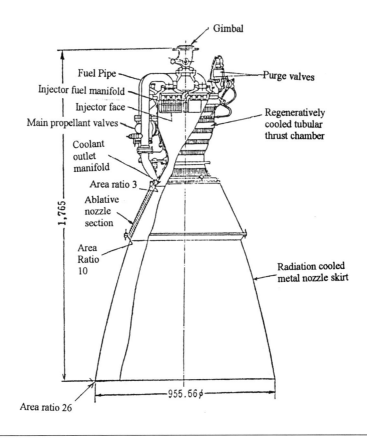

Fig. 11-1 The LE-3 thrust chamber was used for the second stage of the N-1 SLV and had three different cooling schemes (dimensions in mm). Courtesy of MHI and Rocketdyne, adapted from drawings obtained through personal communications from K. Kishimoto, retired chief engineer, MHI, and G. Suzuki, retired manager, Rocketdyne).

nickel material. The nozzle extension is dump cooled by a small amount of hydrogen, and it has 650 small tubes of 286 high-temperature alloy. The LE-5 injector is shown Fig. 11-4. It and the LE-7 injectors (both for TCs and GGs) use coaxial injection elements. These elements are arranged in circles around the central igniter passage. The LE-5 TC has 24 acoustic absorbers in the TC, and they are tuned to about 4800 Hz. The turbopumps for all of the LOX/LH$_2$ engines (beginning with the LE-5) and many of the GGs were developed and built by IHI. They are mentioned in the next section of this chapter. The TC was gimbaled and could provide pitch and yaw flight corrections. The attitude control was by two separate small hydrazine monopropellant modules, also built by IHI and described in the next section.

Japan's Liquid Propellant Rocket-Engine History **823**

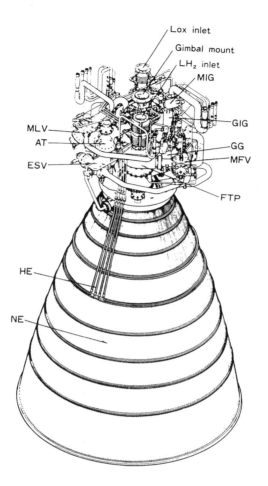

Fig. 11-2 LE-5 was the first Japanese LPRE with the energetic cryogenic propellants of LOX/LH$_2$. It propelled the second stage of the H-I SLV. The abbreviations are explained in Fig. 11-3. Courtesy of MHI; copied with AIAA permission from Ref. 9.

The startup method of the LE-5 engine is given here in some detail because it was unique and was novel at the time.[8–10,*] The Japanese called it a "coolant bleed start." Prior to the start both propellants are drained or bled out through the engine in order to chill down the hardware. Figure 11-5 shows the start and shutoff sequences. The igniters of the TC and GG are started using gaseous oxygen and gaseous hydrogen at a fuel-rich mixture. A small flow of propel-

---

*Personal communications, K. Kishimoto, retired chief engineer, Mitsubishi Heavy Industries, 2000–2005.

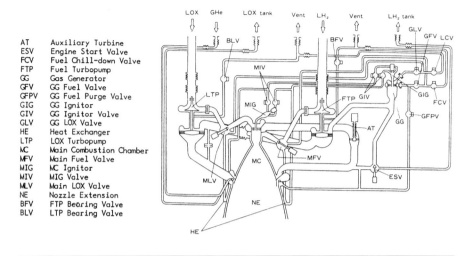

Fig. 11-3 Simplified schematic flow diagram of the LE-5. Courtesy of MHI; copied with AIAA permission from Ref. 9.

lant is tapped off from the main propellant valves and heated in the nozzle. The propellants, flowing under tank head pressure, begin to burn in the TC. About 25% of the gasified fuel flow is tapped off the flow from the cooling jacket; this then flows through the hydrogen engine start valve (ESV) to drive the two turbines in series. When the vehicle reaches an appreciable level of acceleration (at 50% thrust), the vehicle controller gives a second-stage engine lock-in signal (SELI in the figure). With this SELI signal the engine controller causes the gas generator to start and also to close the ESV. The turbopumps are then accelerated further, and the engine soon reaches its full thrust level. Three precalibrated orifices (two in the propellant lines to the GG and one in the turbine bypass of the oxidizer TP's turbine) give the desired 100% thrust level and the desired mixture ratio. This is a relatively complex start sequence that was simplified in subsequent versions of this engine.

The LE-5A is an uprated improved version.[7,11–14,*] It has a higher chamber pressure, higher thrust, a larger throat area, and a hydrogen nozzle coolant bleed engine cycle, instead of a conventional gas-generator engine cycle. The engine is shown in Fig. 11-6, and a flow sheet with this nozzle coolant bleed cycle is shown in Fig. 11-7. The turbines are driven by tapping gaseous hydrogen from the nozzle coolant loop, thereby reducing complexity and mass. It has a 5% thrust idle mode for more precise orbit control and for an easier restart. The engine then

---

*Personal communications, K. Kishimoto, retired chief engineer, Mitsubishi Heavy Industries, 2000–2005.

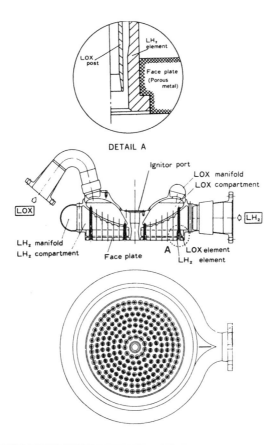

Fig. 11-4 View of injector face and two sections of the LE-5 injector. It has 208 coaxial injection elements arranged in circular patterns around the igniter opening. The oxidizer inner posts or sleeves are recessed, and their exit end is chamfered. Courtesy of MHI; copied with AIAA permission from Ref. 9.

operates at low thrust on tank pressure without having to run the pumps. This idle stage is a unique contribution to the state of the art of LPREs, but it has some similarities to the idle of the German V-2 engine. The TC construction is similar to the LE-5, and it has no acoustic cavities. The first flight was in February 1994. Both the LE-5A and the LE-5B have a throttling capability down to 88.1 kN, but this lower thrust was not used in flight.

The LE-5B was aimed at reducing the cost of the engine and at further increasing the thrust.[1,7,*] This engine is seen in Fig. 11-6, and the flow

---

*Personal communications, K. Kishimoto, retired chief engineer, Mitsubishi Heavy Industries, 2000–2005.

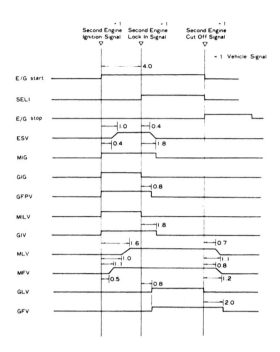

Fig. 11-5 LE-5 engine sequence. The abbreviations are explained in Fig. 11-3 and in the text. Courtesy of MHI; copied with AIAA permission from MHI through Ref. 9.

diagram is in Fig. 11-7. The turbine inlet temperature was reduced, which made the turbine-blade stresses less critical. The nozzle area ratio and the chamber pressure were reduced, and the specific impulse was reduced slightly. The engine cycle, called a chamber expander bleed cycle, is a further modification as can be seen by comparing the two flow diagrams. The turbine gases are bled off the cooling jacket, which goes around the chamber and the nozzle-throat region; the nozzle extension is dump cooled. It has a different startup sequence. The injector elements are reduced in number. For the first time in Japan, a milled channel cooling-jacket design was used with a copper alloy (instead of tubes) in the chamber and nozzle-throat area.

The LE-7 is the first large LOX/LH$_2$ LPRE and the first Japanese engine to use a staged combustion cycle with a fuel-rich preburner.[1,7,11-15] It is used to propel the booster or first stage for the H-II SLV. The LE-5A or -5B engine drive the second stage. The LE-7 is shown in Fig. 11-8, its flow diagram in Fig. 11-9, and some data in Table 11-2. The preburner gas temperature is 810 K or about 1000°F. The inner wall is copper with a small amount of chromium to improve the metal strength at elevated temperatures. There were problems during the

Fig. 11-6 LE-5A and LE-5B rocket engines. Courtesy of MHI (from personal communications, K. Kishimoto, retired chief engineer, MHI, 2004).

development of the engine, which delayed the first launch by a year. For example, this included a safe start sequence under a variety of different conditions. It takes about 5 to 6 s from start signal to full power, as shown in Fig. 11-10. Engine development was between 1984 and 1992. Several technical problems, encountered during development, were successfully overcome, and some of them are listed here: turbine manifold burst (change manufacturing process), engine start troubles (new start sequence), TC cooling (coolant passage redesign, film cooling added), external fire caused by leaks (joint redesign), nozzle flow separation (nozzle-exit contour shape), high turbine inlet temperature caused by low pump performance (increase throat diameter), turbine manifold and vane cracks (change welding process, vane removed, external support ring added), main injector burst (change welding process, manifold reshape), and fuel TP second impeller crack (reduce stress by increasing radius of impeller edge). The auxiliary propulsion in roll control is provided by four small thrusters of 150 kg each. This extra roll control is provided to allow a fine

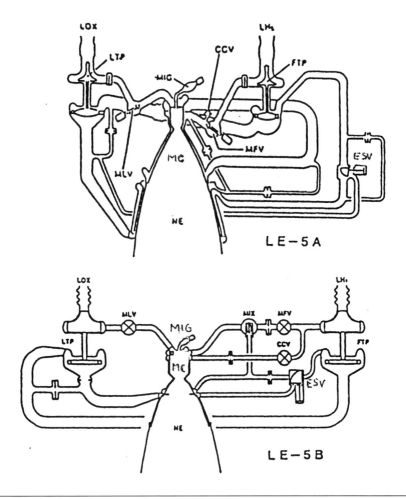

Fig. 11-7 Simplified flow diagrams of the LE-5A and LE-5B LPREs. Abbreviations are explained in Fig. 11-3. Courtesy of MHI.

tuning in roll, because the solid rocket booster roll control, was too coarse. The propellant for the auxiliary thrusters is gaseous, tapped off the preburner of the LE-7, but diluted with hydrogen to achieve a lower temperature. These roll thrusters were fixed (not hinge mounted), operated in a pulse mode, and were developed at MHI.

Although the H-II SLV was a great technical achievement of the Japanese space program, it had not been a commercial success. The cost of a launch in the 1990s reached $190 million, which at that time was about twice the cost of a U.S. Atlas SLV or the French Ariane SLV. As a result, several Japanese commercial satellites were launched by foreign SLVs, and the H-II was limited to a few Japanese government payloads. To overcome this cost disadvantage,

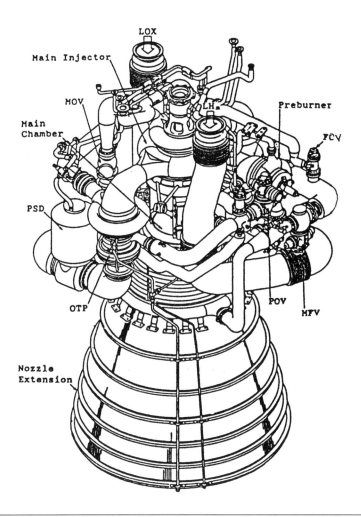

Fig. 11-8 Simplified sketch of the LE-7 rocket engine. The abbreviations are explained in Fig. 11-9. Courtesy of MHI.

NASDA initiated the H-IIA development with the objective of cutting costs in half. Depending on the mass of the payload, the new H-IIA could fly with two or four solid-propellant strap-on boosters or alternatively without strap-on boosters.

The new LE-7A engine was more robust and aimed at lower cost and alleviating some of the problems uncovered in the development of the LE-7, as outlined in the preceding list.[13,14,*] It had the same thrust, a 10% lower

---

*Personal communications, K. Kishimoto, retired chief engineer, Mitsubishi Heavy Industries, 2000–2005.

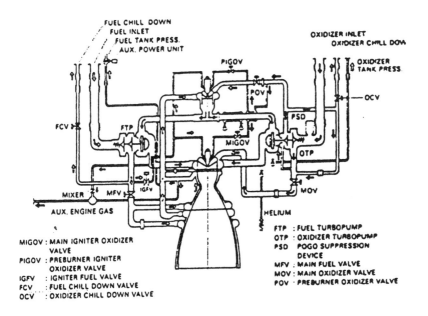

Fig. 11-9 Simplified flow diagram of the LE-7 LPRE. Courtesy of MHI; from Ref. 11.

chamber pressure, a 10% larger throat area, lower pump speeds, slightly less specific impulse, a slightly higher nozzle area ratio, and a lower preburner gas temperature (reduced by about 9% from 810 to 740 K). All of these changes made the engine less critical and more reliable than its predecessor, the LE-7. Stresses were lower because of reduced pressures in the turbopump, cooling jackets or valves, and the heat transfer was reduced. Also there were fewer parts. For example, the baffles and acoustic cavities were eliminated (the injector was found to be stable without them), and the number of tubes in the nozzle cooling jacket was reduced. Fig. 11-11 shows this engine. It can be throttled to 72% of rated thrust, but this feature was not used in flight. It still used the same staged combustion engine cycle.

All of the large Japanese LOX/LH$_2$ engines have provisions for furnishing heated hydrogen gas to pressurize the LH$_2$ tank and helium or heated oxygen to pressurize the LOX tank. In some vehicles the helium spherical containers are immersed in the LH$_2$ tank. This saves inert mass by avoiding storage at ambient temperatures.

A joint project between MHI and Rocketdyne has been underway for the past several years to develop a new upper-stage engine of about 60,000 lbf thrust using LOX/LH$_2$ propellants. Identified as MB-60, it has progressed by the end of 2003 through the design phase into component development.

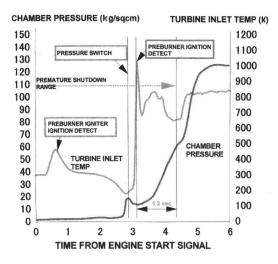

Fig. 11-10 Start characteristics of the LE-7 engine. Courtesy of MHI (from personal communications, K. Kishimoto, retired chief engineer, MHI, 2004).

Ishikawajima Harima Heavy Industries (IHI)

This company has been active in the LPRE field since 1964. It started by developing some small monopropellant thrusters both with hydrogen peroxide and later hydrazine.[7] A small experimental LOX turbopump and a small engine with storable bipropellants were early projects, and they provided a learning experience.[1,7] They obtained the hydrazine monopropellant technology through a license from TRW (today Northrop Grumman) and built LPRE with these thrusters in Japan beginning about 1969.

In 1971 IHI began to build copies of the main thrust chamber and several minor components of the Rocketdyne MB-3J engine.* These TCs were tested and put into a limited production and delivered to MHI for assembly. The small cooled pair of roll control vernier TCs at 1000 lbf thrust each were part of the engine, and were also copied by IHI. They were supplied with propellants from the main engines's TP. They can be seen in Figs. 4.6-3 and 4.6-4.

All of the turbopumps for all of the Japanese LOX/LH$_2$ engines were developed and produced by IHI.[16-20] In the many years of work on TPs, this company developed a well-known capability for their design and fabrication. As a sample, two turbopumps for the LE-5A are shown in Fig. 11-12 and 11-13.

---

*Personal communications, K. Kishimoto, retired chief engineer, Mitsubishi Heavy Industries, 2000–2005.

832 History of Liquid Propellant Rocket Engines

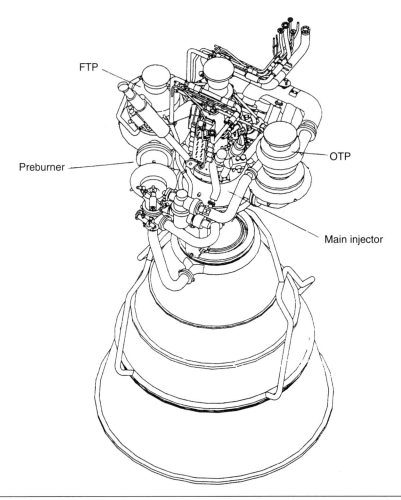

Fig. 11-11 Sketch of the LE-7A LPRE. Courtesy of MHI.

Their ball bearings are cooled and lubricated by propellants. They have been redesigned for a larger flow than the TPs for the LE-5, with a new set of inducers, and precision castings for reduced cost. The LH_2 pump has external thermal insulation. The larger TP for the LE-7 is shown in Fig. 11-14. IHI has designed, built, and tested about a dozen different TPs and has became a respected TP developer. A few years ago Pratt & Whitney in their work on an experimental upper-stage engine of a nominal thrust of 60,000 lb (vac) obtained their hydrogen TP by contract from IHI. The company designed, built, tested, and delivered this advanced hydrogen TP to Pratt and Whitney recently. This is also mentioned in Chapter 7.10.

IHI developed a series of bipropellant and monopropellant thrusters and their systems for use in satellites and spacecraft for maneuvers and attitude

Japan's Liquid Propellant Rocket-Engine History 833

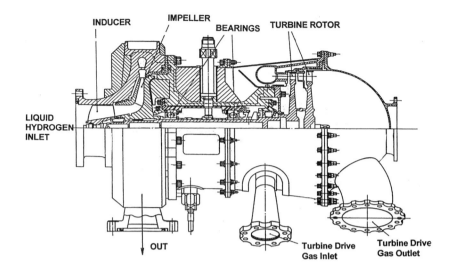

Fig. 11-12 Simplified half-section of the fuel turbopump for the LE-5A. Courtesy of IHI; obtained from Ref. 7.

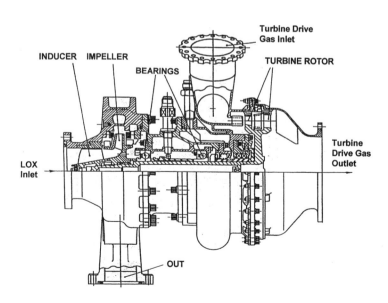

Fig. 11-13 Simplified half-section of the oxygen turbopump of the LE-5A. Courtesy of IHI; obtained from Ref. 7.

Assembly of first-stage engine (LE-7) LH₂ turbopump

Fig. 11-14 External view of an early version of the liquid-hydrogen turbopump of the LE-7 engine. Courtesy of IHI from Ref. 7.

control.[7] One example is IHI's unified propulsion system for a GEO mission, and it is shown in Fig. 11-15. It has a regulated pressurized feed system, an apogee 1.7-kN bipropellant TC (it has a specific impulse of 321 s at area ratio of 240 with almost 1-h duration), and 16 monopropellant hydrazine catalytic decomposition thrusters (4 at 50 N and 12 at 0.86 N thrust). The small thruster had to demonstrate a capability for a minimum of operating for 270,000 cycles. IHI developed a number of complete liquid propellant propulsion systems with multiple small thrusters, and they have flown in Japanese satellites.

Another example of small thrusters is the reaction control systems or auxiliary propulsion systems for the upper stages of the H-I and H-II spaceflight vehicles. They were assembled in two modules of six thrusters each, used monopropellant hydrazine, a helium gas-pressurizing subsystem, and were integrated with their own propellant tanks into a package or module. They were attached to the vehicle skirt around the main engine. They are shown (without their cover sheets) in Fig. 11-16 for the H-I upper stage (1970s) and in Fig. 11-17 for the H-II upper stage.[7] The six monopropellant hydrazine thrusters in each module are four at 50 N and two at 18 N thrust. These modules were self-contained and independent of the main engine, which means they could operate when the main engine was or was not running.

Another example is a propulsion system for a lunar exploration satellite developed for ISAS. It has a bipropellant main thruster (500 N, NTO/hydrazine, with 321 s) and 10 small monopropellant thrusters for attitude control and a pressurized-helium-gas feed system.[7] It is shown in a perspective sketch in

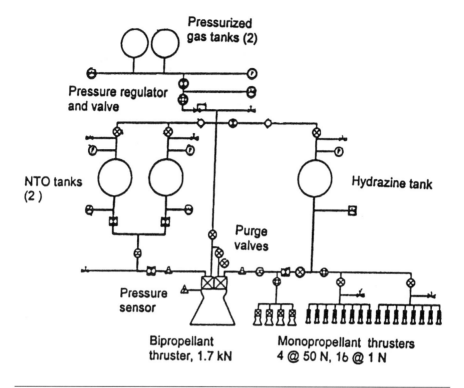

Fig. 11-15 Flow diagram of IHI Unified Propulsion System. It provides space maneuvers and attitude control for the Comets satellites. Courtesy of IHI from Ref. 7.

Fig. 11-18. The bipropellant engine was started in 1983 and has been successful in putting several satellites into orbit. A bipropellant thrust chamber, similar to the one used in the lunar module, is shown in a ground test with a shortened nozzle in Fig. 11-19.

IHI also performed the checkout, testing, servicing, and installation of the second stage of the N-2 stage. The design and fabrication of this stage and its Aerojet AJ10-118FJ engine were done by Aerojet in the United States, including the cold-gas-pressurizing system, gimbal, and gimbal-actuating mechanism. Licenses were obtained from the Douglas Aircraft Company for the stage and from Aerojet Propulsion Company for the engine. The engine had about 10,000 lbf thrust and used NTO/A-50 as propellants. Roll control was provided by cold-gas jets. Eight sets of this upper stage were built at Aerojet between 1970 and 1983 and were delivered by Aerojet to IHI for further processing. All eight of these upper stages flew satisfactorily between 1981 and 1991.

IHI in coordination with JAXA's predecessor has also worked on a 10-kN (2240-lbf) experimental LOX/LH$_2$ engine for orbit insertion and orbit transfer

Fig. 11-16 One of two modules of the reaction control system with its own pressurized gas feed system for the upper stage of the H-I SLV. It has six small hydrazine pulsing monopropellant thrusters, which are also shown in an enlarged detail. Courtesy of IHI; copied from Ref. 7.

H-II rocket second-stage gas jet system

Fig. 11-17 One of four modules for the reaction control of the upper stage of the H-II SLV. It has six small hydrazine monopropellant thrusters. Some thrusters are redundant. Courtesy of IHI from Ref. 7.

of a future planned satellite using an expander cycle.[21] This was the first LPRE in Japan with this engine cycle. Some component tests and engine ground tests were performed, but this engine was apparently not selected for flight.

Facilities

The facilities used for making and testing LPREs are spread over Japan. The MHI engineering and manufacture plant is at Komaki, north of Nagoya. This plant also makes jet engines. Rocket-engine test firings were at MHI facilities in Nagasaki on Kyushu Island and at Tashiro on the northwest side of Honshu. The Tashiro facility also has the capability to test engines mounted in complete stages. IHI produces components at its Tanashi plant in Tokyo and cleans, assembles, and factory tests at the Mizuho plant in Tokyo. IHI has a test facility for ground firings and hot component tests at Aioi City in Hyogo prefecture.

Two facilities for simulated high-altitude testing are at the JAXA Laboratory at Kakuda, near Sendai in northern Honshu.[22] A simplified diagram for one of these high-altitude test facilities at Kakuda is shown in Fig. 11-20. It can operate at a vacuum of 6 torr during operation and 13 torr prior to start. The operating duration of about 10 min is limited by the capacity of the high-pressure steam storage. The smaller vacuum facility was used for altitude testing of the LE-3 engine, and the larger facility was used for the simulated altitude tests of the LE-5 series of LPREs. Government R&D on experimental LPRE components is also largely accomplished in the NAXA laboratories at Kakuda.

The JAXA facility at Tanegashima, an island south of Kyushu, has multiple capabilities. It is equipped with a large test stand for ground testing of the large LE-7 engine in a vertically downfiring position, and another large stand for

838 History of Liquid Propellant Rocket Engines

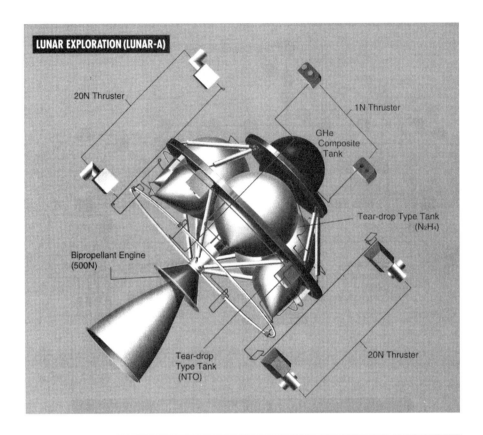

Fig. 11-18 Artist's sketch of Lunar-A propulsion system with a single axial bipropellant thruster of 500 N thrust, 10 ACS monopropellant thrusters (20 N each), and several 1 N thrusters. It has six tear-shaped propellant tanks. Courtesy of IHI from Ref. 7.

testing engines mounted in their captive large vehicle stages and for launching (in a southwest direction) all space launches of Japanese SLVs over the Pacific Ocean. It includes assembly buildings, propellant supplies, security personnel, control stations, and appropriate off-site tracking stations.

Concluding Comments

The early Japanese LPRE efforts before and during World War II had little direct influence on later engine developments, but they helped to build a capability. This included the experience in copying the German Me 163 fighter aircraft and its engine. An orderly, deliberate, and ambitious space program was planned by the Japanese in 1970, about 25 years after the war had ended. It proposed successively larger SLVs with new larger rocket engines, with the second generation using LOX/LH_2 propellants for top performance. They also planned and

Japan's Liquid Propellant Rocket-Engine History   839

Fig. 11-19 Ground test on an apogee thrust chamber with a shortened ground-test nozzle. Courtesy of IHI from Ref. 7.

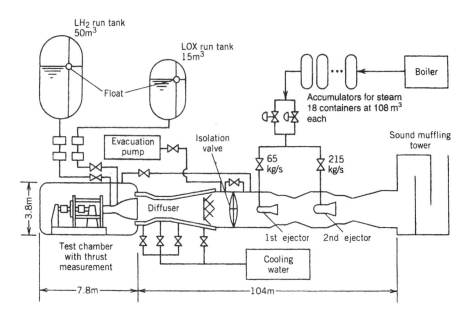

Fig. 11-20 Simplified diagram of an altitude test facility for horizontal firing at very low ambient pressure. Courtesy of JAXA; copied from Ref. 22.

later built the appropriate test and launch facilities. They deliberately obtained licenses to copy engines from Rocketdyne (Thor, first stage), Aerojet (Delta, second stage) and from TRW (small monopropellant thrusters) and/or bought a few of these engines for their initial SLVs. They cautiously then developed their own engines, came up with a novel start sequence, adopted a staged combustion engine cycle, and improved fabrication techniques. Beginning with their H-II SLV, all main engines used high performance LOX/LH_2 propellants. As costs were high in Japan, they modified and simplified their latest SLV; this included simplifications and part elimination on their two main engines. They made good use of their LPRE resources and did not develop, qualify, or produce engines that were later not needed. The Japanese divided the work between two contractors IHI and MHI and assigned some R&D to Government Laboratories. MHI did the final engine assembly and large thrust chambers, while IHI did all the turbopumps, all small thrusters and assembled some second stages. In 2004 they had good, reliable, high performance LPREs and were continuing their space program with further launches of satellites.

References

[1]*Space in Japan*, Aeronautics and Space Sec. Science and Technology Agency, Government of Japan, March 1967.

[2]Matogawa, Y., "Japanese Liquid Rockets in World War II," International Academy of Astronauties, Paper 97-2.3.02, Oct. 1997.

[3]Yamada, Y., Katata, H., and Sekita, R., "Historical Review of Japan's Liquid Rocket Propulsion," AIAA Paper 93-2598, 1993.

[4]Makino, I., and Mochida, Y., "The Only Japanese Rocket Fighter Airplane '{Shusui}' from Its Birth to Extinction," Vol. 34, No. 428, 1995, pp. 67–85 (in Japanese).

[5]Takenaka, Y., "Overview of Space Activities in Japan," *Proceedings of the Pacific Basin International Symposium on Advances in Space Science Technology and its Application*, June 1987.

[6]Takenaka, Y.," Development History of the N-1 Launch Vehicle," *Journal of the Japanese Institute of Aeronautics and Astronautics*, Vol. 32, No. 362, March 1984.

[7]Ishikawajima Harima and Mitsubishi Heavy Industries, Company Data Sheets, Brochures, or Illustrations, circa 2003.

[8]Hirata, A., Denda, Y., Yamada, A., Kochiyama, J., Fujita, T., and Katsuta, H., "Development Tests of the LE-5 Rocket Engine," *13th International Symposium on Space Technology and Science Proceedings*, AGNE Publishing, Tokyo, 1982, pp. 237–242.

[9]Yanagawa, K., Fujita, T., and Miyajima, H., "Development of the LOX/LH_2 Engine LE-5," AIAA Paper 84-1223, June 1984.

[10]Sogame, E., Yanegawa, K., Taniguchi, H., and Katsuta, H., "Liquid Hydrogen and the Japanese Space Program," *Proceedings of the Third World Hydrogen Energy Conference*, Vol. 3, Pergamon, New York, 1981.

[11]Fujita, M,. and Fukushima, K., "Improvements in the LE-5A and LE-7 Engines," AIAA Paper 96-2847, July 1996.

[12]Fukushima, Y., Nakatsuzi, T., Koganezawa, T., and Warashina, S., "Development Status of the LE-7A and LE-5A Engines for the H-IIA Family," International Astronautical Federation, Paper 97-S102, 1997.

[13]Davis, N. W., "Japan's Space Hopes Ride on H-II," *Aerospace America*, March 1994.

[14]Fukushima, Y., Watanabe, Y., Hasegawa, K., and Warashina, S., "Development Status of the H-IIA Rocket First Stage Propulsion System," AIAA Paper 98-35075, July 1998.

[15]Tamura, H., Yatsuyanagi, N., Gomi, H., Niino, M., and Kumakawa, A., "Experimental Study in $LO2/LH2$ Staged Combustion Cycle Rocket Engines, (Preburner and Main Burner),"

Proceedings of the 15th International Symposium of Space Technology and Science, AGNE Publishing, Tokyo, 1986, pp. 383–388.

[16]Shimura, T., Kamijo, K., and Yamada, A., "Dynamic Response of the Liquid Oxygen Pump of the LE-5 Rocket Engine," *Proceedings of the 13th International Symposium on Space Technology and Science*, AGNE Publishing, Tokyo, 1982.

[17]Kamijo, K., Yoshida, M., and Tsujimoto, Y., "Hydraulic and Mechanical Performance of the LE-7 LOX Pump Inducer," *Journal of Propulsion and Power*, Vol. 9, No. 6, Nov.-Dec. 1993, pp. 819–924.

[18]Shimura, T., and Kamijo, K., "Dynamic Response of the LE-5 Rocket Engine Oxygen Pump," *Journal of Spacecraft and Rockets*, Vol. 22, No. 2, 1985.

[19]Kamijo, K., Sogame, E., and Okayasu, A., "Development of the Liquid Oxygen and Hydrogen Turbopumps for the LE-5 Rocket Engine," *Journal of Spacecraft and Rockets*, Vol. 19, No. 3, 1982, pp. 226–231.

[20]Kamijo, K., Yamada, H., Hashimoto, R., and Tsujimoto, Y., "Performance of the LE-7 LOX Pump Inducer," *18th International Symposium on Space Technology and Science*, AGNE Publishing, Tokyo, 1992.

[21]Tanatsuga, N., and Suzuki K., "The Study of High Pressure Expander Cycle Engines with Advanced Concept Combustion Chamber," *Acta Astronautica*, Vol. 13, No. 1, 1986, pp.1–7.

[22]Yanagawa, K., Fujita, T., Miyajima, H., and Kishimoto, K., "High Altitude Simulation Tests of the LOX/LH2 Engine LE-5," *Journal of Propulsion and Power*, Vol. 1, No. 3, 1985, pp. 180–186.

Liquid Propellant Rocket Engines in the United Kingdom or Britain

Britain had a vigorous and extensive LPRE program between 1947 and 1971, but since that time the LPRE work was limited to engines for spacecraft maneuvers and attitude control. In the late 1940s the impetus to the British development of LPREs came in part from urgent government priorities and from examining the accomplishments of the Germans (V-2 and hydrogen-peroxide aircraft rocket engines), which became known to them in detail only at the end of World War II.[1,2] As a learning experience, the U.K. military people decided to launch several V-2 missiles. They collected V-2 hardware in Germany and selected an old German artillery range near Cuxhaven in Germany as the assembly and launch site.[1] Four V-2 missiles were assembled and launched into the North Sea. The team consisted of about 2500 British military personnel and about 1000 German manufacturing and launch personnel. High-level visitors were invited to attend the launches, including Theodore von Kármán of Cal Tech, William Pickering, director of the U.S. JPL, Valentin P. Glushko, the most experienced LPRE engineer at that time in the Soviet Union, and Sergei P. Korolev, who later became the top man in the USSR missile and space programs.

In particular, the British concentrated on using high-strength hydrogen peroxide as the oxidizer and extended this LPRE technology, which was originated and developed at the German Hellmuth Walter Company. This propellant relates to much of the U.K. history of LPREs between 1945 and 1965. They obtained the services and the help in the United Kingdom of a number of German rocket experts, including three years from Hellmuth Walter, the founder and boss of the H. Walter Company in Kiel, Germany, and Ulrich Barske, a famous pump expert, who had a special low-flow pump named after him and who helped with turbopumps. The Germans did make some contributions to the development of LPREs in the United Kingdom.

Early Efforts

There was some work with LPREs in the United Kingdom prior to 1945 during World War II. The first British LPRE work, intended to become a JATO, was started in 1941 and was given the nickname "Lizzy". It is shown in Fig. 12-1, and it was intended to provide 1000 lb thrust for 20 s (Refs. 1 and 3). The first tests in 1941 were at much lower thrust and duration, used a blow torch for ignition, and were beset by hardware failures. Lizzy was then modified several times and retested until 1944. Lizzy used a pressurized-nitrogen-gas feed system, an uncooled ceramic TC, LOX/gasoline propellants diluted with water, and a pyrotechnic igniter. This LPRE was developed by a small team (11 men) of the Asiatic Petroleum Company (now Shell International Petroleum Company). At that time little was known about prior LPRE work in Germany, the Soviet Union, or the United States, and this group had to tackle many of the basic issues, such as heat-resistant chamber materials, injector design, adequate and safe ignition, or purging propellants. The nozzle was inclined by 18 deg to prevent the accumulation of unburned propellants in the chamber. Lizzy never was used as a JATO. But it provided the background for an upper-stage LPRE for a ground-to-air projectile, which was flown successfully. This sustainer LPRE had a thrust of 454 kg or 1000 lbf for a duration of 25 s. Its propellants were LOX and a fuel consisting of 60% methyl alcohol and 40% water. It was a model from which the LPRE for a subsequent rocket test vehicle (RTV-1 described next) was patterned.

One of the early activities in the United Kingdom after the war was to hot fire a few German LPREs using hydrogen peroxide. This included the firing of a 500-kg thrust German JATO, and it was a good learning experience for U.K. rocket personnel. The firing of a recently developed larger Walter prototype JATO was also planned. It had thrusts of between 1000 and 1500 kg (2220 and 3300 lbf) and used 80% hydrogen peroxide, an aqueous liquid permanganate

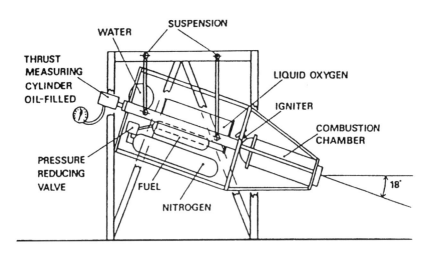

Fig. 12-1 Simplified schematic layout of one version of the first British experimental JATO rocket engine, nicknamed "Lizzy" mounted in a firing test fixture. Copied with permission of the British Interplanetary Society from Ref. 3.

catalyst solution for its decomposition, and gasoline for fuel. This JATO unit had been satisfactorily ground tested many times, and it even had undergone some flight tests before the end of the war. The test in Britain resulted in an unexpected explosion and the death of three people, namely, a certain Dr. Schmidt, a former German designer of the Walter company, and two Englishmen. This accident was reported by Hellmuth Walter.[4] He suspected an improper starting sequence of the three liquids as one potential cause of the accident. Another likely cause was a leak and an accumulation of the hydrogen peroxide in a pocket of the sheet metal engine cover around the JATO; the unclean cover metal caused the peroxide to be catalyzed, and it exploded. The accident caused the British investigators to tighten operational procedures.

The *Rocket Propulsion Establishment*, which was later identified by other names, was set up in 1946 by the Ministry of Supply at Westcott at a former Royal Airforce training base.[5-7] Its mission was to undertake R&D work for the propulsion of British flying rocket vehicles. This government organization actually did a lot of the original LPRE development work with high-strength hydrogen peroxide, which was of 90% concentration. It was tested with a number of different fuels and with compatible materials of construction. They later also investigated LOX/alcohol and LOX/alcohol mixed with hydrazine and water. Later, when Britain worked with the RZ2 LPRE (described later in this chapter) they worked with LOX/kerosene. In the 1960s some exploratory work was done with LOX/LH$_2$, but this work was then redirected toward storable monopropellants and storable hypergolic bipropellants. This included RFNA with mixed amine fuels, NTO, or mixed oxides of nitrogen with hydrazine and MMH. They worked closely with those companies of the British industry, where the LPRE developments were continued and production of several LPREs was undertaken. One of the technology demonstration projects undertaken at Westcott is shown in Fig. 12-2. It depicts an experimental packaged LPRE with a solid-propellant gas-pressurizing system that moves two curved pistons in the propellant tanks. It was a clever design and might have been

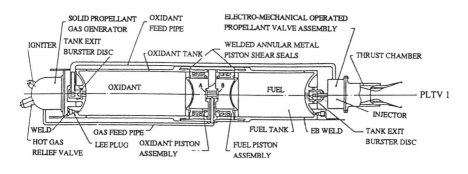

Fig. 12-2 Experimental packaged LPRE with piston expulsion and a solid-propellant pressurization system developed by the U.K.'s Rocket Propulsion Establishment. Copied with permission of the British Interplanetary Society from Ref. 5.

ahead of its time This author visited this U.K. government development and test facility at Westcott in the 1950s.

In 1985 the Royal Ordnance, plc, a commercial company, was formed by the government with personnel from the government establishment at Westcott; several facilities were transferred to the company. The abbreviation "plc" stands for public limited company, which in the United States would be a private commercial company. Some of the Westcott facilities and the relevant technology were transferred to this new company. Royal Ordnance has concentrated on small LPRES with storable bipropellants and has developed, built, and sold new small rocket engines. This is discussed again later in this chapter.

The *Alpha* LPRE was developed at the Westcott government organization and built and flight tested by Vickers Armstrong in the late 1940s. It propelled a flying U.K. experimental vehicle (18 in. diam) aimed at transonic aerodynamic studies.[8] Figure 12-3 shows that the engine was housed in a cylindrical body with an ogive-shaped nose. It had 800 lbf of thrust in a single TC. Propellants were (for this as well as several other U.K. LPREs) 80% hydrogen peroxide and a hypergolic fuel mixture of 57% methyl alcohol, 30% hydrazine hydrate, and 13% water (a copy of the German C-Stoff). The liquid catalyst was an aqueous solution of 17% potassium cuprocyanide, and it was added at 13 cc per liter of fuel. A blowdown system with 4000 psi air was used to expel the propellants. The loaded mass of 234 kg or 471 lb (including all tanks and piping) gave a propellant fraction of 42.5%, which is reasonable for that period of time and for a pressure feed system. The forward end of the Alpha TC had three separate injection elements or nozzles (similar to the Walter TC) with a face shutoff valve using a spring-loaded pintle design. One of the injection elements is shown in Fig. 12-4. It has some similarities to later pintle designs discussed in Chapter 7.9. The first Alpha flight occurred in 1947.

The *Beta* LPRE and its test vehicle were started to support a planned vertical takeoff interceptor.[5,8,9] The TC development was performed by the Rocket Propulsion Establishment at Wescott, and the feed system was developed by

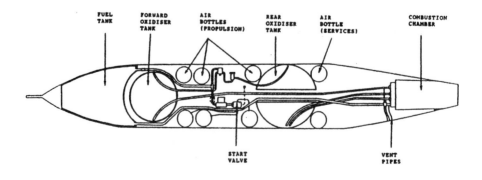

Fig. 12-3 Simplified diagram of the Alpha LPRE. Copied with permission from Ref. 6.

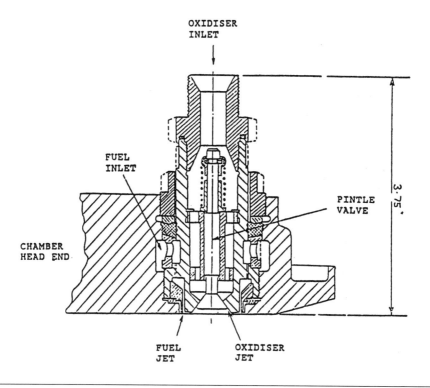

Fig. 12-4 The Alpha LPRE had an advanced pintle injection element. Copied with permission from Ref. 6.

Fairey Aircraft Company.[8] An early version is sketched in Fig. 12-5. Design thrust with two TCs was 1800 lbf, and the nominal chamber pressure was 250 psia. It had a relatively large combustion chamber (L^* of 90 in.) to achieve good performance. The TCs were regeneratively cooled by the oxidizer flowing through spiral cooling passages. Later versions had hinged TCs, one to permit pitch control and the other (with its rotary axis at 90 deg to the first one) for yaw control. This was the first U.K. LPRE to use a hinge-mounted TC and a turbopump (TP); the final version of the TP is sketched in Fig. 12-6. Earlier TP designs featured straight vane pump impellers, and the static seals were plastic diaphragms supported by radial leaf springs. The dynamic seals relied on an impeller creating sufficient centrifugal pressure to balance the pump inlet pressure. The shaft extended beyond the TP's external envelope for a possible auxiliary drive. The TP development had problems with static and dynamic seals, bearing seizures, axial thrust bearings, and failures of diffuser vanes and impeller vanes. After some design modifications the TP was run satisfactorily. When the liquid catalyst of prior engines was replaced by a solid catalyst, it allowed a simplification of the system and eliminated a heavy pressurized tank and some valves. The first beta hydrogen-peroxide GGs used a pellet-type solid catalyst, but it was soon replaced by a silver-plated multiple screen-type catalyst. The first flight was in 1949.

848 History of Liquid Propellant Rocket Engines

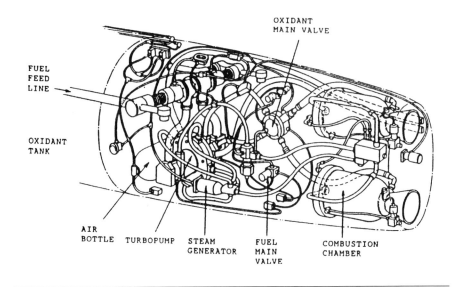

Fig. 12-5 Simplified view of the Beta LPRE with two thrust chambers. This version does not show hinged TCs. Copied with permission from Ref. 6.

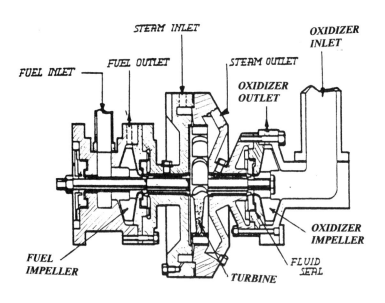

Fig. 12-6 Simplified cross-section sketch of the Beta 1 turbopump designed with suggestions from U. Barske (German pump advisor). Copied with permission from Ref. 6.

RTV-1 (research test vehicle # 1) was small-scale model for testing antiaircraft missile parameters and operation.[8] The sustainer stage of this vehicle was originally intended to use LOX/gasoline, but the fuel was switched to a methyl alcohol/water mixture (60/40), in part to not attenuate the radio signals (which had to travel through the plume) for vehicle control and in part because the gasoline cooled TC had burnouts after 15 s of operation. As shown in Fig. 12-7, it had a nearly spherical combustion chamber, 900 lbf thrust, a regeneratively fuel cooled TC (with film-cooling augmentation at the converging section of the nozzle), dual pyrotechnic igniters, and an expansion provision (bulge in outer wall). The RTV-1 was boosted by an array of solid-propellant rocket motors burning for less than half a second, but this was sufficient to give a vehicle velocity allowing aerodynamic stability. Experimental vehicles flew between 1948 and 1952.

Aircraft Rocket Engines

One of the early JATO flight tests was performed in a modified Avro Lancaster aircraft in 1951 using a German Walter JATO unit at the Wescott airfield of the Rocket Propulsion Establishment.

The De Havilland Engine Company built and flew a series of *Sprite JATO and aircraft-assist LPREs* between 1948 and about 1960 (Refs. 2 and 9). The Sprite 1 had 5000 lb thrust for 15 s with a dry mass of 350 lbm. As seen in Fig. 12-8, it used a liquid permanganate catalyst solution and was basically a monopropellant hydrogen-peroxide rocket engine. It has been called an improved version of a German Walter hydrogen-peroxide monopropellant JATO unit

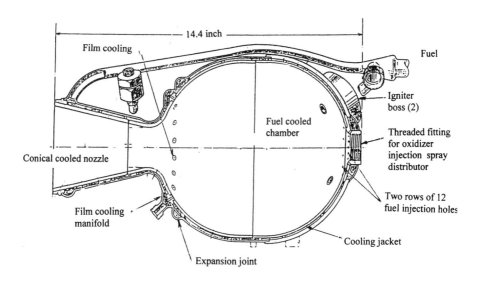

Fig. 12-7 Section of RTV-1 thrust chamber. Copied with permission from Ref. 5.

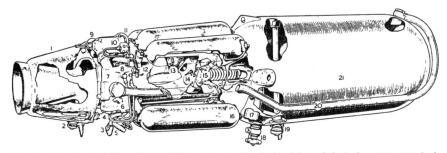

1, Catalyst tank; 2, catalyst filling point; 3, air filling point; 4, catalyst feed to injector (14); 5, air pressure gage; 6, air distributor valve for catalyst; 7, air distributor valve for hydrogen peroxide; 8, starting valve; 9, air-reducing valve; 10, check thrust valve; 11, air feed pipe to catalyst tank (1); 12, air manifold; 13, reaction chamber; 14, catalyst injector; 15, hydrogen peroxide injector; 16, compressed-air bottles, nine in number; 17, hydrogen peroxide collector pipe; 18, hydrogen peroxide dump valve; 19, hydrogen peroxide filling point; 20, air feed pipe to peroxide tank (21); 21, hydrogen peroxide tank.

Fig. 12-8 Partial cutaway drawing of the Sprite 1 JATO monopropellant rocket engine. Courtesy of De Havilland Engine Company, from company release.

discussed in Chapter 9.3. It was flight tested as a JATO with a British Valiant bomber. The Sprite 2 had a catalyst bed with silver-plated nickel screen packs designed for optimum bed loading and bed flow. As mentioned before, the solid catalyst allowed a simpler engine system. Static and flight tests were in 1952. It was used experimentally as a JATO on the Comet airliner with the aim to facilitate takeoff with a heavy load on hot days or at high-altitude airports. This is the only known JATO flight test on a commercial airliner. However it was not put into commercial service.

Improved and uprated Sprite versions were developed with higher thrust, better packaging, and higher performance by adding fuel and making it a bipropellant engine. This fuel had a composition similar to the German C-Stoff and was hypergolic. Figure 12-9 shows two partially sectioned drawings of the Sprite 4. It also used compressed nitrogen gas for tank pressurization, had a regeneratively cooled TC, and a screen-type catalyst bed in the TC. The fuel tank was wrapped around the nozzle.

De Havilland engines then developed the bipropellant Sprite 5 (also known as *Spectre*) LPRE for augmenting fighter aircraft climbing and altitude performance.[2,9,10] It had a turbopump with unusual pump impellers, a low-loss turbine, screen pack catalyst beds for the GG and the TC, and regenerative cooling. Static tests began in 1953, and flight approval tests were completed in 1956. It was flown as an experimental auxiliary engine in a modified Saunders Roe SR 53 interceptor aircraft, which was normally equipped with a turbojet engine for takeoff, slow climb, and cruise or normal flight operations. This LPRE flew 39 times in this aircraft and flew at Mach number of 1.33 and altitudes up to 55,000 ft. In 1958 the aircraft was damaged in a takeoff accident (unrelated to the LPRE) resulting in the death of the test pilot. There was a planned follow-on with the SR177 aircraft. A version of the Spectre LPRE was used to assist with the takeoff of the Valiant and Victor bomber aircraft and

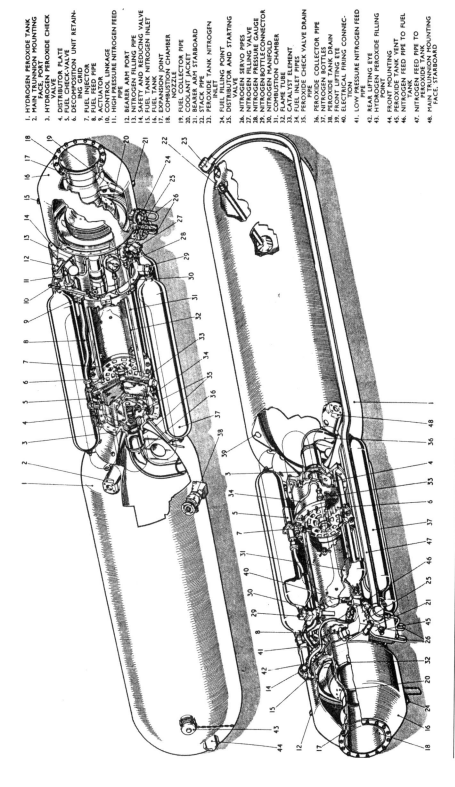

Fig. 12-9 Partial cutaway drawings of the De Havilland Sprite 4 bipropellant JATO rocket engine. The fuel tank surrounds the nozzle of the TC. Courtesy of De Havilland Engine Company, from company release.

were flight tested. Both the Spectre engine and the SR-177 aircraft became victims of a U.K. government decision, and shortly thereafter the programs were canceled.

To create competition in the hydrogen-peroxide LPRE field (not only used for aircraft, but also for launch vehicles), the U.K. government brought additional companies into this business. For example, in the early 1950s Napiers provided such LPREs for missile tests based on Beta engine technology. In 1956 Napiers completed testing of its Scorpion JATO LPRE, which was flight tested on two Canberra military aircraft. Another company was Armstrong Siddeley Motors; they had already started in the LPRE business in 1947 and had developed two JATO engines using LOX as the oxidizer. Their Snarler JATO unit used LOX with water-diluted methyl alcohol, had 900 kg or 1980 lbf thrust, and was flight tested in a Hawker P.1072 aircraft. This company also developed a larger JATO unit called Screamer using LOX/kerosene with a thrust of 3600 kg or 7920 lbf, but it did not fly, and the project was canceled. LOX was found to be impractical for quick response military use. In 1959 this company was reorganized, absorbed the de Havilland rocket engine organization, and became known as Armstrong Siddeley, Ltd. It developed several noteworthy LPREs described later in this chapter. Around 1967 Rolls–Royce, Britain's foremost aircraft engine company, acquired Bristol-Siddeley. The rocket department was reinforced by technical personnel from Rolls–Royce, and they developed the larger LPREs discussed in the following section.

Launch Vehicles and Missiles

The technology and the propellant experience from the Beta and Sprite engines were used by Bristol-Siddeley, the successor to De Havilland, to develop new LPREs for an air-launched missile and a sounding rocket. This company then developed the Gamma Mk 201 engine for the *Black Knight Rocket*, a research vehicle and sounding rocket.[2,9,11,12] This vehicle was conceived in 1955 to give the United Kingdom an experimental capability (in anticipation of a larger ballistic missile) for testing reentry of model warheads, guidance systems, and associated equipment. The Gamma Mk 201 LPRE design was actually started by the predecessor Armstrong Siddeley Motors, but the development was completed, and the engines were built and serviced by Bristol-Siddeley. This engine used hydrogen peroxide and kerosene as propellants and had four hinged TCs of 1857 kg or 4100 lbf thrust each, for a total nominal thrust of about 16,000 lbf. Each TC was regeneratively cooled by the oxidizer and was supplied with propellants from its own turbopump. Figure 12-10 shows a section of this engine. The oxidizer flow was decomposed in a sliver screen catalyst pack, and the resulting hot steam with oxygen gas flowed through the injector into the chamber, where it was mixed and burned with liquid fuel. Visible in this figure are the hinge joints, through which the pressurized propellants flowed to the hinged TC, the turbopump with one roller bearing (for radial loads) and one ball bearing (for radial and axial loads), the purge gas connections, and the valves for the oxidizer, fuel, and GG supply. The

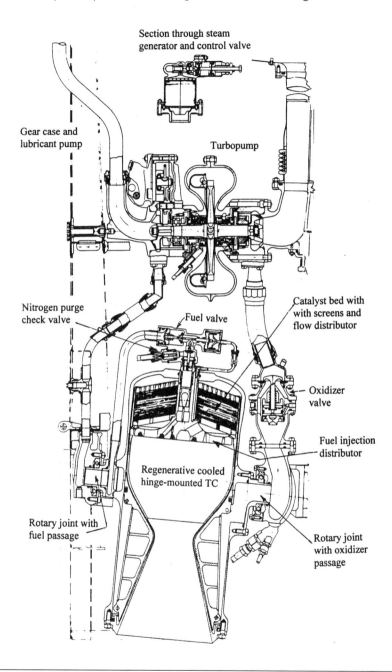

Fig. 12-10 Section through one of four Gamma Mk 201 engines for the Black Knight test vehicle. The thrust chamber is hinge mounted. Copied with permission from Ref. 11.

injector contained passages and injection holes for the fuel (fuel entry through the top) and holes for the decomposed heated oxidizer gas from the catalyst pack. The liquid oxidizer enters into the TC catalyst bed from the top of the cooling jacket.

The Mk 201 LPRE for the Black Knight rocket vehicle was started in an unusual way. A gas-pressurized ground supply of oxidizer entered into the screen catalyst bed of the TC and simultaneously entered into the catalyst bed of the gas generator. This allowed a relatively fast start without using any of the oxidizer in the vehicles own propellant tanks. In the GG the oxidizer was decomposed into steam and hot oxygen, and this hot gas was supplied to the turbine. As the TP accelerated, the pump discharge pressures began to rise, and the chamber pressure also began to rise. When the fuel pressure reached a predetermined value, the fuel valve is automatically opened admitting fuel to the combustion chamber, where it ignites spontaneously with the hot oxidizer decomposition gases. Simple check valves are used for control of the flow direction. The flow of hydrogen peroxide to the GG is then supplied from the oxidizer pump. Once the pump discharge pressure exceeds the pressure of the ground supply, feeding from the ground stops. The TP continues to accelerate, and the pump pressures and the chamber pressure continue to increase rapidly. Once the pressures exceed a predetermined value, restriction orifices slide into place, finding an equilibrium position so as to maintain a steady flow condition at approximately the desired mixture ratio. Once the chamber pressure in all four TCs reaches the rated value, the vehicle hold-down device is released, and the vehicle begins to lift off of its launchpad, and the connections for the pneumatic supply (actuation and purge), oxidizer (pressurized start charge), and electrical (power and control signals) are severed. The Mk201 LPRE did not have an automatic mixture-ratio control, and therefore cutoff usually occurred when one of the propellants ran out and the remainder of the other propellant was at times excessive.

Shortly after starting the work on the Gamma Mk201 LPRE in 1956, Bristol-Siddeley was given the job for the *Stentor engine*, which was developed for the *Blue Steel* standoff guided bomb.[9] This air-to-surface weapon was capable of flying at Mach 1.6 and was originally intended to carry a nuclear warhead. The Blue Steel air-launched missile became a successful supersonic vehicle, and it entered service with the Royal Air Force in 1963 and was decommissioned in 1971. Over 120 Stentor engines were delivered.

Figure 12-11 shows the Stentor engine; its development schedule allowed enough time for several engine innovations.[1,2,9] It had two TCs, one at 24.2 kN or 5500 lbf with variable thrust for cruising and a larger one at 88.9 kN or 20,200 lbf with constant propellant flow for the initial acceleration, namely, for the boost and initial climb maneuvers. It had about 25% more chamber pressure than the Mk201 and a larger nozzle area ratio (60 to 1). The TP was also unusual; it had a single-stage turbine, two oxidizer pumps with spiral inducers (one for the larger TC and one for the small TC), a single fuel pump, an auxiliary gear case, and an oil pump for cooling and lubricating bearings and the small gear case. The large oxidizer pump ran dry when only the small

Fig. 12-11 Stentor LPRE for an air-to-surface missile with a large and a small tubular thrust chamber. Engine height is approximately 5 ft. Copied with permission from Refs. 9 and 11.

TC was operating. It is the only TP known to the author to run dry during engine operation.

The Stentor engine was the first in the United Kingdom to have a mixture-ratio control using a single throttle valve activated by pressure signals from two venturi flowmeters, one in each propellant line. In addition thrust control was provided by throttling the flow of oxidizer to the GG; a chamber pressure signal activated the throttle valve, so that constant chamber pressure was maintained. Starting was at the altitude of the launching bomber aircraft, and the engine start was about 5 s after release from the bomber. This delay prevented the bomber aircraft from being exposed directly to the hot flame of the rocket engine. It was the first British LPRE with a tubular cooling jacket. The tubular concept was basically copied from prior U.S. TCs, except that welding was used to join the tubes together instead of brazing. It was a relatively complex and high-performance LPRE. Components and technology from the Stentor engine were used later in the development of the Gamma Mk301 engine.

In 1958 work started on a more sophisticated, more controllable and somewhat larger Bristol-Siddeley LPRE, called Mk301. It was an upgraded version of the Mk201 and was used for an enhanced version of the same application, namely, the Black Knight.[1,9,11–13] Shown in Fig. 12-12, it also had four hinge-mounted thrust chambers, whose design was based on the Stentor engine. The flow of oxidizer coolant was in opposite directions in adjacent tubes. This

TC is shown in Fig. 12-13. Its injector had been improved to allow a modest performance increase when compared to the Mk201 engine. The engine thrust could be adjusted between 17,300 and 21,800 lbf depending on the flight need. At maximum thrust the chamber pressure was 4.1 MPa or about 595

Fig. 12-12 Gamma Mk301 LPRE with four thrust chambers in an assembly fixture. Copied with permission from Refs. 11 and 12.

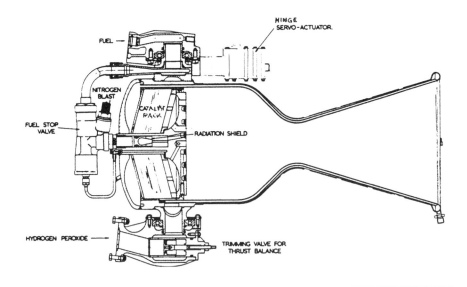

Fig. 12-13 Section of one of the four thrust chambers of the Gamma Mk301 rocket engine. Copied with permission from Ref. 11.

psi. This Gamma 301 had a larger turbopump feeding all four thrust chambers and also had a simpler control of the propellant mixture ratio. Changes to the design included a large nozzle area ratio (60 to 1) for better altitude performance (higher specific impulse and higher thrust), features to improve life, automatic mixture-ratio control, and transistorized engine controls.

Between 1958 and 1965 a total of 22 flights of Black Knight were made out of the Woomera launch range in Australia, and the LPREs worked well during these flights. Of these 14 flights were with the Mk201 LPRE and eight were with the Mk301 LPRE. Some of these were two-stage vehicles with a solid-propellant motor for the other stage. The test flights permitted the investigation of various atmosphere reentry phenomena and various design features for a bigger vehicle, the Blue Streak; it is described later in this chapter.

In 1953 Rolls–Royce, the automobile and aircraft engine company, started to work on LPREs and became the most important supplier of LPREs for British missiles and SLVs. Rolls–Royce inherited the experience and key rocket personnel from Bristol-Siddeley and its predecessors. Rolls–Royce developed the LPREs for the first and second stages for the *Black Arrow* space launch vehicle, which had been considered by some people to be an uprated version of Black Knight. It used a Rolls–Royce Gamma 8 engine (with eight TCs) in the first stage, a Rolls–Royce Gamma 2 engine (with two TCs) in the second stage, both running with high-strength hydrogen peroxide (90%) and kerosene.[1,9,14,15,*] The Gamma 8 LPRE can be seen in Fig. 12-14. The eight TCS are hinge mounted in a cruciform pattern, and two adjacent TCs are mechanically linked together, so that they will move or rotate on their hinges in unison. This feature provides full pitch, yaw, and roll control of the powered flight. The regeneratively cooled thrust chambers are essentially identical, except that the second-stage TCs had a radiation-cooled nozzle extension with a large nozzle-exit-area ratio of 350 to 1. A new solid-propellant motor provided the propulsion for the third stage, which also had a cold-nitrogen-gas attitude control system. The two TPs supplied eight TCs (four TCs for each TP) in the first stage, and each regeneratively cooled TC gives about 7000 lbf (or about 31 kN) thrust each. The second-stage engine assembly was called Gamma 2. It is shown in Fig. 12-15. Its two TCs were gimbaled, produced about 15,600 lbf or 70 kN of total thrust and were supplied from a single TP.[11,12] The mixture ratio was controlled to achieve oxidizer exhaustion, and the nominal mixture ratio was 8.2:1 for both stages. Black Arrow launches started in 1968, and four vehicles were launched. The last one successfully placed a satellite into orbit on 28 October 1971. With this satellite, named Prospero, the United Kingdom became the sixth nation to have successfully launched a payload into an Earth orbit. It was the first and only British satellite to be launched by a British launch vehicle. Ironically the launch took place approximately three months after the program was canceled.

---

*Personal communications, V. Cleaver, Rolls–Royce, 1955–1958, and D. Millard, Senior curator, Science Museum, London, 2004.

Fig. 12-14 View of the Rolls–Royce Gamma 8 engine showing four sets of coupled dual-hinged thrust chambers. Two turbopumps supply propellants to the eight thrust chambers. Author's photo from Science Museum display, London, 2004. Courtesy of Rolls–Royce.

The British government decided around 1954 to build and deploy a medium-range ballistic missile to counteract the perceived threat of the Soviet long-range surface-to-surface missile. It was called the *Blue Streak*, and Rolls–Royce was given the job of providing the LPRE for it. The U.S. government agreed to help. Accordingly in 1955 Rolls–Royce obtained a license from Rocketdyne to acquire the then best available technology of large LPREs using LOX and RP-1, which was a type of kerosene. This was the technology of one version of the Thor LPRE described in Chapter 7.8. The nozzle-exit sections were then conical with a 15-deg half-angle. This author remembers making trips on behalf of Rocketdyne to Rolls–Royce in Derby, England, in connection with technical discussions related to this agreement.

Rolls–Royce integrated two of the modified Thor rocket engines together for the first stage of Blue Streak and identified the engine as RZ2.[16,17,*] To put two engines together, Rolls–Royce changed the location of the TP relative to the TC and modified the turbine exhaust ducts and heat exchangers. The

---

*Personal communications, V. Cleaver, Rolls–Royce, 1955–1958.

Liquid Propellant Rocket Engines in the United Kingdom or Britain   859

Fig. 12-15 View of the Gamma 2 engine developed by Rolls–Royce. A central TP supplies two gimbaled TCs. Author's photo taken at Science Museum display, London, 2004. Courtesy of Rolls–Royce.

rocket engine for the Blue Streak is shown in Fig. 12-16, and the Blue Streak first stage is shown in Fig. 12-17. It had two tubular TCs with conical nozzle exits, each with its own geared turbopump feed system using a gas-generator engine cycle. The TC of this missile is shown in detail in Fig. 4.3-5, and the TP is essentially identical to the one shown in Fig. 4.4-2. The GG in Fig. 12-18 had a close resemblance to the original Thor GG and generated a flow with reasonably uniform gas composition and gas temperature. The total takeoff thrust was 1350 kN or more than 300,000 lbf, which makes it the highest-thrust LPRE in Britain. Propellants were LOX/kerosene. About 24 engine sets were built, and there were 11 satisfactory flights of this engine in the booster stage with various upper-stage configurations. In one of the early flights, there was excessive sloshing of the propellants in the tanks of the Blue Streak stage, but this was remedied by additional baffles.[1]

In 1960 the British government canceled the Blue Streak program as a military missile, relying instead on U.S. Thor and Polaris missiles, which were being deployed in Britain at that time. However Blue Streak continued its life

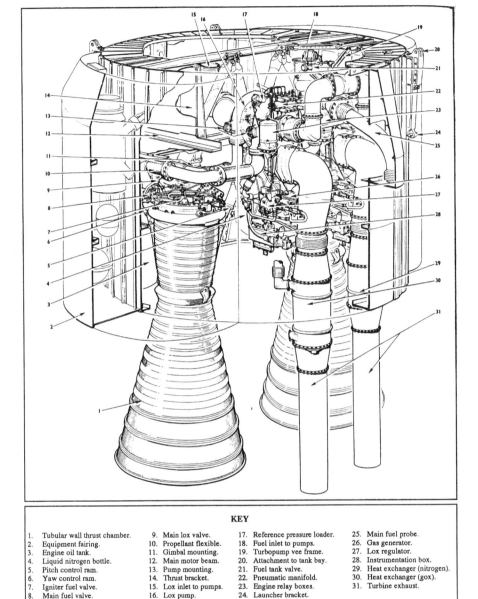

Fig. 12-16 Blue Streak RZ2 engine using two LPREs with gimbaled thrust chambers running on liquid oxygen and kerosene fuel. Courtesy of Rolls-Royce (from personal communications, V. Cleaver, Rolls-Royce, circa 1958).

KEY

1. Tubular wall thrust chamber.
2. Equipment fairing.
3. Engine oil tank.
4. Liquid nitrogen bottle.
5. Pitch control ram.
6. Yaw control ram.
7. Igniter fuel valve.
8. Main fuel valve.
9. Main lox valve.
10. Propellant flexible.
11. Gimbal mounting.
12. Main motor beam.
13. Pump mounting.
14. Thrust bracket.
15. Lox inlet to pumps.
16. Lox pump.
17. Reference pressure loader.
18. Fuel inlet to pumps.
19. Turbopump vee frame.
20. Attachment to tank bay.
21. Fuel tank valve.
22. Pneumatic manifold.
23. Engine relay boxes.
24. Launcher bracket.
25. Main fuel probe.
26. Gas generator.
27. Lox regulator.
28. Instrumentation box.
29. Heat exchanger (nitrogen).
30. Heat exchanger (gox).
31. Turbine exhaust.

Fig. 12-17 Blue Streak booster stage with the RZ2 engine. Courtesy of Roll-Royce (from personal communications, V. Cleaver, Rolls-Royce, circa 1958).

for a few more years as the first or booster stage for an Europa SLV, which at the time was supported by several European countries. The second stage of the Europa was based on the French Coralie rocket vehicle, and the third had a new German Astris rocket engine. They both had gas-pressurized feed systems and are described in Chapters 9 and 10, respectively. These two chapters also contain discussions about the Europa SLV.

The Blue Streak first stage generally performed very well as a booster for 11 launches (1964 to 1971) of the Europa SLV, but there were several failures of the French and German LPREs in upper stages and in other vehicle subsystems. None of the 11 Europa flights was really completely successful. There were many discussions and negotiations with other European countries about continuing a joint European effort using the Blue Streak as a first stage, but they did not result in an agreement. With the lack of a successful three-stage Europa flight, the failure to reach an agreement, and after realizing how much the cost would be, Britain decided not to be in the space launch field. Further negotiations for Britain to participate in a European space launch vehicle stopped. The Blue Streak program and related efforts were terminated in 1971. The work on LPREs in Britain diminished drastically, and the capable LPRE development organization at Rolls-Royce was disbanded.

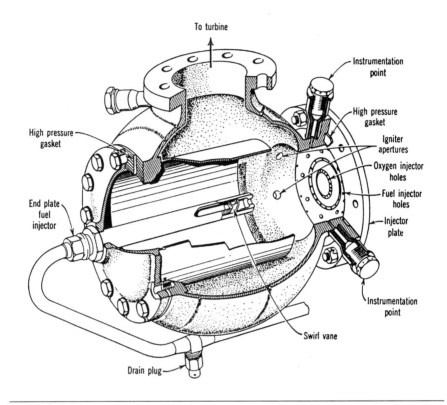

Fig. 12-18 Gas generator of the Blue Streak booster engine. Courtesy of Rolls–Royce; Copied from Ref. 18.

Small Thrusters

Some good work on small LPREs for reaction control systems did continue in Britain since the demise of the Blue Streak. Today the Royal Ordnance, plc, a 1985 commercial offshoot of the Westcott government rocket organization, is the only LPRE company in the United Kingdom. It still builds small thrusters and occasionally also a LPRE system for attitude control and space trajectory maneuvers for European satellites.[5] This small company is a leader in using pure hydrazine with NTO (or NTO with NO or nitrogen oxide additive) in bipropellant thrusters. Hydrazine fuel gives a slightly higher specific impulse and propellant density than the customary fuels, such as UDMH or MMH. This extra margin is needed for demanding spaceflight missions. These thrusters from Royal Ordnance are made of niobium with a disilicide internal coating as oxidation protection and with titanium injectors. For example, the development of the Leros 1 thruster began in 1986. It is an apogee engine with 500 N or 114 lbf thrust using mixed oxides of nitrogen (which is N_2O_4 with some NO) and hydrazine. Leros 20 is a small attitude control thruster at 22 N thrust with MMH as a fuel.

This Royal Ordnance company was acquired by ARC of the United States in 1997, and the work on small thrusters has continued in Britain.* This acquisition is discussed in Chapter 7.11. The acquisition has allowed Royal Ordnance to build and sell other ARC products in Europe, such as hydrazine monopropellant thrusters or positive expulsion propellant tanks.

Concluding Remarks

The United Kingdom of Britain started their LPRE developments with JATO units and experimental flight test vehicles, which helped to build their expertise. They learned from the Germans and adopted technology from the Hellmuth Walter Company's use of concentrated or high strength hydrogen peroxide (80%) with a hypergolic fuel (German C-Stoff). They used this oxidizer in most of their engines, such as those of the Alpha, Beta, Sprite, Black Knight, Black Arrow vehicles. The last two of these test vehicles were used to explore technical features of future MRBMs and SLVs. Britain became the world leader in the application of high strength hydrogen peroxide and they improved the performance by going to kerosene as a fuel or going in some cases to 85% hydrogen peroxide. When Britain decided to develop a medium-range ballistic missile (MRBM) in the mid-1950s, they obtained engine technology from Rocketdyne (U.S.) and developed a reliable dual engine (LOX/kerosene) with 300,000 lbf thrust for the booster stage of its new Blue Streak MRBM. In 1960 the British government decided to cancel further development of the MRBM, but they approved continued work on the Blue Streak booster for an SLV and space exploration. The booster performed well in all 11 launches of the Europa space launch vehicle, but the vehicle did not achieve its flight objectives in any of these flights. In 1971 the British government canceled the Blue Streak program and with it they lost most of their LPRE capability. Limited LPRE work continued on a few small thrusters for satellites.

References

[1] *British Rockets and Missiles, Part I, Space List Number 14*, Rocket Services, Wareham, Dorset, England, U.K., 2001.

[2] Stokes, P. R., "Hydrogen Peroxide for Power and Propulsion," paper presented and distributed at the Science Museum, London, 14 Jan. 1998.

[3] Griffiths, J., "Lizzy—The First British Liquid Propellant Rocket Motor," *Journal of the British Interplanetary Society*, Vol. 38, Dec. 1985, pp. 390–394.

[4] Walter, H., "Experience with the Application of Hydrogen Peroxide for Production of Power," *Jet Propulsion*, May–June 1954, pp. 166–171.

[5] Klepping, A. H., "Liquid Space Engine Development at Royal Ordnance Rocket Motors Division," *Journal of the British Interplanetary Society*, Vol. 44, May 1991, pp. 195–201.

[6] Maxwell, W. R., "A Note on the History of the Westcott Establishment, 1946-1977," *Journal of the British Interplanetary Society*, Vol. 46, 1993, pp. 286–288.

---

*Personal communications, M. Marin, ARC, Niagara Falls, NY, 2002–2004.

[7]Rothmund, C., and Harlow, J., "A History of European Liquid Propellant Rocket Engines for Aircraft," AIAA Paper 99-2901, June 1999.

[8]Harlow, J., "Alpha, Beta, and RTV-1: The Development of Early British Liquid Propellant Rocket Engines," *History of Rocketry and Astronautics*, American Astronautical Society History Series, Vol. 22, Univelt, San Diego, CA, 1998, Chap. 8, pp. 173–193.

[9]Millard, D., *The Black Arrow Rocket, a History of a Satellite Launch Vehicle and Its Engines*, NMSI Trading, Ltd., London, 2004.

[10]Watts, R. V., "Throttling of the Spectre Engine," *Rocket Propulsion Technology*, Plenum, New York, 1961, pp. 63–78.

[11]Andrews, D., and Sunley, H., "The Gamma Rocket Engines for Black Knight," *Journal of the British Interplanetary Society*, Vol. 43, 1990, pp. 301–310.

[12]"The British Black Knight Rocket," *History of Rocketry and Astronautics*, edited by J. Backlake, American Astronautical Society History Series, Vol. 17, Univelt, San Diego, CA, 1995.

[13]Robinson, H. G. R., "Black Knight Control," *Rocket Propulsion Technology*, Plenum, New York, 1961, pp. 49–62.

[14]Gould, R. D., and Harlow, J., "Black Arrow: the First British Satellite Launcher," *History of Rocketry and Astronautics*, American Astronautical Society History Series, Vol. 20, Univelt, San Diego, CA, 1997, Chap. 8, pp. 257–270.

[15]Andrews, D., "Gamma Type 2 Rocket Engine for Black Arrow," *Spaceflight*, Vol. 10 1969, pp. 390–394.

[16]Cleaver, A. V., "The Blue Streak Propulsion System," European Symposium on Space Technology, June 1961.

[17]Samson, D. R. (ed.), *Development of the Blue Streak Satellite Launcher, Proceedings of the Second Space Engineering Symposium*, Pergamon, New York, 1963.

[18]Sutton, G. P., *Rocket Propulsion Elements*, 3rd ed. Wiley, New York, 1963, p. 186.

Chapter 13

Liquid Propellant Rocket Engines in the People's Republic of China

The Chinese LPREs were developed more quickly than in most of the other nations, and they are described in this chapter. This country has six centuries of background in solid-propellant rocket motors, but has been working intensively in large LPREs for only about 50 years. The author found no information about earlier efforts, which might have preceded the intensive work on large LPREs in the 1950s. The Chinese program to develop rocket vehicles and thus also LPREs for military weapons and space applications was focused in 1956 under the leadership of Hsuh Shen Tsien, who had just returned from the United States to China a year before. He was an outstanding graduate student of Theodore von Kármán, who was the chairman of the Aeronautical Engineering Department at the California Institute of Technology in Pasadena, California.[1] Tsien became a very young professor and participated in a few classified U.S. R&D programs. This author, as a graduate student, met Tsien in the 1940s and had several student–faculty interactions with him. In 1950 Tsien attracted the unfavorable attention of the U.S. authorities when he refused to testify in a perjury case involving a colleague with alleged communist sympathies. The consequence was the withdrawal of Tsien's U.S. security clearance. The pleading by von Kármán and several other prominent people was not effective to appeal the government decision. So when Tsien saw little future for himself in the United States, he decided to go home to China, but he was arrested, searched, and detained by U.S. authorities for two weeks.[1] He tried to take 1800 lb of papers and notes with him; these were searched, but no classified data were found. He then was prohibited from leaving America for five years. In 1955 he returned to China and within a year he became the leading scientist and top official in China's aerospace programs.

In the initial period of rocket development, the Chinese received considerable help from the Soviet Union, namely, design information, missile hardware, and advisory technical personnel.[2] The Chinese even built a duplicate of a Soviet ballistic missile, the R-2. However this assistance was stopped in 1960 when the relationship between these two countries deteriorated and the

Soviets withdrew more than 1000 advisors. During the 1950s, with the help of the Soviet Union the Chinese had duplicated the Soviet R-2 missile, which used LOX/alcohol propellants. They produced R-2s and launched several. This experience helped China to develop their own missiles beginning in the 1950s. By 1960 they had built their first own medium range ballistic missile (MRBM) and launched it.[2-4] It was called the Dong Feng 1 (DF-1), which is translated as East Wind 1. In 1960 official announcements were made of the launch of a sounding rocket and of an experimental version of an MRBM. More complex and larger ballistic missiles followed. The DF-2 (940-mile range) appeared in 1964, the DF-3 (2200-mile range) in December of 1966, the DF-4 (5000-mile range) in 1970, and the DF-5 (range of 7500 miles) in 1980.[2]

The Chinese, being late entrants into the LPRE field, did not have to struggle with some key issues faced by other countries that started earlier. For example, they learned from the literature, exhibits, technical meetings, patents, and news releases that LPREs and their components could be developed and would fly. With a desire to build up an indigenous capability and with a high government priority, they were able to develop the technology for LPREs rather quickly. They learned how to develop large and small LPREs with storable propellants and later cryogenic propellants They solved problems on bearings or seals of their turbopumps, restarts in zero gravity and at low temperature, on accurate propellant filling, thrust vector control, thrust magnitude control, mixture-ratio control, and combustion stability. The Chinese have been open about some of their failures and technical problems.[2] Some are described later. Historically it appears that the Chinese made no big mistakes with their LPREs and did not develop or produce LPREs that had to be canceled, as happened in other countries.

The China Academy of Launch Vehicle Technology (CALT, formerly the Beijing Wan Yuan Industry Corporation) near the town of Nan Yuan (15 km south of Beijing) develops and builds the cryogenic engines. The Shanxi Liquid Rocket Engine Company works on the storable propellant engines. Static firing tests are conducted at the Beijing Rocket Test Center 50 km west of the capitol.

Chinese Liquid Propellant Rocket Engines

Table 13-1 shows some data on several large LPREs developed in the People's Republic of China.[2,5-8]* In this table The SLVs are identified as Long March or LM and the engines as YF, which is for Yei-ti Fadong-ji or liquid-state engine. All large Chinese engines seem to use a gas-generator engine cycle.

The first two engines in the table are the oldest, and they used IRFNA/ UDMH or inhibited red-fuming nitric acid and unsymmetrical dimethyl hydrazine. The next two engines in the table and most small attitude control engines and

---

*Personal communications about China's LPREs came from J. Morehart, The Aerospace Corporation, M. Coleman and T. Moore, Chemical Propulsion Information Agency, M. Gu, Shanghai Bureau of Astronautics, and F. Winter, Smithsonian Institute, 2001.

Table 13-1 Some Parameters for Several Chinese LPREs

LPRE designation	Application Military	Application Space missions	Stage	Oxidizer	Fuel	Thrust, kN	Thrust, klbf	Specific impulse, s	No. of TCs	First flown as a SLV
YF-2	IRBM	LM-1	First	IRFNA	UDMH	4 × 255 SL	4 × 57.3 SL	241 SL	4	1970
YF-3	IRBM	LM-1	Second	IRFNA	UDMH	294 vac	66 vac	287 vac	1	1970
YF-21	ICBM	LM-2-3	First	NTO	UDMH	4 × 742 SL	4 × 167 SL	286 vac	4	1974
YF-22*	ICBM	LM-2-3	Second	NTO	UDMH	742 vac	167 vac	298 vac	1	1974
YF-20	—	LM-4	Booster	LOX	LH_2	4 × 790 SL	4 × 166 SL	NA	4	2003
YF-73	—	LM-3	Third	LOX	LH_2	44.1 vac	9.9 vac	420 vac	4	1984
YF-75	—	LM-3A/3B	Third	LOX	LH_2	2 × 78.4 vac	2 × 17.6 vac	442 vac	1	1994

Some engines were built in two or more models and the performance data listed here may have different values in some models.
*Thrust includes the YF-23 vernier rocket engine at 47 kN vac (10,600 lbf) with 4 TCs.
Data taken from Refs. 2,4,5,8–10.

orbit maneuver engines use NTO/UDMH; NTO does not deteriorate during storage. The last two engines use LOX/LH$_2$ propellant because a high performance is needed in the third stage space launch application to achieve the desired orbit payloads. The first two engines YF-2 and YF-3 were developed for China's IRBM, and this vehicle (historically with the first large truly Chinese LPREs) was then adapted into their first SLV the Long March 1 (LM-1) with a 300-kg low-Earth-Orbit (LEO) payload. The YF-2 and YF-3 use an injector with cooled baffles and spray injection nozzles similar to the one described for the YF-1 below. The thrust chamber and nozzle are cooled and use a corrugated intermediate metal sheet between the outer and inner walls, similar to the Soviet design, which was originally suggested by Alexei Isayev (Chapter 8.6 and Fig. 4.3-4). The turbine of the turbopump is placed between the two propellant pumps. The rotation of the turbine is started by a solid propellant start cartridge. The tanks are pressurized by inert nitrogen gas which has been heated in a heat exchanger by turbine exhaust gas. The forces needed to control the flight during this first stage operation are obtained from rotating graphite jet vanes placed in the nozzle exhaust similar to the Soviet R-2 missile, which had been copied by the Chinese. The same basic engine is used in a cluster of four in the booster stage and in a single engine the second stage, but the latter has a larger nozzle-exit-area ratio (48:1). The oxidizer tank was pressurized by warmed gasified NTO, which had been heated in a heat exchanger by turbine exhaust gas. The fuel tank is pressurized by gas tapped off the fuel rich gas from the turbine exhaust duct and this gas flow is cooled by propellant.[8] Cavitating venturis and calibration orifices are used to control the propellant flow to the gas generator and they control the engine thrust and the GG mixture ratio and gas temperature.[8] It first was launched for a military application in 1960 and as a space launch vehicle 10 years later.[2] After two successful launches (1970 and 1971) this LM-1 SLV was superseded by a larger SLV with a larger payload and with larger LPREs.[3-5]

Severe combustion vibrations were encountered in the YF-1 engine, which seems to have been the predecessor of the YF-2 and is similar to the YF-2. A number of different approaches were tried, and the problem was eventually remedied.[9] The major changes were these:

1) The injector elements and the distribution of propellants over the face of the flat injector were modified. The spray elements of this injector are shown in Fig. 13-1. There were three types of injection elements: a) coaxial spray and swirl elements with tangentially directed inlet holes, with two versions, high flow (larger holes) and low flow; b) a division-type injector element, which had a conical fuel spray induced by a spiral swirler and a set of inclined impinging holes for the injection of the oxidizer. Typical development changes in the distribution patterns of these elements are shown in Fig. 13-2. These changes improved the combustion stability, but there were still incidences of instability.

2) The propellants were changed from IRFNA with 27% NTO and MA-50 fuel (mixture of 50% triethylamine and 50% triethyl benzamine) to IRFNA with 20% NTO and UDMH. The ignition delay was much shorter with

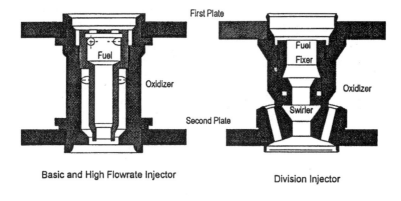

Fig. 13-1 Cross sections of two types of injection elements used in the YF-1 injector. Copied with AIAA permission from Ref. 9.

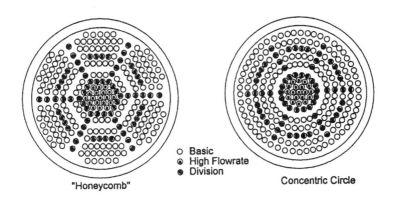

Fig. 13-2 YF-1 experimental patterns and distributions of injector elements across the injector face. Copied with AIAA permission from Ref. 9.

the UDMH, but there were still combustion stability problems during transients.

3) Cooled baffles were added. The final configuration shown in Fig. 13-3 was satisfactory under all operating conditions.

There was an orderly progression in the development of the engines for the LM-2, LM-3, and LM-4 SLVs with some variations of each of these models. The engines for the first and second stage of these SLV are essentially similar, but differ in nozzle exit area ratio.[3-7] The YF-20 LPRE is the backbone of the Chinese long-range launch vehicles, and a cluster of four (used in the first stage) is designated as the YF-21 engine. The four first stage engines were hinge mounted and could be rotated by ± 10 degrees. The nozzles each had a nozzle exit area ratio of 12.69.[8] The YF-22 for the second vehicle stage is a high altitude version of the YF-20 with a larger nozzle-exit diameter (area ratio

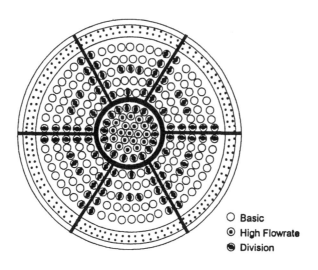

○ Basic
◉ High Flowrate
◉ Division

Fig. 13-3 Final version of YF-1 injector with a seven-compartment baffle. Copied with AIAA permission from Ref. 9.

in one version was 24.2:1) and an altitude start capability.[5–8] In order to properly cool the upper stage engine, which appears to have a higher chamber pressure than the YF-21, the cooling jacket uses an inner wall with a milled channel design, which is brazed and welded to the outer wall. The same geared turbopump, injector, controls, or principal types of valves are used in three versions of the same engine, namely, the YF 20, YF-22, and strap-on boosters (mentioned later). A flow sheet is seen in Fig. 13-4. The second stage with the fixed YF-22 also has four hinged smaller vernier TCs with a combined thrust of approximately 46 kN (10,360 lbf) using the same propellants; the vernier engine has its own TP and GG and has been designated as the YF-23 engine. The hinge-mounted vernier TCs could swing through +/− 60 deg. The vernier TC's nozzle exit section was radiation cooled and made of niobium with a protective coating on the inside surface. The Chinese also installed a propellant utilization system on the main engine and it is intended to minimize the amount of residual propellant remaining in the tanks and engine after engine cut-off. The control for changing the mixture ratio during flight for this propellant utilization system is obtained by a throttling valve in a by-pass line around the fuel pump.

These first and second stages form the core for a series of Chinese SLVs with different durations, different tank lengths, and different third stages. The first and second stage and the YF-20 and YF 22 engines were originally designed for the Chinese ICBM, and these were later adapted for a series of SLVs. The first SLV flight with these engines was in 1974. For heavy payloads, four strap-on boosters each with a TF-20 type LPRE are added. Four of these strap-on boosters first flew on the LM-3B SLV in 1990. There seem to be pro-

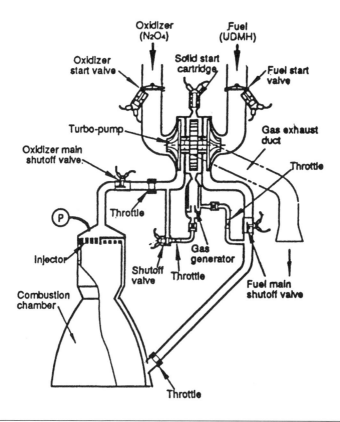

Fig. 13-4 Simplified flow diagram of the YF-20 or YF-22 LPREs. The call-out "throttle" in four places means a calibrated throttling orifice. Courtesy of China Academy of Launch Vehicle Technology (CALT), obtained through personal communications, J. Morehart, The Aerospace Corporation, M. Coleman and T. Moore, Chemical Propulsion Information Agency, M. Gu, Shanghai Bureau of Astronautics, and F. Winter, Smithsonian Institute, 2001).

visions for the pressurization of the propellant tanks, by evaporation and heating of propellants.

The YF-20 experienced incidences of combustion instability during the early development phases.[9] Vibration measurements as high as 600 g were recorded at the injector. The remedy was again a change of the propellant flow density distribution over the face area of the injector by the judicious placement of different types of injector elements with different flows and by installing cooled baffles.

According to Ref. 4, a "hot separation" is used in separating the first and second stages in flight. This would result in a somewhat more efficient flight trajectory compared to one with a conventional pause during separation. In a customary stage separation it takes some time between the termination of the first stage and the starting of the second stage in order to have a reasonable separation distance between the two stages. Otherwise the high-velocity exhaust gas from the upper-stage engine can cause not only damage to the lower expended stage, but the potential blowback of the hot fast gas could also damage the upper stage. During this coast period, the upward velocity of the vehicle is diminished by the pull of Earth's gravity. In the Chinese hot separation there is essentially no time interval with zero thrust. When the thrust of the first-stage engine reaches a reduced level during the normal engine shutdown, the second-stage engine is started, and the explosive bolts, holding the two stages together, are activated or fired. The exhaust gas of the second-stage engine pushes on the dome of the forward propellant tank of the lower stage and in effect pushes the two stages apart. The interstage structure has vents that prevent an excessive buildup of pressure of the exhaust gas inside of the interstage structure. This event has to happen fast enough so that there is no heat damage to parts of the second stage.

Third Stage and LOX/LH$_2$ LPREs

The two LOX/LH$_2$ engines in Table 13-1 are for different versions of a third stage of a SLV. The YF-73 was developed first and has a single TP (its bearings are cooled by LH$_2$) and four hinged TCs.[2,10,11] The engine is shown in Fig. 13-5, and a flow diagram of it is in Fig. 13-6. Cavitating venturi sections are used in four places to control propellant flows. The TCs are hinge mounted and allow full flight-path control. The YF-73 used pyrotechnic igniters in the TC and the GG, and the design provided for restart. The nozzle area ratio of 40 is relatively small for an upper-stage engine. A portion of the high-pressure hydrogen and the oxygen flow is sent through heat exchangers in the hot turbine exhaust duct, and these gases are used to pressurize the propellant tanks. Engine development started in 1976. The thrust might have been too low for some of the missions. The YF-73 engine apparently was the cause for three early flight malfunctions, when the second burn and restart of the engine was flawed.[2]

The YF-75 came about 10 years later. Two gimbal-mounted engines are used in the third stage of LM-3A and LM-3B to allow flight-path control of this stage during powered flight. Each LPRE has two separate turbopumps, one for the fuel and the other for the oxidizer, and the thrust was much higher than the earlier YF-73 LOX/LH$_2$ engine.[3,11] It was based on the experience gathered with the YF-73. A flow diagram of the YF-75 is shown in Fig. 13-7, and an external view in Fig. 13-8. Engine development was started in 1986. The thrust chamber uses a milled channel cooling-jacket design, and the inner wall is made of a copper alloy with some zirconium added for strength at high temperature. The nozzle has an area ratio of 80, and the nozzle extension has rectangular cooling passages, which are bent in spiral shapes and are welded together. The fuel-rich gas from the GG drives the two turbines in series. The YF-75 includes some special technical features, such as the cooling of gear cases with hydrogen, controlling mixture ratio in step changes, using high-pressure hydrogen for actuating the gimbal motions, or a restart capability. The flow diagram indicates that cooling and lubricating of one of the bearings

Fig. 13-5 View of YF-73 LPRE with four hinged TCs and a single TP. It is used on the third stage for LM-3. Courtesy of CALT (from personal communications, J. Morehart, The Aerospace Corporation, M. Coleman and T. Moore, Chemical Propulsion Information Agency, M. Gu, Shanghai Bureau of Astronautics, and F. Winter, Smithsonsan Institute, 2001).

near the turbine of the oxygen TP was done with gaseous hydrogen, and this may have been different. The state of the art of the technology of this YF-75 LPRE seems to be advanced.

Little information was available on the optional storable-propellant engine identified as YF-40 for an optional third stage of LM-4.[3–5] The total vacuum thrust is given as 98.1 kN or 22,100 lbf; it uses NTO/UDMH and has a vacuum specific impulse of 303 s. There are two TCs, and the engines operate with a gas-generator cycle. This third stage was the first in China with a common bulkhead between the fuel and the oxidizer tanks, and this allowed a shorter vehicle length.

A separate hydrazine monopropellant engine with multiple thrusters using a gas-pressurized feed system is also part of some version of the third stage.[3,4] It has two 300-N (67-lbf) thrusters, two at 45 N (10 lbf), and 12 more small attitude control thrusters at a smaller thrust level. It is used for propellant management

874 History of Liquid Propellant Rocket Engines

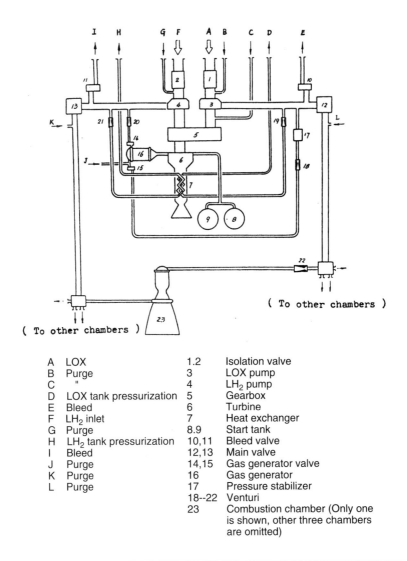

A	LOX	1,2	Isolation valve
B	Purge	3	LOX pump
C	"	4	LH$_2$ pump
D	LOX tank pressurization	5	Gearbox
E	Bleed	6	Turbine
F	LH$_2$ inlet	7	Heat exchanger
G	Purge	8,9	Start tank
H	LH$_2$ tank pressurization	10,11	Bleed valve
I	Bleed	12,13	Main valve
J	Purge	14,15	Gas generator valve
K	Purge	16	Gas generator
L	Purge	17	Pressure stabilizer
		18--22	Venturi
		23	Combustion chamber (Only one is shown, other three chambers are omitted)

Fig. 13-6 Simplified flow diagram of the YF-73 LPRE. Only one of the four TCs is shown here. Courtesy of CALT, (from personal communications, J. Morehart, Aerospace Corporation, M. Coleman and T. Moore, Chemical Propulsion Information Agency, M. Gu, Shanghai Bureau of Astronautics, and F. Winter, Smithsonsan Institute, 2001).

or positioning of the propellants in the tanks after second-stage shutoff and prior to third stage start, for payload orientation after third-stage cutoff, and also for three-axes attitude control maneuvers during that portion of the flight when the third-stage engine is not operating. Little information was available on these small thrusters or their engine systems.

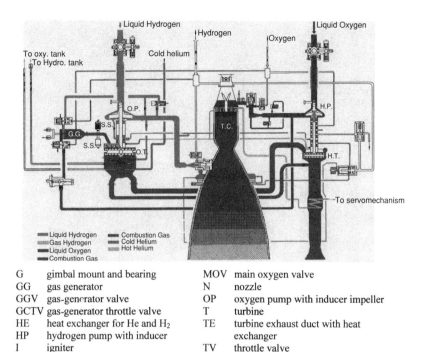

G	gimbal mount and bearing	MOV	main oxygen valve
GG	gas generator	N	nozzle
GGV	gas-generator valve	OP	oxygen pump with inducer impeller
GCTV	gas-generator throttle valve	T	turbine
HE	heat exchanger for He and H_2	TE	turbine exhaust duct with heat exchanger
HP	hydrogen pump with inducer		
I	igniter	TV	throttle valve
IV	inlet valve	V	venturi flow section
MFV	main fuel valve		

Fig. 13-7 Simplified flow diagram of the YF-75 LPRE. Courtesy of CALT (annotated by the author from personal communications, J. Morehart, The Aerospace Corporation, M. Coleman and T. Moore, Chemical Propulsion Information Agency, M. Gu, Shanghai Bureau of Astronautics, and F. Winter, Smithsonsan Institute, 2001; also adapted from Ref. 11, 2002).

Other Engines

To further increase the payloads in LEO and GEO, China decided to use strap-on boosters.[3,4,6] They modified the SLV to allow two or four liquid propellant boosters to be attached and thereby to augment the takeoff thrust and the payloads. Each booster has a version of a YF-20 engine, and the boosters are dropped off when their propellants are consumed. One version of the LM-2 with four strap-on boosters was modified to carry astronauts. The modifications included a space crew capsule with an emergency escape tower (similar to the ones used in the Apollo or the Soyuz) and additional redundancies and safety provisions in the vehicle and the engines. Several unmanned flights of the LM-2F were made with an unmanned version. The first launch with an astronaut onboard was on 14 October 2003. The People's Republic of China thus became the third nation to send an astronaut into a space orbit. No information was found on the rocket engines for orbit injection, attitude control, reentry alignment, retro action, or orbit adjustments.

Fig. 13-8 View of dual-engine cluster of YF-75 LPREs used in third stage of LM-3A and LM-3B. Courtesy of CALT (from personal communications, J. Morehart, The Aerospace Corporation, M. Coleman and T. Moore, Chemical Propulsion Information Agency, M. Gu, Shanghai Bureau of Astronautics, and F. Winter, Smithsonsan Institute, 2001–2002).

In 1968 the People's Republic of China decided to become a power in the world's space endeavors and started to modify their military ballistic missiles for spaceflight applications.[3,4] An Academy of Space Technology was established in that year, and it became responsible for all of the space technology and the management of all satellites. Tsien was its first director.[2] The first space launch (Long March 1) took place in 1970 based on IRBM military launch vehicle and its LPREs. The development of the intercontinental ballistic missile DF-5 was undertaken in parallel with the LM-2 space launch vehicle, and they basically use the same launch vehicle and the same engines. In 1985 China started to market its space launch capability to the world and has offered a series of SLVs for different payloads, low Earth orbits, and geosynchronous-orbit missions. They were successful to attract several foreign customers, whose payload they have launched. There were two failures to reach the intended orbit on two of these commercial flights, and this caused concern in the international community of potential customers.

China also has developed and produced small maneuvering and attitude control LPREs.[5,8] Work on *hydrazine monopropellant thrusters* and engines started in China in the 1970s, approximately 10 years after other nations had done so. All had pressurized gas feed systems, usually with compressed nitrogen. Some used an electrically actuated explosive valve to admit the gas to the propellant tank, which was isolated during storage by burst diaphragms, upstream and down stream of the tank. The early versions had nominal thrust levels of 5 N (1.1 lbf) and 45 N (10.1 lbf) and used two different catalysts, which were probably based on iridium. Because of the high freezing point of the hydrazine (1.4°C), the propellant tanks and the key components had to be heated. The Chinese also developed a hydrazine monopropellant with a freezing point depressant as an additive. It had a freezing point of −30°C and it could use a catalyst that had a relatively small amount of the costly iridium ingredient. However the specific impulse with the additive was considerably lower than pure hydrazine. The Chinese have since developed hydrazine monopropellant thrusters in other sizes up to 1,000 N or 225 lbf.[8]

The Chinese have developed a *variety of different bipropellant thrusters* with thrusts between 25 N (5.6 lbf) and 2,500 N (562 lbf).[4,8,12] All used pressurized gas feed systems. They had different propellants, including NTO/UDMH, nitric acid/UDMH, NTO/A-50, mixed oxides of nitrogen with UDMH, or NTO/MMH. Several of these bipropellant thrusters had acoustic cavities for maintaining stable combustion. Many of the propellant tanks used flexible bladders or surface tension devices to assure positive expulsion of propellants in a zero gravity environment and no gas bubbles in the propellant.

The F-25 apogee bipropellant engine of 490 N (11 lbf) thrust has operated for a cumulative duration of more than 30 min, and it has been flown more than once. One engine system for a satellite had hydrazine monopropellant engines (used with a single larger engine in an upper stage) with four thrusters at 45 N or 10 lbf (pitch), two of 45 N (yaw), and four at 11 N or 2.2 lbf each (roll). However little detail about small thrusters was found.

Technical R&D Base

One of the strength of the People's Republic of China in the LPRE field is in their universities and research institutes. In the last 40 years a good number of these have undertaken R&D projects related to LPREs, as revealed by searches of technical publications. Many of their analytical and experimental investigations have been published, mostly in domestic professional magazines or journals. A number of their technical papers in English have appeared in American and European publications. Many technical papers on the technology of LPREs were prepared by staff members, faculty, or graduate students and have been presented at learned technical meetings all over the world. The literature on combustion (fundamental investigations, combustion vibrations, injector behavior) and on engine condition monitoring (fault detection, sensors, neural networks, algorithms, or flaw verification) have been particularly noteworthy. Some of the analytical work shows novel approaches, but some of it relates to work done earlier in other countries. Here is a partial list of typical Chinese universities and research institutions engaged in R&D related to LPREs: National University of Defense Technology, Changsha; Beijing University of Aeronautics and Astronautics; Tsinghua University, Beijing; Jiaotong University, Shanghai; Naval Aviation Engineering Academy, Yantai, Shandong; Northwestern Polytechnical University, Xian; Shaangxi Engine Design Institute, Changsha; Lanzhou Institute of Physics; Shanghai Institute of Space Power Machinery; Fourth Institute and Eleventh Institute, Beijing; Harbin Institute of Technology; Aerospace Research Institute of Materials and Processing Technology, Beijing; and Institute of (Equipment) Command and Technology, Beijing.

The following is a partial list of sample R&D subjects investigated[12-16] in the last 35 years: flexible support of turbopump rotor; combustion fundamentals, instability, with different injector elements; radiation with a cluster of four thrust chambers; finite element analysis for liquid transients in propellant lines; diffusion of toxic gas after a liquid propellant explosion; neural networks for fault detection in LPRE; critical speeds of rotating machinery including cyclic radial forces from pumps; microexplosions of small droplets; experimental studies of cavitating venturi in liquid hydrogen; and recession analysis of carbon-carbon composite nozzles for LPREs.

Concluding Comments

The People's Republic of China has developed a very respectable set of LPREs in a relatively short time. They have demonstrated their capability to produce reliable engines for military and space launch vehicle, and they have successfully competed in the world's launch-vehicle market. In 2003 they launched and recovered the first manned Chinese space module. Their industry is supported by an extensive network of universities and R&D institutes. All of their larger LPREs have used reliable gas-generator engine cycles, and so far they seem not to have developed LPREs with the somewhat higher-performing closed engine cycles. The latest LOX/LH$_2$ engine seems to have a mature state of the art. The hot stage separation appears to be effective in China. Being a late starter in the LPRE field, the Chinese had the advantage of seeing many key design features, such as TPs, TCs, propellant utilization schemes, or hot separation, that were flown earlier by either the USA or the USSR. Even if they may not have known about the specific details of foreign designs, they did know that it was a practical scheme and that it would be successful. The Chinese seem to have managed their LPRE programs and their R&D resources well because the author could not find a major LPRE program that was canceled, as has happened in other countries. The author did not find much published information on the early LPREs for the MRBMs or on small thrusters.

References

[1]von Kármán, T., and Edsen, L., *The Wind and Beyond, Theodore von Kármán*, Little, Brown and Company, Boston, 1967, pp. 308–315.

[2]Chang, I-S., Toda, S., and Kibe, S., "Chinese Space Launch Failures," Paper ISTS 2000-g-08, 22nd International Symposium on Space Technology and Science, May–June 2000.

[3]Chen, X. "Launch Vehicle Technology in China," *Space Forum*, Vol. 2, 1998, pp. 65–76.

[4]Isakowitz, S. J., Hopkins, J. B., and Hopkins, J. P., Jr. (eds.), "Long March," *International Reference Guide to Space Launch Systems*, 4th ed., AIAA, Reston, VA, 2004, pp. 229–262.

[5]*"Launch Services and Space Technology*," Great Wall Industry Corp., Brochure, China, circa 1998.

[6]*Space Directory*, Jane's Information Group, Inc., Coulsdon, Surrey, England, U.K., 1996–1997, 2000–2001.

[7]Zhu, N., "A Personal Viewpoint on the Development of China's Liquid Propellant Rocket Engines," (in Chinese); translated by Air Force Inst. of Technology. FTD-ID(RS) T-0545-91, 1990.

[8]Sun, H-M., "The Development of Chinese Liquid Rocket Engines," International Astronautical Federation, Paper 96-5.1.01, Oct. 1996.

[9]Hurlbert, E. A., Sun, J. L., and Zhang, B., "Instability Phenomena in Storable Bipropellant Rocket Engines," *Liquid Rocket Engine Combustion Instabilities*, edited by V. Yang, and W. E. Anderson, Progress in Astronautics and Aeronautics, Vol. 169, AIAA, Reston, VA, 1995, Chap. 5, pp. 136–140.

[10]Wang, Z., and Gu, M., "Oxygen-Hydrogen Rocket Engines for CZ-3," International Astronautical Federation, Paper 89-299, Oct. 1989.

[11]Gu, M., and Liu, G., "The Oxygen/Hydrogen Engines for the Long March Vehicle," AIAA Paper 95-2838, July 1995.

[12]Zhu, L., and Zhang, M., "Research on N_2O_4/MMH Thrusters," *Journal of Propulsion Technology*, Aug. 1999, pp. 20–32 (in Chinese).

[13]Liu, W., and Qizhi, C., "Recession Analysis for Carbon-Carbon Composite Nozzles of Liquid Propellant Rocket Engines," AIAA Paper 96-3214, 1996.

[14]Zhu, S., "An Experimental Study of the Cavitating Venturi in Liquid Hydrogen," *16th International Symposium of Space Technology and Science Proceedings*, Vol. 1, AGNE Publishing, Tokyo, 1988, pp. 827–829.

[15]Zhang, Y., Wu, J., Huang, M., Zhu, H., and Chen, Q., "Liquid Propellant Rocket Engine Health Monitoring Techniques," *Journal of Propulsion and Power*, Vol. 14, No. 5, Sept./Oct. 1998, pp. 789–796.

[16]Liang, H., and Chen, Z., "Characteristic Analysis of Flow-Induced Vibrations in Inducers of Oxygen Pumps," *Journal of Propulsion and Power*, Vol. 18, No. 2, March-April 2002, pp. 289–294.

Chapter 14

Liquid Propellant Rocket Engines in India

In 1963 this country started solid-propellant sounding rockets for investigating the upper atmosphere and continued these flights for several decades. Some were two-stage vehicles. This was a prelude to future space exploration. On 18 July 1980 a solid-propellant flight vehicle put a small satellite into a low Earth orbit, and India became the eighth nation to have accomplished this feat. Its first few low-Earth-orbit or LEO satellites (with small payloads of about 40 kg or 88 lbm) were flown up to 1983 and were based on solid-propellant rocket motors for a multistage launch vehicle. Later strap-on stages, driven by solid-propellant motors, were added to increase payload. The author did not see information on some small LPREs for trajectory corrections and attitude control of these early vehicles.

Early work on LPREs started with the Prithvi vehicle, a possible sounding rocket and potential short-range missile.[1] Its LPRE was an adaptation of the engine of the Soviet SA-2 antiaircraft missile; its engine is described in Chapter 8.6. At that time they obtained some help from the Soviet Union. Efforts toward large LPREs began in the early 1980s, when India made the decision to develop a low-polar-orbit satellite system and a geosynchronous-Earth-orbit or GEO satellite capability of their own with remote sensing payloads. They wanted to be independent of foreign SLVs and not rely on information from foreign satellite observation or communication systems.

India had the capability of developing sizeable solid-propellant motors for their early SLVs, and they were able to engineer and build large solid-propellant rocket motors for larger payloads. However, it would take several stages of very large motors in a launch vehicle to place a useful payload into a higher orbit. They knew that they needed LPREs, which provided higher performance than their solid-propellant motors, for putting a meaningful payload (2000 to 5000 lb) into a LEO or at least 1000 lb into a GEO.[1-3] The Indian Space Research Organization (ISRO) then planned a most unusual spaceflight launch vehicle. It had a large segmented solid-propellant rocket motor in the first stage and a single LPRE in the second stage. The third stage had initially a solid-propellant motor, but it would be replaced by a LPRE for more advanced and later versions with the higher payload missions. The fourth stage had a LPRE, initially a simple one with storable propellants and a pressurized-gas feed system. ISRO planned on two kinds of missions. One is aimed at a low-altitude polar Earth orbit and called polar satellite launch vehicle (PSLV) and the other for a stationary orbit or a geosynchronous satellite launch vehicle (GSLV). No other nation has this unusual combination of solid-propellant motors for stages 1 and 3 and LPREs for stages 2 and 4 (Refs. 1–3).*

The PSLV was the first to be pursued. It has a large segmented solid-propellant rocket motor in the first stage with liquid side injection into the nozzle for pitch and yaw control. For roll control they have a pair of hinged liquid propellant thrust chambers of 6.4 kN or 1400 lbf thrust, and they operate continuously while the solid motor is running. They use NTO/MMH propellants and have a separate gas-pressurized feed system.

For the second-stage LPRE ISRO obtained a license from SEP (Société Européene de la Propulsion, Vernon, France) for one of the early versions of the French Viking engine with about 155,000 lbf SL thrust. This French engine is described in Fig. 10-7 and 10-8. They preferred this route of licensing over developing such a large engine themselves because it would save time and resources. The Indians have given the name Vikas to their version of the Viking engine. They were able to copy and build the Vikas in India. They used some components from France during the early manufacture, but now all of the Vikas engine components have domestic sources including the turbopump and the GG.

A single Vikas engine is used for the second stage of their SLV.[1,3] It uses a gas generator engine cycle, NTO/UDMH as propellants, has a thrust of 725,000 kN or 163,000 lb (vac), a chamber pressure of 51.9 atm or 763 psia, and a vacuum specific impulse of 295 s at a nozzle-exit-area ratio of 31:1.0. It is shown in Fig. 14-1. The TC is a copy of the French Viking IV, is gimbal mounted, film cooled, and has a large composite nozzle-throat insert and still has a conical nozzle exit. The Indians developed their own source of its silica-phenolic resin throat insert. It was first flown in 1993. The early French Viking engines used UDMH as the fuel, but the French switched to a mixture of 75% UDMH and 25% hydrazine because of combustion instability problems. The information sources indicate that the Indians have not switched the Vikas fuel to a hydrazine mixture. Two small solid-propellant rocket motors are fired just before the engine start in order to orient the ullage in the propellant tanks and ensure that propellants are covering their tank outlets. The second stage has two sets of two roll control nozzles at 300 N (67 lbf) thrust each; they are supplied with "warm" gas, directly from the Vikas gas generator. The nozzles are fixed and point into a tangential direction. Hot-gas valves provide an on/off control of roll moments applied to the stage. The initial third stage was a solid-propellant rocket motor with a swiveled nozzle.

The fourth stage had two gimbal-mounted bipropellant TCs for further increasing the flight velocity and for controlling pitch, yaw, and roll during of the flight of this stage. The LPRE of this fourth stage has a helium-gas-pressure feed system that supplies the two TCs at 7.5 kN (1689 lbf) thrust each, at a

---

*Personal communication, R. Nagappa, India, 2001.

Fig. 14-1 Vikas second-stage LPRE in a static test facility. This LPRE has an old style conical nozzle exit. Height is about 10 ft. Courtesy of ISRO-LPC.

specific impulse of 305 s (vac) at a nozzle expansion ratio of 60 to 1. It uses a modified NTO with MMH.[4-7] It can provide propulsion for trajectory or orbit maneuvers (e.g., apogee propulsion or orbit insertion), and it was able to restart. This fourth stage also has six small thrusters at 50 N (11 lb) each for attitude control during the coasting period or the unpowered part of the spaceflight of the fourth stage.[4] Initially all types of small thrusters were investigated, including ablatively cooled and regeneratively cooled TCs. For the attitude control of the SLVs, the Indians selected radiation-cooled versions with film cooling using refractory metals.[1,5] For the flight-path control of satellites, they developed some low-thrust hydrazine monopropellant thrusters. The sizes are 1 N (0.2 lbf) and 10 N (2.2lbf) with a solid catalyst beds.

The ISRO is responsible for all spaceflight efforts including the satellites and LPREs.[1-3] Its Liquid Propulsion Center is headquartered at Trivandrum near the southern tip of the country. The Center has a branch in Bangalore, and its static test facility is at Mahendriagiri, southeast of Trivandrum.[1,*] The Vikas engine's first full-duration firing (150 s) was at this facility in January 1988.

The geosynchronous satellite launch vehicle or GSLV uses the same first stage (solid rocket motors) and same second stage (Vikas), but a new LOX/LH_2 engine for the third stage. The higher-energy liquid propellants were needed

---

*Personal communication, R. Nagappa, India, 2001.

to achieve the desired payloads. They requested and received some help from the Soviet Union. Prolonged negotiations with the Soviets to obtain hardware and a license for a LOX/LH$_2$ engine failed in 1993, but resumed thereafter. ISRO agreed to obtain a license and buy seven KVD-1 LPREs.[8,*] This engine can be seen in Fig. 8.6-9, is discussed briefly in Chapter 8.6, and was originally developed by the Isayev Design Bureau in the Soviet Union for an upper stage of an advanced version of the Soviet N-1 moon vehicle. This engine was technically advanced, operated with a staged combustion engine cycle and an estimated specific impulse of 363s (vac). This KVD-1 engine was never flown by the Soviets. It was rated at 75 kN or 17,000 lbf thrust, with a specific impulse of 460 at a nozzle area ratio of 360:1.0. It used a gas-generator engine cycle, the thrust chamber was gimbal mounted, and the engine would need to run for more than 10 min and could be restarted in space. For roll control some cold-gas jets were used. This purchased engine, the KDV-1, flew for the first time in an upper stage of an Indian space launch vehicle in April of 2001, and it has flown again since then.

As a backup and a future replacement of the Russian engine, India started to develop such a LOX/LH$_2$ engine on their own beginning in 1993. It reportedly was aimed at a thrust level of 76 kN (17,100 lbf) and a gas-generator cycle. Work on this propellant combination actually started in India in 1989 at a small scale. In 1993, shortly after they installed a liquid-hydrogen manufacturing plant, they had an explosive failure of a 10-kN (2270-lbf) subscale model engine during a static test. Work on the new Indian LOX/LH$_2$ engine was reported to be in progress with a possible first flight in 2006. No details about this engine were available.

To increase the orbital payload, ISRO has developed two kinds of strap-on boosters.* Initially they used six solid-propellant rocket motors, each similar in size to the first-stage motor, which put the first Indian satellite into orbit. They have a relatively short duration and are dropped off, before the first stage completes its operation. They also developed a strap-on booster with a LPRE; this booster was larger in size and could deliver more performance. Four such boosters can be fastened to the first stage, and they also started at about the same time as the large solid-propellant first-stage motor. Each liquid booster is propelled by a modified Vikas engine. Whereas the solid-propellant strap-on boosters are dropped off after their propellant is consumed, the liquid boosters burn longer and stay with the empty first-stage motor case (after thrust termination of the first stage) and are dropped off together with the empty motor case. This seems to be a unique arrangement, peculiar to their particular set of solid-propellant and liquid propellant stages. No information was found on the methods of pressurizing the propellant tanks in the various stages.

The planned geosynchronous satellite or spacecraft called Insat 2 was launched by their GSLV and had its own propulsion system.[4] It has a bipropellant (NTO/MMH) apogee maneuver thruster of 440 N (100 lbf) thrust and

---

*Personal communication, R. Nagappa, India, 2001.

eight bipropellant 22-N (5-lbf) thrusters for attitude control and stationkeeping. These and the other small thrusters already mentioned have radiation-cooled chambers and nozzles using niobium (sometimes still called columbium) walls coated on the inside with disilicide. The injector is made of titanium. All use pressurized-inert-gas feed systems. The design and materials are similar to thrusters developed earlier in other countries. The 22-N thruster has been endurance tested for more than 250,000 cycles (Ref. 4).

India's ISRO has been offering orbital launch services on their SLVs for payloads developed by other nations.[8,*] After some failures to reach the intended orbit, several commercial payloads have been placed into a GEO. With the various LPREs they now offer low Earth orbits, polar orbits, or geosynchronous orbits for small- and medium-sized payloads.

The Indians have been conducting studies and experiments related to their LPRE efforts in their ISRO propulsion research establishments and also in some universities. A survey of databases of technical publications showed quite a few entries on a variety of subjects related to LPREs. Some appeared to be repetitions of investigations done earlier in other countries, but they must have been useful as a learning experience. A listing of typical investigations conducted in India between 1988 and 2002 is given here: impingement atomization of gelled propellants; risk analysis for land impact of a liquid-propellant upper stage at or near a launch facility; analytical simulations (computer modeling) of rocket engines, including transients; analysis of potential POGO vibrations in a vehicle second stage; processing systems for the manufacture of UDMH or MMH; spark torch igniters for cryogenic propellants; combustion, atomization, and ignition of liquid propellants; coaxial swirl injection elements for a tripropellant rocket engine; investigation of various materials compatible with different propellants; vortex formation in a propellant tank being emptied; experimental investigations of heat transfer with storable and cryogenic propellants; combustion instability analysis including possible damping of the vibrations; analyses of alternate schemes for simulating altitude in ground-test facilities; erosion of nozzle-throat inserts; morphology of silica-phenolic ablative liners; effect of geometry on the heat conduction in coolant channels of cooling jackets; and kinetics of the oxygen/UDMH reaction.

Concluding Comments

India has a remarkable, but limited capability in LPREs in the areas of storable propellant engines for SLVs/missiles and small thrusters. Their space launchers all use a large segmented solid propellant rocket motor as the power source for the first stage. There has been criticism about the effect of the exhaust on the upper atmosphere. They are working on a croygenic engine for an upper-stage application. They now have entered the space-launch-vehicle business

---

*Personal communication, R. Nagappa, India, 2001.

and are able to provide transportation of small- and medium-sized payloads for various orbital space missions. They have flown missions with foreign payloads and their own missions with their own payloads. India started late in the LPRE business, and their approach has been conservative by licensing foreign engines or by developing engines similar to those, which have worked well for other countries. As far as the author can determine, India has not released any information up to 2003 indicating a significant major technical contribution to the state of the art of the technology of LPREs. India seemed to have planned their LPRE activity properly and did not put into production a major LPRE that subsequently had its program canceled, as happened in other countries.

References

[1]"Launch Vehicles and Their Propulsion," *Jane's Space Directory*, 12th and 16th eds., Jane's Information Group Limited, Coulsdon, Surrey, England U.K., 1996–1997, 2000–2001.

[2]Raj G., *Reach for the Stars: The Evolution of India's Rocket Programme*, Viking by Penguin Books India, New Delhi, India, 2000.

[3]Isakowitz, S. J., Hopkins, J. B., and Hopkins, J. P., Jr., *International Reference Guide to Space Launch Systems*, 4th ed., AIAA, Reston, VA, 2004.

[4]Dhas, N. S., Valliappan, K. L., Joseph, E. S., and Balan, C. G., "Low Thrust Liquid Engines at ISRO," *Euro-Asian Space Week on Cooperation in Space*, Proceedings, European Space Agency, Noordwijk, The Netherlands, 1999, pp. 579–585.

[5]Muthunayagam, A. E., "Development of Liquid Propulsion Systems at ISRO," International Astronautical Federation, Paper 88–224, Oct. 1988.

[6]Sudhakar, V., Sivaramakrishnan Nair, K., and Ramamurthi, K., "Development of a 7 kN Liquid Propellant Engine," International Astronautical Federation, Paper 88–225, Oct. 1988.

[7]Sivaramkrishnan Nair, K., Thomas, R. P., and Mahesh, G., "High Performance Liquid Engine," *Recent Advances in Aerospace Sciences and Engineering, Proceedings of the International Symposium*, Vol. 2, Interline Publishing, Bangalore, India, 1993.

[8]Clark, P., "India's GSLV Reaches Orbit, But Can It Be a Contender?," www.Jane's.com/aerospace/civil/news/misc, April 2001.

CHAPTER 15

General Findings, Comments, and Conclusions

The history of LPREs really started with the first flight of an awkward-looking experimental flight vehicle in 1926. Before this date several historical technical papers and books on space flight and rocketry were written by several outstanding visionaries in several countries. This helped to popularize the subject. Several pioneers in different countries undertook key experimental work on LPRE since the 1920s, and the progress in the technology was fast. Enormous technical strides have changed and improved the design of these engines. An estimated 1500 different types of LPREs have been developed in this world.

The emphasis on the engine development criteria has changed. The early practitioners of LPREs had repeated TC burnouts and damaging accidents. In the early decades the LPRE investigators were happy if the engine held together, ran for the intended duration, and did not fail. Making it work was the key objective. Soon the emphasis shifted to the needs of the application, reliability and safety, and then to obtaining maximum possible performance. Today the emphasis is still on these same criteria, but special attention is being given to cost, environmental compatibility, and for reusable engines also to long life. Safety in design, facilities, operation, and servicing is today important.

LPREs were developed because their performance and engine characteristics helped flight vehicles perform better than other means of propulsion using chemical energy as a source of power. LPREs allowed more payload, a more ambitious flight regime, such as more range or a higher-altitude orbit, or better pulsing characteristics or restarts. The design could be tailored to fit the flight application. The unique engine features that caused LPREs to be selected for specific missions have been described.

Many of the flight applications, for which LPREs have done well, are today obsolete. For example, there is today no interest in using LPREs for JATO applications, aircraft superperformance, or sounding rocket vehicles. The emphasis today is on applications for space launch vehicles, spacecraft, or reaction control for the steering of these vehicles. There have been other applications, which today are still of interest. These include underwater submarine-launched ballistic missiles in Russia and reusable propulsion for research sleds on rails for test with high acceleration. The engines of this last category did not fly and are not mentioned in this book.

The development of a rocket engine has usually been a team effort requiring different kinds of disciplines and specialties. As knowledge in this field increased and more of the physical and chemical processes were understood, specialists were needed in areas, such as materials behavior, interaction with flight vehicles, test equipment, instrumentation, rotating machinery, heat transfer, stress analysis, cavitation, or propellants. Each decade the technology database has become larger, and the needed skills and expertise have become more specialized, particularly with large engines. With time the engine teams seemed to become bigger, but many experts are needed only for a brief time, particularly for more complex LPREs.

The first major project in essentially all new LPRE development organizations was usually a simple engine with an inert gas pressurizing the propellant tanks, such as an engine for a sounding rocket or a JATO unit. Later improvements by using warm gas helped to reduce the inert mass of the feed system. This type of engine system is still popular today. Turbopumps (TPs) allowed a major reduction of the propellant tank masses and an improved vehicle performance. The first TPs were flown in Germany in 1938 (HWK), in the Soviet Union in 1947 (RNII), and in the United States in the 1950s (RMI). Almost all large (high total impulse) rocket engines have used TPs and a gas-generator engine cycle. More advanced engine cycles offered somewhat more performance, which was needed for ambitious space missions, and they were developed later. The first flight with LPRE using a staged combustion engine cycle took place in the Soviet Union in 1961 (S1.5400 Korolev Design Bureau), in the United States in 1981 (SSME, Rocketdyne), and in Japan in 1993 (LE-7, MHI). An engine with an expander engine cycle was first flown in the United States in 1963 (RL-10, P&W), and variations of this cycle were flown by the Japanese between 1986 and 2001 (LE-5, MHI). However the numbers of different LPRE using these advance engine cycles are relatively small, estimated at perhaps no more than 50.

One can distinguish several types of LPREs, and each has some specific features or characteristics that make it attractive for certain applications. This includes high thrust (1.7 million lbf) vs low thrust (0.005 lbf), with different propellants, for a single start, several restarts or a few or thousands of restarts (pulsing), for manned or unmanned flight vehicles, with or without one of various thrust-vector-control schemes, high or low chamber pressure, sea-level or high-altitude operation, with or without a control of thrust or mixture ratio or condition monitoring, with high or low chamber pressure, with different feed systems and different engine cycles, and with a steady or variable thrust. Each of these types has been discussed in this book.

The majority of all LPREs that have flown or are currently flying have used pressurized feed systems, many with small thrusters. The next most common type was larger engines with a TP feed system and operating with a gas-generator engine cycle. There are relatively few engines with staged combustion engine cycles or with expander engine cycles.

Liquid propellant strap-on boosters were first introduced in the Soviet Union and first flown in the 1950s (RD-107/108, Energomash). Since that time, these strap-on boosters with a LPRE have been developed and flown in France, China, and India. The United States flew its first one in 2005 (RS-68, Rocketdyne).

The progress in the technology of LPREs has been truly remarkable in the last eight decades. The technical milestones and key LPREs have been discussed in this book. The advance is mirrored in the new engine or component features, new designs, clever inventions, amazing innovations, better instrumentation, and requirements of new applications. These advances took place in different countries, with different development organizations, and at different time periods. In this book attempts have been made to identify these technical advances, to give credit to those organizations or people, who were responsible, and to identify the benefits to specific applications. Advances in components, such as TCs, TPs, GGs, TVCs, controls, or tank pressurization systems have been discussed. The rate of innovation, new materials or new designs, was higher in the first few decades of this history than it has been in the more recent decades. Although there are still some areas where further R&D can have beneficial results for LPREs, the technology is today mature. Any one of several existing development organizations can today perform the design and development of a new LPRE with confidence.

Some of the technology seems to be focused in only one development organization or one country. This includes the use of start turbines in the Soviet Union, the development of TCs with aerospike nozzles in the United States, the use of relatively high-thrust vernier TCs or separate vernier engines in the Soviet Union, and operation of a modified bleed cycle in Japan. Examples of clever technology residing in only one organization are the platelet technology for injectors at Aerojet, tank pressurization provided by one engine in a cluster of four engines at CADB in Russia, or the beryllium chambers for small thrusters at Rocketdyne.

The LPRE has been applied to a good number of ballistic missiles, and most of the countries discussed here have developed SRBMs, MRBMs, or ICBMs, some with nuclear warheads. The first experimental ICBM was launched by the Soviets in 1956 and by the United States in 1957. A number of these ICBMs have since been decommissioned. However China and Russia still have such ballistic missiles with LPREs in their current military arsenals. It is indeed remarkable that none of the MRBMs or ICBMs have been fired against another country in anger or used in a war. The exceptions are a few short-range ballistic missiles (SRBM), such some versions of the R-11 (SCUD), which the Soviets had exported to Iraq and to Egypt; they were used in combat in 1973 and 1991.

The highest production of a LPRE with a pressurized feed system was with the RMI Bullpup in the 1950s; more than 50,000 were delivered. The highest production of LPREs with a turbopump feed system was those used with antiaircraft missiles in the Soviet Union in about the 1950s. A total of between 10,000 and 20,000 engines were delivered, and they consisted of at least four different LPREs designed by several Soviet design bureaus.

As stated before, the LPRE field is today essentially mature. The basic engine system and key components had been fairly well defined about 40 to 50 years ago, and truly reliable operations have been achieved by many LPREs perhaps 30 or 40 years ago. Certainly there have been a few new technical ideas in the last few decades, such as better materials for turbopumps, better nozzle configurations, lower cost components, special tank pressurization systems, or a single-piece stiff turbopump rotor. These and other good ideas are still improving LPREs. There have also been a few new potential requirements, such a microminiaturized LPREs or reusable LPRE strap-on boosters in the United States. However, the opportunities for developing a new LPRE are today not nearly as plentiful as they used to be.

Early engine developments overcame three major problems, but only after extensive R&D investigations. The first were occurrences of combustion instability, which destroyed a number of LPREs in less than 1s. Today a lot more is understood about the combustion behavior. A series of techniques for successfully eliminating combustion vibrations have been described, but none seem to apply to all likely TCs. All new engines and modified engines can now be developed to be free of combustion instability problems. However there are still some areas that are not well understood. Secondly, the cooling of thrust chambers was difficult in the first couple of decades, particularly with large TCs and high chamber pressures. Various cooling techniques, materials of construction, and different designs have solved this issue. The third set of problems is related to the start and restart procedures and a successful transient to full thrust. This too has been discussed in this book.

In the first 50 years a large number, perhaps more than 2500 different liquid propellant combinations, have been considered, and many of these have been evaluated in laboratories. The more interesting chemicals have been investigated or evaluated for their properties, storability, ignition behavior, safety, compatibility with material of construction, and performance. For several decades an intensive effort was focused on evaluating and synthesizing high-energy propellants for better performance; most were very toxic and/or hazardous and were abandoned. As many as 300 propellant combinations might have been tested in experimental small TCs and perhaps 40 in actual flying LPREs. In the last 30 or 40 years only a few specific propellant combinations have been found to be practical, each for a specific type of LPRE or particular types of applications. The storable propellants NTO with either MMH, UDMH, hydrazine, or a mix of these fuels, have emerged as the most practical for many applications. LOX/LH$_2$ has been selected as the most practical high-energy propellant mostly for upper stages of larger SLVs. Hydrazine remains the only practical monopropellant for small thrusters.

All liquid propellants are very energetic materials, and their hazards are well known. They can be corrosive, explosive, flammable, and/or toxic. The author was personally involved in engine failures that caused injuries and damage. A lot has been learned about the safety of the design of the LPREs, safety of test equipment and launch facilities, and the appropriate handling of liquid propellants. It is safe to operate any modern proven LPRE, provided the hazards are identified and understood, and the personnel have been properly trained in safety procedures. Nevertheless, there have been accidents and injuries. The Russians, and the Soviets before them, had perhaps more than 160 fatalities in accidents involving liquid propellants. This is far more than any other country. However the rate of fatal accidents appears to be decreasing.

As in many other technical endeavors, the techniques for design, manufacture, and testing have drastically changed. This author remembers having designed and analyzed engines in the 1940s using a slide rule and a notebook, and many of the manufacturing instructions were verbal at that time. The LPRE industry has changed, first to handheld electronic calculators, and then to today's sophisticated computers. Computer-aided-design techniques have become common, often tied into a computer-aided manufacturing system, an accounting system, or a data-management system. Test recording, data analysis, and data displays are now largely computerized. Engineering analysis, such as heat transfer, plume behavior, or combustion vibration, can all be helped by sophisticated computer programs. The control of vendor performance, material/parts tracing, and product quality is routinely assisted by computers. Computer simulation has reduced the amount of testing and manufacturing that used to be required and allows a comparison of predicted engine parameters and measured engine parameters in condition monitoring systems. The way business is being conducted in the LPRE field has changed.

One of the deterrents to the development of new LPREs has been the high cost of development and flight qualification. Although these costs and the number of required tests have come down greatly over the last 50 years, they are still high. The governments (or prime vehicle contractors), who pay for these costs, are reluctant to spend money on new engines, when existing proven engines or a modification of one will do the job, albeit not as well as a new engine.

The number of countries that have a capability in LPREs has proliferated. There are separate chapters for each of the eight countries discussed in this book. Approximately 10 other countries have a more limited LPRE capability and either are or were active in a narrow part of the field. An additional group of countries has had some genuine interest, formed organizations to investigate, did some LPRE experiments, or have rocket societies and amateur organizations. Altogether perhaps 30 countries have been involved, and this is a small number of all of the countries in the world. The majority of countries have no interest in LPREs, do not have the resources needed to be in this business, and are preoccupied by their own country's urgent issues.

The developers and manufacturers of LPREs are usually second-tier contractors with a government agency or a vehicle organizations usually being in the prime-contractor role. This means that the LPRE organizations are subject to the deliberations, priorities, budgets, and whims of government agencies as well as prime contractors. The LPREs are just one of several key subsystems of a rocket vehicle, and often these other subsystems get a lot more attention. Typical problems of LPRE programs have been vehicle program cancellations, program reorientations, stretching out the schedule, effects of major accidents, overruns of the budget, or lack of available funds. This has made the management of LPRE programs difficult at times.

The LPRE business in the United States has seen its peak in the late 1950s to the early 1970s. This was the period when LPRE employment and sales in the United States were at their highest. The peak activity in Germany was between 1935 and 1945, and in Russia (formerly the Soviet Union) it was between 1946 and 1975. Since then, the need for new SLVs and LPREs has decreased, in part because there are fewer than expected satellite launches and the satellites have a longer life in orbit. Although the volume of business has greatly diminished, there is today still a lot of activity worldwide. This decrease has brought about mergers, acquisitions, cooperative agreements, consolidations, reorganizations, and some companies have just abandoned the field. With these changes often came name changes for the organization. A number of these organization changes are described in this book. There are fewer U.S. companies and fewer Russian design bureaus engaged with LPRE today, and the staffing levels for LPRE work have also been decreased. Competition of other types of rocket propulsion for the same application and the competition of other urgent government programs (nonpropulsion) have diminished the resources available for LPREs

Although there is no known listing, the number of technical articles dealing with subjects related to LPREs seems to have been increasing between 1940 and 1995. The number of professional journals, technical magazines, or newsletters dealing with this subject also seems to be larger. There never has been a magazine or scientific journal exclusively devoted to LPREs. The LPRE subjects must usually share the printed space with analyses and discussions of other types of propulsion or with topics on space technology and military efforts. Today there are no national or international professional societies devoted exclusively to LPREs.

There is a symbionic relationship in all of the eight countries between their LPRE industry and certain university professors and government laboratories in tackling some technical LPRE issues. To the best of the author's knowledge, there are no universities that have an academic department devoted exclusively to LPREs, but there are many who perform research and teach about LPREs, usually in their mechanical engineering or aerospace departments. There are more than four dozen universities and technical organizations worldwide that have recently given courses in LPRE technology, and sometimes these course cover also other propulsion technologies. Judging from the sale of the author's book *Rocket Propulsion Elements* (the book has been translated into Chinese, Russian, and Japanese), more than 40 universities worldwide have used it in a course or as a reference in the last couple of decades.

In the last few decades several of the countries working on LPREs have learned to cooperate on an international level. Examples are the European Space Agency, which relies on work by approximately 15 European countries. A few years ago the United States obtained licenses and some engines from Russia and some French technology on composite materials. Pratt & Whitney recently developed a new upper-stage LPRE with key components purchased from organizations in Russia, Japan, or Belgium.

It takes a few years to become proficient in this business. Each country and each organization has had to learn about LPREs from scratch, acquire a suitable set of computer programs, build its own facilities, and laboratories, and train its own personnel. Each usually has had to repeat some developments or tests, which have already been done by others.

The engines, propulsion concepts, and the LPRE technology in one country or one organization often seem to be very similar to that of some of the others. For example, most countries have used very similar niobium (columbium) chambers, and nozzles for their small bipropellant thrusters or injector patterns successfully developed in one country would find their way into other countries. There have been duplicate efforts in certain aspects of LPRE technology between different countries, between different companies, or government laboratories. The engine concepts and often the detail features and materials are not all that different. Some duplicate R&D effort might be appropriate. It is very expensive to enter into the LPRE field, and it takes years to become proficient. However several countries have seen a benefit being in the LPRE business and have made this investment in recent decades.

The work done on most engines by the LPRE industry has become more efficient. This is in part because of the learning curve (better educated and more experienced personnel) and major advances in computers, databases, and software programs for engineering, manufacturing, or management. It is only recently that one U.S. company designed a new LPRE completely by computer. Efficiency measurements, such as sales per employee, seem to have risen.

In the last several decades the amount of annual available business with LPREs has decreased. There are now perhaps still a dozen organizations that can develop and produce with confidence most kinds of LPREs. However, this capability is diminishing, as experienced craftsmen and engineers are retiring, as government contracting becomes more complex, and as the backlog for new business is declining. The number of employees working directly with LPREs has been drastically reduced by a factor between 2 and 10 when compared to peak employment a few decades ago. To compensate for this loss, several of the Russian design bureaus have undertaken work of a civilian nature unrelated to LPREs, such as commercial pumps, oil refining equipment, or medical devices.

The author started in the LPRE business in 1943, 62 years ago. It has been an exciting time to observe the progress and evolution of this cutting-edge technology. The first flight beyond the speed of sound, the harnessing of enormous kinetic energy, the routine handling of highly combustible hazardous propellants, the propelling of man to the moon and back, or the dispensing of precise small impulse bits have been fabulous achievements. The space age could not have happened without the new technology of LPREs. It has been a satisfaction to the author to have been a part of it.

ABOUT THE AUTHOR

George P. Sutton has been active in the design, research, development, testing, teaching, installation, and management of rocket propulsion since 1943. He was personally involved in several early historic liquid propellant rocket engine programs and solid propellant rocket motor projects as well. In the aerospace industry he worked for three years at Aerojet Engineering Company and for more than 25 years at Rocketdyne (now a part of The Boeing Company) where he held several positions, including Executive Director of Engineering and Director of Long Range Planning. His book *Rocket Propulsion Elements* (currently in its 7th edition) is the classic text on this subject, has been translated into three other languages, and is used by more than 40 colleges worldwide. First published in 1949, it has been in print longer than any aerospace text. For 11 years he served as a member of the propulsion panel of the U.S. Air Force Scientific Advisory Board. In academia he was the Hunsaker Professor of Aeronautical Engineering at the Massachusetts Institute of Technology and has been on the faculty of the California Institute of Technology. He has worked for the U.S. Governments as Chief Scientist of the Department of Defense Advanced Research Projects Agency (ARPA), where he started major propulsion programs, and as a project leader at the Lawrence Livermore National Laboratory. For several years he was an officer in two commercial companies and has been a member of the board of directors of two industrial private companies. He is an AIAA fellow, a past president of the American Rocket Society (predecessor society of the AIAA), an author of 50 technical articles, the recipient of several professional society awards, and is listed in "Who's Who in America".

INDEX

A

ablative cooling, also known as ablative liners or ablative thrusters, 76, 86, 87, 165–167, 304, 374, 375, 377, 455–457, 477, 478, 484–487, *see also* cooling of thrust chambers
acoustic cavities, 233, 235, 236, 459, 460, *see* combustion and combustion vibrations
Aerojet Propulsion Company, 168, 269, 297, 298, 308, 359–403
  adapting Soviet NK-33 engine, 400
  Aerobee sounding rocket, 372
  Aerotojet, 368, 369
  aircraft rocket engine, 9, 10, 367–371
  Bomarc LPRE, 377–379
  JATO, 317, 360–367
  liquid hydrogen engines, 380
  small thrusters (acquired from Rocket Research Co., and Marquardt Co.), 388, 399
  Titan hot stage separation, 389, 390
  Titan LPREs, 380–385
  upper stage engines, 371–377
aerospike, *see* nozzles
Aerozine 50 or A-50 (50% hydrazine with 50% UDMH), 373, 381–383, 425, 436, 445
Aestus LPRE, 774, 776–778
  Agena LPRE 9,59, 518–525
Agena, 2000
aircraft rocket engines, 9–11, 12, 311–317, 346–350, 553, 559–576, 629–632, 662, 730, 734, 761, 793–797, 817, *see also* JATO
altitude test facility, 837, 838
American Rocket Society, 269, 283–286
amateur rocket societies, 281–288
  Britain, 287, 288
  Germany, 281, 282
  others, 288
  Soviet Union, 281, 282
  United States, 282–287

American Pacific Corp., on 1 Oct. 2004, 297, 298, 308, 510–529
  ACS for Mercury manned space capsule, 512
  bipropellant thrusters, 168, 169, 518–525
  Chevaline missile system, 517
  hydrazine monopropellant thrusters (originally from Hamilton Standard), 510, 515
  hydrogen peroxide monopropellant thrusters, 510–515
  Minuteman post boost control system, 516, 517
  positive expulsion tanks, 525, 526
  Royal Ordnance, 510, 514, 863
  self-renewing silica thermal insulation coating, 523
analysis of engine systems, 70–72
Apollo/Saturn V LPREs, 16
  booster engines, *see* F-1 LPRE
  command module ACS, 167
  lunar ascent engine, 435–437, 527, 528
  lunar descent and landing, 484, 485
  lunar excursion module, 436
  service module LPRE, 376
  sketch of propulsion systems, 425
application of LPREs, *see also* JATO, aircraft rocket engines, flywheel unloading
  ballistic missiles, 13–15
  post boost control, 20
  propellant settling prior to in-flight start, 19, 455
  reaction control, 19, 20
  sounding rockets, 6
  space launch vehicles, 5,16–18
  spacecraft maneuvers, 18, 19
  tactical missiles, 11, 13
  terminal maneuvers of defensive missiles, 20, 460
Ariane SLV's LPREs, 775–777, 791, 797–801
Astris LPRE (propelled Ariane third stage), 773, 774

898 Index

Atlantic Research Corporation, Liquid Rocket Division, was renamed Ambac-ISP and bought by
Atlas ballistic missile's LPREs, 14, 408, 416–420, 422, 423
attitude control systems (ACS), 10, 20, 147, 148
Australia, 288, 857
Austria, 241, 288, 807

B
baffles, 232, 233, selected examples
　China's YF-1 cooled baffle, 869
　Chinese LPRE, 868
　consumable baffle, RD-0110 (CADB) 639, 641
　F-1 LPRE, 234
　Isayev's cooled cruciform baffle, 663
　Lunar ascent engine, 435, 437
　SSME (early version), 233
ballistic missiles, 13–15, 728, 889, selected LPREs for
　Belgium, 807
　Britain, 858
　China, Peoples Republic of, 867
　France, 792, 793
　Soviet Union
　　R-1, R-2, R-5, R-7; R-9A, R-12, R-14, R-16, R-36 (Energomash), 589–591
　　R-16, RS-20B, RS-20V, RS-22 (Youzhnoye), 711
　　UR-200, RS-10, RS-20A, RS-18, RS-19, RS-20V, RSM-54, (CADB), 631
　United States
　　Atlas ICBM, 416–423
　　Redstone SRBM, 411–413
　　Titan ICBM, 380–388
beryllium, 455, 457–460 see also small thrusters, cooling,
bladder, 56, 451, 668, see also positive expulsion tanks or devices
inflatable dual bladder, 60, 668
Blue Streak missile and its LPRE, see Britain
Bavarian Motor Works (BMW), 297, 763–768
　JATOs, 763
　Taifun missile LPRE, 764, 766
　Wasserfall LPRE, 763–766
　X-4 air-launched missile, 767
The Boeing Company, 404–473 see also Rocketdyne Propulsion and Power
　Boeing B-29 bomber, JATO for, 364
　Boeing B-47 bomber JATO for, 364–357
Bölkow, Ludwig and Messerschmitt Bölkow Blohm Co., 769, 771–773, 775
Bomarc missile booster LPRE, 53, 377–379
booster pump or booster turbopump, 121, 122, 304, 408–410, 437–442, 608, 612, 652, see also pumps for propellants

dual concentric shaft oxygen booster turbopump, 652, 653
booster rocket engines for multi-stage flight vehicles (examples)
　Apollo/Saturn SLV, 16, 425, see F-1 LPRE (Rocketdyne)
　Bomarc missile, 377–379
　RD-107, RD-108 (R-7 ICBM and several SLVs, Energomash), 599–604
　Vanguard SLV, 355
　Viking engine for Ariane SLV, 791, 797–801
　YF-21 LPRE for China's DF-4 ICBM or LM-2 SLV, 866–872
booster strap-on stages with LPREs, 408, 442, 600–604, 870
BORD project, 448, 467, 773
boron hydride fuels, 343, 626
Brazil, 301, 302
Britain, also called Great Britain or United Kingdom, 9, 12, 241, 305, 843–863
　aircraft rocket engines, 849–852
　Black Arrow, 857
　Blue Streak booster stage of IRBM and SLV, 858–861
　Europa SLV, 790–792, 797, 861
　Royal Ordnance, 299, 846, 863
　Rolls Royce, 297, 857, 858, 861
　satellite launch, 300, 301, 857
　small thrusters, 862, 863
British Interplanetary Society, 287, 288
Bulgaria, 288
Bullpup missile and LPRE, 11, 53, 321–324, 889
burst diaphragm or burst disks, 192, 193, 766

C
C-Stoff (German hypergolic fuel), 49, 738, 757, 846
catalysts for monopropellant thrusters or gas generators
　hydrazine, 161–163, 304
　hydrogen peroxide, 159–161
cavitating venturi, see controls for LPREs
ceramic liner or ceramic nozzle insert, 251, 455, 539
chamber pressure, 25, 496–502, 729
Chemical Automatics Design Bureau (CADB) or KB Khimautomatiki, 629–660
　aircraft engines, 632
　booster engine for submarine launched missile, 648–650
　dual concentric shaft oxidizer booster turbopump, 653
　engines for missiles and space flights (mostly for upper stages), 634–647

Index **899**

engines for surface-to-air missiles, 632–634
listing of key CADB LPREs, 630, 631
LOX/LH$_2$ LPRE, 650–655
LOX turbopump for Pratt & Whitney's RL60, 503
LPRE with expansion/deflection nozzle, 655–657
chemically generated (warm) gas, 56, 58, *see also* tank pressurization
China, People's Republic of, 15, 299, 865–879
   China Academy of Launch Vehicle Technology, 297, 866, 876
   combustion vibrations, 868, 869
   first satellite flight, 301
   listing of Chinese LPREs, 867
   LPREs with storable propellants, 866–872
   LPREs with LOX/LH$_2$ propellants, 872–874
   manned space flight, 875
   Shanxi Liquid Rocket Engine Company, 297, 866
   small bipropellant and monopropellant thrusters, 299, 874, 877
   universities and research institutes, 877, 878
cleaning of oxidizer flow passages, 198–200
cold gas propulsion systems, 19, 158, 159, 474, 475, 691, 720
combustion and combustion vibrations, 228–238, *see also* acoustic cavities, baffles
   combustion behavior, 228–230
   examples of remedying combustion instability
     General Electric experimental LPREs, 331, 332
     Soviet RD-0110 LPRE, 636–639
     Space Shuttle Orbiter reaction control thruster, 236, 237
     United States F-1 LPRE, 232–234
   POGO instability, 383
   RD-170 baffle and injector spray element pattern, 608, 612, 613, 614
   rating technique for confirming combustion stability, 231
   techniques for eliminating combustion instability, 231–238
   three modes/types of combustion vibrations, 230
companies, *see* LPRE organizations/companies
composite materials (France), 808, 809
computer applications and software for LPREs, 305
   computer assisted design of LPRE, 442, 443
   computer simulation, 891
controls for LPREs, *see also* engine-out control, pressure regulator, and thrust vector control

cavitating venturi flow control, 184, 185, 483, 868, 973
condition monitoring or engine health monitoring, 16, 31, 72, 182, 189, 190, 438, 440, 654
differential throttling, 226, 227
items to be controlled, 182, 266, 267
mechanical linkages, 183
mixture ratio control or propellant utilization control, 30, 182, 189, 197, 872
pneumatic control, 184
thrust control, 182, 186
timer control and step ladder control, 183, 495
cooling of thrust chambers, 75–87, 730, 890
   see also ablative cooling
   double wall cooling jacket, 80, 539, 552, 562, 567, 568, 573
   dump cooling, 76, 92, 427, 428
   evaporating cryogenic coolant, 316–319
   fabrication of cooling jacket, 81–86, 596, 634
   film cooling, 75–77, 256, 498, 631, 746
   inter-regenerative or Interregen cooling, 175, 176, 456–458, 779–781
   milled channel design, 84, 85, 772, 773
   nozzle extension, cooling of, 92, 93
   radiation cooling, 76, 92, 167–171
   regenerative cooling, 77–80, 171, 172, 686, 688
   table of alternate cooling jacket designs, 76
   transpiration or sweat cooling, 77, 493, 498–500
   tubular cooling jacket or tubular thrust chamber, 81–85, 304, 413–418
Coralie second stage LPRE, 773, 790, 792, 861
costs of LPREs, 28, 891
Crocco, Luigi, 276, 277
Curtiss Wright Corporation, 268, 309, 346–350
   XLR-25 LPRE for research airplane, X-2, 346–350

D

Delta SLVs LPREs, 17
   Aerojet upper stage engines for, 374–376
   booster engine (modified Thor engine) for Delta I and II, 408, 420–422
   Northrop Gruimman upper stage engines, 485
   RL10 for upper stage, 491–496
   Rocketdyne RS-68 for Delta IV, 442–444
Denmark, 807

Index

Diamant A and B (France), 791–793
differential throttling, see controls for LPREs
Dushkin, Leonid S., 112,139, 556, 559, 561, 573

E

EADS, 162
  Bölkow Design Bureau, Messerschmitt, Bölkow, Blohm; DASA, Astrium, 172,177, 298
  SEP, SNECMA, 298, 786
  SEPR, 785, 786, 788, 793–797
early team efforts, 289, 290
Egypt, 680
electric propulsion systems, 148
Energomash, see NPO Energomash
engine complexity, 28
engine cycles, 63–68
  engine failures, 32
  expander cycle, 56, 64, 65, 304, 491–494, 658, 659 821, 828, 888
  gas generator cycle, 56, 63, 866, 889
  modified expander cycle or coolant bleed cycle, 67 821, 828, 888
  staged combustion cycle, 56, 66, 67, 69, 304, 438, 440, 448, 612, 621, 643, 649, 652, 681, 696, 699, 700, 710–713, 719, 722–724, 729, 769–773, 826, 888
  TC tap-off or topping cycle, 63, 304, 435
engine families, 27, 28
engine-out control, 6, 31, 182, 191, 434, 701
engine systems
  engine system analyses, 70–72
  for primary LPREs, 54, 73
  for small thrusters, 68–70
environmentally friendly propellants and exhaust gases, 6
Esnault-Pelterie, Robert, 268, 280, 785
Europa (European SLV), 790–792, 861
expansion/deflection nozzle, 304, 657, see nozzles
exhaust gas plume or flame, 6, 305
export of LPREs (Russia), 736
extendible nozzle, see nozzles
extra functions of LPREs, 28

F

F-1 LPRE, 23, 303, 408, 427–431
  start sequence, 429, 430
  uprated F-1A
  facilities for testing, 305–306, 837, 838
feed systems, 56–63, 151–154, 257, 258, 304, 786–793, 798
  first feed system, 251, 304
  gas pressurized with regulated pressure or blow-down, 56–58, 191

pump pressurized 59–61, with turbopump, 61–63
  with positive expulsion tanks, 56, 59–61
first vehicle flight with a LPRE, 253–255
flow diagrams for selected LPREs
  Blue Streak, 860
  Bomarc, 379
  D-57, 695
  Gamma Mark, 201, 853
  H-1, 65
  Hermes A-1, 329
  HM-7, 804
  J-2, 433
  KTDU-35, 668
  LE-5, 824
  LE-5A and LE-5B, 828
  LE-7, 830
  NK33, 704
  OR-2, 544
  postboost control system, 178
  RD-0120, 653
  RD-0212, 645
  RD-107, 600
  RD-119, 608
  RD-170, 611
  RD-1Kh, 565
  RD-253, 69
  RL10, 66
  Sprite 4, 851
  SSME, 440
  submarine LPRE, 674
  V-2, 745
  Vanguard booster, 337
  various satellites, 71, 174, 835
  Veronique, 788
  Viking, 800
  Wasserfall, 766
  YF-20, 871
  YF-73, 874
  YF-75, 875
fluorine, liquid, 44 and as used in these engines:
  with ammonia in ARC Agena LPRE, 525
  with ammonia in Energomash RD-301 LPRE, 622–624
  with German small TC, 780
  with hydrazine in Rocketdyne NOMAD, 445
  with LH_2 in Pratt & Whitney RL10, 504
flywheel unloading or desaturation, 19, 148
France's LPREs, 9,12, 241, 302, 785–813, see also Coralie second stage, LPRE Veronique sounding rocket, Vulcain LPRE,
  aircraft superperformance LPREs, 793–797
  amateur society, 288

EADS/SEPR/SEP, 298, 785, 786, 788, 794, 795, 796
EADS/SNECMA, 297, 785
first satellite launch, 301
LOX/LH$_2$ LPREs, 791, 801–807
ONERA, 812
pressurized gas LPREs, 786–793
small thrusters, 809–811
Furaline, French rocket fuel, 787

G

GALCIT, 264, 267, 268, 269, 289, 290, 360–362
JATO, 362
predecessor of Aerojet, 290, 360–362
red fuming nitric acid, 289
thrust coefficient, 289
Gamma LPRE (Britain), 853, 857–859
Gas Dynamics Laboratory (Soviet Union), 532, 533–542
gas generators (GG) and preburners (PB), 131–146, 260–263
chemical tank pressurization by GG, 133–135
definition of preburner, 131
early Soviet GG at GDL, 138,141
fuel rich and oxidizer rich GGs, 132, 262
GG for selected LPREs, 137, 140, 142–146
Goddard's GG, 139, 141
monopropellant hydrogen peroxide GG, 56, 135–138
Veronique tri-propellant GG, 788
water dilution and cooling of gas from GG, 138, 139, 141, 262, 548, 549, 572, 788–790
gasoline, 41
gear pumps, 56, 112, 113, 125, 260, 524, 564–566
Gemini manned spacecraft, 455
General Electric Co., 268, 269, 297, 298, 308, 327–345
flame holder for combustion stability, 331
hydrogen peroxide monopropellant GG and TCs, 343, 344
major LPREs, 332–339
plug nozzle (aerospike), 240–243
Project Hermes, 327–331
self renewing thermal insulation coating, 340, 342
Germany
see also V-2 LPRE, Wasserfall anti-aircraft missile LPRE
LPRE, amateur rocket society
Army Research Station at Peenemünde, 205, 210, 297 737, 740–753

BMW (Bavarian Motor Works), 763–768
Hellmuth Walter Company, 10, 747, 755–762
LPRE of V-2 missile, 24, 740–750
V-2 engine unique feature, 750–752
LPREs, 9, 12, 294, 295, 297, 298, 299, 737–782
LPREs since 1945, 769–782
P-111 LPRE (Bölkow), 769, 770–772
propellant investigations, 738, 739
GIRD (Gruppa Isutcheniya Reaktivnovo Dvisheniya), 532, 533, 542–545, 546
Glushko, Valentin P., 61, 66, 139, 241, 280, 721, 557, 585, 586, 588, 592, 603, 605, 606, 620, 622, 721, 843
Goddard, Robert Hutchins, 6, 61, 139, 141, 184, 185, 213, 241
first flight with a LPRE, 253–255
first igniters for LPREs, 258, 259
first thrust chamber tests, 256, 257
first turbopumps, 260–263
first U.S. bi-propellant gas generator, 139, 141, 260, 262
flight control by air vanes and jet vanes, 263, 264
flight stabilization with gyroscope, 263–265
patents, 249, 250, 267
significant contributions to the technology of LPREs, 256–267
valves, pressure regulator, 258

H

Hamilton Standard product line, now owned by ARC, 161, 298, 510, 515
heat transfer, 71, 79, see cooling of thrust chambers
Heeresversuchsanstalt Peenemünde (Army Experimental Station), see Germany
Heinkel aircraft with LPREs, 10
Heinkel He111 bomber (JATO), 754
Heinkel He112, 756
Heinkel Kadett He176, 10, 11, 756
Hellmuth Walter Company, see Germany
high energy propellants or high performance propellants, 26, 44–46, see also fluorine, liquid
hydrogen peroxide with pentaborane, 343, 625
hot separation or hot stage separation, 386–388, 706, 871, 872
hover test facility, 460
Hughes Aircraft Co., 298, 309

902 Index

Hydrazine, 47, 48, 304, 890, *see also* propellants, liquid; small thrusters; and catalysts for monopropellant thrusters or gas generators
   bipropellant thrusters, 510, 514, 862, 863
   dual mode thruster, 480
   monopropellant thrusters, 359, 392–397, 663–665, 809, 810, 831–837
   ultrapure, 305
hydrogen, liquid, 50–52
hydrogen peroxide, 37, 46, 47, 49, 50, 356–358, *see* catalysts for monopropellant thrusters or gas generators, T-Stoff
   bipropellant thrusters, 754–761, 846–857
   experimental LPRE with pentaborane, 49, 343, 625
   monopropellant gas generators, 139, 313, 332, 335, 337, 344, 412, 572, 592, 599, 600, 738, 746, 749
   monopropellant thrusters, 34, 343, 344, 356–358, 510–513, 843, 844
hydroxylammonia nitrate, 393
hypergolic propellants, *see* propellants, liquid

I
ignition of LPREs, 211–216
   dual electronic spark igniter, 433
   hypergolic ignition, *see* propellants, liquid
   table of common ignition mechanisms, 211
impeller, *see* pumps for propellants
India, 15, 288, 299, 302, 881–886
   first satellite launch, 301
   geosynchronous satellite launch vehicle, 882
   Indian Space Research Organization (ISRO), 881, 882
   Liquid Propulsion Center, 883
   LOX/LH$_2$ LPRE, 883, 884
   polar satellite launch vehicle, 882
   solid propellant first stage, 881
   Vikas LPRE, 882, 883
inducer impeller, *see* pumps for propellants
injector, 97–107, *see* thrust chamber
   coaxial bipropellant injector elements, 438, 773, 774, 776, 777
   for small thrusters, 155, 156
   injector elements with conical sprays, 99, 493, 603, 618, 638–640, 776, 779, 807
   injector elements with hole pattern, 98, 235–237, 774, 811
   injector structure, 103–106
   pintle injector, 101, 480–483, 484, 846, 847
   platelet injector, 377
   side injection (Viking, Valois; France), 789, 793, 800
injuries, deaths, accidents, 735, 891

interconnecting components, 196–198
intercontinental ballistic missile (ICBM), 13–15, 589–591, 630, 631, *see* Atlas ballistic missile's LPREs, Titan ICBM and SLV
internal cleaning of propellant hardware and components, 198–200
interregenerative cooling or interregen, *see* cooling of thrust chambers
Iran, 680
Iraq, 680, 681
Isayev, Alexei Mikhailovich, 661–663, 724, *see* KB Khimmash
Ishikawa Harima Heavy Industries (IHI), 297, 503, 815, 831–838
Israel, 298, 300, 301, 305
Italy, 241, 288, 305, 807

J
J-2 LOX/LH$_2$ LPRE, 408, 431–435
Japan's LPREs, 301, 302, 815–840, *see* Mitsubishi Heavy Industries, Ishikawa Harima Heavy Industries
   first satellite launch, 300, 301
   LOX/LH$_2$ LPREs, 820–830
   LPRE organizations, 815, 816
   small thrusters, 834–837
   space launch vehicle, 818
JATO (Jet Assisted Take-Off), 7, 8, 267, 351–355, 361–367, 404, 754–756, 763, 786, 844, 849–852, 887
JAXA (Japan Aerospace Exploration Agency), 815, 816, 837, 839
Jet Propulsion Laboratory, 6, 299
Jupiter LPRE, 420, 421–423

K
Katorgin, Boris (Energomash), 588
Keldish Research Center (Russia), 579, 582, 726
M. W. Kellogg Company, 309, 351–355
Kerosene, 41, *see* liquid oxygen/kerosene LPRE
Walter Kidde & Co., 298, 309, 356–358
   hydrogen peroxide monopropellant thrusters, 356–358
KB Khimmash also known as Isayev's Design Bureau, OKB-154, 297–299, 578, 661–683, *also see* Isayev
   LOX/LH$_2$ LPRE, 681, 883, 884
   LPREs for submarine launched missiles (immersed in propellant tank), 669–677
   small thrusters, 677–681
   space rocket engines, 666–669
   tactical missile engines, 677–681
Kistler Aerospace Corporation, 402

Korolev, Sergei, 139, 568, 569, 578, 585, 662, 679, 698, 721, 843
Korolev's Design Bureau also known as NPO Energiya or OKB-1, 297, 578, 721–724
  first flight with a LPRE using a stage combustion engine cycle, 722, 723
  vernier TC for RD-107, 722
Kuznetsov Design Bureau, now NPO Samara, 23, 297, 578, 698–709, 721
  N-1 SLV, 702–704, 706–709
  NK-15 LPRE for N-1 SLV, 23, 699, 703, 704
  NK-33 LPRE, 399, 400, 699, 700

L

Lance missile rocket engine, 452–454
liquid hydrogen, see hydrogen, liquid
liquid oxygen (LOX), 50
liquid oxygen/alcohol LPREs, 41, 253, 408, 411, 413, 550, 740
liquid oxygen/kerosene LPRE
  Soviet Union, 590–592, 597–616, 626, 630, 631, 634–639, 655, 722, 723
  United States, 38, 408, 409, 416–431
liquid oxygen/liquid hydrogen (LOX/LH$_2$) rocket engines, 890
  China, People's Republic of, 867, 872–874
  France, 791, 801–808
  Germany, 775–777
  India, 883, 884
  Japan, 819–831
  Russia/Soviet Union, 650–655, 681, 693–697
  United States, 304, 383, 408, 409, 424, 425, 431–435, 442–449
Lockheed NF104A fighter aircraft with LPRE, 463
liquid propellants, 25–27, 33–52 also listed with each specific LPRE
LPRE organizations/companies
  aquisitions, mergers, consolidations, 892
  first satellite launches, 300–302
  government budget, 293
  list of organizations, world wide, 296, 299, 300
  number of employees, 893
  project cancellations, 294
  proliferation of LPREs, 294
Liquid Propellant Rocket Engine Systems, 25–27, 33–53, 55–73
  clusters of multiple engines, 604, 700–702, 733
  merits of, 5
LRBR (Laboratoire de Recherches Ballistic et Aéronautique), 785
lunar ascent engine, 424, 435, 436, 527, 528
lunar descent/landing engine, 484
  ablative liner, 485
  throttling, 848
Lybia, 9, 796, 860
OKB Lyulka, now NPO Saturn, 297, 578, 693–697

M

man rating of an engine, 31
manned space flight, LPREs for, 424, 455, 511–513, 601, 728, 875
Marquardt Company, now part of Aerojet, 28, 168, 308, 396–399, see small thrusters
NII Mashinostroeniya (abbreviated as NII Mash) R&D Institute for Mechanical Engineering, 170, 298, 578, 684–692
mature technology, 889, 890
Messerschmitt Me 163 aircraft, 11, 13, 367, 757, 760, 761, 767, 769, 817
methane, 41, 43, 657
Minuteman III post boost control system, 54, 456, 457, 516, 517
Mirage, French rocket assisted aircraft, 795, 796
Mitsubishi Heavy Industries (MHI), 65, 66, 297, 415, 423, 815–830
  LOX/LH$_2$ LPREs, 820–830
mixture ratio control or propellant utilization control, see control for LPREs
monomethyl hydrazine (MMH), 19, 44, 165, 398, 399, 401, 455, 457
monopropellants, 46–48, see also hydrazine, hydrogen peroxide
monopropellant thrusters, see hydrazine, hydrogen peroxide

N

N-1 large Soviet lunar exploration SLV, 16, 23, 387, 388, 693, 698, 702, 703, 704, 706–709, 724, 728, 733
NASA Reseach Center, Cleveland Ohio, 106
Navaho ramjet powered missile, booster LPRE for, 413–416
The Netherlands, 807
Nike-Ajax anti-aircraft missile, 14, 305, 371, 372, 401
niobium also called columbium, 167, 168, 170, 172, 179, 304, 686, see small thrusters, radiation cooled
nitric acids, 18, 37, 39
  red fuming nitric acid, 37
  inhibited red fuming nitric acid, 39
  selected LPREs using nitric acid, 540, 548, 550, 551, 561, 564, 867, 868
nitrogen tetroxide, see propellants, liquid
NOMAD upper stage with fluorine/hydrazine propellants, 445

North American Aviation, 10, 404, see Rocketdyne Propulsion and Power
P-51 fighter aircraft with LPRE assist, 369
F-86 fighter aircraft with LPRE assist, 371, 463
North Korea, 680
Northrop flying wing aircraft, 369, 370
  small scale model MX 324 aircraft (first U.S. aircraft rocket flight), 10, 12, 369
Northrup Grumman Corporation, Propulsion Products Center (formerly TRW), 160, 164, 166, 297–299, 308
  bipropellant thruster, 477–480
  gaseous propellants, 474, 475
  gelled propellants, 487–490
  hydrazine monopropellant thrusters, 475–477
  pintle injectors, 155, 480–483
  secondary combustion augmented thruster, 480
Norway, 807
nozzles
  aerospike or plug nozzle, 94–96, 304, 464–470, 889
  cooling of nozzle, see cooling of thrust chambers
  different nozzle diverging section configurations, 90
  expansion/deflection nozzle, 90, 94, 95, 304, 449, 450, 655–657
  extendible nozzle, 95, 96, 496, 497, 696
  flow separation in diverging section, 93, 94
  nozzle extension approaches, 92, 93
  over-expanded, under-expanded, 88
  supersonic, 88–96
NPO Energomash (formerly OKB 456), 297, 578, 585–628
  cooling jackets, 596
  experimental TCs, 595, 596
  gas generator for RD-111, 139, 140
  high energy propellant engines, 626
  LOX/kerosene engines (incl. RD-107, RD-108, RD-170, RD-180), 597–616
  LPREs with storable propellants (incl. RD-216, RD-253, RD-270), 616–622
  propellant valves, 195–197
  tri-propellant experimental engines, 626
  upgraded version of German V-2, 592, 593

O
Oberth, Hermann, 241, 245, 268, 271–275, 281, 282, 738
  his launch vehicle concepts, 271
  small cone-shaped thrust chamber, 272, 273

OKB Fakel, 725
organizations and companies, see LPRE organizations/companies
orbital maneuver engine, see Space Shuttle, orbiting maneuver engine
oxygen, liquid, see propellants, liquid

P
P-111 LPRE, 769–773, see staged combustion engine cycle
Pakistan, 9, 796
performance of different propellants, 40, 41
pintle injector, see injector
piston pump, see pumps for propellants
pioneers of LPREs, 241–280, 887
platelet injector, 304, 377, 378, 889
platinum, 175, 176, 781, see also cooling of thrust chambers, inter-regenerative
plug nozzle, see nozzle, aerospike
POGO combustion instability, see combustion and combustion vibrations
Poland, amateur society, 288
positive expulsion tanks or devices, 59, 60, 525–528, 732, see bladder, surface tension propellant orienting device; rolling, peeling diaphragm
postboost control systems, 20, 176–179, 456, 457, 516, 517, 715–717
Pratt & Whitney, a United Technologies Company, 308, 491–508
  engines of the RL10 family of LPREs, 491–496
  first LOX/LH$_2$ engine flight, 491
  high chamber pressure efforts, 496–502
  first flying extendible nozzle, 496, 497
  Russian joint venture (RD-180), 505–508
preburners, see gas generators and preburners
prepackaged storable LPREs, 304, 321–324, 450–452, 667, 668
pressure regulator, 58, 134, 151, 182, 191, 197, 203, 258
pressurized gas feed system also called gas expulsion feed system, 56–59, 151–154, 304, 321, 329, 356, 512, 887, 888
propellants, liquid, 25–27, 33–52, 730, see also hydrogen, hydrogen peroxide, monomethyl hydrazine, nitric acids, unsymmetrical dimethyl hydrazine
  additives, 39, 40
  alcohols, 34, 41
  ammonia, 49, 314, 622
  cryogenic propellants, see liquid oxygen/liquid hydrogen rocket engines, methane
  gelled propellants, 48, 487–490

Index **905**

high performance or high energy
  propellants, 44–46, 890
hydrazine, 27, 46–48
hydrocarbons, kerosene, RP-1, methane,
  41–43
methane, 41, 43
monopropellants, 19, 46–48
nitrogen tetroxide, 18, 27, 41, 686
oxygen, liquid (LOX), 26, 27, 41, 50
propellant research organizations, 34–39,
  325, 728–735, 738, 739
self-igniting or hypergolic propellants, 24,
  36, 39, 215, 536, 566, 730
silica thermal insulation, 48, 49, 523, 524
theoretical propellant performance, 40, 41
tripropellants, 51
Propulsion Products Center, see Northrop
  Grumman Corporation
propulsion systems for Apollo/Saturn V lunar
  landing space vehicle, 425
pressurized gas feed system or gas pressure
  expulsion system, 56–58, 190, 191
  first pressurized gas feed system, 257
  for small thrusters, 151
  high pressure gas tank, 54
  typical systems, 152, 164, 174
pumps for propellants, see also gear pumps,
  turbopump
  axial flow pump, 118, 119
  booster pumps or booster turbopumps,
    121, 367, 607, 608, 610, 611, 652, 653
  centrifugal impeller pump, 56, 108, 118,
    260–263
  high discharge pressure, 731
  inducer impellers, 56, 110, 111, 120, 121
  piston pumps, 56, 60, 127, 245, 250, 260
  positive displacement pumps, 125
  pumps for small thrusters, 30, 56, 152
  vane pump, 847
quick restart, fast pulsing of small thrusters, 5,
  149

R

R-7 ICBM and modified for several SLVs, its
  LPREs, 300, 303, 589, 600, 728
Rachuk, Vladimir, 629, 631, 632, 636, 640,
  642, 644, 646, 648, 650
reaction control system, 19, 20, 148, 887, see
  small thrusters
Reaction Motors, Inc. (RMI), 6, 53, 268, 297,
  299, 308, 311–326
  aircraft engines, 311–317
  Bullpup missile, 321–324
  Lark LPRE, 320, 321
  small thrusters, 324

  turbopumps, 313, 317–319
  Viking sounding rocket LPRE, 319, 320
RD-0110 upper stage LPRE (CADB), 636–640
RD-107, RD-108 LPREs for R-7 ICBM and
  several SLVs (Energomash), 589,
  597–605
RD-170 highest thrust booster LPRE for SLV
  (Energomash), 303, 507, 589, 608–616
RD-180 booster LPRE for U.S. SLV, 505–508,
  590, 616
RDA-1-150 aircraft rocket engine, the first to
  fly in USSR, 559–561
Redstone LPRE (Rocketdyne), 137, 408,
  411–413, 420
regenerative cooling, see cooling of thrust
  chambers
rhenium material for TCs, 170, 171
reliability, 6, 156–158, 691, 887
RL10 first LOX/LH$_2$ engine to fly, 491–496,
  504, 508
RL10B-2 engine with highest specific impulse,
  304, 496, 497,
Republic Aircraft Co, F-84 fighter airplane
  with rocket assist, 366
restart in gravity free space, 30,182, 888
RNII (Soviet Propulsion Research Institute),
  547, 548, 554
Rocketdyne Propulsion and Power, The Boeing
  Co., 297–299, 308, 317, 404–471
  aircraft rocket engines, 461–463
  axial flow tubopump, 119
  Lance surface to air missile LPRE, 452–454
  large LPREs, 413–449
  prepackaged LPREs, 450–452,
  small thrusters, 152, 167, 175, 178,
    454–461
  spark igniter, 213
  special nozzles, aerospike, 464–470
  SSME Space Shuttle Main Engine and
    RS-68 LPREs (Delta), 437–444
  table of selected large LPREs, 408–410
  thrust chambers (TC), 83, 85, 234, 235
  tubular TC, 81–84, 415
  injectors, 234, 235–237
rocket propelled aircraft, see aircraft rocket
  engines
Rocket Propulsion Establishment (Britain) and
  some of their LPREs, 845–849
rolling, peeling diaphragm, 60, see positive
  expulsion tanks or devices
Rocket Research Company, now part of
  Aerojet, 308, 392–395
rocket societies, see amateur rocket societies
Rolls Royce, 297, 423, 857–861
Romania, 241, 288

Royal Ordnance (Britain), acquired by ARC, 510, 514, 862, 863
RP-1, or Rocket Propellant #1, 42, 43
RP-318 glider airplane, propelled by LPRE, 10, 550
RS-68 LOX/LH$_2$ booster rocket engine for Delta IV SLV, 442–444
Russia, formerly the Soviet Union, LPREs, 8, 16, 531–736
 amateur society, 289
 experimental thrust chambers, 592–596
 list of LPRE Design Bureaus and related organizations, 578, 579
 list of typical LPRE production plants, 582
 rocket propelled aircraft, 559–576
 summary of efforts in LPREs, 728–736
Russian Scientific Center for Applied Chemistry, 726, 727
RZ2 LPRE (Britain) for Blue Streak missile and SLV, 845, 660–861

S

safety, 268, 887, 891
Sänger, Eugen, 79, 241, 277–279, 285, 286, 786
satellite propulsion systems, 18–20, 71, 174, 397, 834
satellite launches, initial, 300–302
Saturn V, moon landing SLV, LPREs, 303, 408, 424–436
Saudi Arabia, 680
selection of LPREs, 5
SEPR or Société d'Études de la Propulsion par Reaction, 794, 796, 785, 792, 797
Shuttle, see Space Shuttle
silcon dioxide thermal insulation or self-renewing wall insulation coating, 48, 49, 340, 523, 524
Slovakia, 680
small thrusters, 147–180, see hydrogen peroxide; hydrazine, monopropellant thrusters, postboost control systems, terminal maneuver system
 applications, 16–20, 147–149, see also flywheel unloading or desaturation
 bipropellant, 165–177, 396–399, 478–480, 510–515, 664–666, 684–690, 779, 781, 810, 811, 832–837, 838, 877, 884, 885
 carbon and composites fiber material, 460, 461, 808, 809
 feed systems, 57, 151–154
 injectors, 155, 156, 237
 interregen cooling, 172, 173, 175–177, 456–460, 779, 781
 list of manufacturers, 298, 299
 radiation cooled, 169, 396–399, 663–666, 684–692
 reliability/redundancy, 156–158, 691
 requirements, 149, 150
 with ablative liner, see ablative cooling
 with regenerative cooling, 153, 171, 172, 686–688, 781
SNECMA or Société National d'Études et de Construction de Moteurs d'Avion, 297, 785
solid carbon formation layer, 43
sounding rockets, 6, 402, 734, 787–789, 866
South Africa, 9, 796
Soviet Union or United Soviet Socialist Republics (USSR), LPREs, 531–736, see specific Design Bureau or specific rocket engine
 All-Union Society for Study of Interplanetary Travel, 281
 first satellite launch, 301
 list of key LPRE organizations, 577–583
 Soviet experimental military aircraft for testing auxiliary aircraft rocket engines, 561,562, 569, 574, 575, 579,
 summary of LPRE efforts, 728–736
Soviet N-1 SLV for moon travel, see N-1 large Soviet lunar exploration SLV
space flight, LPREs, 16–20, 30, 242, 244, 300–302, 511–513, 728, 865, 866, 867, see also manned space flight
space launch vehicle (SLV), 16–18, 887
space maneuvers, 18–20, 147, 219–222
space shuttle, 16,18
 man rated and reusable, 438
 Orbiting maneuver engine (OME), 375–377, 811
 Space Shuttle main engine (SSME), 408, 437–442, 444
 staged combustion engine cycle, 437, see engine cycles
 turbopumps from Pratt & Whitney and powerhead, 441
Spain, 288, 796, 807
specific impulse or I_s, 41, 304, 358, 382, 408–410, 447, 476, 496, 551, 574, 791, 821, 867
 highest known I_s, value achieved 304, 496
Sprite 1 monopropellant JATO, 850
Sprite 4 bipropellant JATO, 851
staged combustion engine cycle, selected specific engines, 437–442, 444, 447, 498, 608–616, 618–622, 652, 653, 769–773, see engine cycles
starting of LPREs, 201–210
 start sequences for selected LPREs

Index **907**

H-1 LPRE, 206–208
F-1 LPRE, 429, 430
LE-5 LPRE, 823–826, 828
LE-7 LPRE, 826–831
  with a separate start turbine, 210, 714, 715
steering with LPREs, *see* thrust vector control
strap-on booster stage (droppable) with LPREs, 729, 875, 884
submarine launched missile engines (submerged in propellant), 648–650, 659, 669–677, 682, 731, 887
successful LPREs, definition, 2
superperformance aircraft rocket engine, *see* aircraft rocket engines
surface contamination from exhaust plume, 51
surface tension propellant orienting device, 59, 60
Sweden, 807
Switzerland, 9, 241, 288, 796
Syntin (also spelled Sintin) a synthetic type of kerosene, 42, 724
Syria, 680

T
T-Stoff (literally T-substance) or 80% hydrogen peroxide, 737, 754, hydrogen peroxide
tactical missiles propelled by a LPRE, 11–13
  Beta engine for British surface-to-air missile, 847
  Bomarc booster stage, Aerojet engine for air-to-surface missile (operational), 377–379
  Bullpup air-to-surface missile (operational) RMI LPRE, 321–324
  HS 293 glide bomb with a Walter LPRE (operational), 761
  NALAR unguided air-to-surface and air-to-air missile, Rocketdyne LPRE, 450–452
  Nike-Ajax anti-aircraft missile (operational) with an Aerojet LPRE, 11, 14, 371
  R-11 USSR SRBM (operational) with a KB Khimmash LPRE, 679, 680
  SE 4300 French surface-to-air missile, 787
  Taifun BMW LPRE for surface to air missile, 11, 764, 766
  Wasserfall German surface-to-air missile, 763–766
tank pressurization, 133–135, 888
  by cold gas, 19, 133
  by direct propellant injection into tank, 734
  by water diluted reaction gas, 788, 790, *see* gas generators and preburners
  by warm gas from decomposition of monopropellant, 134
  by warm gas from gas generators, 58, 133–135, 304
terminal maneuvering system, 20, 178, 459, 460
thermo-chemical analysis of combustion reaction, 71
Thiokol Corporation, 311
Thor LPRE and its upgraded versions for Delta series of SLVs, 153, 408–410, 420–423
thrust or thrust force, *see also* small thrusters
  control of thrust magnitude, 182, 186–188
  initiation of thrust, *see* starting of LPREs
  range of values, 23, 24
  termination, 30
  thrust-to-engine weight ratio, 28
  variable thrust, 5, 30, 452, 483
thrust chamber (TC), 74–107, *see also* cooling of thrust chambers
  configuration, 74
  dual concentric (Lance LPRE), 452–454
  first TC firing, 251
  life prediction, 87
  thruster, *see* small thrusters
  tubular TC, *see* cooling of thrust chambers
  vernier TC, 219, 225–227, 602, 603, 636, 644, 645, 674, 710, 733
thrust vector control (TVC), 30, 218–227
  actuators for rotating TC, 224, 872
  alignment of thrust direction, 147, 222, 223
  by differential throttling of multiple TCs, 226, 700, 706
  by gimbal mounting of TC, 220, 221, 222, 224, 265, 335, 379, 872
  by hinge mounting of TCs, 220, 225, 416, 847, 872
  by jet vanes, 218, 219, 263, 264, 732, 746, 748
  by side injection into nozzle, 226, 713, 714, 733
  by using turbine exhaust gas, 224, 424, 606–608
Titan ICBM and SLV, 300, 380–388
  engines for Titan I, II, III, and IV, 382
  Titan booster engine, 380–385
  Titan LPRE family, 28
  Titan sustainer engine, 380, 386
topping engine cycle, *see* engine cycles
trends, technical and historical, 23–32
  expanding the range of thrust values, 23, 24
  increasing chamber pressure, 25
  liquid propellants, 25–27
Trident, French fighter aircraft with auxiliary LPRE, 9, 794, 795

Tri-ethyl aluminum (hypergolic start fuel), 38
tripropellant experimental rocket engines, 591, 655
Truax, Robert C., 277
Tsander, Fridrikh A., 6, 241, 532, 542, 543, 544, 546
Tsien, Hsuh Shen, 865, 876
Tsiolkowsky, Konstantin E., 241–246, 280
  Tsiolkwsky equation, 243
tubular cooling jacket, *see* cooling of thrust chambers
OKB Tumansky (small thruster), 299
turbine, 116, *see also* turbopump
  radial inflow turbine, 350, 502
  start turbines (USSR), 128, 704, 705, 714, 715, 722
turbopump (TP), 108–130, 260, 261, 888, *see also* turbine; pumps for propellants; booster pumps
  advanced experimental TP, 502, 503
  alternate arrangement of key components, 108,109
  earliest turbopumps (H. Walter, Germany), 110, 111, 756
  for selected aircraft rocket engines, 313, 364, 465
  gear cases, 110, 122–124, 494
  high pressure TPs, 499, 500, 501, 731
  largest production of TPs, 732
  selected TP from Britain, 847, 848
  selected TP from Russia/USSR, 115, 572, 601, 605, 653, 704, 705, 729
  selected TPs from Japan, 831–835
  selected U.S. turbopumps, 110–112, 123, 260, 261, 313, 334, 375, 494, 501, 503
  TP of German V-2 LPRE, 113, 114, 750
  with two in-line shafts, 108, 109, 114, 115
Turkey, 305

U

unsymmetrical dimethyl hydrazine (UDMH), 6, 19, 27, 44, 165, 727
  LPREs with UDMH, 590, 591, 630–632, 685, 711, 712
United Kingdom, *see* Britain
United States LPREs, 9, 10, 303–528
  first aircraft flight with a LPRE, 370
  first flight of a LPRE (1926), 253–255
  first satellite launch (Explorer 1), 300, 301
  LPREs developers and manufacturers, 306–309
  summary of key accomplishments and innovations, 303–306

universities and government laboratories, 309, 310
use of computers and software, *see* computer applications and software for LPREs
USSR or United Soviet Socialist Republics or Soviet Union, *see* Russia

V

V-2 LPRE also known as A-4 LPRE, 24, 737, 740–750
  copy by Soviet Union, 592, 593, 752
  copy by United States, 406, 407, 752
  launch demonstration by Britain, 843
  unique or novel characteristics, 750–752
valves for LPREs, 192–196, 198, 304, *see also* burst diaphragm
  electromagnetic valves for small thrusters, 196, 198, 686
  face shut-off valve, 156, 481
Vanguard three stage SLV, 335, 372
  its booster stage engine, 336, 337
vehicle mass ratio, 6
vehicle flight performance, 887
Verein für Raumschiffahrt (VfR) (German amateur rocket society), 282, 283, 738
Veronique sounding rocket, 6, 787–789
vernier thrust chamber, *see* thrust chamber
Viking LPRE (France), 797–801
Viking upper atmosphere sounding rocket vehicle (USA), 319, 320
Vikas LPRE (India), 882, 883
Volvo Aero (Sweden), 504
von Braun, Wernher, 283, 289, 740, 742, 752
von Kármán, Theodore, 269, 289, 290, 359, 865
Vulcain LPRE, 802, 803, 805, 806
  injector, 776
  thrust chamber, 775–777, 807
  turbopumps, 808, 809

W

WAC Corporal sounding rocket, 7, 371
  Bumper WAC, first two stage flight tests, 327, 328
warm gas, *see* tank pressurization, gas generators
Wasserfall anti-aircraft missile LPRE, 11, 328, 330, 677, 678, 763–766
water hammer, 67, 68
water dilution and gas cooling, *see* gas generators
Winkler, Johannes, 282

X

X-15 experimental supersonic aircraft and its LPRE, 10, 12, 314–317

Y

NPO Youzhnoye now Southern Machine Building and Production Association, 297, 299, 577, 578, 710–720
 list of their key rocket engines, 711, 712

Z

zink diethyl (hypergolic start liquid), 38
zirconium oxide coating, 696

SUPPORTING MATERIALS

A complete listing of AIAA publications is available at http://www.aiaa.org.